Seafood Enzymes

FOOD SCIENCE AND TECHNOLOGY

A Series of Monographs, Textbooks, and Reference Books

Seafood Enzymes

Utilization and Influence on Postharvest Seafood Quality

edited by

Norman F. Haard

University of California
Davis, California

Benjamin K. Simpson

McGill University
Ste. Anne de Bellevue, Quebec, Canada

CRC Press
Taylor & Francis Group
Boca Raton London New York

CRC Press is an imprint of the
Taylor & Francis Group, an **informa** business

First published 2000 by Marcel Dekker, Inc.

Published 2018 by CRC Press
Taylor & Francis Group
6000 Broken Sound Parkway NW, Suite 300
Boca Raton, FL 33487-2742

ISBN-13: 978-0-8247-0326-4 (hbk)
ISBN-13: 978-0-367-39888-0 (pbk)

Visit the Taylor & Francis Web site at
http://www.taylorandfrancis.com

and the CRC Press Web site at
http://www.crcpress.com

Foreword

Gathering food from the oceans represents the last major hunting effort of humans. As has happened in other fields, however, our technological abilities have outstripped our capacity to deal with the social, cultural, and economic consequences of these technologies. Where national and international regulations, exhortations, and pleas have failed, the common property status of sea resources had led to overexploitation, which has reduced the number of animals available to be used as foods. This decline has been accompanied by an increase in human populations and, more importantly, rising expectations for improved diets in newly developing areas of the world. These factors have increased demands on fish and shellfish stocks and increased strains on the ability of the fish and shellfish to maintain their numbers. Quite naturally, this has had the most impact on those species that have the greatest attraction as food for humans.

Traditionally, the fisheries have been a wasteful industry, in that often no more than 50% of the flesh of even desirable species is converted to high-value human food, and often much high-value seafood protein and lipid is dumped back, unused, into the oceans. No one denies the great need to utilize our aquatic resources more efficiently. It not only makes economic sense, but on a planet with limited resources, there is a moral imperative not to needlessly destroy valuable raw materials. First, we must use traditional commercial species to the fullest extent. This means adding value by utilizing processing byproducts to produce high-value materials (e.g., enzymes) and developing techniques to re-cover all the flesh and use it for human food.

Approximately half of all the species caught in the world today go into the production of fish meal and oil. Development of procedures that will make these resources directly available for human food would greatly improve their efficiency of use. The chemical instability of both the protein and lipid fractions, the presence of high concentrations of unstable dark muscle, seasonal fluctuations in catch, unfavorable sizes and shapes, strong flavors, and skeletal structure that does not permit easy removal of bones are all factors limiting the use of these species. Advances in our knowledge of the chemistry and biochemistry of the unstable components and improvement in processing procedures will be necessary to adapt these species for human food. We must face and accept this challenge.

A thorough understanding of the nature of the critical components of seafood tissues and how they respond to processing, storage, and handling procedures is an absolute necessity to achieve these goals. No class of components of seafood tissues is more important than their enzymic systems. The number of books or reviews devoted exclusively to seafoods is small compared to those dealing with the muscle tissue of land animals. Indeed, in many cases land animals or birds have been used as models to discuss fishery problems. Although there are many similarities between the muscles of fish and land animals, some important considerations for specific seafoods are often overlooked or downplayed. One point often overlooked in discussions of enzymes in foods is that most of the important commercial species are caught in cold water. In fact, 95% of the ocean has a temperature of less that 5°C year-round, leading to species with lipids that contain a large percentage of the highly unsaturated fatty acids eicosapentaenoic and docosahexaenoic acids (which maintain fluidity of the lipids at low temperature). This makes the lipid fraction susceptible to oxidation. Likewise, proteins (enzymes) of seafoods must have a greater inherent flexibility to be able to function at low temperatures—a flexibility that also makes them less stable. Thus, seafood enzymes function well at low temperatures and refrigeration might not have the same inhibitory effect on postmortem changes that it would for warm-blooded land animals.

Seafood Enzymes covers a myriad of topics on how enzymes are important to improve uses of seafood raw materials. These topics include the nature of the enzymes themselves and biological factors that affect them; the role of native enzymes in postmortem effects on quality attributes such as texture, flavor, and color; the use of products of enzymic breakdown as quality indices; control of enzymic activity by modification of environmental conditions, processing, or use of inhibitors; and the use of enzymes isolated from fish processing byproducts as processing aids. These discussions of the special roles of seafood enzymes in postmortem fish metabolism and the quality changes they effect are critical pieces of knowledge in achieving the goal of obtaining maximum value from the available species. We have a strong obligation to use seafood resources wisely and responsibly so future generations may also enjoy their benefits.

Herbert O. Hultin, Professor
Department of Food Science
University of Massachusetts/Amherst
Gloucester Marine Station
Gloucester, Massachusetts

Preface

It has been over 100 years since Eduard Büchner showed that molecules with catalytic activity could be isolated from yeast cells. Following decades of intensive research on these molecules (enzymes), we now know they are ubiquitous in living systems and are the agents that make chemical reactions possible in a diversity of life forms, albeit sometimes under vastly different environmental conditions. In 1961, when the International Commission on Enzymes of the International Union of Biochemistry established a system to classify enzymes, the committee listed only 712 enzymes. The total number of enzymes identified has since grown to more than 3000.

Most of the known enzymes have been extensively studied in land mammals, such as rats, and microorganisms, such as *Escherichia coli*. So why devote a book to enzymes from aquatic animals? Although most of the enzymes discussed in this book are also found in terrestrial life forms, homologous enzymes from different sources, which have the same name and Enzyme Commission (EC) number, may exhibit vastly different properties with respect to stability, temperature optimum, secondary substrate specificity, and others. These differences are based on adaptation and are magnified when the cellular milieu of the source organisms varies because of habitat conditions or other reasons. Aquatic organisms occupy unique and often extreme environments, such as the deep ocean where pressure is high and light is absent; temperature ranges from –2°C in Polar saline gradients to 103°C at thermal vents; salinity ranges from very low to saturated; and daily fluctuations in oxygen availability for tidal organisms. For this reason, the topic of seafood enzymes is of interest to the comparative biochemist.

There are also practical reasons for publishing a book on seafood enzymes. As a general rule, fish and shellfish are more perishable than other food myosystems, e.g., beef and poultry. Several studies have shown that the initial rate of food quality deterioration in microbiologically sterile and nonsterile fish is the same. Thus, it is evident that so-called "autolysis" or endogenous biochemical reactions, rather than spoilage microorganisms, are responsible for the loss of prime quality in seafood. Furthermore, it has been observed that tropical fish, coming from a habitat temperature similar to the body temperature of mammals, keep much better than seafood from the more typical cold (<5°C) environment of the oceans. Fish are poikilotherms, which means that their body temperature is, in most cases, a direct reflection of the water temperature and their enzymes are accordingly cold adapted. Moreover, methods developed to extend the shelf life of chilled seafood, such as low-dose ionizing radiation, modified atmospheres, and most chemical treatments, have been ineffective in preserving prime quality—and in some cases accelerate loss of prime quality shelf life. This is because these methods target spoilage microorganisms rather than endogenous tissue enzymes. Accordingly, future development of methods to preserve the quality of chilled fish must evolve from an understanding of seafood enzymes and aim to target key enzyme-catalyzed reactions that contribute to loss of appearance, flavor, texture, nutrition, and functional properties.

Seafood enzymes are also becoming more important as industrial processing aids. Although enzymes were inadvertently used as processing aids for centuries prior to Büchner's discovery, the wide-ranging use of enzymes as seafood processing aids described in this book is relatively new. Enzymes recovered from fishery products' wastes have recently been obtained as "value added" products for use as industrial enzymes. The ideal enzyme for any application should have a unique combination of properties, such as thermal stability, temperature optimum, pH stability, inhibitor sensitivity, and catalytic specificity. Thus, given the unusual and wide-ranging comparative biochemistry of seafood enzymes stemming from the taxonomic and environmental diversity of aquatic animals, enzymes from seafood wastes that more adequately fit processing niches have been identified and used in food and other industrial sectors.

Seafood Enzymes is organized into five parts. Chapter 1 serves as an introduction discussing how the properties of enzymes from aquatic organisms are related to inter- and intraspecific factors. Part II (Chapers 2–11) reviews specific enzymes or enzyme groups that are known to be important to the seafood technologist and have been studied in recent years. The third part (Chapters 12–15) delves into the relationship between enzymes and seafood quality. Part IV (Chapters 16–20) focuses on the control of enzyme activity in seafood products.

Finally, Part V (Chapters 21–23) covers the application of enzymes as seafood processing aids and the recovery of useful enzymes as byproducts from seafood wastes.

This book contains contributions from international experts from countries around the globe, including Canada, Germany, India, Japan, Mexico, Norway, Poland, Spain, Taiwan, and the United States. While some duplication of information in the chapters in unavoidable, we have made an effort to provide comprehensive and timely coverage of the subject matter.

Norman F. Haard
Benjamin K. Simpson

Contents

Part III. Enzymes and Seafood Quality

Part IV. Control of Enzyme Activity in Seafoods

Part V. Industrial Application of Enzymes

Contributors

Haejung An Department of Food Science and Technology/Coastal Oregon Marine Experimental Station, Oregon State University, Astoria, Oregon

Isaac N. A. Ashie Novo Nordisk BioChem, North America, Inc., Franklinton, North Carolina

Keith R. Cadwallader Department of Food Science and Technology, Mississippi State University, Mississippi State, Mississippi

Chau-Jen Chow Department of Seafood Science, National Kaohsiung Institute of Marine Technology, Kaohsiung, Taiwan, Republic of China

Manuel Díaz-López Department of Applied Biology, University of Almería, Almería, Spain

Marilyn Erickson Center for Food Safety and Quality Enhancement, Department of Food Science and Technology, University of Georgia, Griffin, Georgia

Taichiro Fujimura Laboratory of Flavor Substances, Shiono Koryo Kaisha, Ltd., Osaka, Japan

Fernando Luis García-Carreño Department of Biochemistry, CIBNOR, La Paz, BCS, Mexico

Asbjørn Gildberg Norwegian Institute of Fisheries and Aquaculture, Ltd., Tromsø, Norway

Tom Gill Canadian Institute of Fisheries Technology, DalTech/Dalhousie University, Halifax, Nova Scotia, Canada

K. Gopakumar Fisheries Division, Indian Council of Agricultural Research, New Delhi, India

Norman F. Haard Department of Food Science and Technology, Institute of Marine Resources, University of California, Davis, California

Patricia Hernández-Cortés Department of Biochemistry, CIBNOR, La Paz, BCS, Mexico

Shann-Tzong Jiang Department of Food Science, College of Fisheries Science, National Taiwan Ocean University, Keelung, Taiwan, Republic of China

Tetsuo Kawai Laboratory of Flavor Substances, Shiono Koryo Kaisha, Ltd., Osaka, Japan

Jeongmok Kim Department of Food Science and Human Nutrition, University of Florida, Gainesville, Florida

Edward Kołakowski Department of Food Science and Technology, Agricultural University of Szczecin, Szczecin, Poland

Jen-Min Kuo Department of Food and Health, Chia-Nan College of Pharmacy, Tainan, Taiwan, Republic of China

Tyre C. Lanier Department of Food Science, North Carolina State University, Raleigh, North Carolina

Clara López-Amaya Department of Food Science, University of Guelph, Guelph, Ontario, Canada

Rune Male Department of Molecular Biology, University of Bergen, Bergen, Norway

Alejandro G. Marangoni Department of Food Science, University of Guelph, Guelph, Ontario, Canada

M. R. Marshall Department of Food Science and Human Nutrition, University of Florida, Gainesville, Florida

Yoshihiro Ochiai Food Science Laboratory, Faculty of Education, Ibaraki University, Bunkyo, Mito, Japan

Bonnie Sun Pan Department of Food Science, College of Fisheries Science, National Taiwan Ocean University, Keelung, Taiwan, Republic of China

Hartmut Rehbein Department of Biochemistry and Technology, Federal Research Centre for Fisheries, Hamburg, Germany

Krisna Rungruangsak Torrissen Department of Aquaculture, Institute of Marine Research, Bergen, Norway

Zdzisław E. Sikorski Department of Food Chemistry and Technology, Technical University of Gdańsk, Gdańsk, Poland

Benjamin K. Simpson Department of Food Science and Agricultural Chemistry, McGill University, Ste. Anne de Bellevue, Quebec, Canada

Carmen G. Sotelo Department of Chemistry and Biochemistry of Seafood Products, Instituto de Investigaciones Marinas (CSIC), Vigo, Spain

Wonnop Visessanguan Food Biotechnology Laboratory, BIOTEC Yothi-Research Unit, National Center for Genetic Engineering and Biotechnology, Bangkok, Thailand

Cheng-i Wei Nutrition and Food Science Department, Auburn University, Auburn, Alabama

Seafood Enzymes

1
Seafood Enzymes: The Role of Adaptation and Other Intraspecific Factors

Norman F. Haard
University of California, Davis, California

I. INTRODUCTION

The enzyme makeup of aquatic organisms is directly related to growth and ultimately to the tissue composition, nutritive value, shelflife, and edible characteristics of seafood. Intraspecific factors (e.g., biological age, diet, water quality and habitat temperature, etc.) are known to influence the composition, postmortem durability, and edible characteristics of seafood (1, 2). The primary importance of enzymes rather than microorganisms to the loss of prime quality in seafood has been reviewed elsewhere (3). The seafood technologist is thus concerned with the influence of adaptation and intraspecific factors on the comparative biochemistry of fish tissues, notably muscle tissues.

A. Metabolic Plasticity

The enzymology of fish muscle and other food myosystems have been compared (4). There are many references to how a homologous enzyme from fish A differs, quantitatively or qualitatively, from that found in fish specie B or from a terrestrial animal. For example, fish A contains more of a particular enzyme than B; the thermal stability of homologues from A and B differ; or the Michaelis constant of homologue A is greater than homologue B. In some cases, the genetic makeup of the animal is the raison d'être for such differences in homologous enzymes. For example, the expression of glycolytic enzymes among 15 taxa of the fish genus *Fundus* indicates that some of the variation in enzyme concentration may be nonadaptive (5). However, most of such differences can be explained by ontogenetic and adaptation mechanisms related to ontogeny and environment rather than strictly to genetic variation.

This review summarizes some recent information pertaining to the metabolic plasticity of fish that reflects their incredibly adaptive genetic structures. The introduction of rDNA technology and gene cloning methods has already had an enormous impact on our understanding of this metabolic plasticity. For example, in searching for "the" gene for Atlantic salmon trypsin, no fewer than five distinctive clones were identified in the gene library of this organism (6). One of the genes identified codes for a trypsin isoform that resembles the physical and catalytic properties of homologous trypsins from mammals. It is therefore possible to explain why feeding salmon diets containing soybean meal led to an adaptation whereby the animal began secreting a trypsin with lowered sensitivity to soybean trypsin inhibitor, (i.e., a trypsin homologue comparable to human trypsin) (7). Knowledge of the biochemical adaptability of fish, and the relationship of enzyme chemistry to the growth of fish and utilization of seafood has important implications, especially when applied to the cultivation of fish.

B. Interspecies Differences in Enzymes

The biochemical pathways of animal cells are for the most part very similar (8). The comparative enzymology of food myosystems has been reviewed (4). When comparing homologous enzyme(s) of different species it is important to normalize all intraspecific factors; in most cases this is not done. To normalize intraspecific factors means that species A and B should have the same age, size, sex, spawning phase, spawning history, diet composition, feed intake, habitat temperature and depth, water quality, stress history, etc. This is rarely possible.

However, there appear to be some enzymes present in a given species, that are not at all represented in the genome of other species (Table 1). Whether the gene for a deficient enzyme is absent or merely suppressed is difficult to know without probing for the gene using molecular biology techniques. Some apparent qualitative interspecific differences in enzymes from aquatic organisms that have an impact on seafood quality are briefly discussed below.

1. Ascorbate Biosynthesis

Some, but not all, fish species have the ability to synthesize ascorbic acid. The ability of animals to make vitamin C appears to depend on the presence of gulonolactone oxidase, an enzyme that catalyzes the final step in ascorbate biosynthesis. Primitive fish, such as sea lamprey, contain the enzyme gulonolactone

Table 1 Interspecific Variation in Enzyme Content Affecting Food Quality Indices

Quality index	Enzyme(s)	Comments
Nutrition	Gulonolactone oxidase	Primitive sp. of fish have the capacity to synthesize vitamin C.
	Thiaminase	Some fish(raw) contribute to vitamin B deficiency.
Color	Polyphenol oxidase	The integument of crustaceans may undergo enzymic browning.
	Various enzymes involved with carotenoid synthesis	"Innovator" species have the capacity to synthesize xanthophyll pigments such as astaxanthin.
Texture	Amylose-like enzyme	Time of rigor onset is influenced by the rate of glycogen degradation postmortem.
	TMAO demethylase	Gadoid sp. contain an enzyme that forms formaldehyde, leading to myosin cross-linkage in frozen fish.
	Proteinases	Specific proteinases contribute to unusual texture softening in some fish sp.
Flavor	Various antioxidant enzymes	Lipid peroxidation occurs faster in some species of fish.
	Inosine hydrolase and inosine phosphorylase	Hypoxanthine (bitter) accumulates in some species of fish.
	Various enzymes involved with pyruvate metabolism	The flavor of some marine invertebrates is influenced by metbolites (e.g., succinate, alanine) that accumulate postharvest.
	Urease	Urea is rapidly converted to ammonia in elasmobranch sp.

oxidase and the capacity to renew 4–5% of the body's ascorbate pool daily (9). The gene for L-gulonolactone oxidase has been introduced into rainbow trout, a fish that does not normally have the capacity to synthesize vitamin C (10).

2. Oxycarotenoid Biosynthesis

While no known animal has the metabolic equipment for de novo biosynthesis of carotenoids, several organisms have the ability to convert one form of carotenoid to another. Some aquatic organisms have the ability to assimilate specific types of dietary carotenoids and others the ability to modify the structure of dietary carotenoid (11). So-called "innovator" species such as goldfish and shrimp contain enzymes needed to convert carotenes (e.g., yellow β-carotene) to xanthophylls (e.g., red astaxanthin).

3. Conversion of Inosine to Hypoxanthine

About one-third of fish species are so-called inosine accumulators; they do not form hypoxanthine from inosine in situ in postmortem muscle (12) (see also Chap. 2). This suggests the possibility that the enzymes responsible for converting inosine to hypoxanthine, inosine hydrolase (products are ribose and hypoxanthine) or inosine phosphorylase (products are ribose-1-phosphate and hypoxanthine) are absent in the muscle tissue of "inosine accumulators."

4. Antioxidant Enzymes

The activities of hepatic catalase, glutathione peroxidase, glutathione-S-transferase, and malondialdehyde levels vary considerably among species of fish (13). Major differences are observed in freshwater species, although differences are also observed among marine fish.

The content of antioxidant enzymes (superoxide dismutase and catalase) in elasmobranchs (shark and ray) is lower than in teleosts and seems to follow the overall metabolic oxygen consumption and oxidative activity for each major fish taxonomic group (14). Liver peroxidation, as measured by malondialdehyde concentration, is approximately one order of magnitude higher in elasmobranchs than in mammals.

However, it is difficult to establish phylogenetic-based differences in the presence of antioxidant enzymes since their presence or absence is very sensitive to intraspecific factors (see Sec IV.B). For example, the differences in antioxidant enzymes in tissues of various fish may reflect the antioxidant levels, derived from diet, in the tissue rather than distinctive genetic variation (15).

5. Glycogenolysis and Glycolysis

Glycogen degradation in muscle is normally initiated by the enzyme phosphorylase. The contribution of amylase-like enzymes as an alternative route to glycogenolysis in fish muscle has been known for some time (16, 17) and was recently reported in five species of fish (18). There do not appear to be reports of amylase-like enzymes in the muscle of terrestrials and glycogenolysis is well studied in land animals. Hence, the hydrolytic pathway may indeed be peculiar to the muscle tissue of fish and not present (or expressed) in terrestrial myosystems.

The muscles of some mollusks, notably tidal organisms, accumulate metabolites other than lactate (e.g., pyruvate, succinate, alanine, octopine, during anaerobic glycolysis) (19), reflecting a different biochemical makeup adapted to the daily change in oxygen availability to the animal.

6. Other Enzymes

A novel cysteine protease was reported to be present at high concentration in the dark muscle of several fish and absent from mammalian muscles (20). The enzyme (95–100 kDa) specifically hydrolyzes Z-Arg-Arg-Nmec at pH 5.5–6.0.

Examples of other phenotypic enzymes that are noteworthy to seafood technologists include phenolase in crustacean species, thiaminase in some fish and shellfish, carnosinase in Anguillidae fish, and trimethylamine oxide demethylase in gadoid fish (4).

II. ONTOGENIC FACTORS

Factors such as size, changing diet, metamorphosis, sexual maturation, and spawning may be associated with changes in tissue enzymes and seafood quality (Table 2). Older fish differ in a number of ways from younger members of the population. The influence of growth and aging on the intrinsic variation in food

Table 2 Ontogenic Factors: Impact on Enzymology and Seafood Quality

Parameter	Enzyme(s)	Impact on seafood quality
Sexual maturation	Protein redistribution	Depletion of muscle protein content and increased proteolytic activity.
	Glycogenolysis pathway	Rate and extent of postmortem glycolysis is central to conversion of muscle to meat.
Biological age	Digestive enzymes	Appropriate feed formulation for farmed fish varies with age.
	Protein depletion	Meat of older members of species may develop "jellied" condition.
	Glycogenolysis pathway may increase	Rate and extent of postmortem glycolysis is central to conversion of muscle to meat.
	Phospholipase A	Increased phospholipid hydrolysis contributes to quality loss during frozen storage.
Postmortem	Antioxidant enzymes	Decreased antioxidant enzymes is associated with increased formation of lipid hydroperoxides.
	5′-nucleotidase	Increased IMP degradation is associated with loss of fish freshness.
	Lysosomal enzymes	Leakage of proteolytic enzymes can cause texture deterioration.

quality of fish has been reviewed elsewhere (2). Developmental changes may also be associated with migration from fresh water to the sea or vice versa. It is not always possible to differentiate between the influence of developmental stage and such other factors. Food deprivation and inhabitation of deep water might explain, for example, the unusual chemical composition and poor food quality of older members of the American plaice population, where there are differences in temperature, pressure, and illumination (21).

Changes in relative proportions of enzymes during development sometimes reflect altered metabolic capacity and metabolic patterns related to biological function. *Sebastalobus altivelis* larvae have much higher activities of malate dehydrogenase, pyruvate kinase, and citrate synthase per gram of muscle tissue than do juvenile and adult stages (22). Such differences in enzyme activity may reflect the transition from pelagic to demersal behavior as the fish undergo a vertical migration in the water column of up to 1300 m.

A. Sexual Maturation

The seasonal dynamics of lactate dehydrogenase, alanine transferase, aspartate aminotransferase, glutamate dehydrogenase, and creatine kinase concentration were studied in several species of fish (23). Major changes in all enzyme activities occur during spawning and spawning is the only notable seasonal factor responsible for peak creatine kinase and lactate dehydrogenase activities. On the other hand, there are no differences in serum aspartate aminotransferase, glutamate dehydrogenase, creatine kinase, cholinesterase, and lactate dehydrogenase of nase fish due to sex (24).

Female seabass have greater liver fructose 1,6 bisphosphatase activity than males, especially at spawning time (25). These differences show that glucose requirements peak during spawning and that females have larger glucose requirements than males over the reproductive cycle. The adaptive capacity of fructose 1,6 bisphosphatase is also influenced by broodstock diet.

Spawning induces a shift in several enzymes involved with energy metabolism in female rainbow trout (26). Phosphofructokinase activity was measured as an index of glycolytic capacity, 3-hydroxyacyl-CoA dehydrogenase for β-oxidation, citrate synthase for citric acid cycle, and cytochrome oxidase for oxidative capacity. Large changes in white muscle enzyme activity were observed during the spawning period reflecting a 50% reduction in glycolytic capacity and a fourfold increase in oxidative capacity, notably for fatty acid utilization.

The brain cytochrome P450 aromatase activity of goldfish undergoes a 50-fold range of activity during different seasons of the year (27). Activity peaks with the onset of gonadal regrowth and the synthesis of this enzyme in the natural environment is mediated by aromatization of androgen to estrogen. Likewise, lipoxygenase activity appears to be greater in spawning than in nonspawning saltwater salmon (28).

B. Age and Size

1. Digestive Enzymes

In sea bass larva, brush border enzymes (aminopeptidase, maltase, and γ-glutamyl transpeptidase) increase with age while the cytosolic enzyme leu-ala peptidase decreases concurrently (29). These changes are associated with the capacity of larva to adapt to weaning diets.

Amylase and proteolytic activities were determined in different stages of ontogenetic development of carp (30). The normal increase in digestive tract enzyme activities of this species is dependent on feed intake rather than a programmed developmental process. In several freshwater fish (pike, bream, perch, loach), carbohydrase activity varies more so with age than proteinase and phosphatase activity (31). The changes in carbohydrase activity parallel changes in feeding preference from initial planktophages to predators to benthophages.

2. Energy Metabolism

In white skeletal muscle of conspecific pelagic fishes, the activity of enzymes associated with anaerobic glycolysis usually increase with body size, while the opposite pattern is seen for enzymes of aerobic metabolism. In barred sand bass, the increase in lactate dehydrogenase (LDH) activity, as a function of body size, is not governed by LDH mRNA concentration (32). It appears that other mechanisms, such as rate of mRNA translation or reduced rate of LDH turnover, explain the higher levels of LDH in large fish.

The influence of morphometric maturity and body size on the metabolic capacity of muscle tissue was studied in snow crab (33). Metabolic capacity was estimated using the enzyme markers cytochrome c oxidase, citrate synthase, and lactate dehydrogenase. Metabolic capacity of muscle tissue, but not the digestive gland, is greater in adults than in adolescents. Indicators of protein synthesis do not differ in the muscle of adults and adolescents. Larger individuals in populations appear to have enhanced anaerobic capacity for burst swimming speed in order to counteract their increased friction due to larger size.

3. Growth Rate

Glycolytic enzyme markers are strongly related to the growing rate of Atlantic cod when expressed as units per gram wet mass, units per gram dry mass, or with respect to muscle protein or DNA (34). The most sensitive index of growth rate is the activity of pyruvate kinase or lactate dehydrogenase expressed as units per microgram DNA. On the other hand, enzymes reflecting the aerobic capacity of the white muscle (e.g., cytochrome oxidase) are not specifically related to growth rate.

Ornithine decarboxylase activity has been used as an index of growth rate in Atlantic salmon (35). However, unlike the use of RNA concentration as an index of growth, feeding strategy also influences ornithine decarboxylase activity independent of growth rate. The influence of growth rate on enzyme indices of growth may be compounded by water temperature, since the enzyme used may also increase if water temperature is decreased (36).

4. Other Enzyme Changes

An increase in phospholipase A activity occurs during development of Atlantic halibut (37). The activity change appears to be related to the restructuring of membrane phospholipids and decreased membrane fluidity corresponding with upward movement of larvae in the water column.

Brain catalase activity of *Channa punctatus* fish remains constant during the animal's lifespan, but the enzyme activity becomes increasingly sensitive to Mn^{+2} inhibition with advancing age (38). Increasing susceptibility to Mn^{+2} during aging suggests changes in the regulation of catalase.

C. Postmortem Time

While postmortem history does not fit the biologists definition of ontogeny, changes in the amount or properties of enzymes that occur after death are certainly relevant to seafood technology. Accordingly, some examples of changes in activity of enzymes after death will be reviewed at this juncture.

Mackerel and tuna glutathione peroxidase activity decreases during storage of the fish at 4°C (39). Loss of this activity, and of the substrate glutathione, is associated with the formation of lipid hydroperoxides.

The 5′-nucleotidase activity of black rockfish muscle shows a more complex change during postmortem storage at 4°C (40). Activity decreases from its initial value to a minimum at 1–2 days and then gradually rises to a maximum after 4 days. These changes influence the rate of inosine monophosphate (IMP) degradation and the freshness of the fish.

Lysosomal marker enzymes leak into the cell interstitial fluid from muscle tissue of rainbow trout after only 1 day of ice storage (41). Enzyme leakage is also accelerated by damage caused by slow freezing (see Chap. 12).

III. SEASONAL CHANGES IN ENZYMES

It is difficult to differentiate between seasonal and ontogenetic (e.g., spawning) or environmental factors (e.g., water temperature or feed availability). The seasonal variation of plasma enzymes of nase fish was studied and related to physiological and environmental parameters (24). Enzymes monitored were aspartate aminotransferase, glutamate dehydrogenase, creatine kinase, cholinesterase, and lactate dehydrogenase. The activities of cholinesterase and glutamate dehydrogenase varied with water temperature. Spawning influenced the activities of all the enzymes monitored.

However, in some cases the cause of seasonal changes in enzymes is multifactorial and/or incompletely understood. None the less, the fishing crew and the marketplace perhaps better understand the relationship between season of catch and seafood quality (Table 3).

A. Energy Metabolism

Seasonal changes of lactate dehydrogenase and succinate dehydrogenase activities in several Black Sea fish were best revealed by assaying the enzyme at the habitat temperature of the season to remove temperature anomalies (42).

The activities of channel catfish hepatic and muscle glucose-6-phosphate dehydrogenase, 6-phosphogluconate dehydrogenase, lactate dehydrogenase, and malate dehydrogenase were measured to examine seasonal variation in the temperature acclimation response (43). Enzyme indices showed different acclimation patterns in fall and winter than in spring and summer fish.

B. Digestive Enzymes

The influence of fishing season on several carbohydrases of the intestinal mucosa of six fresh water species was investigated (44). Total amylolytic and invertase activities increase during times of intense feeding and due to the presence of carbohydrate in the feed. The kinetic properties of these enzymes also vary as a

Table 3 Seasonal Factors: Impact on Enzymology and Seafood Quality

Parameter	Enzyme(s)	Impact on seafood quality
Feed availability	Digestive enzymes	"Feedy" fish may exhibit greater autolysis and poor quality.
	Glycogenolysis	Feed deprivation leads to low muscle glycogen and high ultimate muscle pH.
Water temperature	Various	Tissue enzyme activities tend to increase as a low temperature compensation
Spawning	Various	Sexual maturation is frequently associated with poor flesh quality.
	Cathepsin D	Peak protease activity is associated with spawning.

function of the seasonal water temperature. Evidence is provided that the lipid matrix of the enterocyte membranes plays a role in altering the kinetic properties of these enzymes.

C. Other Enzyme Changes

The seasonal change in the cathepsin D activity of sardine dark and white muscle revealed maximum activity in April and October (45). The authors suggested the peak in activity is associated with the spawning period of this species and that cathepsin D contributes to the ripening process of salted sardine.

IV. ENVIRONMENTAL POLLUTION

There is a plethora of information on the influence of environmental contaminants on the enzyme content of fish tissues. Enzyme activities or the presence of specific enzyme isoforms are often used as indices for environmental pollution of fishing grounds. A small sample of reports on this topic is provided herein. This information has relevance to the safety of seafood and, moreover, gives insight into how adaptation of fish to environmental stressors may lead to enzymes that otherwise have an impact on the quality attributes of seafood (Table 4). A good example of the latter is the induction of antioxidant enzymes in fish subjected to oxidative stressors.

The induction of cytochrome P450 isozymes in fish has been specifically reviewed elsewhere (46). Long-term dietary exposure of rainbow trout to 2,2′,4,4′,5,5′-hexachlorobiphenyl induces liver cytochrome P450 activities, in-

Table 4 Environmental Pollutants: Impact on Enzymology and Seafood Quality

Parameter	Enzyme(s)	Impact on seafood quality
Chlorinated hydrocarbons, pesticides, etc.	Cytochrome P450 enzymes, etc.	Increase in indicator enzymes is evidence of food safety concerns.
Oxidative stress	Antioxidant enzymes	As above; tissues may be less susceptible to oxidative rancidity postmortem.
Heavy metals	Phosphatase	Indice of sublethal Cd pollution in shellfish.
Organotin	Chymotrypsinogen, trypsinogen	Direct inhibition; may influence feed efficiency.
Chromate	ATPase	Osmoregulation impaired in fish.

cluding arylhydrocarbon hydroxylase and ethoxyresorufin-O-deethylase (47). Ethoxylresorufin-O-deethylase activity is a much more sensitive and rapid indicator of dioxin and metal pollution than glutathione S-transferase in carp (48). Similar results have been obtained with eel and trout exposed to polluted water (49).

A. Precautions

Caution must be taken, in choice of diet and feeding strategy, when testing the biochemical response of trout to xenobiotic metabolizing enzyme (50). Hepatopancreatic parameters of xenobiotic metabolism include aminopyrine demethylase, ethoxyresorufin-O-deethylase, benzo α pyrene hydroxylase, UDP glucuronyltransferase, glutathione reductase, glucose-6-phosphate dehydrogenase, glutathione-S-transferase, glutathione peroxidase, and glucophosphogluconate dehydrogenase.

Commercial feeds contain both inducers and inhibitors of several of these indicator enzymes. Hence, it is advisable to fast fish prior to testing the effect of environmental pollutants on indicator enzymes. Moreover, moderate temperature elevation and wounding can cause similar changes as chemical stressors such as heavy metals and acid water (51).

B. Antioxidant Enzymes

An animal subjected to oxidative stress from contaminates, including polyaromatic hydrocarbons, polychlorinated biphenyls, and pesticides, often shows an increase in antioxidant enzymes. These include glutathione peroxidase, superoxide dismutase, catalase, glucose-6-phosphate dehydrogenase, and glutathione reductase and they have been recommended as useful indices to biomonitor environmental pollution (52). Specific Cu, Zn-superoxide dismutase isoforms have been recommended as biomarkers of oxidative stress caused by transition metals or organic xenobiotics (53).

Peroxidase activity of various tissues, especially kidney, appears to decrease when *Channa gachua* fish inhabit polluted water (54).

C. Phosphatases

Increased acid phosphatase activity and changes in molecular isoforms appear to be a general marker for sublethal cadmium pollution in the water of various fresh water mollusks, crustaceans, and fish (55). Serum and gill, but not liver, acid phosphatase activity also increases in carp exposed to vegetable oil factory effluent (56).

Of various enzymes/tissues tested as an index of copper stress in carp, gill and serum alkaline phosphatase were the most sensitive indicators (57).

D. Digestive Enzymes

Red Sea bream were exposed to the organotin compound triphenyltin at 0.13–3.23 ppm for up to 8 weeks (58). The liver chymotrypsinogen and trypsinogen contents, but not those of chymotrypsin and trypsin, were greatly depressed by exposure to the antifouling compound. The changes were attributed to a direct influence of the triphenyltin rather than to an indirect result from changes in feeding activity.

E. Energy Metabolism

Endosulfan, an organochlorine pesticide, specifically reduces the activity of cytoplasmic and mitochondrial malate dehydrogenase in catfish (59). Similar results were obtained in vivo and in vitro indicating endosulfan forms an inhibitor complex with the enzyme.

Exposure of mudskipper to sublethal doses of potassium chromate causes a decrease in total ATPase activity in various tissues (60). Chromium may thereby act by interfering with the osmoregulatory mechanism of fish.

V. INJURY

Catfish harvested by a turbine pump have more reddened cutaneous abrasions and reddened fins than those harvested by net. Although activities of serum aspartate aminotransferase and lactate dehydrogenase tend to increase during harvest, there was no consistent difference in these enzyme activities as a function of harvest method (61).

VI. EXERCISE

Changes in locomotory behavior may be associated with the ontogenetic characteristics of the organism (22) or with water movement in cultured fish. The latter can have an impact on the edible characteristics of farm-raised fish (1) (Table 5). An increase in the swimming effort of the fish, due to water movement, enlarges the quantity of dark muscle at moderate flow and increases the quantity of white muscle at rapid velocity (62). Chronic exercise can also induce the activity of certain antioxidant enzymes in the muscle of animals that could influence lipid peroxidation after harvest (4).

Table 5 Exercise: Impact on Enzymology and Seafood Quality

Parameter	Enzyme(s)	Impact on seafood quality
Long-term exercise	Aerobic metabolism	Increased proportion of dark muscle.
	Antioxidant enzymes	Decreased propensity for lipid oxidation postmortem.
Short-term exercise	Anaerobic metabolism	Rate of postmortem pH decline may increase; lower pH decline.
	AMP deaminase	K value may increase more rapidly.

A. Energy Metabolism

Key glycolytic enzymes, glycogen phosphorylase and glycogen phosphorylase kinase, are maintained in active readiness in fish that make a powerful movement for a short period of time (63). This provides for fast activation of the glycogenolysis enzyme system by Ca^{+2} and cellular metabolites. The complete cascade for activating glycogen phosphorylase (glycogen phosphorylase kinase, cAMP-dependent protein kinase, as well as levels of cAMP levels) is activated in trout skeletal muscle and liver during exhaustive exercise (52). Glycogen phosphorylase kinase activity increases 60 % in liver and 40% in white muscle but is not affected in other organs.

Phosphofructokinase binding is a key mechanism to control glycolysis in muscle tissue. Exhaustive exercise of goldfish leads to much higher levels of PFK binding in muscle of warm- than in cold-acclimated fish (64).

Exercise-trained leopard shark exhibit a 36 % increase in the activities of lactate dehydrogenase and citrate synthase and a 34 % increase in white fiber diameter (65). It is not clear whether the changes in enzyme activity reflect the higher growth rates or the increased energetic demands of the training activity.

B. Nucleotide Metabolism

Exercise also results in modification of rainbow trout muscle AMP-deaminase, possibly by a reversible phosphorylation (66). The modified form of the enzyme is less sensitive to feedback inhibition by IMP and has a lower Arrhenius energy of activation. The phosphorylation of the enzyme during exercise results in an enzyme that more effectively converts AMP to IMP under physiological conditions. AMP isoform changes are also associated with ontogenetic changes in the muscle of several animals (4). These results are of particular interest to the seafood technologist, since maintaining high levels of muscle AMP or IMP has a direct bearing on the consumer acceptability of seafood (see Chaps. 2 and 12).

VII. TEMPERATURE

A plethora of information shows that increased enzyme activity or changes in the kinetic properties of enzymes are a form of compensation in cold acclimated fish. The subject was reviewed more than a decade ago (67). Since then, a mounting body of evidence indicates that the expression of cold-adapted enzymes by organisms is determined more by habitat temperature (adaptation) than by intrinsic phenotypic expression. Acclimation to temperature change has traditionally been classified as short-term adaptation, whereas differences in phenotypic expression are classified as long-term adaptation (68).

A survey of data on mitochondrial enzymes in cold-acclimated fish indicates that the increase in enzyme content and/or molecular activity does not always compensate for differences in habitat temperature (69). Adaptation to the cold may also be tissue/species-specific. For example, acclimation to cold increases the Ca^{+2}-ATPase activity of trout heart muscle and has the opposite effect on carp heart (70). These changes may influence seafood quality in a number of ways (Table 6) (see also Chap. 12).

A. Habitat Temperature

Three enzymes from Antarctic fish (blood glucose-6-phosphate dehydrogenase, liver L-glutamate dehydrogenase, and muscle glycogen phosphorylase b) are more efficient catalysts at low temperature than the homologous enzymes from mesophiles (71). All three enzymes are irreversibly inactivated at temperatures much higher than the physiological temperature and only slightly lower than those of homologous mesophilic enzymes. However, heat inactivation occurs very abruptly over a narrow temperature range. Similar results were reported for erythrocyte glucose-6-phosphate dehydrogenase from two Antarctic fish (72). The cold-adapted homologues have a slightly higher subunit mass than mam-

Table 6 Temperature: Impact on Enzymology and Seafood Quality

Parameter	Enzyme(s)	Impact on seafood quality
Low temperature	Digestive proteinases, etc.	Increased autolytic activity at ice temperature.
	Lipogenic enzymes	Increased carcass fat content.
	Decreased anaerobic metabolism	Rate of postmortem pH decline may be influenced.
	Nucleotide-degrading enzymes	K value increases more rapidly.

malian enzyme. The influence of temperature on the kinetic properties of LDH from an Antarctic fish has also been recently studied (73).

Many similar observations have been reported in the literature for a variety of enzymes (67) and will not be detailed in this review.

B. Temperature Adaptation

Carp reared at the same feeding rate at low temperature (17°C) have more body fat and hepatopancreas glycogen than those reared at high temperature (27°C) (74). Most hepatopancreatic enzymes were more abundant in the cold acclimated fish. In particular, lipogenic enzymes such as glucose-6-phosphate dehydrogenase, NADP-malate dehydrogenase, and the rate of fatty acid synthesis from glucose and amino acid in the hepatopancreas were activated during cold acclimation. This would appear to explain the higher body fat content of carp reared at lower temperature. Hepatopancreatic-activities of glucosephosphate isomerase, glucose-6-phosphate dehydrogenase, phosphogluconate dehydrogenase, and glutamic pyruvic transaminase, as well as body lipid content, also increase with feeding rate (75). It appears that acclimation temperature and feeding rate affect carbohydrate-metabolizing activities and lipogenesis by separate or cooperative routes.

Adaptation of loach to low temperature (5°C) for 25 days gives rises to changes in the inactivation kinetics of lactate dehydrogenase from skeletal muscles (76). It is curious that the enzyme from fish adapted to low temperature has greater thermal stability, although it has the same temperature optimum and greater specific activity. The temperature at which the Km value of loach LDH is minimal decreases as the acclimation temperature is decreased (77). Acclimation of tilapia to lower temperature (26°C vs. 20°C) results in decreased white muscle LDH activity, but increased citrate synthase and creatine phosphokinase activities (78). Oxygen consumption rate is greater in the cold-acclimated fish than those held at 26°C.

Channel catfish were reared at temperatures of 7, 15, and 25°C to evaluate enzymatic acclimation to habitat temperature (79). No temperature-induced changes in isozyme expression were found for glucose-6-phosphate dehydrogenase, 6-phoshogluconate dehydrogenase, lactate dehydrogenase, or malate dehydrogenase in liver and muscle. Increased activity of glucose-6-phosphate dehydrogenase during cold acclimation was quantitative and not due to gene expression for different isoform(s). No temperature-induced changes in LDH were observed in this study.

Cold acclimation of striped bass increases red muscle citrate synthase activity 1.6-fold and carnitine palmitoyltransferase activity two-fold (80). Carnitine palmitoyltransferase is the rate-limiting enzyme for β-oxidation of long-chain fatty acids. The authors concluded that capacity for fatty acid oxidation increases during cold acclimation due primarily to proliferation of mitochondria, without altering the kinetic properties of carnitine palmitoyltransferase

C. Mechanisms of Enzymatic Cold Adaptation

1. Gene Expression and Isozymes

Skeletal muscle fibers undergo modifications in mitochondrial content and oxidative metabolism in response to low-temperature acclimation. Mitochondrial enzyme markers are higher in both red and white muscle of cold-acclimated fish (81). Study of the mitochondrial mRNA and DNA copy numbers in cold-acclimated fish (4°C vs. 18°C) indicates that response to the physiological stimuli of temperature likely results from the regulation of nuclear-encoded genes.

The activity of muscle M-4 lactate dehydrogenase was studied in loach acclimated to different temperatures (82). "Cold" and "warm" isoforms of LDH were purified and characterized. The "cold" LDH has greater specific activity and stability in the pH range 5.0–6.5.

The cytosolic form of malate dehydrogenase from longjaw mudsucker (an eurythermal fish) has two gene loci (83). One isozyme ('warm") has a higher thermal stability, and a lower Michaelis-Menten constant for NADH, than the other ("cold") isozyme. Winter-acclimatized fish contain more "cold" isozyme while the reverse is true for fish collected in the summer. Similar results were obtained when the fish were acclimatized at 10°C and 30°C in the laboratory. The isozymes serve to conserve Km value of the enzyme at high and low temperatures. Similar results have been reported for myofibril ATPase from other eurythermal fish species such as carp (84). "Cold" isoforms of ATPase have relatively high activity at low temperature at the expense of thermal stability. The physiological result of the switch in gene expression is that muscle of fish acclimated to low temperature develops more force and power during contraction at low body temperature than muscle from warm-acclimated fish. For more details on the structure of myosin isoforms, see Chapter 3.

Atlantic salmon trypsin has a large number of gene loci (6). Some of the trypsin clones are anionic and others are cationic; the differences between the two forms are comparable to those between salmon and mammalian trypsin. All the residues, which differ in charge between the two forms, are located on the surface of the protein.

2. Membrane Lipids

In general, the rate of muscle ATP catabolism is greater in cold-water fish than in warm-water fish. The reader is referred to Chapter 2 for details of this pathway.

The mechanism by which the rate of ATP conversion to hypoxanthine varies in fish from different habitat temperatures is not completely understood. The rate of IMP dephosphorylation is one of the steps in this pathway. The K_m values of microsomal 5'-nucleosidase from several fish increase at the temperature range below the breakpoint on the Arrhenius plots of enzyme activity (85). The breakpoint temperature is correlated with the content of saturated and unsaturated fatty acids in phospholipids of muscle microsome (86). However, contribution of membrane lipids does not appear to explain differences in adaptation of several intestinal cell membrane enzymes (leucine aminopeptidase, alkaline phosphatase, and maltase) in Antarctic and temperate-water fishes (87).

3. Protein Structure

Differences in the thermal and urea stability of cold-adapted cod trypsin and bovine trypsin are related to the lower conformational stability of cod trypsin (88). Fluorescence spectrophotometry of phenylmethylsulphonyl fluoride derivatives was used to show that the free energy change of unfolding was 9 kJ/mol for cod trypsin and 19 kJ/mol for bovine trypsin.

The mechanism of cold adaptation for trypsin from an Antarctic fish (*Paranotothenia magellanica*) was studied by developing a three-dimensional model of the enzyme from the amino acid sequence (89). The enzyme has a relatively high molecular activity at low and moderate temperatures and a relatively low thermal stability compared to mesophilic homologues. Among the key structural differences between the cold-adapted trypsin and bovine trypsin were lack of Tyr-151 in the substrate-binding pocket, an overall decrease in the number of salt bridges and hydrophobicity, and an increase in surface hydrophilicity. The basic structural features and the amino acid sequence of the active site peptides of glucose-6-phosphate dehydrogenase and glutamate dehydrogenase from an Antarctic fish and the enzymes from mesophilic organisms are very similar, despite differences in some catalytic and thermodynamic properties (90). The genes for two psychrophilic elastases from Atlantic cod have been cloned (91). The two cod elastases have only 50–64% amino acid sequence homology with mammalian elastases and a relatively high content of methionine. The high methionine content is believed to be characteristic of psychrophilic fish enzymes.

Pepsins from cold water fish such as Atlantic cod exhibit relatively high molecular activity compared to homologous mesophilic enzymes. Cod pepsin was recently crystallized and sequenced (92). Sequence alignment with other aspartate proteinases indicates that the enzyme is more related to mammalian pepsins than to mammalian gastricsins and chymosins, lysosomal cathepsin Ds, and pepsin from tuna fish.

VIII. HYPOXIA

The activities of several glycolytic enzymes and their association with subcellular structures are influenced by oxygen deprivation in sea scorpion (93). Although total activity of several glycolytic enzymes increased in brain tissue, in other tissues the distribution rather than the amount of these enzymes is mostly affected by oxygen deprivation. In muscle, phosphofructokinase is reduced from 57.4 to 41.7% bound, while the proportion of bound aldolase and triosephosphate isomerase increases during hypoxia. A new bound form of fructose-1,6-bisphosphatase appears in white muscle of anoxic animals. These changes are likely to affect glycogen metabolism, e.g., in postmortem muscle, since binding of glycolytic enzymes to subcellular structures is an important regulatory mechanism of carbohydrate metabolism (Table 7).

In vivo exposure of sea scorpions to hypoxia increased the proportion of muscle AMP deaminase bound to cellular substructures (94). AMP-deaminase binding has also been reported in other species of fish (95, 96). The data collected by these authors suggest that those combined effects of allosteric modifiers and enzyme binding with cellular structure control AMP-deaminase activity

Table 7   Hypoxia, Pressure, and Salinity: Impact on Enzymology and Seafood Quality

Parameter	Enzyme(s)	Impact on seafood quality
Oxygen deprivation	Anaerobic glycogen metabolism	Short-term adaptation may increase rate of postmortem pH decline. Long-term adaptation in tidal organisms may lead to alternate routes of pyruvate utilization and end products that have an impact on flavor.
	AMP-deaminase	K value may increase more rapidly.
Hyperbaric habitat	Anaerobic glycogen metabolism	Rate of postmortem pH decline may be decreased; kinetic properties of key control enzymes may be altered.
	Superoxide dismutase, etc.	Thermal stability of enzymes may be lower.
Habitat salinity	ATPase, etc.	Osmoregulation influences intracellular solute concentrations that impact flavor and stability.
	Lipogenic enzymes	A metabolic shift may increase lipid content of muscle.
	Various	Stability and/or activation of enzymes by salt may affect brined/salted fermentations.

in the cell. The importance of the reaction catalyzed by AMP-deaminase and nucleotide metabolism in general in postharvest metabolism of fish is discussed in Chapter 2.

IX. WATER DEPTH

A notable characteristic of fish living at great depths is a distinctive lipid composition, including more saturated fatty acids and wax esters (2). Deep-water fish often have a relatively watery flesh for the species (2, 21). Relatively little research has been done on the relationship between enzymes and the food quality characteristics of deep-water fish (Table 7).

The LDH homologues from shallow-living species are more susceptible to proteolytic inactivation than the enzymes from deep-living species (97). This appears to be related to the greater stability of enzymes from organisms that live at depths as great as 4000 m. Unlike shallow-living species, the Km values for LDH from deep-living fish are conserved as assay pressure is increased (98).

Superoxide dismutase from the bathophile fish *Lampanyctus crocodilus* has some unique properties, including higher sensitivity to thermal treatment, a higher molecular mass, and interesting substitutions in amino acid sequence that may explain its habitat adaptation (99).

Indices of glycolytic potential (lactate dehydrogenase, pyruvate kinase) and the citric acid cycle (malate dehydrogenase, citrate synthase) are lower in deep-living short-spined thornyhead than the cofamiliar shallow-living spotted scorpionfish. Laboratory rearing under different environmental conditions and feeding strategies revealed that lower muscle enzyme activities in the deep-living fish can be explained by the lower abundance of feed, low temperature, low ambient oxygen, and darkness in deep water (100).

X. WATER SALINITY

The range of mineral composition in fish is fairly narrow and fish will osmoregulate to keep the internal mineral balance reasonably constant (2). Osmoregulation requires energy, and results in the accumulation of solutes such as amino acids, urea, and trimethylamine oxide in muscle tissue. These metabolites influence seafood quality in a variety of ways. Hence, it would be expected that varying the water salinity of a fish would influence the food quality attributes of the flesh (Table 7).

Euryhaline species, such as *Pomacanthus imperator,* undergo a fundamental reorganization of metabolism to favor carbohydrate and lipid retention in response to isoosmotic salinity (101). Acclimation to isoosmotic saltiness (15–22 ppm NaCl) for 1 month results in elevated liver glycogen, glucose-6-phosphate dehydrogenase activity, and muscle lipid. Under hyperosmotic conditions (22–33 ppm) brachial Na-K-ATPase activity is generally lowered.

Gill Na^+-K^+ ATPase activity plays a role in maintaining hydromineral balance during seawater adaptation. Elevated plasma ion levels are associated with resistance to smolting in juvenile. Nonanadromous Atlantic salmon has a lower level of hypo-osmoregulatory ability than anadromous salmon (102). The increase in Na^+-K^+ ATPase activity is slower in nonanadromous than in anadromous salmon and mortality approaches 100% in spite of increased activity. Hypo-osmoregulatory inability and a suppression of the smolting process may be associated with sexual maturation at a young age.

The activities of marine invertebrate (scallop, shrimp, squid) muscle transglutaminases are salt activated (e.g., scallop enzyme is activated 11-fold by 0.5 N NaCl)(103). The osmotic pressure of these invertebrate muscles is isotonic to seawater. There are no other reports of TGase whose activities are stimulated by salt. The reader is referred to Chapter 6 for more information on TGase. In contrast to porcine pepsins, the activities of pepsin isoforms from Greenland cod are likewise stimulated by seawater concentrations of sodium chloride (104, 105). There is also evidence for qualitative differences in the lipoxygenase-catalyzed formation of aromatic carbonyls in fresh and saltwater-reared salmon (28).

XI. DIET AND NUTRITION

The impact of nutritional status on metabolic changes and qualitative responses of fish is not completely understood and is the subject of considerable controversy in the scientific literature. It is clear that dietary alteration causes changes in enzyme activity. Changes may be mediated by induction of more enzymes (normally slow response) as well as by changing the concentration of substrate or effector molecules (normally rapid response) as, for example, occurs with carp hepatopancreas phosphofructokinase (106). There is, however, an increasing amount of information showing that the diet consumed by fish can influence digestive and intracellular enzyme activities. These changes are important to the aquaculturist and seafood technologist because they can influence the body composition of the fish and the postharvest stability of the catch (Table 8).

A. Dietary Protein

The enzyme cysteine dioxygenase is involved with taurine biosynthesis in fish. Specific activity of cysteine dioxygenase activity in rainbow trout hepatopancreas increases exponentially as dietary egg albumin is increased (107). On the other hand, no clear effect of protein level on this enzyme was observed with di-

Table 8. Diet and Nutrition: Impact on Enzymology and Seafood Quality

Parameter	Enzyme(s)	Impact on seafood quality
Protein	Digestive proteinases and peptidases	Quality and quantity of feed protein can alter quantity of digestive proteinases that may leach to flesh and cause autolysis postmortem.
	Cysteine dioxygenase	Increased activity can lead to an increase in intracellular taurine, an important precursor of browning in dried fish.
	Muscle protein catabolism	Restricted protein intake can lead to depletion of muscle protein.
Carbohydrate	Lipogenic enzymes	Some forms of dietary carbohydrate may increase lipid content of flesh.
Lipid	Lipogenic enzymes	A metabolic shift may increase lipid content of muscle.
	Digestive lipases & proteinases	Increased activity may favor autolytic reactions postmortem.
Protein hydrolysates, live feed	Digestive enzymes	The development of the digestive process decreases weaning time of larval fish.
Antioxidant vitamins	Antioxidant enzymes and lipoxygenase	Lower indices of oxidative stress associated with lower propensity for lipid oxidation postmortem.
Starvation	Lipogenic enzymes	Decreased activity leads to lower lipid content of flesh.
Carnitine	Pyruvate carboxylase	Increased protein and decreased lipid in muscle.

etary casein. It was concluded that cysteine dioxygenase activity responds to dietary sulfur amino acid levels more than to protein levels.

The long-term effects of endogenous or exogenous protein, generated respectively by prolonged starvation or by feeding a high-protein diet, was examined in rainbow trout (108). Two important amino acid-metabolizing enzymes, glutamate dehydrogenase (GDH) and alanine aminotransferase (AAT), were monitored in this study. A strong adaptive response was observed in the kinetic properties of these amino acid-metabolizing enzymes after both dietary manipulations. Feeding fish with a high-protein diet significantly increases the liver and kidney GDH Vmax and catalytic efficiency and liver AAT Vmax without changing Km. Starvation has a similar effect on the liver enzymes, but kidney enzymes are not altered.

Common carp were fed a commercial diet for 30 days at seven different feeding rates ranging from satiation to starvation (109). The activities of hepatopancreatic glucose-phosphate isomerase, pyruvate kinase, glucose-6-phosphate dehydrogenase, phosphogluconate dehydrogenase, NADP-isocitrate dehydrogenase, and malate dehydrogenase significantly decreased with feeding rate. Glucose-6-phosphatase and alanine aminotransferase activities remained constant. The results were interpreted as indicative that feed restriction stimulates fatty acid and glycogen mobilization, depresses glycolysis and lipogenesis, and maintains gluconeogenesis and amino acid degradation. Amino acids are the major energy source for the hepatopancreas regardless of feeding rate (110).

Increasing the dietary protein (casein, gelatin, or squid meal) content of shrimp diets increased the hepatopancreas trypsin content, but did not alter trypsin isoforms, over the protein range of 25–48% (111). On the other hand, amylase content and isoform numbers decrease as casein content in the diet is lowered. Varying the casein content of the diets does not influence chymotrypsin levels and isoforms. However, animals fed isonitrogenous diets with gelatin contained less chymotrypsin than those fed casein or squid meal. It was concluded that changes in shrimp digestive enzymes are the specific result of dietary impact rather than to growth rate (112). A relatively small amount (15%) of replacer protein in white shrimp increased the hepatopancreas proteinase activity (113). Total azocasein hydrolytic activity was higher for shrimp fed tuna waste meal as replacer. Trypsin and chymotrypsin activities were higher in animals fed menhaden meal replacer. The aminopeptidase activities of white shrimp hepatopancreas are also influenced by 15% replacer protein (114). Principal component analysis showed that 10 aminopeptidase activities were greater for animals fed menhaden fish meal replacer and less for shrimp fed soybean meal replacer. The results indicate that activities of hepatopancreas aminopeptidases and proteinases increase when shrimp are fed diets with small amounts of poor quality replacer protein. Autolysis of muscle protein in crustacean seafood occurs rapidly after harvest and has been attributed to leaching of digestive proteases into the flesh after capture (115).

B. Dietary Carbohydrate

The influence of starch and glucose on growth and hepatic enzymes was studied with tilapia juveniles (116). Fish fed starch gain more weight and have higher protein and lipid deposition than those fed glucose. The type of carbohydrate in the diet does not affect hepatic hexokinase, phosphofructokinase, or glucose-6-phosphatase activities. Malic enzyme, glucose-6-phosphate dehydrogenase, and phosphogluconate dehydrogenase activities are higher in fish fed starch than fed glucose. The results indicate that lipogenic activity can adapt to dietary carbohydrates in the fish liver.

Rainbow trout were fed diets containing 32% glucose, maltose, dextrin, or raw corn starch using different feeding strategies (117). Liver lipogenic enzyme activities are generally higher in continuously fed fish than in meal-fed fish. Trout fed glucose and maltose have higher glucose-6-phosphate dehydrogenase and 6-phosphogluconate dehydrogenase activities than those fed starch. It was concluded that continuous feeding of carbohydrate improves carbohydrate utilization and enhances fat deposition by promoting lipogenesis.

Carp fed diets supplemented with 30% starch, glucose, fructose, or galactose exhibit different metabolic responses of hepatopancreatic enzyme activity (118). Galactose causes a large growth reduction along with the amounts of hepatopancreatic enzyme activities. Starch, glucose, and fructose depress the activities of gluconeogenic and amino acid-degrading enzymes and raise the activity of phosphofructokinase. Starch and glucose raise lipogenic enzyme activities but not fructose. Salmon contain hepatic lipogenic enzymes including NADP-isocitrate dehydrogenase, malic enzyme, glucose-6-phosphate dehydrogenase, and phosphogluconate dehydrogenase. However, the activities are not influenced by absorbed dietary starch (119).

Fructose 1,6 bisphosphatase activity differs in the liver of sea bass fed different diets: standard diet > high carbohydrate diet > deficient ω-3 fatty acid diet (25). These differences suggest that gluconeogenic activity of sea bass may be influenced by diet.

Soluble NADP-isocitric dehydrogenase activity increases in the liver of carp and tilapia fed a high-carbohydrate diet or in those fed at a high feeding rate (120). The soluble form of the enzyme appears to contribute to hepatic fatty acid biogenesis by supplying NADPH.

C. Dietary Lipid

Juvenile tilapias were fed isoenergetic and isonitrogenous diets containing different amounts of corn oil/cod liver oil/lard (1:1:1) (121). Liver malic enzyme, glucose-6-phosphate dehydrogenase, and phosphogaluconate dehydrogenase activities decreased as the lipid content of the diet was increased. Body lipid composition is greater in animals fed diets with up to 20% lipid, although maximum growth occurs at 12% dietary lipid. In Atlantic salmon, absorbed fat is negatively correlated with the activities of lipogenic enzyme activities: including NADP-isocitrate dehydrogenase, malic enzyme, glucose-6-phosphate dehydrogenase, and phosphogluconate dehydrogenase (119). However, dietary absorbed fat does not influence the levels of plasma triacylglyceride nonesterified fatty acids.

Mahseer fingerlings fed increasing levels of sardine oil (3–12%) with 40% fish meal form higher levels of digestive lipase and proteinase activities (122). Dietary lipid may not only spare protein for growth but also increase feed efficiency at intermediate concentrations.

Juvenile yellowtail fed different amounts of carbohydrate (3–24%) and lipid (16–8%) in isoenergetic and isonitrogenous diets show adaptation to utilize dietary lipid and protein sparing (123). In fish fed high-lipid diets, there is an increase in the activities of hepatic glucose-6-phosphate dehydrogenase, phosphogluconate dehydrogenase, malic enzyme, glucose-6-phosphatase, and alanine aminotransferase. In effect, dietary lipid depressed hepatic lipogenesis, gluconeogenesis, and amino acid degradation, making lipid more effectively used as an energy source and protein-sparing component of the diet. Likewise, dietary supplementation of lipid in carp diets depresses lipogenesis, glycogen synthesis, gluconeogenesis, and amino acid degradation in the hepatopancreas (124).

D. Nature of Feed

1. Dry and Moist Pellets

The activity of digestive enzymes was measured in sea bream juveniles under intensive rearing (125). Moist and dry diets were tested for their effects on digestive enzyme pattern. Changes in enzyme patterns were observed in relation to growth and diet composition.

2. Protein Hydrolysates

Addition of protein hydrolysate to the larval diet of sea bass appears to stimulate development of the digestive system (126), characterized by increased trypsin activity and decreased amylase activity. Protein hydrolysate (35%) increases the activity of brush border indicator enzymes and thereby stimulates the maturation process during weaning. Sea bass larva fed diets containing di- and tripeptides derived from fish meal showed better growth and survival than those fed isonitrogenous and isoenergetic control diets with fish meal (29). The growth improvement was associated with stimulation of pancreatic chymotrypsin activity. Intestinal brush border activities corresponding to normal development of the intestine, including aminopeptidase, maltase, and γ-glutamyl transpeptidase, also increase at a greater rate in larva fed peptides.

3. Live Feed

Larval goldfish fed artemia grow better than those fed dry diet. The growth performance of larvae fed different diets was positively related to the trypsin activity and the integrity of liver hepatocyte structures (127).

Supplementation of sea bass larvae diets with digestive enzymes did not have the same positive effect on growth as artemia (128). However, the combination of artemia and digestive enzyme enhanced the assimilation and deposition of dietary nutrients in the larval body.

An increase in the growth and activity of digestive enzymes occurs in shrimp fed on live clams (129).

E. Diet and Ontogenetic Changes

The development of digestive amylase and proteolytic activities was studied in carp larvae (30). Ontogenetic development of digestive enzyme activities in the larvae occurred in 12 days for a diet group continuously fed zooplankton, but sharply declined when natural feed was withdrawn. The authors concluded that exogenous feeding of zooplankton regulates the development of digestive enzymes in carp larvae. Soy protein concentrate (SPC) was used as a major protein source (casein replacer) in carp larvae diets (130). Incorporation of SPC levels of up to 40 % in the diet does not affect growth or survival of carp larvae. However, larvae groups fed 60 and 70 % SPC causes growth retardation and significant decrease in trypsin-specific activity. Purified soybean trypsin inhibitor has a similar influence on trypsin-specific activity but does not adversely affect growth. Amylase specific activity does not differ between diet groups fed SPC replacer. It was concluded that antinutritional factors, other than soybean trypsin inhibitor, are responsible for the growth impairment of carp fed SPC.

The digestive enzyme capacity of sea bass larvae fed different diets was assessed in the pancreas and the enterocyte brush border membrane and cytosol (131). Larvae fed artemia (control group) grew better than those fed a compound diet from day 15 or day 25 until day 40. Pancreatic amylase and trypsin synthesis is similar in larvae fed artemia or compound diet. However, pancreatic secretion is lower in the group fed compound diet. A marked decrease in cytosolic leucine–alanine peptidase and concurrent increase in this enzyme in the brush border occurs similarly in both groups. It was concluded that low enzyme production does not explain the poor growth of larvae fed compound diet compared to those fed artemia.

F. Other Dietary Factors Influencing Enzymes

1. Antioxidant Vitamins

The concentration of tissue antioxidants such as vitamins A, E, and K, and carotenoids (presumably of diet origin) is correlated with the activities of the prooxidant enzyme lipoxygenase and antioxidant enzymes (superoxide dismutase, peroxidase, and glutathione reductase) in muscle and liver of several species of fish (15). Glutathione peroxidase activities in the liver and erythrocytes and hepatic superoxide dismutase activity of wild fish are significantly lower than in farmed yellowtail (132). Hepatic 2-thiobarbituric acid reactive substances are higher in farmed fish. The results suggest that wild fish are less susceptible to oxidative stress than cultured fish because they receive more dietary antioxidants (tocopherol and ascorbate). Vitamin A deficiency has also been found to compromise the integrity of skeletal muscle and skin lysosomes, causing release of acid hydrolases (133).

2. Carnitine

Atlantic salmon fed diets supplemented with L-carnitine have increased body protein and decreased body fat. It was shown that dietary carnitine increases fatty acid oxidation by increasing the flux through pyruvate carboxylase and not by decreasing flux through the branched chain α-keto acid dehydrogenase complex (134). This change would be expected to provide more carbon for amino acid synthesis, as would preventing the oxidative loss of the branched-chain amino acids. Metabolic flux through mitochondrial pyruvate carboxylase is increased 230%, although the dehydrogenase complex is unchanged by dietary carnitine.

3. Lactate

Dietary sodium lactate enhances growth of Arctic char, but not Atlantic salmon (135). The presence of lactate in the diet increases the intestinal cholytaurin hydrolase activity of char but not salmon. The authors concluded that the positive influence of lactate on char growth is indirect by stimulating a change of the intestinal microbiota to contain lactic acid-tolerant bacteria.

4. Hormones

Feeding dietary hormones to Indian major carp increases growth by improving protein and lipid digestibility, presumably by stimulating secretion of digestive enzymes (136).

G. Starvation

Starvation of fish influences the food quality of seafood in a number of ways (1, 2). A general impact of long-term feed deprivation is reduced muscle glycogen and other aspects of energy metabolism, with consequent alteration of postharvest pH decline and rigor onset. Some species, like cod, become less susceptible to gaping during seasons of the year that feed is not available.

Feed deprivation was shown to increase γ-amylase activity of various tissues of *Heteropneustes fossilis* including muscle (137). The amylase catalyzes an alternate route of glycogenolysis (see Sec. I.B.5.) Refeeding fish caused a reduction of tissue amylase activity, suggesting a shift to the phosphorylase pathway in well fed animals.

Starvation decreases the hepatic activities of glucose-6-phosphate dehydrogenase, 6-phosphogluconate dehydrogenase, and malic enzyme in rainbow trout by almost 60% (138). The Michaelis constants of these enzymes are not influenced by starvation/refeeding. It was concluded that starvation causes hepatic protein repression, while refeeding leads to induction of enzyme synthesis.

Fasting of rainbow trout has a similar effect to feeding a high-protein diet in increasing activity of hepatic enzymes involved with amino acid metabolism: glutamate dehydrogenase and alanine aminotransferase (108). The response appears to be an adaptive reaction to the availability of amino acids.

Starvation of carp and tilapia results in low amounts of soluble NADP-isocitric dehydrogenase, an enzyme associated with lipogenesis (120). On the other hand, the mitochondrial (bound) form of NAD-isocitric dehydrogenase decreases slowly during prolonged starvation. The latter appears to function in oxidative phosphorylation.

XII. CONCLUSIONS

Seafood is obtained from a broad taxonomic array of vertebrate and invertebrate organisms. In addition to the tens of thousands of taxonomic species, genetic variation through natural selection in the wild and genetic crossing by the fish farmer gives rise to even greater genetic diversity (strains and stocks). Given the considerable qualitative and quantitative variation of enzymes in fish due to adaptation, relatively few illustrations of interspecific differences can be made with certainty. On the other hand, there is abundantly more information in the literature indicating that the enzyme makeup of aquatic organisms is a function of intraspecific factors. This review is intended to illustrate this point. Intraspecific differences in the enzyme make-up of fishery products are important to the seafood technologist because they are often causally related to the food quality indices that influence consumer acceptance of seafood products. Moreover, culture conditions used by the fish farmer can have an important impact on metabolism and thus the chemical composition, quality, utility, and postharvest durability of seafood products.

REFERENCES

1. NF Haard. Control of chemical composition and food quality attributes of cultured fish. Food Res Intl 25: 289–307, 1992.
2. RM Love. Growth, metamorphosis and aging. In: The Food Fishes. Their Intrinsic Variation and Practical Implications. New York: Van Nostrand Reinhold Company, 1988, pp 25–42.
3. NF Haard. Extending prime quality of seafood. J Aquatic Food Prod Technol 1: 9–28, 1992.
4. NF Haard. Enzymes from food myosystems. J Muscle Foods 1: 293–338, 1990.

5. VA Pierce, DL Crawford. Phylogenetic analysis of glycolytic enzyme expression. Science 276: 256–259, 1997.

6. R Male, JB Lorens, AO Smalas, KR Torrissen. Molecular cloning and characterization of anionic and cationic variants of trypsin from Atlantic salmon. Eur J Biochem 232: 677–685, 1995.

7. NF Haard, R Arndt, FM Dong. Estimation of protein digestibility IV. Properties of pyloric caeca enzymes from Coho salmon fed soybean meal. Comp Biochem Physiol B 115: 533–540, 1996.

8. M Dixon, EC Webb. Enzymes. New York: Academic Press, 1979.

9. R Moreau, K Dabrowski. Body pool and synthesis of ascorbic acid in adult lamprey *Petromyzon marinus:* An agnathan fish with gulonolactone oxidase activity. Proc Natl Acad Sci USA 95: 10279–10282, 1988.

10. A Krasnov, M Reinisalo, M Nishikimi, H Molsa. Expression of rat gene for L-gulono-gamma-lactone oxidase, the key enzyme of L-ascorbic acid biosynthesis in guinea pig cells and in teleost fish rainbow trout (*Oncorhynchus mykiss*). Biochim Biophys Acta 1381: 241–248, 1998.

11. NF Haard. Biochemistry of color and color changes in seafoods. In: R Martin, R Ory, G. Flick, ed. Seafood Biochemistry: Composition and Spoilage. Lancaster. PA: Technomic Publishing Co., 1992, pp 305–361.

12. T Saito, K Arai, and D Matuyoshi. A new method for estimating the freshness of fish. Bull Jpn Soc Sci Fish 24(a):749–750, 1959.

13. S Kolayi, M Arikan, D Uzonosmanoglu, B Vanizor, E Kiran, R Sagban. Comparative studies on antioxidant enzyme activities and lipid peroxidation in different fish species. Turk J Zool 21: 171–173, 1997.

14. D Wilhelm Filho, A Boveris. Antioxdant defences in marine fish: II. Elasmobranchs. Comp Biochem Physiol C 106:415–418, 1993.

15. II Rudneeva-Titova. Ecological and phytogenetic features of some antioxidant enzyme activity and antioxidant content in Black sea cartilaginous and teleost fish. Z Evol Biokh Fiziol 33: 29–37, 1997.

16. Y Konishi. Characteristics of the lysosomal hyrolytic pathway of glycogen in striated muscle of diploid and triploid masu salmon. Nippon Suisan Gakk 57: 1147–1149, 1991.

17. F Nagayama. Mechanisms of breakdown and synthesis of glycogen in tissues of marine animals. Nippon Suisan Gakk 32: 188, 1966.

18. T Nakagawa, M Ando, Y Makinodan. Properties of glycogenolytic enzymes from fish muscle. Nippon Suisan Gakk 62: 434–438, 1996.

19. YB Lee. Postmortem biochemical changes and muscle properties in surf clam (*Spisula solidissima*). J Food Sci 43: 35–37, 1978.

20. M Matsumiya, A Mochizuki. Distribution and characteristics of new type acid cysteine protease in muscle of a few fishes. Bulletin of the College of Agriculture and Veterinary Medicine, Nihon University 50: 139–144, 1993.

21. NF Haard. Protein and non-protein constituents in jellied American plaice, *Hippoglossoides platessoides*. Can Inst Food Sci Technol J 20: 98–101, 1987.

22. JF Siebenaller. Analysis of the biochemical consequences of ontogenic vertical migration in a deep-living teleost fish. Physiol Zool 57: 598–608, 1984.

23. V Luskova. Influence of spawning on enzyme activity in the blood plasma of fish. Polsk Arch Hydrobiol 44: 57–66, 1997.

24. V Luskova, K Halacka, S Lusk. Yearly dynamics of enzyme activities and metabolite concentrations in blood plasma of *Chondrostoma nasus*. Folia Zool 44: 75–82, 1995.

25. L Garcia-Rejon, MJ Sanchez-Muros, J Cerda, M De La Higuera. Fructose 1,6 bisphosphatase activity in liver and gonads of sea bass *Dicentrarchus labrax*. Influence of diet composition and stage of the reproductive cycle. Fish Physiol Biochem 16: 93–105, 1997.

26. A Kiessling, L Larsson, K-H Kiessling, PB Lutes, T Storebakken, SSS Hung. Spawning induces a shift in energy metabolism from glucose to lipid in rainbow trout white muscle. Fish Physiol Biochem 14: 439–448, 1995.

27. D Gelinas, GA Pitoc, GV Callard. Isolation of a goldfish brain cytochrome P450 aromatase cDNA: mRNA expression during the seasonal cycle and after steroid treatment. Mol Cell Endocrinol 138: 81–93, 1998.

28. DB Josephson, RC Lindsay, DA Stuiber. Influence of maturity on the volatile aroma compounds from Pacific and Great Lakes salmon. J. Food Sci 56: 1576–1579;1585, 1991.

29. JLZ Infante, CL Cahu, A Peres. Partial substitution of di and tripeptides for native proteins in sea bass diet improves *Dicentrarchus labrax* larval development. J Nutr 127: 608–614, 1997.

30. R Chakrabarti, J Sharma. Ontogenic changes of amylases and proteolytic enzyme activity of Indian major carp, *Catla catla* Ham. in relation to natural diet. Ind J Animal Sc 67: 932–934, 1997.

31. VV Kuzmina. Influence of age on digestive enzyme activity in some freshwater teleosts. Aquaculture 148: 25–37, 1996.

32. TH Yang, GN Somero. Activity of lactate dehydrogenase but not its concentration of messenger RNA increases with body size in barred sand bass, *Paralabrax nebulifer* (Teleostei). Biol Bull 191: 155–158, 1996.

33. E Mayrand, H Guderley, J-D Dutil. Effect of morphometric maturity and size on enzyme activities and nucleic acid ratios in the snow crab *Chionocetes opilio*. J Crust Biol 18: 232–242, 1998.

34. D Pelletier, PU Blier, J-D Dutil, H Guderley. How should enzyme activities be used in fish growth studies? J Exp Biol 198: 1493–1497, 1995.

35. S Arndt, T Benfey, R Cunjak. A comparison of RNA concentrations and ornithine decarboxylase activity in Atlantic salmon *Salmo salar* muscle tissue with respect to specific growth rates and diet. Fish Physiol Biochem 13: 463–471, 1994.

36. H Gurderley, BA Lavoie, N Dubois. The interaction among age, thermal acclimation and growth rate in determining muscle metabolic capacities and tissue masses in the threespine stickleback, *Gasterosteus aculeatus*. Fish Physiol Biochem 13: 419–431, 1994.

37. RP Evans, CC Parrish, P Zhu, JA Brown, PJ Davis. Changes in the phospholipase A2 activity and lipid content during early development of Atlantic halibut *Hippoglossus hippoglossus*. Marine Biol Berl 130: 369–376, 1998.

38. H Ogata, F Aranishi, K Hara, K Osatomi, T Ishihara. Age-related changes in cata-

lase activity and its inhibition by magnesium II chloride in the brain of two species of poikilothermic vertebrates. Arch Gerontol Geriatrics 26: 119–129, 1998.

39. F Watanabe, M Goto, K Abe, Y Nakano. Glutathione peroxidase activity during storage of fish muscle. J Food Sci 61: 734–735, 1996.

40. DW Marseno, K Hori, K Mayazawa. Comparison of membrane-bound and cytosol 5′-nucleosidases from black rockfish *Sebastes inermis* muscle and their influence on freshness of fish. Fish Sci Tokyo 60: 115–121, 1994.

41. K Nilsson, B Ekstrand. The effect of storage on ice and various freezing treatments on enzyme leakage in muscle tissue of raibow trout *Oncorhynchus mykiss*. Z Lebensm-Unters Forsc 197: 3–7, 1993.

42. IV Emeretli. Effects of annual cycle temperatures on the activity of energy metabolism enzymes in Black Sea fishes. Vopr Ikhtiol 34: 395–399, 1994.

43. WL Seddon, CL Prosser. Seasonal variations in the temperature acclimation response of the channel catfish *Ictaluras punctatus*. Physiol Zool 70: 33–44, 1997.

44. VV Kuzmina, IL Golovanova, GI Izvekova. Influence of temperature and season on some characteristics of intestinal mucosa carbohydrases in six freshwater fishes. Comp Biochem Physiol B 113: 255–260, 1996.

45. MC Gomez-Guillen, I Batista. Seasonal changes and preliminary characterization of cathepsin D-like activity in sardine *Sardina pilchardus* muscle. Int J Food Sci Technol 32: 255–260, 1997.

46. G Iurovitskii. Cytochrome P450-dependent monooxygenase in fish tissues. Izv Akad Nauk Ser Biol 1: 20–29, 1998.

47. EG Da Costa, LR Curtis. Bioaccumulation of 2,2′, 4,4′,5,5′-hexachlorobiphenyl and induction of hepatic aryl hydrocarbon hydroxylase in rainbow trout *Oncorhynchus mykiss*. Environ Toxicol 14: 1711–1717, 1995.

48. G Chen, L Xu, Y Zheng, KW Schramm, A Kettrup. Influence of dioxin and metal-contaminated sediment on phase I and II biotransformation enzymes in silver crucian carp. Ecotoxicol Environ Safety 40: 234–238, 1998.

49. H Fenet, C Casellas, J Bontoux. Labotatory and field-caging studies on hepatic enzymatic activities in European eel and rainbow trout. Ecotoxicol Environ Safety 40: 137–143, 1998.

50. L Vigano, A Arillo, M Bagnasco, C Bennicelli, F Melodia. Xenobiotic metabolizing enzymes in uninduced and induced rainbow trout *Oncorhynchus mykiss*. Effects of diet and food deprivation. Comp Biochem Physiol C. Pharmacol Toxicol Endocrinol 104: 51–55, 1993.

51. Y Iger, HA Jenner, SEW Bonga. Cellular responses in the skin of the trout *Oncorhynchus mykiss* exposed to temperature elevation. J Fish Biol 44:921–935, 1994.

52. A Rodriguez-Ariza, J Peinado, C Pueyo, J Lopez-Barea. Biochemical indicators of oxidative stress in fish from polluted littoral areas. Can J Fish Aquatic Sci 50: 2568–2573, 1993.

53. JR Pedrajas, J Peinado, J Lopez-Barea. Oxidative stress in fish exposed to model xenobiotics. Oxidatively modified forms of Cu,Zn-superoxide dismutase as potential biomarkers. Chem Biol Int 98:267–282, 1995.

54. KY Chitra, and NSR Kumar. Effect of water pollution on peroxidase activity in fish, *Channa gachua*. J Environ Biol 18:191–194, 1997.

55. IL Tsvetkov, SL Zarrubin, GA Urvantseva, AS Konichev, B Filippovich Yu. Acid phosphatase of hydrobionts as an enzymatic marker of biochemical adaptation to the action of toxicants. Izvest Akad Nauk Ser Biol a Moscow 5:539–545, 1997.

56. M Ramesh, G Palanivel, K Sivakumari, MK Kanagaraj, R Manavalaramanujam. Changes in acid phosphatase activity in a freshwater teleost, *Cyprinus carpio* when treated with vegetable oil factory effluent. Uttar Pradesh J Zool 13:160–162, 1993.

57. V Karan, S Vitorovic, V Tutundzic, V Poleksic. Functional enzymes activity and gill histology of carp after copper sulfate exposure and recovery. Ecotoxicol Environ Safety 40:49–55, 1998.

58. R Kuroshima, A Kakuna, J Koyama. Effects of triphenyltin on the potential activities of trypsinogen and chymotrypsinogen of red sea bream. Nippon Suisan Gakk 63: 85–89, 1997.

59. R Mishra, SP Shukla. Impact of endosulfan on cytoplasmic and mitochondrial liver malate dehydrogenase from the freshwater catfish (*Clarius battrachus*). Comp Biochem Physiol C 117:7–18, 1997.

60. J Thayer, J Chhaya, R Mittal, AP Mansuri, R Kundu. Effects of chromium (VI) on some ion-dependent ATPases in gills, kidney and intestine of a coastal teleost *Periophthalmus dipes*. Toxicology 112: 237–244, 1996.

61. JM Grizzle, LL Lovshin. Injuries and serum enzyme activities of fingerling channel catfish *Icalurus punctatus* harvested with a turbine pump. Agric Engin 15: 349–357, 1996.

62. M Greer-Walker, GA Pull. Skeletal muscle function and sustained swimming speeds in the coalfish *Gadus virens* L. Comp Biochem Physiol A 44:495–501, 1973.

63. TP Serebrenikova, VK Shmelev, VP Nesterov. The particularities of glycolytic enzyme asctivation in muscle of fishes with a rush type swimming. Z Evolyuts Biokhim Fiziol 32:44–49, 1996.

64. M Huber, H Guderley. The effect of thermal acclimation and exercise upon the binding of glycolytic enzymes in muscle of the goldfish *Carassius auratus*. J Exp Biol 175:195–209, 1993.

65. SJ Gruber, KA Dickson. Effects of endurance training in the leopard shark, *Triakis semifasciata*. Physiol/Zoo 70:481–492, 1997.

66. VI Lushchak, KB Storey. Effect of exercise on the properties of AMP-deaminase from trout white muscle. Int J Biochem 26:1305–1312, 1994.

67. BK Simpson, NF Haard. Cold adapted enzymes from fish. In: D Knorr, ed. Food Biotechnology. New York: Marcel Dekker, Inc., 1987, pp 495–527.

68. IA Johnston, J Dunn. Temperature acclimation and metabolism in ectotherms with particular reference to teleost fish. Symp Soc Exp Biol 41:67–93, 1987.

69. C Nathanailides. Are changes in enzyme activities of fish muscle during cold acclimation significant? Can J Fish Aquatic Sci 53: 2333–2336, 1996.

70. E Aho, M Vornanen. Ca^{+2}-ATPase activity and Ca^{+2} uptake by sarcoplasmic reticulum in fish heart: effects of thermal acclimation. J Exp Biol 201:525–532, 1998.

71. MA Ciardiello, L Camardello, G Di Prisco. Enzymes of Antarctic fishes: Effect of temperature on catalysis. Cybium 21:443–450, 1997.

72. MA Ciardiello, L Camardello, G Di Prisco. Glucose-6-phosphate dehydrogenase from the blood cells of two Antarctic teleosts: Correlation with cold adaptation. Biochim Biophys Acta 1250: 76–82, 1995.

73. E Rodriques, R Rosa, M Bacila. The effect of temperature on the kinetic properties of lactate dehydrogenase from the epaxial muscle of the Antarctic fish *Notothenia gibberifrons* Lonneberg. Arq Biol Tecnol Curitiba 38:1231–1236, 1995.

74. T Shikata, S Iwanga, S Shimeno. Metabolic response to acclimation temperature in carp. Fish Sci Tokyo 61:512–516, 1995.

75. S Shimeno, T Shikata. Effects of acclimation temperature and feeding rate on carbohydrate-metabolizing enzyme activity and lipid content of common carp. Nippon Suisan Gakk 59:661–666, 1993.

76. OS Klyachko, ES Polosukhina, AV Persikova, ND Ozernyuk. Kinetic differences of lactate dehydrogenase from muscles of fish at temperature adaptation. Biofizika 40:513–517, 1995.

77. ND Ozernyuk, OS Klyachko, ES Polosukhina. Acclimation temperature affects the functional and structural properties of lactate dehydrogenase from fish *Misgurnus fossilis*. Comp Biochem Physiol B 107:141–145, 1994.

78. DM Mwangangi, G Mutungi. The effects of temperature acclimation on the oxygen consumption and enzyme activity of red and white muscle fibers isolated from the tropical freshwater fish *Oreochromis niloticus*. J Fish Biol 44: 1033–1043, 1994.

79. WL Seddon. Mechanism of temperature acclimation of the channel catfish *Ictalurus punctatus:* Isozymes and quantitative changes. Comp Biochem Physiol A 118: 813–820, 1997.

80. KJ Rodnick, BD Sidell. Cold acclimation increases carnitine palmitoyltransferase I activity in oxidative muscle of stripped bass. Am J Physiol 266: 405–412, 1994.

81. BJ Battersby, CD Moyes. Influence of acclimation temperature on mitochondrial DNA, RNA and enzymes in skeletal muscle. Am J Physiol 275:905–912, 1998.

82. OS Klyachko, ES Polosukhina, VA Klyachko, ND Ozernyuk. The effect of pH on the functional properties of M-4-lactate dehydrogenase from skeletal muscles of fish at thermal adaptations. Biofizika 42: 334–337, 1997.

83. JJ Lin, GN Somero. Temperature dependent changes in expression of thermostable and thermolabile isozymes of cytosolic malate dehydrogenase in the eurythermal goby fish *Gillichthys mirabilis*. Physiol Zool 68: 114–128, 1995.

84. G Goldspink. Adaptation of fish to different environmental temperature by qualitative and quantitative changes in gene expression. J Therm Biol 20: 167–174, 1995.

85. SA Yang, H Kawai, K Endo. Temperature dependency of 5′-nucleosidase and acid phosphatase in muscle microsomes of fish. Comp Biochem Physiol A 119:765–771, 1998.

86. SA Yang, K Endo. Acclimation temperature affects of activities of 5′-nucleosidase and acid phosphatase and lipid fatty acid composition in carp muscle microsomes. J Food Sci 59: 1009–1012, 1994.

87. M Maffia, R Acierno, G Deceglie, S Vilella, C Storelli. Adaptation of intestinal cell membrane enzymes to low temperatures in the Antarctic teleost *Pagothenia bernacchii*. J Comp Physiol B 163: 265–270, 1993.

88. M Amiza, R Owusu Aspenten. Urea and heat unfolding of cold-adapted Atlantic cod *Gadus morhua* trypsin and bovine trypsin. J Sci Food Agric 70: 1–10, 1996.

89. S Genicot, F Rentier-Delrue, D Edwards, J Vanbeeumen, C Gerday. Trypsin and

trypsinogen from an Antarctic fish: Molecular basis of cold adaptation. Biochim Biophys Acta 1298: 45–57, 1996.

90. MA Ciardiello, L Camardello, V Carratore, G di Prisco. Enzymes in Antarctic fish: glucose-6-phosphate dehydrogenase and glutamate dehydrogenase. Comp Biochem Physiol A 118: 1031–1036, 1997.

91. E Gudmundsdottir, R Spilliaret, Q Yang, CS Craik, JB Bjarnason, A Gudmunds-dottir. Isolation and characterization of two cDNAs from Atlantic cod encoding two distinct psychrophilic elastases. Comp Biochem Mol Biol 113: 795–801, 1996.

92. S Karlsen, E Hough, RL Olsen. Structure and proposed amino acid sequence of a pepsin from Atlantic cod *Gadus morhua*. Acta Crystallogr D 54: 32–46, 1998.

93. VI Lushchak, TV Bahnjukova, KB Storey. Effect of hypoxia on the activity and binding of glycolytic and associated enzymes in sea scorpion tissues. Braz J Med Biol Res 31: 1059–1067, 1998.

94. VI Lushchak, YD Smirnova, KB Storey. AMP-deaminase from sea scorpion white muscle: Properties and redistribution under hypoxia. Comp Biochem Physiol B 119: 611–618, 1998.

95. D Smirnova Yu, VI Lushchak. Solubilization of AMP-deaminase in the muscles of teleost fish sea scorpaena. Ukr Biokhim Z 68: 63–68, 1996.

96. VI Lushchak, YD Smirnova, KB Storey. Unusual AMP-deaminase solubilization from teleost fish white muscle. Biochem Mol Biol Int 43: 685–694, 1997.

97. JR Hennessey, JF Siebenaller. Pressure-adaptive differences in proteolytic inacti-vation of M_4-lactate dehydrogenase homoloques from marine fishes. J. Exp Zool 241: 9–15, 1987.

98. JF Siebenaller, and GN Somero. Biochemical adaption to the deep sea. Rev Aquatic Sci 1: 1–25, 1989.

99. C Capo, ME Stroppolo, A Galtieri, A Lania, S Costanzo, R Petruzzelli, L Cal-abrese, F Polticelli, A Desideri. Characterization of Cu, Zn superoxide dismutase from the bathophile fish, *Lampanyctus crocodilus*. Comp Biochem Physiol B 117: 403–407, 1997.

100. T-H Yang, and GN Somero. Effects of feeding and food deprivation on oxgen con-sumption, muscle protein concentration and activities of energy metabolism en-zymes in muscle and brain of shallow-living *Scorpaena guttata* and deep-living *Sebastlobus alascanus* scorpaenid fishes. J Exp Biol 181: 213–232, 1993.

101. NYS Woo, KC Chung. Tolerance of *Pomacanthus imperator* to hypoosmotic salinities: Changes in body composition and hepatic enzyme activities. J Fish Biol 47: 70–81, 1995.

102. TP Birt, JM Green. Acclimation to seawater of dwarf nonanadromous Atlantic salmon, *Salmo salar.* Can J Zool 71: 1912–1916, 1993.

103. H Nozawa, S Mamegoshi, N Seki. Partial purification and characterization of six transglutaminases from ordinary muscles of various fishes and marine inverte-brates. Comp Biochem Physiol B 118: 313–317, 1997.

104. J Squires, NF Haard, LAW Feltham. Pepsin isozymcs from Greenland cod, *Gadus ojac*. 2. Substrate specificity and kinetic properties. Can J Biochem Cell Biol 65: 210–214, 1986.

105. J Squires, NF Haard, LAW Feltham. Pepsin isozymes from Greenland cod, *Gadus ojac.* 1. Purification and physical properties. Can J Biochem. Cell Biol 65: 205–209, 1986.

106. S Shimeno, T Shikata, S Iwanaga, T Sugita. Concentrations of metabolic intermediates and their relation to phosphofructokinase activity in carp hepatopancreas. Nippon Suisan Gakk 64: 110–115, 1998.

107. M Yokoyama, M Udagawa, J-C Nakazoe. Influence of dietary protein levels on hepatic cysteine dioxygenase activity in rainbow trout. Fish Sci Tokyo 60: 229–233, 1994.

108. MJ Sanchez-Muros, L Garcia-Rejon, L Garcia-Salguero, M DeLahiguera, JA Lupianez. Long-term nutritional effects on the primary liver and kidney metabolism in rainbow trout: Adaptive response to starvation and a high protein, carbohydrate-free diet on glutamate dehydrogenase and alanine aminotransferase kinetics. Int J Biochem Cell Biol 30: 55–63, 1998.

109. S Shimeno, T Shikata, H Hosokawa, T Matsumoto, D Kheyyali. Metabolic response to feeding rates in common carp, *Cyprinus carpio.* Aquaculture 151: 371–377, 1997.

110. T Shikata, S Shimeno. Effects of feed restriction and starvation on fatty acid synthesis and oxidation of glucose and alanine in carp hepatopancreas. Fish Sci Tokyo 63: 301–303, 1997.

111. G Le Moullac, B Klein, D Sellos, A Van Wormhoudt. Adaptation of trypsin, chymotrypsin and a-amylase to casein casein level and protein source in *Penaeus vannamei* Crustacea Decapoda. J Exp Marine Biol Ecol 208: 107–125, 1997.

112. G Le Moullac, A Van Wormhoudt, A Aquacop. Adaptation of digestive enzymes to dietary protein, carbohydrate and fiber levels and influence of protein and carbohydrate quality in *Penaeus vannamei* larvae. Aquacult Living Resources 7: 203–210, 1994.

113. JM Ezquerra, FL Garcia-Carreno, NF Haard. Effects of feed diets on digestive proteases from the hepatopancreas of white shrimp (*Penaeus vannamei*). J Food Biochem 21: 401–419, 1997.

114. JM Ezquerra, FL Garcia-Carreno, G Arteaga, NF Haard. Effect of diet on aminopeptidase activities from hepatopancreas of white shrimp (*Penaeus vannamei*). J Food Biochem 23, 1999.

115. H Salem, AM Youssef, AMN El-Nakkadi, DA Jones. Proteolytic decomposition of shellfish proteins under different conditions. Alex J Agr Res 18: 61–66, 1970.

116. J-H Lin, S-Y Shiau. Hepatic enzyme adaptation to different dietary carbohydrates in juvenile tilapia *Oreochromis niloticus* x O. aureus. Fish Physiol Biochem 14: 165–170, 1995.

117. SSO Hung, T Storebakken. Carbohydrate utilization by rainbow trout is affected by feeding strategy. J Nutr 124: 223–230, 1994.

118. T Shikata, S Iwanaga, S Shimeno. Effects of dietary glucose, fructose, and galactose on hepatopancreatic enzyme activities and body composition in carp. Fisheries Sci Tokyo 60: 613–617, 1994.

119. P Arneson, A Krogdhal, I Kristiansen. Lipogenic enzyme activities in the liver of Atlantic salmon *Salmo salar* L. Comp Biochem Physiol B 105: 541–546, 1993.

120. S Shimeno, Y Saida, T Tabata. Response of hepatic NAD- and NADP-isocitric dehydrogenase activities to several dietary conditions in fishes. Nippon Suisan Gakk 62: 642–646, 1996.

121. B-S Chou, S-Y Shiau. Optimal dietary lipid level for growth of juvenile hybrid tilapia, *Oreochromis nilticus* x *Oreochromis aureus*. Aquaculture 143: 185–195, 1996.

122. MM Bazaz, P Keshavanath. Effect of feeding different levels of sardine oil on growth, muscle composition and digestive enzyme activities of *Mahseer torkhudree*. Aquaculture 115: 111–119, 1993.

123. S Shimeno, H Hosokawa, M Takeda. Metabolic responses of juvenile yellowtail to dietary carbohydrate to lipid ratios. Fisheries Sci Tokyo 62: 945–949, 1996.

124. S Shimeno, D Kheyyali, T Shikata. Metabolic responses to dietary lipid to protein ratios in common carp. Fisheries Sci Tokyo 61: 977–980, 1995.

125. G Caruso, L Genovese, S Greco. Preliminary investigations on the digestive enzymes of reared *Pagellus acarne* Risson, 1826 juveniles in relation to two different diets. Oebalia 22: 3–13, 1996.

126. CL Cahu, JL Zambonini Infante. Maturation of the pancreatic and intestinal digestive functions in sea bass, *Dicentrarchus labrax:* Effect of weaning with different protein sources. Fish Physiol Biochem 14: 431–437, 1995.

127. A Abi-Ayad, P Kestemont. Comparison of the status of goldfish *Carassius auratus* larvae fed with live, mixed and dry diet. Aquaculture 128: 163–176, 1994.

128. S Kolkovski, A Tandler, MS Izquierdo. Effects of live food and dietary digestive enzymes on the efficiency of microdiets for seabass *Dicentrachus labrax* larvae. Aquaculture 148: 313–322, 1997.

129. PD Maugle, D Deshimaru, T Katayama, KL Simpson. Effect of short-necked clam diets on shrimp growth and digestive enzyme activities. Bull Jpn. Soc Sci Fish 48: 1759–1764, 1982.

130. AM Escaffre, I Zambonino, CL Cahu, M Mambrini, P Bergot, SJ Kaushik. Nutritional value of soy protein concentrate for larvae of common carp *Cyprinus carpio* based on growth performance and digestive enzyme activities. Aquaculture 153: 63–80, 1997.

131. CL Cahu, JLZ Infante. Is the digestive capacity of marine fish larvae sufficient for compound diet feeding? Aquaculture Int 5: 151–160, 1997.

132. H Murata, T Sakai, K Yamauchi, T Ito, T Tsuda, T Yoshida, M Fukudome. In vivo lipid peroxidation levels and antioxidant activities of cultured and wild yellowtail. Fisheries Sci Tokyo 62: 64–68, 1996.

133. P Harikumar, R Kakati, UC Goswami. Vitamin A deficiency and its effects on the lysosomal enzymes of fish. Int J Vit Nutr Res 66: 97–100, 1996.

134. H Ji, TM Bradley, GC Trembly. Atlantic salmon *Salmo salar* fed L-carnitine exhibit altered intermediary metabolism and reduced tissue lipid, but no change in growth rate. J Nutr 126: 1937–1950, 1996.

135. G Gislason, RE Olsen, E Ringo. Comparative effects of dietary sodium lactate on Arctic char, *Salvelinus alpinus* L. and Atlantic salmon, *Salmo salar* L. Aquaculture Res 27: 429–435, 1996.

136. V Jayaprakas, and BS Sindhu. Effect of hormones on growth and food utilization of the Indian major carp, *Cirrhinus mrigala*. Fishery Technol 33: 21–27, 1996.

137. DK Sharma, S Sengupta. Gamma amylase activity an alternate pathway of carbo-hydrate metabolism in animals. Curr Sci Bangalore 64: 325–327, 1993.
138. JB Barroso, J Peragon, C Contreras-Jurado, L Garcia-Salguera, FJ Corpas, FJ Esteban, MA Peinado, M De La Higuera, JA Lupianez. Impact of starvation-refeeding on kinetics and protein expression of trout liver NADPH-production systems. Am J of Physiol 274: 1578–1587, 1998.

2
Nucleotide-Degrading Enzymes

Tom Gill
DalTech/Dalhousie University, Halifax, Nova Scotia, Canada

I. INTRODUCTION

A. Postmortem Changes in Fish and Shellfish

Upon death, there are pronounced changes in the appearance, texture, chemistry, and redox potential of the muscle. It is perhaps not uncommon to see the terms muscle and meat used interchangeably in the literature and although meat is derived from living muscle, there are many differences between the two. Perhaps the most notable change occurring upon death is the cessation of the heart beat, which supplies blood and oxygen to the muscle. The lack of oxygen initiates most of the deteriorative changes that ensue immediately after death.

In its struggle to supply energy to the anabolic processes carried out in the living animal, the postmortem muscle must produce adenosine-5'-triphosphate (ATP), the major source of readily available contractile energy, in the absence of oxygen. In the meantime, ATP levels are quickly depleted as the muscle enters the rigor process.

Rigor or rigor mortis is the term describing the contractile process that takes place in postmortem muscle and results in both shortening and rigidity that persists for several hours or days, depending upon a number of factors. A detailed discussion of the biochemical basis and physical aspects of rigor is beyond the scope of this chapter but is well documented in a number of review articles, including that of Hultin (1).

Instead, this chapter deals with the enzymatic catabolism of ATP and related compounds as it pertains to the ultimate eating quality of the fish. In addition, a section will be devoted to the measurement of the major nucleotide catabolites and the applicability of using individual compounds and combination of compounds as quality indicators in chilled and processed fish.

B. ATP-Related Compounds and the Factors Affecting Breakdown

Under physiological conditions, the ATP content of rested fish muscle averages between 7 and 10 μmoles/g tissue (Fig. 1).

Early studies pointed to the fact that the disappearance of ATP after death for several fish species involved the multistep process in which uric acid was the terminal catabolic product (2–5). Much of the early work on nucleotide-degrading enzymes and phosphorus compounds in fish was carried out and reviewed by Tarr (6).

In live fish, muscle contraction is powered by the release of energy resulting from the breakdown of ATP. The enzyme involved in this rapid process is calcium-activated ATPase and is controlled by the influx and efflux of Ca^{+2} into the cell liquid (sarcoplasm). When intracellular Ca^{+2} levels are $> 1 \mu M$, Ca^{+2}-activated ATPase reduces the amount of free ATP in the sarcoplasm and the muscle begins to contract due to the cyclic nature of the formation of actin/myosin noncovalent cross bridges. When the Ca^{+2} concentration of the sarcoplasm drops below about 0.5 μM, ATP can no longer be hydrolyzed because of the inactivation of Ca^{+2}-activated ATPase, and the ATP then functions as a plasticizing agent.

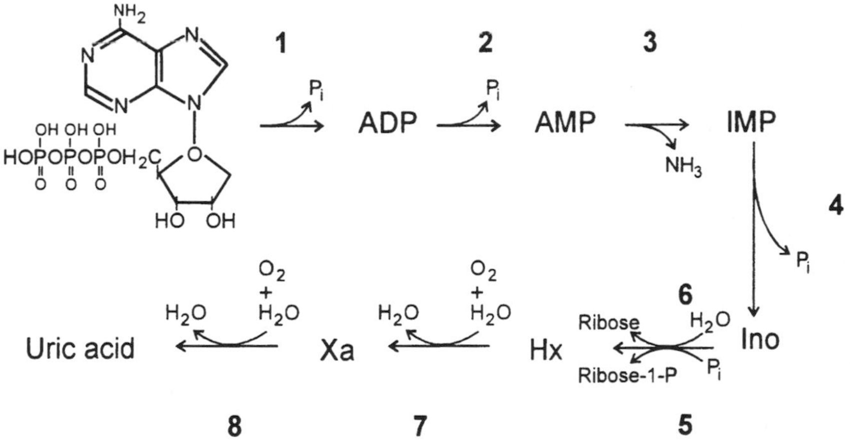

Figure 1 Postmortem ATP degradation in fish. Enzymes include: (1) ATPase; (2) Myokinase; (3) AMP deaminase; (4) 5′ nucleotidase; (5) Nucleoside phosphorylase; (6) Inosine nucleosidase; 7, 8. Xanthine oxidase. ADP, adenosine diphosphate; AMP, adenosine monophosphate; IMP, inosine monophosphate; Ino, inosine; Hx, hypoxanthine; Xa, xanthine, Pi, inorganic phosphate.

In postmortem muscle, the conversion of ATP to ADP, ADP to AMP, and AMP to IMP usually takes place within 24 h or less. These changes are thought to be totally autolytic since, in most instances, insufficient time has elapsed to allow the proliferation of spoilage microorganisms. Of course several factors can affect the rate of IMP accumulation, including temperature, species, and handling.

1. Temperature

It is perhaps logical to expect that the enzymatic breakdown of ATP and related compounds would occur at a faster rate at higher postmortem storage temperatures. This is indeed true for most fish harvested from temperate waters including salmon, herring, and Pacific cod (7) as well as Atlantic cod, to name a few. This is clearly shown in a study done by Fraser et al. (8) in which Atlantic cod was stored at three different temperatures and the degradation of ATP monitored in the muscle tissue (Fig. 2).

However, for fish harvested from tropical and subtropical waters, some examples have been found in which iced storage actually enhanced nucleotide

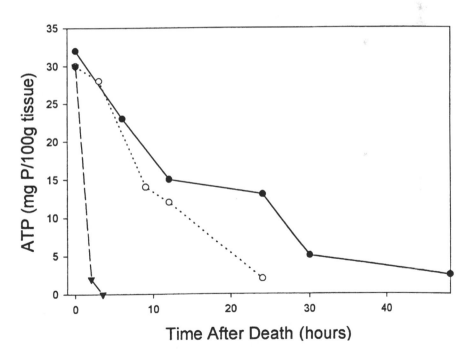

Figure 2 ATP degradation in cod muscle at different storage temperatures. ATP units are expressed as acid labile phosphorus per 100 g tissue. • ——— •, 0°C; ○ ——— ○, 9°C; ▼ — — ▼, 25°C. (Adapted from Ref. 8.)

breakdown compared to fish stored at much higher temperatures. One notable example was published by Iwamoto et al. (9), who found that for Japanese plaice, ATP degradation as well as the onset of rigor was far faster at 0°C than at 20°C. It was also interesting to note that in this case the formation of inosine and hypoxanthine as measured by the K-index (10) was quickest at 20°C compared to the other temperatures tested including 0, 5, 10 and 15°C. That is, although the onset of rigor was accelerated by chilling to 0°C, the fish stored at this temperature remained in rigor far longer than those stored at higher temperatures. Although the onset of rigor was delayed at 20°C, the breakdown of IMP to Ino and Hx was much faster as was the onset of spoilage at the higher temperatures. The authors suggested that one explanation for the accelerated ATP-ase activity at low temperatures was the release of Ca^{+2} from the mitochondria and because the ability of sarcoplasmic reticulum to absorb Ca^{+2} is inhibited at low temperatures.

Another study by Iwamoto et al. (11) on the striated adductor muscle of Japanese scallops showed that not only did the depletion of ATP increase and onset of rigor shorten, but the K-index (10) also increased more quickly for tissue stored at -3 and 0°C than it did at 5 and 10°C. The use of nucleotides as an index of quality was discouraged for this particular species.

A study on the effects of environmental temperature on the progress of rigor was presented by Abe and Okuma in 1991 (12). Rigor mortis in carp proceeded more quickly and more vigorously with increasing differences between acclimation temperatures before death and storage temperatures after death (Fig. 3). The authors also pointed out that after transferring live carp from 14 to 30°C water, it took 4 weeks of acclimation at 30°C to change the progress of rigor to a typical 30°C pattern, suggesting that protein or lipid biosynthesis is involved in the changes of ATP enzymatic breakdown resulting from changes in environmental temperature.

In addition to acclimation temperature of premortem fish and the storage temperature of postmortem fish, there are obviously species-to-species variations in the temperature dependence of nucleotide degradation. Tomioka et al. (13) compared the IMP degradation of three different species: carp, cod, and yellowtail. The dephosphorylation of IMP is carried out by 5'-nucleotidase (5'-Ntase) and was used in this study as an indicator of freshness loss since IMP degradation is often the rate-limiting step in the overall breakdown of nucleotides. This enzyme (5'-Ntase) is a membrane-bound microsomal enzyme and for the three species studied, exhibited zero-order kinetics. That is, the degradation of IMP in stored fish was linear (at least at initial stages of spoilage) with time of storage, and in all three species IMP degradation increased with storage temperature in the range -5 to +20°C.

The authors plotted the rates of formation of Ino and Hx in the flesh

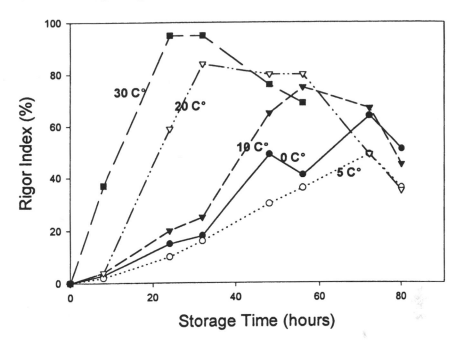

Figure 3 Progress of rigor mortis in carp muscle as affected by environmental temperature prior to slaughter. The numbers on the graph represent the differences between premortem acclimation temperature and postmortem storage temperature. (Adapted from Ref. 12.)

against 1/temperature for carp, cod, and yellowtail (Fig. 4, solid lines) and found that the rate of degradation changed abruptly at approximately 5°C for carp ($1/T = 3.6 \times 10^{-3}$) and -2°C for both cod and yellowtail. Although it is perhaps not surprising to find enzyme kinetics to be Arrhenius-like in nature, it is interesting that the temperature dependence could be represented with two nearly straight lines. Figure 4 also illustrates species differences in activation energies for the degradation of IMP as determined from the slopes of the Arrhenius plots. The plot for carp 5′-Ntase showed only one break point and the activation energies calculated to be 14 kcal/mole and 46 kcal/mole for reactions taking place above and below 5°C, respectively. The shape of the Arrhenius plot for cod was similar to the plot for carp, except that the break point in cod appeared to take place at about the freezing point for the muscle (-2°C). A similar break in linearity appeared to be present at -2°C in yellowtail muscle; the authors suggested this was due to the change of state of water in the muscle. This was substantiated by the

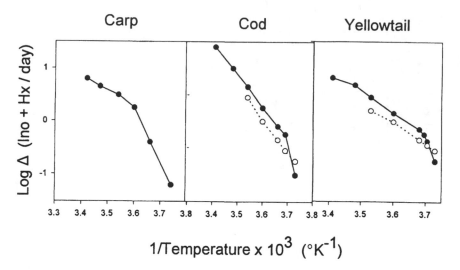

Figure 4 Arrhenius plots of nucleotide degradation (IMP → Ino + Hx) for three different fish species. Dotted lines indicate the effect of glycerin added to chopped fish as a cryoprotectant. (Adapted from Ref. 13.)

fact that perfusion of the tissue with glycerol to depress the freezing point (dotted lines, Fig. 4) resulted in unbroken Arrhenius kinetics even at temperatures below the freezing point of muscle.

The changes in slopes above the freezing point as observed for both carp and yellowtail were explained as possibly being due to temperature-dependent microsomal membrane lipid phase separation, which in turn results in a conformational change of the 5′-Ntase activity is bound to the membrane. In addition to the work reported on IMP dephosphorylation within the muscle, Tomioka et al. (13) also prepared and tested the 5′-Ntase activity in the microsomes purified from carp, cod, and yellowtail. Arrhenius plots for these samples were very much like the data plotted for intact muscle.

Although Fraser et al. (8) demonstrated that, at least for prerigor cod, the rate of ATP degradation decreased with temperature above the freezing point of muscle, further reduction of temperature can actually lead to enhanced nucleotide breakdown. This was clearly demonstrated by Nowlan and Dyer (14) and is perhaps of greater significance today because of the growth in demand by the consumer for very fresh (prerigor) fish. Temperatures in the range of -0.8–-5°C have a dramatic effect on the degradation of nucleotides in fish muscle, presumably because a considerable amount of unfrozen water remains in the muscle at these temperatures. Figure 5 was adapted from work published by Nowlan and

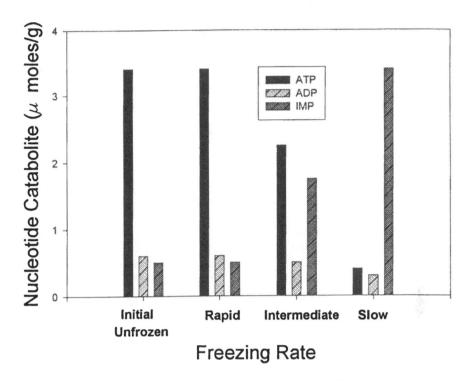

Figure 5 Effect of freezing rate on prerigor cod muscle nucleotides. Rates refer to time for the centre of a 2.8 cm thick block to pass through the critical zone (-0.8°C to -5°C). Rapid = 0.22 h; intermediate = 5.4 h; slow = 19.9 h. (Adapted from Ref. 14.)

Dyer (14) and suggests that the nucleotide pattern for rapidly frozen cod is essentially the same as that for fresh fish. Intermediate and slow freezing (as measured by the time required to pass through the so-called critical zone: –0.8 to -5.0°C) result in far greater ATP breakdown than that occurring during rapid freezing.

For scallops, it is common commercial practice to quick freeze. In many instances although the scallops are quickly frozen shortly after death, retail distribution often involves temperatures within the range of -8 C to 0°C. Hiltz, and colleagues (15) found that partial freezing in this temperature range could lead to accelerated rates of nucleotide degradation. In this species, AMP is the major accumulated catabolite in the initial stages of spoilage. ATP was rapidly (85%) broken down between -8 and -6°C in about 2 h, resulting in the accumulation of AMP below the freezing point of scallop muscle (-1.4°C).

More recently, Iwamoto et al. (11) found similar results while studying nucleotide degradation in scallop muscle. Above the freezing point, nucleotide degradation breakdown increased with storage temperature of the adductor muscle. At and below freezing, (0 and -3°C) the degradation of ATP related compounds was greater than at 5 or 10°C. In this instance, however, the sensory evaluation indicated that the 10°C storage treatment yielded poorest quality. Nucleotide degradation cannot therefore be used as a universal indicator of seafood quality.

2. Species

The enzymatic degradation of ATP and related compounds has been used for seafood quality evaluation since at least 1959. However, one of the most compelling reasons for not using the change in ATP catabolites as a universal seafood quality indicator is the species-to-species differences in degradation patterns.

Dingle and Hines (16) compared the rates of breakdown of IMP and Ino in Atlantic cod, haddock, pollock, halibut, plaice, and winter flounder stored on ice. Haddock and winter flounder exhibited a very slow rate of IMP dephosphorylation (low 5'-Ntase activity) compared to the other four species. The data presented in this study clearly indicate the problem in making generalizations using nucleotide degradation as an indicator of quality without first determining the species-specific patterns.

In most species, with a few exceptions, the degradation of either IMP or Ino is a rate-limiting step during chilled storage. One such exception is squid (*Illex argentinus*)(17), in which the degradation of AMP is clearly the rate-limiting step. In this particular species it took nearly 12 days for AMP to disappear completely from the mantle muscle. Also important was the fact that the rates of nucleotide degradation changed in squid with degree of sexual maturation. The data on Argentine squid were in sharp contrast to those published for Japanese common squid, *Todarodes pacificus*, (18). In this study, Ino and Hx were produced quickly upon death and together represented nearly 50% of the nucleotide pool after 7 days of storage at 0°C.

Short-finned squid (*Illex illecebrosus*) harvested in the north Atlantic were different from both Argentine and Japanese squid. AMP was accumulated after death and was converted almost directly into Hx without accumulation of much IMP or Ino as intermediate steps (19).

Much of the species-related work on nucleotide degradation has been published on vertebrate fish, but a limited amount of work has been published on crustaceans and molluscs. Hiltz and Dyer (20) found that sea scallops (*Placopecten magellanicus*) dephosphorylated much more slowly than most vertebrate postmortem muscle. It took 2–9 days for complete disappearance of ATP compared to finfish, which normally degrade ATP within the first 24 h after death. In scallops harvested from the North Atlantic, AMP was the major nucleotide catabolite that accumulated to a maximum level by the third day postmortem at 0°C. To date, the detailed analysis of nucleotide catabolism in scallop has not been thoroughly examined. The major route for conversion of AMP to Ino via adenosine rather than IMP has been proposed but never confirmed (21).

The nucleotide degradation patterns vary not only among related species of shellfish, such as the Japanese oyster, hard shell clam, the ark-shell, and abalone (22), but also among different tissues such as the mantle, gill, body, and adductor muscle of the oyster *Crassostrea gigas* reported by Yokoyama et al. (23). The postmortem degradation of lobster ATP, to form ADP, AMP, IMP, Ino, and Hx, was reported by Dingle et al. (24) and later confirmed by Gill and Harrison (25). However, it has been the experience of the Canadian Institute of Fisheries Technology (25) that there is a high degree of individual variability from one lobster to another with regard to the rate of nucleotide degradation. This variability was probably due to premortem stress differences and general physiological conditions as a result of live holding of lobster prior to death.

Nucleotide degradation patterns in kuruma prawns were studied (26) at different temperatures. Although ATP was found to degrade more quickly at -1°C than at 0 and 5°C, the shelf life was greater and degradation of all other nucleotide catabolites was retarded at -1°C.

3. Handling

A great deal has been published on the effects of handling fish including the various stages of processing fresh fish, frozen fish, and canned fish. In addition, a great deal of attention is currently given to the handling of aquaculture fish. In most instances, fish farmers have an important advantage over traditional fishermen since they have the opportunity to control every stage of the processing and handling including the harvest conditions. Virtually all steps in the processing and handling of fish, whether farmed or caught in the wild, have an effect on nucleotide degradation, perhaps with the exception of quick freezing.

Early studies recognized the fact that ATP was the most predominant nucleotide derivative found in the muscle of living fish. Jones and Murray (27) showed that in trawled codling ATP was nearly completely converted to IMP prior to landing due to struggle in the net and probably also because of the enormous pressure placed upon the flesh during the trawling operation. Of course, the duration of the trawl as well as the size of the catch both affect the damage done to the catch during harvest. Fraser et al. (28) found that there was little or no ATP or glycogen remaining in trawled cod regardless of subsequent handling on board or in the processing plant. They also found that there was little difference among all of the fish landed by otter trawl, confirming that struggle resulted in exhaustion of the energy reserves in commercially handled fish in comparison to fish that were dip-netted live from cod traps in Newfoundland. The authors pointed out, however, that the processing of very fresh fish with significant levels of ATP in the muscle would no doubt lead to the formation of thaw rigor if frozen immediately upon landing. This condition of thaw rigor is due to the effects of ice crystal formation in prerigor fish, causing the rupture of the sarcoplasmic reticulum membranes and resulting in the rapid and complete release of Ca^{+2} ions into the sarcoplasm, causing rapid shortening and contortion of the muscle upon thawing (1).

Stress of capture and transportation of live farmed fish causes rapid loss of ATP, rapid onset of rigor, and concomitant disruption and softening of the edible flesh (29–32). In the latter study, farmed salmon were stressed by being placed into small seawater tanks for 10 min before CO_2 stunning. Control fish were handled the same as stressed fish except they were placed directly in CO_2 stun tanks prior to slaughter. The stressed group underwent vigorous muscular activity for 10 min prior to slaughter. This significantly accelerated the onset of rigor and loss of ATP (producing IMP). This phenomenon has perhaps a parallel in the pork-processing industry, caused by the rapid onset of rigor and unusually rapid drop in postmortem pH. Pale, soft, exudative (PSE) pork can result from premortem stress in pigs and results in significant economic loss in the meat industry (1). Softening in teleost fish is due to weakening of the myocomomata or connective tissue that separates the myotomes or flakes. The rapid onset of isometric tension generated in stressed fish cause the muscle to pull itself apart. The onset of rigor shortening along with protein denaturation associated with the rapid generation of lactic acid both contribute to the softening and gaping observed in stressed fish.

A number of authors have found that immediate destruction of the brain upon capture reduces the negative impacts of stress and in many instances brain destruction ("iki jimi" in Japanese) delays the onset of spoilage. In some studies, rested harvest coupled with iki jimi resulted in a reduction in the isometric tension developed within the muscle during rigor (29, 33, 34) whereas in other instances, no differences in maximum rigor tension between stressed and nonstressed fish could be detected (35). These discrepancies are perhaps due to species-to-species differences or to the differences among the various methods of measurement of rigor tension. Table 1 illustrates a number of studies performed on a variety of fish species in which rested harvest and immediate destruction of the brain have had beneficial effects on quality as related to the retardation of the enzymatic degradation of ATP and its associated delay of the rigor process. It is believed that the rapid deterioration in stressed fish is under hormonal control; stressed fish have far higher blood cortisol levels and lower ATP levels than their unstressed counterparts (31).

Bleeding fish subsequent to harvest is also important for a number of reasons. The most obvious reason for bleeding fish at harvest is to reduce the incidence of blood spots or bruises in the flesh. However, as Chiba et al. (37) pointed out, bleeding significantly retards the degradation of ATP and related compounds. This effect appears to be more pronounced if the fish is properly chilled after death.

Table 1 Examples of Studies Claiming Beneficial Effects of "Rested Harvest" and
Destruction of the Brain

Species studied	Beneficial effects	References
Tilapia	Retards ATP depletion Retards rigor onset Retards drop in postmortem pH	36
Japanese loach	Retards depletion of ATP and creatine phosphate Retards drop in postmortem pH	37
Carp	Retards ATP depletion Retards rigor onset Reduces magnitude of muscle shortening Retards ATP depletion	29
Snapper	Retards ATP depletion Lowers K-value Retards rigor onset Reduces blood cortisol levels	38
Horse mackerel	Retards ATP depletion Retards creatine phosphate depletion Retards rate of lactic acid production Increases muscle break strength Retards rigor onset and resolution	33
Chub mackerel and round scad	Retards ATP depletion Retards creatine phosphate depletion Retards rigor onset and resolution Increases muscle break strength	34

Handling during the filleting of fish or crushing of the tissue with ice or other fish can accelerate nucleotide degradation in fresh fish. Surette et al. (39) showed that for cod, manipulation of the flesh during the filleting operation advanced the rate of inosine production by at least 4 days at 3°C even when the fillets were removed aseptically. It has been known for some time that physical handling procedures such as filleting can result in a dramatic increase in bacterial numbers and therefore promote the rate of spoilage. However, physical handling has a detrimental effect on autolytic spoilage as measured by nucleotide catabolism. This is perhaps not surprising since most of the enzymes involved in ATP catabolism are believed to be membrane bound or tightly bound to structural proteins in the living fish. Physical handling, therefore, in postmortem fish could lead to the release of bound enzymes and a concomitant increase in nucleotide degradation.

Although there is little evidence that low-temperature processing (at least $\leq -20°C$) has an effect on nucleotide decomposition in fish flesh, there is evidence that nucleotide degradation products such as hypoxanthine and inosine actually promote protein denaturation in frozen stored fish (40). These authors found that ADP, AMP, and IMP actually exerted a protective effect on denaturation in frozen stored model systems containing milkfish actomyosin.

Thermal processing of fish, such as is commonly carried out in the canning process, usually results in significant degradation of ATP-related compounds. Gill et al. (41) spiked tuna with authentic standards prior to processing. The nucleotide recoveries after processing at 122°C for 67 min were on average 50%, 75%, 64%, and 92% for AMP, IMP, Ino, and Hx, respectively. The significance of these data is that the individual rates of thermal destruction are different among the various catabolites. The use of nucleotides for the evaluation of canned fish quality has been proposed for sardines (42), tuna (43), and herring (44). With the exception of Hx, all of the ATP-related compounds are thermally labile, thus making it difficult or impossible to use nucleotide levels in canned products to reflect the quality of raw material prior to processing.

II. ENZYMES INVOLVED IN THE POSTMORTEM DEGRADATION OF ATP AND ITS CATABOLITES

Figure 1 illustrates the ATP breakdown products most commonly formed in fish after death. The individual enzymes are also indicated.

A. ATPase

ATPase activity is controlled within the muscle through the modulation of calcium. The most important form of ATPase in the muscle is actually a structural component of the major contractile protein, myosin. It is part of the globular head of the myosin molecule (called heavy meromyosin) and is only active when the sarcoplasm contains sufficiently high levels of soluble calcium. In the living animal, Ca^{+2}-activated ATPase is responsible for the breakdown of ATP with the release of energy used for in the contractile process. For more information about myosin the reader is referred to Chapter 3.

B. AMP Deaminase (EC 3.5.4.6)/AMP Aminohydrolase and Myokinase

In 1957, Jones and Murray (27) discovered that IMP levels in postmortem cod muscle first increased and then subsequently decreased during iced storage. IMP has been recognized for many years as having flavor-enhancing properties. The production of ammonia as a spoilage compound also gives special significance to AMP deaminase. AMP deaminase was first reported in mammals during the 1920s (45) but much of the work on this enzyme in the skeletal muscle of fish was reported during the 1960s.

Dingle and Hines (46) studied the AMP deaminase in both pre- and postrigor cod muscle. Although found as a water-soluble sarcoplasmic protein in prerigor muscle, it was thought to be somehow associated with myosin under certain conditions. In postrigor muscle, the enzyme became tightly bound to the myofibrillar proteins. Apparently, the enzyme has a high degree of specificity for AMP but was allosterically activated by ATP and K^+ ions, at least in the thornback ray (47). The enzyme is inhibited by inorganic phosphate and the inhibition was also dependent upon the amount of 5'-AMP present during the reaction. 3'-AMP was found to be a competitive inhibitor.

The optimum in vitro pH for catalysis ranges from pH 6.6 to 7.0 and the Michaelis-Menten constant (Km) was found to be 1.4 to 1.6 $\times$ 10^{-3} moles/L in cod (46).

In the living fish, this enzyme is extremely important for the regulation of energy production via the breakdown of ATP. In rapidly contracting muscle during intense swimming activity, ADP builds up as a result of the breakdown of ATP through the action of ATPase. Here, myokinase acts to remove ADP, which is a potential inhibitor of ATPase. The myokinase reaction results in the production of AMP during muscular activity.

$$2\,ADP \leftrightarrow ATP + AMP$$

The AMP deaminase reaction is almost nonexistent in the resting muscle due to the relative absence of AMP. The continuous removal of AMP keeps the myokinase reaction moving in a forward direction.

The end products for the AMP deaminase reaction are IMP and ammonia (NH_3). The IMP may be converted to Ino and then Hx, whereas the ammonia is transported to the gills through the blood. Excretion of the potentially toxic NH_3

is carried out mainly in the gills with minor amounts being excreted in the urine. In marine elasmobranch fish such as sharks, which manufacture large quantities of urea for osmotic regulation, ammonia enters the urea cycle.

For a complete description of the nitrogen metabolism of fish, readers are refered to work by Watts and Watts (48).

C. 5′ Nucleotidase (EC 3.1.3.5)

There are in fact three possible enzymes responsible for the conversion of IMP to Ino in fish muscle: 5′ nucleotidase, alkaline phosphatase, and acid phosphatase. Of these three, 5′-Ntase has received the most attention and is probably the most important of the three in chilled postmortem fish. Perhaps the most remarkable feature of the 5′-Ntase reaction:

$$IMP \leftrightarrow Ino + Pi$$

is that its rate is highly variable among different fish species. This was clearly demonstrated by Dingle and Hines (16), who showed, for example, that both haddock and halibut had little 5′-Ntase whereas Atlantic cod, pollack, American plaice, and winter flounder lost most of the IMP within 4 days of postmortem storage at 0° C.

The rate of IMP dephosphorylation is highly dependent upon temperature. The 5′-Ntase enzyme is membrane bound in some species but is also found in the cytosol (sarcoplasmic fluid) in other fish. In some cases, both bound and free 5′-Ntases are found in the same fish but the membrane-bound type appears to be most important. Table 2 illustrates the wide variability in 5′-Ntase properties among different species. However, the following generalizations can be made:

1. 5′-Ntase has greatest affinity (lowest Km) and highest specificity (Vmax/Km) for 5′-AMP rather than 5′-IMP. In addition to 5′-IMP and 5′-AMP, the enzyme dephosphorylates 5′-UMP and 5′-CMP but does not react with the 2′ or 3′ phosphates.
2. Both soluble and insoluble 5′-Ntase enzymes are inhibited by ADP, ATP, and high concentrations of phosphocreatine.
3. 5′-Ntase enzyme levels have a great influence on fish freshness as defined by the K_i value defined by Karube et al. (49) where:

$$K_i = \frac{[Ino] + [Hx]}{[IMP] + [Ino] + [Hx]} \times 100$$

This relationship between freshness and 5′-Ntase activity was demonstrated for a variety of species by Tomioka and Endo (50).
4. It would appear that most of the 5′-Ntase enzymes studied in fish to date have been activated by a variety of divalent cations and inhibited

by the presence of EDTA, which acts as a chelator (51). The phenolic antioxidant BHA was also found inhibitory at the millimolar concentration range (52) but BHT, another phenolic antioxidant, was not inhibitory.

5. Most of the 5′-Ntase enzymes isolated from fish tissue to date have been glycoproteins with two or more subunits that may or may not be identical.

D.   Nucleotide Phosphorylase (EC 2.4.2.1) and Inosine Nucleosidase (EC 3.2.2.2)

The degradation of Ino to Hx is a critical step in the overall breakdown of ATP and its catabolites. This is because the formation of Hx is often an indication of the last stage in edible fish quality. Tarr (59, 60) suggested that this reaction was catalyzed by two enzymes, nucleoside phosphorylase (NP) and nucleoside hydrolase (sometimes referred to as inosine nucleosidase, IN, Fig. 1). Tarr reported that both enzymes were found in Pacific lingcod, but this is apparently not the case in all fish species. LeBlanc (61) was able to demonstrate the presence of only minute quantities of NP or IN in the muscle tissues of freshly killed Atlantic cod (*Gadus morhua*). In fact, the levels of IN were more than 1000-fold less than the levels previously reported for Pacific lingcod by Tarr (59).

LeBlanc (61) found that although levels of IN remained low for whole gutted cod stored on ice over a 16 day storage study, the NP activity rose from nondetectable levels on day 0 to progressively higher levels on days 1 and 3, remained constant until day 11, and then began to drop by day 15 (Fig. 6), at which time much of the Ino had been degraded. These results suggested that, at least for very fresh Atlantic cod, the degradation of Ino to Hx was negligible at least as a result of autolysis by IN.

In the same study, LeBlanc stored fresh cod fillets in order to promote spoilage and found not only that the activities for both enzymes were higher upon storage, but also that the marked increase in enzyme activities for NP preceded those for IN by approximately 2 days (Fig. 7). Subsequently, a mixed bacterial culture was isolated from spoiling cod and was found to have substantial IN and NP activity.

The work by LeBlanc (61) and subsequent work reported by Surette (39) established the following, at least for Atlantic cod:

1. Most of the degradation of Ino to Hx is due to the presence of both autolytic and bacterial IN and NP.
2. The presence of spoilage bacteria greatly enhances the production of Hx from Ino, although the reaction does occur in sterile muscle (39).

Table 2 Properties of 5′-Nucleotidase Enzymes in Different Fish Species

Fish	Membrane-bound-B-soluble-S	Km ($\times 10^3$ M) for IMP	Molecular weight ($\times 10^{-5}$)	Subunit structure	Inhibitors (I)/activators (A)	Optimal pH	References
Carp	B	0.093	2.40		ADP-I ATP-I EDTA-I	7.5	51, 53
Black rockfish	S	62	2.42	Heterodimer 148,000 94,000	PMSF-I DFP-I BHA-I EDTA-I 2-ME-I DTT-I Glut-I AA-I	8.1	54
Black rockfish	B	61	2.65	Homotetramer 67,000 $\times$ 4	PMSF-I DFP-I EDTA-I BHA-I ADP-I ATP-I Taurine-A Glutamate-A 2-ME-A AA-A Glutathione-A 20 mM $MgCl_2$-A	8.3	52

Species					Effectors	pH	Ref.
Snapper	B	0.03	3.6	Homotetramer 89,000 × 4	ADP-I; ATP-I; 1 mM Cu^{+2}-I; 1 mM Zn^{+2}-I; EDTA - I; 1 mM Mn^{+2}-A; 1 m M Co^{+2}-A	8.5	55
Pacific cod	S	800			Mn^{+2}, 0.5 M - A; EDTA - I; pyrophosphate - I; KF - I; Zn^{+2} - I	7.6	56
Bonito	S	108	1.15	-	Mg^{+2}-A; Fe^{+2}-A; Ca^{+2}-I; Mn^{+2}-I; Fe^{+3}-I; Zn^{+2}-I; Ba^{+2}-I	5.5	57, 58

EDTA, ethylene diamine tetraacetic acid; PMSF, phenylmethyl sulfonyl fluoride; DFP, diisopropyl fluorophosphate; BHA, butylated hydroxy anisole; 2-ME, 2-mercaptoethanol; DTT, dithiothreitol; AA, ascorbic acid; Glut, glutathione.

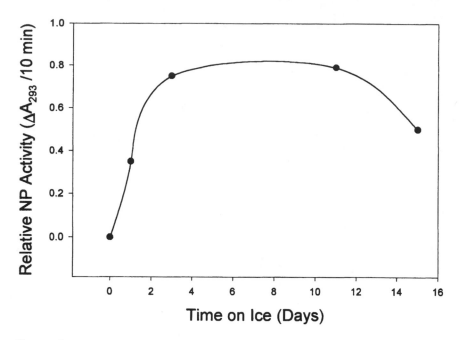

Figure 6   Nucleoside phosphorylase activity in iced, gutted Atlantic cod. (Adapted from Ref. 61.)

3.   NP activity is more important in early stages of spoilage, whereas IN is more important in later stages of spoilage.
4.   The bacterial NP was isolated from cell free extracts of *Proteus* but could only be demonstrated if bacteria were first ultrasonically disrupted (62).
5.   Although spoilage bacteria produce IN, this contribution to the purine degradation process in chilled Atlantic cod appeared to be far less than that from NP. Extensive dialysis of the Ino-degrading cell-free bacterial extracts to remove phosphate (a requirement for NP activity but not IN activity) completely eliminated the ability to degrade Ino (63).

NP has been isolated from a variety of sources including bacteria (63–67), erythrocytes (68–70), fish (60) and chicken (71). All carry out the reaction:

Ino + Pi ↔ Hx + Ribose – 1 – phosphate

Most have been found to exhibit maximum activity between pH 6.5 and 8.0. The NP purified from cod spoilage bacteria by Surette et al. (63) has an apparent mol-

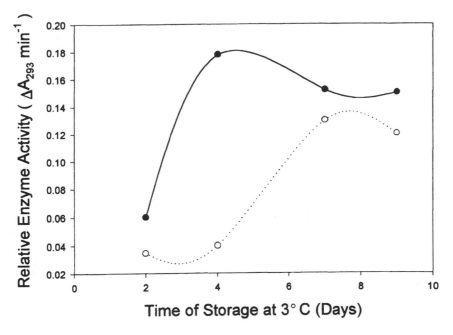

Figure 7 Nucleoside phosphorylase (solid line) and inosine nucleosidase (dotted line) activities in Atlantic cod fillets stored at 3°C. (Adapted from Ref. 61.)

ecular weight of 120,000 Da, a Km for Ino of 3.9×10^{-5}, and an isoelectric point of 6.8. It showed broad specificity, also catalyzing the breakdown of guanosine, and is competitively inhibited by adenosine. It is interesting to note that the autolytic NP activity isolated from Pacific lingcod (60) and that isolated from *P. vulgaris* cultured from spoiling Atlantic cod (63), have similar affinities for Ino, with Km values of 3.2×10^{-5} and 3.9×10^{-5}, respectively.

IN catalyzes the reaction:

$$\text{Ino} + \text{H}_2\text{O} \rightarrow \text{Hx} + \text{D-ribose}$$

and has been reported in a variety of sources including bacteria, fungi, protozoa, yeast, plants, and fish. The IN found in Pacific ling cod (*Ophiodon elongatus*) has an optimum pH of 5.5 and has broad specificity for guanosine, adenosine, xanthosine, and cytidine (59). Like NP, its importance in the autolytic degradation of Ino in fish is probably dependent on species. It is interesting to note that abundant activity was found in the muscle of Pacific lingcod but only small amounts in Atlantic cod.

E. Xanthine Oxidase (EC 1.1.3.22)

Xanthine oxidase (XO) has been found in a variety of species and catalyzes the final step in the nucleotide breakdown:

$$Hx + H_2O + O_2 \rightarrow Xa + H_2O_2$$
$$Xa + H_2O + O_2 \rightarrow \text{Uric acid} + H_2O_2$$

Hypoxanthine is not only an objective indication of spoilage in chilled fish but also contributes the off flavors typical of spoiled fish. Hx is itself bitter and contributes bitterness, but only under certain circumstances (44). The taste threshold for Hx is much lower in aqueous solutions than when Hx is present in fish products. Hx can only be detected when the population of spoilage bacteria exceed 10^6 colony-forming units (cfu)/g tissue. It is believed that fresh fish contain substances that mask the bitter off-flavor imparted by Hx, whereas bacterial spoilage either removes such masking agents or produces compounds that enhance the Hx flavor.

From a practical perspective, the activity of bacterial xanthine oxidase is perhaps less important than the other enzymes in the catabolic breakdown of nucleotides, since Hx often begins to appear only after the fish quality has been judged unacceptable by trained panelists. Nevertheless, xanthine oxidase is important for other reasons, one of which is the use of the enzyme in immobilized form as part of a biosensor.

Xanthine oxidase (XO) is also called the type O form of xanthine oxidoreductase, whereas xanthine dehydrogenase (XD) is called the type D form. The former is derived from the latter in nature. That is, proteolytic cleavage of xanthine dehydrogenase by trypsin-like enzymes results in the conversion of XD with a molecular weight of 140,000 Da into XO with a molecular weight of approximately 91,000 Da (72).

XD has been isolated from a number of sources (mostly mammalian) but little has been published on the enzyme causing or associated with spoiling fish. Like NP and IN, it is probably present in both the muscle of fish and in the bacteria that cause fish spoilage. In mammals, its physiological importance is still under study. Apart from the metabolism of purines, XO has some toxicological relevance because of its ability to generate superoxide anions that can lead to certain pathological conditions (73). The isoelectric point and Km for XO extracted from mouse liver are 6.7 and 3.4 μM, respectively.

XD is similar to XO in that it is responsible for the oxidation of Hx to xanthine, but does not use molecular oxygen as does the XO reaction. Instead, XD transfers electrons from Hx to NAD$^+$:

$$Hx + NAD^+ \rightarrow Xa + NADH + H^+$$

thus no hydrogen peroxide is produced.

Hultin (74) has suggested that the disappearance of ATP and ADP from postmortem fish muscle with a concomitant increase in XO activity contributes to the production of hydroxyl radicals ($\cdot$OH), through the generation of superoxide and hydrogen peroxide in the presence of reduced iron Fe^{+2}.

$$Hx + H_2O + O_2 \rightarrow Xa + O^-_2 + H_2O_2$$
$$Fe^{+2} + H_2O_2 \rightarrow Fe^{+3} + \cdot OH + OH^-$$

The formation of hydroxyl radicals, which are extremely reactive, leads to postmortem oxidation of lipids and destruction of membranes. The oxygen for the first reaction can arise from filleting, slicing, or mincing the fish tissue; the reduced iron increases in stored fish muscle as it is gradually released from proteins such as myoglobin, ferritin, and membrane lipids (75).

Not only does the disappearance of ATP and ADP lead to the production of Hx (which can in turn lead to lipid oxidation) but the anaerobic conditions present in the postmortem muscle promote the production of XO from its precursor, XD.

In summary, because of the presence of XO, the reintroduction of oxygen to postmortem fish muscle can activate the production of H_2O_2 and $\cdot$OH which produce oxidized flavors arising from the lipids. These reactions are best controlled through elimination of oxygen and either freezing or consumption of the fish before significant levels of Hx are formed, since the disappearance of ATP and ADP during the spoilage process, promotes the enzymatic formation of pro-oxidants.

III. USE OF NUCLEOTIDE CATABOLITES AND ENZYMES TO MEASURE SEAFOOD QUALITY

A. The K index

Figure 8 illustrates the postmortem changes in the ATP-related catabolites in chilled Atlantic cod muscle. It is a chain reaction in which metabolites typically rise and then fall as the next catabolite in the degradative chain begins to rise. Saito et al. (10) were the first to report that since the degradation of ATP and its catabolites occurred in a more or less predictable pattern, a formula was developed for fish freshness based upon the changes in the nucleotides. The K-value or freshness index

$$K\% = \frac{[Ino] + [Hx]}{[ATP] + [ADP] + [AMP] + [IMP] + [Ino] + [Hx]} \times 100$$

represents the relative degree of decomposition as measured by conversion to the end-products of the spoilage process, Ino and Hx. The values for each of these compounds in the formula were expressed as molar concentrations and determined by the use of ion exchange chromatography on Amberlite IRA-400 to separate Ino and Hx from the rest of the ATP-derived catabolites. Concentrations of the compounds

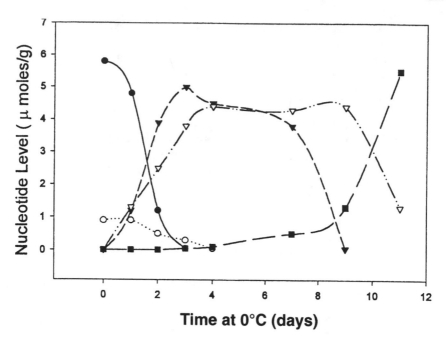

Figure 8 Nucleotide degradation pattern of relaxed cod muscle at 0°C. Onset of rigor occurred at 3 h. • —— •, ATP; ○ — — ○ , ADP; ▼- - - ▼, IMP; ▽ — - - ▽ , Ino; ■ — — ■, Hx. (Adapted from Ref. 76.)

in the numerator were determined after fractionation using absorbance readings at 250 nm and compared with readings of the unfractionated extracts.

The higher the K-value, the poorer the quality of the fish. For many pelagic species, the K-value correlates well with the degree of spoilage as judged by trained panelists. However, as indicated in Section I, the degradation patterns as seen in Figure 8 vary a great deal from one species to another. Also, as explained previously, K-values may be very different even within the same fish species, depending upon the differences between postharvest and acclimation temperature and postmortem storage temperature.

Nevertheless, K-value has given good information on fish quality, provided that the data from one species are not compared with data from another. One advantage to using nucleotide catabolites or their molar ratios to measure quality is that they measure both bacterial spoilage and autolytic degradation in an integrated manner over time.

The K_i value was proposed by Karube et al. (49) as an approximation of the K = freshness index, where:

$$K_i = \frac{[Ino] + [Hx]}{[IMP] + [Ino] + [Hx]} \times 100$$

The K_i value often approximates the K index for commercial (wild stock) fish landed by traditional methods since, in most instances, the amounts of ATP, ADP, and AMP in the muscle are negligible compared to IMP, Ino, and Hx. For practical purposes, the K_i value is simpler to obtain since only three analytes are required.

B. Analytical Methods

There have been many different approaches to the measurement of nucleotide catabolites as an objective measurement of fish freshness. These methods have been reviewed previously by Gill (77), Gill (78) and Botta (80), and Gill (79), to name a few. The methods can be broadly categorized as either enzymatic or nonenzymatic. The nonenzymatic methods include ion exchange, thin-layer and high-performance chromatographies (HPLC), the latter being the most common analytical method at the time of writing. A review of these and other nonenzymatic methods of nucleotide analysis is beyond the scope of this chapter.

However, a variety of enzymatic methods for the analysis of nucleotide catabolites in fish have been reported. These include test paper strips that change color in the presence of certain nucleotides. The first such strip was introduced by Jahns et al. (81) for the determination of Hx with the use of immobilized xanthine oxidase and an indicator dye. The most sophisticated strips have the ability of detecting more than one analyte and in some cases can be used for the calculation of Ki value.

Still others have used conventional enzyme technologies in which the analytes have been degraded to uric acid and hydrogen peroxide through a series of steps carried out in the presence of soluble enzymes such as 5'Ntase, NP and XO.

The third broad category of enzymatic analysis of nucleotides is biosensor technology, in which one or more enzymes are immobilized on a semipermeable membrane in the vicinity of a Clark-type oxygen electrode (82). Upon the introduction of the analyte (usually found in a soluble fish extract), the reaction converting Hx to uric acid results in the generation of oxygen, which is quantified using the calibrated electrode. Such biosensors have become popular topics in the literature during the past 10–15 years. Some of these technologies have become commercially available while others have not been nearly so successful.

Table 3 summarizes much of the work presented in the literature over the years on the enzymatic determination of fish quality using the nucleotide catabolites as objective indicators. Most of the assay procedures are specific and easy to perform. However, perhaps one of the disadvantages of these (in comparison

Table 3 Enzymatic Determination of Nucleotide Catabolites in Fish

Method type	Analyte(s)	Enzymes	Detection method	References
Test paper strips	Hx	XO	Color (resazurin)	81
Freshness testing paper	Hx, Ino, Imp, Ki value	NP, XO, 5'Ntase	Color	83
Test paper strips	IMP	IMP dehydrogenase, diaphorase	Color (tetrazolium salt)	84
Chromogenic dip stick	Ki value	5'Ntase, NP, XO	Color	85
Simple enzyme assay	Hx	XO	A_{290}	86
Simple enzyme assay	Ino, Hx	NP, XO	Color (2,6 dichlorophenol indophenol)	87
Simple enzyme assay	Ino, Hx	NP, XO	A_{293}	88
Simple enzyme assay	AMP, IMP, Ino, Hx, K index	Adenosine deaminase, Alkaline phosphatase, NP, XO, peroxidase	Color; H_2O_2 determined using peroxidase, phenol and 4-aminoantipyrine	89
Simple enzyme assay	K index	Adenosine deaminase, Alkaline phosphatase, NP, XO	A_{290}	90
Simple enzyme assay	Hx	XO	Clark oxygen electrode	91
Simple enzyme assay	Ki value	NP, Alkaline phosphatase, Nucleoside oxidase	Color (4-aminoantipyrine)	92
Enzyme electrode	Hx	XO	Clark oxygen electrode	93
Enzyme electrode	Ki value	5'Ntase, NP, XO	Clark oxygen electrode	49
Enzyme electrode	IMP	5'Ntase, NP, XO	Clark oxygen electrode	94
Enzyme electrode	AMP, IMP, Ino, Hx	AMP deaminase, 5'Ntase, NP, XO	Clark oxygen electrode	95

Enzyme electrode	Hx, Ino	NP, XO	Clark oxygen electrode	96
Oxygen electrode and soluble enzymes	K index	Soluble enzymes: Alkaline phosphatase, NP, XO	Clark oxygen electrode	97
Enzyme electrode	Ki value	5'Ntase, NP, XO	Clark oxygen electrode	98
Oxygen electrode and soluble enzymes	Ki value	5'Ntase, NP, XO	Clark oxygen electrode (Pegasus Freshness Meter)	99
Enzyme electrode	Hx	XO	Clark oxygen microelectrode	100
Enzyme electrode	Hx	XO	Clark oxygen electrode	101
Enzyme electrode	IMP, Ino, Hx	5'Ntase, NP, XO	Clark oxygen electrode	102
Enzyme electrode	Hx	XO	Platinum electrode with immobilized XO	103
Enzyme electrode	Hx/IMP+Ino+Hx	XO, 5'Ntase and soluble NP	Clark oxygen electrode	104
Enzyme electrode and capillary electrophoresis	Hx/IMP+Ino+Hx	XO, catalase, NP, 5'Ntase	Clark oxygen electrode and capillary electrophoresis	105
Immobilized enzyme bioreactor	Hx	XO	A_{290}	106

NP, nucleoside phosphorylase; XO, xanthine oxdiase; 5'Ntase, 5'nucleotidase; A_{290}, spectrophotometic detection at 290 nm.

to HPLC methods) is that the equipment is often specialized and cannot be used for the analysis of other objective quality indicators such as amines, aldehydes, organic acids, and alcohols. Also, many of the procedures listed in Table 3 are restricted to particular nucleotide catabolites such as Hx or combinations of catabolites such as the Ki value. By comparison, most of the HPLC techniques currently employed for seafood nucleotides are capable of resolving all of the nucleotides in a single analysis, thus making it possible for the analyst to obtain a more complete overview of the degradation patterns.

IV. CONCLUSIONS

ATP degradation patterns in fish, shellfish, crustaceans, and cephalopods can be used effectively to estimate eating quality. However, many factors can lead to false or misleading conclusions. Most important are the effects of temperature, species differences, and handling and the interactions among these variables. Although the nucleotide degradation patterns are of great value in assessing fish quality objectively, there is no universal nucleotide or combination of nucleotides that may be used under all circumstances.

REFERENCES

1. HO Hultin. Postmortem biochemistry of meat and fish. J Chem Ed 61(4):289–298, 1984.
2. JM Shewan, NR Jones. Chemical changes occurring in cod muscle during chill storage and their possible use as objective indices of quality. J Sci Food Agric 8:491–498, 1957.
3. J Saito, K Arai. Slow freezing of carp muscle and inosine formation. Nature 179:820–822, 1957.
4. VM Creelman, N Tomlinson. Inosine in the muscle of Pacific salmon stored in ice. J Fish Res Bd Can 17:449–451, 1960.
5. B Kassemsarn, B Perez, J Murray, NR Jones. Nucleotide degradation in the muscle of iced haddock, lemon sole and plaice. J Food Sci 28:28–37, 1963.
6. HLA Tarr. Phosphorus compounds in fish. Bull Jpn Soc Sci Fish 32(2):213–223, 1966.
7. MD Huynh, J Mackey, R Gawley. Freshness assessment of Pacific fish species using K-value. In: E G Bligh, ed. Seafood Science and Technology. Oxford: Fishing News Books, 1992, pp 258–268.
8. D Fraser, S Punjamapirom, WJ Dyer. Temperature and the biochemical processes occurring during *rigor mortis* in cod muscle. J Fish Res Bd Can 18(4):641–644, 1961.
9. M Iwamoto, H Yamanaka, S Watabe, K Hashimoto. Effect of storage temperature

on *rigor-mortis* and ATP degradation in plaice *Paralichthys olivaceus* muscle. J Food Sci 52(6):1514–1517, 1987.

10. T Saito, K Arai, M Matsuyoshi. A new method for estimating the freshness of fish. Bull Jpn Soc Sci Fish 24(9):749–750, 1959.

11. M Iwamoto, H Yamanaka, S Watabe, K Hashimoto. Changes in ATP and related breakdown compounds in the adductor muscle of Itayagai scallop *Pecten albicans* during storage at various temperatures. Nippon Suisan Gakk 57(1):153–156, 1991.

12. H Abe, E Okuma. Rigor-mortis progress of carp acclimated to different water temperatures. Nippon Suisan Gakk 57(11):2095–2100, 1991.

13. K Tomioka, T Kuragano, H Yamamoto, K Endo. Effect of storage temperature on the dephosphorylation of nucleotides in fish muscle. Nippon Suisan Gakk 53(3):503–507, 1987.

14. SS Nowlan, WJ Dyer. Glycolytic and nucleotide changes in the critical freezing zone, -0.8 to -5C, in pre-rigor cod muscle frozen at various rates. J Fish Res Bd Can 26:2621–2632, 1969.

15. DF Hiltz, LJ Bishop, WJ Dyer. Accelerated nucleotide degradation and glycolysis during warming to and subsequent storage at -5C of pre-rigor, quick-frozen adductor muscle of the sea scallop (*Placopecten magellanicus*). J Fish Res Bd Can 31(7):1181–1187, 1973.

16. JR Dingle, JA Hines. Degradation of inosine 5′-monophosphate in the skeletal muscle of several North Atlantic fishes. J Fish Res Bd Can 28(8):1125–1131, 1971.

17. A Sagedhal, JP Brusalmen, HA Roldan, ME Paredi, M Crupkin. Post-mortem changes in adenosine triphosphate and related compounds in mantle of squid (*Illex argentinus*) at different stages of sexual maturation. J Aquat Food Prod Technol 6(4):43–56, 1997.

18. E Ohashi, M Okamoto, A Ozawa, T Fujita. Characterization of common squid using several freshness indicators. J Food Sci 56(1):161–163, 174, 1991.

19. SM Langille, TA Gill. Postmortem metabolism of short-finned squid muscle (*Illex illecebrosus*). Comp Biochem Physiol 79B:361–367, 1984.

20. D Hiltz, WJ Dyer. Principal acid-soluble nucleotides in adductor muscle of the scallop *Placopecten magellanicus* and their degradation during postmortem storage on ice. J Fish Res Bd Can 27:83–92, 1970.

21. N DeVido de Mattio, ME Paredi, M Crupkin. Post mortem changes in glycogen, ATP, hypoxanthine and 260/250 absorbance ratio in extracts of adductor muscles from *Aulacomya ater ater (Molina)* at different biological conditions. Comp Biochem Physiol 103A(3):605–608, 1992.

22. Y Yokoyama, M Sakaguchi, F Kawai, M Kanamoir. Chemical indices for assessing freshness of shellfish during storage. Fisheries Science 60(3):329–333, 1994.

23. Y Yokoyama, M Sakaguchi, F Kawai, M Kanamori. Changes in concentration of ATP-related compounds in various tissues of oyster during ice storage. Nippon Suisan Gakk 58(11):2125–2136, 1992.

24. JR Dingle, JA Hines, DI Fraser. Post-mortem degradation of adenine nucleotides in muscle of the lobster, *Homarus americanus*. J Food Sci 33:100–103, 1968.

25. TA Gill, KE Harrison. Premortem physiology and postmortem biochemistry of the

American lobster (*Homarus americanus*) and related product quality. Progress report prepared for the Natural Sciences and Engineering Research Council of Canada (Cooperative Research and Development Grant Program), 1989. Project No. 0039115, 29 pp.

26. M Matsumoto, H Yamanaka. Post-mortem biochemical changes in the muscle of kuruma prawn during storage and evaluation of freshness. Nippon Suisan Gakk 56(7):1145–1149, 1990.

27. NR Jones, J Murray. Nucleotides in the skeletal muscle of codling (*Gadus callarias*). Biochem J. 66:5–6, 1957.

28. DI Fraser, HM Weinstein, WJ Dyer. Post-mortem glycolytic and associated changes in the muscle of trap- and trawl-caught cod. J Fish Res Bd Can 22(1):83–100, 1965.

29. T Nakayama, L Da-Jia, A Ooi. Tension changes of stressed and unstressed carp muscle in isometric rigor contraction and resolution. Nippon Suisan Gakk 58:1517–1522, 1992.

30. ML Izquierdo-Pulido, K Hatae, NF Haard. Nucleotide catabolism and changes in texture indices during ice storage of cultured sturgeon, *Acipenser transmontanus*. J Food Biochem 16:173–192, 1992.

31. T Lowe, JM Ryder, JF Carrager, RMG Wells. Flesh quality in snapper, *Pagrus auratus,* affected by capture stress. J Food Sci 58:770–773, 796, 1993.

32. T Sigholt, U Erikson, T Rustad, S Johansen, TS Nordtvedt, A Seland. Handling stress and storage temperature affect meat quality of fanned-raised Atlantic salmon (*Salmo salar*). J Food Sci 62(4):898–905, 1997.

33. S Mochizuki, A Sato. Effects of various killing procedures and storage temperature on post-mortem changes in the muscle of horse mackerel. Nippon Suisan Gakk 60(1):125–130, 1994.

34. S Mochizuki, A Sato. Effects of various killing procedures on post-mortem changes in the muscle of chub mackerel and round scad. Nippon Suisan Gakk 62(3):453–457, 1996.

35. T Berg, U Erikson, TS Nordtvedt. Rigor mortis assessment of Atlantic salmon (*Salmo salar*) and effects of stress. J Food Sci 62(3):439–446, 1997.

36. RW Korhonen, TC Lanier, F Gilsbrect. An evaluation of simple methods for following rigor development in fish. J Food Sci 55(2):346–348, 368, 1990.

37. A Chiba, M Hamaguchi, M Kosaka, T Tokuno, T Asai, S Chichibu. Quality evaluation of fish meat by[31] Phosphorus nuclear magnetic resonance. J Food Sci 56(3):660–664, 1991.

38. T Nakayama, T Toyoda, A Ooi. Physical property of carp muscle during rigor tension generation. Fisheries Sci 60(6):717–721, 1994.

39. M Surette, TA Gill, PJ LeBlanc. Biochemical basis of postmortem nucleotide catabolism in cod (*Gadus morhua*) and its relationship to spoilage. J Agric Food Chem 36(1):19–22, 1988.

40. ST Jiang, BS Hwang, CY Tsao. Effect of adenosine nucleotides and their derivatives on the denaturation of myofibrillar proteins in vitro during frozen storage at -20C. J Food Sci 52(1):117–123, 1987.

41. TA Gill, JW Thompson, S Gould, D Sherwood. Characterization of quality deterioration in yellowfin tuna. J Food Sci 52(3):580–583, 1987.

42. CJ Rodriguez, V Villar-Estalote, I Besteiro, C Pascual. Biochemical indices of freshness during processing of sardine (*Sardina pilchardus* (walb.)) for canning. In: J B Luten, T Borresen, J Oehlinschlager, eds. Seafood from Producer to Consumer. Amsterdam: El sevier, 1997, pp 203–210.

43. Y Fujii, K Shudo, K Nakamura, S Ishikawa, M Okada. Relation between the quality of canned fish and its content of ATP breakdown—III. ATP breakdowns in canned albacore and skipjack in relation to the organoleptic inspection. Bull Jpn Soc Sci Fisheries 39(1):69–84, 1973.

44. RB Hughes, NR Jones. Measurement of hypoxanthine concentration in canned herring as an index of the freshness of the raw material with a comment on the flavour relations. J Sci Food Agric 17:434–436, 1966.

45. YP Lee. Adenylic deaminase. In: P D Boyer, H Hardy, K Myrback, eds. The Enzymes, 2nd ed., vol. 4. New York: Academic Press, 1960, pp. 279–83.

46. JR Dingle, JA Hines. Extraction and some properties of adenosine 5'-monophosphate aminohydrolase from prerigor and postrigor muscle of cod. J Fish Res Bd Can 24(8): 1717–1730, 1967.

47. W Makarewicz. AMP-aminohydrolase in muscle of elasmobranch fish. Purification procedure and properties of the purified enzyme. Comp Biochem Physiol 29:1–26, 1969.

48. RL Watts, DC Watts. Nitrogen metabolism in fisheries. In: M Florkin, B T Scheer, eds., Chemical Zoology, vol. 8. New York: Academic Press, 1974, pp 369–445.

49. Karube, H Matsuoka, S Suzuki, E Watanabe, K Toyama. Determination of fish freshness with an enzyme sensor. J Agric Food Chem 32:314–319, 1984.

50. K Tomioka, K Endo. K-value increasing rates and IMP-degrading activities in various fish muscles. Bull Jpn Soc Sci Fish 50:889–92, 1984a.

51. K Tomioka, K Endo. Properties of 5'-nucleotidase from carp skeletal muscle. Bull Jpn Soc Sci Fish 50:1739–44, 1984b.

52. DW Marseno, K Hori, K Miyasawa. Purification and properties of membrane-bound 5'-nucleotidase from black rockfish (*Sebastes inermis*) muscle. J Agric Food Chem 41:863–69, 1993a.

53. K Tomioka, K Endo. Purification of 5'-nucleotidase from carp muscle. Bull Jpn Soc Sci Fish 50(6):1077–81, 1984c.

54. DW Marseno, K Hori, K Miyazawa. Purification and properties of cytosol 5'-nucleotidase from black rockfish (*Sebastes inermis*) muscle. J Agric Food Chem 41:1208–1212, 1993b.

55. K Nedachi, N Hirota. Purification and properties of 5'-nucleotidase from snapper muscle. Nippon Suisan Gakk 58(10):1905–1911, 1992.

56. HLA Tarr, LJ Gardner, P Ingram. Pacific cod 5'-nucleotidase. J Food Sci 34:637–640, 1969.

57. N Hirota. Studies on 5'-nucleotidase of a bonito muscle—I. Isolation and purification of 5'-nucleotidase. Bull Jpn Soc Sci Fish 39(12):1271–1278, 1973a.

58. N Hirota. Studies on 5'-nucleotidase of bonito muscle—II. Some properties of purified enzyme. Bull Jpn Soc Sci Fish 39(12):1279–1283, 1973b.

59. HLA Tarr. Fish muscle riboside hydrolases. Biochem J 59:386–391, 1955.

60. HLA Tarr. Lingcod muscle purine nucleotide phosphorylase. Can J Biochem Physiol 36:517–530, 1958.

61. PJ LeBlanc. Approaches to the study of nucleotide catabolism for fish freshness evaluation. M Sc thesis, Technical University of Nova Scotia, 1987.

62. ME Surette. Isolation and immobilization of nucleotide catabolic enzymes for evaluation of fish freshness. M Sc thesis, Technical University of Nova Scotia, 1987.

63. ME Surette, T Gill, S MacLean. Purification and characterization of purine nucleoside phosphorylase from *Proteus vulgaris*. Appl Environ Microbiol 56(5):1435–39, 1990.

64. R Gardner, A Kornberg. Purine nucleoside phosphorylase of vegetative cells and spores of *Bacillus cereus*. J Biol Chem 242:2383–88, 1967.

65. A Imada, S Igarasi. Ribosyl and deoxyribosyl transfer by bacterial enzyme systems. J Bacteriol 94:1551–59, 1967.

66. KF Jensen. Purine nucleoside phosphorylase from *Salmonella typhimurium* and *Escherichia coli*. Initial velocity kinetics, ligand binding and reaction mechanisms. Eur J Biochem 61:377–86, 1976.

67. Y Machida, T Nakanishi. Properties of purine nucleoside phosphorylase from *Enterobacter cloacae*. Agric Biol Chem 45:1801–1807, 1981.

68. RP Agarwal, RE Parks Jr. Purine nucleoside phosphorylase from human erythrocytes. Crystallization and some properties. J Biol Chem 244:644–47, 1969.

69. BK Kim, S Cha, RE Parks Jr. Purine nucleoside phosphorylase from human erythrocytes. Purification and properties. J Biol Chem 243:1763–1770, 1968.

70. BK Kim, S Cha, RE Parks Jr. Purine nucleoside phosphorylase from human erythrocytes. Kinetic analysis and substrate binding studies. J Biol Chem 243:1771–1776, 1968.

71. K Murakami, K Tsushima. Molecular properties and nonidentical trimeric structure of purine nucleoside phosphorylase from chicken liver. Biochim Biophys Acta 453:205–210, 1976.

72. K Stark, P Seubert, G Lynch, M Baudry. Proteolytic conversion of xanthine dehydrogenase to xanthine oxidase:evidence against a role for calcium-activated protease (calpain). Biochem Biophys Res Commun 165(2):858–864, 1989.

73. G Carpani, M Racchi, P Ghezzi, M Terao, E Garattini. Purification and characterization of mouse liver xanthine oxidase. Arch Biochem Biophys 279(2):237–241, 1990.

74. HO Hultin. Biochemical deterioration of fish flesh. In: H.H. Huss, M. Jakobsen, J. Liston, eds. Quality Assurance in the Fish Industry. Oxford:Elsevier, 1992, pp 125–138.

75. E Dekker, Z Xu. Minimizing rancidity in muscle foods. Food Technol 52(10):54–59, 1998.

76. DI Fraser, JR Dingle, JA Hines, SC Nowlan, WJ Dyer. Nucleotide degradation monitored by thin-layer chromatography and associated postmortem changes in relaxed cod muscle. J Fish Res Bd Can 24(8):1837–1841, 1967.

77. TA Gill. Objective analysis of seafood quality. Food Rev Int 6(4):681–714, 1990.

78. TA Gill. Biochemical and chemical indices of seafood quality. In: H.H. Huss, M. Jakobsen, J. Liston, eds. Quality Assurance in the Fish Industry, Amsterdam:Elsevier, 1992, pp 377–388.

79. TA Gill. Advanced analytical tools in seafood science. In: JB Luten, T. Borresin, J. Oehlenschlager, eds. Seafood from Producer to Consumer, Integrated Approach to Quality, Amsterdam:Elsevier, 1997, pp. 479–490.

80. JR Botta. Freshness quality of seafoods: a review. In: F. Shahidi, J.R. Botta, eds. Seafoods—Chemistry, Processing, Technology and Quality. London:Blackie Academic and Professional, 994, pp. 140–167.

81. FD Jahns, JL Howe, RJ Coduri, AG Rand Jr. A rapid visual enzyme test to assess fish freshness. Food Technol 30:27–30, 1976.

82. LC Clark Jr. The hydrogen peroxide sensing platinum anode as an analytical enzyme electrode. In. S. Fleischer, L Packer, eds., Methods in Enzymology, Vol. 56. New York: Academic Press, 1979, pp 448–479.

83. EAC (Environmental Analysis Center. Technical manuscript of freshness test paper (FTP)), 1985. 1-1, Higashi-Ikebukuro 3-chome, Toshima, Japan.

84. S Negishi, I Karube. An enzymatic assay method for IMP determination using IMP dehydrogenase as an application of the principle to a test paper method. Nippon Suisan Gakk 55(9):1591–1597, 1989.

85. Diffchamb AB. Transia Fresh Tester, Baka Bergogata 7, 422 46 Hisings Baka, Sweden, *http://www.diffchamb.com/products/other/freshtest.html.* 1998.

86. NR Jones, J Murray, EI Livingston, CK Murray. Rapid estimations of hypoxanthine concentrations as indices of the freshness of chill-stored fish. J Sci Fd Agric 15:763–774, 1964.

87. JR Burt, GD Stroud, NR Jones. Estimation of hypoxanthine concentrations in fish muscle by a rapid, visual modification of the enzymatic assay procedure. In: R. Kreuzer, ed. Freezing and Irradiation of Fish. London:Fishing News Books., Ltd., 1969, pp. 367–370.

88. S Ehira, H Uchiyama. Rapid estimation of freshness of fish by nucleoside phosphorylase and xanthine oxidase. Bull Jpn Soc Sci Fish 35(11):1080–1095, 1969.

89. F Uda, E Hayashi, H Uchyama, K Kakuda. Colorimetric method for measuring K value, an index for evaluating freshness of fish. Bull Tokai Reg Fish Res Lab. No. 111:55–60, 1983.

90. K Nakamura, S Ishikawa. An enzymatic method of measuring K value. Bull Tokai Reg Fish Res Lab. No. 118:39–43, 1985.

91. JM Kim, M Suzuki, RD Schmid. A novel amplified enzymatic assay method for hypoxanthine. Biocatalysis 3:269–275, 1990.

92. Y Isono. A new colorimetric measurement of fish freshness using nucleoside oxidase. Agric Biol Chem 54(11):2827–2832, 1990.

93. E Watanabe, K Ando, I Karube, H Matsuoka, S Suzuki. Determination of hypoxanthine in fish meat with an enzyme sensor. J Food Sci 48:496–500, 1983.

94. E Watanabe, K Toyama, I Karube, H Matsuoka, S Suzuki. Determination of inosine-5-monophosphate in fish tissue with an enzyme sensor. J Food Sci 49:114–116, 1984.

95. E Watanabe, S Tokimatsu, K Toyama, I Karube, H Matsuoka, S Suzuki. Simultaneous determination of hypoxanthine, inosine, inosine-5'-phosphate and adenosine-5'-phosphate with a multielectrode sensor. Anal Chim Acta 164:139–146, 1984.

96. E Watanabe, K Toyama, I Karube, H Matsuoka, S Suzuki. Enzyme sensor for hypoxanthine and inosine determination in edible fish. Appl Microbiol Biotechnol 19:18–22, 1984c.

97. M Ohashi, N Arakawa, T Ashahara, S Sakamoto. Method for determing index of freshness of fish and mollusks. European Patent Application 84305761.3, 1985.

98. E Watanabe, H Endo, N Takeuchi, T Hayashi, K Toyama. Determination of fish freshness with a multielectrode enzyme sensor system. Bull Jpn Soc Sci Fish 52(3):489–495, 1986.

99. J Luong, K Male, A Nguyen. Development of a fish freshness sensor. Am Biotechnol Lab 6(8): 38–41, 1988.

100. M Suzuki, H Suzuki, I Karube, R Schmid. A disposible hypoxanthine sensor based on a micro oxygen electrode. Anal Lett 22(15):2915–2927, 1989.

101. A Mulchandani, J Luong, K Male. Development and application of a biosensor for hypoxanthine in fish extract. Anal Chim Acta 221:215–222, 1989.

102. A Mulchandani, K Male, J Luong. Development of a biosensor for assaying postmortem nucleotide degradation in fish tissues. Biotechnol Bioeng 35:739–745, 1990.

103. A Nguyen, J Luong, AM Yacynych. Retention of enzyme by electropolymerized film: a new approach in developing a hypoxanthine biosensor. Biotechnol and Bioeng 37:729–735, 1991.

104. J Luong, K Male. Development of a new biosensor system for the determination of the hypoxanthine ratio, an indicator of fish freshness. Enzyme Micro Technol 14:125–130, 1992.

105. J Luong, K Male, C Masson, A Nguyen. Hypoxanthine ratio determination in fish extract using capillary electrophoresis and immobilized enzymes. J Food Sci 57(1):77–81, 1992.

106. D Balladin, D Narinesingh, V Stoute, T Ngo. Immobilization of xanthine oxidase and its use in the quantition of hypoxanthine in fish muscle tissue extracts using a flow injection method. Appl Biochem Biotechnol 62:317–329, 1997.

3
Myosin ATPase

Yoshihiro Ochiai
Ibaraki University, Bunkyo, Mito, Japan

Chau-Jen Chow
National Kaohsiung Institute of Marine Technology, Kaohsiung,Taiwan, Republic of China

I. INTRODUCTION

The activities of animals are largely supported by muscle contraction. The most abundant protein in muscle is myosin, which behaves as a molecular motor. Therefore, myosin had long been considered to be a protein specific to muscle, but later it was established that myosins are also involved in other biological activities (such as cytokinesis and organelle motility). The molecular shape of myosins differ with respect to function (1,2). Therefore, myosins form a broad family of proteins, and those found in muscle are classified as myosin II (conventional myosin). myosin II is characterized by two head portions and a long tail (3,4) as described later (Sec. II.C). From the viewpoint of food, the edible part of myosystems is mostly the myofibrillar protein and predominantly myosin.

A. Muscle Proteins

Muscle is composed of numerous proteins. Based on solubility, muscle proteins can be classified into three groups. The sarcoplasmic proteins (water-soluble proteins) are soluble in low-ionic strength buffer (ionic strength I=0.05–0.15, pH 6.5–7.5). This fraction is composed of glycolytic enzymes, parvalbumin, myoglobin, and others. The second group, myofibrillar proteins, are soluble in high ionic strength buffer (about I=0.5–1.0, pH 6.5–7.5), and thus are called salt-soluble proteins. The third group, stroma proteins, are insoluble, irrespective of salt concentration. The major component of the stroma fraction is collagen. The myofibril fraction occupies 40–70% of muscle protein, and the most abundant protein (50%) in the myofibril is myosin. Thus, the properties of myosin largely affects the function of muscle as well as the characteristics of muscle as food. Since fish muscle normally contains a lower proportion of stroma protein than the myosystems of land animals, it normally contains more myofibrillar protein than other meats. The importance of myosin ATPase to seafood quality is discussed in detail in Chapters 1 and 12.

B. Muscle Classification

Muscles of vertebrates, including fish, are classified into striated muscle and smooth muscle. The former is further classified into skeletal muscle and cardiac muscle. Skeletal muscle is classified into fast-twitch and slow-twitch muscles, based on the shortening characteristics. Fast- and slow-twitch muscles are respectively called "white" and "red" muscles due to the difference in myoglobin content. White muscle is glycolytic and suitable for a short duration of powerful movement, whereas red muscle is oxidative (contains many mitochondria), and is suitable for sustained contraction. In the case of fish, these muscles are often referred ιo as "ordinary" and "dark" muscles, respectively. The latter muscle is normally distributed along the lateral lines of fish, and is used for constant cruising. Cruising fish such as tuna and sardine have a relatively large amount of dark muscle: as much as 30% of the total musculature (5). Invertebrate muscles are classified into three types: striated, smooth, and obliquely striated muscles. All these muscles are driven by the two contractile proteins, myosin and actin, at the expense of chemical energy from ATP (about 30 kJ/mol) supplied via hydrolysis by myosin ATPase activity. Using this chemical energy, thin filaments can slide into thick filaments (6,7).

C. Muscle Structure

Skeletal muscle is composed of many muscle cells (so-called muscle fibers) 10–160 μm in diameter. Along the longitudinal direction of muscle fiber run many myofibrils (about 1 μm in diameter) (Fig. 1). The sarcomere (sectioned by Z lines) is the basic unit of the myofibril contractile system. The length of sarcomere is about 2 μm. Under the resolution of light microscopy, the myofibril is striated by the repetition of isotropic bands (I bands) that look light, and anisotropic bands (A bands) that look dark. The I and A bands consist of thin filaments (5–7 nm in diameter) and thick filaments (10–11 nm in diameter), respectively. The former filament is mainly composed of actin and the latter is mainly composed of myosin. Actin is a globular protein of a single polypeptide chain, whose molecular weight is about 42,000 Da (G-actin). Under physiological con-

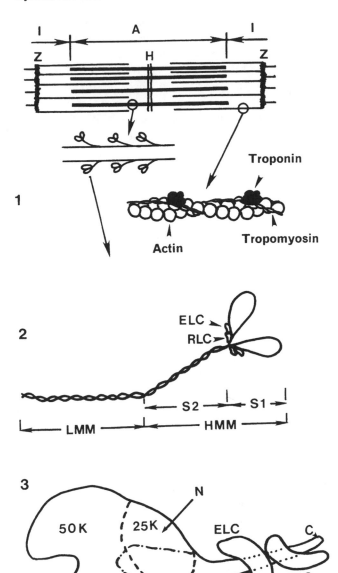

Figure 1 Schematic structures of myofibril, myosin, and S1. (1) Myofibril, thick and thin filaments; (2) myosin molecule; (3) S1. A, A band; I, I band; H, H zone; Z, Z line in scheme 1; ELC, essential light chain, RLC, regulatory light chain in schemes 2 and 3. 25K, 50K, and 20K, 25 kDa, 50 kDa, and 20 kDa domains of S1 heavy chain; N and C, *N*- and *C*- termini of S1 heavy chain, respectively.

ditions, G-actin associates into a double helical thin filament (F-actin). Binding of F-actin to myosin results in the formation of actomyosin and remarkable activation of Mg^{2+}-ATPase of myosin, as described in Section II.A.1.

Thick filaments of invertebrate muscles have a core formed by paramyosin, which does not show ATPase activity but occupies up to about 40% of myofibrillar protein. The paramyosin core is overlaid with myosin. Paramyosin appears to be involved in the catch mechanism of bivalve shellfish, which can continue to close the shells for a long time with low energy expenditure.

D. Muscle Contraction

In resting muscle, thick filaments (myosin crossbridge) cannot attach to thin filaments because the myosin binding site on actin is blocked by tropomyosin. Ca^{2+} concentration in resting muscle is about 10^{-7} M. A nerve impulse releases Ca^{2+} from the sarcoplasmic reticulum up to the 10^{-5} M level, and Ca^{2+}, bound to the troponin C subunit, causes conformational changes in the thin filament, and thus the myosin binding site on actin is unblocked. This system is called "actin-linked" regulation. The myosin crossbridge then can attach to an actin filament, where it swivels to push the actin filament toward the center of the sarcomere. On binding another ATP molecule, it is detached and reoriented to the resting position, ready to repeat the cycle (8). Tropomyosin, together with troponin, binds to F-actin, and offers Ca^{2+} sensitivity to Mg^{2+}-ATPase of actomyosin. On the other hand, in the case of some mollusks (such as scallops, clams), Ca^{2+} binds to the regulatory light chain of myosin, which facilitates actin–myosin interaction. This mechanism is called "myosin-linked" regulation (9,10). Muscle contraction of some species (squid and others) is considered to be regulated by both systems.

ATP hydrolysis provides energy for muscle contraction. ATP also prevents actin–myosin interaction and dissociates the actin–myosin complex during relaxation of muscle. The cyclic series of contraction and relaxation continues uninterrupted until Ca^{2+} is sequestered by the sarcoplasmic reticulum, and the thin filament is turned off. As a result, actin is no longer available to bind the myosin crossbridge (physiological relaxation) or until muscle ATP is exhausted and no ATP is available to bind to the cross bridge and dissociate it from actin (as occurs in rigor mortis after the death of an animal).

E. Myosin Isolation

As described above, myosin is soluble at high ionic strength, and elutes from the myofibril under high ionic conditions. When the extraction time is short (up to 10 min), myosin is the major component in the extract. When extraction time is prolonged, the extract contains many other myofibrillar proteins (actin, tropomyosin, troponin, etc.). Fish myosin tends to be contaminated with actin even if extraction time is short. The fraction thus obtained (usually overnight extraction) is called myosin B, which is actually natural actomyosin. Actomyosin can also be artificially prepared by mixing isolated myosin and actin at a molar ratio of about 1:2. This product is called reconstituted actomyosin, and differs from myosin B in that it lacks regulatory proteins (tropomyosin and troponin). Actomyosin solution (at high salt concentration, about I=0.5) shows high viscosity, because actin and myosin form complexes. However, upon addition of ATP, the viscosity decreases due to the dissociation of actomyosin as follows:

$$\text{Actomyosin} + \text{ATP} + \text{H}_2\text{O} \rightarrow \text{Actin} + \text{Myosin} + \text{ADP} + \text{H}_3\text{PO}_4$$

Intact myofibrils are obtained when muscle is homogenized, and all the water-soluble proteins and stroma protein are removed by washing with low-ionic strength buffer and selective filtration and centrifugation (in order to remove ATPase from sarcoplasmic reticulum and mitochondria). Myofibrils thus prepared maintain the tertiary structure of intact myofibrils in muscle. The properties are very close to those of muscle, unlike solubilized actomyosin. "Myofibrils" are often used in assays for the properties of myosin, because they are much more stable compared to myosin itself, and are easy to prepare and handle.

1. Fish Myosin Isolation

High-purity fish myosin was first isolated from cod by Connell in 1954. Fish myosin is generally obtained as follows: muscle mince is repeatedly washed with low-ionic-strength solution to remove water-soluble components, and extracted with a high-ionic-strength buffer for 5–10 min. The extract is diluted with water to precipitate myosin under low ionic strength. However, this precipitate is usually contaminated with actin. Therefore, actomyosin complex is removed, in the presence of Mg^{2+}-ATP, by centrifugation or by ammonium sulfate fractionation. Fish myosin is quite unstable (11), and thus is susceptible to rapid inactivation of ATPase and proteolysis by endogenous proteinases with concomitant formation of aggregates. To avoid such deterioration, antidenaturants (such as sugars and sodium glutamate) and protease inhibitors are often added throughout preparation procedures.

Reports on skeletal muscle myosin ATPase profiles are available for several fish species such as cod(12), pollack (13, 14), tunas (11, 15, 16), tilapia (17), carp (17), trout (18), mackerel (19), yellowtail (16, 20), milkfish (21), shark (22), etc. The properties of dark muscle myosin have also been reported (19, 20, 23).

II. MOLECULAR STRUCTURE IN ASSOCIATION WITH FUNCTION AND PROPERTIES

Myosin is a hexamer of two heavy-chain subunits ($\approx$200,000 Da) and four light-chain subunits ($\approx$20,000 Da), and therefore, the mass molecular weight is about 480,000 (24, 25). As shown in Figure 1, the myosin molecule is composed of two globular (ellipsoidal) heads and a fibrous tail. Myosin heads protrude from the thick filament surface and form a crossbridge with the thin filament. The myosin molecule is about 150 nm in length, and the head portion is about 20 nm in length and 9 nm in width, while the tail (rod) is 135 nm in length and 1.5 nm in width. The rod has a coiled-coil structure of two parallel α-helices wound about each other. In the rod, alternate clusters of positively and negatively charged amino acids are recognized, with the major repetition of 28 amino acids formed by the minor repetition unit of 7 residues (26). The molecular structure of fish myosins is essentially the same as that of higher vertebrates (25).

Myosin is rich in negatively charged amino acids (glutamic and aspartic acids) and poor in positively charged ones (histidine, lysine, arginine), and thus is negatively charged under physiological ionic strength and pH. The isoelectric point in the absence of divalent cations (Mg^{2+}, Ca^{2+}) is 5.4, but in their presence, the isoelectric point shifts to an alkaline side as a result of their binding to the molecule. Fifty to 55% of the myosin polypeptide chain is in the form of α-helices. Myosin heavy chain has about 40 cysteine residues (12–13 residues are located in S1), but has no cystine (27). Physicochemical properties (amino acid composition, intrinsic viscosity, sedimentation coefficient, etc.) and molecular structures of myosins from fish and shellfish resemble those of myosins from higher vertebrates (17, 19, 27, 28).

A. General Characteristics

Striated muscle (or sarcomeric) myosins have three physiologically important characteristics: ATPase activity, actin binding, and thick filament formation (29). Each property is described below.

1. ATPase

Myosin (EC 3.6.1.3) hydrolyzes ATP into ADP and inorganic phosphate. ATPase activity of myosin was first discovered by Engelhardt and Ljubimova in 1939. In the absence of actin, myosin is activated by Ca^{2+}, but inhibited by Mg^{2+}. However, in the presence of actin, myosin is activated by both Mg^{2+} and Ca^{2+}. Specificity to substrate is relatively low, because inorganic triphosphate, ribose triphosphate, and many other nucleotide triphosphates can be the substrate of myosin. Pyrophosphate is a competitive inhibitor of myosin ATPase activity.

2. Actin Binding

The dissociation constant for actin and myosin is 10^{-8} to 10^{-7} M. The actin–myosin complex (actomyosin) specifically dissociates with ATP, pyrophosphate, and a few other polyanions. The higher the ionic strength, the less ATP is required. Mg^{2+} is required for the dissociation.

3. Filament Formation

Myosin aggregates to form anisotropic (bipolar) filaments. Aggregation is due to the repetitive pattern of charged amino acids distributed in the rod domain of the molecule (26), and is inhibited by high ionic strength. The length of the filament is about 1.5 μm, and one thick filament consists of about 300 myosin molecules.

B. Myosin ATPase

Myosin ATPase activity is stimulated by Ca^{2+}. The activity reaches its maximum with 3–5 mM Ca^{2+}. This activity is solely due to myosin alone, and thus is not essentially affected by the presence of actin. Ca^{2+}-ATPase activity is a good parameter to estimate the quality or the extent of deterioration of protein in muscle food, as described in Section III.A. and Chapter 12. Ca^{2+}-ATPase activity of fish ordinary muscle myosin shows optima at around pH 6 and 9 (12, 20, 22). Myosin ATPase activity is inhibited by Mg^{2+} at low (physiological) ionic strength, but in the presence of actin it is highly enhanced. On the other hand, Mg^{2+}-ATPase of actomyosin (or actin-activated Mg^{2+}-ATPase of myosin) is the activity directly involved in muscle contraction. The physiological level of Mg^{2+} in the muscle fiber is 2–3 mM. In the presence of K^+ and EDTA (to remove divalent cations Mg^{2+}, Ca^{2+}), myosin ATPase is also activated. This ATPase activity is high at alkaline pH, and the maximum activity is observed at around pH 9 (22). The extent of filament formation (self-associating ability) of myosin also largely affects ATPase activity. Filament formation of fish myosin is generally affected by the concentrations of Mg^{2+} and ATP. Filaments of fish myosin, formed in vitro under physiological conditions (5 mM Mg^{2+}, 1 mM ATP, I = 0.12), are shorter than native thick filaments and, therefore, the activation level by actin is generally low (30).

Myosin ATPase is also largely affected by chemical modification of reactive SH residues (SH1, SH2) with p-chloromercuribenzoate (pCMB) and N-ethylmaleimide (NEM) or modification of the reactive lysine with 2,4,6-trinitrobenzenesulfonate (TNBS) (31). When the most reactive cysteine, SH1, is alkylated, K^+-EDTA-ATPase is inactivated, but Ca^{2+}-ATPase is activated. Fish myosin behaves in a similar way (32). Modification of another thiol, SH2, results in inactivation of Ca^{2+}-ATPase.

C. Substructure of Myosin

The myosin molecule has two sites very susceptible to proteolysis. When myosin is treated with trypsin, the molecule is cut into the major part of the rod and the two heads with a short tail (29). The former and the latter are called light meromyosin (LMM) and heavy meromyosin (HMM), respectively (Fig. 1). When myosin is digested with chymotrypsin in the presence of EDTA, the head is released from the rod, and is called subfragment-1 (S1). A short piece of the rod in HMM is called subfragment-2 (S2). S1 thus prepared consists of one heavy-chain fragment and one alkali light chain, and is a mixture of two iso-forms based on associating light chain. S1 contains the ATPase activity and actin-binding ability from myosin. As shown in Figure 1, S1 is a very asymmetrical molecule, and is characterized by several deep clefts and pockets where the nucleotide binds and S1 is split into two major domains (33). S1 is water-soluble, and no longer forms a filament. The sizes of HMM, LMM, S1, S2, and rod are about 350 kDa, 140–150 kDa, 95–115 kDa, 120–140 kDa, and 220–240 kDa, respectively. HMM and S1 from fish myosins show similar properties to those from higher vertebrates (34–36).

Further digestion of S1 gives rise to three fragments, i.e., (25 kDa, 50 kDa, and 20 kDa domains), which are connected in this order from the N-terminus (37). These fragments are produced because the junctions (corresponding to loop I and loop II in Fig. 2) of these domains are susceptible to proteolytic digestion. Many studies have been done to determine the functions and specific sites of S1 as follows: actin-binding site resides on both 50 kDa and 20 kDa domains; ATP binds to 25 kDa and 50 kDa domains; reactive sulfhydryl residues (SH1 and SH2) are on the 20 kDa where light chains bind to this domain; and a reactive lysine residue is on the 25 kDa domain. These facts were further confirmed by the later understanding of the tertiary structure of S1 revealed by x-ray diffraction studies on the crystal of S1 (38).

The 50 kDa domain is the most fragile part of the three segments of S1, and is susceptible to conformational change by heat treatment. Tryptic decay of the 50 kDa domain and inactivation of Ca^{2+}-ATPase follow a first-order reaction rate, and thus both factors are in good relationship (39). Binding of ATP to S1 causes a remarkable stabilization of the S1 structure (37). The 25 kDa and 20 kDa domains of fish S1s are generally sensitive to proteolysis after heat treatment, unlike the S1 of rabbit (40). In the presence of actin, cleavage of the 50 kDa and 20 kDa domains is inhibited, because actin binds to this region. Even after proteolytic cleavage, S1 maintains its tertiary structure, and shows full ATPase activity (41). As a result of heat denaturation, fish S1s give rise to turbidity and light chain is released from S1 heavy chain, but the extent is dependent on species (40).

```
pollack   MST-DAEMAI YGAAAIYLRK PERERLEAQS TPFDAKAAAY VSDVKELYVK CTMTKRDAGK   59
carp10    .G--.G..EC F.P...... ......S.T I.....T.FS .T.AA.M.L. S.LISIE...   58
carp30    .G--.G..EC F.P..V... T....I...N ......T.FF .V.PD.M.L. G.LVSKEG..   58
chicken   .ASP....A F.E..P.... S.K..I...N K.....SSVF .VHP..SF.. G.IQSKEG..   60

pollack   VTVTILATKE ERTVKEDDVY PMNPPKYDKI EDMAMMTHLN EASVLYNLAE RYAAWMIYTY  119
carp10    A..KTHCG.T V-.....EIF ......F..M ......L.... .P...F..KD ..........  117
carp30    A..KTHSG.T V-.....EIF ......F.... .......... .PA..F..K. ..........  117
chicken   V..KTEGGET L-.....QVF S......... .......... .H .PA....K. ..........  119

pollack   SGLFCATVNP YKWLPVYDQS CVNAYRGKKR MEAPPHIFSV SDNAFQFMLT DRENQSVLIT  179
carp10    .......... ........AV V.AG...... I........I ....Y..... ..........  177
carp30    .....V.... ........AV V.GG...... I........I ....Y..... ..........  177
chicken   .....V.... ........NPE V.L....... Q........I ....Y..... ....I....  179
```

P-loop   **loop 1**
```
pollack   GESGAGKTVN TKRVIQYFAT IAVGGGGEKA DVGAGKIKGS LEDQIIAANP LLEAYGNAKT  239
carp10    .......... .......... .-AMA.PK.A EAVP..MQ.. .......... ..........  236
carp30    .......... .......... . VGAMS.PK.P EPVP..MQ.. .....V.... ..........  237
chicken   .......... .......... ..A-S.EK.K EEQS..MQ.T ......S... .......F...  238
```

switch I loop
```
pollack   VRNDNSSRFG KFIRIHFHAN GKLSSADIET YLLEKSRVSF QLPDERGYHI FFQMMTNHKP  299
carp10    I......... ......SGT ...AK..... .......T. ..SA..S... .Y.L..G...  296
carp30    .......... ......GTT ...A...... .......T. ..SA..S... .Y.L..G...  297
chicken   .......G.T ...A...... .......T. ..A..S... .Y.I.S.K..  298

pollack   EIIEMTLITT NPYDFPMCSQ GQITVASIDD KEELDATDAA IDILGFTSED KVAIFKFTGA  359
carp10    ....AL.... ....Y..I.. .E...K.... V..FI...T. .......ADE .IS.Y.L...  396
carp30    ....AL.... ....Y..I.. .E...K..N. V..FI...T. .......ADE .IS.Y.L...  397
chicken   ...D.L.... ....YHYV.. .E...P.... Q...M...S. ......SADE .T..Y.L...  398

pollack   VLHHGNMKFK QKQREEQAEP DGNEEADKIC YLLSLNSADM LKALCYPRVK VGNEYVTKGQ  419
carp10    .M...A.... ......... ..A....A ..MGI..... .......... ....M.....  416
carp30    .M........ ......... ...T.V...A ..MG...... .......F.... .....M....  417
chicken   .M.Y..L... ......... ...T.V...AA ..MG....EL ......F.... .........F....  418

pollack   TVPQVNNSVS ALAKSIYERL FLWMVIRINT MLDTKQARQF YIGVLDIAGF EIFDYNSMEQ  479
carp10    .......A.. ..C..V..KM ........V..E ..N.TNP.EY .......F.... .....F..L..  476
carp30    .......A.. ..S..V..KM .........E .......P...F .......F.... .....F..L..  477
chicken   ..S..H...G ....AV..KM .........Q Q....P..Y F........ .......F..F..  478
```

switch II
```
pollack   LCINFTNEKL QQFFNHTMFV LEQEEYKKEG IIWEFIDFGM DLAACIELIE KPMGIFSILE  539
carp10    .......... .......... ......... ..E.A..... .......... ...........  536
carp30    .......... ......H... ......... ..E....... .......... ...........  537
chicken   .......... ......H... ......... ..E....... .......... ...........  538

pollack   EECMFPKASD VTFKNKLFDQ HLGKNRAFEK PKPAKDKAEA HFSLVHYAGT VDYNVTGWLD  599
carp10    .......T. TS.....H... ...CS..Q... ...G.G..... .......... ...IN...E  596
carp30    .......T. TS.....H... ...TA..Q... ...G.G..... .......... ...IV....  597
chicken   .......T. TS.....Y... ...SNN.Q... .......... .......... ...IS...E  598
```

loop 2
```
pollack   KNKDPLNDSV IGLYQKSSNK LLPVLYHPVV EEV--GGAKK GGKKKGGSMQ TVSSQFRENL  657
carp10    .......... VQ.....AL. V.AL..VA-. P.AEAA.K.G .-.....F. ...AV.....  654
carp30    .......... VQ......L. V.AF..ATHG A.AEG..G.. .-.....F. ...AL.....  656
chicken   .......ET. .......V. T.AL.FATYG G.AEG..G.. .......S.F. ...AL.....  658

pollack   GKLMTNLRST HPHFVRCLIP NESKTPGLME NHLVIHQLRC NGVLEGIRIC TKGFPSRIIY  717
carp10    .......... .......... ......F.... .......V.... .......... ........H.  714
carp30    .......... .......... .......Y.... .......... .......... R......L.  716
chicken   N..A..... .......I.. ..T....A.. HE..L..... .......... R......L.  718

pollack   ADFKQRYKVL NASVIPDGQF IDNKKASEKL LGSIDVPHDE YKFGHTKVFF KAGLLGTLEE  777
carp10    G......... .....E.... ......T.... ....D.NQ ......... ..........  774
carp30    G......... .....E.... .......... ....D.TQ ......... .......A...  776
chicken   ......R.. A..E... M.S....... ....D.TQ .R....... .......L...  778
```

ELC binding site   **RLC binding site**
```
pollack   MRDEKLAALV GMIQAAGRGY VMRKEYVKMT ERREAVYTIQ YNIRSFMNVK HWPWMKVYYK  837
carp10    .....[.SH.. T.T..LA... .....F...M ...I.S.... .......... ......F..  834
carp30    .....[.L.. T.T..LC... .....F...M ...SI.S.... .......... ......L.F.  836
chicken   ...D.[..EII TRT..RC... L..V..RR.V ...SIFC...V.......... .......LFF.  838

pollack   IKP                                                               840
carp10    ...                                                               837
carp30    ...                                                               839
chicken   ...                                                               841
```

Figure 2 Amino acid sequences of myosin heavy chain (heavy meromyosin portion) from Alaska pollack, carp, and chicken. Underlined parts correspond to actin-binding sites. Carp 10 and carp 30, carp acclimated to 10 and 30°C, respectively. (64); loops 1 and 2, junctions of 25–50 kDa and 50–20 kDa domains; dots and bars, the same residues and deletions, respectively. Sequences are from Ref. 65 for pollack, Ref. 64 for carp, and Ref. 62 for chicken. (From Y Ochiai, 1998, courtesy of Asakura Publishing Co., Tokyo, Japan.)

1. Light Chains of Myosin

The light chains are noncovalently bound to the vicinity of the junction of head and tail portions (33, 42, 43). There are two categories of light chain: essential light chain and regulatory light chain. In the case of skeletal muscle myosin, the former is usually referred to as alkali light chain, because it is dissociated from myosin heavy chain under alkaline condition with concomitant loss of ATPase activity. However, Sivaramakrishnan and Burke (44) reported for the first time that S1 heavy chain retaining full ATPase activity could be prepared by the thermal treatment of S1 in the presence of Mg-ATP. Recent experiments suggest that light chains are necessary for shortening and might serve to transduce, amplify, and moderate the ATP-induced conformational change of myosin (45). The atomic structure of S1 shows that the light chains wrap the α-helical domain of S1 C-terminus (38). Alkali light chain is generally detected as two different molecular species (designated A1 and A2, in the order of decreasing molecular weight). A1 and A2 light chains have basically the same primary structure, except that A1 has an additional peptide at its N-terminus (46). This region is called a "difference peptide," which is rich in alanine and proline, and shows affinity for actin. With respect to alkali light chain combination, three isomers of myosins are possible: two homodimers associating two moles of A1 or A2 per molecule, and one heterodimer associating one each of A1 and A2 per molecule. In the case of higher vertebrates, A1 and A2 light chains are produced from one gene by alternative splicing (47–49). However, this does not seem the case with fish so far: two light chains are translated from different genes (50, 51). Alkali light chains appear to be necessary for stabilization of myosin head (52).

The other category, regulatory light chain, is often referred to as DTNB light chain, because it is dissociated from heavy chain by treatment with a kind of sulfhydryl reagent, 5,5′-dithobis(2-nitrobenzoic acid) (DTNB). This class of light chain is involved in the regulation of actin–myosin interaction in molluscan muscle, but the function in skeletal muscle is not yet clear. In the case of molluscan myosin, regulatory light chain is reversibly removed by treatment with EDTA, and is thus called EDTA light chain. Upon removal of EDTA light chain, molluscan myosin loses Ca^{2+}-sensitivity: Mg^{2+}-ATPase is not activated even in the presence of Ca^{2+} (desensitization). When this light chain is added back to desensitized myosin, Ca^{2+}-sensitivity is restored (resensitization). EDTA light chain is essential for myosin-linked regulation of muscle contraction in molluscs (10). Xie et al. (43) revealed by structural studies on the regulatory domain of scallop myosin that two light chains and heavy chain moiety are essential for the formation of the Ca^{2+}-binding site of myosin. DTNB light chain from fish skeletal myosin as well as those of higher vertebrate binds to the desensitized molluscan myosin, but does not offer Ca^{2+}-sensitivity, unlike EDTA light chain. Mg^{2+}-ATPase activity of myosin shows Ca^{2+}-sensitivity even in the absence of actin, and the changes through desensitization and resensitization are similar to those of actin-activated Mg^{2+}-ATPase. Molluscan myosin, devoid of regulatory light chains, irreversibly lose both superprecipitation activity and Ca^{2+}-sensitivity. Ca^{2+}-sensitivity of Mg^{2+}-ATPase activity of molluscan myosin decreases when measured at the temperature over 20°C (53). This results from the release of regulatory light chain, in other words, thermal desensitization.

The composition of light chains shows remarkable species specificity due to molecular weight, isoelectric point, and molar ratio (54–56). In most animals, the molecular weight of A2 is lower than that of DTNB light chain, but, in the case of some fish species (such as mackerel, sardine, tuna), A2 exceeds DTNB light chain in molecular weight (55, 56). To some extent, it is possible to identify the animal species of origin of muscle or meat products by analyzing the light chain pattern on sodium dodecyl sulfate (SDS)–polyacrylamide gel electrophoresis (57). On the other hand, dark and cardiac myosins of fish have only two kinds of light chain, similar to the slow muscle of higher vertebrates (23, 58).

S1 prepared by chymotryptic treatment of myosin in the presence of EDTA is devoid of DTNB light chain, and is a mixture of two isoforms with respect to associating alkali light chain: S1(A1) and S1(A2). DTNB light chain is lost through the digestion, because it is susceptible to proteolysis in the absence of divalent cations. S1 isoforms were used to study the role of alkali light chain, because they have the same heavy chain fragment (36, 52, 59). The ATPase activity of the isoforms does not practically differ in the absence of actin, but S1(A2) is more thermostable than S1(A1), as observed by the inactivation rate of ATPase and turbidity increase profiles by heat treatment (52, 60).

Differences between the S1 isoforms were also observed in actin activation under low ionic strength (both in apparent dissociation constant for actin and maximum velocity), but become very small under physiological conditions (36, 59). The pH-dependency of ATPase activity in the absence of actin was essentially the same for the two isoforms (36).

D. Primary Structure

The primary structures of myosin heavy-chain and light-chain subunits have been determined by sequencing the proteins and recently by cloning cDNA. For heavy chain, the primary structures have been reported so far for rabbit (61), chicken (62), scallop (63), carp (64), Alaska pollack (65), and others (66). The sequences for the HMM portions (S1 and S2) of Alaska pollack and carp are compared with that of chicken in Figure 2. Fish S1s have similar functional domains to those of rabbit and chicken. The genome size encoding heavy chain of carp is about 12 kbp. The size is about half of the counterpart of rabbit and chicken. In the case of scallop, it was shown that heavy chains of striated and smooth muscle are produced from one gene by alternative splicing (63). Data are also available on the sequence of light chains (46, 50, 51, 64, 67, 68).

Similarity in the sequence of each portion of the myosin molecule is shown in Table 1. The table suggests that there are not so many substitutions of amino acids between fish and higher vertebrates, especially, in the regions essential for functions of myosin. On the other hand, the junctions of the three domains are not very similar. The substitutions are considered to be the result of adaptation to body temperature (ambient temperature for poikilothermal animals) to facilitate ATP hydrolysis and interaction with actin at lower body temperature (i.e., to acquire flexibility of the molecule). These substitutions result in the differences in the thermal stability of myosin or myofibril. Stability of myosin exhibits a good relationship to the body temperature of the animal as described later. Expressions of different isoforms of heavy chain and light chain are observed in carp as a result of temperature acclimation, causing differences in activity and stability of myosin or myofibrillar ATPase (64,69). Studies on the re-

Table 1 Comparison of Amino Acid Sequences of Heavy Chain of Heavy Meromyosin against Pollack Counterpart

Position	Homology(%)		
	Carp[d]	Chicken	Rabbit
Total (1–1287)[a]	83	79	79
S1 (1–840)	79	76	76
ATP-binding site-I (116–136)	100	95	95
ATP-binding site-II (163–193)	97	94	94
ATP-binding site-III (230–256)	96	96	93
Actin-binding site-I (405–419)	93	93	100
Actin-binding site-II (531–560)	87	87	83
Actin-binding site-III (569–580)	83	92	92
RLR region 2[b] (78–93)	82	100	100
SH1-SH2 region 3[c] (687–717)	95	95	95
RLC-binding site (817–834)	100	94	94
ELC-binding site (789–811)	78	57	52
25K-50K junction (204–218)	27	33	33
50K-20K junction (621–646)	60	57	63
S2 (841–1287)	90	83	83

[a] Residue numbers.
[b] 11 Residues including RLR in the middle.
[c] 31 Residues including SH1 and SH2 in the middle.
[d] Acclimated to 10°C (64).
RLR, reactive lysyl residue; RLC, regulatory light chain; ELC, essential light chain.
Source: Ref. 65.

lationship between instability of fish myosin and amino acid substitutions are now in progress (65).

E. Reaction Mechanism

The substrate of myosin is actually Mg-ATP. The reaction mechanism is considered as follows (2, 70):

$$M + ATP \leftrightarrow M1ATP \rightarrow M2ATP \leftrightarrow M. ADP.Pi \rightarrow M.ADP + Pi \rightarrow$$
$$M + ADP + Pi$$

where, M represents myosin and Pi is inorganic phosphate.

First, myosin forms a loose complex with ATP (M1ATP), and then turns into a tight complex (M2ATP). These steps proceed very quickly. Through this change, there is an increase in the intrinsic fluorescence of a tryptophan residue in the S1. The next step (splitting of ATP to ADP and Pi on the myosin molecule without release of products) is a rate-limiting process and, especially in the absence of actin, the rate is very low. Actually, ATPase activity of pure myosin is very weak in the presence of Mg^{2+} (as described above). M.ADP.Pi is in a metastable state, and is unstable in the presence of trichloroacetic acid. Thus, when the reaction is stopped by this reagent, Pi is easily released, resulting in the apparent activity enhancement (referred to as initial burst). Actin increases specific activity of myosin Mg^{2+}-ATPase by 100–200-fold by promoting the decomposition of M. ADP.Pi. In resting muscle, almost all the crossbridges contain 1 mole each of ADP and Pi (hydrolyzed products of ATP). In the "energized" state, they still store the chemical energy of ATP. In the absence of Ca^{2+}, the energized head cannot interact with actin, because tropomyosin blocks the myosin-binding site on actin as described above. Energy transduction proceeds during product release, and thus hydrolysis of ATP does not synchronize with force generation. Pi is released prior to ADP.

III. IMPLICATIONS IN THE PROCESSING AND STORAGE OF MARINE PRODUCTS

A. Instability of Fish Myosin

Deterioration of fish muscle during processing and storage is greatly influenced by denaturation of myosin. Fish myosins are generally unstable, and easily and rapidly form aggregates with concomitant decrease of ATPase activity. Many researchers have reported the instability of fish myosin (11, 12, 16, 27, 28, 32, 71, 72). By mild heat treatment, the ATPase activity of myosin, actomyosin, and myofibrils decreases gradually. However, thermal aggregation of myosin seems to start from the

S2 portion by forming hydrophobic bonding, S2 is considered to be the most susceptible part of myosin to thermal conformational change (73). Through thermal denaturation, S–S bonding is first formed at S1 and S2 portions. Heat-induced formation of myosin oligomer precedes inactivation of myosin ATPase (74). Furthermore, formation of oligomer (the product as a result of noncovalent intermolecular bonding) with concomitant loss of filament formation ability proceeds independently of inactivation of Ca^{2+}-ATPase, suggesting that the responsible part for oligomer formation is the rod portion.

When Ca^{2+}-ATPase is taken as a parameter, inactivation follows a first-order rate reaction against heating time, provided the heating condition is moderate (18, 75). In the case of Mg^{2+}-ATPase activity, inactivation is complicated and follows a two-phase change (initial increase followed by subsequent decrease) (76, 77). The rate constant of inactivation (K_D) is species-specific (75). Such species-specific differences in thermostability are also observed in the case of actomyosin and myofibril (75, 78, 79). Stability of myosin is highly dependent on the body temperature of the animal (in the case of fish, it depends on the temperature of inhabiting waters, except for cruising thermogenic fish such as tuna). When the free energy of denaturation is plotted against habitat temperature, a linear relationship is observed for myofibril and myosin ATPase. Direct measurement of the respective thermal absorption of myosin and actin in muscle by differential scanning calorimetry revealed that the stabilities of both proteins depended on the habitat temperature (80).

Incidentally, as a result of acclimation to different ambient temperatures, changes in myofibrillar ATPase are observed for carp, roach, *etc.* (69, 76, 81, 82). Carp expresses different molecular species (isoforms) of myosin heavy chain as shown in Figure 2 (64). Myosin isoforms thus produced show substantial differences in ATPase activity and thermal stability.

Activation enthalpy and activation entropy of myofibrillar ATPase by thermal inactivation exhibit proportional relationships (74, 83). The enthalpy of activation ($\Delta H^{\neq}$), the entropy of activation ($\Delta S^{\neq}$), and the free energy of activation ($\Delta G^{\neq}$) at 25°C estimated by Arrhenius plots, are 80–100 kcal/mol, 170–240 e.u., and 25–27 kcal/mol for warm-water fish; and 50–70 kcal/mol, 100–160 e.u., and 23–25 kcal/mol for cold-water fish. Positive correlation between $\Delta G^{\neq}$, $\Delta H^{\neq}$, or $\Delta S^{\neq}$ and environment temperature is observed. This is the result of interspecies compensatory adaptation to environmental temperature at the molecular level. Changes in the thermal characteristics of myosin subunits during ice and frozen storage suggest that subunits of cold-water fish deteriorate more rapidly than subunits of tropical fish (80). Such differences in the stability of actin have also been established, but the extent is much less (about 5-fold at maximum). On the other hand, medium pH also greatly affects the stability of fish myofibrils: K_D of myofibrillar Ca^{2+}-ATPase is low (pH 7–8.6)(79).

As for the dark muscle myosin, it is much more stable than ordinary muscle myosin under neutral pH in the case of mackerel (19), but shows almost no difference for yellowtail (20). An Arrhenius plot of K_D value revealed that myofibril is more stable than myosin alone, implying that actin is involved in the stabilization of myosin (84). On the other hand, myosin from shark, which accumulates urea in the body, shows resistance to physiological levels of urea (22).

By taking myofibrillar ATPase as a parameter, many researchers estimated the extent of denaturation or deterioration of muscle foods (see Chap. 12). For example, total Ca^{2+}-ATPase activity (specific activity multiplied by protein concentration) in fish meat paste (surimi) is in good relationship with its gel-forming ability (85). Heat-induced inactivation of Ca^{2+}-ATPase is intensified as neutral salt concentration increases, and the effect is in the order of LiCl>NaCl>KCl. This order is in accordance with the hydration level of salt. On the other hand, many substances were found to prevent or delay the denaturation of muscle protein through this kind of experiment.

B. Prevention of Denaturation

The reagents effective for stabilization of myosin are classified into two groups: ATP and its derivatives in the presence of Mg^{2+}; and sugars, carboxylic acids, and amino acids (86–88). In the case of ATP and its derivatives, the effect is the order of ATP>ADP>pyrophosphate>AMP. ATP shows a similar extent of protection as actin. These substances directly bind to myosin head to cause conformational change, resulting in stabilization of this protein. Actually, S1 structure is considerably stabilized in the presence of Mg-ATP through conformational change. S1 heavy chain (free of light chain) was successfully isolated in the presence of ATP (44). The conformational changes can be monitored by tryptic digestion pattern of S1, the distance between SH1 and SH2, reactivity of lysine residue with TNBS, and ultraviolet spectrum. Such inhibitory effects of ATP and its derivatives are stoichiometrically saturated when S1 binds one molecule per mole. However, the effect of pyrophosphate is complicated and differs depending on the origin of S1 (fish species) and the condition (especially salt concentration) of the assay mixture (89). Decreasing the rate of S1 formation from myofibrils by proteolytic treatment is in good accordance with inactivation rate of ATPase.

On the other hand, the substances belonging to the second group (sorbitol, lactitol, citrate, gluconate, lactate, sodium glutamate, etc.) stabilize proteins by changing the property of environmental water (e.g., dielectric constant) (86). Unlike the substances in category 1, there is a proportional relationship between the molar concentration of additives and the K_D values of Ca^{2+}-ATPase (86, 90). The protective effect of sugars depends on the numbers of hydroxyl groups. On the other hand, basic amino acids promote the denaturation of proteins. The myofibrillar protein of Antarctic krill is remarkably unstable, and its Ca^{2+}-ATPase is rapidly inactivated even during ice storage. To prevent denaturation, the protein should be stored in the presence of sugar (more than 20%) (91).

Sensitivity of myofibrillar Ca^{2+}-ATPase to thermal denaturation is in good relationship with the loss of ATPase activity during frozen storage. The inactivation of myofibrillar Ca^{2+}-ATPase follows basically a first-order rate reaction as a function of frozen storage time (90). Resistance of myofibrillar proteins against thermal and freezing denaturation is also in good accordance. However, the structural change of myosin caused by freezing is considered to be different from that caused by heating (92). In particular, the rod portion is believed to remain undamaged during frozen storage. On the other hand, the inactivation profile of myofibrillar Ca^{2+}-ATPase under high pressure (1–2 kbar) is somewhat different from thermal denaturation in that the former does not follow a first-order rate reaction.

IV. CONCLUSIONS

Studies on fish and shellfish myosins have been limited in comparison to study of myosin from higher vertebrates mainly because of the former's instability. However, tremendous efforts have been made to overcome such handicaps, and a lot of information is now available on fish and shellfish myosins. Myosin ATPase, its inactivation profile in fish muscle, and methods to prevent or retard denaturation have been established by the use of model systems of myofibrillar ATPase activity. During the coming decade, it is probable that effective techniques will be developed to understand the instability of fish myosins and the denaturation mechanism of myofibrillar proteins from the molecular point of view. This will facilitate further effective utilization of fish and shellfish. On the other hand, the relationship between adaptation to habitat environment and the biological strategy of acquiring the best-designed myosin molecule will also be further understood.

REFERENCES

1. HC Mannherz, RS Goody. Proteins of contractile systems. Annu Rev Biochem 45: 427–465, 1976.
2. RS Adelstein, E Eisenberg. Regulation and kinetics of actin-myosin-ATP interaction. Annu Rev Biochem 49: 921–956, 1980.
3. HV Goodson, JA Spudich. Molecular evolution of the myosin family: relationships derived from comparisons of amino acid sequences. Proc Natl Acad Sci USA 90: 659–663, 1993.
4. MS Mooseker, RE Cheney. Unconventional myosins. Annu Rev Cell Dev Biol 11: 633–675, 1995.

5. T Suzuki, S Watabe. New processing technology of small pelagic fish protein. Food Rev Int 2: 271–307, 1987.

6. HE Huxley. The mechanism of muscular contraction. Science 164: 1365–1366, 1969.

7. AF Huxley, R Simmons. Proposed mechanism of force generation in striated muscle. Nature 233: 533–538, 1971.

8. JM Murray, A Weber. The cooperative action of muscle proteins. Sci Am 230: 58–71, 1974.

9. J Kendrick-Jones, E M Szentkiralyi, AG Szent-Gyorgyi. Regulatory light chains in myosins. J Mol Biol 104: 747–775, 1976.

10. J Kendrick-Jones, JM Scholey. Myosin-linked regulatory systems. J Muscle Res Cell Motil 2: 347–372, 1981.

11. CS Chung, EG Richards, HS Olcott. Purification and properties of tuna myosin. Biochemistry 6:3154–3161, 1967.

12. JJ Connell. Studies on the proteins of fish skeletal muscle. 3. Cod myosin and cod actin. Biochem J 58: 360–367, 1954.

13. I Kimura, M Takahashi, E Nagahisa, T Fujita. Preparation of monomeric myosin from walleye pollack surimi. Bull Jpn Soc Sci Fish 48: 251, 1982.

14. T Ojima, S Yoshikawa, K Nishita. Isolation and characterization of myosin from Alaska pollack surimi. Fisheries Sci (Tokyo) 63:811–815, 1997.

15. T Murozuka. Comparative studies on the properties of myosins from the frozen muscle of bigeye tuna and yellowfin tuna. Bull Jpn Soc Sci Fish 42: 163–169, 1976.

16. I Kimura, T Murozuka, K Arai. Comparative studies on biochemical properties of myosins from frozen muscle of marine fishes. Bull Jpn Soc Sci Fish 43: 315–321, 1977.

17. R Takashi. Studies on muscular proteins of fish. VIII. Comparative studies on the biochemical properties of highly purified myosin from fish dorsal and rabbit skeletal muscles. Bull Jpn Soc Sci Fish 39: 197–205, 1973.

18. H Buttkus. Preparation and properties of trout myosin. J Fish Res Bd Canada 23: 563–573, 1966.

19. S Watabe, K Hashimoto. Myosins from white and dark muscles of mackerel—some physico-chemical and enzymatic properties. J Biochem (Tokyo) 82: 1491–1499, 1980.

20. S Watabe, K Hashimoto, S Watanabe. The pH-dependency of myosin ATPases from yellowtail ordinary and dark muscles. J Biochem (Tokyo) 94: 1867–1875, 1983.

21. C S Chen, D C Hwang, S T Jiang. Purification and characterization of milkfish myosin. Nippon Suisan Gakkaishi 54:1423–1427, 1988.

22. S Kanoh, S Watabe, K Hashimoto. ATPase activity of requiem shark myosin. Bull Jpn Soc Sci Fish 51: 973–977, 1985.

23. S Watabe, TNL Dinh, Y Ochiai, K Hashimoto. Immunochemical specificity of myosin light chains from mackerel ordinary and dark muscles. J Biochem (Tokyo) 94: 1409–1419, 1983.

24. WF Harrington and ME Rodgers. Myosin. Annu Rev Biochem 53:35–73, 1984.

25. PJR Barton, M E Buckingham. The myosin alkali light chain proteins and their genes. Biochem J 231: 249–261, 1985.

26. AD McLachlan, J Karn. Periodic charge distributions in the myosin rod amino acid sequence match cross-bridge spacings in muscle. Nature 299: 226–231, 1982.

27. H Buttkus. The sulfhydryl content of rabbit and trout myosins in relation to protein stability. Can J Biochem 49:97–107, 1971.

28. G Hamoir, HA McKenzie, MB Smith. The isolation and properties of fish myosin. Biochim Biophys Acta 40: 141–149, 1960.

29. S Lowey, HS Slayter, AG Weeds, H Baker. Structures of the myosin molecule. I. Subfragments of myosin by enzymatic degradation. J Mol Biol 42: 1–29, 1969.

30. M Matsuura, K Konno, K Arai. Thick filaments of fish myosin and its actin-activated Mg-ATPase activity. Comp Biochem Physiol 90B: 803–808, 1988.

31. E Reisler, M Burke, WF Harrington. Cooperative role of two sulfhydryl groups in myosin adenosine triphosphatase. Biochemistry 13: 2014–2022, 1974.

32. T Murozuka. Comparative studies on the enzymatic properties and thermal stabilities of N-ethylmaleimide modified myosins. Bull Jpn Soc Sci Fish 45: 1503–1512, 1979.

33. I Rayment, WR Rypeniewski, K Schmidt-Base, R Smith, DR Tomchick, MM Benning, DA Winkelmann, G Wesenberg, HM Holden. Three-dimensional structure of myosin subfragment-1: a molecular motor. Science 261: 50–58, 1993.

34. I Kimura, K Arai, S Watanabe. Heavy meromyosin from skipjack tuna, *Euthynnus pelamis*—preparation and enzymatic properties. J Biochem (Tokyo) 86:1629–1638, 1979.

35. T Ikariya, I Kimura, K Arai. Preparation and characterization of myosin subfragment-1 from carp. Bull Jpn Soc Sci Fish 47: 947–955, 1981.

36. Y Ochiai, A Handa, T Kobayashi, S Watabe, K Hashimoto. Isolation and enzymatic properties of myosin subfragment-1 isozymes from the ordinary muscle of tilapia *Oreochromis niloticus*. Nippon Suisan Gakkaishi 55:2143–2149, 1989.

37. A Setton, A Muhlrad. Effect of mild heat treatment on the ATPase activity and proteolytic sensitivity of myosin subfragment-1. Arch Biochem Biophys 235: 411–417, 1984.

38. I Rayment. The structural basis of the myosin ATPase activity. J Biol Chem 271:15850–15853, 1996.

39. M Burke, S Zaager, J Bliss. Substructure of skeletal myosin subfragment-1 revealed by thermal denaturation. Biochemistry 26: 1492–1496, 1987.

40. M Hamai, K Konno. Structural stability of fish myosin subfragment-1. Comp Biochem Physiol 95B: 255–259, 1990.

41. M Burke, V Kamalakannan. Effect of tryptic cleavage on the stability of myosin subfragment 1. Isolation and properties of the severed heavy-chain subunit. Biochemistry 24: 846–852, 1985.

42. P F Flicker, T Wallimann, P Vibert. Electron microscopy of scallop myosin. Location of regulatory light chains. J Mol Biol 169: 723–741, 1983.

43. X Xie, DH Harrison, I Schlichting, RM Sweet, VN Kalabokis, AG Szent-Györgyi, C Cohen. Structure of the regulatory domain of scallop myosin at 2.8Å resolution. Nature 368: 306–312, 1994.

44. M Sivaramakrishnan, M Burke. The free heavy chain of vertebrate skeletal myosin subfragment 1 shows full enzymatic activity. J Biol Chem 257: 1102–1105, 1982.

45. S Lowey, GS Waller, KM Trybus. Skeletal muscle myosin light chains are essential for physiological speeds of shortening. Nature 365: 454–456, 1993.

46. G Frank, AG Weeds. The amino-acid sequence of the alkali light chains of rabbit skeletal-muscle myosin. Eur J Biochem 44: 317–334, 1974.

47. Y Nabeshima, Y Fujii-Kuriyama, M Muramatsu, K Ogata. Alternative transcription and two modes of splicing result in two myosin light chains from one gene. Nature 308: 333–338, 1984.

48. M Periassamy, EE Strehler, LI Garfinkel, RM Gutbis, N Ruiz-Opazo, B Nadel-Ginard. Fast skeletal muscle myosin light chains 1 and 3 are produced from a single gene by a combined process of differential RNA transcription and splicing. J Biol Chem 259: 13595–13604, 1984.

49. P Emerson, SI Bernstein. Molecular genetic of myosin. Annu Rev Biochem 56:695–726, 1987.

50. LD Libera, E Carpine, J Theibert, JH Collins. Fish myosin alkali light chains originate from two different genes. J Muscle Res Cell Motil 12: 366–371, 1991.

51. Y Hirayama, S Kanoh, M Nakaya, S Watabe. The two essential light chains of carp fast skeletal myosin, LC1 and LC3, are encoded by distinct genes and change their molar ratio following temperature acclimation. J Exp Biol 200, 693–701, 1997.

52. Y Ochiai, A Handa, S Watabe, K Hashimoto. Alkali light chains are involved in stabilization of myosin head. Int J Biochem 22:1097–1103, 1990.

53. T Ojima, K Nishita, S Watanabe. Reversible changes in the ATPase activity and in the regulatory light chain content upon heat (30°C)-treatment of "akazara" striated adductor myosin. J Biochem (Tokyo) 89: 1333–1335, 1981.

54. B Focant, F Huriaux, IA Johnston. Subunit composition of fish myofibrils: the light chains of myosin. Int J Biochem 7, 129–133, 1976.

55. S Watabe, Y Ochiai, K Hashimoto. Identification of 5, 5′-dithio-bis-2-nitro-benzoic acid (DTNB) and alkali light chains of piscine myosin. Bull Jpn Soc Sci Fish 48: 827–832, 1982.

56. Y Ochiai, T Kobayashi, S Watabe, K Hashimoto. Mapping of fish myosin light chains by two-dimensional gel electrophoresis. Comp Biochem Physiol 95B: 341–345, 1990.

57. N Seki. Identification of fish species by SDS-polyacrylamide gel electrophoresis of the myofibrillar proteins. Bull Jpn Soc Sci Fish 42: 1169–1176, 1976.

58. TNL Dinh, S Watabe, Y Ochiai, K Hashimoto. Myosin light chains from the cardiac muscle of mackerel, *Pneumatophorus japonicus japonicus*. Comp Biochem Physiol 80B: 203–207, 1985.

59. PD Wagner, AG Weeds. Studies on the role of myosin alkali light chains. Recombination and hybridization of light chains and heavy chains in subfragment-1 preparations. J Mol Biol 109: 455–469, 1977.

60. A Mrakovcic-Zenic, C Oriol-Audit, E Reisler. On the alkali light chains of vertebrate skeletal myosin. Nucleotide binding and salt-induced conformational changes. Eur J Biochem 115: 565–570, 1981.

61. J R Sellers, HV Goodson. Motor protein 2: myosin. In: P Sheterline, ed. Protein Profile, vol.2. London: Academic Press, 1995, pp. 1323–1423.

62. T Maita, E Yajima, S Nagata, T Miyanishi, S Nakayama, G Matsuda. The primary

structure of skeletal muscle myosin heavy chain: IV. Sequence of the rod, and the complete 1,938-residue sequence of the heavy chain. J Biochem (Tokyo) 110: 75–87, 1991.

63.   L Nyitray, A Jancso, Y Ochiai, L Graf, AG Szent-Györgyi. Scallop striated and smooth muscle myosin heavy-chain isoforms are produced by alternative RNA splicing from a single gene. Proc Natl Acad Sci USA 91: 12686–12690, 1994

64.   S Watabe, Y Hirayama, M Nakaya, M Kakinuma, K Kikuchi, XF Guo, S Kanoh, S Chaen, T Ooi. Carp expresses fast skeletal myosin isoforms with altered motor functions and structural stabilities to compensate for changes in environmental temperature. J Therm Biol 22: 375–390, 1998.

65.   T Ojima, N Kawashima, A Inoue, A Amauchi, M Togashi, S Watabe, K Nishita. Determination of primary structure of heavy meromyosin region of walleye pollack myosin heavy chain by cDNA cloning. Fisheries Sci (Tokyo) 64:812–819, 1998.

66.   HM Warrick, JA Spudich. Myosin structure and function in cell motility. Annu Rev Cell Biol 3: 379–421, 1987.

67.   JH Collins. Myosin light chains and troponin C: Structural and evolutionary relationships revealed by amino acid sequence comparisons. J Muscle Res Cell Motil 12:3–25, 1991.

68.   Y Hirayama, A Kobiyama, Y Ochiai, S Watabe. Two types of mRNA encoding myosin regulatory light chain in carp fast skeletal muscle differ in their 3′ non-coding regions and expression patterns following temperature acclimation. J Exp Biol 201: 2815–2820, 1998.

69.   G Goldspink. Adaptation of fish to different environmental temperature by qualitative and quantitative changes in gene expression. J Therm Biol 20: 167–174, 1995.

70.   CR Bagshaw, JF Eccleston, F Eckstein, RS Goody, H Gutfreund, DR Trentham. The magnesium ion-dependent adenosine triphophatase of myosin. Two-step processes of adenosine triphosphatase association and adenosine diphosphate dissociation. Biochem J 141: 331–349, 1974.

71.   JJ Connell. The relative stabilities of the skeletal-muscle myosins of some animals. Biochem J 80: 503–509, 1961.

72.   T Murozuka, R Takashi, K Arai. Relative thermo-stabilities of Ca^{2+}-ATPase of myosin and actomyosin from tilapia and rabbit. Bull Jpn Soc Sci Fish 42:57–63, 1976.

73.   A Bertazzon, TY Tsong. High-resolution differential scanning calorimetric studies of myosin, functional domains, and supramolecular structures. Biochemistry 28: 9784–9790, 1989.

74.   I Kimura, K Arai, K Takahashi, S Watanabe. Earliest event in the heat denaturation of myosin. J Biochem (Tokyo) 88: 1703–1713, 1980.

75.   A Hashimoto, A Kobayashi, K Arai. Thermostability of fish myofibrillar Ca-ATPase and adaptation to environmental temperature. Bull Jpn Soc Sci Fish 48: 671–684, 1982.

76.   I A Johnston, N Frearson, G Goldspink. The effects of environmental temperature on the properties of adenosine triphosphatase from various species of fish. Biochem J 133: 735–738, 1973.

77.   T Taguchi, M Tanaka, Y Nagashima, K Amano. Thermal activation of actomyosin

Mg^{2+}-ATPase from flying fish and black marlin muscles. J Food Sci 51: 1407–1410, 1986.

78. K Arai, K Kawamura, C Hayashi. The relative thermostabilities of actomyosin AT-Pase from the dorsal muscles of various fish species. Bull Jpn Soc Sci Fish 39: 1077–1085, 1973.

79. A Hashimoto, K Arai. The effects of pH and temperature on the stability of my-ofibrillar Ca-ATPase from some fish species. Bull Jpn Soc Sci Fish 44: 1389–1393, 1978.

80. BK Howell, AD Matthews, AD Donnelly. Thermal stability of fish myofibrils: a differential scanning calorimetric study. Int J Food Technol 26: 283–295, 1991.

81. IA Johnston, W Davidson, G Goldspink. Adaptation in Mg^{2+}-activated myofibrillar ATPase activity induced by temperature acclimation. FEBS Let 50: 193–195, 1975.

82. RK Penny, G Goldspink. Regulatory proteins and thermostability of myofibrillar ATPase in acclimated goldfish. Comp Biochem Physiol 69B: 577-538, 1981.

83. IA Johnston, G Goldspink. Thermodynamic activation parameters of fish myofib-rillar ATPase enzyme and evolutionary adaptation to temperature. Nature 257: 620–622, 1975.

84. K Yamashita, K Arai, K Nishita. Thermo-stabilities of synthetic actomyosins in various combinations of myosin and actin from fish, scallop, and rabbit muscles. Bull Jap Soc Sci Fish 44: 485–489, 1978.

85. N Katoh, W Nozaki, I Komatsu, K Arai. A new method for evaluation of the qual-ity of frozen surimi from Alaska pollack. Bull Jpn Soc Sci Fish 45: 1027–1032, 1979.

86. T Ooizumi, K Hashimoto, J Ogura, K Arai. Quantitative aspect for protective ef-fect of sugar and sugar alcohol against denaturation of fish myofibril. Bull Jpn Soc Sci Fish 47, 901–908, 1981.

87. T Ooizumi, Y Nara, K Arai. Protective effect of carboxylic acids, sorbitol, and Na-glutamate on heat denaturation of chub mackerel myofibrils. Bull Jpn Soc Sci Fish 50: 875–882, 1984.

88. G A Mac Donald, TC Lanier. Actomyosin stabilization to freeze-thaw and heat de-naturation by lactate salts. J Food Sci 59:101–105, 1994.

89. K Konno. Suppression of thermal denaturation of myosin subfragment-1 of Alaska pollack (*Theragra chalcogramma*) by sorbitol and accelerated inactivation by py-rophosphate. J Food Sci 57: 261–264, 1992.

90. Y Matsumoto, T Ooizumi, K Arai. Protective effect of sugar on freeze-denatura-tion of carp myofibrillar protein. Bull Jpn Soc Sci Fish 51: 833–839, 1985.

91. T Ooizumi, M Nakamura, A Hashimoto, K Arai. Thermal denaturation of myofib-rillar protein of Antarctic krill and protective effect of sugar. Bull Jpn Soc Sci Fish 49: 967–974, 1983.

92. Y Azuma, K Konno. Freeze denaturation of carp myofibrils compared with ther-mal denaturation. Fisheries Sci (Tokyo) 64: 287–290, 1998.

4
Phospholipases

Clara López-Amaya and Alejandro G. Marangoni
University of Guelph, Guelph, Ontario, Canada

I. GENERAL CONSIDERATIONS

Lipases, the lipolytic enzymes able to degrade phospholipids, are grouped into two categories: acylhydrolases and phosphodiesterases. The nomenclature of phospholipases is based on the position of their attack on the phospholipid. Phospholipase A_1 splits the sn-1 fatty-acyl ester bond; phospholipase A_2 splits the sn-2 fatty-acyl ester bond; phospholipase B splits both or the remaining ester in the lysophospholipid; phospholipase C attacks the diglyceride-phosphate link; and phospholipase D, the choline–phosphate link (Fig. 1).

Phospho(lipases) differ from classic esterases in that their natural substrates are insoluble in water and their activity is maximum only when the enzyme is adsorbed onto a lipid–water interface. Several hypotheses have been proposed to explain this phenomenon referred to as interfacial activation, and they can be organized into two groups: the enzyme models and the substrate models (1).

Enzyme models relate interfacial activation to a regulation of the catalytic power of the enzyme by the lipid–water interface. At present, the model postulated by Verger (2) is the most commonly accepted. A conformational change of the lipolytic enzyme upon interaction with interfaces (3) would induce a modification in the active site, triggering the high catalytic rates observed with substrate aggregates. In this model, the existence of an additional site in the enzyme (topographically and functionally distinct from the active site) responsible for the interaction between the enzyme and the substrate aggregate (2, 4) is assumed. This region has been identified as the "penetration

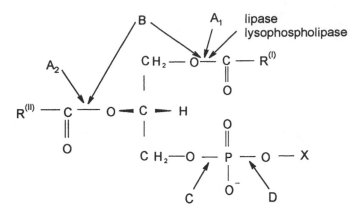

Figure 1 Attack sites of phospholipases. $R^{(I)}$, $R^{(II)}$ are alkyl chains; X denotes any of the moieties (choline, serine, ethanolamine, etc.) present in 3-sn-phosphoglycerides. (From Ref. 7.)

site" (2), "supersubstrate binding site" (4), and "interface recognition site" (IRS) (5).

Substrate models describe interfacial activation as the result of changes in the substrate itself, caused by a surface created by an aggregation process. Changes in substrate concentration and conformation/orientation may be involved in the process (6). When a micelle is formed in a phospholipid solution, the effective concentration of a phospholipid can increase by at least three orders of magnitude depending on the critical micellar concentration (CMC) (7). It has been demonstrated for pancreatic lipase and for phospholipase A_2 that the hydrolysis rate for a substrate aggregate can be 10^3–10^4 times higher than that obtained for a monomeric state substrate (6). While substrate molecules in the interface have an orientation that favors their interaction with soluble enzymes, free monomers tumble free in solution (5). In phospholipid aggregates the orientation of the glycerol backbone relative to the interface changes (8, 9). As a result of this conformational change, the two-acyl esters become nonequivalent, with the ester at position 1 located deeper in the hydrophobic region. The orientation of the esters at positions 1 and 2 of the glycerol is crucial during phospholipase–substrate interactions (10).

The origin, properties, and functions of phospholipases A_1, A_2, and C are briefly described in the following sections. Other phospholipases are not included either because very little is known about them or because their significance in aquatic organisms is not relevant.

A. Phospholipases A$_1$ and A$_2$

Phospholipases A$_1$ (PLA$_1$s) are widely distributed in nature, have broad specificity, and can act on lysophospholipids. Microsomal and lysosomal are the two types of PLA$_1$s present in mammalian intracellular organelles (11). Ca^{+2} and charged amphipaths alter the surface charge on the substrate micelle or membrane influencing hydrolysis; however, Ca^{+2} is not required for activity (7). PLA$_1$s from rat liver plasma membrane (10) and from the muscle of bonito (*Euthynnus pelamis*) (12) catalyze hydrolysis and transacylation.

PLA$_1$ seems to be the predominant bacterial lypolitic enzyme and is perhaps the best characterized PLA$_1$. *E. coli* has a detergent-resistant PLA$_1$ associated with its outer membrane and a detergent-sensitive PLA$_1$ associated with the cytoplasmic membrane and soluble cytosolic fractions, both of which are Ca^{+2}-dependent (11).

Phospholipase A$_2$ (PLA$_2$) is ubiquitous in nature and probably the most studied species of all lipolytic enzymes. PLA$_2$s have been classified into three main groups and several subgroups (13, 14) and more recently a fourth group has been added (15). Groups I, II, and III are extracellular enzymes, also termed secretory PLA$_2$s, with a high disulfide bond content, Ca^{+2} requirement for catalysis, and low molecular weight (14 kDa). Group IA includes PLA$_2$s from cobras and kraits; group IB includes porcine and human pancreatic enzymes; group IIA includes rattlesnakes, vipers, and human synovial/platelet phospholipases; group IIB includes gaboon viper phospholipase; and group III includes bee and lizard enzyme phospholipases. Although PLAs of these three groups are mainly extracellular, enzymes of groups I and II have recently been found in human tissues (15). Group IV comprises the Ca^{+2}-dependent intracellular PLA$_2$ with an approximate molecular weight of 85 kDa and specificity for arachidonic acid. Ca^{+2}-independent intracellular PLA$_2$s could constitute another group. A potential member of this group is a 40 kDa cytosolic PLA$_2$, isolated from canine myocardium, with a preference for arachidonyl-containing phospholipids (13, 14). PLA$_2$s with similar characteristics have been found in P388D1 cells (16) and in cod (*Gadus morhua*) muscle. The latter has a molecular weight of over 50 kDa (17). Determination of the sequences of these PLA$_2$s will help one to decide to which group they belong. In addition to these groups, several Ca^{+2}-dependent and Ca^{+2}-independent PLA$_2$s with distinct properties are currently being characterized; therefore, it is assumed that in the future a wider classification will have to be made (15). Some of the properties of PLA$_2$ are summarized in Table 1.

Intracellular PLA$_2$s are present in almost every mammalian cell in low concentrations (10), playing diverse and important roles. They participate in basal phospholipid metabolism and some of them play a digestive role in phagocytosed materials (15). In many tissues, PLA$_2$s are involved in the retailoring of the molecular species of membrane lipids (7). Furthermore, Ca^{+2}-independent

Table 1 Comparison of Properties of PLA$_2$ Groups

Characteristics	Group		
	I/II/III	IV	?[a]
Localization	secreted	cytosolic	cytosolic
Molecular mass	~14 kDa	~85 kDa	~40 kDa
Amino acids	~125	~750	
Cysteines	10–14	9	
Disulfides	5–7	0	
Dithiothreitol sensitivity	+	–	–
Arachinodate preference	–	+	+
Ca^{+2} required	mM	μM, not absolute[b]	none
Ca^{+2} role	catalysis	memb. assoc.	none
Regulatory protein	–	–	PFK[c,d]
Regulatory cofactors	–	–	ATP
Regulatory phosphorylation	–	+	–
Lyso-PLA activity	–	high	+
PLA$_1$ activity	–	+	–
Transacylase activity	–	+	---
Fatty acyl-CoA hydrolase	–	–	+

[a] Canine myocardial PLA$_2$. Group number has not been determined yet.
[b] Ca^{+2} can be substituted by metals, salts in vitro.
[c] Phosphofructokinase.
[d] To be confirmed.
Source: Ref. 15.

PLA$_2$s participate in eicosanoid synthesis and group IV PLA$_2$s are involved in signal transduction (15).

Pancreatic phospholipases are stored in secretory granules as inactive precursors (proPLA$_2$s). Once they are secreted into the intestine, trypsinolysis occurs, releasing a small peptide and the activated enzyme, which digests emulsified phospholipids from the diet or from bile juice (10, 18, 19). In the mechanism proposed by Pieterson et al. (20) for pancreatic PLA$_2$s, monomeric lecithins can be hydrolyzed by both the proenzyme and the activated enzyme, while micellar substrates can only be hydrolyzed by the activated enzyme. This suggests that the active site is already functional in the proenzyme. When tryptic activation occurs, the zymogen loses its activation heptapeptide and experiences a conformational change with the formation of an ion pair between the -NH$_3^+$ group of the N-terminal alanine and a buried carboxylate. Parallel to the formation of this salt bridge, a supersubstrate-binding site is created. This site allows enzyme interaction with micelles (20). Figure 2 presents a schematic representation of the hypothetical series of

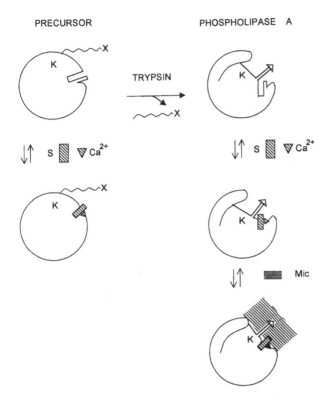

Figure 2 Schematic representation of PLA$_2$ zymogen conversion to the activated enzyme, and its interaction with the substrate. X, the zymogen activation heptapeptide; S, monomer substrate; K, catalytic region that forms the monomer substrate-binding site, Ca^{+2} binding site; arrow, supersubstrate binding site. (From Ref. 20.)

events that occur during zymogen activation of pancreatic PLA$_2$. Venom PLA$_2$s are involved in neurotoxic, miotoxic, anticoagulant, or hemolytic and digestive functions (21–23). A host defense function, suggested by their expression in inflammatory diseases, a role in signal transduction, and their participation in eicosanoid production are physiological functions that have also been attributed to group II PLA$_2$s (24).

Numerous PLA$_2$'s three-dimensional (3D) structures, determined by x-ray crystallography, have been published but none of them are from fish origin. The first 3D structure obtained from a PLA$_2$ was from the mature bovine phospholipase (19). Recently, a 1.5 Å orthorombic form of the bovine recombinant PLA$_2$ was obtained (25). Of the amino acid residues, 44.7% are in α helical conformation and 6.5% are a two-stranded element of β structure (Fig. 3). This bovine

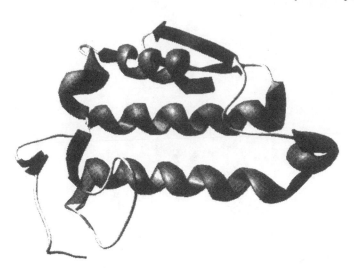

Figure 3   Schematic representation of the three-dimensional structure of the orthorombic form of the bovine recombinant PLA$_2$ (PDB entry: 1UNE).

PLA$_2$ has 123 residues, five alpha helices, two 3,10 helices, nine beta turns, one beta hairpin, and seven disulfide bonds that maintain together all the elements of the secondary structure (25).

Over 80 pancreatic, snake venom, and intracellular PLA$_2$ species have been identified and sequenced. The high degree of homology among these phospholipases, with the exception of bee-venom PLA$_2$, indicates that they may have evolved from a common ancestor (19). A comparison of some PLA$_2$s sequences is illustrated in Figure 4. In snake venom and in pancreatic PLA$_2$s, the catalytic triad (His-48, Asp-99, H$_2$O) is conserved (26). The bee-venom PLA$_2$ sequence shows less amino-acid identity; however, its three-dimensional structure reveals an active site topology very similar to that of the other PLA$_2$s. The stretch between residues 62 and 66 is the most notable difference between the pancreatic and snake venom enzymes (19).

The fact that the active site of PLA$_2$ has a similar configuration to that of serine proteases, including the presence of an His–Asp couple, suggests that they might also have a similar catalytic process. Three steps are involved in the serine proteases mechanism:

1.   The serine OH group performs nucleophilic attack.
2.   The imidazole ring of histidine transfers protons.
3.   Slightly positively charged NH groups fix and stabilize the substrate's carbonyl oxygen atom.

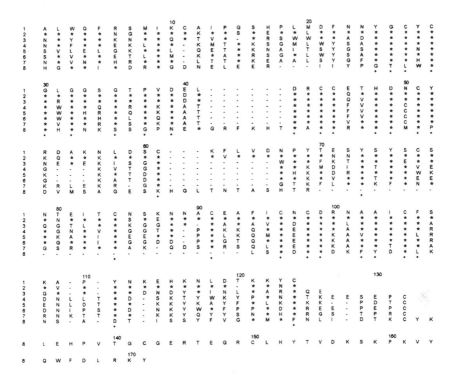

Figure 4 Amino acid sequence comparison of different PLA₂s. Lane 1, pig; lane 2, cow; lane 3, *Naja naja atra;* lane 4, *Agkistrodon piscivorus piscivorus,* Asp-49 enzyme; lane 5, *Agkistrodon piscivorus piscivorus,* Lys-49 PLA₂ homologue; lane 6, *Crotalus atrox;* lane 7, human synovial fluid; lane 8, bee venom. Asterisks represent identity with the sequence of the porcine pancreatic PLA₂. Dashes were introduced for alignment purposes. Diamonds indicate conserved residues in 95% of the active phospholipases. (From Ref. 19.)

In the deacylation step of this proton-relay type reaction, the nucleophilic function performed by the serine may be fulfilled in PLA₂s by the immobilized water molecule (19). The proton relay system of all the PLA₂s is buried in the interior of the hydrophobic active-site wall (10).

Based on x-ray crystallographic studies of PLA–substrate analog or –transition state analog complexes, the catalytic mechanism of PLA₂ hydrolysis of monomeric substrates has been proposed (19, 10, 15). In this model a substrate molecule binds to the enzyme's active site with the participation of hydrophobic forces and polar interactions between phosphate and calcium, and phosphate and the side chain of residue 69. A water molecule, hydrogen-

bonded to His-48, attacks the carbonyl carbon atoms of the substrate. A tetrahedral intermediate is formed and stabilized by both the calcium ion and the
peptide NH group of residue 30 (19). Ca^{+2} is fixed in the carbonyl oxygens of
Tyr-28, Gly-30, Gly-32, two water molecules, and Asp-49 (26). At the end, a
fatty acid and a lysophospholipid are released as a result of the breakdown of
the tetrahedral intermediate and a proton is transferred to His-48. An
overview of the interaction of a mutant porcine pancreatic phospholipase with
an amide inhibitor is shown in Figure 5. A strong hydrogen bond between the
$N^{\delta 1}$ atom and the nitrogen of the inhibitor is formed. Pancreatic and snake
venom PLA_2s have either equivalent functional groups or identical residues
involved in binding. The conformation of the inhibitor in the active site corre-

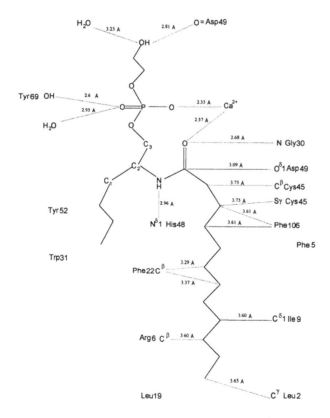

Figure 5   Schematic synopsis of the interaction of (R)-2-dodecanoyl-amino-1-hexa-
nol-phosphoglycol with the active site of a mutant porcine pancreatic PLA_2. (From
Ref. 19.)

sponds to that of aggregated phospholipids. Binding of the inhibitor does not affect the active-site residue positions. Therefore, the active site conformation is preformed to bind a lipid molecule in the conformation determined by the substrate aggregate (19).

X-ray crystal structures have been obtained only with monomeric substrate analogs; consequently, a direct picture of PLA_2–substrate aggregate complexes is not available. More information is required to explain the PLAs' high activity in the presence of aggregated substrates. Interesting results from biochemical and x-ray studies on residues involved in aggregate binding can be found in work by Wooley (19).

B. Phospholipases C and Signal Transduction

As a result of the cleavage of membrane phospholipids at the phosphate moiety by phospholipase C (PLC), the polar head group is liberated and diacylglycerols or ceramides are produced. PLCs have a wide substrate specificity, ranging from the highly specific PLCs such as phosphatidylinositol (PI)-PLC and sphingomyelinase (Smase) C, which hydrolyze PI and sphingomyelin SM, respectively; to the less specific such as phosphatidylcholine (PC)-PLC, which hydrolyzes PC and also phosphatidylserine (PS) and phosphatidylethanolamine (PE) (19). PLC is usually associated with secretions from pathogenic bacteria. For instance, *Clostridium welcheii* PLC and its toxins are responsible for the damage caused during gas gangrene (7). Bacterial PLCs are small enzymes (Mr= 20–30 kDa) that usually require zinc for activation. Enzymes that require zinc also need divalent cations, particularly Ca^{+2}. This additional requirement is probably related to the necessity of PLCs for a positive ξ potential. Therefore, the role of Ca^{+2} and other cations may be to supply an appropriate charge to the supersubstrate–water interface (11). *Bacillus cereus* PLC has been successfully used as a mimicking agent of the action of mammalian PC-PLC (19).

PLC from mammalian tissues has been associated with the agonist-induced generation of second messengers in signal transduction pathways. The hydrolysis of phosphatidylinositol-4-5-bisphosphate (PIP_2), catalyzed by PLC to liberate inositol-1,4,5-trisphosphate (IP_3) and diacylglycerol (DAG), is stimulated by more than 40 hormones, growth factors, and other agents. Signal transduction also seems to occur either via agonist-induced breakdown of PC by PLC to yield DAG, or via phospholipid hydrolysis, catalyzed by phospholipase D (PLD), to produce phosphatidic acid. Phosphatidic acid phosphohydrolase can attack phosphatidic acid, yielding DAG. PC-PLC activation is an important step in the mitogenic response to platelet-derived growth factor, as is the involvement of DAG, derived from PC, in the atypical regulation of protein kinase C in cells transformed by ras and src oncogenes. Furthermore, PC has been identified as a major source of the eicosanoid precursor arachidonic acid (19 and references therein).

II. FISH PHOSPHOLIPASES

A. Introduction

Lipids in fish are utilized for two purposes: as a supply of metabolic energy and as essential components of cellular membranes. It is known that juvenile and adult fish are able to synthesize phospholipids (27). Dietary sources of phospholipids such as phosphatidylcholine and phosphatidylinositol, rich in ω3 polyunsaturated fatty acids, are commonly required by fish (28). For example, the growth rate and the food efficiency of the red sea bream *Pagrus major* dramatically improve with a diet rich in ω3 polyunsaturated fatty acids. It has also been demonstrated that diets rich in phospholipids are essential for growth of red sea bream and larval ayu *Plecoglossus altivelis* (29), and that semipurified diets supplemented with phospholipids significantly improve survival and growth of carp *Cyprinus carpio* larvae (27). The nature of the requirement has not been elucidated but it is assumed that, in larval growing stages, the demanding processes of membrane building and renewal possibly exceed the capacity of phospholipid synthesis (27).

Parallel to the importance of phospholipids is the role of phospholipases in a variety of biological functions in fish. Phospholipases in fish muscle may be responsible for membrane integrity and processes ranging from the production of eicosanoids to the deterioration of fish muscle upon freezing. Despite their importance, documentation about fish muscle phospholipases is poor due to difficulties in their isolation and characterization (17). For the same reason, there is also a lack of information about the enzymology of digestive lipolysis in fish (30). Purification of PLA_2s from pollock, *Pollachius virens* (31); pyloric caeca (32) and hepatopancreas of red sea bream, *Pagrus major* (30); liver of trout, *Salmo gairdneri* (33); and cod, *Gadus morhua* muscle (17), has been reported. A summary of the main characteristics of these phospholipases is presented in Table 2.

A compilation of the most relevant information available about fish phospholipases including general properties, function, and importance for aquatic organisms is presented in this section. Due to extensive knowledge of mammalian phospholipases, mainly phospholipase A_2, and to their similarity with fish phospholipases, comparisons with these enzymes are made throughout the chapter.

B. Phospholipase A_1 from Bonito (*Euthynnus pelamis*)

Living cells have a system to get rid of lysophosphatidylcholine (LPC), which would otherwise cause cellular damage due to its lytic activity. LPC has been detected after cell death, in cases of myocardial ischemia, and in fish muscle

Table 2 Main Properties of Purified Fish Phospholipases

Origin	Pollock (Pollachius virens)[a] Muscle	Red sea bream (Pagrus major)[b] Pyloric caeca	Red sea bream (Pagrus major)[c] DE-1 Hepatopancreas	Red sea bream (Pagrus major)[c] DE-2 Hepatopancreas	Cod (Gadus morhua)[d] Possibly lysosomal cytosolic	Rainbow trout (Salmo gairdneri)[e] Liver	Carp Cyprinus carpio[f] PC hydrolysis Hepatopancreas	Carp Cyprinus carpio[f] Wax ester synthesis Hepatopancreas
MW (kDA)	~13	~14	~13.5	~13.5	Above 50	~73		
pH optimum for activity	8.5–9	8–9	10[g]	8–9	4[h]	20°C: 8–8.5 5°C: 9	5	4–7
Temperature (°C) optimum for activity	37–42				40		35–45	30–40
Ca^{+2} requirement	Absolute	Absolute	Absolute[g]	Absolute[g]	None	ma	ma	
EDTA inhibition	Yes	Yes	No	No	No		No	No
Sodium deoxycholate requirement	ma	ma	ma[g]	PC: ma PE: inhibits activity			ma	None

DE-1, isoform 1; DE-2, isoform 2. ma; for maximal activity.

Source: [a] Ref. 31; [b] Ref. 32; [c] Ref. 30; [d] Ref. 17; [e] Ref. 33; 40; [f] Ref. 49.

[g] Substrates: PC, PE (2mM).

[h] For PC at 25°C.

during frozen storage. In fish, the activity of lysophospholipase is higher than the activity of phospholipase; therefore, the storage of lysophospholipids is a rare event. Recently, almost 10% of the phospholipid found in the fresh muscle of bonito (*Euthynnus pelamis*) was LPC rich in highly unsaturated fatty acids such as docosahexanoic acid (DHA) and eicosapentanoic acid (EPA). Furthermore, it was found that during frozen storage of bonito muscle almost twice the initial content of LPC had accumulated. A great proportion of the LPC was 1-lyso-2-acyl-sn-glycero-3-phosphocholine (2-acyl LPC). The presence of PLA_1 activity for PC in the muscle of bonito has been connected with the accumulation of LPC (34).

PLA_1 from the muscle of bonito has an optimum activity at pHs between 6.5 and 7.0 and temperatures from 20 to 30°C. About 50% of the PLA_1 activity at 20°C remains at 4°C. Moreover, the activity of this PLA_1 is independent of Ca^{+2} and unaffected by EDTA. Involvement of this PLA_1 in wax ester synthesis and transacylation has been reported (12). Using 1-stearoyl-2-arachidonoyl PC as a donor and oleyl alcohol as an acceptor molecule, in the absence of Ca^{+2}, a high percentage of the wax fraction obtained is stearic acid. It seems that PLA_1 transfers the sn-1 fatty acid to the alcohol. Bonito muscle has a high proportion of 1-palmitoyl-2-DHA PC. When 1,2 dipalmitoyl PC is used as a donor and 2-DHA LPC as acceptor in the presence of PLA_1, 1-palmitoyl-2-DHA PC is produced. However, when the donor or the acceptor or the PLA_1 is omitted, 1-palmitoyl-2-DHA PC is not detected. Transacylation of fatty acid from the sn-1 position of the donor's PC to LPC occurs and strict specificity for the sn-1 position is observed for both the donor and the acceptor (12).

C. Phospholipase A_2 from Red Sea Bream (*Pagrus major*)

Lysophosphatidylcholine (LPC) and free fatty acids (FFA) are the products of the hydrolysis of the acid moiety from the sn-2 position of the diacylglycerophospholipid catalyzed by phospholipase A_2. Phospholipase A_2 hydrolyzes essential dietary phospholipids in marine teleosts. Information about the organs that produce and secret digestive PLA_2 and the ontogeny of the PLA_2 cells would help to improve the composition and quality of diets used in cultured fish (35).

Pagrus major, as with many fish species, instead of having a pancreas, has the pancreatic tissue in the liver, an organ referred to as the hepatopancreas (36). It has been shown that the anti-*Naja naja* venom phospholipase A_2 antibody immunoreacts with phospholipase A_2 purified from fish hepatopancreas and intestine (35). Using this antiserum phospholipase A_2 of adult *Pagrus major* was found to be stored in zymogen granules in the pancreatic acinar cells as for mammalian pancreatic PLA_2s (35, 37, 38). It is assumed that the prophospholipase of red sea bream is secreted into the intestinal lumen from the hepatopancreas and then is activated by tryptic hydrolysis (35) following a similar process to that of pancreatic PLA_2s.

A Ca^{+2}-dependent phospholipase A_2 from the pyloric caecum (32) and two phospholipase A_2 isoforms DE-1 and DE-2, related to group I PLA_2, from the hepatopancreas of red sea bream *Pagrus major* (30) have been purified and characterized. After secretion into the intestinal tract, the PLA_2 isoforms from the hepatopancreas along with pyloric caeca PLA_2 probably hydrolyze dietary and biliary phospholipids. The molecular weight of PLA_2 from the hepatopancreas of red sea bream is 13,500 Da (30). Moreover, this PLA_2 is Ca^{+2} dependent, has an optimum pH in the alkaline region, is stable against acid, heat treatment, and 0.1% TFA-49.5% acetonitrile (32), and has a cystein content of 10. All the above-mentioned properties of red sea bream PLA_2 are very similar to those of mammalian pancreatic PLA_2s (32 and references therein).

A comparison of the sequences of red sea bream hepatopancreas DE-1 and DE-2 PLA_2 isoforms, mammalian pancreatic PLA_2, *Naja naja* venom, the pancreatic PLAs of pigs, cattle, and human beings is illustrated in Figure 6. It is worth noting the striking homology among the compared sequences. Residues at positions 4, 5, 8, 9, 25–30 are conserved in all these PLA_2s (30).

PLA_2 is also located in hepatic tissue cells with iron in their cytoplasm. These cells possibly correspond to phagocytic leukocytes (35). It is assumed that intestinal digestive PLA_2 is produced and secreted by cells located in the bottom of the intestinal crypt of the pyloric caeca and the upper intestine where PLA_2 like immunoreactivity can be detected. The cells located in this region are similar to the Paneth cells of certain mammals and may complement the digestive function of the acinar pancreatic cells (35).

Red Sea bream larvae begin to feed on the fourth or fifth day after hatching.

```
      1       5         10          15          20        25        30
1) A L W Q F G N M I Q C A Q P G V N P F L     Y N D Y G C Y C G L G G
2) A L N Q F R Q M I L C V M P D S W P I F D Y A D Y G C Y C G L G G
3) A L W Q F R S M I K C A I P G S H P L M D F N N Y G C Y C G L G G
4) A L W Q F N G M I K C T I P S S E P L M D F N N Y G C Y C G L G G
5) A V W Q F R K M I K C V I P G S D P F L E Y N N Y G C Y C G L G G
6) N L Y Q F K N M I K C T V   P S R S W W D F A D Y G C Y C G R G G
```

Figure 6 N-terminal amino acid sequence comparison of red sea bream hepatopancreas DE-1 (1), DE-2 (2), porcine pancreatic (3), bovine pancreatic (4), human pancreatic (5), *Naja naja naja* venom (6) PLA_2. The boxes contain the conserved residues. (From Ref. 30.)

How do they digest phospholipids if zymogen granules in the pancreas do not appear until the 13th day and larvae reserves have been exhausted by this time? It has been proposed that either the enzyme formation is triggered by food intake or enzymes are provided by the food organism (35). On the other hand, PLA_2 in intestinal cells is not detected until after the 85th day. Although the hypotheses mentioned previously could apply in this case, perhaps this phospholipase functions only as a complement of pancreatic PLA_2 or its role is to control intestinal bacterial flora (35).

D. PLA_2 and Early Development in Atlantic Halibut

The role of PLA_2 during early development in fish is regulation of membrane lipid modifications as a response to environmental changes, provision of fatty acids for metabolic energy substrates, and biosynthesis of prostaglandin. Although PLA_2 activity has not been detected in eggs of Atlantic halibut, a 122% increase in PLA_2 activity between days 10 and 17 of embryo development at 20°C has been observed. The highest PLA_2 activity in 17-day embryos was detected at 37°C and pH 8. Halibut embryo development temperature is between 5 and 7°C. Surprisingly, the optimum temperature for various phospholipases usually is much higher than the temperatures at which embryos develop (39). Trout liver PLA_2 activity is higher at 20°C for trout acclimated at 5°C than for trout acclimated at 20°C (40). The finding of an optimum enzyme activity between 34°C and 37°C led Evans et al. (39) to conclude that there was not a relationship between enzyme activity and lipid phase transition. It should not be forgotten that in addition to the PC's transition temperature, there is a pretransition temperature of 35°C at which the hydrocarbon chains adopt a perpendicular orientation to the bilayer facilitating enzyme binding. Therefore, the highest PLA_2 activity at temperatures around 35°C may be associated with this process.

In Atlantic halibut a decrease in the total lipids and in the unsaturated fatty acid/saturated fatty acid ratio and unsaturation index in PC indicate that egg membranes are becoming less fluid (39). The presence of elevated cholesterol levels in the period between the unfertilized egg and the hatching and the increase in the PC/PE ratio are also indicators of less membrane fluidity (41). Conversely, a decrease in the PC/PE ratio by 13% between hatching and 7 days posthatch of Atlantic halibut (preferential retention of PE during hatching) has been observed (39). It is known that accumulation of PE allows fish membranes to remain fluid during cold adaptation (41). During embryo development, 32% of PUFA is probably removed through oxidation to products such as prostaglandins. PLA_2 plays an essential role in the alteration of acyl chains for phospholipid restructuring during the early stages of temperature acclimation (42).

Additional lipid synthesis would be required as a response to the decline in embryo membrane fluidity. Developing embryos can use two sources of precursor molecules for lipid synthesis: the free amino acids (FAA) large pool and amino acids from protein degradation. Evidence of FAA pool depletion in turbot *Scophthalamus maximus,* lemon sole *Microstomus kitt,* and gilthead sea bream *Sparus aurata* has been shown (39 and references therein).

Eicosanoids are 20 carbon compounds derived from essential fatty acids, with very strong and interacting biological functions that affect almost every tissue in mammals. Arachidonic acid is converted to a cyclic endoperoxide that can form prostaglandins and other compounds (7). Prostaglandins affect many cellular and tissue functions by regulating the synthesis of the intracellular messenger molecule 3', 5'-cyclic AMP (cAMP) which, in turn, mediates the action of many hormones (43).

Precursors for 2-series prostaglandins can be obtained through removal of arachidonic acid from PE and triglycerides (TG) in Atlantic halibut developing embryos. Arachidonic acid seems to be the main precursor of prostaglandins in fish, as well as in mammals (44, 45). A relationship between an excess of AA-derived prostaglandins and atherothrombotic and chronic inflammatory disorders in humans (46) and cardiac histopathology in salmon has been established (47).

E.   Effects of Diet on PLA$_2$ and Prostaglandins in Atlantic Salmon

Diet influences the development of cardiac lesions in Atlantic salmon, *Salmo salar,* as well as cardiac phospholipase A activity. Diets supplemented with fish oil (FO), sunflower oil (SFO), or linseed oil (LO) as lipid sources were given to Atlantic salmon for 12 weeks. After this period, a local degeneration of the myocardium with associated leukocyte accumulation in SFO-fed fish and minor tissue damage in FO-fed fish were observed. There were no lesions in LO-fed fish. SFO-fed fish incorporated more AA heart and leukocyte phospholipids, while LO-fed fish accumulated less. In the latter case, there was a higher proportion of 18 and 20-carbon ω -3 PUFAs than ω -6 PUFAs. Competition at the Δ-6-desaturase between 18:2ω -6 and 18:3ω -3 and at the Δ-5-desaturase between 20:3ω -6 and 20:4ω -3 results in a low production of AA (47). ω -3 Fatty acids are precursors of the 1- and 3-series prostaglandins, act as competitive inhibitors for enzyme binding sites with arachidonic acid, and lead to a decreased prostaglandins synthesis from AA (48).

Fatty acid composition can influence enzyme activities in membranes (47). In the above study it was also found that fish fed with SFO (rich in 18:2ω -6) had the highest cardiac phospholipase A activity (Table 3). A high phospholipase activity on endogenous phospholipids induces fatty acid turnover releasing substrate for eicosanoid synthesis, and may compromise membrane integrity (47).

Table 3   Phospholipase A Activities in Hearts of Salmon Fed with Diets
in Which the Lipid Source was Fish Oil (FO), Sunflower Oil (SFO), or
Linseed Oil (LO)

Phospholipase A activity	Diets		
	FO	SFO	LO
nmol fatty acid liberated/h per mg protein	53.1±4.9[b]	96.4±6.9[a]	59.9±13.4[b]
nmol PC, PE hydrolized/h per mg protein	11.2±1.4[b]	26.3±7.0[a]	13.9±304[b]

Values are means ± SD, n=4.
Values in the same row with different superscript letters are significantly different
(p<0.05).
Source: Ref. 47.

F.   Carp Hepatopancreas PLA$_2$ and Wax Ester Synthesis

Wax esters, esters of long chain fatty acids and fatty alcohols, are an alternative
form of energy reserve to triacylglycerols in the oil of sperm whales, the flesh oils
of several deep sea fishes such as orange roughy, and in zooplankton. Marine fish,
like all marine organisms, use carbohydrate and amino acid precursors to synthesize wax esters (7). Mankura et al. (49) studied the relationship between PLA$_2$ hydrolysis and wax ester synthesis using radiotracer techniques. About 10% of the
radioactivity   liberated   from   L-1-palmitoyl-2-[1-^{14}C]arachidonyl-3sn-glyc-
erophosphatidylcholine (^{14}C-PC) by PLA$_2$ was found in wax esters. Sodium de-
oxycolate (5 mM), Triton X-100 (10 mM), and Ca^{+2} (10 mM) increased PLA$_2$
activity but did not alter wax ester synthesis. However, both PLA$_2$ activity and
wax ester synthesis from ^{14}C-PC increased with the addition of oleyl alcohol.
Consequently, oleoyl alcohol seems to be a limiting factor for the synthesis of
wax esters. Wax synthesis was detected in the hepatopancreas, kidneys, gills,
spleen, and intestine. The highest wax synthesis was detected in the intestine.
Arachidonic acid in position 2 of PC is an efficient precursor for prostaglandins
type 2 and for wax ester synthesis in carp hepatopancreas preparations. In carp
hepatopancreas, FFA released by PLA$_2$ from the β-position of phospholipids are
used in wax ester synthesis. It has been hypothesized that wax esters derived from
AA may serve to provide the precursors for prostaglandins synthesis in carp (49).

G.   A Cellular Acidic PLA$_2$ from Cod (*Gadus morhua*) Muscle

PLA$_2$ from cod muscle is an exceptional fish PLA$_2$ with regard to pH requirements. Usually fish PLA$_2$ optimal activity is observed at alkaline pH (17). This
phospholipase exhibits maximal activity at pH 4.0, in a pH range from 4 to 8.6 at

25°C. High PLA$_2$ activity from cod muscle is only present in approximately one out of four cod. The activity of this 50 kDa PLA$_2$ displays an optimum at 40°C and has a PI of 5.2. A cod muscle PLA$_2$ elution profile obtained with gel filtration shows two peaks: a major peak with a shoulder and a minor peak. This profile is identical to the gel filtration profiles obtained from red sea bream hepatopancreas (50) and from pollock muscle (31). While a major activity peak is associated with the shoulder of the first peak obtained from the Red Sea bream PLA$_2$, a little peak near the shoulder and a major peak associated with the second protein peak appear in the profile from pollock muscle PLA$_2$ (17).

The activity of this enzyme is not dependent on Ca^{+2} and this may be related to its pH requirements. At low pH values the role that has been attributed to Ca^{+2} in the polarization of the ester bond of fatty acids in the sn-2 position of glycerol during catalysis, disappears because the amino acid acidic groups and the phosphate are protonated (17). Perhaps catalysis Ca^{+2} independent PLA$_2$s proceeds through an acyl-enzyme intermediate (15). The origin of this enzyme could be lysosomal, which would explain its acidic pH requirements (17).

III. PHOSPHOLIPASES AND TEMPERATURE ADAPTATION

Saturated fatty acids (SFA) in the sn-1 and polyunsaturated fatty acids (PUFA) in the sn-2 are the main components of phospholipids in marine fish tissues (51). Remodeling is the most common mechanism used by poikilotherms organisms to restructure their biological membranes as a response to changes in temperature (40, 52). Highly unsaturated fatty acids accumulate in poikilotherms upon exposure to low temperatures (53, 54). Polyunsaturated fatty acids presumably compensate for the ordering effects induced by low temperatures by establishing a balanced degree of membrane order. Membrane unsaturation also facilitates reduction by 4–8°C in transition temperatures helping with function and survival in cold temperatures (52). Among the presumed biosynthetic adjustments that contribute to acyl chain remodeling during cold acclimation are high activity in the desaturation and elongation processes, increased synthesis of phospholipids with a high proportion of long-chain PUFA, and increased phospholipid biosynthesis by minor pathways to produce unsaturated species of phospholipids (52).

Catabolic processes, although less well documented than biosynthetic capacities, have also been postulated where membrane restructuring is a response to low temperatures. In mammals, through the deacylation/reacylation (D/R)-cycle, products of de novo synthesis are modified to unique molecular species of phospholipids and long-chain PUFAs become constituents of biological membranes (52). A relationship between the modification of the sn-1 position fatty acids and the physiological properties of biomembranes as a response to temperature adaptation has been demonstrated (55). In carp liver an increase in PC and

PE 1-oleoyl-2-DHA species, coupled with a decrease in 1-palmytoil-2-DHA and 1-stearoyl-2DHA species, occurs upon cooling from 25 to 5°C. Fatty acid remodeling through deacylation/reacylation is responsible for the formation of these compounds. In bonito muscle about 10% of PC molecular species is 1,2-diDHA PC (12). Fatty acid remodeling in the bonito muscle likely results from the participation of PLA_1 and acyltransferase and/or a transacylase (12). On the other hand, PLA_2 participates in the production of new membrane phospholipids through cleavage of the sn-2 fatty acid moiety followed by reacylation (7). Remodeling of membrane lipid composition of trout liver has been attributed to the D/R cycle (56).

Phospholipase activity is the rate-limiting step in the D/R cycle; therefore, it may play an important role in metabolic regulation. The increases in phospholipase activity upon exposure to cold temperatures both in vivo and in vitro suggest this regulatory role (52 and references therein). The PLA_2 of trout liver shows maximal activity towards molecular species of PC with monoenoic fatty acids at sn-2 position, but these species are not present at cold temperatures. This observation suggests that substrate specificity, along with enzyme activity, could contribute to the molecular species pattern observed at low temperatures. However, using a complex distribution of molecular species of PC similar to that found in trout microsomes, Hazel (52) found that rates of PC molecular species catabolism were not significantly influenced by temperature. Consequently, the favored incorporation of unsaturated fatty acids during cold acclimation may be derived from the substrate specificity of the biosynthetic rather than the catabolic limb of the D/R cycle (52).

IV. PHOSPHOLIPASES C AND OLFACTORY SIGNAL TRANSDUCTION

The binding of an odorant molecule to odorant receptors present in the cilia of olfactory receptor neurons initiates olfactory signal transduction (57). Upon binding, one or several cascades may be activated. The receptor-mediated phosphoinositide turnover plays an essential role in transduction in many cellular systems. Once agonists occupy appropriate receptors, PLC hydrolyzes PIP_2 with the production of IP_3 and diacylglycerol. These products are potential second messengers. There is liberation of calcium, possibly due to the interaction of IP_3 with a specific receptor in the endoplasmic reticulum, and diacylglycerol activates protein kinase C that subsequently phosphorylates tissue-specific proteins (58).

In several species the signaling process is mediated by GTP-binding proteins (59 and references therein). The second messengers used by the transduction pathways mediated by GTP are either cyclic nucleotides or IP_3 (60 and references therein). GTP-binding proteins (G proteins) regulate the activity of adenylcyclase or phospholipase C or both depending on the nature of the odorant (59 and references therein). Following this, the second messengers cAMP and IP$_3$ open ion channels in the ciliary membrane and, consequently, depolarization of the receptor neuron takes place (61). The proteins that could mediate olfactory transduction include the putative receptors G$_{olf}$, the G$_i$ proteins, type III adenylyl cyclase, and the cyclic nucleotide-gated channels. A description of an additional ciliar IP$_3$ receptor protein that accumulates in response to exposure to some odorants and probably has identical properties to the calcium channel gated by IP$_3$ has also been reported (59 and references therein).

The two types of olfactory receptor cells, ciliated and microvillus, present in the teleost olfactory epithelium seem to have amino acid binding sites (60). The olfactory PLC of several fish is stimulated by odorants. Zebrafish respond by electrophysiological and behavioral measurements to the most common odorants detected by other fish species. For this reason zebrafish are widely used as a model for studies of the olfactory system (61). This fish detects amino acids and bile odorants with relatively independent odorant receptors; however, the transduction cascades activated as a result of the interaction have not been elucidated. In zebrafish, neomycin (50 µM), a PLC inhibitor, reduces responses evoked by amino acids by 80–90%, while responses to bile salts are reduced by just 20–50%. This finding demonstrates that PLC activation is essential to odor transduction and is consistent with the presence of the PLC/IP$_3$ cascade, which is activated by amino acids but not by bile salts (61). In vivo neurophysiological studies and biochemical studies of isolated cilia preparations of the channel catfish *Ictalurus punctatus* revealed that its olfactory system exhibits odorant-stimulated phosphoinositide turnover (58). The stimulation of the phosphoinositide turnover by GTP and analogues in catfish preparations suggests the involvement of a GTP-binding regulatory protein in the activation of this pathway; this process has similar characteristics to that of the agonist-stimulated phosphoinositide metabolism of other cells (58). The G-protein subunits G$_s$/G$_{olf}$, G$_{i1}$, G$_{i2}$, G$_q$, and G$_\beta$ have been identified in channel catfish olfactory epithelium. They are members of three subfamilies and are expressed in the dendrita and cilia of olfactory receptor neurons. It appears that G$_s$/G$_{olf}$ mediate odorant activation of adenyl cyclase, while G$_i$ and G$_\beta$ mediate odorant activation of PLC. PLC of the cilia is stimulated by the odorant amino acid L-alanine and by guanine nucleotides (59).

The properties of PLC and the molecular forms differ depending on the cells. The properties of olfactory phospholipase C from catfish have been studied and a linear relationship between production of inositol phosphate products and PLC has been found. The half-maximal and the maximal hydrolysis of PIP$_2$ is observed at about 2 and 10 min of incubation, respectively. Of the total inositol phosphate products, 91% is IP$_3$. A peak of enzyme maximal activity appears at pH 6.7 and a smaller peak is found at pH 7.0. The presence of two pH optima has also been observed in rat brains and bovine myocardial PLCs, and it has been attributed to heterogeneity (58 and references therein). Low levels of intracellular free Ca^{+2} are required for the appropriate functioning of PLC in odorant responses; for catfish such levels are around 29 $\oplus$ 12nM (62). Boyle (58) found maximal PLC activity in the absence of added Ca^{+2}. Presumably the Ca^{+2} used during the treatment of the tissue to separate the cilia was retained by the cilia. Moreover, 1 mM EDTA inhibited the production of inositol phosphates, while IP_2 was independent of exogenous Ca^{+2} and inhibited at 1 mM added Ca^{+2}. The enzyme of the cilia is produced in very low amounts; therefore, it has not been possible to determine its molecular weight. However, from the cytosolic fraction a peak of 100,000 Da and two small peaks of 82,000 and 60,000 have been detected (58). The cytosolic cilia enzyme may be or may not be different from the olfactory PLC. It has been hypothesized that enzyme activity in the cilia may correspond to cytosolic enzyme trapped during resealing of the membranes (58).

Amino acids and bile salts are potent olfactory stimuli for Atlantic salmon (60). PIP_2 breakdown has been detected in a plasma rich fraction from Atlantic salmon olfactory rosettes (63). The threshold of detection of bile acids by the olfactory system of Atlantic salmon is more than an order of magnitude lower than that for amino acids. Taurocholic acid (TChA) is a potent olfactory stimulus. In the presence of GTPγS, TChA stimulates PIP_2 breakdown in a dose dependent manner; however, neither TChA alone nor GTPγS alone can stimulate PIP_2 breakdown (60).

AlF_4^- and TChA stimulate PLC activity but TChA is only effective in the presence of GTPγS; in contrast, PLC activity is inhibited by preincubation with GDPγS. These findings are consistent with PLC regulation mediated by G protein reported for the PLC-β family. GTPγS is more effective than GTP stimulating PLC activity in other systems. The fact that binding of alanine is not affected by the presence of TChA indicates that olfactory discrimination between bile salts and amino acids may occur at the receptor level (60).

Ca^{+2} also stimulates olfactory PLC activity. A high-affinity Ca^{+2}-ATPase activity present in salmon olfactory cilia and other plasma preparations might be responsible for maintaining the low Ca^{+2} levels required to sensitize PLC to stimulation by TChA and amino acids. In salmon, stimulus-dependent PLC activity is maximal at or below 10 nM free Ca^{+2}. When, because of high concentrations of Ca^{+2}, amino acids do not stimulate PLC, an interesting alternative mechanism takes place. Ca^{+2} stimulates PLC activity regardless of the presence of GTPγS or GDPβS (60).

V. PHOSPHOLIPASE ACTIVITY IN POSTHARVEST SEAFOOD

Fish quality is negatively affected by nonesterified fatty acid (FFA) accumulation during frozen storage. FFA may be responsible for the formation of off flavors or off odors, the destruction of some vitamins and amino acids, the changes in texture and water-holding capacity of muscle proteins (31 and references therein), and the alteration of lipid oxidation processes (64).

Fish species, particularly the members of the lean-fleshed Gadoid family, develop myotomal tissue toughening during frozen storage. Textural changes in frozen tissue may result from the accumulation of FFA, which stimulate the formation of protein–lipid cross-linkages producing myofibrillar denaturation. Research on protein denaturation and lipid hydrolysis in frozen cod fillets was first published in 1959 when extractability of myofibrillar proteins was detected following FFA formation (65 and references therein). In the myotomal tissue of Atlantic cod, 90% of the lipid fraction corresponds to phospholipids (66). During the first 8 weeks of cod storage at -12°C, a rapid increase in FFA levels and a proportional decrease in phospholipid content have been observed. After this period, accumulation of FFA diminishes. The depletion of phospholipids with a corresponding accumulation of FFA is due to phospholipid hydrolysis that results from the activity of endogenous phospholipases. When Atlantic cod myotomal tissue is stored at -30°C, the activities of microsomal enzyme-bound enzymes and other cytosolic enzymes decrease over a 12-week storage period. Conversely, enhancement of the microsomal phospholipase hydrolytic capacity is detected in the first 8 weeks of storage. During the following 4 weeks, phospholipase activity returns to its initial value. These findings on phospholipase activity are consistent with the phospholipid and FFA levels observed at -12°C (66 and references therein). Frozen storage induces changes in membrane permeability that trigger increases in phospholipase activity, facilitating phospholipid hydrolysis. There is evidence that ice crystals liberate phospholipase complexes from microsomal and mitochondrial membranes. Hanaoka et al. (67) found that PC's decomposition is accelerated during freezing of carp ordinary muscle. Glycerylphosphorylcholine (GPC), used as indicator of decomposition of PC, increased with the dehydration rate. After rapid freezing, a proces that avoids tissue disruption, GPC levels were unexpectedly higher than after slow freezing (67). Therefore, the process of dehydration that occurs during freezing may also contribute to the promotion of phospholipid hydrolysis.

Lipid oxidative reactions in seafood lead to generation of both fresh flavors and off flavors (68; see Chaps. 13, 14), the latter, particularly due to lipid peroxidation (69). Numerous studies (64, 69, and references therein) have reported an association between lipid oxidation and the liberation of FFA from lipolytic activity during frozen storage of fish. Conflicting results have been obtained from these studies, perhaps due to the use of different fish species and different experimental conditions. Enzymatic lipid hydrolysis has been identified either as a promoter or as an inhibitor of lipid oxidation; and the opposite, lipid oxidation as a stimulator of phospholipid hydrolysis, has also been reported. Enzymatic phospholipid hydrolysis has been detected in frozen cod, lemon sole, halibut, trout, herring, whitefish, flounder, salmon, silver hake, carp, red sea bream, and capelin (69 and references therein). Rancid odors and flavors produced from oxidation of unsaturated fatty acids may be the main factors contributing to the short shelf life of frozen catfish (70). Oxidative rancidity of mullet has been related to the presence of phospholipid-free fatty acids (31 and references therein). In horse mackerel most of the docosahexanoic acid produced by enzymatic lipolysis is subsequently auto-oxidized during storage at -10°C (64 and references therein). In sardines, refrigerated after a freeze–thawing process, acceleration of enzymatic lipolysis without quality deterioration has been detected. In contrast, in horse mackerel, treated under the same conditions, both enzymatic lipolysis and lipid oxidation have been observed, accompanied by quality deterioration (71). The relationship between FFA produced by hydrolysis of phospholipids and lipid oxidation was studied in carp *Caprynus carpio* and Pacific mackerel *Scomber japonicus* stored at -5°C for 4 weeks. After treating the homogenates from both fish with PLA_2, TBA and $TBA_{37°C-2h}$ compounds, whose presence is associated-with lipid oxidation, increased in Pacific mackerel while in carp TBA compounds increase slightly and $TBA_{37°C-2h}$ compounds decrease. Depression of antioxygenic activities was also found. The presence of myoglobin in mackerel may be responsible for the lipid oxidation observed. Consequently, FFA released by the action of PLA_2 did not inhibit lipid oxidation (64).

The rate of oxidation of FFA may be faster than that of esterified fatty acids (64 and references therein, 69 and references therein). However, a negative correlation between FFA content and lipid oxidation has been reported (69). Reduction in lipid oxidation in cod and mullet has been associated with increased phospholipid hydrolysis. Moreover, in cod and trout, added PLA reduces lipid oxidation (69 and references therein). Shewfelt et al. (69) carried out a study to establish the role of phospholipid hydrolysis on lipid peroxidation of the microsomal fraction of flounder (*Pseudopleuronectes americanus*) muscle. They found that preincubation of the microsomes with PLA lowers both enzymatic and nonenzymatic oxidation processes. With the addition of BSA, which removes FFA from the membrane, to PLA-treated samples, a partial recovery of lipid oxidation was observed. This result is good evidence that FFA have an effect on both enzymatic and nonenzymatic oxidation. The mechanisms of enzymatic and nonenzymatic oxidation are different; so presumably are the hydrolysis points of inhibition in lipid oxidation. Deoxycholate, a detergent that affects lipid–lipid interactions, inhibits both oxidation systems while phospholipase C inhibits only enzymatic oxidation (69). Production of thiobarbituric acid-reactive substances (TBARS) and lipid hydroperoxides decline upon preincubation of the sarcoplasmic reticulum of flounder with phospholipase A_2 in both enzymatic and nonenzymatic systems (72). Why is lipid oxidation inhibited by phospholipid hydrolysis? The opposite should be true since it has been proposed that free fatty acids oxidize faster than esterified fatty acids. It is possible that this model works only for TAG but not for phospholipids (69). Shewfelt et al. (72) proposed a model to explain the inhibitory effect of phospholipid hydrolysis on both enzymatic and nonenzymatic peroxidation. During phospholipid hydrolysis, there is a structural rearrangement of the membrane fatty acids that decreases intractions between the polyunsaturated regions of the fatty acids. Fewer fatty acid interactions, in turn, decrease the free-radical chain propagation and alter the reaction kinetics. It appears that enzymatic oxidation is inhibited by enzymatic destabilization stimulated by membrane disruption (72). On the other hand, it has been demonstrated in vitro that enzymatic and nonenzymatic lipid peroxidation of either rainbow trout microsomes or whole muscle may trigger phospholipid hydrolysis by phospholipase A_2 muscle (73).

Lipid hydroperoxides (ROOH), intermediates of peroxidation that accumulate in peroxidized lipids, react with transition metals producing RO· and ROO· radicals. The latter radicals can be responsible for protein structural changes such as cross-linking and fragmentation. The resulting denatured proteins may alter cellular function or radicals may produce tissue degeneration. Compounds used by organisms to reduce accumulation of lipid hydroperoxides are tocopherols, carotenoids, and different enzymes. One such enzyme is glutathione peroxidase, which transforms the hydroperoxide group to a less toxic moiety but only acts on nonesterified fatty acid substrates. At this stage PLA_2 plays an important role, releasing peroxidized fatty acids from membranes. An interesting fact is that PLA_2 has a higher activity for these acids than for normal acids. PLA_2 can protect membranes from lipid peroxidation damage by providing the appropriate substrate for glutathione peroxidase to remove the toxic ROOH moiety, and through reacylation, allowing the formation of a normal lipid molecule in the membrane (7).

Pro- and antioxidant levels and the lipids' microenvironment are the main factors involved in flavor generation, the latter being the most important factor for lipid oxidation. The nonhomogeneous distribution of lipid oxidation compounds creates multiple environments within a tissue. For example, the ease of O_2 access to external environments facilitates oxidation of external compared to internal environments in channel catfish (68 and references therein).

Lipid susceptibility depends on oxidant stress. When emulsions of fatty acids and phospholipid liposomes are the oxidative substrate, cobalt ions are effective catalysts whereas cooper ions are ineffective catalysts. The influence of initiation systems on lipid susceptibility also illustrates the dependence of lipid susceptibility on oxidant stress. When the enzymatic initiation system of NADPH, ADP, and iron is applied to liposomes 20% of endogenous tocopherol is depleted before peroxidation begins. In contrast, when either the water-soluble AAPH or the lipid soluble AMVN azo compounds are applied, a depletion of approximately 70% of the antioxidant precedes the propagation phase (68 and references therein).

Relative susceptibilities of lipid classes have been assessed in minced muscle tissue from two strains (LSU and AQUA) of catfish stored at fluctuating temperatures between -6 and -18°C for 9 months (70). While hydroperoxides and conjugated dienes, the primary products of lipid oxidation, did not increase during the first 6 months, a dramatic increase was detected in the next 3 months. This result differs from the traditional trend observed in the early periods of oxidation where hydroperoxide levels increase (70 and references therein). TBARS gradually increased during the first 6 months, after which a sharp increase was observed. Interaction of hydroperoxides with iron ions, favored by the process of mincing, could have caused the inability to detect primary products during the early stages of storage. Levels of hydroperoxides, conjugated dienes, and TBARS were higher in AQUA than in LSU. On the other hand, tocopherol degradation was not statistically significant during the first 3 months and 75–90% of the original tocopherol had been degraded after 9 months. In the final period of storage, oxidation of tocopherol increased in such a manner that this compound was no longer sufficient to protect membrane lipids. While phospholipid fatty acid concentration decreased with storage time, FFA increased. The decrease of phospholipid PUFA, along with the increase of FFA fraction of PUFA concentrations, detected after 3, 6, and 9 months of storage, may indicate phospholipase participation. Increased phospholipid hydrolysis in LSU samples may be related to reduced levels of TBARS in comparison to AQUA samples. Decrease in phospholipid fatty acid quantities from membranes did not correspond to the increase in FFA produced. This indicates that a proportion of the FFA is generated by TGA hydrolysis. As LSU samples had greater FFA concentrations, it is assumed that these samples had higher lipase activity and that lipases may also be involved in alteration of lipid oxidation processes (70).

VI. CONCLUSIONS

Fish phospholipases are lipolytic enzymes of great importance because of their role in diverse physiological functions, ranging from the biosynthesis of eicosanoids to the deterioration of fish during frozen storage. Unfortunately, at present, not enough information on fish phospholipases is available and much research needs to be undertaken. Results from such studies would help establish evolutionary trees for fish phospholipases, develop appropriate culture conditions, and improve composition and quality of diets. The determination of suitable manipulation and storage conditions for postharvest fish would allow for the supply of nutritious products of excellent quality. Some of the main areas that require attention are as follows:

Methods for fish phospholipase isolation and purification
Amino acid sequences and tertiary structure of fish phospholipases

Enzymology of digestive lipolysis

Organs that produce and secrete digestive PLA_2 and the ontogeny of the phospholipase cells

Muscle phospholipases

Fish-muscle composition and oxidative susceptibility

Physicochemical changes that occur in fish during the freezing process and their relationship with phospholipase activity, phospholipid hydrolysis, and lipid oxidation

REFERENCES

1. R Verger, GH de Haas. Interfacial enzyme kinetics of lipolysis. Annu Rev Biophys Bioeng 5:77–117, 1976.
2. R Verger, MCE Mieras, GH de Haas. Action of phospholipase A at interfaces. J Biol Chem 248:4023–34, 1973.
3. P Desnuelle, L Sarda, G Ailhaud. Inhibition de la lipasa pancréatique par le diéthyl-p-nitrophenyl phosphate en emulsion. Biochim Biophys Acta 37:570–1, 1960
4. H Brockerhoff. A model of pancreatic lipase and the orientation of enzymes at interfaces. Chem Phys Lipids 10:215–22, 1973.
5. E Verger. Enzyme kinetics of lipolysis. Methods Enzymol 64:340–392, 1980.
6. JJ Volwerk, G De Haas. Pancreatic phospholipase A_2: a model for membrane-bound enzymes? In: Lipid–Protein Interactions, Volume 1. Toronto: John Wiley & Sons, 1982, pp 70–113.
7. M Gurr, JL Harwood. Lipid Biochemistry: An Introduction, 4th ed. New York: Chapman and Hall, 1991, pp 307–314.
8. MF Roberts, AA Bothner-By, EA Dennis. Magnetic non-equivalence within the fatty acyl chains of phospholipids in membrane models: ^{1}H nuclear magnetic resonance studies of the α-methylene groups. Biochemistry 17:935, 1978.
9. J DeBony, EA Dennis. Magnetic nonequivalence of the two fatty acyl chains in phospholipids of small unilamellar vesicles and mixed micelles. Biochemistry 20:5256, 1981.
10. M Waite. The Phospholipases. Handbook of Lipid Research, volume 5. New York: Plenum Press, 1987, pp 155–242.
11. H Brockerhoff, RG Jensen. Lipolytic Enzymes. New York: Academic Press, Inc, 1974, pp 197–282.
12. K Hirano, A Tanaka, K Yoshizumi, T Tanaka, K Satouchi. Properties of phospholipase A_1/transacylase in the white muscle of bonito Euthynnus pelamis (Linnaeus). J Biochem 122:1160–1166, 1997.
13. RA Wolf, RW Gross. Identification of neutral active phospholipase which hydrolyzes choline glycerophospholipids and plasmalogen selective phospholipase A_2 in canine myocardium. J Biol Chem 260:7295–7303, 1986.
14. SL Hazen, RJ Stuppy, RW Gross. Purification and characterization of canine myocardial cytosolic phospholipase A_2. J Biol Chem 265:10622–10630, 1990.

15. EA Dennis. Diversity of group types, regulation, and function of phospholipase A_2. J Biol Chem 269:13057–13060, 1994.

16. EJ Ackermann, ES Kempner, EA Dennis. Ca^{+2}-independent cytosolic phospholipase A_2 from macrophage-like $P388D_1$ cells. Isolation and characterization. J Biol Chem 269:9227–9233, 1994.

17. B Aaen, J Flemming, J Benny. Partial purification and characterization of a cellular acidic phospholipase A_2 from cod (*Gadus morhua*) muscle. Comp Biochem Physiol 110B:547–554, 1995.

18. HM Verheij, AJ Slotboom, GH de Haas. Structure and function of phospholipase A_2. Rev Physiol Biochem Pharmacol 91:91–203, 1981.

19. P Wooley, SB Petersen. Lipases: Their Structure, Biochemistry and Application. Cambridge: Cambridge University Press, 1994, pp 119–134.

20. WA Pieterson, JC Vidal, JJ Volwerk, GH de Haas. Interaction of phospholipase A_2 and its zymogen with divalent metal ions. Biochemistry 13:1439–1445, 1974.

21. PY-K Wong, EA Dennis. Phospholipase A_2: Role and Function in Inflammation. New York: Plenum Press, 1990.

22. M Waite. Handbook of Lipid Research. New York: Plenum Press, 1987.

23. BD Howard, CB Gundersen Jr. Effects and mechanisms of polypeptide neurotoxins that act presynaptically. Annu Rev Pharmacol Toxicol 20:307–326, 1980.

24. S Barbour, EA Dennis. Antisense inhibition of group II phospholipase A_2 expression blocks the production of prostaglandin E_2 by $P388D_1$ cells. J Biol Chem 268:21875–21882, 1993.

25. K Sekar, M Sundaralingam. Carboxylic ester hydrolase, 1.5 Angstrom orthorhombic form of the bovine recombinant PLA_2. Protein Data Bank (PDB), Brookhaven National Laboratory.

26. C van den Berg, AJ Slotboom, HM Verheij, GH de Haas. The role of Asp-49 and other conserved amino acids in phospholipases A_2 and their importance for enzymatic activity. J Cell Biochem 39:379–390, 1989.

27. I Geurden, J Radunz-Neto, P Bergot. Essentiality of dietary phospholipids for carp (*Cyprinus carpio* L.) larvae. Aquaculture 131:303–314, 1995.

28. CB Cowey, JR Sargent. Bioenergetics and growth. In: WS Hoar, DJ Randall, JR Brett, eds. Fish Physiology. New York: Academic Press, 1979.

29. A Kanazawa, S-T Teshima, M Sakamoto. Effects of dietary bonito-egg phospholipids and some phospholipids on growth and survival of the larval ayu *Plecoglossus altivelis*. Z Angew Ichthyol 4:165–170, 1985.

30. H Ono, N Iijima. Purification and characterization of phospholipase A_2 isoforms from the hepatopancreas of red sea bream, *Pagrus major*. Fish Physiol Biochem 18:135–147, 1998.

31. MA Audley, KJ Shetty, JE Kinsella. Isolation and properties of phospholipase A from pollock muscle. J Food Sci 43:1771–1775, 1978.

32. N Iijima, S Chosa, M Uematsu, T Goto, T Hoshita, M Kayama. Purification and characterization of phospholipase A_2 from the pyloric caeca of red sea bream *Pagrus major*. Fish Physiol Biochem 16:487–498, 1997.

33. NP Neas, JR Hazel. Partial purification and kinetic characterization of the microso-

mal phospholipase A_2 from thermally acclimated rainbow trout (*Salmo gairdneri*). J Comp Physiol B 155:461–469, 1985.

34. K Satouchi, M Sakaguchi, M Shirakawa, K Hirano, T Tanaka. Lysophosphatidylcholine from white muscle of bonito *Euthynnus pelamis* (Linnaeus): involvement of phospholipase A_1 activity for its production. Biochim Biophys Acta 1214:303–308, 1994.

35. K Uematsu, M Kitano, M Morita, N Iijima. Presence and ontogeny of intestinal and pancreatic phospholipase A_2-like proteins in the red sea bream, *Pagrus major.* An immunocytochemical study. Fish Physiol Biochem 9:427–438, 1992.

36. R Fange, D Grove. Digestion. In: WS Hoar, DJ Randall, JR Brett, ed. Fish Physiology. New York: Academic Press, 1979, pp 161–260.

37. MC Carey, DM Small, CM Bliss. Lipid digestion and absorption. Annu Rev Physiol 45:651–677, 1983.

38. GH De Haas, NM Postema, W Nieuwenhuizen, LLM Van Deenen. Purification and properties of an anionic zymogen of phospholipase A from porcine pancreas. Biochim Biophys Acta 159:118–129, 1968.

39. RP Evans, CC Parrish, P Zhu, JA Brown, PJ Davis. Changes in phospholipase A_2 activity and lipid content during early development of Atlantic halibut (*Hippoglossus hippoglossus*). Mar Biol 130:369–376, 1998.

40. NP Neas, JR Hazel. Phospholipase A_2 from liver microsomal membranes of thermally acclimated rainbow trout. J Exp Zool 233:51–60, 1985.

41. JR Hazel. Cold adaptation in ectotherms: regulation of membrane function and cellular metabolism. In: LCH Wang, ed. Advances in Comparative and Environmental Physiology. Berlin: Springer-Verlag, 1989, pp 1–44.

42. EE Williams, JR Hazel. Restructuring of plasma membrane phospholipids in isolated hepatocytes of rainbow trout during brief in vitro cold exposure. J Comp Physiol (B) 164:600–608, 1995.

43. A Lehninger, DL Nelson, MM Cox. Principles of Biochemistry With an Extended Discussion of Oxygen-Binding Proteins, 2nd ed. New York: Worth Publishers, 1993, pp 258.

44. RJ Henderson, DR Tocher. The lipid composition and biochemistry of freshwater fish. Prog Lipid Res 26:281–347, 1987.

45. JG Bell, DR Tocher, JR Sargent. Effects of supplementation with 20:3 (n-6), 20:4 (n-6) and 20:5 (n-3) on the production of prostaglandins E and F of the 1-, 2-, and 3- series in turbot (*Scophthalmus maximus*) brain astroglial cells in primary culture. Biochim Biophys Acta 211:335–342, 1994.

46. PC Weber. The modification of the arachidonic acid cascade by n-3 fatty acids. In: B Samuelsson, S-E Dahlen, J Fritsch, P Hedqvist, eds. Advances in Prostaglandin, Tromboxane and Leukotriene Research, Vol 20. New York: Raven Press, 1990, pp 232.

47. JG Bell, JR Dick, AH Mc Vicar, JR Sargent, KD Thompson. Dietary sunflower, linseed and fish oils affect phospholipid fatty acid composition, development of cardiac lesions, phospholipase activity and eicosanoid production in Atlantic salmon (*Salmo salar*). Prostaglandins Leukot Essent Fatty Acids 49:665–673, 1993.

48.   T Terano, JA Salmon, GA Higgs, S Moncada. Eicosapentanoic acid as a modulador of inflammation. Effect on prostaglandin and leukotriene synthesis. Biochem Pharmacol 35:779–785, 1986.

49.   M Mankura, M Kayama, N Iijima. The role of phospholipase A_2 on wax ester synthesis in carp hepatopancreas preparations. Bull Jpn Soc Sci Fish 2:2107–2114, 1986.

50.   N Iijima, M Nakamura, K Uematsu, M Kayama. Partial purification and characterization of phospholipase A_2 from the hepatopancreas of Red Sea Bream. Nippon Suisan Gakk 56:1331–1339, 1990.

51.   MV Bell, JR Dick. Molecular species composition of the major diacyl glycerophospholipids from muscle, liver, retina and brain of cod (Gadus morhua). Lipids 26:565–573, 1991.

52.   JR Hazel. Role of molecular species catabolism in the temperature-induced restructuring of phosphatidylcholines in liver microsomes of thermally-acclimated rainbow trout (Oncorhynchus mykiss). Fish Physiol Biochem 15:195–204, 1996.

53.   C Buda, I Dey, N Balogh, LI Horvath, K Maderspach, M Juhasz, YK Yeo, T Farkas. Structural order of membranes and composition of phospholipids in fish brain cells during thermal acclimatization. Proc Natl Acad Sci USA 81:8234–8238, 1994.

54.   C Wallaert, PJ Babin. Thermal adaptation affects the fatty acid composition of plasma phospholipids in trout. Lipids 29:373–376, 1994.

55.   E Fodor, RH Jones, C Buda, K Kitajka, I Dey, T Farkas. Molecular architecture and biophysical properties of phospholipids during thermal adaptation in fish: an experimental and model study. Lipids 30:1119–1126, 1995.

56.   JR Hazel, AF Hagar, NN Pruitt. The temperature dependence of phospholipid deacylation/reacylation in isolated hepatocytes of thermally-acclimated rainbow trout. Biochim Biophys Acta 918:149–158, 1987.

57.   D Lancet. Vertebrate olfactory reception. Ann Rev Neurosci 9:329–355, 1986.

58.   AG Boyle, YS Park, T Huque, RC Bruch. Properties of phospholipase C in isolated olfactory cilia from the channel catfish (Ictalurus punctatus). Comp Biochem Physiol 88B:767–775, 1987.

59.   FC Abogadie, RC Bruch, AI Farbman. G-protein subunits expressed in catfish olfactory receptor neurons. Chem Senses 20:199–206, 1995.

60.   YH Lo, SL Bellis, L-J Cheng, J Pang, TM Bradley, DE Rhoads. Signal transduction for taurocholic acid in the olfactory system of Atlantic salmon. Chem Senses 5:371–380, 1994.

61.   L Ma, WC Michel. Drugs affecting phospholipase C-mediated signal transduction block the olfactory cyclic nucleotide-gated current of adult zebrafish. J Neurophysiol 79:1183–1192, 1998.

62.   D Restrepo, T Miyamoto, BP Bryant, JH Teeter. Odor stimuli trigger influx of calcium into olfactory neurons of the channel catfish. Science 249:1166–1168, 1990.

63.   YH Lo, TM Bradley, DE Rhoads. Stimulation of Ca^{+2} regulated olfactory phospholipase C by amino acids. Biochemistry 32:12358–12363, 1993.

64.   M Toyomizu, K Hanaoka, K Yamaguchi. Effect of release of free fatty acids by enzymatic hydrolysis of phospholipids on lipid oxidation during storage of fish muscle at -5°C. Bull Jpn Soc Sci Fish 47:615–620, 1981.

65. P Chawla, RF Ablett. Detection of microsomal phospholipase activity in myotomal tissue of Atlantic cod (*Gadus morhua*). J Food Sci 52:1194–1196, 1987.

66. P Chawla, B MacKeigan, SP Gould, RF Ablett. Influence of frozen storage on microsomal phospholipase activity in myotomal tissue of Atlantic cod (*Gadus morhua*). Can Inst Food Sci Technol 21:399–402, 1988.

67. K Hanaoka, M Toyomizu. Acceleration of phospholipid decomposition in fish muscle by freezing. Bull Jpn Soc Sci Fish 45:465–468, 1979.

68. MC Erickson, RV Sista. Influence of microenvironment on oxidative susceptibility of seafood lipids. In: F Shahidi, KR Cadwallader, eds. Flavor and Lipid Chemistry of Seafoods. Washington, DC: American Chemical Society Symposium Series 674, 1997, pp 175–185.

69. RL Shewfelt, RE Mcdonald, HO Hultin. Effect of phospholipid hydrolysis on lipid oxidation in flounder muscle microsomes. J Food Sci 46:1297–1301, 1981.

70. MC Erickson. Compositional parameters and their relationship to oxidative stability of channel catfish. J Agric Food Chem 41:1213–1218, 1993.

71. C Kuizumi, C-M Chang, T Ohshima, S Wada. Influences of freeze-thawing process on the quality of sardine and horse mackerel during refrigerated storage. Nippon Suisan Gakk 54:2203–2210, 1988.

72. RL Shewfelt, HO Hultin. Inhibition of enzymic, non-enzymic lipid peroxidation of flounder muscle sarcoplasmic reticulum by pretreatment with phospholipase A_2. Biochim Biophys Acta 751, 432–438, 1983.

73. T-J Han, J Liston. Lipid peroxidation, phospholipid hydrolysis in fish muscle microsomes, frozen fish. J Food Sci 52:294–296, 1987.

5
Lipases

Clara López-Amaya and Alejandro G. Marangoni
University of Guelph, Guelph, Ontario, Canada

I. GENERAL CONSIDERATIONS

Lipolytic reactions play an important role in the quality deterioration of chilled, frozen, and processed seafood. The importance of these reactions in seafood is discussed in Chapter 4, Sec. V; Chapter 5, Sec. II.H; and Chapter 16, Sec. II.E. and III.D. of this book. Moreover, with the increased use of lipases as food-processing aids, there is growing interest in identifying lipases with appropriate specificity (e.g., for transesterification reactions that enhance the content of long-chain, highly unsaturated fatty acids in fish oil) (see Chapter 21).

A. The Lipase Gen Family

The vertebrate genes for lipoprotein lipase (LPL), hepatic lipase (HL), and pancreatic lipase (PL), derived from a common ancestral gene, are members of the lipase gene family (1). Hide et al. (1) performed homology studies on all sequences of LPL, HL, and PL published until 1992 and deduced the phylogeny of the lipase family. PL may have branched off earlier than LPL and HL, who share a more recent ancestor. These enzymes hydrolyze circulating and dietary triglycerides that will be assimilated and distributed to the central and peripheral tissues (1).

Homology among the lipase members has consistently been a topic of study because lipases involve a wide spectrum of enzymes and have structural similarities with other proteins. One of the structurally conserved features among the lipase groups, serine esterases and proteases, is the consensus sequence

Glycine(Gly)-X-Serine(Ser)-X-Gly, where X stands for any amino acid residue (1). In esterases this site has been recognized as the substrate-binding site (2).

Studies of the structure and function of human pancreatic lipase (hPL) have improved our knowledge of the mechanism for interfacial activation and catalysis. The active site of the enzyme is a trypsin-like catalytic triad, Aspartate (Asp) . . . Histidine (His) . . . Ser, that includes Ser 152. Ser 152 is essential to triglyceride hydrolysis. The so-called oxyanion hole is an essential feature of the active site of serine proteases. In lipases the two hydrogen-bond donors that stabilize the oxyanion of the tetrahedral intermediate come from the main chain nitrogen atoms of residues 77 and 153 (3).

The main stereochemical features of the hydrolytic activity of hPL and other lipases may be similar to those of serine proteases of the trypsin family. This assumption is based on the presence of a common catalytic triad Asp . . . His . . . Ser. With this similarity in mind, the catalytic reaction has been described in the following steps (3) (Fig. 1):

1.   Creation of the noncovalent Michaelis Menten complex.
2.   Formation of the tetrahedral, hemiacetal intermediate by the attack of the nucleophilic serine O^γ.
3.   Cleavage of the substrate ester bond, breakdown of the tetrahedral intermediate, protonation and release of the diglyceride.
4.   Attack on the serine ester by an activated water molecule and formation of the deacylation intermediate.
5.   Cleavage of the acyl-enzyme complex, protonation of the remaining group and fatty acid release.

Based on available models for the Michaelis complex and for the tetrahedral intermediate of serine proteases and inverting the chirality of the tetrahedral intermediate, the tetrahedral intermediate for triacylglycerol hydrolysis was modeled. The following steps correspond to the construction of a working model of the lipase catalytic domain and the reaction steps of substrate cleavage (3):

1.   The 238–260 loop was omitted from the model, considering that in the catalytically active state the tip of this loop could not remain in the immediate vicinity of the active site. The C domain was also omitted because of its distant location from the catalytic center.
2.   Two aromatic side-chains, tyrosine (Tyr) 114 and phenylalanine (Phe) 215, located near the active site and making contact with the flap residues, were reoriented into alternative conformations. This reorientation provided enough room for the substrate.
3.   The chain segment around Phe 77 was moved about 2 Å to enlarge the cavity even further. This segment is on the surface of the protein and, once removed, the flap only interacts weakly with the rest of the protein.

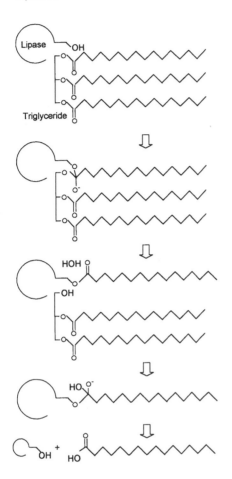

Figure 1 Basic steps of the cleavage of a trilaurin substrate by lipase. (From Ref. 3.)

4.   Ethyl butyrate was attached covalently to O^γ of Ser 152 in the hemiac-
     etal form and the two hydrogen bonds, believed responsible for the
     oxyanion hole stabilization, were enforced (Fig. 2). The resultant
     model was called the "activated" lipase model.

Using the "activated" lipase model and the x-ray structure of trilaurin, stereo-
chemically satisfactory models of the Michaelis complex, hemiacetal- and acyl-enzyme, water attack, and product were obtained. Moreover, the proposed inverse
handedness of the hemiacetal intermediate has been detected in the complex of *Rhizomucor miehei* lipase (RmL) with a covalently attached phosphonate inhibitor (3).

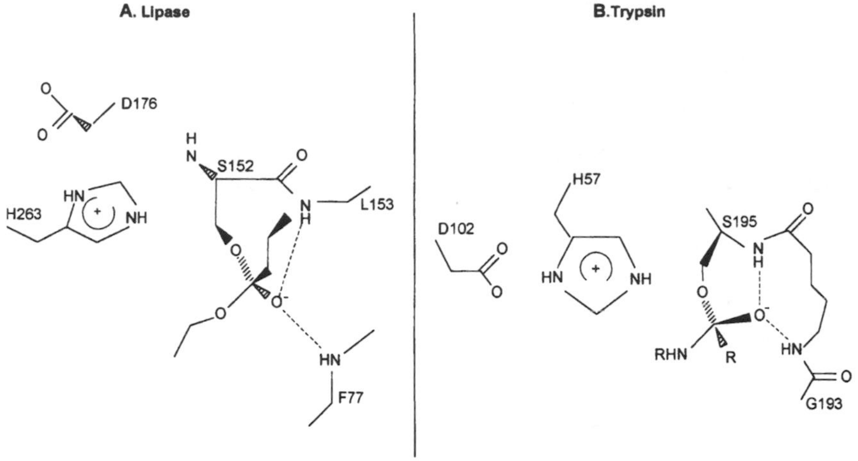

Figure 2 Models of the tetrahedral intermediates of (A) lipase with an ethyl butyrate model substrate and (B) trypsin with a generic amide substrate. The tetrahedral centers of these enzymes have opposite handedness. (From Ref. 3.)

When the critical micellar concentration of partly soluble substrate molecules is surpassed, (phospho)lipase activity experiences a dramatic increase. This phenomenon is called interfacial activation. It has been postulated, among numerous other hypotheses, that the conformational changes of the enzyme upon binding are intimately related to this phenomenon (4). In phospholipases, when comparing the structures of apoenzymes with those of apoenzyme–inhibitor complexes, only small changes can be detected in the backbone structure. Along with these changes some adjustments to the side chains occur. The hydrophobic channel of the apoenzymes, which binds to the aliphatic chains of the inhibitors, is preformed and has two large openings at the surface of the protein. One opening, which may allow entry of the substrate and exit of the products, becomes sealed after enzyme adsorption to the interface. The other opening is solvent-accessible and leads to the binding site of the phospholipid head group. In contrast, the triglyceride lipase apoenzyme structures have sealed cavities filled with solvent and active sites that are deeply buried and solvent-inaccessible. Moreover, to create a substrate-binding site, important backbone changes have to take place because hydrophobic side-chains of the protein fill the space around the active site. Studies of covalent inhibitor–RmL complexes have shed light on the nature of the conformational changes of the enzyme upon substrate binding. A short α-helical lid blocks the entrance to the active site. Upon substrate binding, the lid unhinges and exposes the hydrophobic patch that surrounds the entrance to the active site cleft to the substrate interface (3,4). Although hPL has a short helix similarly positioned to the RmL helix, for this enzyme the presence of a colipase is required in order to bind to the bile salt covered interface. Procolipase binds to the lid of the lipase. The open lid and the extremities of the procolipase fingers form an extended hydrophobic surface that interacts with the lipid water interface covered by bile salts. As a result of the enzyme conformational changes, a lipid-binding site is created, the active site is restructured and becomes accessible, the oxyanion loop is stabilized, and the interaction with procolipase is strengthened (4). Different features have been observed in different lipolytic enzymes. The active site of *Candida rugosa* lipase is covered by more than one loop (5). Cutinase, an enzyme that hydrolyzes both insoluble and soluble substrates, does not have a flap and does not experience interfacial activation either. It may be that the flap's first role is not to protect the active site but to avoid the hydrolysis of soluble esters by the lipase (3, 6).

1. Lipoprotein-Lipase

Lipoproteins are high-molecular-weight aggregates of triglycerides, phospholipids, cholesterol, cholesterol esters, and proteins, which circulate in the blood (7). The triglycerides from these lipoproteins are hydrolyzed by LPLs releasing fatty acids, which will provide energy on the peripheral tissues (8). The blood lipoproteins, in order of increasing density and decreasing triacylglycerol (TAG) content, are chylomicrons, very-low-density lipoproteins (VLDL), low-density lipoproteins (LDL), and high-density lipoproteins (HDL) (7).

The parenchymal cells of tissues such as adipose, cardiac muscle, red skeletal muscle, lung, kidney, medulla, aorta, and mammary gland synthesize LPL. Macrophages can secrete LPLs (9). After synthesis, LPL is secreted into the capillary endothelium and is bound to the cell surface by sulfated glycosaminoglycans (10). LPLs are essential for processes such as catabolism of chylomicron and VLDL and transfer of phospholipids, cholesterol, and apolipoproteins among lipoprotein particles (1).

Heparine stimulates the liberation of LPL from the tissues to the blood and apolipoprotein C-II (apoC-II), a protein component of plasma lipoproteins, enhances LPL activity (9). ApoC-II may facilitate lipoprotein interface–enzyme interaction and modulate the specificity of LPL so that the hydrolysis of long-chain fatty acids is preferred over the hydrolysis of short-chain fatty acids (10). Other characteristics of LPLs are their inhibition by sodium chloride and protamine and a pH optimum of 8.0–8.5. Although heparin does not seem to be immediately involved in catalysis, it may play a role in enzyme activation. An allosteric mechanism of activation and the involvement of heparin in the enzyme–supersubstrate binding process have both been suggested (7 and references therein). Along with LPL, another TAG hydrolase from liver is stimulated by heparin.

Unlike LPL, this enzyme, referred to as hepatic (endothelial) lipase, is not inactivated by protamine or by NaCl and its role seems to be related to the catabolism of remnants of VLDL and HDL_2. Disorders associated with hypertriglyceridemia are coronary heart disease, peripheral occlusive artherosclerosis, cerebral vascular disease, fatty liver, and pancreatitis. Elevated plasma triglyceride and decreased LPL levels have also been associated with diabetes mellitus, endogenous hypertriglyceridemia, and renal failure (9).

LPL appears to be a serine esterase with a catalytic mechanism similar to that of histidine–serine proteases. LPL is specific for primary esters but does not have absolute stereospecificity (9).

2. Hepatic Lipase

The enzyme referred to as hepatic lipase has alkaline pH requirements and can be found in cytosolic, microsomal, and plasma membranes (11). The other lipolytic enzyme synthesized in the liver, the lysosomal acidic enzyme, is of lysosomal origin and has an acidic pH optimum. Heparin stimulates membrane release of HL. The plasma membranes of liver endothelial cells are probably the physiological action site of HL. HL hydrolyzes tri-, di-, and monoacylglycerols and has phospholipase activity towards the acyl sn-1 position. HL is inhibited by apolipoproteins but the inhibition mechanism is not clear (12). HL also participates in lipoprotein metabolism. It is involved in metabolism of high-density lipoproteins, through which it may participate in the delivery of cholesterol from peripheral tissues to the liver, and in the metabolism of intermediate-density lipoprotein to low-density lipoproteins in the liver (1).

3. Pancreatic Lipase

This enzyme plays an important role in dietary fat absorption. It binds to the lipid interface in the presence of bile salts, colipase, and calcium to hydrolyze the ester bonds of TAGs (1). PL has an approximate molecular weight of 50kDa and has been isolated from numerous mammals. Inhibition of PL hydrolysis by bile salts can be overcome by a cofactor called colipase. It seems that colipase anchors PL to lipid micelles, which contain bile salts, in the intestine. In medicine the topic of inhibition of human pancreatic lipase is very important for the treatment of obesity. This enzyme has four potential glycosylation sites and all cysteins are completely conserved (3).

B. Bile Salt-Activated Lipases

In addition to PL, the pancreas synthesizes another lipolytic enzyme that catalyzes not only the hydrolysis of acylglycerols but also those of phospholipid, cholesterol esters, and vitamin esters. This enzyme, which has also been found in mammary glands (13) and liver (14), has other names such as nonspecific lipase, lipase A, lysophospholipase, cholesterol esterase, carboxylesterase, carboxyl ester hydrolase, carboxyl ester lipase, bile salt-stimulated lipase, bile salt-dependent lipase, and bile salt-activated lipase (BAL). Bile salts activate and directly interact with BAL, the major physiological substrates of which are dietary long-chain acylglycerols (15). Amino acid sequences of human, rat, rabbit, and bovine pancreatic BAL place these enzymes in the esterase B or cholinesterase family (16). The crystal structure of bovine BAL–taurocholate complex determined at 2.8Å reveals the presence of a BAL dimer. Two bile salt-binding sites are present in each BAL molecule within the BAL–taurocholate complex. One site is located close to a loop near the active site and the other site is located far from the active site. The former loop changes its conformation upon binding of taurocholate. This conformational change may lead to the formation of the substrate-binding site (17). The three-dimensional structure of the BAL–taurocholate complex is illustrated in Figure 3. The available structures of members of the esterase B family

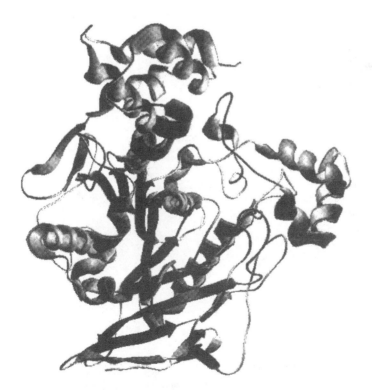

Figure 3 Three-dimensional structure of the bovine bile salt-activated lipase–taurocholate complex. One of the two BAL molecules is shown (PDB entry: 1AQL).

show dramatic similarities even when their sequence identity is low (5, 18). The esterase B enzymes belong to the α/β hydrolase fold structure family (19). The secondary structure of *C. rugosa* is illustrated as an example of this family (Fig. 4). There is no sequence identity between these enzymes and pancreatic triacylglycerol hydrolases other than the consensus sequence Gly-X-Ser-X-Gly present in all serine esterases. While BAL's hydrolytic activity on soluble and partially soluble substrates is bile salt-independent, on insoluble lipid substrates it is strictly bile salt-dependent (16).

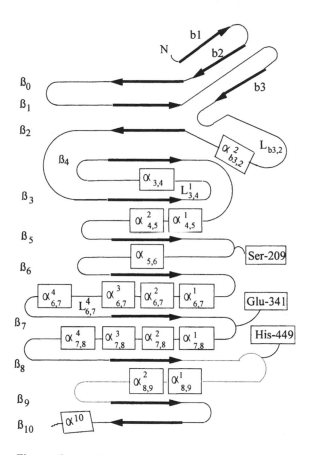

Figure 4 Topology and stereo view of the *Candida rugosa* lipase polypeptide chain. Strands of the small β-sheet are denoted by b_j where j ranges from 1 to 3. Large sheet strands are denoted by β. α-Helices are marked as α^k_{ij}, where the subscripts refer to the loop in which the helix is embedded and k represents the number of the helix within the loop. (From Ref. 5.)

The participation of both PL and BAL in dietary lipid digestion shows the importance of the fat digestion process in the evolution of the development of digestive organs in vertebrates. In rat pancreatic juice, the lipolytic activity of PL is 10–60-fold higher than that of BAL. In contrast, in leopard shark (20), cod (21), salmon (16), and red sea bream (22) pancreas, PL is not present in the pancreatic juice and fat digestion is restricted to BAL. It remains to be established whether this property is common to all fish species.

The fat absorption capacity of the poorly developed pancreas of premature infants improves when formula is supplemented with human milk as a source of BAL. This lipase is present in cat's milk, suggesting kittens' dependence on BAL for fat digestion. Kittens fed with formula supplemented with human milk BAL grew twice as fast as those fed only with formula. On the other hand, the growth rate of breast-fed kittens was similar to that of kittens fed with formula supplemented with BAL. Therefore, BAL involvement in the digestion of triacylglycerols and minor dietary fats may be of special importance in early development (15).

The amino acid sequences of BALs from the human pancreas and mammary glands are identical. However, the human BAL amino acid sequence is longer than those of the rat and the bovine enzymes in 130 and 143 residues, respectively. An extension of a proline (Pro)-rich region near the C terminus of these enzymes is responsible for sequence size differences among different species. The Pro-rich structure has repeating units of 11 amino acid residues (PVPPTGDSGAP), with minor substitutions. These repeating units seem to be unique among lipolytic enzymes; consequently, the development of BAL must have been late in evolution. With the development of BAL, the development of other enzymes able to transform cholesterol to bile salts and of the pancreatic digestive organ was required (15). Because BAL is the only lipase detected in leopard shark (20) and cod pancreas (21), BAL may have appeared earlier than PL in the evolution of the pancreas of vertebrates (15). The recent findings of salmon (16) and red sea bream (22) BALs, as the apparently sole lipolytic enzymes in the pancreas of these fish, support this hypothesis.

C. Gastric Lipases

It has long been thought that the hydrolysis of dietary triglycerides began in the intestinal lumen with the exclusive participation of PL. However, recently the presence of preduodenal lipolysis was indirectly demonstrated. In premature infants, where PL and bile secretions are minimal, and in other cases of PL deficiency such as cystic fibrosis and chronic pancreatitis, approximately 70% of the fat was absorbed. In a different study performed with patients who had congenital PL deficiency, more than 50% of the ingested fat was absorbed. These findings suggested the existence of a second lipolytic enzyme. Although the exact physiological role of preduodenal lipase in the lipolysis process has not yet been elucidated, it has been suggested that this enzyme may cooperate with PL (23 and references therein).

In mammals preduodenal lipase activity has been found in lingual, pharyngeal, and gastric tissues. All of these enzymes have high stability and activity under acidic conditions, unlike PL (23). Human gastric lipase (hGL) is a single 379-amino-acid-protein with a molecular weight of 43kDa. hGL, rabbit gastric lipase (rGL), and dog gastric lipase (dGL) are glycoproteins with 15% carbohydrate. There is up to an 80% sequence identity between hGL and lingual lipase. Based on the molecular cloning of cDNA for human lysosomal acid lipase (hLAL), it has been demonstrated that preduodenal and lysosomal lipases belong to a new lipase gene family. Amino acid sequences of hLAL, hGL, and rat lingual lipases do not share identity with other mammalian or microbial lipases; however, the amino acid sequence of hLAL has a 58% homology with that of hGL and a 57% homology with that of rat lingual lipases. These three groups of lipases have the Gly-X-Ser-X-Gly sequence common to all lipase and serine proteases. The conserved Ser within this pentapeptide, Ser 153 in hGL and Ser 152 in porcine and human pancreatic lipases, is probably the essential Ser for catalysis within the catalytic triad Asp.. His.. Ser. The single disulfide bridge present in gastric lipases such as hGL, rGL, and dGL is essential for catalytic activity (23).

II. FISH LIPASES

A. Introduction

For carnivorous fishes, carbohydrates are not easily available; therefore, lipids are important sources of energy (21, 22, and references therein). On the other hand, marine fishes require polyunsaturated n-3 fatty acids, namely eicosapentanoic acid (20:5n-3) (EPA) and docosahexanoic acid (22:6n-3) (DHA), as essential fatty acids for normal growth (24). In mammals with two major digestive lipolytic enzymes, PL and BAL, polyunsaturated fatty acids (PUFA) are more resistant to hydrolysis by PL. Although information on the enzymology of digestive lipolysis in fish is scarce, some findings suggest that fish may possess a mammalian type PL and colipase system and homologues of BAL. At present, mammalian-type PLs have only been isolated from the pancreas of rainbow trout (*Oncorhynchus mykiss*) (25) and from sardine (*Sardinella longiceps*) hepatopancreas (26). However, purity or identity of the former has been questioned (21) and colipase dependency from the sardine enzyme has not been detected. Colipase has been found in the pancreas of elasmobranchs and purified from dogfish (*Squalus acanthias*) pancreas. Activation of porcine PL by dogfish colipase and of rainbow trout PL by porcine pancreatic colipase have been determined (22 and references therein). The presence of homologues of BAL involved in triglycerides hydrolysis has been reported in fish species such as the anchovy (*Engraulis mordax*), striped bass (*Morone saxatilis*), pink salmon (*Oncorhynchus gorbuscha*) (27), leopard shark (*Triacus semifasciata*) (20), rainbow trout (*Salmo gaidnerii*) (28), and dogfish (29). The properties of cod (*Gadus morhua*) lipase have been compared with those of mammalian BALs (30). In addition, mammalian-type BAL has been isolated from cod pyloric caeca (21), red sea bream (*Pagrus major*) hepatopancreas (22), and Atlantic salmon (*Salmo salar*) pancreas (16). Properties of purified fish lipases are summarized in Table 1.

In the following sections we will review the current knowledge on the enzymology of fish triglyceride and bile salt-activated lipases. The main topics covered include classification, general biochemical properties, factors that influence enzyme activity, and the amino acid sequences of fish lipases. These are placed in the context of our more extensive knowledge of the properties and activity of vertebrate (mammalian) and microbial lipases.

B. Fish Lipid Digestion and Absorption

Fish have slower rates of digestion and absorption than mammals (31). The pyloric caecum organ in fish not only helps retain food, facilitating enzyme-substrate exposure, but also provides an increased surface area for the absorption of the products of digestion. In rainbow trout *Oncorhynchus mykiss* and in Atlantic cod *Gadus morhua* the pyloric caecum epithelium corresponds to 70 and 69%, respectively, of the postgastric surface area available for nutrient absorption. Similarly, the foregut of the Arctic chart has various pyloric caeca and the hindgut, an extension of the midgut, has diminished digestive and absorption functions (31).

Both sn-1,3-specific and nonspecific lipases have been found in fish (21, 27, 32). Lipase preparations (acetone–ether powders) of scallop *Patinopecten yessoensis* hepatopancreas have high specificity for the sn-1(3)-position of long-chain triacylglycerol molecules; therefore, scallop hepatopancreas represents a potential source of 1,3-specific lipases (33). A nonspecific lipase predominates in the intestine of Arctic charr (*Salvelinus alpinus* L.). Interestingly, lipolytic activity has been detected along the whole intestinal tract of Arctic charr, cod (34), *Chanos chanos* (35), and turbot (36).

Fatty acid digestibility in fish decreases with increasing chain length and increases with increasing degree of unsaturation (21, 36, 37). PUFA followed by monounsaturated fatty acids (MUFAs) and then saturated fatty acids are good substrates for fish digestive lipases (30, 37). The lipid absorption process in fish is presumed to be similar to that of mammals (38), in which the products of lipolysis embedded in bile salt micelles travel from the intestinal lumen to the intestinal mucosa to be absorbed as monomers. Free fatty acid absorption rates increase with unsaturation and decrease with chain length in Atlantic salmon

Table 1 Main Biochemical Properties of Purified Fish Lipases

	Red sea bream (*Pagrus major*)[a]	Cod (*Gadus morhua*)[b]	Leopard shark (*Triacus semifasciata*)[c]	Dogfish (*Squalus ancanthias*)[d]	Rainbow trout (*Salmo gairdnerii*)[e]	Oil sardine (*Sardinella longiceps*)[f]
Origin	Hepatopancreas	Pyloric caecum		Pancreas	Intercaecal pancreatic tissue	Hepatopancreas
MW (kDA)		~60				~54–57
pH optimum for activity	8–9	4 Nytrophenyl-myristate:6.5–7.5		8.5		8
Temperature (°C) optimum for activity		4 Nytrophenyl-myristate: 25	Olive oil: 36	Olive oil: 35		Tributyrine: 37
Specificity		sn-1,3 specific				α-Ester bond of triglycerides
Ca^{+2} role on lipolysis	None		Required		Triolein: required Tributyin: none	Tributyrine: required Short-chain TAG: none
Bile salt dependency	Sodium taurocholate/cholate: yes Sodium deoxycholate: no	Absolute on insoluble substrates	Absolute	Yes	Yes	

Sources: [a] Ref. 22, [b] Ref. 21, [c] Ref. 20, [d] Ref. 29, [e] Ref. 25, [f] Ref. 26.

(*Salmo salar*). The rate of absorption of individual free fatty acids (FFA) seems to be inversely related to their melting point (39). Therefore, triglycerides containing unsaturated and short chain fatty acids are preferentially hydrolyzed, and the released free fatty acids are preferentially absorbed, over saturated and long-chain fatty acids (31). Rates of digestion and absorption of individual fatty acids in Arctic charr have been studied. Arctic charr were fed with diets containing coconut oil and PUFA and MUFA combinations either as FFA or as triacylglycerol. Lipid digestion in Arctic charr takes place mainly in the pyloric caecum. While PUFAs (18:2n-6 or 18:3n-3) were almost completely absorbed, MUFAs (20:1n-9 and 22:1n-9) were less efficiently absorbed. 22:1n-9 accumulated in the digesta, indicating that its absorption is limited by its rate of release from TAG. The best fatty acid substrate for intestinal lipase was 12:0 followed by 14:0 and 16:0. In contrast, 18:0 was resistant to digestion and its absorption was very low (31).

In both Indian oil sardines (*Sardinella longiceps*), a planktonivorous pelagic fish, and ribbon fish (*Trichiurus sp.*) a carnivore bottom feeder, lipase activity is mainly detected in the muscle, gut, and liver. Within the gut, the hindgut (intestine portion) has higher lipolytic activity than the foregut (stomach portion) in both species. On the other hand, the liver has the highest lipase activity. In liver and muscle, oil sardine lipase activity is three times higher than that in ribbon fish. While seasonal changes in lipase activity have been detected in the oil sardine, in ribbon fish no significant differences have been observed. Seasonal lipase activity variation in the gut, muscle, and hepatopancreas has also been reported in catfish, catla, and silverfish (40 and references therein). Enzyme activity may vary seasonally due to factors such as water chemistry, pH, temperature, age of fish, feeding habitat, diet, filling of gut, and spawning season. During the period of the study of lipase activity in oil sardines and ribbon fish, fluctuations in temperature were minimal; therefore, other factors probably were responsible for the observed changes in lipase activity (40 and references therein). It has also been suggested that fish may have endogenous rhythms that control the enzymatic activities during different seasons (41).

Knowledge of the physiology of herbivorous marine fish is minimal and no solid evidence exists for the existence of enzymes that degrade cell wall structural polysaccharides (42). The presence of volatile fatty acids in the guts of reef fish has been associated with microbial symbionts that may degrade the algae's carbohydrates (43). The participation of microbial symbionts in the digestion of herbivorous could be similar to that observed in terrestrial insects and ungulates; however, studies on the microorganisms present in the gut of fish have only been performed on rudderfish, mullets, rabbitfish, and surgeonfish (42 and references therein). Flagellates, ciliates, bacteria, and members of a recently recognized group (44) that includes the largest known bacteria (600 µm long) *Epulopiscium fishelsoni* (epulos) have been found in the gut of herbivorous fish (42). High densities of epulos are harbored in the middle intestine of the brown

Table 2 Intestinal Lipase Activity of the Brown Surgeon fish, *Acanthurus nigrofuscus*, with (+) and without (-) the Gut Symbiont, *Epulopiscium fishelsoni*

	pH 6		pH 7		pH 8	
	+	−	+	−	+	−
Mean (Sigma-Tietz units/ml)	3.48	1.79	2.30	2.89	3.90	4.96
Standard deviation	2.82	0.58	1.05	0.60	1.13	1.75
(N)	10	5	10	5	10	5
Variation coefficient (%)	81.0	32.4	45.7	20.8	29.0	35.3

Source: Ref. 42.

surgeonfish, *Acanthurus nigrofuscus,* from the Red Sea. At pH 8.0 fish lipase activity, in the absence of epulos, is 2.8 times higher than that at pH 6.0. Epulos reduces the pH of the host's middle intestine to ~ 6.5. The anterior and posterior intestinal pH remains at 7.2–7.8. From these measurements it would appear that epulos indirectly inhibits the host lipase activity by reducing pH. However, at pH 6 lipase activity is surprisingly high (Table 2). This indicates that the activity detected probably originates from the symbiont rather than from the host. Winter is virtually the only period of the year when brown surgeonfish gain weight and deposit lipids, possibly as a result of lipid intake from the digestion of diatoms. The suppression of lipolytic activity caused by microbes in fish would appear to be maladaptive. However, this is not the case because in what it seems to be a mutualistic relationship, epulos enhances its own lipolytic activity, compensating for the deficiency of the host (42).

C. Bile Salt-Activated Lipase from Red Sea Bream

The activities of both the crude enzyme extract of the delipidated powder and the purified enzyme from the hepatopancreas of red sea bream are increased by the addition of sodium cholate or taurocholate. This increase in activity has been detected using either ρ-nitro-phenyl myristate or triolein as substrates (22). The activation of BAL by bile salts is consistent with findings reported on other fish species (20, 21, 25, 28, 30, 31). Optimal activity of BAL toward ρ-nitro-phenyl myristate and triolein is in a pH range of 7–9. BAL activity is not stimulated by sodium deoxycholate and is Ca^{+2}-independent. Moreover, BAL appears to be the only or at least the predominant enzyme present in the extract. Red sea bream hepatopancreas lipase hydrolyzes 20:4 n-6 and 20:5 n-3 ethyl esters faster than 18:1 n-6, 18:2 n-6 and 18:3 n-3 ethyl esters. However, the hydrolytic rate of 22:6 n-3 is slower than that of 20:4 n-6 and 20:5 n-3 ethyl esters. The purified enzyme from the hepatopancreas of red sea bream is approximately 64 kDa. This molecular mass is comparable to that of cod (21) and mammalian pancreatic BALs, namely porcine, bovine, and rat (22 and references therein). The molecular mass of human milk and pancreatic BALs is larger, approximately 100 kDa. Differences in molecular size between species have been attributed to different degrees of glycosylation (45) and to the presence of proline-rich sequences near the C-terminus. The number of repetitions of these sequences varies depending on the species (15).

Red sea bream lipase binds tightly to a cholate–Sepharose column (22). The same type of interaction has been observed in mammalian BALs that have been purified by affinity chromatography using cholate as a ligand (22 and references therein). These findings agree with the BAL property of direct interaction with bile salts (15). On the other hand, as occurs with human milk and human pancreatic BALs (13), red sea bream BAL is highly activated by the 7 α-hydroxyl group of bile salts. From the findings described above it is concluded that red sea bream hepatopancreas lipase is homologous to the mammalian BAL (22).

D. Amino Acid Sequence from Atlantic Salmon BAL

The first nonmammalian BAL sequence determined was that from salmon (16). This sequence had a 57–59% homology with those from human, rat, and bovine pancreas BALs. The authors assumed that the mature salmon BAL sequence started at alanine (Ala) 14, as is the case of mammalian BALs, and defined this position as residue number 1. The mature salmon BAL has 527 amino acids and does not have the proline-rich C-terminus repeats found in human and rat BALs (Fig. 5) (16 and references therein). Salmon BAL has an active serine in the Gly-Asp-Ser–Ala-Gly sequence that is characteristic of esterase B class enzymes. While the amino acids of the central β sheet structure of acetylcholine esterases are well conserved in the BAL salmon sequence, the α-helical elements are less conserved (16). Sequence identity of the salmon BAL with other members of the esterase B enzymes ranges from 27% with *Candida rugosa* lipase to 35% with mouse acetylcholine esterase. *G. candidum* (46) and *C. rugosa* (5) lipases have a catalytic triad similar to that of other serine esterases. The salmon triad motif Ser190, His433 and Asp315 was identified by alignment of the salmon BAL sequence to mammalian BAL and to other esterase B enzyme sequences. The neighboring C-terminal amino acid to the His residue is a conserved Ala in BALs while in esterase B enzymes is a conserved Gly. On the other hand, the BALs' catalytic triad has an Asp that in esterase B enzymes, with the exception of some from Drosophila, is replaced by glutamate (Glu). In *Candida rugosa* and *Geotrichum candidum* lipases, the loop region between cysteine (Cys) 64 and Cys80 functions as the active site lid. In BALs only a short section that may not

```
                         1                      2 0
                         ↓                      ↓
                       L V A S A L F L  G S A S A A T L G V V Y T E G G M V  Q G K K V N S D G L - L R T  M D V F K G I  P Y A D K P
SALBAL
RATBAL  M G R L E V L  F L G L T C C L A A A C A A K . . A . . . . . . F . E . V N . K L . L L G G D S V . I . . . . . F . T A .
HUMBAL  M G R L Q L V  V L G L T C C W A V A S A A K . . A . . . . . . F . E . V N K K L G L . Q - D S V . I . . . . . F . A P T
BOVBAL    L G A S R L G P S P G C L A V A S A A K . . S . . . . . . F . E . V N K K L S L F G - D S V . I . . . . . F . A P .
RABBAL  M G R L E L T  V L G L A C L L D S G A C G D . . P . . . . . . F . E . E N . K L . L L G A T S V . I . . . . . F . T P A

           4 0                      6 0                      8 0
           ↓                        ↓                        ↓
SALBAL  G V F E K P K  R H P G W D G V L K A T E F K P R C M Q L N L L Q S D T R  G Q E D C L Y L N I  W V P Q G L S - V S T G L P
RATBAL  K T L . N . Q . . . . . Q . T . . . D . . K . L A T I T . D . . Y . . . . . . . . . . . . R K Q . . H D . .
HUMBAL  K A L . N . Q P . . . . Q . T . . . K N . . K . L . A T I T . D S . Y . D . . . . . . . . . . . R K Q . . R D . .
BOVBAL  K A L . . . E . . . . . Q . T . . . K S . . K . L . A T . T . D S . Y . N . . . . . . . . . . . R K E . . H D .
RABBAL  - I L . N . Q . . . . . Q . T . . . K D . . K . L . A T . T . D S . F . D Q . . . . . . . . . . R K E . . H N . .

           1 0 0                    1 2 0                    1 4 0
           ↓                        ↓                        ↓
SALBAL  V M V W I  F G G G Y L V G G S M G A N F L D N Y L Y D G E E I  A N R G K  V I V V T L  G Y R V G  T L G F L S S E M H D - -
RATBAL  . . . . . Y . . A F . M . S G Q . . . . K . . . . . . . . . . . . . . . T . . N . . . . F N . . . . P . . . . T G D A N L P
HUMBAL  . . I . . Y . . A F . M . S G H . . . . N . . . . . . . . . . . . T . . N . . . . F N . . . . P . . . . T G D A N L P
BOVBAL  . . I . . Y . . A F . M . A . Q . . . . S . . . . . . . . . . . . T . . N . . . . F N . . . . P . . . . T G D S N L P
RABBAL  . . I . . Y . . A F . M . S . Q . . . . S . . . . . . . . . . . . T . . N . . . . F N . . . . P . . . . T G D A N L P

           1 6 0            ⊗        1 8 0          *         2 0 0
           ↓                        ↓                        ↓
SALBAL  C N Y V V - D Q H A A I  A W V N R N I  R D F G G D P N N I  T V F G E S A  G P A S V S F Q T L S  P H N K G L I  R R A I  S Q
RATBAL  G . F G L R . . . M . . . . A A . . . . . . . A A . . . . . . . I . . . . . G . . . . . L . . . . . Y . . . . . . . . .
HUMBAL  G . . G L R . . . M . . . . K . . . A A . . . . . . . . L . . . . . G . . . . L . . . . Y . . . . . . . . .
BOVBAL  G . . G L W . . . M . . . . K . . E A . . . . . D . . . . L . . . . . G . . . . L . . . . Y . . . . K . . . .
RABBAL  G . . G L R . P . M . . . . K A . A A . . . . D . . . I . . . . . . . . . L . . . . . Y . . . . . . . .

           2 2 0                    2 4 0                    2 6 0
           ↓                        ↓                        ↓
SALBAL  S G V A L C P W A I  N H N P R A F A E M V A G K V G C P T D D - - Q M M A C L K L I  N A K E L  T L A G T L S L A G S P S
RATBAL  . . . . . S . . . . Q E . . L F W . K T I . K . . . . . E . T A K . A G . . . I T D P R A . . . . Y R . P . . K . Q E
HUMBAL  . . . . . S . . V . Q K . . L F W . K K . . . E . . . . V G . A A R . A G . . . V T D P R A . . . . Y R . . . . A G L E
BOVBAL  . . . G . . . . . Q Q D . L F W . K R I . E . . . . V . . T S K . A G . . . I T D P R A . . . . Y K . P . . . . T E
RABBAL  . . . . . S . . D . Q K . . L F W . K K I . . . . . . L . Y T A T . A Q . V . I T D P H S . . . . Y N F P . . A G L A

           2 8 0                    3 0 0          *         3 2 0
           ↓                        ↓                        ↓
SALBAL  T P I V D N L  A L S P V I  D G D F L P D H P G N L  F H N A A D I  D Y L A G V N S M D A H L F T  G L D L P A V N K P I  A N
RATBAL  Y . . . H Y . . F I . . V . . . . . D . I . . Y D . . . . . . . . . I . D . G . . . A T V . V . . I D . A K Q D
HUMBAL  Y . M L H Y . V G F V . . . . . . . I . A D . I . . Y A . . . . . . . . I . . T . N . . G . I . A S I . M . . I . G N K K
BOVBAL  Y . K L H Y . . S F V . . . . . . . I . . D . V . . Y A . . . . . . I . . . T . D . G . . . V . M . V . . I . S N K Q D
RABBAL  Y . M . H Y . . G F I . . V . . . . E D . I I . Y A . . . . . . . . . . T . D . G . . . A T V . M . . I D . S Y K D

           3 4 0                    3 6 0                    3 8 0
           ↓                        ↓                        ↓
SALBAL  L P L S D V - K L L L G S L  T K K G E A S I  N S R F A E Y T A D W G D K  P S Q E T I  K K T V V  M I  E T D Y V F L V P T Q
RATBAL  V T E E . F Y R . . V S . H T V A . . L K G T Q A T . D I . . E S . A Q D . . . . N M . . . . . A F . . . I L . . I . . E
HUMBAL  V T E E . F Y K . V S E F T I . . L R G A K T T . D V . . E S . A Q D . . . . N K . . . . . D F . . . V L . . . . . E
BOVBAL  V T E E . F Y K . V S . . . V T . . L R G A . A T Y E V . . E P . A Q D S . . . . R . . . M . D L . . . I L . . I . . E
RABBAL  I S D Q . F Y . . V S . M T V T . . S . G A Q A T Y S I . . E S . A Q D S . . Q N K . . . . . D L . . . I L . . I . . E

           4 0 0                    4 2 0          *         4 4 0
           ↓                        ↓                        ↓
SALBAL  A A L Y L H A  S N A Q S A R T Y S Y L F S E P S R M S G V V L P F  P S W M E A D H A E D L Q F  V F G K P F T T P L A Y W
RATBAL  M . . A Q . R A H . K . . K . . . . . . H . . . . . . P I Y . K . . G . . . . D . . Y . . . . . . A . . . . G . R
HUMBAL  I . . A Q . K . . H . K . . N . T . . . . H . . . . . . P V Y . K . V G . . . . D . I . Y . . . . . A . . . T G . R
BOVBAL  I . V A Q . R A . . H . K . . N . . T . . . Q . . . . . . P I Y . K . . G . . . . D . I . Y . . . . . A . . . G . R
RABBAL  M . . A Q . R A . S S T . K . . . . . . H . . . . . . P I Y . . . . G . . . . D . . Y I . . . . . A . . T G . R

           4 6 0                    4 8 0                    5 0 0
           ↓                        ↓                        ↓
SALBAL  P K H R N V S K Y F I  A Y W T N F  A R T G D P N K G E S N V P V T  W P A Y T T S G Q K Y L E I  N A K M N R N S V H E K M
RATBAL  A Q D . T . . . A M . . . . . . . . K S . . . . M . N P . . T H . . Y P . . . E N G N . . D . . K . I T S T . M K . H L
HUMBAL  . Q D . T . . . A M . . . . . . . . K . . . . M . D A . . T H . E P . . E N S G . . . . T K . G S S . M K R S L
BOVBAL  A Q D . T . . . A M . . . . . . . . T . H . T . . A N . D P . L E D D N . . . . . K Q . D S . . M K L H L
RABBAL  . Q D . T . . . T L . . . . . . . . T . F G S . A R H . E P . . L E N G S . . . . . K . . S Q D . T K S H L

           5 2 0
           ↓
SALBAL  R V R F V N W W S N T - - - L P S I
RATBAL  . E K . L K F . A V . F E M . . . T V V »
HUMBAL  . T N . L R Y . T L . Y L A . . T V T »
BOVBAL  . T N Y L Q F . T Q . Y Q A . . T V T »
RABBAL  . N S . L Q F . A G P T K A . . V V L »
```

Figure 5 Alignment of salmon BAL amino acid sequence to rat BAL (RATBAL), human BAL (HUMBAL), bovine BAL (BOVBAL), and rabbit BAL (RABBAL) sequences. In salmon BAL Ala14 was set as amino acid position 1. Residues identical to those of the salmon sequence (.), insert regions (-), residues in the catalytic triad (★), and a putative N-glycosilation site (⊗) are shown. (From Ref. 16.)

be enough to play the lid function role has been conserved. The presence of a lid in lipases has been associated with substrate specificity (47). Moreover, interfacial activation has not been detected in BALs (16).

Bile salts may aid in the binding of BALs to the lipid–bile–salt interface. BALs are the only enzymes that bind to bile salts; therefore, residues involved in the binding process should be conserved in all BALs and be unique for these enzymes. From multiple alignments of BALs to some esterase B sequences only a few sequence elements are exclusively present in BALs (Fig. 6). The salmon BAL sequence formed by the amino acids 115–128, GANFLXNYLYDGEE, is conserved in all BALs (16). When comparing BAL sequences to esterase B enzyme sequences, the N-terminal section of this sequence, GANFLXNY, is an insertion. This unique, highly conserved insertion may be associated with the process of bile-salt binding. The presence of some arginines (Arg) such as Arg63, Arg132, Arg173, Arg417, and Arg456 in BALs has also been found. In the human BAL molecule one or more Arg side chains bind the anionic head groups of bile salts (48).

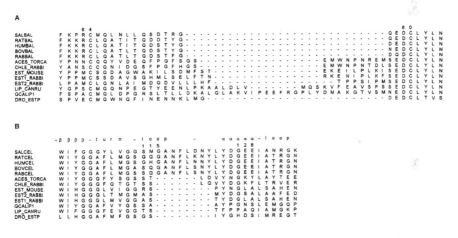

Figure 6 Sequence comparison of salmon BAL (SALBAL) with mammalian BALs and other esterase B and P enzymes. (A) Disulfide-stabilized loop between positions 64 and 80. (B) BALs conserved sequence showing the insert (region between positions 115 and 124). Insert regions are denoted by (-). Sequences used are rat BAL (RATBAL), human BAL (HUMBAL), bovine BAL (BOVBAL), rabbit BAL (RABBAL), *T. californica* (ACES_TORCA), rabbit butyrilcholine esterase (CHLE_RABBI), mouse liver BAL (EST_MOUSE), rabbit liver esterase 1 (EST1_RABBI) and 2 (EST2_RABBI), *Geotrichum candidum* lipase 1 (GCALIP), *Candida rugosa* lipase (LIP_CANRU), and *Drosophila melanogaster* esterase P (DRO_ESTP). (From Ref. 16.)

E. Cod Digestive Lipolysis

The first purified BAL of nonmammalian origin was that of the pyloric caecum from cod. This BAL is the only enzyme present in the cod's pyloric caecum that hydrolyzes TAGs, 4-nitrophenyl esters, and cholesteryl esters. It has a molecular weight of 50 kDa as determined by sodium dodecyl sulfate–polyacrylamide gel electrophoresis (SDS-PAGE) and of 56 kDa as determined by gel filtration chromatography. Multimers of more than 200 kDa appear after exposure of the enzyme to bile salts (21). The amino acid composition of cod BAL is similar to the amino acid composition of some mammalian BALs (Table 3). Cod BAL has a high Ser and glutamate/glutamine (Glx) content and a low lysine (Lys) content. Although all the compared enzymes have a high content of Gly, the Gly content of the cod enzyme is exceptionally high. In human milk BAL, where Pro-rich repeats were not included in the calculations, the Pro content was similar to those of cod and rat. This finding is consistent with the attribution of differences in molecular mass between species to the presence of repetitive Pro sequences. Incubation of the cod BAL crude extract for 24 h at 4°C results in 95% loss in enzyme activity; in contrast, only 20% of the activity is lost when sodium taurocholate is added. These results indicate that bile salts protect cod BAL from denaturation, as has been reported for BALs from other sources (21).

Table 3   Amino Acid Composition of Cod BAL and Selected Mammalian BALs

	Cod BAL	Human BAL	Rat BAL	Dog BAL	Human N-term Milk BAL
Cysteine	0.4				0.8
Aspartate/asparagine	9.8	11.1	10.9	12.8	11.7
Threonine	5.0	8.2	5.5	5.8	7.7
Serine	10.5	5.4	8.6	7.1	5.4
Glutamate/glutamine	11.8	6.0	9.6	7.6	7.1
Proline	3.9	13.4	6.0	10.7	6.5
Glycine	19.0	10.4	10.2	10.5	9.6
Alanine	8.3	9.7	10.8	9.9	10.0
Valine	4.8	8.4	7.5	6.0	7.1
Methionine	1.2	1.6	1.5	1.8	2.1
Isoleucine	3.3	3.5	4.5	4.4	4.0
Leucine	7.8	6.2	7.5	6.8	8.2
Tyrosine	3.5	3.5	0.3	3.6	5.0
Phenylalanine	3.8	3.6	4.0	3.3	4.8
Histidine	1.4	1.5	1.8	1.9	1.6
Lysine	1.6	4.8	6.4	4.4	6.3
Arginine	4.0	2.8	4.7	3.5	3.6

Source: Ref. 21.

Gjellesvik et al. (21), based on Wang's et al. approaches (49), calculated rate constants of cod BAL-catalyzed glyceride hydrolysis. Diolein was hydrolyzed 10 times faster than triolein and mono-olein hydrolysis was slow. Cod BAL was expected to display higher catalytic activities at low temperatures, as generally occurs with enzymes from cold adapted poikilotherms. However, the activity of cod BAL at temperatures between 0°C and 8°C, the normal habitat range temperature of cod, compared to that of human pancreatic BAL was not significantly higher. The average hydrophobicity calculated, based on the amino acid composition, was 3.44 kJ/mol. This value is notably lower than that of the human BAL (4.83 kJ/mol) indicating that in cod, as in other poikilotherms, the participation of hydrophobic forces in protein stabilization is not strong (21).

Generally the catalytic efficiency of poikilotherms is higher than that of their homologous homeotherms. However, the catalytic efficiency of the cod BAL was lower than that of the human BAL (21). BALs have been shown to be able to hydrolyze esters that PL does not hydrolyze (50). Cod BAL hydrolyzed glyceryl esters of C20:4 and C20:5 at high rates, but others, such as the glyceryl esters of 22 carbon series fatty acids, were poorly hydrolyzed. In this regard, it is important to note that hydrolysis rates may be altered by assay conditions and that the different bile salts mixtures may affect fatty acid specificity (21).

Cod BAL may be homologous to mammalian BALs despite the differences found in catalytic efficiencies. These differences could be expected due to the evolutionary distance between teleosts and mammals. It also appears that cod BAL is the only enzyme responsible for lipid digestion in cod. The lower specific activity of this BAL for triacylglycerols compared with that of PL might be compensated by the prolonged exposure of lipid nutrients to lipolysis facilitated by the pyloric caecum organization (21).

F. Human, Zebrafish, and Rainbow Trout Lipoprotein Lipases

In mammals, the cDNA and amino acid sequences of LPLs have a high degree of homology (51). The hLPL gene contains 10 exons and nine introns (52). A couple of interesting and unique characteristics of this gene have been identified. The length of exon 10 is greater than that of the sum of the other nine exons and intron 9 splits TGA into TG/A. The cDNA from zebrafish *Brachidanio rerio* has been successfully aligned to the known mammalian cDNA. In both zebrafish and rainbow trout *Oncorhynchus mykiss,* the codon spanning the last two exons is AGA while in mammals is TGA. Moreover, intron 9 is 0.128kb and 0.127kb long in zebrafish and rainbow trout, respectively, compared to 3.090kb long in mammals. Other fish genes with smaller introns than those found in mammals have been reported (53, 54). The more compact fish genome is due to the presence of very few repetitive DNA sequences, smaller intergenic sequences, smaller introns, and the absence of pseudogenes. The C-terminal of zebrafish is

17 amino acids longer that that of mammals. The amino acids encoded by the beginning of exon 10 in zebrafish and trout may participate in the binding of LPL to heparine sulfate proteoglycan of endothelial cell membranes. From these findings it has been postulated that mutations may have converted the AG/A sense codon present in fish, into the stop codon TG/A of mammals. Whether the C-terminal end loss in LPL of mammals represents an advantage, regarding enzyme activity, or a disadvantage, related to susceptibility to diseases such as atherosclerosis or diabetes, has not been elucidated yet (51).

G. Rainbow Trout Liver and Adipose Tissue Lipases

Lipids in fish are stored in different compartments including mesenteric fat, liver, and dark muscle. Hydrolysis of long-chain TAGs has been detected in the dark muscle of rainbow trout. Two TAG lipases have been isolated and partially purified from rainbow trout liver (55) and adipose tissue (56), respectively. The biochemical properties of these enzymes are comparable with those of rat and chicken adipose tissue "hormone-sensitive" lipases. Insulin and glucagon alter the phosphorylation state of TAG hepatic lipase from rainbow trout (57). The reversibility of this activation has been proven; consequently, the importance of phosphorylation to the activation of rainbow trout TAG hepatic lipase has been confirmed. Michelsen et al. (58) studied the activation of adipose tissue lipolysis in rainbow trout. cAMP/ATP activated rainbow trout adipose tissue TAG lipase as previously observed in rainbow trout liver lipase (55). Activation of the enzyme was not enhanced by the addition to the trout preparation of the protein kinase catalytic subunit. This result indicated that there was sufficient endogenous protein kinase in the preparation. The purest adipose tissue TAG lipase preparation contained a phosphoprotein with a molecular weight of 45 kDa. ^{32}P from [^{32}P] ATP was found to be associated with a 45kDa protein; therefore, phosphorylation of the enzyme occurs during the process of activation (58 and references therein).

H. Lipases in Postharvest Seafood

Lipolysis, a process that has been associated with quality deterioration of frozen seafood, occurs in many fish species. Fish muscle tissue has red or dark and white fibers. Lipolysis predominates in red muscle tissue, which can constitute from 0 to 30% of the total muscle tissue and in most fish is used for normal swimming. In contrast to white muscle, red muscle has a higher concentration of total lipid and phospholipid. As most of the studies on fish lipolysis have been performed on lean fish, whose predominant lipid is phospholipid, little information is available regarding triglyceride hydrolysis. Triglycerides in live fish are a source of energy reserve and their hydrolysis is restricted to hormonal control. Triglycerides are essential for red muscle metabolism (59).

Accumulation of free fatty acids in frozen fish has been associated with quality deterioration. The main indirect effects of free fatty acid accumulation in frozen fish are alterations of textural properties by stimulation of protein denaturation and production of off flavors by promotion of lipid oxidation. Triglycerides are hydrolyzed in both red and white muscle in frozen Baltic herring (59). Senthivel et al. 92 (60) followed lipase activity of oil sardine and ribbon fish stored at -18 ⊕ 2°C at pH 8.0 during 6 months. They found that after 60 days of frozen storage, lipase activity of both species decreased reaching the least activity at 180 days (60). Temperature fluctuations from -12°C to -35°C for 96 h and slow freezing can activate lysosomal lipase of the dark lateral line muscle of rainbow trout. However, intermediate and fast freezing do not trigger lipase release from lysosomes (61). This is consistent with the results of Partman (62), who found that temperature reduction to the freezing range should be performed in less than 30 min in order to minimize quality alterations caused by enzymes. The production of free fatty acids from phospholipids and triglycerides in Cape hake mince stored at -5°C, -18°C, and -40°C has been reported (63). During frozen storage free fatty acids are produced from both phospholipids and neutral lipids and the relative contribution of each depends on the lipid composition, storage temperature, and storage time. At -5°C for example, the phospholipid contribution to the total FFA approximately doubles that of neutral lipids. At -40°C, on the other hand, both the phospholipid and the neutral lipid contributions to the total FFA are similar. Above -12°C phospholipids hydrolyze faster than neutral lipids and below that temperature-neutral lipids hydrolyze faster than phospholipids (63).

Triglyceride hydrolysis may promote lipid oxidation. In contrast, phospholipid hydrolysis may inhibit lipid oxidation (59, 64). In ground beef, lipase and lipoxygenase stimulate pigment oxidation, while phospholipase reduces it (65). The importance of lipid oxidation in fish quality deterioration and the relationship between lipolysis and lipid oxidation is described in Chapter 4, Sec. V.

III. CONCLUSIONS

Lipids are essential sources of energy for marine carnivorous fish. On the other hand, bile salt-activated lipases (BALs) are very important because they are the predominant, if not the sole, lipolytic enzymes present in the pancreas of marine fish. This means that the role of BALs involves not only hydrolysis of cholesterol esters or vitamin esters but also digestion of triacylglyceride and wax esters. In mammals these functions are shared by pancreatic lipase and BAL. The information available on fish digestive lipolysis is scarce. Although some important advances have been made with studies performed on purified fish BALs (21, 16, 22, and references therein), much more investigation into lipid digestion is required. More studies of lipolytic enzymes in fish pancreas are required to establish whether the absence of pancreatic lipase is a common characteristic of fish. Purification, sequencing, and three-dimensional structure resolution of fish lipases would help to inform a wide range of important topics from evolution to enzyme function and lipase-substrate binding and catalysis. Other specific topics related to fish lipases that need investigation are the ontogeny and mechanisms of intestinal function in larval fishes, the digestive physiology of herbivorous marine fish, the gut content and the relationship between gut microbes and lipid digestion in herbivorous fish, and lipid absorption. Although lipase activity has been detected during fish frozen storage, a direct relationship between lipase activity and quality deterioration needs to be established.

REFERENCES

1. WA Hide, L Chan, W-H Li. Structure and evolution of the lipase superfamily. J Lipid Res 33:167–178, 1992.
2. A Svendsen. Sequence comparisons within the lipase family. In: P Wooley, SB Petersen, eds. Lipases—Their Structure, Biochemistry and Application. Cambridge: Cambridge University Press, 1994, pp 1–21.
3. FK Winkler, K Gubernator. Structure and mechanism of human pancreatic lipase. In: P Wooley, SB Petersen, eds. Lipases—Their Structure, Biochemistry and Application. Cambridge: Cambridge University Press, 1994, pp 139–157.
4. H van Tilbeurgh, MP Egloff, C Martinez, N Rugani, R Verger, C Cambillau. Interfacial activation of the lipase-procolipase complex by mixed micelles revealed by x-ray crystallography. Nature 362:814–820, 1993.
5. P Grochulski, YG Li, JD Schrag, F Bouthillier, P Smith, D Harrison, B Rubin, M Cygler. Insights into interfacial activation from an open structure of *Candida rugosa* lipase. J Biol Chem 268: 12843–12847, 1993.
6. KA Dugi, HL Dichek, GD Talley, HB Brewer, Jr., S Santamaria-Fojo. Human lipoprotein lipase: the loop covering the catalytic site is essential for interaction with lipid substrates. J Biol Chem 35:25086–25091, 1992.
7. H Brockerhoff, RG Jensen. Lipolytic Enzymes. New York: Academic Press, 1974.
8. M Bownes. Why is there sequence similarity between insect yolk proteins and vertebrate lipases? J Lipid Res 33:777–790, 1992.
9. LC Smith, HJ Pownall. Lipoprotein lipase. In: B Borgstrom, HL Brockman eds. Lipases. New York: Elsevier Science Publishing, 1984, pp 263–270.
10. M Gurr, JL Harwood. Lipid Biochemistry, an Introduction, 4th ed. New York: Chapman and Hall, 1991.
11. GL Jensen, B Daggi, A Bensadoun. Triacylglycerol lipase, monoacylglycerol lipase and phospholipase activities of highly purified rat hepatic lipase. Biochim Biophys Acta 710: 464–470, 1982.
12. PKJ Kinnunen. Hepatic endothelial lipase. Isolation, some characteristics, and physiological role. In: B Borgstrom, HL Brockman eds. Lipases. New York: Elsevier Science Publishing, 1984, pp 307–321.

13. L Blackberg, D Lombardo, O Hernell, O Guy, T Olivecrona. Bile salt-stimulated lipase in human milk and carboxyl ester hydrolase in pancreatic juice. FEBS Letter 136:284–2288, 1981.

14. ED Camulli, MJ Linke, HL Brockman, DY Hui. Identity of a cytosolic neutral cholesterol esterase in rat liver with the bile salt-stimulated cholesterol esterase in pancreas. Biochim Biophys Acta 1005:177–188, 1989.

15. C-S Wang, JA Hartsuck. Bile salt-activated lipase. A multiple function lipolytic enzyme. Biochim Biophys Acta 1166: 1–19, 1993.

16. DR Gjellesvik, JB Lorens, R Male. Pancreatic carboxylester lipase from Atlantic salmon (*Salmo salar*) cDNA sequence and computer-assisted modelling of tertiary structure. Eur J Biochem 226: 603–612, 1994.

17. XQ Wang, CS Wang, J Tang, F Dyda, XJC Zhang. The crystal structure of bovine bile salt activated lipase: insights into the bile salt activation mechanism. Structure 5: 1209–1218, 1997.

18. M Cygler, JD Schrag, JL Sussman, M Harel, I Silman, MK Gentry, BP Doctor. Relationship between sequence conservation and 3dimensional structure in a large family of esterases, lipases and related proteins. Protein Sci 2:366–382, 1993.

19. DL Ollis, E Cheah, M Cygler, B Dijkstra, F Frolow, SM Franken, M Harel, SJ Remington, I Silman, JD Schrag, JL Sussman, KHG Verschueren, A Goldman. The α,β-hydrolase fold. Protein Eng 5:197–211, 1992.

20. JS Patton, TG Warner, AA Benson. Partial characterization of the bile salt-dependent triacylglycerol lipase from the leopard shark pancreas. Biochim Biophys Acta 486: 322–330, 1977.

21. DR Gjellesvik, D Lombardo, BT Walther. Pancreatic bile salt dependent lipase from cod (*Gadus morhua*): purification and properties. Biochim Biophys Acta 1124:123–134, 1992.

22. N Iijima, S Tanaka, Y Ota. Purification and characterization of bile salt-activated lipase from the hepatopancreas of red sea bream, *Pagrus major*. Fish Physiol Biochem 18:59–69, 1998.

23. F Carriere, Y Gargouri, H Moreau, S Ransac, E Rogalska, R Verger. Gastric lipases: cellular, biochemical and kinetic aspects. In: P Wooley, SB Petersen, eds. Lipases—Their Structure, Biochemistry and Application. Cambridge: Cambridge University Press, 1994, pp 181–205.

24. MV Bell, RJ Anderson, JR Sargent. The role of polyunsaturated fatty acids in fish. Comp Biochem Physiol 83B: 711–719, 1987.

25. C Leger, D Bauchart, J Flanzy. Some properties of pancreatic lipase in *Salmo gairdnerii* Rich.: Effects of bile salts and Ca^{+2}, gel filtration. Comp Biochem Physiol 57: 359–363, 1977.

26. MK Mukundan, K Gopakumar, MR Nair. Purification of a lipase from the hepatopancreas of oil sardine (*Sardinella longiceps* Linnaceus) and its characteristics and properties. J Sci Food Agric 36:191–203, 1986.

27. C Leger. Digestion, absorption, and transport of lipids. In: CV Cowey, AM Mackie, JG Bell, eds. Nutrition and Feeding in Fish. London: Academic Press, 1985, pp 229–231.

28. DR Tocher JR Sargent. Studies of triacylglycerol, wax ester and sterol ester hydrolases in intestinal caeca of rainbow trout (*Salmo gairdneri*) fed diet rich in triacylglycerols and wax esters. Comp Biochem Physiol 77B:561–571, 1984.

29. BA Raso, HO Hultin. A comparison of dogfish and porcine pancreatic lipase. Comp Biochem Physiol 89B:671–677, 1988.

30. DR Gjellesvik. Fatty acid specificity of bile salt-dependent lipase: enzyme recognition and super-substrate effects. Biochim Biophys Acta 1086: 167–172, 1991.

31. RE Olsen, RJ Henderson, E Ringo. The digestion and selective absorption of dietary fatty acids in Arctic charr, *Salvelinus alpinus*. Aquaculture Nutrition 4:13–21, 1998.

32. DR Gjellesvik, AJ Raae, BT Walther. Partial purification and characterization of a triacylglycerol lipase from cod (*Gadus morhua*). Aquaculture 79:77–184, 1989.

33. Y Itabashi, T Ota. Lipase activity in scallop hepatopancreas. Fish Sci 60:347, 1994.

34. O Lie, E Lied, G Lambersten. Lipid digestion in cod (*Gadus morhua*). Comp Biochem Physiol 98A: 159–163, 1987.

35. IG Borlongan. Studies on the digestive lipases of milkfish, *Chanos chanos*. Aquaculture 89:315–325, 1990.

36. WM Koven, RJ Henderson, JR Sargent. Lipid digestion in turbot (*Scophthalmus maximus*). II: Lipolysis in vitro of ^{14}C-labelled triacylglycerol, cholesterol ester and phosphatidylcholine by digesta from different segments of the digestive tract. Fish Physiol Biochem 13:275–283, 1994.

37. O Lie, G Lambertsen. Digestive lipolytic enzymes in cod (*Gadus morrhua*): fatty acid specificity. Comp Biochem Physiol 80B: 447–450, 1985.

38. MV Ostos Garrido, MV Nuñez Torres, MA Abaurrea Equisoain. Lipid absorption by enterocytes of the rainbow trout, *Oncorhynchus mykiss:* diet induced changes in the endomembranous system. Aquaculture 110:161–171, 1993.

39. S Sigurgisladottir, SP Lall, CC Parrish, RG Ackman. Cholestane as a digestibility marker in the absorption of polyunsaturated fatty acid ethyl esters in Atlantic salmon. Lipids 27:418–424, 1992.

40. A Senthilvel, LN Srikar, G Vidya Sagar Reddy. Lipase activity in oil sardine and ribbon fish during different seasons. Environ Ecol 12:371–373, 1994.

41. R Hofer. The adaptation of digestive enzymes to temperature, season and diet in roach, *Rutilus rutilus* L and rudd, *Scardinus erythrophthalmus* L. 2. Proteases. J Fish Physiol 15:373–379, 1979.

42. PE Pollak, WL Montgomery. Giant bacterium (*Epulopiscium fishelsoni*) influences digestive enzyme activity of an herbivorous surgeonfish (*Acanthurus nigrofuscus*). Comp Biochem Physiol 108A:657–662, 1994.

43. DW Rimmer, WJ Wiebe. Fermentative microbial digestion in herbivorous fishes. J Fish Biol 31: 229–236, 1987.

44. WL Montgomery, PE Pollak. *Epulopiscium fishelsoni* n.g., nsp., a protist of uncertain affinities from the gut of an herbivorous reef fish. J Protozool 35:565–569, 1988.

45. N Abouakil, E Rogalska, J Bonicel, D Lombardo. Purification of pancreatic carboxylic ester hydrolase by immunoaffinity ant its application to the human bile salt-stimulated lipase. Biochim Biophys Acta 961:299–308, 1988.

46. JD Schrag, Y Li, S Wu, M Cygler. Ser-His-Glu forms the catalytic site of the lipase from *Geotrichum candidum*. Nature 351:761–764, 1991.

47. A Hjorth, F Carriere, C Cudrey, H Woldike, E Boel, DM Lawson, F Ferrato, C Cambillau, GG Dodson, L Thim R Verger. A structural domain (the lid) found in pancreatic lipases is absent in the guinea pig (phospho)lipase. Biochemistry, 32:4702–4707, 1993.

48. L Blackberg, O Hernell. Bile salt-stimulated lipase in human milk. Evidence that bile salt induces lipid binding and activation via binding to different sites. FEBS Lett 323:207–210, 1993.

49. CS Wang, JA Hartsuck, D Downs. Kinetics of acylglycerol sequential hydrolysis by human milk bile salt activated lipase and effect of taurocholate as fatty acid acceptor. Biochemistry 27:4834–4840, 1988.

50. Q Chen, B Sternby, B Akesson, A Nilsson. Effects on human pancreatic lipase-colipase and carboxyl ester lipase on eicosapentaenoic and arachidonic acid ester bonds of triacylglycerols rish in fish oil fatty acids. Biochim Biophys Acta 1044:111–117, 1990.

51. F Arnault, J Etienne, L Noé, A Raisonnier, D Brault, JW Harney, MJ Berry, C Tse, C Fromental-Ramain, J Hamelin, F Galibert. Human lipoprotein lipase last exon is not translated, in contrast to lower vertebrates. J Mol Evol 43:109–115, 1996.

52. TG Kirchgessner, JC Chuat, C Heinzmann, J Etienne, S Guilhot, K Svenson, D Ameis, C Pilon, ALD Andalibi, MC Schotz, F Galibert, AJ Lusis. Organization of the human lipoprotein lipase gene and evolution of the lipase gene family. Proc Natl Acad Sci USA 86:9647–9651, 1989.

53. M Tanaka, S Fukada, M Matsuyama, Y Nagahama. Structure and promoter analysis of the cytochrome P-450 aromatase gen of the teleost fish, medaka (*Oryzias latipes*). J Biochem (Tokyo) 117:719–725, 1995.

54. I Hirono, T Aoki. Characteristics and genetic analysis of fish transferrin. Fish Pathol 30:167–174, 1995.

55. JS Harmon, KG Michelsen, MA Sheridan. Purification and characterization of hepatic triacylglycerol lipase isolated from rainbow trout, *Oncorhynchus mykiss*. Fish Physiol Biochem 9:361–368, 1991.

56. MA Sheridan, WA Allen. Partial purification of a triacylglycerol lipase isolated from steelhead trout (*Salmo gairdneri*) adipose tissue. Lipids 19:347–352, 1984.

57. JS Harmon, LM Rieniets, MA Sheridan. Glucagon and insulin regulate lipolysis in trout liver by altering phosphorylation of triacylglycerol lipase. Am J Physiol 265:R255–R265, 1993.

58. KG Michelsen, JS Harmon, MA Sheridan. Adipose tissue lipolysis in rainbow trout, Oncorhynchus mykiss, is modulated by phosphorylation of triacylglycerol lipase. Comp Biochem Physiol 107B:509–513, 1994.

59. RL Shewfelt. Fish muscle lipolysis—a review. J Food Biochem 5:79–100, 1981.

60. A Senthivel, LN Srikar, GV Sagar Reddy. Effect of frozen storage on protease and lipase activities of oil sardine and ribbon fish. J Fd Sci Technol 29:392–394, 1992.

61. EJ Geromel, MW Montgomery. Lipase release from lysosomes of rainbow trout (*Salmo gairdneri*) muscle subjected to low temperatures. J Food Sci. 45:412–415, 419, 1980.

62. W Partman. The effects of freezing and thawing on food quality. In: RG Duckworth, ed. Water Relations in Food. London: Academic Press, 1975, p 505.

63. AJ de Koning, TH Mol. Rates of free fatty acid formation from phospholipids and neutral lipids in frozen cape hake (*Merluccius spp*) mince at various temperatures. J Sci Food Agric 50:391–398, 1990.

64. RL Shewfelt, RE Mcdonald, HO Hultin. Effect of phospholipid hydrolysis on lipid oxidation in flounder muscle microsomes. J Food Sci 46:1297–1301, 1981.

65. S Govindarajan, HO Hultin. Myoglobin oxidation in ground beef: mechanistic studies. J Food Sci 42:571–577, 582, 1977.

6

Transglutaminases in Seafood Processing

Isaac N. A. Ashie
Novo Nordisk BioChem, North America, Inc., Franklinton, North Carolina

Tyre C. Lanier
North Carolina State University, Raleigh, North Carolina

I. INTRODUCTION

Most common enzymes used in food industry processes, such as amylases, proteases, and lipases, contribute to the breakdown of food components. The unique ability of transglutaminase (TGase) to modify protein functionality by covalent crosslinking has thus generated enormous interest. Less than a decade ago, studies on surimi derived from some fish species (*Theragra chalcogramma, Sardinops melanostictus,* and *Micropogan undulatus*) led to the discovery that an endogenous TGase is responsible for the spontaneous gelation of surimi pastes at low temperature (5–40°C); this leads to a strengthening of the "set" gel upon cooking (1–3). This discovery has led to further studies of the content and nature of TGase endogenous to seafood, as well as the evaluation of TGase from other sources in seafood applications as a processing aid for quality improvement and product development.

TGase is a transferase, having the systematic name protein-glutamine γ-glutamyltransferase (EC 2.3.2.13). It catalyzes the acyl transfer reaction between γ-carboxyamide groups of glutamine residues in proteins, peptides, and various primary amines. When the ε-amino group of lysine acts as acyl acceptor, it results in polymerization and inter- or intra-molecular crosslinking of proteins via formation of ε-(γ-glutamyl) lysine linkages. This occurs through exchange of the

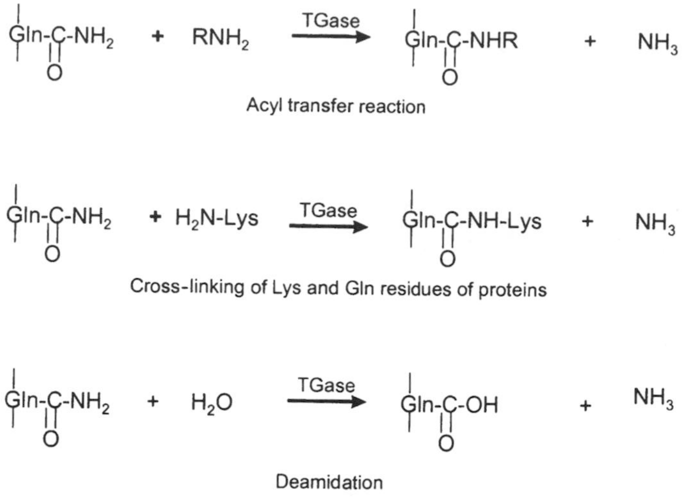

$$\text{Gln-}\underset{O}{\overset{|}{C}}\text{-NH}_2 \;+\; \text{RNH}_2 \;\xrightarrow{\;\text{TGase}\;}\; \text{Gln-}\underset{O}{\overset{|}{C}}\text{-NHR} \;+\; \text{NH}_3$$

Acyl transfer reaction

$$\text{Gln-}\underset{O}{\overset{|}{C}}\text{-NH}_2 \;+\; \text{H}_2\text{N-Lys} \;\xrightarrow{\;\text{TGase}\;}\; \text{Gln-}\underset{O}{\overset{|}{C}}\text{-NH-Lys} \;+\; \text{NH}_3$$

Cross-linking of Lys and Gln residues of proteins

$$\text{Gln-}\underset{O}{\overset{|}{C}}\text{-NH}_2 \;+\; \text{H}_2\text{O} \;\xrightarrow{\;\text{TGase}\;}\; \text{Gln-}\underset{O}{\overset{|}{C}}\text{-OH} \;+\; \text{NH}_3$$

Deamidation

Figure 1   Reactions catalyzed by TGases.

ε-amino group of the lysine residue for ammonia at the carboxyamide group of a glutamine residue in the protein molecule(s) (Fig. 1). In the absence of primary amines, water may act as the acyl acceptor, resulting in deamination of γ-carboxyamide groups of glutamine to form glutamic acid. Formation of covalent crosslinks between proteins is the basis of the ability of TGase to modify the physical properties of protein foods.

TGase activity in tissues may be determined by one of several mechanisms including: amine incorporation into substrates using monodansylcadaverine, hydroxamate or radioactive putrescine (4–6); disappearance of amino groups by trinitrobenzenesulfonate or fluorescence intensity methods (7); increase in molecular weight of substrate by Sodium dodecylsulfate–polyacrylamide gel electrophoresis (SDS-PAGE) (7); release of ammonia (8); or measurement of functional effects such as viscosity and gel strength (9). More recently, Ohtsuka et al. (10) have also developed an enzyme-linked immunosorbent assay (ELISA) for specifically estimating microbial TGase activity in surimi-based products.

II.   SOURCES AND CHARACTERISTICS

A.   Physiological Function

The physiological role of the plasma TGase of higher animals (also referred to as fibrinoligase or factor XIIIa) has been well established as crosslinking of the fibrin clot during hemostasis. The zymogen factor XIII is activated by the protease thrombin, another component of plasma that also induces clotting of fibrinogen. A commercial preparation comprised primarily of beef thrombin and plasma is presently used as an adhering agent for the manufacture of cold-restructured meat, poultry, and seafoods (11). Similarly, in the hemocyte lysate of the Japanese horseshoe crab (*Tachypleus tridentatus*), endotoxins from invading bacteria activate an endotoxin-sensitive zymogen, setting off a cascade of reactions that eventually activate the proclotting TGase. The activated TGase subsequently converts a soluble coagulogen to an insoluble coagulin gel that may act as a barrier against invading microorganisms (12). In plant leaves, the higher energy availability or higher uptake of polyamines into the chloroplasts by light appears to stimulate TGase activity. The enzyme is therefore thought to be involved in photosynthesis, probably by modifying ribulose-bis-phosphate carboxylase or oxygenase (13, 14). The physiological roles of TGase from most other sources, however, still remain to be elucidated.

B. Extraction and Purification

TGase has been identified in tissues of various species including mammals, birds, fish and shellfish, micro-organisms, and plants (Table 1). Purification from mammalian sources has mainly involved the use of guinea pig liver (15, 16). The unfrozen liver, perfused with 0.15 M NaCl, is homogenized, centrifuged, and the soluble fraction subjected to chromatography on DEAE-cellulose. The enzyme is precipitated with protamine and then selectively extracted from its complex with protamine by ammonium sulfate and gel filtration on agarose. This process commonly achieves a 250-fold purification and 15–30% recovery. In a later modified procedure by Brookhart et al. (16), the soluble fraction obtained after centrifugation is subjected to QAE–Sephadex ion exchange followed by adsorption to a hydroxylapatite column and subsequent elution of various fractions. The partially purified enzyme is exhaustively dialyzed and finally subjected to affinity chromatography on a phenylalanine-sepharose 4B column to obtain the pure enzyme. Brookhart et al. (16) further suggested a shorter procedure, which gives about 70% recovery by proceeding directly to the affinity chromatography step after centrifugation and filtration of the homogenate.

Purification of the microbial enzyme from *Streptoverticillium* species has been reviewed by Zhu et al., (17) based on the methods of Ando et al. (18) and Gerber et al. (19). It involves aerobic fermentation for 2–4 days at temperatures ranging between 25 and 35°C for growth and enzyme production. Being an extracellular enzyme, it is dissolved in the fermentation broth and can then be separated using conventional methods of enzyme purification. For example, in the method of Ando et al. (18), the microorganisms were separated from the fermentation broth by centrifugation followed by concentration of the supernatant by

Table 1 Sources and Characteristics of Some TGases

Source	Mol. Wt. (kDa)	Optimum Temp. (°C)	Optimum pH	Reference
Mammal				
Human plasma factor XIII	300–350[t]			27
Bovine factor XIIIa				
Guinea pig liver	75–85[m]		8.0	27
Rabbit liver	80[m]			93
Plant				
Pea seedlings				23
Alfalfa	39[m]			13
Microbial				
S. mobaraense	40[m]	50	6.0–7.0	18
Physarum polycephalum	77[d]			28
S. ladakanum	37.5	50	6.0	32
Seafoods				
Red sea bream liver	78	55	9.0–9.5	21
Carp muscle	80			20
Walleye pollack liver	77	50	9.0	33
Lobster muscle	200			94
Japanese oyster	84/90	40/25	8.0	34
Limulus hemocyte	86			12
Scallop	80			35
Botan shrimp	80			"
Squid	80			"
Rainbow trout	80			"
Atka mackerel	80			"

[t]tetramer; [m]monomer; [d]dimer.

ultrafiltration. The concentrate was subjected to ion exchange and gel filtration chromatography to obtain the pure enzyme, which is often mixed with stabilizers such as salts, sugars, and polyols. In the modification by Gerber et al. (19), the ultrafiltration and gel filtration steps were omitted. The total recovery of TGase activity in the two procedures was 42% and 40%, respectively.

The fish enzyme is generally prepared by homogenization of the tissue followed by centrifugation and filtration to obtain a TGase-rich supernatant that is subjected to a series of chromatographic purification and dialysis procedures depending on the extent of purification required (20, 21). For example, partial purification of the enzyme from carp (*Cyprinus carpio*) muscle required both DEAE-cellulose and Sephacryl S-300 column chromatographies, while complete purification from red sea bream (*Pagrus major*) liver required three separate chromatographic procedures (20, 21). In the latter procedure (21), the purification steps involved anion exchange on DEAE–Sephacel followed by cation exchange on a CM–Sepharose column after elution and dialysis of the TGase fraction. Further purification on a heparin–Sepharose column resulted in a total yield of 14%.

The recent cloning and expression of red sea bream TGase activity in *Escherichia coli* is a novel approach that makes the prospects for commercial production of fish TGases rather encouraging (22). The process involves transformation of *E. coli* by cloning an expression plasmid for production of red sea bream TGase into *E. coli* cells. The transformed cells are then grown under aerobic conditions and the synthesis of TGase induced when the level of tryptophan is depleted in the growth medium. The growth and synthesis of TGase is continued for about 20 h and monitored by optical density measurements at 660 nm. Since the enzyme is produced intracellularly, the cells are harvested by centrifugation, suspended in an extraction buffer, and subjected to ultrasonication at 4°C for cell lysis to release the enzyme. The cell lysates are further centrifuged at 4°C to separate soluble and insoluble fractions. While these workers observed an increase in production levels up to a maximum at 32–37°C culture temperature, most of the enzyme produced at temperatures beyond 32°C formed aggregates and thus was inactive. Construction of producer strains containing both heat shock chaperones and expression plasmids enabled them to suppress polypeptide aggregation and increase the amount of soluble TGase from 10% to 50% and a corresponding four-fold increase in activity (22).

TGase activity has been demonstrated in several plant tissues (13, 14, 23). Falcone et al. (14) extracted the enzyme from etiolated sprout apices, slices of tuber medullary parenchyma, leaves, and flower buds of *Helianthus tuberosus*. The tissues were extracted in an ice-cold mortar, filtered through layers of cheesecloth, and centrifuged. The supernatants were further purified by chromatography on a Sephadex column.

C. Properties of the Enzyme

TGase is thought to be either a cytosolic/soluble enzyme (type II) (15) or, in some organisms, partially bound to membranes (type I) in the lysosomes or mitochondria (24–26). Depending on the source, it may exist as a monomer, dimer, or tetramer. For example, TGase from guinea pig liver is a monomer (27), while that isolated from the mold *Physarum polycephalum* is a dimer (28). Plasma factor XIII exists as a tetrameric zymogen that is dissociable to two dimeric forms, with TGase activity residing in one of these dimers (27). TGase isolated from muscle of various seafood species are generally monomeric proteins. Such TGase activity has been demonstrated in several fish species including carp (*Cyprius carpio*), rainbow trout (*Oncorhynchus mykiss*), chum salmon (*Oncorhynchus keta*), atka mackerel (*Pleurogrammus azonus*), and white croaker

(*Argyrosomus argentatus*) (Table 1). Araki and Seki (29) have further shown that TGase activities vary between different fish species. In a study of TGase activities in some fish species, they found white croaker (*Argyrosomus argentatus*) to have the highest activity (2.41 units/g wet weight) followed in descending order by carp (*Cyprinus carpio*), walleye pollock (*Theragra chalcogramma*), chum salmon (*Oncorhynchus keta*), atka mackerel (*Pleurogrammus azonus*), and rainbow trout (*Oncorhynchus mykiss*) with the least activity at 0.1 units/g wet weight.

TGase is a sulfhydryl enzyme with a seemingly conserved pentapeptide (Tyr-Gly-Gln-Cys-Trp) active site sequence and is readily inactivated by sulfhydryl reagents (such as *N*-ethylmaleimide and p-chloromercuribenzoate), which alkylate free sulfhydryl groups (27, 28). Cations such as Cu^{2+}, Zn^{2+}, and Pb^{2+} also inhibit TGase activity by binding to the thiol group of the active site cysteine residue (30). In mammalian TGases, the sulfhydryl group responsible for covalent binding with the acyl portions of substrates is located at the apex of a hydrophobic crevice on the enzyme surface into which the γ- and β-methylene groups of peptide-bound glutamine residues or α- and β-methylene groups of aliphatic amides must fit for catalysis (27).

Until recently, the enzyme had been widely described as a Ca^{2+}-dependent enzyme, but isolation of TGase from new sources indicates that the requirement of Ca^{2+} ions is not universal. This requirement may even vary depending on the nature or conformation of the substrate. For instance, pea seedling TGase was stimulated by Ca^{2+} ions when *N-N*-dimethyl casein was used as substrate, but was neither affected by high Ca^{2+}-ion concentrations nor by Ca^{2+}-ion chelators such as ethylenediaminetetraacetic acid (EDTA) and ethyleneglycoltetraacetic acid (EGTA) during TGase-catalyzed putrecine incorporation into proteins (23). It has also been reported that while the enzyme from *Streptoverticillium mobaraense* is Ca^{2+}-independent in a hydroxamate assay, crosslinking of caseinate is Ca^{2+}-dependent (31). Still other TGases isolated from some microbial and plant sources have been shown to be entirely Ca^{2+}-independent (13, 18, 32). In addition to nonreliance on Ca^{2+} ions for activity, TGase from *Streptoverticillium* also differs from mammalian TGase in molecular weight, thermal stability, isoelectric point, and substrate specificity (18, 32). For instance, its isoelectric point (pI) is about 8.9 compared to 4.5 for TGase of guinea pig liver. Like mammalian TGases, all forms of TGase isolated thus far from seafoods are Ca^{2+}-dependent. Sr^{2+} can replace the activation by Ca^{2+}, which is thought to function by orienting an essential sulfhydryl group into a suitable relationship with other groups in the active site during catalysis (27). However, the seafood enzymes also show differences in their sensitivity to Ca^{2+}-ions. For example, the concentration of Ca^{2+} required to express maximum activity in the walleye pollock enzyme is 3 mM while TGase from red sea bream liver and carp muscle required 0.5 mM and 5 mM Ca^{2+}, respectively, for full activity (20, 21, 33). The corresponding optimal

Ca^{2+} concentrations for limulus hemocyte,guinea pig liver, and Japanese oyster were found to be 8 mM, 10 mM, and greater than 25 mM, respectively (12, 18, 34). Nozawa et al. (35) also observed similar differences in the Ca^{2+} ion requirement of various species studied.

TGase isolated from some seafood sources has shown significant sequence homology when compared to that from other sources (36). For example, limulus TGase sequence homologies with guinea pig liver TGase, human factor XIIIa, and human keratinocyte TGase, were 32.7%, 34.7%, and 37.6%, respectively. However, the active site sequence (Tyr-Gly-Gln-Cys-Trp), which is presumed to be conserved in TGases, was not observed in the enzyme isolated from red sea bream (*Pagrus major*) liver (21).

GTP (guanosine triphosphate) in micromolar concentrations inhibits guinea pig liver TGase, yet the enzyme purified from limulus hemocyte was not affected by nucleoside 5'-triphosphates including ATP, GTP, UTP, and CTP (12). Salt (NaCl) levels tested had no effect on both red sea bream and guinea pig liver TGases as well as carp dorsal muscle TGase (12, 20). Limulus hemocyte TGase was inhibited by increasing NaCl concentration (up to 1 M) while two isozymes isolated from Japanese oyster (*Crassostrea gigas*) were either inhibited or activated by the same conditions (12, 34). Similarly, the activities of TGase partially purified from some marine invertebrates such as scallop (*Patinopecten yessoensis*), botan shrimp (*Pandalus nipponensis*), and squid (*Todardes pacificus*) were enhanced by increasing NaCl concentration (35). Additionally, these enzymes showed different affinities for monodansylcadaverine and succinylated casein (35).

Differences may also exist between isozymes isolated from the same species. In Japanese oyster, an 84 kDa isozyme (TG-1) was optimally active at 40°C compared to 25°C for the 90 kDa (TG-2) isozyme (34).

III. CROSSLINKING OF SEAFOOD MUSCLE PROTEINS

The spontaneous gelation of fish meat paste containing NaCl (2–4%) and incubated at low temperatures (5–40°C) is a well-known occurrence in the manufacture of surimi-based products. This phenomenon, referred to as "setting" or "suwari," involves network formation of myosin due to crosslinking by endogenous TGase (1–3). Gels formed in this way develop much greater strength once cooked than gels formed directly by cooking of unset pastes at high (>80°C) temperature (37).

Setting therefore is commonly used to strengthen the texture of many surimi seafood products. For example, in composite-type imitation crab products composed of both fibers and an enmeshing matrix (both made from surimi), the fibers may be engineered to be stronger than the enmeshing matrix by employing a setting process prior to cooking (38). Tani et al. (39) also reported the production of imitation seafoods using shark-fins prepared by TGase-catalyzed crosslinking of gelatin, collagen, or their mixture. TGase activity may also be involved in the manufacture of "himono," a dried fish product of widespread consumption in Japan made by curing of fish fillets in concentrated salt solutions followed by drying. Ito et al. (40) reported that the changes in physical characteristics of the product during drying involves crosslinking of the myosin heavy chain. Based on high-performance liquid chromatography (HPLC) and SDS-PAGE analyses, Kumazawa et al. (41) were able to show an increase in ε-(γ-glutamyl)lysine crosslink formation in blocks of horse mackerel meat during the drying process, indicating possible TGase involvement in the process. In a recent report, Fukuda et al. (42) have attributed the unique chewy texture of fish eggs (caviar) developed during preservation in salt solutions to TGase activity. After washing the eggs of different fish species in isotonic saline and preserving in 7% sodium chloride solution for 3 days, subsequent analysis of the eggs showed an increase in breaking strength, ε-(γ-glutamyl)lysine content, and TGase activity. It was presumed that the higher TGase activity at the elevated salt concentration was responsible for these changes.

Gel filtration chromatography and SDS-PAGE analyses were used to demonstrate that, in crosslinking of muscle proteins, the myosin heavy chains, specifically the rod and heavy meromyosin portions, constitute the main site of enzymic crosslinking with no involvement by the light chains (43). Funatsu et al. (44) showed that actomyosin complexed with all other protein components of muscle to form large aggregates during the setting reaction.

The rate of crosslinking of myosin heavy chains varies significantly among fish species (45). Using carp muscle TGase, Maruyama et al. (45), showed that carp (*Cyprinus carpio*) myosin heavy chain forms dimers at a rate of 8 μM/h while rainbow trout (*Oncorhynchus mykiss*) and atka mackerel (*Pleurogrammus azonus*) myosin form dimers and polymers at identical rates (55 μM/h). The higher rate of crosslinking in rainbow trout and atka mackerel than in carp myosin was attributed to the limited number of reactive glutamine residues in the latter. When myosins from trout and carp were mixed, interspecies crosslinking was very minimal and the crosslinking rate increased with increasing proportions of trout to carp suggesting a prevalent intra-species crosslink formation. However, the crosslinking rate of rainbow trout and atka mackerel myosin mixtures was independent of the relative proportions and formed large polymers at the same rate, leading the authors to suggest the possible formation of interspecies crosslinks (45).

When the fish is of prime quality the level of endogenous TGase activity may be adequate to ensure good setting ability in surimi prepared from this fish. The endogenous enzyme activity in low-quality fish can be significantly reduced (2, 46). In such cases, the addition of exogenous TGase could improve the gelling properties. Mammalian TGase was previously the sole source of commercial TGase, but the high price for the enzyme (about U.S. $80.00 per unit) prevented its use in foods (17). The recent development of microbially derived TGase will likely broaden applications for TGase as an additive in food products. A recombinant form of factor XIIIa has been produced by fermentation of *Saccaromyces cerevisiae*. Variants of *Streptoverticillium mobaraense* and *Aspergillus spp.* have also been used for TGase production (31, 47, 48). Since the TGases produced by these variants are extracellular, purification is less demanding, making it feasible for production on a commercial scale.

Addition of microbial TGase to surimi significantly increases its gel strength, particularly when the surimi has lower natural setting ability (presumably due to lower endogenous TGase activity) (47, 49). Using SDS-PAGE and HPLC analyses, Lee et al. (47) observed an increase in non-disulfide polymerization and formation of ε-(γ-glutamyl)lysine dipeptides with increase in setting time and microbial TGase concentration. When microbial TGase was added, the rate of increase in gel stress with time of holding at 25°C was greater. At equal levels of ε-(γ-glutamyl)lysine content, gels prepared with added microbial TGase displayed higher gel stress. Since these gels achieved this level of isopeptide content more rapidly than pastes gelled with endogenous TGase only, it was concluded that the rate of myosin polymerization may also be a factor influencing gel strength, not isopeptide content alone. Yasunaga et al. (50) observed increasing isopeptide content concomitant with increased gel strength as microbial TGase was increasingly added. They also noted that the relatively higher gel strength with an increasingly lower gel strain therefore produced gels that differed in texture from those produced by setting without additives (51).

IV. FACTORS AFFECTING TGASE ACTIVITY IN MUSCLE TISSUE

The setting temperature and duration may be critical for surimi derived from certain species. For example, while incubation of surimi sols from Alaska pollock (*Theragra chalcogramma*) and Atlantic croaker (*Micropogan undulatus*) surimi at 40°C for 2–3 h increased gel strength, further incubation beyond this period resulted in reduced gel strength (1). When the setting temperature was raised to 50°C, the gels were progressively weakened, showing no initial increase in strength. The gel weakening phenomenon, referred to as "modori," has been attributed to the activity of thermostable proteases that are optimally active at 50–60°C (52). Thus the gel-weakening activity of these proteases may compete with gel strengthening due to protein crosslinking by TGase. Addition of exogenous TGase could possibly offset somewhat the effects of the heat-stable proteases because the isopeptide bonds formed by TGase are supposedly resistant to proteolytic attack (53). Generally, however, a protease inhibitor substance, such as beef plasma or egg white, is added effectively to eliminate gel weakening by proteases and thus maximize the impact of any TGase crosslinking on gel texture. Imai et al. (54) showed that while myosin isopeptide crosslinking and gel strength initially increase during incubation of pollock surimi at 40–50°C, just as occurs at lower incubation temperatures, measurable isopeptide bonds do not increase in number once gel strength begins to decline, presumably as a result of proteolysis.

While it is not clear how TGase activity in live fish species is regulated in vivo, there is evidence that some muscle components may control the activity of the enzyme in vitro. A deproteinized extract from chum salmon muscle was found to contain large amounts of anserine that inhibited TGase activity, preventing crosslinking and low-temperature gelation of actomyosin (55). In spite of its abundant supply, chum salmon is not utilized for surimi due to its poor gelforming ability. Several other fish species, such as yellowfin tuna, bluefin tuna, little tuna, skipjack, and black marlin, have been identified as having poor gelling qualities; these also contain large amounts of anserine (56, 57). This potential regulatory activity of anserine may not be present in all species as some fish have good gelling ability without TGase.

The presence of certain ingredients in the meat paste may also influence the characteristics of the gels formed by TGase catalysis. When wheat and soy proteins were added to salted Alaska pollock surimi paste and allowed to set at 10°C for up to 72 h, the presence of these proteins decreased the setting response of the gels (58). Similar reductions of the setting response have been observed with the addition of 10% whole egg, albumen, or yolk to walleye pollock surimi (59). Addition of L-lysine also suppresses myosin crosslinking by TGase, as evidenced by reduced breaking force and strain of kamaboko prepared from Alaska pollock surimi (60).

Torley and Lanier (61) found that increasing admixture of comminuted beef with surimi led to diminishing of the low-temperature setting ability. Since such admixture also led to a decrease in pH, they measured pH effects on setting of surimi and found that at corresponding pH values to those of beef–surimi mixtures, similar setting responses were obtained. Of course, beef paste displays no low-temperature setting ability, so that dilution alone could account for some portion of the reduced setting ability in these mixtures.

Kataoka et al. (62) noted that chitosan, a naturally-occuring polymer of β-1, 4-linked glucosamine residues, influences TGase catalyzed crosslinking of myosin heavy chains in seafoods. The breaking strength of gels formed from salted Alaska pollock surimi pastes increased with increasing chitosan concentration. The mechanism by which chitosan enhances TGase crosslinking of myosin heavy chains was not explained.

Because the tissue TGase is calcium dependent, added ingredients that affect calcium availability can also affect the cold setting reaction. Matsukawa et al. (63) showed that addition of pyrophosphate diminished the setting response of pollock surimi. However, addition of a small amount of calcium salts to a surimi paste seems to remove any influence of phosphate on crosslinking and gel strengthening, and crosslinking activity of microbial TGase, which is insensitive to calcium level, is not affected by added phosphates.

Addition of beef plasma to surimi-based products may also enhance isopeptide crosslinking, as evidenced by the proportionately greater increase in strength of Pacific whiting surimi gels containing plasma when subjected to low temperature setting, as compared to that obtained by other process schedules (Fig. 2) (64). Plasma TGase is normally found in close association with fibrinogen, such that this fibrinogen-rich plasma may also augment that TGase activity endogenous to the surimi (65). Alternatively, the added fibrinogen might enhance substrate reactivity (66). Another component of beef plasma, $\alpha_2$macroglobulin (α_2M) can form $\epsilon(\gamma$-glutamyl)lysine bonds between γ-carboxyamide groups of glutamine residues in proteins and a variety of primary amines (53, 67). Seymour et al. (68) found considerable monodansyl cadaverine (MDC)-incorporation in liquid plasma. More than 85% of this activity was not calcium dependent, indicating α_2M involvement rather than plasma TGase activity. In addition to its possible involvement in the crosslinking process, α_2M had also been shown to inhibit a broad spectrum of proteases in fish muscle, thus contributing to the reduction of proteolytic activity during surimi manufacture (69).

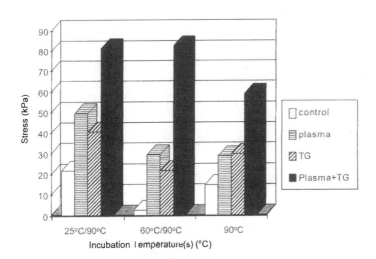

Figure 2 Effects of added plasma, TGase, or their mixture on strength of Pacific whiting surimi gels. (From Ref. 64.)

V. SPECIFICITY AND SUBSTRATE MODIFICATION

TGase shows a high degree of specificity in the crosslinking of proteins. Unlike other enzymes that act on the γ-carboxyamide group of glutamine, it requires protein- or peptide-bound glutamine, in which both α-amino and α-carboxyl groups of glutamine are in peptide linkage, for activity. A hydrophobic methylene group of peptide-bound glutamine is suggested to be essential for interaction with a complementary hydrophobic region in close proximity to the enzyme active site (27). The enzyme also has an absolute stereospecificity for the L-form of the amino acid and shows different levels of reactivity with different substrates (27). According to Folk (27), the basis for the differences in reactivity may reside in the nature of amino acid residues flanking the susceptible glutamine residues in the substrate. The proposed mechanism of enzyme–substrate interaction requires that only amides with unbranched carbon atoms α and β to their carboxamide groups are able to interact productively with the active site, whereas branched groups in either of these positions prevent productive interaction (27). Charged groups in close proximity to the susceptible glutamine residue also tend to hinder enzyme–substrate interaction. For TGase from some sources, it has been suggested that the glutamine residue should be located in a flexible region, in a surface loop, or in regions of reverse turns for catalysis (70–72). Thus a high content of glutaminyl residues does not necessarily ensure a high level of reactivity with TGase, as evidenced by the 11S and 7S seed storage proteins of pea legumin, which are rather poor substrates in spite of the high glutaminyl and lysyl content (73). Similarly, out of six glutamine and 12 lysine residues in α-lactalbumin, only one glutamine and three lysine residues are available for TGase reaction even in the molten globule state (74). In another study, Grootjan et al. (75) also found that Gly, Asp, Pro, His, and Trp preceding the acyl acceptor lysine have adverse effects on substrate reactivity while Val, Arg, Phe, Ser, Ala, Leu, Tyr, and Asn tend to have an enhancing effect. Such selectivity for the lysine residues is a curious observation, since the commonly held notion is that the enzyme is less selective toward the acyl acceptor than it is to the donor.

The high degree of specificity of TGase is evidenced by its relative reactivity with various substrates. Some workers have shown by SDS-PAGE analysis that myosin is the most susceptible among muscle proteins to TGase activity while actin is unaffected by the enzyme (1, 76). Tropomyosin and troponin-T have also been suggested as possible substrates, although to a limited extent (77). When carp muscle TGase was used to crosslink actomyosins from different fish species, different rates of polymerization were observed, giving strong indication that the crosslinking process is substrate-dependent (29). Joseph et al. (78) observed that while addition of guinea pig liver TGase to Alaska pollock (*Theragra chalcogramma*) and Atlantic croaker (*Micropogan undulatus*) surimi re-

sulted in increased gel strength, the pH and temperature optima for the setting reaction remained the same as that for the naturally occurring setting reaction in
surimi of those species, which is induced by TGase endogenous to those species.
This indicates that the temperature and pH optima for setting may be determined
by the availability of reactive points on the myosin surface (conformation of the
substrate) rather than by the enzyme source. Also, the optimum setting temperature correlates with the habitat temperatures of the live fish (79), which further
suggests that TGase activity is dependent on the stability of the myosin. For example, myosin heavy chain of fish is polymerized by TGase during low-temperature setting of the salted paste, whereas beef paste is unaffected due to the
greater stability of beef myosin (61).

Substrate conformation may be altered to increase susceptibility to TGase
activity. Pea legumin, a poor substrate for TGase activity, is suggested to have 37
glutaminyl and 22 lysyl residues per $\alpha\beta$-subunit, of which only two glutaminyl
residues per $\alpha\beta$-subunit will normally form ϵ-(γ-glutamyl)lysine bonds (73).
Chemical modification of pea legumin by citraconylation induced conforma-
tional changes in the protein, resulting in increased TGase activity (73). Addition
of reductants such as dithiothreitol, cysteine, or glutathione may also be used to
modify substrates by partially unfolding protein molecules to unmask lysyl
and/or reactive glutaminyl residues to improve their accessibility to TGase as
has been demonstrated in α-lactalbumin and β-lactoglobulin (80, 81) and actomyosin (78). Similarly, surimi does not set into a gel until its proteins are destabilized and solubilized by the addition of salt (82). Conversion of α-lactalbumin
to its molten globule state (characterized by a compact globular structure with
native-like secondary structure and disordered tertiary structure) by using EDTA
to chelate bound Ca^{2+}-ions resulted in a 10-fold increase in TGase activity when
compared to the native state (74). Similarly, reaction at an oil–water interface
was also shown to enhance TGase crosslinking of various proteins due to increased exposure of the hydrophobic protein core to the enzyme (81).

Enzymic modification of substrates might be more acceptable for the food
industry. Using trypsin concentrations as low as 0.02–0.05% in the presence of
Ca^{2+}-ions, Kolakowski et al. (83) demonstrated that mild proteolysis of bream
(*Abramis brama*) mince greatly enhanced TGase-mediated crosslinking. It was
suggested that tryptic hydrolysis of the native myosin molecule to heavy
meromyosin (HMM), tail, and S-1 fragments unfolds the helical structure to a
random coil that is more readily accessible to TGase catalysis. Hamada (84) also
reported that peptidoglutaminase activity was substantially increased when protein hydrolysates rather than native proteins were used as substrates. Thus when
deamidation is preceded by proteolysis, molecular size and/or conformation of
substrate may be altered, thereby increasing accessibility to the enzyme. An investigation of peptidoglutaminase activity toward soy proteins had also shown
that heat treatment alone or in combination with proteolysis increased activity of

the enzyme. For example, peptidoglutaminase activity was increased 27 times at 20% hydrolysis when heat treatment was combined with proteolysis (85).

High pressure might also be utilized by the food industry for substrate modifications (76). Pressurizing surimi pastes at 300 MPa prior to TGase-induced setting resulted in about a threefold increase in gel strength compared to gels set without prior pressure treatment. SDS-PAGE revealed that myosin was crosslinked while actin was not, even after pressurization. Thus while myosin is generally a very good substrate for TGase activity, unfolding of myosin induced by high pressure may further increase accessibility of previously hidden glutamine and lysine residues to the enzyme, resulting in increased gel strength. Gilleland et al. (86) similarly observed a sixfold increase in the stress value of cooked gels from Alaska pollock surimi subjected to 300 MPa of pressure for 30 min prior to setting at 25°C for 120 min. Pressurization of other food proteins (e.g., bovine serum albumin, ovalbumin, lysozyme, and γ-globulin) at 200–600 MPa also enhanced their reactivity with MTGase (87). Ammonium salts, which inactivate TGase, suppressed crosslinking of myosin heavy chain and ε-(γ-gltamyl) lysine formation resulting from pressure treatment of pollock surimi (88).

The degree to which high pressure induces unfolding of muscle proteins also varies depending on their relative stability, which is a function of habitat/body temperature (79). Thus pressure treatment did not enhance subsequent TGase-induced gelation of lean turkey meat paste as had occurred for surimi pastes, unless pressurization was carried out at a higher temperature to further destabilize the myosin (76).

VI. CONCLUSIONS

The endogenous TGase activity of fish muscle has long been utilized to build strong texture in surimi-based gelled foods by allowing the enzyme to crosslink myosin during a lower temperature preincubation (setting treatment) prior to cooking. More recently we have isolated the natural TGase activity of blood plasma for cold restructuring of foods, and further extended the applications of TGase into other protein foods through isolation and commercial production of microbial TGase. The latter enzyme is not calcium sensitive like its tissue counterparts, thus allowing application in seafoods that might contain high amounts of phosphates or other chelating agents.

Thus far the primary applications of TGase in seafood processing have been for cold restructuring, cold gelation of pastes, or gel-strength enhancement through myosin crosslinking. The enzyme may be used in the future to modify the functionality of marine-derived proteins for food, pharmaceutical, or industrial use (89). TGase activity may be potentially utilized to enrich the nutritional value of various foods by covalently crosslinking proteins containing complementary limiting essential amino acids. For example, the enzyme may be used to incorporate methionyllysine or arginyllysine to correct methionine and arginine deficiencies in caseins. An issue likely to be raised by such crosslinked peptides, especially the glutamyl-lysine bonds, is their digestibility and bioavailability. Recent evidence does indicate the existence of a kidney enzyme, γ-glutamylamine cyclotransferase, and γ-glutamyltransferase also found in kidney, blood, and the intestinal mucosal walls, which are capable of cleaving the γ-glutamyllysine bonds to release lysine (90, 91). As with the changes in nutritional value, complementary protein crosslinking may also be applied to modify or improve protein functionality.

Marine-derived forms of the enzyme could possibly have utility for specific applications where either low temperature must prevail or heat-inactivation at low temperature is desired, since most enzymes derived from cold-water organisms are less heat-stable than those derived from warm blooded animals or plants (92). Likewise, seafood muscle proteins are less stable to heat and pressure than those of homeotherms, and therefore better substrates for TGase crosslinking. However, both seafood and other proteins can be modified by chemical, enzymatic, or physical means to improve their reactivity with TGase. Several approaches appear promising for commercial exploitation. This can reduce the cost of TGase use as well as improve process efficiency and enzyme performance.

REFERENCES

1. GG Kamath, TC Lanier, EA Foegeding, DD Hamann. Nondisulfide covalent crosslinking of myosin heavy chain in setting of Alaska pollock and Atlantic croaker surimi. J Food Biochem 16:151–172, 1992.
2. N Seki, H Uno, NH Lee, I Kimura, K Toyoda, T Fujita, K Arai. Transglutaminase activity in Alaska pollack muscle and surimi, and its reaction with myosin B. Nippon Suisan Gakk 56:125–132, 1990.
3. Y Tsukamasa, K Sato, Y Shimizu, C Imai, M Sugiyama, Y Minegishi, M Kawabata. ε-(γ—Glutamyl) lysine crosslink formation in sardine myofibril sol during setting at 25°C. J Food Sci 58:785–787, 1993.
4. JE Folk, PW Cole. Mechanism of action of guinea pig liver transglutaminase. J Biol Chem 241:5518–5525, 1966.
5. JE Folk, SI Chung. Transglutaminase. Methods Enzymol 113:358–375, 1985.
6. L Lorand, T Urayama, JWC Kiewiet, HL Nossel. Diagnostic and genetic studies on fibrin-stabilizing factor with a new assay based on amine incorporation. J Clin Invest 48:1054–1064, 1969.
7. K Ikura, T Kometani, M Yoshikawa, R Sasaki, H Chiba. Crosslinking of casein components by transglutaminase. Agric Biol Chem 44:1567–1573, 1980.

8.   C Backer-Royer, F Traore, J-C Meunier. Polymerization of meat and soybean pro-
     teins by human placental calcium-activated factor XIII. J Agric Food Chem
     40:2052–2056, 1992.
9.   H Sakamoto, Y Kumazawa, M Motoki. Strength of protein gels prepared with mi-
     crobial transglutaminase as related to reaction conditions. J Food Sci 59:866–871,
     1994.
10. T Ohtsuka, K Seguro, M Motoki. 1996. Microbial transglutaminase estimation in
     enzyme-treated surimi-based products by enzyme immunosorbent assay. J Food
     Sci 61:81–84.
11. EJC Pardekooper. Composite meat product and method for the manufacture
     thereof. 1988. U.S. Patent 4,741,906.
12. F Tokunaga, M Yamada, T Miyata, YL Ding, M Hiranaga, T Muta, S Iwanaga.
     Limulus hemocyte transglutaminase. Its purification and characterization, and
     identification of the intracellular substrates. J Biol Chem 268:252–261, 1993.
13. GD Kuehn, M Sotelo, T Morales, MR Bruce-Carver, E Guzman, SA Margosiak.
     Purification and properties of transglutaminase from *Medicago sativa* L. (alfalfa).
     FASEB J 5:A1510, 1991.
14. P Falcone, D Serafini-Fracassini, S Del Duca. Comparative studies of transglutam-
     inase activity and substrates in different organs of *Helianthus tuberosus*. J Plant
     Physiol 142:265–273, 1993.
15. JE Folk, SI Chung. Molecular and catalytic properties of transglutaminase. Adv
     Enzymol Rel Areas Mol Biol 38:109–191, 1973.
16. PP Brookhart, PL McMahon, M Takahashi. Purification of guinea pig liver transg-
     lutaminase using a phenylalanine-sepharose 4B affinity column. Anal Biochem
     128:202–205, 1983.
17. Y Zhu, A Rinzema, J Tramper, J Bol. Microbial transglutaminase—a review of its
     production and application in food processing. Appl Microbiol Biotechnol
     44:277–282, 1995.
18. H Ando, M Adachi, K Umeda, A Matsuura, M Nonaka, R Uchio, H Tanaka, M
     Motoki. Purification and characteristics of a novel transglutaminase derived from
     microorganisms. Agric Biol Chem 53:2613–2617, 1989.
19. U Gerber, U Jucknischke, S Putzien, H L Fuchsbauer. A rapid and simple method
     for the purification of transglutaminase from *Streptoverticillium mobaraense*.
     Biochem J 299:825–829, 1994.
20. H Kishi, H Nozawa, N Seki. Reactivity of muscle transglutaminase on carp my-
     ofibrils and myosin B. Nippon Suisan Gakk 57:1203–1210, 1991.
21. H Yasueda, H Kumazawa, M Motoki. Purification and characterization of a tissue-
     type transglutaminase from red sea bream (*Pagrus major*). Biosci Biotech
     Biochem 58:2041–2045, 1994.
22. K Yokoyama, Y Kikuchi, H Yasueda. Overproduction of DnaJ in *Escherichia coli*
     improves in vivo solubility of the recombinant fish-derived transglutaminase.
     Biosci Biotechnol Biochem 62(6):1205–1210, 1998.
23. I Icekson, A Apelbaum. Evidence for transglutaminase activity in plant tissue. J
     Plant Physiol 84:972–974, 1987.
24. A Elgavish, R W Wallace, DJ Pillion, E Meezan. Polyamines stimulate d-glucose

transport in isolated renal brush-border membrane vesicles. Biochim Biophys Acta 777:1–8, 1984.

25. CW Slife, DM Dorsett, GT Bouguett, E Taylor, S Coroy. Subcellular localization of a membrane-associated transglutaminase activity in rat liver. Arch Biochem Biophys 241:329–336, 1985.

26. M Juprelle-Soret, WC Simone, R Wattiaux. Presence of transglutaminase activity in rat liver lysosomes. Eur J Cell Biol 34:271–280, 1984.

27. JE Folk. Mechanism and basis for specificity of transglutaminase-catalyzed ε-(γ-glutamyl)lysine bond formation. Adv Enzymol 54:1–56, 1983.

28. JD Klein, E Guzman, GD Kuehn. Purification and partial characterization of trans-glutaminase from *Physarum polycephalum*. J Bacteriol 174:2599–2605, 1992.

29. H Araki, N Seki. Comparison of reactivity of transglutaminase to various fish acto-myosins. Nippon Suisan Gakk 59:711–716, 1993.

30. K Seguro, NNio, M Motoki. Some characteristics of a microbial crosslinking en-zyme:Transglutaminase. In Macromolecular Interactions In Food Technology. ACS Sympsium Series 650, American Chemical Society, 1996, pp: 271–280,

31. PM Nielsen. Reactions and potential industrial applications. Review of literature and patents. Food Biotechnol 9:119–156, 1995.

32. G-J Tsai, S-M Lin, S-T Jiang. Transglutaminase from *Streptoverticillium ladakanum* and application to minced fish product. J Food Sci 61:1234–1238, 1996.

33. Y Kumazawa, K Nakanishi, H Yasueda, M Motoki. Purification and characteriza-tion of transglutaminase from walleye pollack liver. 62:959–964. 1996.

34. Y Kumazawa, K Sano, K Seguro, H Yasueda, N Nio, M Motoki. Purification and characterization of transglutaminase from Japanese oyster (*Crassostrea gigas*). J Agric Food Chem 45:604–610, 1997.

35. H Nozawa, S Mamagoshi, N Seki. Partial purification and characterization of six transglutaminases from ordinary muscles of various fishes and marine inverte-brates. Comp Biochem Physiol B: Biochem Mol Biol 118B:313–317, 1997.

36. F Tokunaga, T Muta, S Iwanaga, A Ichinose, EW Davie, K Kuma, T Miyata. Limu-lus hemocyte transglutaminase: cDNA cloning, amino acid seguence, and tissue localization. J Biol Chem 268:262–268, 1993b.

37. TC Lanier. Measurement of surimi composition and functional properties. In: T C Lanier, C M Lee, eds. Surimi Technology. New York: Marcel Dekker, 1992, pp 123–166.

38. MC Wu. Manufacture of surimi-based products. In TC Lanier, and C M Lee, eds. Surimi Technology. NY, Marcel Dekker, 1992, pp 245–272.

39. T Tani, K Iwamoto, M Motoki, S Toiguchi. Manufacture of shark-fin imitation food. Jpn Kokkai Tokkyo Koho 02171160, 1990.

40. T Ito, N Kitada, N Yamada, N Seki, K Arai. Biochemical changes in myofibrillar protein of cured Walleye pollack meat induced by dehydration. Nippon Suisan Gakk 56:999–1006, 1990.

41. Y Kumazawa, K Seguro, M Takamura, M Motoki. Formation of ε-(γ-glutamyl)ly-sine cross-link in cured horse mackerel meat induced by drying. J Food Sci 58:1062–1064, 1083, 1993.

42.   A Fukuda, N Kanzawa, T Tamiya, K Seguro, T Ohtsuka, T Tsuchiya. Transglutaminase activity correlates to the chorion hardening of fish eggs. J Agric Food Chem 46:2151–2152, 1998.
43.   N Seki, C Nakahara, H Takeda, N Maruyama, H Nozawa. Dimerization site of carp myosin heavy chains by the endogenous transglutaminase. Fish Sci 64:314–319, 1998.
44.   Y Funatsu, N Katoh, K Arai. Aggregate formation of salt-soluble proteins in salt-ground meat from walleye pollock surimi during setting. Nippon Suisan Gakk 62:112–122, 1996.
45.   N Maruyama, H Nozawa, I Kimura, M Satake, N Seki. Transglutaminase-induced polymerization of a mixture of different fish myosins. Fish Sci 61:495–500, 1995.
46.   Y Kumazawa, T Numazawa, K Seguro, M Motoki. Suppression of surimi gel setting by transglutaminase inhibitors. J Food Sci 60:715–717, 726, 1995.
47.   HG Lee, TC Lanier, DD Hamann, JA Knopp. Transglutaminase effects on low temperature gelation of fish protein sols. J Food Sci 62:20–24, 1997.
48.   SH Kim, JA Carpenter, TC Lanier, L Wicker. Polymerization of beef actomyosin induced by transglutaminase. J Food Sci 58:473–474, 491, 1993.
49.   K Seguro, Y Kumazawa, T Ohtsuka, S Toiguchi, M Motoki. Microbial transglutaminase and ε-(γ-glutamyl)lysine crosslink effects on elastic properties of kamaboko gels. J Food Sci 60:305–311, 1995.
50.   K Yasunaga, Y Abe, M Yamazawa, K Arai. Heat-induced change inmyosin heavy chains in salt-ground meat with a food additive containing transglutaminase. Nippon Suisan Gaikk 62:659–668, 1996.
51.   Y Abe. Quality of kamaboko gel prepared from walleye pollock surimi with an additive containing transglutaminase. Nippon Suisan Gakk 60:381–387, 1994.
52.   H Saeki, Z Iseya, S Sugiura, N Seki. Gel forming characteristics of frozen surimi from chum salmon in the presence of protease inhibitors. J Food Sci 60:917–921, 928, 1995.
53.   L Lorand. Post-translational pathways for generating ε-(γ-glutamyl)lysine crosslinks. Ann NY Acad Sci 421:10–27, 1983.
54.   C Imai, Y Tsukamasa, M Sugiyama, Y Minegishi, Y Shimizu. The effect of setting temperature on the relationship between ε-glyamy) lysine crosslink content and breaking strength in salt-ground meat of sardine and Alaska pollock. Nippon Suisan Gakk 62(1):104–111, 1996.
55.   J Wan, I Kimura, N Seki. Inhibitory factors of transglutaminase in salted salmon meat paste. Fish Sci 61:968–972, 1995.
56.   M Suyama, T Suzuki, M Maruyama, K Saito. Determination of carnosine, anserine, and balenine in the muscle of animal. Nippon Suissan Gakk 36:1048–1053, 1970.
57.   T Shirai, S Fuke, Y Yamaguchi, S Konosu. Studies on extractive components of salmonids-II. Comparison of amino acids and related compounds in the muscle extracts of four species of salmon. Comp Biochem Physiol 74B:685–689, 1989.
58.   T Yamashita, N Seki. Effects of the addition of soybean and wheat proteins on setting of walleye pollack surimi paste. Nippon Suisan Gakk 62:806–812, 1996.
59.   T Yamashita, N Seki. Effects of the addition of whole egg and its components on

textural properties of kamaboko gel from walleye pollack surimi. Nippon Suisan Gakk 61:580–587, 1995.

60. D Liu, S Kanoh, E Niwa. Effect of L-lysine on the elasticity of kamaboko prepared by setting the paste of Alaska pollock. Nippon Suisan Gakk 61:608–611, 1995.

61. PJ Torley, TC Lanier. Setting ability of salted beef-pollock surimi mixtures. In: EG Bligh, ed. Seafood Science and Technology. Oxford: Fishing News Books, 1992, pp 305–316.

62. J Kataoka, S Ishizaki, M Tanaka. Effects of chitosan on gelling of low quality surimi. J Muscle Foods 9:209–220, 1998.

63. M Matsukawa, F Hirata, S Kimura, K Arai. Effect of sodium pyrophosphate on gelling property and cross-linking of myosin heavy chain in setting-heating gel from walleye pollack surimi. Nippon Suisan Gakk 62(1):94–103, 1996.

64. TC Lanier, IS Kang. Plasma as a source of functional food additives. In: Zu Haque, ed. Physicochemical and Biological Functionality of Food Systems, Vol 1. Lancaster, PA: Technomics Publishing Co., 1999.

65. L Lorand. Activation of blood coagulation factor XIII. Ann. NY Acad. Sci. 485:144–158, 1986.

66. J McDonagh. Biochemistry of fibrin-stabilizing factor (F XIII) *In:* J. McDonagh, R. Seitz, R. Egbring, eds. *Factor XIII.* New York: Schattauer, 1993, pp 2–8.

67. GS Salvesen, CA Sayers, AJ Barrett. Further characterization of the covalent linking reaction of α_2-macroglobulin. Biochem J 195:453–461, 1981.

68. TA Seymour, MY Peters, MT Morrissey, H An. Surimi gel enhancement by bovine plasma proteins. J Agric Food Chem 45:2919–2923, 1997.

69. INA Ashie, BK Simpson. α_2-Macroglobulin inhibition of endogenous proteases in fish muscle. J Food Sci 61(2):357–361, 1996.

70. GAM Berbers, HCM Bentlage, H Bloemendal, WW de Jong. β-Crystallin: endogenous substrate of lens transglutaminase. Characterization of the acyl donor site in the βBp-chain. Eur J Biochem 135:315–320, 1983.

71. R Takashi. A novel actin label: a fluorescent probe at glutamine-41 and its consequences. Biochemistry 27:938–943, 1988.

72. F Wold. Reactions of the amide side chains of glutamine and asparagine in vivo. Trends Biochem Sci 10:4–6, 1985.

73. C Larre, ZM Kedzior, MG Chenu, G Viroben, J Gueguen. Action of transglutaminase on an 11S seed protein (pea legumin): influence of the substrate conformation. J Agric Food Chem 40:1121–1126, 1992.

74. Y Matsumura, Y Chanyongvorakul, Y Kumazawa, T Ohtsuka, T Mori. Enhanced susceptibility to transglutaminase reaction of α-lactalbumin in the molten globule state. Biochim Biophys Acta 1292:69–76, 1996.

75. JJ Grootjans, PJTA Groenen, W de Jong. Substrate requirements for transglutaminases. Influence of the amine acid preceding the amine donor lysine in a native protein. J Biol Chem 270:22855–22858, 1995.

76. INA Ashie, TC Lanier. High pressure effects on gelation of surimi and turkey breast muscle enhanced by microbial transglutaminase. J Food Science 64:704–708, 1999.

77. T Numakura, N Seki, I Kimura, K Toyoda, T Fujita, K Takama, K Arai. Crosslinking reaction of myosin in the fish paste during setting (suwari). Nippon Suisan Gakk 51:1559–1565, 1985.

78. D Joseph, TC Lanier, DD Hamann. Temperature and pH affect transglutaminase catalyzed setting of crude fish actomyosin. J Food Sci 59:1018–1023, 1994.

79. A Hashimoto, A Kobayashi, K Arai. Thermostability of fish myofibrillar Ca^{2+}-ATPase and adaptation to environmental temperature. Bull Jpn Soc Sci Fish 48:671–684, 1982.

80. F Traore, J C. Meunier Cross-linking activity of placental FXIIIa on whey proteins and caseins. J Agric Food Chem 40:399–402, 1992.

81. M Faergemand, BS Murray, E Dickinson. Crosslinking of milk proteins with transglutaminase at the oil-water interface. J Agric Food Chem 45:2514–2519, 1997.

82. T C Lanier. Functional properties of surimi. Food Technol 40:107–112, 1986.

83. E Kolakowski, M Wianecki, G Bortnowska, R Jarosz. Trypsin treatment to improve freeze texturization of minced bream. J Food Sci 62:737–743, 752, 1997.

84. JS Hamada. Peptidoglutaminase deamidation of proteins and protein hydrolysates for improved food use. J Am Oil Chem Soc 68:459–462, 1991.

85. JS Hamada, WE Marshall. Enhancement of peptidoglutaminase deamidation of soy protein by heat treatment and/or proteolysis. J Food Sci 53:1132–1134, 1988.

86. GM Gilleland, TC Lanier, DD Hamann. Covalent bonding in pressure-induced fish protein gels. J Food Sci 62:713–716, 1997.

87. M Nonaka, R Ito, A Sawa, M Motoki, N Nio. Modification of several proteins by using Ca^{2+}-independent microbial transglutaminase with high-pressure treatment. Food Hydrocolloids 11:351–353, 1997.

88. T Shoji, H Saeki, A Wakameda, M Nonaka. Influence of ammonium salt on the formation of pressure-induced gel from walley pollock surimi. Nippon Suisan Gakk 60:101–109, 1994.

89. DA Clare, VW Valentine, HE Swaisgood. Molecular design and characterization of a streptavidin-transglutaminase fusion protein. Abstracts, Annual Meeting of the Institute of Food Technology, Atlanta GA, 1998.

90. ML Fink, SI Chung, JE Folk. γ-Glutamine cyclotransferase: specificity toward ε-(γ-glutamyl)-lysine and related compounds. Proc Natl Acad Sci USA 77:4564–4568, 1980.

91. K Seguro, Y Kumazawa, T Ohtsuka, H Ide, N Nio, M Motoki, K Kubota. ε-(γ-Glutamyl)-lysine: Hydrolysis by γ-glutamyltransferase of different origins, when free or protein bound. J Agric Food Chem 43:1977–1981, 1995.

92. NF Haard. Specialty enzymes from marine organisms. Food Technol 52(7):64–67, 1998.

93. T Abe, SI Chung, RP DiAugustine, JE Folk. Rabbit liver transglutaminase: physical, chemical, and catalytic properties. Biochemistry 16:5495–5501, 1977.

94. R Myhrman, J Bruner-Lorand. Lobster muscle transpeptidase. Methods Enzymol 19:765–770, 1970.

7
TMAO-Degrading Enzymes

Carmen G. Sotelo
Instituto de Investigaciones Marinas (CSIC), Vigo, Spain

Hartmut Rehbein
Federal Research Centre for Fisheries, Hamburg, Germany

I. INTRODUCTION

Some fish species, including hake or cod, have poor stability during frozen storage, which is reflected in a fast texture alteration, usually leading to a tougher and drier product. Much effort has been devoted to clarify the cause of this phenomenon. Almost three decades ago it was discovered that some fish species, especially those with poor freezing properties, accumulated formaldehyde (FA) and dimethylamine (DMA) during frozen storage (1). Some Japanese workers maintained active research on the problem during the 1960s and 1970s. They surveyed a considerable number of fish species and tissues and concluded that the tissues with highest DMA and FA content were usually visceral organs (2–6). They also found that every time DMA was detected, FA was also present. Experiments led to the conclusion that DMA and FA were produced in equimolar amounts and that an enzyme was present in some organisms with the ability to breakdown trimethylamine oxide (TMAO) into DMA and FA (7, 8).

Some debate ensued over the possibility of nonenzymatic TMAO breakdown during frozen storage of gadoids, catalyzed by low-molecular substances present in fish muscle, such as Fe^{2+}/Fe^{3+}, cysteine, or taurine (9), or other compounds such as hemoproteins, that are present in red muscle, blood, or tissues with high blood content (10). The existence of a nonenzymatic mechanism of TMAO breakdown at temperatures above 0°C was described

167

with the finding that TMA is also produced when the catalyst is Fe^{2+} and cysteine (11, 12).

Other scientists working in the field demonstrated that the accumulation of DMA and FA, which takes place during frozen storage of gadoids, is mostly due to the existence of an enzyme or enzymatic system. The fact that adding tissues with high TMAOase activity to white muscle of both gadoids and nongadoids always resulted in DMA production from TMAO during frozen storage (13), whereas either blood or kidney with no TMAOase activity but high hemoprotein content from nongadoids did not produce TMAO breakdown or formation of DMA and FA during frozen storage (14), confirmed the enzymatic nature of TMAO breakdown. Also, the thermolability and high-molecular-weight nature of the compounds catalyzing the reaction gave further evidence for an enzymatic mechanism (15, 16).

In the case of invertebrates, such as the squid *Illex illecebrosus,* it was shown that a nonenzymatic system for breakdown of TMAO is present and activated during cooking with significant production of TMA, DMA, and FA (17).

Several reviews have compiled the work done on enzymatic breakdown of TMAO during frozen storage of fish species and its effects on food quality (18–22).

In this chapter we will review what it is known about the nature of the activity degrading TMAO into DMA and FA in postmortem fish. First, we will deal with the distribution of its substrate, TMAO, and its possible physiological roles, and then we will address the work done on the biochemical characterization of this activity. Finally, we will give some information about some practical ways of inhibiting TMAOase.

II. TRIMETHYLAMINE OXIDE SIGNIFICANCE IN SEAFOOD PRODUCTS

A. Distribution of TMAO in Marine Organisms

Trimethylamine oxide (TMAO) (physicochemical data are given in Table 1) is not only an important compound for maintenance of the physiological functions in fish and shellfish but it is also a key substance in the spoilage of raw or processed seafood. Fish possess the capacity for endogenous synthesis of TMAO (23–26), although they normally take up TMAO by feeding on shrimps or other invertebrate prey (27, 28). Decapoda are rich in TMAO; crustaceans also contain fair amounts of this compound; whereas in oysters, mussels, clams, and scallops usually only small amounts of TMAO have been detected (29).

The content of TMAO in tissues of a large number of aquatic animals has been compiled and discussed in several reviews (2, 28–30). A summary of this earlier work and more recent results of TMAO content in Antarctic fish, deep-sea gadiform teleost, freshwater teleosts, and other fish species are given in Table 2.

Table 1   Physicochemical Properties of TMAO

Property	Reference
Molecular mass: 75.11 Da	
pK$_a$: 4.56	47
Redox potential: +0.130 V (for the half reaction TMAO/TMA)	49
Partial molal volume: 72.67	38
Density (1 M solution at 25°C): 1.000	38

The TMAO content of fish may vary according to the diet available, and is influenced by the age and size of the fish, as well as by the environmental salinity (31), temperature (32), and pressure (33) of the fish.

The highest concentrations of TMAO are found in the muscle of elasmobranchs and deep-sea teleosts: up to 230 mmoles/kg wet weight. The values for many other marine teleosts are between 20 and 100 mmoles/kg wet weight. With a few exceptions (34), freshwater fish contain no TMAO or only small amounts (30).

The distribution of TMAO in different fish tissues is exemplified in Table 1 by the data for Alaska pollack (*T. chalcogramma*). In contrast to earlier findings, TMAO has also been detected in fish blood (Table 2). As an example of the dependence of season (i.e., the water temperature), the TMAO content of rainbow smelt (*Osmerus mordax*) serum varied from about 5 to 55 mM (32). In serum of winter-acclimatized Pacific herring (*Clupea harengus pallasi*), surf smelt (*Hypomesus pretiosus*), and whitespotted greenling (*Hexagrammus stelleri*), TMAO concentration ranged from 47 to 12 mM (32).

In white-fleshed fish, including cod (*Gadus morhua*) or redfish (*Sebastes marinus*), the TMAO content of light muscle is larger than in dark muscle (Table 3), in contrast to dark-fleshed fish (29).

By using [1]H-labeled nuclear magnetic resonance (NMR), the TMAO content of the brain of dogfish (*Squalus acanthias*) was determined to be 58 mmoles/kg wet weight for brain white matter and 99 mmoles/kg for brain grey matter, whereas the body wall of that fish contained 58 mmoles/kg (35).

In summary, varying amounts of TMAO have been detected in all tissues of marine fish, and as well in a few freshwater fish species.

B.   Physiological Significance of TMAO in Marine Organisms

The following biochemical functions have been assigned to TMAO:

1.   Regulation of osmotic pressure in fish tissues (31, 33, 36, 37)
2.   Regulation of buoyancy (38)

Table 2 Occurrence of TMAO in Aquatic Animals

Species and tissue	TMAO content (mmoles/kg wet weight)	Reference
Mollusks		
Octopus, muscle	18	95
Squid, muscle	17–139	95
Elasmobranchs		
Ray, muscle	140–190	95
Shark, muscle	128–231	95
Crustaceans, muscle	9–28	95
Teleosts		
North Atlantic		
Gadoids, light muscle or fillet	62	59
Flatfishes, light muscle	19	59
Sebastes marinus, light muscle	107	59
Clupea harengus, fillet	37	59
Scomber scombrus, fillet	18	59
Antarctic waters		
Notothenoid species, fillet	75	59
Ice-fishes (*Channichthyidae*), fillet	71	59
North Pacific		
Theragra chalcogramma		
Light muscle	86	96
Liver	7	96
Pyloric caeca	14	96
Gall bladder	6	96
Intestine	18	96
Stomach	20	96
Spleen	48	96
Heart	15	96
Testis	87	96
Ovary	26	96
Deep-sea gadiform teleosts		
Antimora microlepis		
Muscle	211	33
Plasma	159	33
Macrouridae		
Muscle	142	33
Plasma	8	33
Freshwater fishes		
Lates niloticus, fillet	25	34
Tilapia spp., fillet	11	34

Table 3 Occurrence of TMAO and Related Amines in Muscle and Blood of North Atlantic Fish

Species and tissue	Content (mmoles/kg wet weight) of		
	TMAO	TMA	DMA
Gadus morhua, cod			
White muscle (n=8)	99.8 (12.5)[a]	2.5 (0.5)	0.1 (0.04)
Red muscle (n=8)	63.6 (13.1)	2.5 (0.4)	0.1 (0.04)
Blood (n=5)	7.7 (4.0)	1.8 (0.7)	0.7 (0.4)
Sebastes marinus, red fish			
White muscle (n=2)	106.7 (11.7)	2.4 (0.3)	0.07 (0.0)
Red muscle (n=2)	70.1 (6.4)	1.4 (0.1)	0.07 (0.0)
Blood (n=2)	1.0 (0.4)	0.5 (0.05)	0.07 (0.0)
Anarhichas lupus, wolf fish			
White muscle (n=2)	63.4 (10.6)	1.9 (0.05)	0.11 (0.05)
Red muscle (n=2)	33.7 (4.3)	2.0 (0.05)	0.07 (0.0)
Blood (n=2)	1.7 (0.4)	1.2 (0.2)	0.18 (0.05)

[a] SD given in parentheses.
Source: Rehbein, unpublished results.

3. Stabilization of proteins against denaturation by urea (39)
4. Protection of proteins against denaturation by freezing (40, 41)
5. Protection of proteins against denaturation by high pressure (33)
6. Buffer (42)
7. Regulatory role in disulfide bond formation in proteins (43)

1. Osmoregulation by TMAO

There is evidence that TMAO is used for osmoregulation in fish. Transfer of flounder (*Pleuronectes flesus*) or three-spined stickleback (*Gasterosteus aculeatus*) from sea water to fresh water resulted in a decrease of TMAO content in the muscle, the latter being correlated to the serum osmolality (31). It was similarly observed that muscle TMAO content showed a linear correlation with the osmotic pressure of the environment, when hagfish (*Myxine glutinosa*) were kept in seawater of different osmolarity (36). Recently, for deep-sea gadiform teleosts the TMAO content of muscle was linearly related to serum osmolatity (33).

Finally, the high content of TMAO in elasmobranchs may be explained by giving the fish osmotic protection, using TMAO as an osmolyte counteracting urea (37).

2. Regulation of Buoyancy

The role of TMAO, other metabolites, and of inorganic ions for buoyancy of elasmobranchs has been studied by measuring concentration of these substances in muscle fluid and plasma, and determining their partial molal volume (38). From the calculation of the contribution of TMAO and urea to the lift of black whaler shark (*Carcharhinus obscurus*), the authors drew the conclusion that a positive buoyancy role must be considered for both compounds in Chondrichthyean fishes.

3. TMAO as an Antagonist of Urea

In the search for the physiological function of the high TMAO concentration in elasmobranchs, the counteraction between urea and TMAO, or other methylamines, was detected. The stabilization of the kinetic properties of enzymes, and of protein structure (44), has been reviewed (39, 45).

   The molar ratio of TMAO and urea in muscle of marine elasmobranchs (shark, ray) was determined to be about 1:2 (46). At this ratio enzymes from elasmobranchs as well as nonelasmobranchs retain their activity (i.e., denaturation by urea is counteracted by TMAO [and other methylamines]). The mechanism of TMAO protection has been studied by physicochemical techniques (47), and was explained by thermodynamic compensation of urea and TMAO interaction with the proteins.

4. TMAO as a Cryoprotectant

Contradictory results have been reported for stabilization of frozen stored fish muscle proteins by TMAO. In one study it was demonstrated that insolubilization of frozen stored (-7°C) red hake (*Urophycis chuss*) muscle proteins was retarded by addition of TMAO (40). However, in a similar study using minced muscle of cod (*Gadus morhua*), addition of TMAO to muscle tissue did not influence the rate or extent of freeze denaturation as measured by protein solubility (41).

5. Protection of Proteins by TMAO Against Denaturation by High Pressure

The elevated levels (up to 211 mmoles/kg wet weight) of TMAO in muscle of deep-sea gadiform teleosts indicate that TMAO may counteract the destabilizing effects of hydrostatic pressure on cellular proteins (33). Lactate dehydrogenase from white muscle of the deep-sea (2000–3900 m) fish *Coryphaenoides leptolepis* was purified, and the effect of TMAO and hydrostatic pressure (300 atm) on NADH K_m was studied. It was found that the increase of K_m induced by high pressure was prevented in the presence of 250 mM TMAO.

6. Buffer Capacity of TMAO

Despite its high concentration in fish muscle, TMAO cannot make a significant contribution to the buffering capacity of fish tissues, since the pK_a of TMAO is 4.56 (47), whereas the pH in muscle or other tissues is about 6.5–7.0. pH values below 5 may be found only within lysosomes.

On the other hand, TMAO may serve as a buffer in processed fish. For example, it is prescribed that the pH value of marinated herring has to be below 4.8 (48), thus being in the range where TMAO may delay the ripening process due to its buffering properties.

7. Regulatory Role of TMAO in S–S Bond Formation

When SH-groups containing amino acids, cysteine, or homocysteine, were reacted with TMAO, disulfides were formed (43). This result suggests that TMAO may play a regulatory role in S–S bond formation in enzymes and other proteins.

III. ENZYMES OF TRIMETHYLAMINE OXIDE METABOLISM

TMAO may be synthesised or degraded by the enzymatic reactions described in the following sections.

A. Trimethylamine Monooxygenase

This enzyme catalyzes the synthesis of TMAO from TMA as follows:

$$TMA + NADPH + H^+ + O_2 \rightarrow TMAO + NADP^+ + H_2O$$

The enzyme has been detected in methylotrophic bacteria (47), in calanoid copepods, and in some marine fish (23, 49, 50). It was partially purified from the marine copepod *Calanus finmarchicus* (51).

The significant TMA monooxygenase activities in the liver of herring (*Clupea harengus*), rainbow smelt (*Osmerus mordax*), and two Antartic fish species were found to be correlated to the serum levels of TMAO in these species (26, 49). Inhibitor studies indicated that the TMAO formation can be largely assigned to a flavin-containing monooxygenase and not to cytochrome P450 monooxygenases.

B. Trimethylamine Dehydrogenase (EC 1.5.99.7)

TMA dehydrogenase catalyses the oxidative demethylation of TMA to DMA and FA:

$$TMA + H_2O \rightarrow DMA + FA + 2H$$

The enzyme has been found in facultative methylotropic bacteria (52), and uses a flavoprotein as an electron acceptor. The enzyme contains a [4Fe-4S] cluster and covalently bound flavin. Up to now existence of TMA dehydrogenase has not been detected in fish spoilage bacteria (50), and DMA or FA production in chilled stored fish is low in the phase of heavy bacterial spoilage of the fish (53).

C. Trimethylamine Oxide Reductase

TMAO is reduced by the action of this enzyme into TMA. This reaction is of great importance for the spoilage of seafood caused by bacteria (50). TMA, being mainly responsible for the off odor of spoilage, has a very low odor threshold compared to DMA or ammonia (55).

$$TMAO + 2H^+ + 2e_- \rightarrow TMA + H_2O$$

In *Shewanella* sp. two TMAO reductases were detected in the periplasmatic fraction. The enzyme of 86 kDa molecular weight was purified and found to be a molybdoenzyme (54).

A large number of bacteria involved in fish spoilage, belonging to the Enterobacteriaceae and the genera *Alteromonas, Photobacterium,* and *Vibrio,* are able to reduce TMAO, which serves as an terminal electron acceptor in anaerobic growth (56). On the other hand, many types of bacteria found in spoiling fish are unable to reduce TMAO (50). Therefore the usefulness of TMA as a quality parameter is dependent on the storage conditions of the product, and the composition of the bacterial flora.

D. Trimethylamine Oxide Demethylase (EC 4.1.2.32)

The demethylation of TMAO rendering DMA and FA is carried out according to the following reaction scheme:

$$TMAO \rightarrow DMA + FA$$

The importance of this endogenous fish activity has been detected in many gadoid fish species, being very important for the quality deterioration of frozen stored fish. Distribution and properties of trimethylamine oxide demethylase (TMAOase) are treated in detail in the next section.

IV. ENDOGENOUS DEMETHYLATION OF TMAO

A. Factors Affecting Distribution of TMAO Demethylase

One of the first questions that comes to mind is whether TMAOase catalyzes the same reaction in vivo as it does postmortem. There have been no experiments to answer this question; more research is needed on the purification of this enzyme or group of enzymes.

Although TMAO is distributed among all kinds of marine fish and invertebrates, and in some fresh water fishes, TMAOase activity has been identified in only 30 species of marine fish belonging to 10 families and 8 species of invertebrates, mainly mollusk (57). However, its real distribution among marine and fresh water organisms is not known because the species studied by different authors are mostly commercial species.

Table 4 compiles most of the available data about detection of postmortem TMAOase in different fish and invertebrates species. Among fish groups, postmortem accumulation of DMA and FA has been described in species belonging to Gadidae, Merluccidae, and Myctophidae (57). Gadiform (families Gadidae and Merluccidae) is the group of fish with more representatives having

Table 4 TMAOase Activity Distribution in Several Marine Organisms

Common name	Latin name	Type of work	Reference
Gadoids			
Cod	*Gadus morhua*	Isolation/characterization	65
		Activity in several tissues	14, 60
Saithe	*Pollachius virens*	Activity in several tissues	14, 60
Haddock	*Melanogramus aeglefinus*	Activity in several tissues	14, 60
Ling	*Molva molva*	Activity in several tissues	60
Blue ling	*Molva dypterygia*	Activity in several tissues	14
Blue whiting	*Micromesistius poutassou*	Activity in several tissues	60
Alaska pollack	*Theragra chalcogramma*	Activity in several tissues	60
Cusk	*Brosme brosme*	Activity in several tissues	60
	Coryphaenoides rupestris	Activity in several tissues	60
Whiting	*Merlangus merlangus*		14
Red hake	*Urophycis chuss*	Microsomal from muscle, isolation, characterization	16, 62, 68, 70
Nongadoids			
Herring	*Clupea harengus*	Activity in kidney	14
Cucumberfish	*Chlorophtalmus nigripinnis*	Production	97
Crustaceans			
Blue crab	*Portunus trituberculatus*	Activity	58
Mollusk			
Bivalve	*Barbatia virescens*	Production	97
Squid	*Illex illecebrosus*	Activity/production	17, 98
	Todarodes pacificus	Activity/production	2
	Sepia esculenta	Activity	2
	Loligo formosa	Production	99
	Thysanoteuthis rombus	Activity	2

TMAOase activity. Also, the level of TMAOase activity in species other than gadiforms is much lower than in gadiforms. The presence of TMAOase in invertebrates is somewhat puzzling. Although there is evidence of DMA and FA accumulation in these organisms, it seems that only viscera contain significant amounts of these compounds (58).

Several groups of fish did not show any TMAOase activity, such as flatfish species, redfishes, Antartic fishes, and clupeids, with the exception of herring (*Clupea harengus*), which has some activity in kidney tissue (14, 59).

B. Biochemical Nature of TMAO Demethylase

The level of TMAOase, in a particular species and tissue, has a large coefficient of variation (60) and seem to be influenced by intraspecific factors (see also Chap. 1). Factors including gender, maturation stage, temperature of the habitat, feeding status, or size can influence the actual level of TMAOase activity. Although systematic measurements of TMAOase activity in fish of different biological condition has not been done, some authors have observed twice the rate of DMA production in large, spawning, summer-caught silver hake than smaller, nonspawning, and winter-caught animals (61).

Some authors have questioned the in vivo substrate of this activity. As mentioned before, TMAO has been postulated as having a different role in fish physiology. However, the link between the regulation of TMAO levels in different fish tissues and organs and the existence of an enzyme breaking-down TMAO has not been established. An alternative hypothesis is that TMAOase does not demethylate in vivo; rather the enzymatic system may have another substrate and therefore catalyze a different reaction (59).

Parkin and Hultin (62, 63) have suggested that enzymatic TMAO breakdown could be related to demethylation reactions that take place in living organisms and are involved in the detoxification of xenobiotic compounds. These reactions have been described as taking place in liver and located in microsomal membranes. The enzymes responsible for such reactions are complexes of electron transfer components, such as flavoproteins and cytochromes (62).

Some workers have studied the anatomical distribution of TMAOase. The organs and tissues with highest TMAOase activity are from the viscera, muscle, and skin. The concentration of DMA is higher in viscera than in muscle (2, 5, 58). Dark muscle usually presents a higher DMA content than white muscle, and in some species with high dark muscle content, such as *Thunnidae* and *Scombridae*, DMA content increased only in dark muscle upon storage at subfreezing temperatures (0° up to -6°C)(64). This is also true for Gadidae, in which dark muscle content correlated with the degree of DMA accumulated during frozen storage (1).

The estimation of TMAOase activity has corroborated the existence of high-TMAOase activity organs such as kidney or spleen of most gadiforms, other organs or biological fluids with moderate TMAOase activity such as pyloric caecum, blood, and others with low or no TMAOase activity, such as liver and muscle (60).

Depending on the source of the enzyme, the activity has been located in the soluble fraction or in the particulate matter of homogenates of organs or tissues. Harada (2) isolated TMAOase from the liver of lizardfish (*Saurida tumbil*). He found a soluble TMAOase, which after a partial purification and an alkaline treatment, could be classified into soluble and insoluble fractions. Only the mixture of fractions possessed TMAO demethylation activity.

Hultin and co-workers have studied the enzyme present in the muscle of red hake (*Urophycis chuss*). These authors found that TMAOase is present both in the soluble and in the particulate fraction of red hake muscle homogenates (16). The insoluble fraction was studied most extensively by these authors.

Gill and Paulson (65) and Joly et al. (66) found that 80% and 80–90% of the activity from cod (*Gadus morhua*) and saithe (*Pollachius virens*) kidney, respectively, was insoluble. Rehbein et al. (67) have found that TMAOase from kidney and spleen of hake (*Merluccius merluccius*) and saithe (*Pollachius virens*) was mostly insoluble (80% in the insoluble fraction). TMAOase from bile of saithe was soluble after centrifugation at $100,000 \times g$.

Much of the work done on isolation, purification, and characterization has focused on insoluble TMAOase, and TMAOase of visceral organs, due to their high activity (65, 66), but there has been some work with the insoluble muscular activity in species such as red hake (16, 62, 68).

1. TMAOase Assay

TMAOase assay provides an estimation of molecules of TMAO converted into DMA and FA in a certain period of time. Several types of TMAOase assay have been described. The main difference is the type of cofactor used for the assay. Other differences include the presence of oxygen during incubation of enzyme extract and substrate, the type of end product measured, and type of activators included in the assay.

The effect of oxygen on the reaction has been a matter of disagreement. Two approaches has been taken to remove oxygen: i.e. use of vacuum (2, 60, 69); and inclusion in the assay of glucose oxidase and glucose, which consumes oxygen during its catalytic activity (62, 70). Some workers have not removed oxygen from the TMAOase assay (65, 68).

The cofactors used in the TMAOase assay have varied. Some authors have used Fe^{2+}, cysteine, and ascorbate (68) and adding reduced methylene blue (65, 66) and excluded oxygen (14, 60). This assay proved to be more sensitive than the same one that excluded reducing agents (not adding cysteine, ascorbate, and methylene blue).

The other type of TMAOase assay has been performed by Hultin and co-workers, who do not add artificial electron donors, such as methylene blue. It has been suggested that a flavonucleotide is the TMAOase cofactor, since addition of $FMNH_2$ or $FADH_2$ stimulates TMAOase activity (2). The cofactor system used includes a flavin (riboflavin, FMN, or FAD) and NAD(P)H (70) or Fe^{2+} with ascorbate or cysteine (16, 68). Reaction is stopped by the addition of an acid (60, 70), followed by centrifugation for removing insoluble matter, and products are determined in supernatant.

There have been also different ways of estimating the concentration of reaction products. Usually FA is measured basically because its determination is easier than determining DMA. FA can be measured with the colorimetric Nash test (71) or can be estimated by the use of coupled assay with a formaldehyde dehydrogenase enzyme (72). Determination of DMA has been performed for the estimation of TMAOase (70) also by using a colorimetric method (73).

2. Purification

Purification of insoluble TMAOase, although it is a membrane protein, has involved methods similar to those employed for water-soluble proteins. Membrane proteins are amphiphiles and need substances that resemble their environment in order to preserve their activity. A certain protein/lipid/detergent ratio is required throughout the purification. If this ratio is altered through the process of purification, the functionality or properties of the protein might be changed (74).

Preparation of Membranes. Subcellular membranes seem to be the most probable localization for insoluble TMAOase. However, the investigations made did not clarify which subcellular organelle or organelles host TMAOase activity. The work done is scarce and with different TMAOase sources. In the case of cod (*Gadus morhua*) kidney, a subcellular fractionation was carried out by isopycnic centrifugation with sucrose medium. TMAOase was found to be related to lysosomal membrane (65). On the other hand, red hake (*Urophycis chuss*) muscle TMAOase was found to be localized in the microsomal fraction (68).

Preparation of membranes has been the main step in the partial purification of TMAOase. This is done by homogenization with a buffer, usually at neutral pH, followed by centrifugation to remove soluble proteins. In most cases, 80–90% of TMAOase is found in the insoluble fraction (65, 66). That could be regarded as a crude preparation of membrane material that is further used for other separation techniques.

Solubilization with Detergents. Solubilization of TMAO demethylase with several detergents has been attempted with different results. Microsomal

TMAOase from red hake muscle was resistant to solubilization with a variety of detergents. Triton X-100 and deoxycholate did not deactivate or solubilize the enzyme, whereas sodium dodecyl sulfate (SDS) was more effective as a solubilizing agent. The activity was resistant to up to 10 mg of SDS per mg of protein (62).

Gill and Paulson (65) were able to solubilize TMAOase with Triton X-100 or SDS. Removal of Triton X-100 detergent resulted in a total insolubilization of the enzyme, whereas in the case of SDS the insolubilization of TMAOase was only 20%. Also, Rehbein and Schreiber (60) employed 0.2% Triton X-100 for preparing TMAOase extracts from fish tissues. Finally, Joly et al. (66) obtained good solubilization yields of TMAOase from saithe kidney using detergents such as sodium deoxycholate, although this detergent increased TMAOase activity.

Chromatographic Separation. The separation of soluble TMAOase has been achieved using different chromatographic techniques. Gill and Paulson (65) applied an SDS-solubilized TMAOase extract to a gel filtration column (Sephadex G-50 or Sephadex G-200). The activity always eluted in the void volume, therefore no purification step was achieved using this methodology.

Joly et al. (66) achieved the separation of three TMAOase fractions from a detergent extract of a membrane fraction using anion exchange chromatography: a DEAE-TSK column equilibrated with 20 mM Tris–HCl pH 7.0 buffer. The fractions eluted with 0.1, 0.2, and 0.25 M NaCl. From this result the authors suggested the possibility of a multienzyme nature of TMAOase. Association with phospholipids and other proteins could be the explanation for the existence of these three TMAOase fractions. The specific activity of one of the fractions, G-0.2, was 8.3 times that of the partially purified extract applied to the column and 83 times that of the crude extract. The yield was 17.5 %.

3. Characterization

Characterization of TMAOase activity has been difficult because of the problems related with its purification. The study of cofactors needed for the reaction has helped investigators to understand the reaction.

It was shown that a dialyzable cofactor was needed for TMAOase to catalyze TMAO breakdown (75). One of the early reports about the nature of TMAOase cofactors is the work done by Harada (2). He described a heat-stable cofactor present in the viscera of several marine organisms that was necessary for the activity of TMAOase from lizardfish liver. He also showed that the reduced forms of NAD or NADP, flavins, and citric acid enhanced TMAOase activity.

Several other substances alter the activity of TMAOase. Some of them enhance TMAOase activity, such as glutathione, Fe^{2+}, ascorbate, flavins, hemoglobin, myoglobin, and methylene blue (76). Others inhibit TMAOase, among them iodoacetamide, cyanide, and azide. Table 5 shows the variety of compounds tested for their ability to activate or inhibit TMAO demethylase. The involvement of nucleotides, especially flavin-type nucleotides, as cofactors was demonstrated by most of the authors. Other compounds, such as methylene blue or phenazine methosulfate, which resemble structurally the flavin core (isoalloxacine), are also strong activators (16).

Later on it was shown that there are at least two physiological cofactor systems. One of them is formed by NAD(P)H and a flavin, in which flavin is an absolute requirement. The other system is composed of Fe^{2+} and ascorbate or cysteine (76). The first system requires anaerobic conditions (62, 68), while the second is not influenced by O_2 (65). Other works have, however, reported that O_2 inhibited activity in the second cofactor system (60).

It has been postulated that the rate-limiting factor for TMAO breakdown to DMA and FA is the concentration of cofactors. Normally they are either below Km or a decrease in concentration can occur prior to frozen storage period (77). The addition to the muscle of compounds that decompose some of the known cofactors, such as NAD or ascorbate, decreased the rate of DMA and FA accumulation during frozen storage (78).

Among metals, it was found that only Fe^{2+} activates the reaction whereas a wide variety of divalent cations did not affect the reaction. Other ions, such as Fe^{3+}, have been shown to be inhibitors at 1 mM concentration (69) or not to have an effect on the reaction at lower concentration (200 ppm) (9).

Inhibition of TMAOase is also promoted by a variety of compounds. However, most of them are general inhibitors of redox reactions, being strong oxidants, such as H_2O_2, which inhibits the reaction (70). The presence of oxygen or oxidants can promote inhibition of enzymatic TMAO breakdown in frozen red hake (*Urophycis chuss*) (79, 80). Azide and cyanide inhibit TMAOase and Parkin and Hultin (68) showed that a hemoprotein linked to the muscle microsomal fraction and that could be involved in the TMAO demethylation (H-444) was not reduced in the presence of azide under anaerobic conditions.

The fact that TMAO demethylation can be stimulated or inhibited by compounds involved in redox reactions has suggested that the type of reaction that takes place is an electron transfer mechanism with the involvement of oxidoreductases. Also, it is quite probable that this reaction proceeds in several stages, since several compounds could function as inhibitors (57).

It has been suggested that in those cases where TMAOase has been found in the particulate fraction, the activity is influenced by lipids, mainly phospholipids. Joly et al. (66) separated and analyzed the phospholipid fraction and found that a different phospholipid/protein ratio could be the explanation for the different chromatographic behavior of the fractions. Also, the phospholipid composition of these fractions was not exactly the same. Some of the fractions were

Table 5   Effect of Different Compounds on TMAOase Activity

Substance	Cofactor	Activator	No effect	Inhibitor	Reference
Nucleotides					
FMN	+				2, 58, 69, 70
NADP(H)	+				2, 58, 69
NAD(H)		+			2, 70
NAD			+		69
RIBOFLAVIN	+				2, 76
FAD				+	69
FADH	+				2
Metals					
Fe^{2+}		+			2, 9, 58, 68, 69, 92
Fe^{3+}				+	69
Fe^{3+}			+		2, 9, 68
Cd^{2+}			+		69
Co^{2+}			+		2, 69
Ni^{2+}			+		69
Zn^{2+}			+		2, 69
Mn^{2+}			+		2, 69
Mg^{2+}			+		2, 69
Ca^{2+}			+		9, 69
Mo^{6+}			+		2
Sn^+		+			9
Amino acids					
Cysteine	+				9, 68
Taurine			+		68
Hypotaurine		+			9
Hypotaurine			+		68
Others					
Methylene blue		+			2, 68
Menadione		+			68
Phenazine methosulfate		+			68
EDTA				+	69
EDTA		+			68, 69
DOC				+	68
Idoacetate				+	9
FA				+	2
Choline				+	68, 69
Betaine		+			69
Betaine			+		68
DMA			+		69
DMA				+	68
TMA				+	68
Dimethylaniline				+	68

rich in very polar phospholipids (although they were not able to identify the compounds). However, these phospholipids did not seem to have a role in TMAOase activity since their digestion with different types of phospholipases did not affect TMAOase activity of the different fractions.

The apparent molecular weight for TMAOase is probably influenced by the presence of lipids and other associated proteins. Therefore, the values given in the literature, which are quite large, could be unreliable. Joly et al. (66) give values from 200 up to 2000 KDa.

Another characteristic of TMAOase is its stability. Although the enzyme is denatured by heat, its cold denaturation temperature seems to be rather low since the system is active at frozen storage temperatures. It has been demonstrated that TMAOase is active down to -29°C (81–83). However, the activity decreases with the temperature at freezing temperatures (14, 84).

The heat denaturation of TMAOase activity has been studied both in experimental fish muscle systems and in partially purified extracts. The results were different depending on the fish species studied. Preheating Alaska pollock muscle at 40°C for 30 min before freezing provoked a 20% decrease in DMA formation after 4 weeks of frozen storage. If the treatment is at 50°C, the production of DMA and FA is totally inhibited (85). However, 60°C preheating treatment of silver hake before freezing had no effect. Only heating up to 80°C produced the desired effect of total inhibition of DMA and FA production in this species (61).

Determination of denaturation temperature in TMAOase extracts produced different results, probably due to different experimental conditions. About 70% of activity in SDS extracts of TMAOase from red hake muscle microsomal fraction is lost upon heating at 40°C for 10 min. The intact microsomal fraction was more heat resistant than this detergent purified fraction (62). An acetone power containing TMAOase from Pollack pyloric caecum, lost most of it activity if the thermal treatment ranged from 40 to 50°C for 5 min (69).

The optimum pH for the reaction in vitro seem to be situated around 5.0 (2, 60, 65, 69, 86). Other authors have found neutral pH to be optimum (e.g., the enzyme purified from Alaska pollack pyloric caecum [10] and the enzyme from red hake muscle microsomal fraction [68]).

Kinetic Characteristics. Several kinetic studies performed in partially purified TMAOase showed that this activity follows Michaelis–Menten kinetics for the degradation of TMAO. The Km for TMAO has been determined and it ranged from 3 mM up to 12 mM (66, 68, 70). This low Km for TMAO suggested that TMAO is not the limiting factor for TMAO breakdown in white muscle, since TMAO concentration in gadoid muscle is higher (59, 76).

Electrophoretic Characteristics. Electrophoretic separation of TMAOase has been described by Gill and Paulson (65). These authors used a freeze-dried, high-molecular-weight fraction from a gel filtration column for a native electrophoretic separation. Using a nonsieving matrix of agarose gels (0.75%) two bands were obtained, one of them with high mobility and the other with very low mobility, both bands gave positive TMAOase-specific stain. When the composition of the separation matrix is changed (0.5% agarose and 3% acrylamide) four bands were obtained, two of them (high-mobility doublet) exhibiting a strong response to TMAOase-specific stain. When acrylamide gradient gels were used (2–16% T) five bands were obtained: a low-mobility band located near the origin that showed a strong TMAOase and lipid stain, and a diffuse and weak TMAOase stain corresponding to four medium-mobility bands. Determination of pI seem to be a difficult task, from the apparent complexity of the TMAOase rich fractions obtained. Gill and Paulson (65) performed isoelectric focusing with the TMAOase fraction obtained by gel filtration and obtained four TMAOase positive bands, the strongest being located in the anodal region of the gel. They concluded that TMAOase activity is related to several high-molecular-weight isoenzymes that also exhibit different pI, although they also suspected that different association among several protein molecules could give a differential pI pattern.

V. INHIBITION OF ENDOGENOUS DEMETHYLATION OF TMAO IN SEAFOOD

TMAOase inhibition has been the main objective for many researchers involved in the study of TMAOase and its consequences in quality loss of frozen fish (1, 87–90).

The inhibition of the TMAO demethylation can be achieved using physical methods, chemical ones, or a combination of both.

Among physical inhibition methods the most important is temperature. Lower temperature and temperature fluctuation during frozen storage decrease the rate of DMA and FA formation (91). However, although the extension of shelf life of frozen gadoids is of primary economic importance, DMA and FA kinetics in situ has scarcely been studied. Also, the mechanisms of FA and DMA production at subzero temperatures and the cold denaturation temperature of the enzymatic system are not fully understood.

The other approach to inhibiting TMAOase is heat denaturation of the enzymatic system. As has been said, a heat treatment of 50°C is needed for any significant reduction of activity (50%). A disadvantage is that this treatment, of course, will lead to some organoleptic changes in the product (61).

Removal of TMAOase rich tissues could be another approach to extend shelf life of frozen fish with TMAOase. The recommended practice is to eviscerate the fish prior to freezing process or not mince the fish muscle with

TMAOase-rich tissues (e.g., red muscle or pieces of kidney), thus preventing the contact of TMAOase-rich tissues with the substrate rich white muscle. However, the localization of kidney and peritoneum makes them difficult to remove when eviscerating. The extensive contact surface of kidney with muscle promotes the production of more DMA and FA in fish where kidney is not removed (92). Also, red muscle removal will likely improve storage characteristic of gadoid fillets.

Other physical methods for removing TMAOase have not been, so far, investigated, such as the use of high-pressure treatments, or heating of localized TMAOase-rich tissues.

Chemical inhibition consists of the addition of substances that can act as inhibitors or can effect a reduction in any of the components needed for the reaction to take place. It has been shown that refrigeration previous to frozen storage, which depletes TMAO by bacterial growth, has decreased the DMA and FA rate formation in frozen gadoids (93). Also, NADH depletion can influence the rate of DMA and FA formation since it is a cofactor of the enzymatic system involved in the reaction (15, 77). Freeze–thaw cycles before the frozen storage period can decrease the cofactor concentration level, thus the rate of TMAO breakdown is lower (77).

Although some research has been devoted to the elucidation of TMAOase inhibitors, the fact that the activity has not been fully purified and characterized hinders the study of the mechanisms of the reaction and the possible inhibitors. TMAOase inhibitors have been described previously and most of them are not of use as food additives (e.g., azide, cyanide, etc.) because of their toxicity. Other substances that should be further tested include sodium citrate, sodium pyruvate, and EDTA (68, 80, 94).

VI. CONCLUSIONS

TMAO is not only an important compound for maintenance of the physiological functions in fish and shellfish but it is also a key substance in the spoilage of raw or processed seafood. In the living animal, TMAO functions in osmoregulation, regulation of buoyancy, protection from urea, cryoprotection, buffering, and regulation of disulfide bond formation. Enzymes involved in TMAO metabolism include trimethylamine monooxygenase, trimethylamine dehydrogenase, trimethylamine oxide reductase, and TMAOase. Formation of formaldehyde by the reaction catalyzed by TMAOase appears to contribute to texture deterioration in frozen fish, especially gadoid species. Characterization of TMAOase activity has been difficult because of the problems related to its purification. The study of cofactors needed for the reaction has helped us to understand its physiological significance and biochemical control.

REFERENCES

1. CH Castell, B Smith, W Neal. Production of dimethylamine in muscle of several species of gadoid fish during frozen storage, especially in relation to presence of dark muscle. J Fish Res Board Can 28:1–5, 1971.

2. K Harada. Studies on enzyme catalyzing the formation of formaldehyde and dimethylamine in tissues of fishes and shells. J Shimonoseki Univ Fish 23:163–241, 1975.

3. K Amano, K Yamada, M Bito. Detection and identification of formaldehyde in gadoid fish. Bull Jpn Soc Sci Fish 29:695–701, 1963.

4. K Amano, K Yamada, M Bito. Content of formaldehyde and volatile in different tissues of gadoid fish. Bull Jpn Soc Fish 29:860–864, 1963.

5. K Amano, K Yamada. The biological formation of formaldehyde in cod flesh. In: R Kreutzer, ed. The Technology of Fish Utilization. London: Fishing News Books Ltd., 1965, pp 73–87.

6. K Yamada, K Amano. Studies on the biological formation of formaldehyde and dimethylamine in fish and shellfish. VII. Effect of methylene blue on the enzymatic formation of formaldehyde and dimethylamine from trimethylamine oxide. Bull Jpn Soc Sci Fish 31:1030–1037, 1965.

7. K Amano, K Yamada. A biological formation of formaldehyde in the muscle tissue of gadoid fish. Bull Jpn Soc Sci Fish 30:430–435, 1964.

8. K Amano, K Yamada. Formaldehyde formation from trimethylamine oxide by the action of pyloric caeca of cod. Bull Jpn Soc Sci Fish 30:639–645, 1964.

9. J Spinelli, BJ Koury. Some new observations on the pathways of formation of dimethylamine in fish muscle and liver. J Agric Food Chem 29:327–331, 1981.

10. T Tokunaga. Biochemical and food scientific study on trimethylamine oxide and its related substances in marine fish. Bull Tokai Reg Fish Res Lab 101:1–5, 1980.

11. EB Vaisey. The non-enzymatic reduction of trimethylamine oxide to trimethylamine, dimethylamine, and formaldehyde. Can J Biochem Physiol 34:1085–1090, 1956.

12. JP Ferris, RD Gerwe, GR Gapski. Detoxication mechanism. II. The iron-catalyzed dealkylation of trimethylamine oxide. J Am Chem Soc 89:5270–5275, 1967.

13. JR Dingle, JH Hines. Protein instability in minced flesh from fillets and frames of several commercial Atlantic fishes during storage at -5°C. J Fish Res Board Can 32:775–783, 1975.

14. H Rehbein. Relevance of trimethylamine oxide demethylase activity and haemoglobin content to formaldehyde production and texture deterioration in frozen stored minced fish muscle. J Sci Food Agric 43:261–276, 1988.

15. RC Lundstrom, FF Correia, KA Wilhelm. Enzymatic dimethylamine and formaldehyde production in minced American plaice and blacback flounder mixed with a red hake TMAOase active fraction. J Food Sci 47:1305–1310, 1982.

16. BQ Phillipy. Characterization of the in situ TMAOase system of red hake muscle. Ph.D. thesis, University of Massachusetts, Amherst, MA, 1984.

17. P Nitisewojo, HO Hultin. Characteristics of TMAO degrading systems in Atlantic short finned *squid (Illex illecebrosus)*. J Food Biochem 10:93–106, 1986.

18. Z Sikorski, J Olley, S Kostuch. Protein changes in frozen fish. Crit Rev Food Sci Technol 8:97–129, 1976.
19. JJ Matsumoto. Chemical deterioration of muscle proteins during frozen storage. In: JR Whitaker, M Fujimaki, eds. Chemical Deterioration of Proteins. Washington: American Chemical Society, 1980, pp 95–124.
20. SYK Shenouda. Theories of protein denaturation during frozen storage of fish flesh. Adv Food Res 26:275–310, 1980.
21. NF Haard. Biochemical reactions in fish muscle during frozen storage. In: EG Bligh, ed. Seafood Science and Technology. Oxford: Fishing News Books, 1992, pp 176–209.
22. CG Sotelo, C Piñeiro, RI Pérez-Martín. Review: Denaturation of fish proteins during frozen storage: formaldehyde role on it. Z Lebensm Unters Forsch 200:14–23, 1995.
23. I Agustsson, AR Strom. Biosynthesis and turnover of trimethylamine oxide in the teleost cod, Gadus morhua. J Biol Chem 256:8045–8049, 1981.
24. RP Charest, M Chenoweth, A Dunn. Metabolism of trimethylamines in kelp bass (Paralabrax clathratus) and marine and freshwater pink salmon. (Oncohynchus gorbuscha). J Comp Physiol B 158:609–619, 1988.
25. T Daikoku, M Murata, M Sakaguchi. Effects of intraperitoneally injected and dietary trimethylamine on the biosynthesis of trimethylamine oxide in relation to seawater adaptation of the eel, Anguilla japonica and the guppy, Poecilia reticulata Comp Biochem Physiol 89A:261–264, 1988.
26. JA Raymond, AL DeVries. Elevated concentrations and synthetic pathways of trimethylamine oxide and urea in some teleost fishes of McMurdo Sound, Antartica. Fish Physiol Biochem 18:387–398, 1998.
27. M Yamagata, LK Low. Banana shrimp, Penaeus merguiensis, quality changes during iced and frozen storage. J Food Sci 60:721–726, 1995.
28. HS Groninger. The occurrence and significance of trimethylamine oxide in marine animals. U.S. Fish Wildl Serv Spec Sci Rep Fish 333:1–22, 1959.
29. CE Hebard, GJ Flick, RE Martin. Occurrence and significance of trimethylamine oxide and its derivatives in fish and shellfish. In: RE Martin, GJ Flick, CE Hebard, DR Ward, eds. Chemistry and Biochemistry of Marine Food Products. Westport, CT: AVI Publ Comp., 1982, pp 149–304
30. JM Shewan. The chemistry and metabolism of the nitrogenous extractives in fish. Biochem Soc Symp No. 6:28–48, 1951.
31. R Lange, K Fugelli. The osmotic adjustment in the euryhaline teleosts, the flounder, Pleuronectes flesus L. and the three-spined stickleback, Gasterosteus aculeatus L. Comp Biochem Physiol. 15:283–292, 1965.
32. JA Raymond. Seasonal variations of trimethylamine oxide and urea in the blood of a cold-adapted marine teleost, the rainbow smelt. Fish Physiol Biochem 13:13–22, 1994.
33. MB Gillet, JR Suko, FJ Santoso, PH Yancey. Elevated levels of trimethylamine oxide in muscles of deep-sea gadiform teleosts: a high-pressure adaption? J Exp Zool 279:386–391, 1997.
34. U Anthoni, T Borresen, C Christophersen, L Gram, PH Nielsen. Is trimethylamine

oxide a reliable indicator for the marine origin of fish? Comp Biochem Physiol 97B:569–571, 1990.

35. JJ Bedford, JL Harper, JP Leader, RA J Smith: Identification and measurement of methylamines in elasmobranch tissues using proton nuclear magnetic resonance (^{1}H-NMR) spectroscopy. J Comp Physiol B. 168:123–131, 1998.

36. C Cholette, A Gagnon. Isoosmotic adaptation in Myxine glutinosa L. II. Variations of the free amino acids, trimethylamine oxide and potassium of the blood and muscle cells. Comp Biochem Physiol 45 A:1008–1921, 1973.

37. PW Hochachka, GN Somero. Biochemical Adaptation. Princeton, NJ: Princeton University Press, 1984, pp 304–354.

38. PC Withers, G Morrison, GT Hefter, TS Pang. Role of urea and methylamines in buoyancy of elasmobranchs. J Exp Biol 188:175–189, 1994.

39. PH Yancey, GN Somero. Counteraction of urea destabilization of protein structure by methylamine osmoregulatory compounds of elasmobranch fishes. Biochem J 183:317–323, 1970.

40. YJ Owuso-Ansah, HO Hultin. Trimethylamine oxide prevents insolubilization of red hake muscle proteins. J Agric Food Chem 32:1032–1035, 1984.

41. G Rodger, R Hastings. Role of trimethylamine oxide in the freeze denaturation of fish muscle—is it simply a precursor of formaldehyde? J Food Sci 49:1640–1641, 1984.

42. M Suyuma, T Shimizu. Buffering capacity and taste of carnosine and its methylated compounds. Bull Jpn Soc Sci Fish 48:89–95, 1982.

43. B Brzezinski, G Zundel. Formation of disulphide bonds in the reaction of SH group-containing amino acids with trimethylamine N-oxide. FEBS Lett 333:331–333, 1993.

44. D L Sackett. Natural osmolyte trimethylamine N-oxide stimulates tubulin polymerization and reverses urea inhibition. Am J Physiol 42:R669–R676, 1997.

45. PM Yancey. Compatible and counteracting solutes. In: K Strange, ed. Cellular and Molecular Physiology of Cell Volume Regulation. Boca Raton, FL: CRC Press, 1994, pp 81–109.

46. PM Yancey, GN Somero. Methylamine osmoregulating solutes of elasmobranch fishes counteract urea inhibition of enzymes. J Exp Zool 212:205–213, 1980.

47. TY Lin, SN Timasheff. Why do some organisms use a urea-methylamine mixture as osmolyte? Thermodynamic compensation of urea and trimethylamine N-oxide interactions with protein. Biochemistry 33:12695–12701, 1994.

48. Guidelines '94 to the German Food Book. Cologne, Germany: Bundesanzeiger Verlagsgesell, 1994, 168.

49. JA Raymond. Trimethylamine oxide and urea synthesis in rainbow smelt and some other northern fishes. Physiol Zool 71:515–523, 1998.

50. JR Baker, A Struempler, S. Chaykin. A comparative study of trimethylamine-N-oxide biosynthesis. Biochim Biophys Acta 71:58–64, 1963.

51. AR Strom. Biosynthesis of trimethylamine oxide in Calanus finmarchicus. Properties of a soluble trimethylamine monooxygenase. Comp Biochem Physiol 65B:243–249, 1980.

51. DJ Steenkamp, K Beinert. Mechanistic studies on the dehydrogenases of methy-

lotrophic bacteria. 1. Influence of substrate binding to reduced trimethylamine dehydrogenase on the intramolecular electron transfer between its prosthetic groups. Biochem J 207:233–239, 1982.

52. EL Barrett, HS Kwan. Bacterial reduction of trimethylamine oxide. Annu Rev Microbiol 39:131–149, 1985.

53. J Oehlenschläger. Evaluation of some well established and some underrated indices for the determination of freshness and/or spoilage of ice-stored wet fish. In: HH Huss, M Jakobsen, J Liston, eds. Quality Assurance in the Fish Industry. Amsterdam: Elsevier, 1992, pp 339–350.

54. JM Regenstein, MA Schlosser, A Samson, M Fey. Chemical changes of trimethylamine oxide during fresh and frozen storage of fish. In: RE Martin, GJ Flick, CE Hebard, DR Ward, eds. Chemistry and Biochemistry of Marine Food Products. Westport, CT: AVI Publ Comp, 137–148, 1982.

55. GJ Clark, FB Ward. Purification and properties of trimethylamine N-oxide reductase from Shewanella sp. NCMB 400. J Gen Microbiol 133:379–386, 1988.

56. HH Huss. Fresh fish—quality and quality changes. FAO Fish Series No 29. Rome: FAO, 1988.

57. Z Sikorski, S Kostuch. Trimethylamine N-oxide demethylase: its occurrence, properties, and role in technological changes in frozen fish. Food Chem 9:213–222, 1982.

58. K Yamada, K Amano. Formaldehyde and dimethylamine in fish and shellfish. Bull Tokai Reg Fish Res Lab 41:89–96, 1965.

59. H Rehbein. Trimethylamine oxide (TMAO) content and TMAO-ase activity in tissues of fish species from the North Atlantic and from Antarctic waters. In: E Tyihak, G Gullner, eds. The Role of Formaldehyde in Biological Systems Budapest: SOTE Press, 1987, pp 237–242.

60. H Rehbein, W Schreiber. TMAOase activity in tissues of fish species from the Northeast Atlantic. Comp Biochem Physiol 79B:447–452, 1984.

61. BS Lall, AR Manzer, DF Hiltz. Preheat treatment for improvement of frozen storage stability at -10°C in fillets and minced flesh of silver hake. J Fish Res Bd Can 32:1450–1454, 1975.

62. KL Parkin, HO Hultin. Partial purification of trimethylamine-N-oxide (TMAO) demethylase from crude fish muscle microsomes. J Biochem 100:87–97, 1986.

63. KL Parkin, HO Hultin. Spectrophotometric evidence for a hemoprotein in fish muscle microsomes: possible involvement in trimethylamine N-oxide (TMAO) demethylase activity. J Agric Food Chem 35:34–41, 1987.

64. T Tokunaga. Trimethylamine oxide and its decomposition in the bloody muscle of fish. II. Production of DMA and TMA during storage. Bull Jpn Soc Sci Fish 36:502–509, 1970.

65. TA Gill, T Paulson. Localization, characterization and partial purification of TMAO-ase. Comp Biochem Physiol 71B:49–56, 1982.

66. A Joly, P Cotin, L Han-Ching, A Ducastaing. Trimethylamine N-oxide demethylase (TMAO-ase) of saithe (*Pollachius virens*) kidney: a study of some physicochemical and enzymatic properties. J Sci Food Agric 59:261–267, 1992.

67. H Rehbein, W Havemeister, MK Nielsen, B Jogersen, F Jessen, CG Sotelo. Purification and characterization of TMAO-ase of saithe (*Pollachius virens*) and hake

(Merluccius merluccius). Annual Report for European Community Project AIR 3-CT94-1921, 1997.

68. KL Parkin, HO Hultin. Fish muscle microsomes catalyze the conversion of trimethylamine oxide to dimethylamine and formaldehyde. FEBS Letts 139: 61–64, 1982.

69. K Tomioka, J Ogushi, K Endo. Studies on dimethylamine in foods. II. Enzymatic formation of dimethylamine from trimethylamine oxide. Bull Jpn Soc Sci Fish 40:1021–1026, 1974.

70. BQ Phillippy, HO Hultin. Distribution and some characteristics of trimethylamine N-Oxide (TMAO) demethylase activity of red hake muscle. J Food Biochem 17:235–250, 1993.

71. T Nash. The colorimetric estimation of formaldehyde by means of the Hantzsch reaction. Biochem J 55:416–421, 1953.

72. H Rehbein. Different methods for measurement of TMAOase activity in extracts of fish tissue. In: JB Luten, T Børrensen, J Oehlenschläger, eds. Seafood from Producer to Consumer. Integrated Approach to Quality. Amsterdam: Elsevier, 1997, pp 521–527.

73. WJ Dyer, YA Mounsey. Amines in fish muscle. II. Development of TMA and other amines. J Fish Res Bd Can 6:359–367, 1945.

74. TC Thomas, MG McNamee. Purification of membrane proteins. Methods Enzymol 182:499–520, 1990.

75. K Yamada, K Harada, K Amano. Biological formation of formaldehyde and dimethylamine in fish and shellfish. VIII. Requirement of cofactor in the enzyme system. Bull Jpn Soc Sci Fish. 35:227–231, 1969.

76. HO Hultin. Trimethylamine-N-oxide (TMAO) demethylation and protein denaturation in fish muscle. In: GJ Flick, RE Martin, eds. Advances in Seafood Biochemistry Composition and Quality. Pennsylvania: Technomic Publishing Co., 1992, pp 25–42.

77. P Reece. The fate of reduced nicotinamide adenine dinucleotide in minced flesh of cod (Gadus morhua) and its association with formaldehyde production during frozen storage. Inst Int Ref, Aberdeen:319–323, 1985.

78. MCM Banda, HO Hultin. Role of cofactors in breakdown of TMAO in frozen red hake muscle. J Food Proc Preserv 7:221–236, 1983.

79. RC Lundstrom, FF Correia, KA Wilhelm. Dimethylamine production in fresh red hake (Urophycis chuss): the effect of packaging material oxygen permeability and cellular damage. J Food Biochem 6:229–241, 1983.

80. Racicot LD, Lundstrom RC, Wilhelm KA, Ravesi EM, Licciardello JJ. Effect of oxidizing and reducing agents on trimethylamine oxide demethylase activity in red hake muscle. J Agric Food Chem. 32:459–464, 1984.

81. JK Babbitt, DL Crawford, DK Law. Decomposition of trimethylamine oxide and changes in protein extractability during frozen storage of minced and intact hake (Merluccius productus) muscle. J Agric Food Chem 20:1052–1054, 1972.

82. C H Castell, B Smith, J Dale. Comparison of changes in trimethylamine, dimethylamine and extractable protein in iced and frozen gadoid fillets. J Fish Res Board Can 30:1246–1248, 1973.

83. DF Hiltz, LB Smith, DW Lemon, WJ Dyer. Deteriorative changes during frozen storage in fillets and minced flesh of silver hake (Merluccius bilinearis) processed

from round fish held in ice and refrigerated sea water. J Fish Res Board Can 33:2560–2567, 1976.

84. CG Sotelo, JM Gallardo, C Piñeiro, RI Pérez-Martín. Trimethylamine and derived compounds changes during frozen storage of hake (*Merluccius merluccius*). Food Chem 50:267–275, 1995.

85. T Tokunaga. Studies in the development of dimethylamine and formaldehyde in Alaska Pollack muscle during frozen storage. Bull Hokk Reg Fish Res Lab 30:90–97, 1964.

86. K Harada, K Yamada. Some properties of a formaldehyde and dimethylamine forming enzyme obtained from *Barbatia virescens*. J Shimonoseki Univ Fish 19:95–103, 1971.

87. JR Dingle, RA Keith, B Lall. Protein instability in frozen storage induced in minced muscle of flatfishes by mixture of muscle of red hake. Can Inst Food Sci Technol J 10:143–146, 1977.

88. DJB DaPonte, DP Rozen, W Pilnik. Effects of additions on the stability of frozen stored minced fillets of whiting. I. Various anionic hydrocolloids. J Food Qual 8:51–68, 1985.

89. YJ Owusu-Ansah, HO Hultin. Effect of in situ formaldehyde production on solubility and cross-linking of proteins of minced red hake muscle during frozen storage. Food Biochem 11:17–39, 1987.

90. DJ Krueger, OR Fennema. Effect of chemical additives on toughening of fillets of frozen Alaska pollack (*Theragra chalcogramma*). J Food Sci 54:1101–1106, 1989.

91. EL LeBLanc, RJ LeBlanc, IE Blum. Prediction of quality in frozen cod (*Gadus morhua*) fillets. J Food Sci 53(2):328–340, 1988.

92. M Rey-Mansilla, M Pérez-Testa, S Aubourg, CG Sotelo. Formaldehyde production and localisation during the frozen storage of European hake (Merluccius merluccius L.). WEFTA meeting, Madrid, 1997.

93. S Kostuch, ZE Sikorski. Interaction of formaldehyde with cod proteins during frozen storage. IIF/IIR Commissions C1 and C2, Karlsruhe, 1977.

94. VD Tran. Solubilization of cod myofibrillar proteins at low ionic strength by sodium pyruvate during frozen storage. J Fish Res Bd Can 32:1629–1632, 1975.

95. S Konosu, K Yamagushi. The flavor components in fish and shellfish. In: RE Martin, GJ Flick, CE Hebard, DR Ward, eds. Chemistry and Biochemistry of Marine Food Products. Westport, CT: AVI Publ Comp, 1982, pp 367–404

96. T Tokunaga. Contents of trimethylamine oxide, trimethylamine and dimethylamine in different tissues of live Alaska Pollack. Bull Tokai Reg Fish Res Lab No 93:79–85, 1978.

97. HA Bremmer. Storage trials on the mechanically separated flesh of three Australian midwater fish species. 1. Analytical test. Food Technol Aust 29:89–93, 1977.

98. DW Stanley, HO Hultin. Amine and formaldehyde production in North American squid and their relation to quality. Can Inst Food Sci Technol J17:157–162, 1984.

99. T Tokunaga. Studies on the development of dimethylamine and formaldehyde in Alaska pollock muscle during frozen storage. Bull Hokkaido Reg Fish Res Lab 29:108–122, 1964.

8

Digestive Proteinases from Marine Animals

Benjamin K. Simpson
McGill University, Ste. Anne de Bellevue,
Quebec, Canada

I. INTRODUCTION

This chapter is on the major proteinases (also known as proteolytic enzymes or proteases) that are produced by the digestive glands of marine animals. Like the proteinases from plants, animals, and microorganisms, digestive proteinases from marine animals are hydrolytic in their action, and catalyze the cleavage of peptide bonds with the participation of water molecules as reactants. In terms of current food industry (and other industrial) applications, proteinases are by far the most important and most widely used group of enzymes (1, 2). They are used to improve product handling characteristics and texture of cereals and baked goods, enhance the drying as well as the quality of egg products, tenderize meat, recover proteins/peptides from bones, and hydrolyze blood proteins. Proteinases are used for the production of protein hydrolysates, reduction of stickwater viscosity, and for roe processing. They are used to make pulses and rennet puddings, and in cheesemaking/cheese ripening. Proteinases are also used for biomedical applications to reduce tissue inflammation, dissolve blood clots, promote wound healing, activate hormones, diagnose candidiasis, and to aid or facilitate digestion (3, 4).

There is great demand for enzymes with the right combination of properties for a plethora of industrial applications. Industrial proteinases are mostly derived from microorganisms, and to a lesser extent from plant and animal sources (3). So far, there is only very limited use of marine proteinases by industry. The reasons for the rather limited use of marine digestive proteinases include the relative paucity of basic information on these enzymes, the cyclical nature of the source material (which precludes supply in a steady manner), and the stereotypical attitude of the general public toward the source material: fish offal. However, marine animals comprise several thousands of very diverse species that subsist under different habitat conditions (5, 6). Some of these differences are in terms of parameters such as temperature, pressure, salinity, light intensity, and aeration. Over the years, marine animals have adapted to different environmental conditions, and these adaptations, together with inter- and intraspecies genetic variations, have resulted in digestive proteinases with certain unique properties compared with their counterpart enzymes from land animals, plants, or microorganisms (6–9). Some of the distinctive features of marine digestive proteinases include a higher catalytic efficiency at low reaction temperatures, lower thermostability, cold stability, and substantial catalytic activity/stability at neutral to alkaline pH (6, 9, 10). Homologous digestive proteinases from marine animals may also differ from one another in their response to specific inhibitors, for example, α-macroglobulin from beef serum inhibited sheephead and bluefish trypsins to different extents (11, 12). Proteinases from the same species may also display season-dependent differences in properties. For example, a thyroid proteinase from winter turbot has different iodoacetate sensitivity than the enzyme from spring turbot. Digestive proteinases from marine animals may also differ in their sensitivity to pressure (12, 13), in their ability to hydrolyze native protein substrates (14–16), or in their sensitivities to pH, acids, alkali, salts, urea, detergents, and other materials.

Worldwide sales of industrial enzymes were estimated at about $1.5 billion for 1998 (17). It is suggested that some of the unique properties of marine enzymes may be exploited in various food applications, and thereby, obtain a share of the lucrative industrial enzymes market to increase profits for the fishing industry. For example, the higher catalytic activity at low reaction temperatures may be used to process foods at low temperature to reduce energy costs and destruction of heat-labile essential food components (9). The lower thermostability of marine digestive proteinases (compared with their homologues from other animals, plants, and microorganisms), would permit their ready inactivation by milder heat treatments, while their ability to denature native protein substrates may be advantageous in fruit juice manufacture, for the inactivation of undesirable endogenous enzymes such as polyphenol oxidases (PPO) or pectin-methyl esterase (PME) (14, 16). Some marine enzymes are currently being used as food process aids (18–21), and this use of marine enzymes is expected to increase due to the possibilities afforded by recombinant DNA technology.

II. CLASSIFICATION OF DIGESTIVE PROTEINASES FROM MARINE ANIMALS

Digestive proteinases from marine animals may be classified by the same criteria used for proteinases from other animals, plants, or microorganisms: on the basis of their similarity to well-characterized proteinases as trypsin-like, chymotrypsin-like, chymosin-like, or cathepsin-like. They may be classified on the basis of their pH sensitivities as acid, neutral, or alkaline proteinases. They may also be described based on their substrate specificities, response to inhibitors, or by their mode of catalysis. The standard method of classification proposed by the Enzyme Commission (EC) of the International Union of Biochemists (IUB), is based on the mode of catalysis. By this approach, digestive proteinases from marine animals are classified into four categories as acid or aspartate proteinases, serine proteinases, thiol or cysteine proteinases, or metalloproteinases. The enzymes in the different classes are differentiated by various criteria, such as the nature of the groups in their catalytic sites, their substrate specificity, their response to inhibitors, or by their activity/stability under acid or alkaline conditions. The EC system for classifying enzymes assigns enzymes four-digit numbers prefaced with the Enzyme Commission's abbreviation "EC." The first number is a group number assigned on the basis of the type of chemical reaction catalyzed. Thus, for the marine digestive proteinases group of enzymes, this first number is "3," signifying that they are all hydrolases (i.e., they all catalyze hydrolytic reactions that require the participation of water as reactant). The second digit in the EC numbering system subdivides the manifold enzymes in the group into subclasses. For the hydrolases, the subdivision into subclasses is made based on the type of bond hydrolyzed. Since the marine digestive proteinases hydrolyze peptide bonds, they all have a common second digit of "4." The third digit in the EC system further subdivides the enzymes in a particular group into susubclasses that narrowly define the types of bonds hydrolyzed, or identify the chemical nature of the essential groups in the catalytic sites of the enzymes. For the hydrolase group of enzymes, at least four subsubclasses have been classified for proteinases based on the EC system (2). Other enzymes that share common first two digits with the proteinases are the peptidases. The peptidases are excluded for the purpose of this chapter. The fourth digit in the EC system is simply a serial number assigned to the enzymes in their subsubclasses. Examples of major proteinases that have been characterized from the digestive glands of marine animals are described in the sections that follow.

A. Acid/Aspartyl Proteinases

The acid or aspartyl proteinases are a group of endoproteinases characterized by high activity and stability at acid pH. This is the basis of their group name "acid"

proteinases. They are referred to as "aspartyl" proteinases (or carboxyl proteinases) because their catalytic sites are composed of the carboxyl groups of
two aspartic acid residues (2, 22). Based on the EC system, all the acid/aspartyl
proteinases from marine animals have the first three digits in common: EC
3.4.23. The three common aspartyl proteinases that have been isolated and characterized from the stomachs of marine animals are pepsin, chymosin, and
gastricsin.

Pepsin is assigned the number EC. 3.4.23.1. It has preferential specificity
for the aromatic amino acids, phenylalanine, tyrosine, and tryptophan. In the EC
system of classification, chymosin (formerly known as rennin) is assigned the
number EC 3.4.23.4. Chymosin has specificity for the aromatic amino acids,
phenylalanine, tryrosine, and tryptophan, similar to pepsin. Gastricsin is assigned a code of EC 3.4.23.3.

B.   Serine Proteinases

The serine proteinases have been described as a group of endoproteinases with a
serine residue in their catalytic site. This family of proteinases is characterized by
the presence of a serine residue, together with an imidazole group and an aspartyl
carboxyl group in their catalytic sites. As a group, they are inhibited by diisopropylphosphofluoridate (DFP), through reaction with the hydroxyl group of the
active site serine residue (22). The proteinases in serine subsubclass all have the
same first three digits: EC 3.4.21. The three major serine proteinases that have
been purified and well characterized from the digestive glands of marine animals
are trypsin, chymotrypsin, and elastase. Trypsin is assigned the code EC 3.4.21.4.
Trypsin has a very narrow specificity for the peptide bonds on the carboxyl side of
arginine and lysine. Chymotrypsin is assigned a code of EC 3.4.21.1., and it has a
much broader specificity than trypsin. It cleaves peptide bonds involving amino
acids with bulky side chains and nonpolar amino acids such as tyrosine, phenylalanine, tryptophan, and leucine. Elastase is designated as EC 3.4.21.11. Elastase
exhibits preferential specificity for alanine, valine, and glycine.

C.   Thiol/Cysteine Proteinases

The thiol or cysteine proteinases are a group of endoproteinases that have cysteine and histidine residues as the essential groups in their catalytic sites. These
enzymes require the thiol (-SH) group furnished by the active site cysteine
residue to be intact, hence the group name "thiol" or "cysteine" proteinases. The
thiol proteinases are inhibited by heavy metal ions and their derivatives, as well as
by alkylating agents and oxidizing agents (22). The first three digits common to
thiol proteinases are EC 3.4.22. An example of a thiol proteinase from the digestive glands of marine animals is cathepsin B, which is designated as EC 3.4.22.1.

D. Metalloproteinases

The metalloproteinases are hydrolytic enzymes whose activity depends on the presence of bound divalent cations. Chemical modification studies suggest that there may be at least one tyrosyl residue and one imidazole residue associated with the catalytic sites of metalloproteinases (2). The metalloproteinases are inhibited by chelating agents such as 1,10-phenanthroline, EDTA, and sometimes by the simple process of dialysis. Most of the metalloproteinases known are exopeptidases. They all have a common first three digit's EC 3.4.24. The metalloproteinases that have been characterized from marine animals (e.g., rockfish, carp, and squid mantle), have not been found in the digestive glands but in the muscle tissues.

III. PREPARATION AND GENERAL PROPERTIES OF DIGESTIVE PROTEINASES FROM MARINE ANIMALS

Digestive proteinases have been isolated from several marine animals using basically the same principles applied to recover similar enzymes from other animal tissues, plant tissues, or from microorganisms. Thus, techniques of purification based on solubility differences (e.g., precipitation with neutral salts or organic solvents); size differences (e.g., size exclusion chromatography or dialysis); charge differences (e.g., ion exchange chromatography, or electrophoresis); and binding to specific ligands (e.g., affinity chromatography, or hydrophobic interaction chromatography) have all been used to purify these enzymes from the digestive glands of various marine animals.

A. Acid/Aspartyl Proteinases

The acid or aspartyl digestive proteinases are found in the stomachs of animals. They are active at acid pH and are unstable/inactive under alkaline conditions. Various researchers have used different approaches in purifying these enzymes from a variety of marine animals. Some of the procedures that have been used to successfully purify some of these enzymes to homogeneity, as well as the common characteristics of these enzymes are briefly described in this sub-section.

1. Pepsin/Pepsinlike Proteinases

Pepsins and pepsinlike enzymes have been prepared from the digestive glands of marine animals such as the Atlantic cod (*Gadus morhua*) (23, 24), capelin (*Mallotus villosus*) (25), Greenland cod (*Gadus ogac*) (26), Polar cod (*Boreogadus saida*) (27), and sardine (*Sardinos melanostica*) (28). Pepsins have also been isolated from the digestive glands of the American smelt (7) and dogfish

(*Squalus acanthias*) (29). Several methods have been described in the literature for preparing pepsins from marine animals. Arunchalam and Haard (27) purified two pepsinogen isozymes from the stomachs of Polar cod by homogenizing the tissue in phosphate buffer (pH 7.3), followed by filtration through a cheesecloth. The filtrate was further clarified by centrifugation and then subjected to affinity chromatography using a CBZ-D-phenylalanine–TETA-Sepharose 4B gel. Two pepsin isozymes were also prepared from the stomachs of capelin by homogenizing the tissues in cold HCl (1 mM) solution, followed by ammonium sulfate fractionation, IEX chromatography on DEAE-cellulose, and size exclusion chromatography on Sephadex G-75 (25). Pepsin was also isolated in the zymogen form from the stomach tissue of rainbow trout by homogenizing the tissue in phosphate buffer (pH 7.1), followed by centrifugation to clarify the homogenate, ammonium sulfate fractionation of the clear extract, and then a series of chromatographic steps involving polylysine-Sepharose, DEAE-Sephacel, Sephadex G-200, and DEAE-Sepharose gels. The pepsins from marine animals were reported to have molecular weights ranging from 27 to 42 kDa (25–28, 30). They had relatively higher pH optima for the hydrolysis of their substrates than mammalian pepsins, and were generally more stable at relatively higher pH values (pH 2.0–4.0) (25–28, 30). While pepsins are unstable under alkaline conditions, their zymogens (pepsinogens) may be stable under these conditions. Pepsins from the digestive glands of marine animals are quite resisant to autolysis at low pH (18). However, marine pepsins were inhibited by pepstatin in a similar fashion to mammalian pepsin (28, 31–32). Pepsins from marine animals display a wide temperature optima range for the hydrolysis of their substrates (37–55°C). In general, the colder-water fish pepsins had lower temperature optima and were more heat-labile than pepsins from animals from the warmer environment (23, 25–27).

2.   Chymosin/Chymosinlike Proteinases

Chymosins have been described as acid proteinases with some characteristics distinct from other acid proteinases. For example, these enzyme are most active and stable around pH 7.0, unlike other acid proteinases. They also have relatively narrower substrate specificity compared with acid proteinases such as pepsin (2). Chymosin is generally found to occur in milk-fed ruminants while pepsins occur in adult ruminants and in animals with a simple stomach. Digestive proteinases with chymosinlike activity were isolated as zymogens from the gastric mucosa of young and adults harp seals (*Pagophilus groenlandicus*) (33–36), by homogenizing freeze-dried powders of the mucosa in phosphate buffer (pH 7.2), followed by centrifugation to separate the clear supernatant. This crude extract was subjected to IEX chromatography on DEAE-Sephadex A-50, size exclusion chromatography on a Sephadex G-100 gel, and then affinity chromatography on a Z-D-Phe-T-Sepharose gel. The distinctive properties of harp seal chymosins compared with pepsins include a relatively high milk clotting to proteolytic activity ratio, a pH optimum of 2.2–3.5 for hemoglobin hydrolysis compared with pH 2.0 for mammalian pepsins, inability to inactivate ribonuclease, low activity on N-acetyl-L-phenylalanyl-3, 5-diiodo-L-tyrosine, instability in 6M urea, and zymogen molecular weight of 33,800 Da. The ratio of chymosinlike isoenzyme to pepsin isoenzyme was also found to be greater in 1-week-old animals than in adult animals. The chymosins from marine animals did not hydrolyze the specific synthetic substrate for pepsin (i.e., *N*-acetyl-L-phenylalanine diiodotyrosine [APD]), appreciably, and were also more susceptible to inactivation by urea.

3. Gastricsin/Gastricsinlike Proteinases

Gastricsins have been described as acid proteinases with enzymatic and chemical resemblance to pepsins (37). However, they differ from pepsins in structure and certain catalytic properties (38). Gastricsins have been prepared and characterized from the gastric juices of salmon (39) and hake (*Merluccius gayi*) (40). Two gastricsin isozymes were isolated in the precursor form from the gastric mucosa of the marine fish, hake, prior to conversion to the active form. The purification process was comprised of homogenization of the tissues in bicarbonate buffer (pH 7.3) followed by centrifugation to clarify the extract. Next, the extract was fractionated with ammonium sulfate, then subjected to two IEX chromatography steps using DEAE–cellulose and DEAE-Sephadex A-50. The zymogens obtained from the IEX chromatography steps were activated to gastricsins by incubation with 0.1 M HCl, and applied to Amberlite IRC-50 columns for further purification. The optimum pH for the hydrolysis of hemoglobin by hake gastricsins was 3.0, which is similar to that of mammalian gastricsins. Hake gastricsins were markedly stable over a broad pH range (up to pH 10.0), unlike mammalian gastricsins that were only stable up to pH 6.7, and were rapidly inactivated at pH ≥ 7.5 (40). The molecular weights of hake gastricsins determined by SDS–PAGE were 32.3 and 33.9 kDa (40).

B. Serine Proteinases

The digestive serine proteinases are found in the pancreatic tissues, pyloric ceca, and intestines of animals. Marine digestive serine proteinases are active at neutral to slightly alkaline pH and are unstable / inactive under acid conditions. Marine serine proteinases have been isolated and characterized from the digestive glands of various marine animals by different methods. Some of the methods successfully used to recover four of the most common marine serine digestive proteinases (i.e., trypsin, chymotrypsin, elastase and crustacea collagenases), as well as the general features of these enzymes are briefly described in this subsection.

1. Trypsins

Trypsins have been isolated from the pyloric ceca, pancreatic tissue, or intestines of several marine animals: anchovy (*Engraulis encrasicholus*) (41), Greenland cod (*Gadus ogac*) (42, 43), Atlantic cod (*Gadus morhua*) (10, 44, 47), capelin (*Mallotus villosus*) (48), mullet (*Mugil cephalus*) (16), sardine (*Sardinos melanostica*) (28), catfish (*Parasilurus asotus*) (49), starfish (50), crayfish (15), and cunner (*Tautogolabrus adspersus*) (14, 44, 51). Several methods have been used to isolate trypsins from marine animals. They invariably involve some fractionation steps with ammonium sulfate or cold acetone, followed by chromatographic methods. For example, cunner trypsins, Greenland cod trypsin and mullet trypsin, were purified to homogeneity by powdering the tissue in liquid nitrogen, making a homogenate of the powdered tissue in Tris–HCl buffer, ammonium sulfate fractionation, acetone precipitation, and affinity chromatography (16, 42, 52). Trypsins were purified from the pyloric ceca of the starfishes *Evasterias trochelii* and *Dermastarias imbricata* by homogenizing the tissue in phosphate buffer (pH 6.5) or Tris-buffer (pH 7.8), followed by fractionating with solid ammonium sulfate, and then chromatography on DEAE-cellulose (53, 54)

Trypsins from marine animals resemble mammalian trypsins with respect to their molecular size (22.5–24 kDa), amino acid composition, and sensitivity to inhibitors such as aprotinin (or trasylol), soybean trypsin inhibitor (SBTI), phenylmethylsulfonyl fluoride (PMSF), diisopropylphosphofluoridate (DFP), benzamidine, and N-tosyl-L-lysine chloromethyl ketone (TLCK). Their pH optima for the hydrolysis of various substrates have been reported to range from 7.5 to 10.0, while their temperature optima for hydrolysis of those substrates ranged from 35 to 45°C (30). Trypsins from marine animals tend to be more stable at alkaline pH, but are unstable at acid pH, unlike mammalian trypsins that are most stable at acid pH.

2. Chymotrypsin

Chymotrypsins have been isolated and characterized from marine species such as the anchovy (*Engraulis japonica*) (55), Atlantic cod (*Gadus morhua*) (55, 57), capelin (*Mallotus villosus*) (58), herring (*Clupeas harengus*) (58), rainbow trout (*Oncorhynchus mykiss*) (59), and spiny dogfish (*Squalus acanthias*) (60, 61). Various methods were used to prepare chymotrypsins from these animals. For example, two isozymes with chymotrypsinlike activities were isolated from the pyloric ceca of Atlantic cod by homogenizing the tissue in cold Tris-HCl buffer (pH 7.5), followed by centrifugation to exclude undissolved tissue, affinity chromatography on *p*-aminobenzamidine Sepharose 4B to remove trypsins, and hydrophobic interaction chromatography (HIC) step with phenyl-Sepharose to obtain the chymotrypsins (57). Two chymotrypsinlike proteinases were isolated from the pyloric ceca of rainbow trout by homogenizing the tissue in Tris-HCl buffer (pH 8.1) followed by ammonium sulfate precipitation, hydrophobic interaction chromatography on phenyl-Sepharose, and then ion-exchange (IEX) chromatography on DEAE-Sepharose. In general, these enzymes are single-polypeptide molecules with molecular weights ranging between 25 and 28 kDa. They are most active within the pH range of 7.5 to 8.5, and are most stable at around pH 9.0 (59). Chymotrypsins from marine animals have a higher catalytic activity and hydrolyzed more peptide bonds in various protein substrates (casein, collagen and bovine serum albumin) at sub-denaturation temperatures than mammalian chymotrypsins. In general, the marine chymotrypsins were more heat-labile than mammalian chymotrypsins (57). Dogfish chymotrypsin was more active toward soy protein isolate from 5 to 35°C than bovine chymotrypsin (61), and could also clot milk. Chymotrypsins from herring and capelin exhibited greater hydrolytic activities toward the synthetic substrate BTEE than the bovine enzyme (58). These findings were similar to those made with Greenland cod trpsin compared with bovine trypsin (9).

3. Elastase

Elastase is a serine proteinase produced by the pancreas to carry out the hydrolysis of elastin under the alkaline conditions that prevail in the intestines. The enzyme has been characterized from Atlantic cod (*Gadus morhua*) (62, 63), bluefin tuna (*Thunnus thynnus*) (64), carp (*Cyprinus carpio*) (65), catfish (*Paracilurus asotus*) (66), Dover sole (*Solea solea*) (67), eel (*Anguila japonica*) (64), monkfish (*Lophius piscatorius*) (68), sea bass (*Lateolabrus japonicus*) (64), stingray (*Dasyatis americana*) (69), teleost tuna (*Thunnus secundodorsalis*) (69), and yellowtail (*Seriola quinqueradiata*) (64). Elastase was prepared from the pancreatic tissue of Atlantic cod by homogenizing the tissue in 0.25 M NaCl, followed by centrifugation to obtain a clear supernatant. The clarified pancreatic extract was then fractionated with solid ammonium sulfate, subjected to IEX chromatography on S-Sepharose gel, and affinity chromatography with SBTI coupled to CNBr activated Sepharose 4B column (70). Cod elastase had optimal activity at 45°C with Suc-(Ala)$_3$-pNA as substrate. Elastases from marine animals are quite stable from pH 5 to 9, and are strongly inhibited by soybean trypsin inhibitor, but only partially inhibited by TPCK. The enzymes were inactivated by elastatinal and denaturing agents such as dimethyl sulfoxide (70, 64). However, the elastases from marine animals were not inhibited by either benzamidine or TLCK.

4. Collagenases

Collagenases have been purified and characterized from the digestive glands of marine animals such as crab, lobster, and prawn (71–76). Collagenase from the fiddler crab (*Uga pugilator*) was purified to homogeneity from the acetone powder by dissolving the powder in tris-HCl buffer (pH 7.5) followed by Gel filtration on Sephadex G-150 and IEX on DEAE cellulose (71). Collagenases from marine animals resemble mammalian collagenases in their action on native collagen. However, marine digestive collagenases differ from mammalian collagenases in exhibiting trypsinlike and chymotrypsinlike specificities. These proteinases are most active from pH 6.5 to 8.0, and are inactivated at pH values lower than 5.0. Fiddler crab collagenase had a molecular weight of 25,000 Da (71), and was inhibited by DFP, TLCK, and PMSF, similarly to other crustacean collagenases (71). Collagenases from hepatopancreas of marine invertebrates have been implicated in postharvest texture deterioration in these animals during storage (75).

C. Cysteine or Thiol Proteinases

Digestive cysteine or thiol proteinases have been found in the hepatopancreatic tissue of marine animals. More than 90% of the proteinase activity in the hepatopancreas of short-finned squid involves cysteine proteinases (77). Marine digestive cysteine or thiol proteinases are most active at acidic pH and inactive at alkaline pH. Various researchers have described different procedures for isolating marine cysteine or thiol proteinases from the digestive glands of marine animals. Some of these methods, as well as the general features of the enzymes, are briefly described in this subsection. A common example of digestive thiol proteinase from marine animals is Cathepsin B. However, not all cathepsins are thiol proteinases. A more extensive treatment of cathepsins is presented in Chapter 19 of this book.

1. Cathepsin B

Cathepsin B has been isolated from a few aquatic animals including the horse clam (*Tresus capax*) (78), mussel (*Perna perna L*) (79), and surf clam (*Spisula solidissima*) (80). Cathepsin B was prepared from the digestive glands of mussel by homogenization in phosphate buffer (pH 6.0) and centrifugation to recover the clear supernatant. The extract was fractionated with ammonium sulfate, and this fraction was further purified by size exclusion chromatography on Sephadex G-75, followed by IEX chromatography on CM-Sephadex C-50. The enzymes were shown to be single polypeptide chains with molecular sizes ranging from 13.6 to 25 kDa. Cathepsins from different species display maximum activity over a broad pH range, from 3.5 in some to 8.0 in others. They are activated by

Cl⁻ions, and require sulfhydryl-reducing agents or metal-chelating agents for activity. The enzymes are inhibited by iodoacetamide, N-ethylmaleimide, N-tosyl-L-phenyl chloromethyl ketone (TPCK), and by heavy metals. Leupeptin or pepstatin does not affect the enzymes. It has been suggested that cathepsins can seep out of the digestive glands of marine animals and cause texture softening (81, 82).

D. Metalloproteinases

Metalloproteinases have been characterized mostly from the muscle tissue of marine animals such as rockfish (83), carp (84), and squid mantle (85). Metalloproteinases do not seem to be common in marine animals.

E. Summary of Some Unique Properties of Digestive Proteinases from Marine Animals

Digestive proteinases from marine animals, particularly those acclimated to the Arctic and sub-Arctic regimes, possess certain unique features distinct from homologous proteinases from tropical waters and homeothermic animals. Some of these unique properties are a low Arrhenius activation energy, relatively lower free energy of activation (ΔG^*), a high apparent Michaelis–Menten constant, cold stability, a low temperature optimum, a low thermostability, and a high pH optimum/pH stability (6, 8, 9, 10, 16, 31). For example, the E_a for the hydrolysis of hemoglobin ranges from as low as 2.9 Kcal/mole for Polar cod pepsin B to 11.2 Kcal/mol for pepsin from warm-blooded animals (27). The structural and functional relationships of enzyme molecules, as judged by their ΔS^* values, suggest that proteinases from cold-adapted marine animals have more flexible structures than their counterparts from land animals (6, 9). The thesis that enzymes from marine animals acclimated to cold habitats have more flexible structures than their counterparts derived from homeothermic species was borne out by CD spectra studies with Greenland cod trypsin and bovine trypsin (9). The more flexible structures are associated with higher catalytic activity and low thermal stability (6, 9, 42). A more flexible enzyme structure is consistent with the observation that the thermal denaturation temperature of pepsins from cold water animals is about 20°C lower than mammalian pepsins (10, 25, 26, 31, 43, 86). Thus, cold-adapted marine pepsins may be more readily inactivated by milder heat treatment, an advantageous property that could be exploited in cheese manufacture, to inactivate the proteinase after the milk clotting reaction is complete (9, 10, 20, 24). Cod fish elastase had about four times more elastolytic activity than the porcine elastase (70). For example, the K_{cat} value and catalytic efficiency (K_{cat}/K_m') for the hydrolysis of Suc-(Ala)3-pNA by cod elastase were 38.5 s⁻¹ and 196 s⁻¹/mM, respectively, at 25°C, while the values for the

Table 1　Comparison of Some Kinetic, Thermodynamic, and Other Physical Properties of Marine Versus Mammalian Digestive Proteinases

Enzyme	Source	Biological temperature (°C)	Mol. wt. (kDa)	pH optimum	Temperature optimum (°C)	K_m' (mM)	V_{max}	Reference
Acid proteinases								
Pepsin	Atlantic cod	0–5	35.5	3.5	40	0.05		25
	Capelin	(-1)–4	23.0; 27.0	3.7; 2.5	38; 43			25,
	Greenland cod	(-2)–2	36.4	3.5	30	0.86		26
	Polar cod	(-1.5)–2	40.0; 42.0	2.0	37	0.06; 1.33	31; 556	27
	Sardine	10	33.0; 37.0	2.0; 4.0	40; 55			28,
	Porcine	37	32.0–35.0	2.0	47	0.04		27, 32, 34
Chymosin			33.8; 44.0	2.2–3.5				34
Gastricin(ogens)	Hake		32.3; 33.9	3.0				
Serine proteinases								
Trypsin	Atlantic cod	0–5	24.2; 24.8	7.5–8.0	40	1.42	0.25	10, 44
	Capelin	(-1)–4	28.0	7.5–8.0	45	0.73; 1.38		42, 44
	Cunner	>5.0	24.0	8.5	45	0.73	0.18	42, 44
	Greenland cod	(-2)–2	23.5	7.5	35	1.67	0.21	16,
	Mullet	15	24.0	8.0	55	0.49	400	28
	Sardine	10	22.9–28.9	10.0	45			
	Bovine	37	23.0	8.2	45	1.02	0.03	44, 86
Chymotrypsin	Atlantic cod	0–5	26.0	7.8	52	0.14; 0.20	207; 214	87
	Dogfish			7.0; 8.0		3.5; 0.27		59
	Rainbow trout	15	28.2–28.8	9.0	55	0.04; 0.02	2.24; 2.52	88
	Bovine	37		7.2	57	0.053	0.85	87, 88
Elastase	Atlantic cod	0–5	28.0	7.5–9.0	43	1.96	38.5	70,
	Porcine	37		7.5–9.0	58	2.42	11.0	70
Thiol proteinase								
Cathepsin B	Mussel		21.0	3.0–4.5				72

porcine elastase-Suc-(Ala)3-pNA hydrolase reaction were 11.0 s^{-1} and 4.5 s^{-1}/mM, respectively, at the same temperature (70). Marine digestive proteinases from the stomachless bonefish, cunner, mullet, and crayfish digested native globular proteins more efficiently than homologous proteinases from higher vertebrates (15, 16, 22, 44, 87). After 4 h incubation with equimolar concentrations of enzyme, approximately 12% of RNAse activity was lost with bovine trypsin whereas cunner trypsin and trypsinlike enzymes caused 58% and 72% inactivation, respectively (44, 51). This unique property of digestive proteinases from stomachless marine animals could be exploited to inactivate enzymes such as PPO and PME that cause undesirable changes in apple juice color and citrus juice cloud stability. Comparisons of some kinetic, thermodynamic, and physical properties of selected marine digestive proteinases are made in Table 1.

Some of the digestive proteinases present in various tissues have been shown to occur in multiple forms, known as isozymes or izoenzymes. Examples of isozymes of digestive proteinases that have been characterized include the two gastricinogens from *Merluccius gayi* (40), two pepsinogens from the Polar cod (27), two chymotrypsins from Atlantic cod with isoelectric points 6.2 and 5.8 (55, 57) and molecular weight of 26,000 Da, and four chymosin isozymes from the gastric mucosa of the harp seal. The properties of the isozyme forms appear to be related to physiological function of the tissue. Furthermore, the amount as well as the level of activity of digestive proteinases appears to be influenced by factors such as season, and the diet. The reader is referred to Chapter 1, Secs. I.C. and I.J. for a more extensive description of the influence of diet and nutrition on marine digestive proteinases.

IV. SOME APPLICATIONS OF DIGESTIVE PROTEINASES FROM MARINE ANIMALS

The applications of (marine) enzymes as industrial processing aids are treated in some depth in Chapters 21 and 22. Thus, this section will briefly discuss some of the specific uses of marine digestive proteinases aimed at taking advantage of some of their unique properties.

The high catalytic activity of Greenland cod trypsin was exploited in the low-temperature curing of herring (matjes) (19, 52) and the fermentation of squid (91). The supplementation of herring ripening with enzyme extracts from herring viscera (92, 93), or trypsin from Greenland cod pyloric ceca (52) or bovine pancreas (94), or the addition of fish trypsin, bovine cathepsin C or a squid hepatopancreas extract rich in cathepsin C to fermenting squid, improves the formation of taste-active amino acids (glutamate, alanine, leucine, serine, lysine, arginine, and proline) and sweet delicious taste (91). Supplementing the low-temperature (10°C) fermentation of matjes with equivalent amounts of pure bovine and Greenland cod trypsins resulted in increased protein solubilization and formation of free amino acids (52). The initial rate of protein solubilization with cod trypsin was almost double that observed with bovine trypsin. Similar results were observed when squid was fermented at 4°C; added cod trypsin increased the formation of free amino acids and sensory scores of the finished product.

Polar cod pepsin has a low temperature coefficient (Q_{10} = 1.2) for the milk clotting reaction compared to pepsins from mammals (Q_{10} = 2.0–2.2). Because of the high molecular activity of cold-adapted pepsins at low reaction temperatures, they are very effective for cold renneting milk (24, 95). There are, however, problems with use of cold-adapted pepsins for cheese making, namely, somewhat lower yields of curd and development of bitter off flavors during aging of Cheddar cheese (24). Atlantic cod pepsin is now produced commercially in Norway (18, 20, 96) and the product appears to be useful for cheese production and as a digestive aid in fish feed.

Cheddar cheese was prepared with seal chymosin (34–36), and the product gave significantly higher sensory scores than cheese made with calf rennet. In particular, Cheddar cheese made with the crude extract from harp seal underwent accelerated aging (34); this was not found with the purified chymosinlike enzyme (36). Moreover, the seal enzyme was able to clot milk at neutral pH because, like chymosin, it is more stable than pepsin at neutral to alkaline pH. The seal enzyme, like calf chymosin and unlike pepsin, has a relatively narrow specificity for hydrolysis of casein. Treatment of milk with bovine trypsin can prevent formation of oxidized flavor (97), however, the trypsin must be inactivated after an appropriate reaction time to avoid excessive proteolysis of milk proteins to affect adversely the taste and texture of the product. Bovine and cod trypsins were similarly effective in preventing oxidation of milk lipids at concentrations greater than $1.3 \times 10^{-3}\%$. However, while the cod trypsin was completely inactivated by the pasteurization process, bovine trypsin retained most of its original activity after the same pasteurization treatment. This demonstrates a process that can take advantage of the more heat-labile nature of Greenland cod trypsin (52).

The ability of trypsins from stomachless marine animals to hydrolyze native proteins was exploited in the inactivation of pectin methyl esterase (PME) and polyphenoloxidases in citrus and apple juice, respectively (98, 16). The heat resistance of citrus fruit PME is such that pasteurization causes deterioration of the product flavor. It seems that the use of industrial enzyme supplementation in production of such products is not widely practiced, perhaps because appropriate enzyme preparations are not available.

Protein molecules carry out diverse functions in biological systems, as biocatalysts, structural elements of cells, mechanical devices, metabolism regulators, transporters, metal chelators, antibodies, protectors, and during storage (99). In food systems, proteins perform critical functions such as thickening, gelation, emulsification, foaming, water uptake and water binding, adhesion, cohesion, and texturization (99). However, food proteins in their native state do not necessarily possess all the optimal functional and nutritional qualities (100) and often require modifications through physical, chemical, or enzymatic treatments to achieve the desired effects (101). One enzymatic approach used to modify the attributes of food proteins is the plastein reaction: resynthesis of high-molecular-weight proteinlike substances from an ill-defined high-concentrate protein hydrolysate. Although the plastein reaction has been extensively studied, its exact mechanism still remains a subject of great controversy. Thus far, plastein formation has been accomplished with proteinases such as trypsin, chymotrypsin, papain, and pepsin, and at temperatures between 37 and 50°C, where hydrophobic interactions are favored. It is to be expected that the products formed from the reaction between 37 and 50°C would be predominantly hydrophobic, and thus have limited solubility in aqueous systems. Hydrophobic plasteins are of limited use in food products. Thus, polar amino acids may be used to make hydrophilic plasteins via an initial chemically modification prior to their incorporation into plasteins (102, 103). The plastein products thus formed are subsequently hydrolyzed to break the ethyl ester linkages to yield hydrophilic plasteins. However, this approach is cumbersome, and advantage may be taken of the high catalytic activity of cold-adapted marine digestive proteinases to circumvent the difficulty. Since hydrophilic interactions are favored at low reaction temperatures, it is expected that these cold-adapted marine protinases (trypsins, chymotrypsins, and pepsins) could facilitate the formation of hydrophilic plastein products for incorporation into food products. The plastein reaction has great potential for application in the food industry: to improve the quality, flavor, solubility, and nutritive value of protein isolates and hydrolysates; prepare gellike products with excellent viscoelastic properties; produce products with a high or low levels of particular amino acids as a dietary supplement to compensate for deficiencies in certain sectors of the population; and to manufacture special types of peptides with important flavor and other properties.

V. BIOTECHNOLOGY AND DIGESTIVE PROTEINASES FROM MARINE ANIMALS

The extraction of enzymes or other useful biochemicals from fish is often not commercially feasible for a variety of reasons such as low yields, and variable supply of raw materials due to seasonal variations and political and environmental restrictions that regulate the harvesting of fish species. Enzymes are used by the food processing industry for a variety of reasons, such as reducing manufacturing cost, preparing products with more consistent properties, minimizing requirements for extensive heat treatments, and improving the quality of end products. Thus, the development of alternative sources of better enzymes would be both relevant and profitable.

The rapid development of genetic engineering has affected the traditional processes for the commercial production of several useful biochemicals, especially drugs, because genetic engineering via recombinant DNA technology allows some of these biological molecules to be cloned and overexpressed in microorganisms. For these and other reasons, research has been encouraged mainly in the biomedical and pharmaceutical sciences, while relatively little work has been done in the area of food science. Nevertheless, food processors are the largest users of industrial enzymes, which can be improved in their functionality and also are produced with more profitability by the application of genetic engineering (104). Although the application of genetic engineering to facilitate food processing is in its infancy, it is almost predictable that many more applications will be forthcoming, to add to notable breakthroughs such as recombinant chymosin, vitamin C, and the genetically modified tomato that are already being commercially exploited (105).

Cloning techniques are well established and have been applied for the production of several thousands of different peptides and proteins by microorganisms. Therefore, by choosing the right techniques and applying appropriate conditions it should be possible to clone and overexpress any fish-enzyme gene in microorganisms. The nucleotide sequences and the structural models of five trypsins variants from Atlantic salmon were elucidated from studies in molecular biology (106). Similarly, the trypsin gene expression in shrimp (*Penaeus vannamei*) was studied (107), and this led to the detection of the cDNAs encoding five isoforms of trypsin. For both studies, the extraction of RNA from the freshly removed fish tissues was performed for the construction of a cDNA library that could be screened with trypsin-specific probes. The probes, which were designed by alignment of known amino acid sequences of trypsins from mammals, insects, and crustaceans, turned out to be trypsin-specific due to the existence of highly conserved regions for this particular enzyme. The cDNAs from Atlantic cod encoding two different forms of trypsinogen were isolated and characterized (108).

In order to understand how psychrophilic enzymes are able to compensate for the low temperature of their environment, Genicot et al. (109) studied the trypsin from an Antarctic fish, *Paranotothenia magellanica*. Determination of the nucleotide sequence was performed after reverse transcription of the poly (A^+) RNA followed by PCR amplification using an oligo $(dt)_{28}$ and a specific 30-mer synthetic oligonucleotide as primers.

The cloning and expression of cathepsin L from shrimp hepatopancreas have been reported (110). Cathepsins are involved in the proteolysis of both endocytosed and endogenous proteins. They are monomeric proteinases with molecular weights of less than 30 kDa.

Cloning of fish enzymes, especially cold-adapted proteinases could allow their overexpression in microorganisms for their use in food processing industry. The competitive advantages of these proteinases over the existing commercial proteinases would make them the enzymes of choice for many food processors if they were commercially available. With regard to the extracellular enzymes such as digestive proteinases, it is expected that there would be a big demand for their recombinant forms by industry. In fact, the possible use of these enzymes in food manufacture has been appreciated for some time, but the enzymes have not been available in commercial quantities.

VI. CONCLUSIONS

The processing of marine animals generates byproducts such as fish guts that can serve as sources of new enzymes. The literature amply demonstrates that certain digestive proteinases from marine animals are more efficient catalysts at low reaction temperature, more heat-labile, or have superior ability to denature native protein substrates (111). The market for industrial enzymes is steadily growing in response to demands for their use in specific applications. Some of these specific applications may best be accomplished by exploiting the unique properties of certain marine enzymes. It is unlikely that relying on the traditional approaches to production would be sufficient to meet the demands for new enzymes. This situation is expected to be improved by the opportunities made possible by recombinant DNA technology (112, 113).

REFERENCES

1. MF Chaplin, C Bucke. Enzyme Technology. Cambridge (UK): Cambridge University Press, 1990, pp 40–79.
2. JR Whitaker. Classification and Nomenclature of Enzymes. In: Principles of Enzymology for the Food Sciences, 2nd ed. New York: Marcel Dekker, 1994, pp 367–385.
3. T Godfrey, J Reichelt. Industrial Enzymology. Surrey (UK): The Nature Press, 1983, pp 582.
4. NF Haard, BK Simpson. Proteases from aquatic organisms and their uses in the seafood industry. In: AM Martin, ed Fish Process: Biotechnological Applications. London (UK): Chapman & Hall, 1994, pp 133–154.
5. NF Haard. A review of proteolytic enzymes from marine organisms and their application in the food industry. J Aquat Food Prod Tech 1: 17–35, 1992.
6. HO Hultin. Enzymes from organisms acclimated to low temperatures. In: JP Danehy and B Wolnak, eds. Enzymes. The Interface Between Technology and Economics. New York: Marcel Dekker, 1980, pp 161–178.

7. NF Haard, N Helbig, LAW Feltham. The temperature characteristics of pepsin from two stocks of American smelt (*Osmerus mordax*). Proceedings of the Workshop on the Labrador Coasting Offshore Region, Newfoundland Institute of Cold Ocean Science, St. Johns (Canada), 1981, pp 174–196.

8. PS Low, JL Bada, GN Somero. Temperature adaptation of enzymes: Roles of free energy, the enthalpy, and the entropy of activation. Proc Natl Acad Sci (USA) 70: 430–432, 1973.

9. BK Simpson, NF Haard. Cold adapted enzymes from fish. In: D Knorr, ed. Food Biotechnology. New York: Marcel Dekker, 1987, pp 495–527.

10. B Asgeirsson, JB Bjarnasson. Purification and characterization of trypsin from the poikilotherm *Gadus morhua*. Eur J Biochem 180: 85–94, 1989.

11. INA Ashie, BK Simpson. Application of high hydrostatic pressure to control enzyme-related fresh seafood texture deterioration. Food Res Int 29: 569–575, 1996.

12. INA Ashie, BK Simpson, H Ramaswamy. Control of endogenous enzyme activity in fish muscle by inhibitors and hydrostatic pressure using RSM. J Food Sci 62: 350–356, 1996.

13. INA Ashie, BK Simpson. α-Macroglobulin inhibition of endogenous proteases in fish muscle. J Food Sci 62: 357–361, 1996.

14. BK Simpson, NF Haard. Characterization of the trypsin fraction from cunner, *Tautogolabrus adspersus*. Comp Biochem Physiol 80B: 475–480, 1985.

15. VG Pfleiderer, R Zwilling, H-H Sonneborn. Eine protease vom molekulargewicht 11,000 und eine trypsinahinche frackion aus *Astacus fluviatilis*. Z Physiol Chem 348: 1319–1331, 1967.

16. N Guizani, RS Rolle, MR Marshall, CI Wei. Isolation and characterization of a trypsin from the pyloric ceca of mullet (*Mugil cephalus*). Comp Biochem Physiol 98: 517–521, 1991.

17. Novo Nordisks. Annual Report, 1998

18. J Raa. Biotechnology in aquaculture and the fish processing industry. A success story in Norway. In: MN Voigt, JR Botta, eds. Advances in Fisheries Technology and Biotechnology for Increased Profitability. Lancaster (PA): Technomic, 1990, pp 509–524.

19. G Stefansson, U Steingrimsdottir. Application of enzymes for fish processing in Iceland—present and future Aspects. In: MN Voigt and JR Botta, eds. Advances in Fisheries Technology and Biotechnology for Increased Profitability. Lancaster (PA): Technomic, 1990, pp 237–250.

20. KA Almas. Utilization of marine biomass for production of microbial growth media and biochemicals. In: MN Voigt, JR Botta, eds. Advances in Fisheries Technology and Biotechnology for Increased Profitability. Lancaster (PA): Technomic, 1990, pp 509–524.

21. T Wray. Fish Processing: New Uses for Enzymes. Food Manufacture. 1988, 63: 48.

22. E. Mihalyi. Application of Proteolytic Enzymes to Protein Structure Studies, 2nd ed. Palm Beach: CRC Press, 1978, pp 215–308.

23. A Martinez, RL Olsen. Characterization of pepsins from cod. U.S. Biochem. Corporation. Editorial Comments. 161: 22–23, 1989.

24. P Brewer, N Helbig, NF Haard. Atlantic cod pepsin. Characterization and use as a rennet substitute. Can Inst Food Sci Technol J 17: 38–43, 1984.

25. A Gildberg, J Raa. Purification and characterization of pepsins from Arctic fish capelin (*Mallotus villosus*). Comp Biochem Physiol 75A: 337–342, 1983.

26. EJ Squires, NF Haard, LAW Feltham. Pepsin isozymes from Greenland cod, *Gadus ogac*. I. Purification and physical properties. Can J Biochem Cell Biol 64: 205–214, 1986.

27. K Arunchalam, NF Haard. Isolation and characterization of pepsin from Polar cod, *Boreogadus saida*. Comp Biochem Physiol 80B: 467–473, 1985.

28. M Noda, K Murakami. Studies on proteinases from the digestive organs of sardine—purification and characterization of three alkaline proteinases from the pyloric ceca. Biochim Biophys Acta 658: 17–26, 1981.

29. TG Merrett, E Bar-Eli, H van Vunakis. Pepsinogens A, C, and D from the smooth dogfish. Biochemistry 8: 3696–3702, 1969.

30. S De Vecchi, Z Coppes. Marine fish digestive proteases—relevance to food industry and the south-west Atlantic region—a review. J Food Biochem 20: 193–214, 1996.

31. EJ Squires, NF Haard, LAW Feltham. Pepsin isozymes from Greenland cod. *Gadus ogac*. 2. Substrate specificity and kinetic properties. Can J Biochem Cell Biol 64: 215–222, 1986.

32. L Sanchez-Chiang, E Cisternas, O Ponce. Partial purification of pepsins from adult and juvenile salmon fish (*Oncorhynchus keta*). Effect of NaCl on proteolytic activities. Comp Biochem Physiol 87B: 793–797, 1987.

33. NF Haard, K Shamsuzzaman, P Brewer, K Arunchalam. Enzymes from marine organisms as rennet substitutes. In: P. Dupuy, ed. Use of Enzymes in Food Technology. Paris: Lavoisier, 1983, pp 237–241.

34. K Shamsuzzaman, NF Haard. Evaluation of harp seal protease as a rennet substitute for Cheddar cheese. J Food Sci 48: 179–183, 1983.

35. K Shamsuzzaman, NF Haard. Characterization of a chymosin-like enzyme from the gastric mucosa of harp seal *Pagophilus groenlandicus*. Can J Biochem Cell Biol 62: 699–708, 1984.

36. K Shamsuzzaman, NF Haard. Milk clotting activity and cheese making properties of a chymosin-like enzyme from harp seal, *Pagophilus groenlandicus*. J Food Biochem 9: 173–192, 1985.

37. JN Mills, J Tang. Molecular weight and amino acid composition of human gastricsin and pepsin. J Biol Chem 242: 3093–3097, 1967.

38. WY Huang, J Tang. On the specificity of human gastricin and pepsin. J Biol Chem 244, 1085–1091, 1969.

39. JS Fruton, M Bergmann. The specificity of salmon pepsin. J Biol Chem 204: 559–560, 1940.

40. L Sanchez-Chiang, O Ponce. Gastricsinogens and Gastricsins from *Merluccius gayi*—purification and properties. Comp Biochem Physiol 68B: 251–257, 1981.

41. A Martinez, RL Olsen, J Serra. Purification and characterization of two trypsin-like enzymes from the digestive tract of anchovy, *Engraulis encrasicholus*. Comp Biochem Physiol 91B: 677–684, 1988.

42. BK Simpson, NF Haard. Trypsin from Greenland cod, *Gadus ogac.* Kinetic and thermodynamic characteristics. Can J Biochem Cell Biol 62: 894–900, 1984.

43. BK Simpson, NF Haard. Trypsin from Greenland cod, *Gadus ogac.* Isolation and comparative properties. Comp Biochem Physiol 79B: 613–622, 1984.

44. BK Simpson, MV Simpson, NF Haard. On the mechanism of enzyme action: digestive proteases from selected marine organisms. Biotech Appl Biochem 11: 226–234, 1989

45. B Asgeirson, JB Bjarnason. Studies on an enzyme preparation from cod viscera. Science Institute, University of Iceland, Reykjavik (Iceland), Report RH-1288, 1988.

46. B Benediktsson. Production of proteolytic enzymes from cod viscera. Icelandic Fisheries Laboratories, Reykjavik (Iceland) Internal Report, 1987.

47. A Gildberg. Aspartic proteinase in fishes and aquatic invertebrates. Comp Biochem Physiol 91B: 425–435, 1988.

48. K Hjelmeland, J Raa. Characteristics of two trypsin type isozymes isolated from the arctic capelin (*Mallotus villosus*). Comp Biochem Physiol 71B: 557–562, 1982.

49. R Yoshinaka, M Sato, T Suzuki, S Ikeda. Enzymatic characterization of anionic trypsin of the catfish *Parasilurus asotus.* Comp Biochem Physiol 77B: 1–6, 1984.

50. C-S Chen, T-R Yan, H-Y Chen. Purification and properties of trypsin-like enzymes and a carboxypeptidase A from *Euphasia superba.* J Food Biochem 21: 349–366, 1978.

51. BK Simpson, NF Haard. Trypsin and trypsin-like enzymes from the stomachless cunner. J Agric Food Chem 35: 652–656, 1987.

52. BK Simpson, NF Haard. Trypsin from Greenland cod as a food processing aid. J Appl Biochem 6: 135–143, 1984.

53. WP Winter, H Neurath. Purification and properties of a trypsin-like enzyme from the starfish *Evasterias trochelii.* Biochemistry. 9: 4673–4679, 1970.

54. Z Camacho, JR Brown, GB Kitto. Purification and properties of trypsin-like protease from the starfish *Desmasteria imbricata.* J Biol Chem 245: 3964–3972, 1980.

55. MS Heu, HR Kim, JH Pyeum. Comparison of trypsin and chymotrypsin from the viscera of anchovy, *Eugraulis japonica.* Comp Biochem Physiol 112B: 557–568, 1995.

56. J Overnell. Digestive enzymes of the pyloric ceca and the associated mesentery in the cod (*Gadus morhua*). Comp Biochem Physiol 46B: 519–531, 1973.

57. B Asgiersson, JB Bjarnasson. Structural and kinetic properties of chymotrypsin from Atlantic cod (*Gadus morhua*). Comparison with bovine chymotrypsin. Comp Biochem Physiol 99B: 327–335, 1991.

58. J Kalac. Studies on herring (*Clupea harengus L.*) and capelin (*Mallotus villosus*) pyloric ceca protease. III. Characterization of anionic fractions of chymotrypsins. Biologia (Bratslava) 33: 939–945, 1978.

59. MM Kristjansson, HH Nielson. Purification and characterization of two chymotrypsin-like proteases from the pyloric ceca of rainbow trout (*Oncorhynchus mykiss*). Comp Biochem Physiol 101B: 247–253, 1992.

60. WF Racicot. A kinetic and thermodynamic comparison of bovine and dogfish chymotrypsins. Ph.D. thesis, University of Massachusetts, Amherst, MA, 1984.

61. M Ramakrishna, HO Hultin, MT Atallah. A comparison of dogfish and bovine chymotrypsins in relation to protein hydrolysis. J Food Sci 52: 1198–1202, 1987.

62. B Asgeirsson, JB Bjarnasson. Properties of elastase from Atlantic cod, a cold adapted proteinase. Biochem Biophys Acta 1164: 91–100, 1993.

63. J Raa, BT Walther. Purification and characterization of chymotrypsin, trypsin and elastase like proteinases from cod (*Gadus morhua* L.). Comp Biochem Physiol 93B: 317–324, 1989.

64. R Yoshinaka, M Sato, M Tanaka, S Ikeda. Distribution of pancreatic elastase and metalloproteins in several species of fish. Comp Biochem Physiol 80B: 227–233, 1985.

65. T Cohen, A Gertler, Y Birk. Pancreatic proteolytic enzymes from carp (*Cyprinus carpio*). I. Purification and physical properties of trypsin, chymotrypsin, elastase and carboxypeptidase B. Comp Biochem Physiol 69B: 639–646, 1981.

66. R Yoshinaka, H Tanaka, M Sato, S Ikeda. Purification and some properties of elastase from the pancreas of catfish. Bull Jpn Soc Sci Fish 48: 573–579, 1982.

67. J Clark, NL Macdonald, JR Stark. Metabolism in marine flatfish. III. Measurement of elastase activity in the digestive tract of Dover sole (*Solea solea* L.). Comp Biochem Physiol 81B: 695–700, 1985.

68. AI Lansing, TB Rosenthal, M Alex. Presence of elastase in teleost islet tissue. Proc Soc Exp Biol Med 84: 689–691, 1953.

69. EN Zendzian, EA Bernard. Distribution of pancreatic ribonuclease, chymotrypsin and trypsin in vertebrates. Arch Biochem Biophys 122: 699–713, 1967.

70. A Gildberg, K Øverbø. Purification and characterization of pancreatic elastase from Atlantic cod (*Gadus morhua*). Comp Biochem Physiol 97B: 775–782, 1990.

71. AZ Eisen, KD Henderson, JJ Jeffrey, RA Bradshaw. A collagenolytic protease from the hepatopancreas of the fiddler crab, *Uca pugilator*. Purification and properties. Biochemistry 12: 1814–1822, 1973.

72. AZ Eisen, JJ Jeffrey. An extractable collagenase from crustacean hepatopancreas. Biochim Biophys Acta 191: 517–526, 1969.

73. GA Grant, JC Sacchettini, HG Welgus. A collagenolytic serine protease with trypsin-like specificity from the fiddler crab, *Uca pugilator*. Biochemistry 22: 354–356, 1983.

74. ES Baranowski, WK Nip, JH Moy. Partial characterization of a crude enzyme extract from freshwater prawn, *Machrobranchium rosenbergii*. J Food Sci 49: 1494–1495, 1984.

75. WK Nip, JH Moy, YY Tzang. Effect of purging on quality changes of ice-chilled freshwater prawn, *Machrobranchium rosenbergii*. J Food Technol 20: 9–15, 1985.

76. Y Chen. Characterization of semi-purified collagenase fraction from lobster hepatopancreas. M.Sc. Thesis, McGill University, Montreal, Canada, 1992.

77. N. Raksakulthai, N.F. Haard. 1999. J. Food Biochem. (In press)

78. RGB Reid, K Rauchert. Catheptic endopeptidases and protein digestion in the horse clam *Tresus capax* (Gould). Comp Biochem Physiol 54B: 467–472, 1976.

79. AH Zeef, C Dennison. A novel cathepsin from the mussel (*Perna perna* Linne). Comp Biochem Physiol 90B: 204–210, 1988.

80. HC Chen, RR Zall. Partial purification and characterization of cathepsin D-like and B-like acid proteases from surf clam viscera. J Food Sci 51: 71–78, 1986.

81. G Siebert, A Schmitt. Fish tissue enzymes and their role in determining changes in fish. In: R. Kreuzer, ed. The Technology of Fish Utilization. London: Fishery News Books, 1965, pp 47–52.

82. SBK Warrier, V Ninjoor, GB Nadkarni. Involvement of hydrolytic enzymes in the spoilage of Bombay duck (*Harpodon nehereus*). In: Technical Report Harvest and Postharvest Technology of Fish. Society of Fisheries Technology (India). 1985, pp 470–472.

83. GE Bracho, NF Haard. Characterization of alkaline metalloproteinases with collagenase activity from the muscle of Pacific rockfish (*Sebastes* sp.) In: SW Otwell, ed. Proceedings of Joint Meeting of the Atlantic Fisheries Technologists and Tropical/Subtropical Fisheries Technologists. Gainesville (FL): Florida Sea Grant College Program, 1991, pp 105–123.

84. Y Makinodan, M Hirotsuka, S Ikeda. 1979. Neutral proteinase of carp muscle. J Food Sci 44: 1110 1979.

85. Y Okamoto, H Otsuka- Fuchino, S Horiuchi, T Tamiyu, JJ Matsumoto, T Tsuchiya. Purification and characterization of two metalloproteinases from squid mantle muscle, myosinase I and myosinase II. Biochim Biohys Acta 1161: 97–104, 1993.

86. NF Haard, LAW Feltham, N Helbig, J Squires. Modification of proteins with proteolytic enzymes from the marine environment. In: RL Feeney, JR Whitaker, eds. Modification of Proteins. Advances in Chemistry Series, Washington (D.C): American Chemical Society, 1982, 198: 223–224.

87. K-D Jany. Studies on the digestive enzymes of the stomachless bonefish *Carassius auratus gibelio* (Bloch): endopeptidase. Comp Biochem Physiol 53B: 31–38, 1976.

88. B Kiel. Trypsin. In: PD Boyer, ed. The Enzymes. New York: Academic Press, 1971, 3: 249–275.

89. B Asgeirsson, JB Bjarnason. Structural and kinetic properties of chymotrypsin from Atlantic cod (*Gadus morhua*). Comparison with bovine chymotrypsin. Comp Biochem Physiol 99B: 27–35, 1991.

90. K Kristjansson, HH Nielsen. Purification and characterization of two chymotrypsin-like proteases from the pyloric ceca of rainbow trout (*Oncorhynchus mykiss*). Comp Biochem Physiol 101B: 247–253, 1992.

91. YZ Lee, BK Simpson, NF Haard. Supplementation of squid fermentation with proteolytic enzymes. J Food Biochem 6: 127–134, 1982.

92. KR Opshaug. Method of enzyme maturing of fish. 1982, PCT patent No 82/03533.

93. C Eriksson. Method for controlling the ripening process of herring. 1975, Canadian patent 969,419.

94. Anonymous. Production process for salted herrings (Matjes). 1975, British patent 1,403,221.

95. NF Haard, N Kariel, G Herzberg, LAW Feltham. Stabilization of protein and oil in fish silage for use as ruminant feed supplement. J Sci Food Agric 36: 229–241, 1985.

96. A Gildberg, KA Almas. Utilization of fish viscera. In: M Le Maguer, P Jelen, eds. Food Engineering and Process Applications. London and New York: Elsevier Applied Science, 1986, 2: 383.

97. D Lim, WF Shipe. Proposed mechanism for the antioxidative action of trypsin in milk. J Dairy Sci 55: 753–758, 1972.

98. M.A Kyei. Digestive proteases from the stomachless cunner fish (*Tautogolabrus adspersus*): preparation and use as food processing aid. M.Sc. Thesis, McGill University, Montreal, Canada. 1997.

99. S Damodaran. Food proteins: an overview. In: S Damodaran, A. Paraf, eds. Food Proteins and their Application. New York: Marcel Dekker, 1997, pp 1–24.

100. KD Schwenke. Enzymes and chemical modification of proteins. In: S Damodaran, A. Paraf, eds. Food Proteins and their Application. New York: Marcel Dekker, 1997, pp 393–423.

101. PM Nielson, Functionality of protein hydrolysates. In: S Damodaran, A. Paraf, eds. Food Proteins and their Application. New York: Marcel Dekker, 1997, pp 443–472.

102. G Hajos. Incorporation of essential amino acids into protein by enzymatic peptide modification (EPM). Nahrung 30: 418–422, 1986.

103. K Aso, H Kimura, M Watanabe, S Arai. Chemical properties of enzymatically modified proteins produced from soy protein by covalent attachment of methionine. Agric Biol Chem 49: 1649–1654, 1985.

104. Y-L Lin. Genetic Engineering and process development for production of food processing enzymes and additives. Food Technol 40: 104–112, 1986.

105. Tucker, G 1996. Biotechnology and enzymes in the food industry. Br Food J 98: 14–19.

106. R Male, JB Lorens, AO Smalas, KR Torrissen. Molecular cloning and characterization of anionic and cationic variants of trypsin from Atlantic salmon. Eur J Biochem 232:677–685, 1995.

107. B Klein. G LeMoullac, D Sllos, A van Wormhaudjt. Molecular cloning and sequencing of trypsin cDNAs from *P. vannamei:* use in assessing gene expression during the moult cycle. Int J Biochem Biol 28: 551–563, 1996.

108. A Gudmundsdottir, E Gudmundsdottir, S Oskarsson, AK Eakin, CS Craik. Isolation and characterization of cDNAs from Atlantic cod encoding two different forms of trypsinogen. Eur J Biochem 217: 1091–1097, 1993.

109. S Genicot, F Rntier-Delrue, D Edwards, J VanBeeumen, C Gerday. Trypsin and trypsinogen from an Antarctic fish: molecular basis of cold adaptation. Biochem Biophys Acta 1298: 45–47, 1996.

110. C LeBoulay, A van Wormhoudt, D Sellos. Cloning and expression of cathepsin L-like proteinases in the hepatopancreas of shrimp *Penaeus vannamei* during inter-molt cycle. J Comp Physiol 166 B: 310–318, 1996.

111. NF Haard. Specialty enzymes from marine organisms. Food Technol. 52(7): 64–67, 1998.

112. BP Wasserman. Evolution of enzyme technology: progress and prospects. Food Technol 44: 118–122, 1990.

113. B Dixon. Cod waste yields useful enzymes. Biotechnology 8: 791, 1990.

9

Trypsin Isozymes: Development, Digestion, and Structure

Krisna Rungruangsak Torrissen
Institute of Marine Research, Bergen, Norway

Rune Male
University of Bergen, Bergen, Norway

I. INTRODUCTION

Trypsin is a major member of the serine proteases, which constitute a large family of biologically important enzymes structurally and functionally conserved from bacteria to mammals. Trypsin consists of a single peptide chain with molecular weight typically 24 kDa. It is synthesized in the pancreas and secreted as an inactive precursor, trypsinogen, from the pancreatic acinar cells together with chymotrypsinogen and proelastase (1, 2). This evidence was observed not only in mammals but also in fish as demontrated in chum salmon, *Oncorhynchus keta* (3), in catfish, *Parasilurus asotus* (4), and in eel, *Anguilla japonica* (5). Trypsinogen is activated by enteropeptidase in the small intestine, by specific cleavage of a unique single peptide-bond between a conserved positively charged residue, usually lysine, and isoleucine. Its N-terminal peptide of six to eight residues, depending on species, is removed (6), and the isoleucine residue at the new N-terminus bends inwards to make several internal contacts forming the catalytically active trypsin (7, 8). The small amount of trypsin activates more trypsinogen and all other pancreatic zymogens: chymotrypsinogen, procarboxypeptidase, and proelastase (1). The active proteolytic enzymes differ markedly in their degree of substrate speci-

Table 1 Pancreatic Proteases

Inactive enzyme	Active enzyme	Site of cleavage
Trypsinogen	Trypsin	Carboxyl side of Lys and Arg
Chymotrypsinogen	Chymotrypsin	Carboxyl side of Phe, Trp, Tyr and of large hydrophobic residues such as Met
Procarboxypeptidase	Carboxypeptidase	Carboxyl-terminal residue with an aromatic or a bulky aliphatic side chain
Proelastase	Elastase	Carboxyl side of residue with small hydrophobic side chain such as Ala

Source: Refs. 1, 9.

ficity, which is usually high and sometimes virtually absolute (Table 1). X-ray studies have shown that these different specificities are due to quite small differences in the binding site. Most proteolytic enzymes also catalyze a different but related reaction, namely, the hydrolysis of an ester bond. The principal protease that is active only in an acid environment of gastric juice is pepsin, while the main proteases that are active in an alkaline medium of the intestine are trypsin and chymotrypsin. Proteins are broken down by these principal proteases into the polypeptides, which are further broken down to amino acids and peptides of smaller molecular size by intestinal peptidases and pancreatic carboxypeptidases. Proteolytic enzymes affect food digestion and supply of amino acid precursors for protein synthesis. Since trypsin is the key enzyme for activating all other pancreatic zymogens, it may play a major role in regulating protein digestion and affect the availability of nutrients for synthesis and growth. Several isoforms of trypsin have been reported in fishes, and the results are summarized in Table 2. Trypsin isozymes are believed to possess differences in their catalytic efficiency (k_{cat}/K_m) as demonstrated in cod, *Gadus morhua,* by Åsgeirsson et al. (18). Isoforms with major differences in the distribution of charged amino acids may have different substrate-binding preferences (15, 22).

　　Since there is limited evidence of the association between different forms of trypsin and their biological characteristics, the aim of this chapter is to review the data that have recently become available on the effect of genetical presence of different trypsin isozymes on growth and physiological changes in fishes. It will also demonstrate that trypsin is the significant enzyme in the digestion process that could be the rate-limiting step for feed utilization and growth in fishes. Studies of protein structure of trypsin and its catalytic activity are also reviewed.

Table 2 Presence of Different Isoforms of Trypsin or Trypsinlike Enzyme in the Digestive System of Different Fishes

Species	Anionic trypsin	Cationic trypsin	Method	Reference
Rainbow trout (*Oncorhynchus mykiss*)	six isozymes (pI 4–6.5)	Exist (pI > 10)	Isoelectric focusing	10, 11
	five isozymes[a]		Disc electrophoresis	12
	two isozymes (pI <4.3–5.6)	two isozymes (pI > 8.4)	Chromatography	13
Atlantic salmon (*Salmo salar*)	six isozymes (pI 4–6.5)	Exist (pI > 10)	Isoelectric focusing	10, 11, 14
	four variants	one variant	cDNA cloning	15
Chum salmon (*Oncorhynchus keta*)	six isozymes (pI 4.1–4.3)	one isozyme (pI 11.1)	Chromatography	16
Arctic charr (*Salvelinus alpinus*)	four isozymes (pI 4–6.5)		Isoelectric focusing	17
Atlantic cod (*Gadus morhua*)	three isozymes (pI 5.5–6.6)		Chromatofocusing	18
Capelin (*Mallotus villosus*)	two isozymes (pI 5.1–5.9)		Chromatography	19
Mackerel (*Scomber japonicus*)	five isozymes (pI <4–6.3)	one isozyme (pI >8.2)	Chromatography	13
Roach (*Rutilus rutilus*)	three isozymes[a]		Disc electrophoresis	12
Sardine (*Sardinops melanosticta*)	one isozyme (pI 4.85)		Chromatography	20
Yellowtail (*Seriora quinqueradiata*)	four isozymes (pI <4.2–5.7)	three isozymes (pI >8.3)	Chromatography	13
Catfish (*Parasilurus asotus*)	one type (pI 5.4)	one type	Chromatography	21

[a] No specification of anionic and cationic forms.

II. IDENTIFICATION OF DIFFERENT FORMS OF TRYPSIN IN SALMONIDS

Trypsin activity is commonly determined by its amidase activity with the substrate benzoyl-DL-arginine-p-nitroanilide (BAPNA), a method basically modified from Erlanger et al. (23). Dahlmann and Jany (24) modified the method for quantitative determination of the enzyme after electrophoresis, by enhancing the sensitivity of trypsin reaction through diazotizing and then coupling with naphthylethylenediamine as described by Bratton and Marshall (25). This method was tried on crude enzyme extracts from the intestine of salmonids, and only a single broad band of trypsinlike activity was detected (K Rungruangsak Torrissen, unpublished result). With further attempt, the method was modified with successful separation of trypsin isozymes by isoelectric focusing on agarose-IEF gel, using flat-bed apparatus FBE 3000 and the instructions for use of agarose-IEF from Pharmacia Fine Chemicals (10). Thus, variations in trypsin isozyme patterns of individuals were first simply identified in Atlantic salmon (26) and Arctic charr, *Salvelinus alpinus* (17). The electrophoretic zymograms of the patterns of trypsin isozymes in the pyloric cecal tissues in comparison between Atlantic salmon and Arctic charr, and their designated names with isoelectric points (pI) are illustrated in Figure 1. Variation in trypsin isozyme patterns was found between the anionic forms of the pI range 4–5 in both Atlantic salmon and Arctic charr (Fig. 1). Three common isozymes (designated TRP-1*100, TRP-2*100, and TRP-3) with two variants in the TRP-1 system (designated TRP-1*75, TRP-1*91) and one variant in the TRP-2 system (designated TRP-2*92) have been detected in Atlantic salmon. Existence of cationic

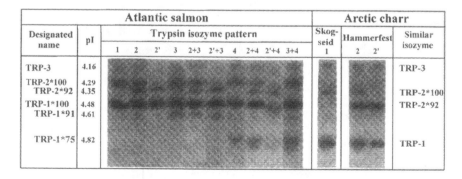

Atlantic salmon											Arctic charr			
Designated name	pI	Trypsin isozyme pattern									Skog-seid	Hammerfest		Similar isozyme
		1	2	2'	3	2+3	2'+3	4	2+4	2'+4	3+4	1	2	2'	
TRP-3	4.16														TRP-3
TRP-2*100	4.29														
TRP-2*92	4.35														TRP-2*100
TRP-1*100	4.48														TRP-2*92
TRP-1*91	4.61														
TRP-1*75	4.82														TRP-1

Figure 1. Electrophoretic zymograms of trypsin isozyme patterns in the pyloric cecum of Atlantic salmon and of two strains of Arctic charr, revealed by isoelectric focusing on Agarose-IEF gel pH 4–6.5. The designated names and isoelectric points (pI) are illustrated. (Data from Ref. 17 and Ref. 27; picture by Øivind Torslett.)

trypsin(s) at pI >10 was observed (11). The anionic trypsin TRP-1*91 was shown to be a major variant in Scottish Atlantic salmon while the TRP-2*92 variant was dominant in Norwegian Atlantic salmon, at a frequency of 0.42 and 0.47, respectively (27). The occurrence of trypsin variant TRP-2*92 in the cultivated Norwegian Atlantic salmon has been at a high frequency of 0.47–0.49 since the first observation in 1985.

In Arctic charr, four trypsin isozymes were detected in the two strains studied (17). Only pattern 1 was observed in slow-growing nonanadromous Skogseid charr, while patterns 2 and 2′ were found in fast-growing anadromous Hammerfest charr (Fig. 1).

Cloning of trypsin genes from pancreatic tissues of Atlantic salmon indicated a large number of gene loci for trypsin (15). Among detectable trypsin genes, five clones containing nearly full-length transcripts were characterized. Three Atlantic salmon trypsins possessed very similar sequences and may represent allelic variants encoded by the same gene locus, while the other two are probably encoded by separate gene loci. Four of them are anionic forms and the fifth clone is cationic trypsin (15). Salmonids still undergo the diploidization process to restore disomic inheritance, and approximately 50% of the duplicated loci are no longer detectable by their protein products (28). However, small differences may not be detectable by conventional protein electrophoresis; thus the number of duplicated active genes may be underestimated. Heredity studies of several salmonid enzymes resulted in offsprings with enzyme patterns that cannot be explained by ordinary disomic (Mendelian) inheritance, and this phenomenon is postulated to be the consequence of a tetraploid event in an ancestral salmonid (28, 29). The occurrence of variations in relative visual intensity between the common trypsin TRP-2*100 and the variant trypsin TRP-2*92 (see Fig. 1) indicated tetrasomic inheritance of duplicated loci (14). Trypsin phenotype frequencies of different salmon families did not clearly reflect Mendel's theory (14). Crossing of individual fish with different trypsin isozyme patterns resulted in offsprings with patterns that cannot be easily explained either by disomic or tetrasomic inheritance (30). This was suggested to be attributed to the experimental method employed, which may underestimate the number of heterozygotes of the *TRP-2* loci if the isozyme alleles exist as a 3:1 ratio.

III. A GENETIC MARKER FOR GROWTH RATE IN SALMONIDS

The association between fish size and different trypsin isozyme patterns in the pyloric cecal tissues were first published in Atlantic salmon fry (14). By using a simple biopsy technique modified from Harvey et al. (31) for the pyloric ceca for isoelectrophoretic evaluation, the individuals were biologically labelled (32). Biopsy can be performed in salmon parr from 30 g with 95–100% survival after

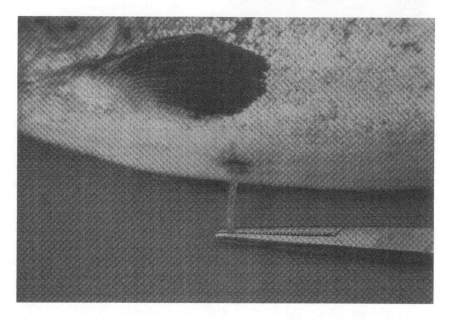

Figure 2. Biopsy technique for pyloric cecum of Atlantic salmon, showing the line cut and the amount of sample removed. (From Ref. 32; photo by Arne Berg.)

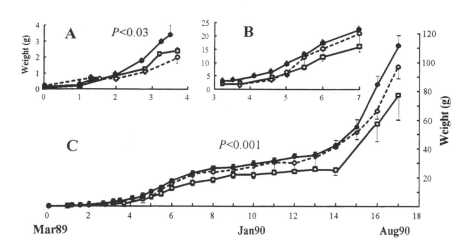

Figure 3. Growth of Atlantic salmon families in freshwater, from first feeding until postsmolts; group according to different frequency (f) of trypsin variant TRP-2*92 at $f \geq$ 0.5 (—●—), $f < 0.5$ (---○---) and $f = 0$ (—□—), during the first 4 months (**A**), from 3 to 7 months (**B**) and during the whole experimental period (**C**). (Data from Ref. 30.)

the operation (Fig. 2). The method allowed for the following of individual performance of genetically known trypsin isozyme pattern. The first evidence on the effect of genetic variation in trypsin isozyme patterns on growth rate was observed in Atlantic salmon from smolts until maturation (32). In Norwegian Atlantic salmon, improved growth has been shown to be associated with the fish possessing the trypsin variant TRP-2*92, during the start-feeding period (14, 30) and during winter of the first sea year (32). Differences in growth rate were manifested 3–5 months after the first feeding, and the differences in weight were maintained without differences in later daily growth rate until smoltification (Fig. 3). Growth rates were manifested again during the first sea winter and the differences in weight were maintained about 30% until harvest during the second sea year (Fig. 4). The two growth periods are enough to affect the size of the fish

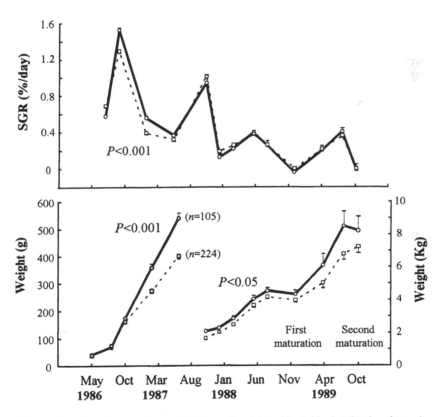

Figure 4. Specific growth rate (SGR) and weight of individual Atlantic salmon in sea water; groups in the presence (—O—) and absence (---□---) of trypsin variant TRP-2*92, from smolts until second maturation at the 4th year of age. (Data from Ref. 32.)

during the whole life cycle. Difference in weight was about 30% on average when comparing fish possessing and lacking the trypsin variant TRP-2*92. The TRP-2*92 salmon became larger without any effect on the second maturation rate (about 75%) when compared to salmon lacking the variant (11). Unfortunately, the first maturation rates were not followed. Increased growth during the first sea winter was observed in two different year classes, and from the size-selected families the average weight was about 5% higher for the fish carrying than the fish lacking the trypsin variant TRP-2*92 (32). If there is an easy and simple method for selection of Atlantic salmon carrying trypsin variant TRP-2*92, the production will be improved in both freshwater and seawater periods, and this will benefit both smolt and food fish producers.

A. Early Development and Temperature Effect on Expression of Different Trypsin Isozymes

Variations in the expression of different trypsin isozymes were observed to be influenced by water temperature during incubation and start-feeding periods (27). Hatching temperatures had no effect on the frequency distribution between salmon parr without (pattern 1: TRP-2*100/100) and with (pattern 2: TRP-2*100/92 and pattern 2′: TRP-2*92/92) the trypsin variant TRP-2*92, but it had an effect on the frequency distribution between the parr with patterns 2 and 2′ (Table 3). The details of trypsin isozyme patterns are shown in Figure 1. The frequency distribution between the fish with patterns 2 and 2′ was changed from 0.55:0.45 at 6°C to 0.68:0.32 at 10°C, regardless of start-feeding temperature (Table 3), indicating an influence of hatching temperature of 10°C on the expression of the common isozyme TRP-2*100. In contrast, start-feeding temperature had no effect on the frequency distribution between the fish of patterns 2 and 2′; the effect was on the frequency between the fish without and with the variant (Table 3). The frequency between the fish without (pattern 1) and with (patterns 2 and 2′) the variant TRP-2*92 was changed from 0.30:0.65 at 6°C to 0.17:0.78 at 12°C, regardless of hatching temperature, indicating an effect of start-feeding temperature of 12°C on induced expression of the variant TRP-2*92. High temperature of 10–12°C, either at hatching or start-feeding period, increased the frequency of the TRP-2*100/92 salmon (pattern 2). Individual characteristics of trypsin isozyme pattern were developed and established during start-feeding and the pattern does not seem to change later in the life cycle (27, 30). Temperature experience during hatching and start-feeding may influence feed utilization and growth at later stage of the life cycle (27). Observations of trypsin isozyme patterns at different times after the first feeding at water temperature of 8°C (14) indicated that food immediately induced trypsin expression, and the enzyme activity was clearly detected by isoelectric focusing on agarose-IEF gel 10 days after feeding (Fig. 5). However,

Table 3 Number of Sampled Fish and Frequency Distribution of Each Trypsin Isozyme Pattern (see Fig. 1) of Norwegian Atlantic Salmon Parr, Hatched and Start-Fed at Different Temperatures and Sampling Times

Temperature (°C)		Sampling time	Trypsin isozyme pattern				Pattern ratio	
Hatching	Start-feeding		1	2	2'	3 and 5	1:2+2'	2:2'
5.9 ± 1.9	12.2 ± 0.5	Nov 94	17 (0.16)	51 (0.47)	34 (0.31)	7 (0.06)		
		Jan 95	23 (0.18)	50 (0.40)	50 (0.40)	3 (0.02)		
Cold–warm group			40 (0.17)	101 (0.43)	84 (0.36)	10 (0.04)	0.17:0.79[a]	0.54:0.46[a]
9.6 ± 1.2	12.2 ± 0.5	Nov 94	27 (0.18)	78 (0.51)	42 (0.27)	7 (0.04)		
		Jan 95	22 (0.17)	68 (0.54)	27 (0.21)	10 (0.08)		
Warm–warm group			49 (0.17)	146 (0.52)	69 (0.25)	17 (0.06)	0.17:0.77[a]	0.68:0.32[b]
5.9 ± 1.3	5.6 ± 1.3	Jan 95	34 (0.28)	49 (0.40)	33 (0.27)	6 (0.05)		
		Aug 95	40 (0.27)	49 (0.33)	48 (0.32)	13 (0.08)		
Cold–cold group			74 (0.27)	98 (0.36)	81 (0.30)	19 (0.07)	0.27:0.66[b]	0.56:0.44[a]
9.6 ± 1.2	5.6 ± 1.3	Jan 95	44 (0.34)	54 (0.41)	23 (0.18)	10 (0.07)		
		Aug 95	37 (0.30)	57 (0.47)	27 (0.22)	1 (0.01)		
Warm–cold group			81 (0.32)	111 (0.44)	50 (0.20)	11 (0.04)	0.32:0.64[b]	0.70:0.30[b]

Within the same column, the ratios with different superscripts are significantly different ($p < 0.01$).
Source: Ref. 27.

Hatching day 1 Day after the first feeding

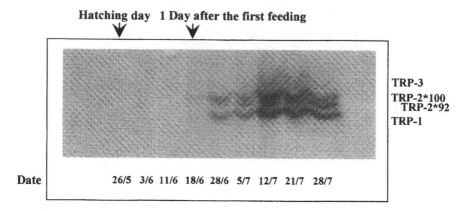

Figure 5. Electrophoretic zymograms of trypsin isozyme expression of Atlantic salmon fry at the early stage of development. (From Ref. 14; picture by Øivind Torslett.)

all isozymes were not expressed at the same time. The trypsin variant TRP-2*92 was detected 1 week later than the three common isozymes (Fig. 5). Each sample came from a mixture of 30–40 fry; the variant would have been detected if it had been produced. These results indicated that the later expression of the variant TRP-2*92 after the first feeding in Figure 5, compared to the other common isozymes, was due to an induction by start-feeding temperature at 8°C. The trypsin variant TRP-2*92 can be induced to express during start-feeding at temperature of 8–12°C.

Trypsin phenotype did not affect egg quality and hatching success but it affected fish growth from the first day of feeding (30). Heredity study with special selection for the presence of the variant TRP-2*92 resulted in offspring with different trypsin isozyme patterns. By following the growth of salmon fry from families with different frequency of the variant TRP-2*92, the families exhibiting a high frequency of the variant ($f \geq 0.5$) had significantly higher growth rate during 3–5 months after the first feeding compared to the families with $f < 0.5$ and lacking ($f = 0$) the variant (Fig. 3). Although the temperature was not reported, the start-feeding temperature at Matre Aquaculture Research Station, Norway, has usually been $\geq 8°C$. A similar experiment was carried out to study the development of trypsin activity in the digestive system of salmon fry from five families with a frequency 0.73–0.85 of the trypsin variant TRP-2*92 and five families with a frequency of 0.57–0.66, 15 fry per family, during the first 4 months of start-feeding period (R Moss and K Rungruangsak-Torrissen, unpublished). A significantly higher increase in trypsin activity was observed in the families with higher frequency of the variant TRP-2*92 after 80 days of feeding at 12°C, while there were no differences in the development of

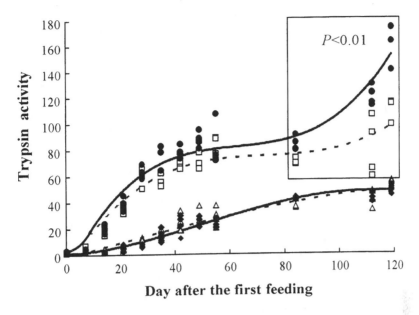

Figure 6. Development of trypsin activity (μmol p-nitroaniline h^{-1} mg protein^{-1}) of Atlantic salmon fry from families with different frequency (f) of trypsin variant TRP-2*92, during 4 months of start-feeding period at different water temperature. Start feeding with warm water of 12°C, families with $f = 0.73$–0.85 (——●——) and with $f = 0.57$–0.66 (---□---). Start feeding with cold water of 6°C, families with $f = 0.73$–0.85 (——◆——) and with $f = 0.57$–0.66 (---Δ---). (From R Moss and K Rungruangsak-Torrissen, unpublished.)

trypsin activity between these families start-fed at 6°C (Fig. 6). The development of protease activities after start-feeding was reported in salmon from five different river strains, showing the maximum increase in the enzyme activities after 2 months of feeding, which became similar to the levels of normal immature fish (33). These results suggest that the time for full development of the digestive system is 2–3 months after the first feeding, when the variation in growth rate can be observed if there are genetic differences in ability for feed utilization.

Growth rates of different trypsin phenotypes from parr to postsmolt during a few months of experimental period have been shown to be different when the water temperature was ≤ 8°C (27, 30, 34), and not different when the temperature > 8°C (27, 34–36). An investigation of three different Norwegian salmon populations indicated a significant association between a higher frequency of the trypsin variant TRP-2*92 and a better growth rate at a low temperature of 2–3°C (Table 4). The differences in weight between the populations were affected by salmon carrying the variant (30). Results from the Norwegian

Table 4   Mean Weight (± SEM) of Atlantic Salmon with and without Trypsin Variant TRP-2*92 from Three Different Populations, during 4 months at Low Water Temperature (2–3 °C)

Salmon strain	Weight (g) in October 1990		Weight (g) in February 1991	
(n=150)	With	Without	With	Without
Dale (f = 0.29)	52.7 ± 2.2^a	42.6 ± 1.2^b	64.3 ± 2.9^a	50.2 ± 1.5^b
Lonevåg (f = 0.10)	47.9 ± 2.8^a	40.7 ± 0.8^b	55.0 ± 4.0*	47.5 ± 1.0^{b*}
Voss (f = 0.09)	40.3 ± 1.7^b	40.5 ± 0.7^b	51.2 ± 2.0^b	49.0 ± 0.7^b

Within the same period, the values with different superscripts or with the sign * are significantly different ($p<0.04$).
f, Frequency of trypsin variant TRP-2*92 in each strain.
Source: Ref. 30.

Sea Project at the Institute of Marine Research in 1995 indicated a higher weight of Atlantic salmon postsmolts possessing trypsin variant TRP-2*92 caught at the area north of latitude 65°N where the temperature could be below 6°C, while the tendency was toward smaller smolts carrying the variant caught below this latitude where temperature was > 10°C, compared to the salmon lacking the variant (27). Spatial distribution in the Norwegian Sea at early migration of wild salmon possessing different trypsin phenotypes indicated that the ambient temperatures (average temperature weighted by fish density using as indicator the average temperature fish experience) of postsmolts possessing patterns 1, 2, and 2′ are 9.3°C, 8.7°C, and 7.7°C, respectively (K Rungruangsak Torrissen, BK Stensholt, and M Holm, to be published). Similar results were observed in Arctic charr (17). A fast-growing anadromous Hammerfest strain in the northern Norway possessed a similar trypsin isozyme TRP-2*92 (see Fig. 1), while this isozyme did not exist in a slow-growing nonanadromous Skogseid landlocked strain living in somewhat higher temperature in the south. Trypsin variant TRP-2*92 has been hypothesized to be manifested at temperature ≤ 8°C, especially below 6°C, while the common isozyme TRP-2*100 is important at the temperature >8°C (27). There is still a question of why the early expression of the variant TRP-2*92 is induced by a high temperature of 8–12°C during start-feeding period (Fig. 5, Table 3). The nature of temperature-related differences in expression of trypsin genes remains to be investigated. The presence of both the common and the variant isozymes in the TRP-2 system of the TRP-2*100/92 fish (pattern 2) will affect good growth at a wide range of water temperatures.

Trypsin isozyme TRP-1*91 (see Fig. 1), the major variant in Scottish Atlantic salmon, has a higher temperature range (>6°C) for effective performance than Norwegian major variant TRP-2*92 (27). This should be an effect of natural selection, since Scottish salmon should be expected to live in the water with temperature somewhat higher than the water where the Norwegian salmon live.

The work demonstrates that gene expressions could be induced by external factor(s) at a very early life stage, and this will affect feed utilization and growth at a later stage of the life cycle. Furthermore, it points out that escaped healthy cultured salmon should not have any adverse effect on the wild population. It is the environmental conditions that have to be controlled to secure gene expressions of the next generations. In addition, rearing conditions of salmon fry for sea-ranching or recruitment and for captive cultured production should be different.

B. Growth Characteristics and Body Composition

Growth of Atlantic salmon with different trypsin phenotypes varied according to rearing temperature (27). An image analysis was studied in Atlantic salmon postsmolts of patterns 1, 2, and 2' reared in seawater at 9°C for 128 days (35). A morphometric component of 12 identifiable landmarks was evaluated using a so-called truss protocol (37), and the body composition was also determined. At the temperature of 9°C, the salmon of pattern 2' (TRP-2*92/92) showed slower growth than the salmon of pattern 1 (TRP-2*100/100) and pattern 2 (TRP-2*100/92), and the fish were shortest due to the absence of the common isozyme TRP-2*100, which is functionally active at temperature > 8°C. When body size was evaluated by centroid size at the end of the experiment, the pattern 2' fish revealed significantly lower values than the other phenotypes, illustrating the first example of morphometric discrimination during growth between fish with different trypsin phenotypes. However, overall evaluation of consensus shape components showed similarity in all landmarks evaluated. Although there were differences in growth and centroid size between different trypsin phenotypes, the proximate composition was still similar between these phenotypes (35). An experiment on an effect of partial prehydrolyzed dietary protein on growth and productive values of protein and lipid (38), and proximate composition (unpublished) of Atlantic salmon with different trypsin phenotypes also indicated that phenotypes did not affect proximate composition on wet weight basis, although differences in growth rate were detected (Table 5). However, fish fed prehydrolyzed dietary protein did have significantly higher moisture (71.6±0.3%) and lower retention of lipid, which caused higher proximate composition ratios of protein to lipid (2.21±0.08%), in comparison to fish fed the diet with nonhydrolyzed protein (69.3 ± 0.1% and 1.74+0.11%, respectively), regardless of phenotypes. Surprisingly the dietary protein quality does not affect protein deposition, on wet weight basis, but it does influence body lipid and water (Table 5).

Table 5 Growth, White Muscle In Vitro Protein Synthesis, and Proximate Composition (on Wet Weight Basis) in Atlantic Salmon Smolts with and without Trypsin Variant TRP-2*92, fed the Feeds containing Either Nonhydrolyzed or Partially Prehydrolyzed Protein by Pepsin, for 7 weeks at a restricted Ration of 0.5% Body Weight Day[-1]

Parameter	Nonhydrolyzed protein		Prehydrolyzed protein	
	With	Without	With	Without
[a]Initial weight (g)	190±12*	125±5*	125±8	110±6
[a]Growth rate (% day[-1])	0.62±0.04	0.69±0.04	0.72±0.07*	0.51±0.06*
[b]White muscle				
rRNA (mg g[-1])	0.86±0.08	0.97±0.06	1.11±0.05*	0.96±0.03*
rRNA activity	20.16±8.29[a]	22.60±4.00[a]	8.61±1.67[b]	10.67±1.92[b]
[b]Proximate composition				
Moisture (%)	68.5±1.0[a]	70.1±0.3[a]	71.5±0.4[b]	71.7±0.5[b]
Protein (%)	16.7±0.3	17.1±0.2	17.1±0.1	16.7±0.2
Lipid (%)	10.9±1.1[a]	9.2±0.4[a]	7.5±0.4[b]	7.9±0.4[b]
Ash (%)	2.3±0.3	2.1±0.1	2.3±0.0	2.2±0.1
Ratio of protein to lipid	1.6±0.2[a]	1.9±0.1[a]	2.3±0.1[b]	2.1±0.1

The fish of the two phenotypes were cultured together in the same tank, at a salinity of 16.9±0.2 ppt and temperature 10.8±0.1°C. Ribosomal RNA (rRNA) activity is presented as pmol ^{14}C-Phe mg rRNA[-1] min[-1].
The values with the sign * or different superscripts are significantly different ($p<0.05$).
Source: [a]Ref. 38; [b]K. Rungruangsak-Torrissen, E. Lied, and M. Espe (unpublished).

C. Effect on In Vivo Utilization of Dietary Protein

1. Feed Consumption and Feed Conversion Efficiency

Feed consumption is an important factor in considering the growth rate of animals. Increased growth rate in Atlantic salmon possessing the trypsin variant TRP-2*92 was due to better food conversion and protein efficiency at water temperature about 6°C (Table 6). Similar results were observed in Arctic charr (Table 6). Unfortunately, there were no differences in apparent digestibility coefficients of protein between the fish without and with the variant, in either Atlantic salmon or Arctic charr. This may be due to the low sensitivity of the chromic oxide method. Increased protein efficiency may be due to protein growth or to increase in water and lipid. Body composition was not analyzed in this experiment. Study of individual consumption at restricted rations by x-radiography method (39–41) did not indicate any differences in weight-specific consumption rates between the two phenotypes of Atlantic salmon postsmolts, with and without the variant TRP-2*92 (36). Surprisingly, consumption rates were similar for fish with more than two times difference in growth rate, indicating

Table 6 Growth and Feed Utilization (mean ± SEM) in Atlantic Salmon Smolts and Two Strains of Arctic Charr, Approximately 1 Year of Age, Group with and without Trypsin Variant TRP-2*92

	Atlantic salmon (55 days experiment)		Arctic charr (36 days experiment)	
	With TRP-2*92	Without TRP-2*92	Hammerfest (with)	Skogseid (without)
Initial weight (g)	58.0 ± 0.7	49.0 ± 0.7	87.7 ± 3.0*	33.0 ± 0.9*
Individual specific growth rate (% day⁻¹)	0.39 ± 0.01*	0.37 ± 0.01*	0.69 ± 0.04*	0.52 ± 0.07*
Feed conversion (dry feed fed per wet weight gain)	0.89 ± 0.01*	0.99 ± 0.02*	0.66 ± 0.04	0.82
Protein efficiency (wet weight gain per crude protein fed)	2.2 ± 0.0*	1.9 ± 0.0*	2.9 ± 0.2	2.3
Apparent digestibility coefficient of protein (%)	81.8	82.7	82.9 ± 1.2	80.6 ± 0.2

Average water temperature was about 6°C. The values without SEM have no replicate and cannot be compared.
* Within the same species, significantly different ($p<0.03$).
Source: Ref. 34.

that differences in growth are not always due to variations in feed consumption but could also be due to differences in feed utilization (36). In addition, a better feed utilization at a restricted ration was observed in TRP-2*92 salmon indicated by a significantly lower maintenance ration (0.11 % body weight day⁻¹) than the fish without the variant (0.13 % body weight day⁻¹) (Fig. 7). No feeding hierarchy was observed between the two phenotypes (36).

An experiment with Scottish Atlantic salmon fry having different trypsin isozyme patterns reared at varying temperature regimens indicated that trypsin phenotype affected feed conversion efficiency (FCE) and growth at different rearing temperatures (27). The Atlantic salmon (pattern 1) lacking the variants TRP-1*91 and TRP-2*92, which effectively function at temperature below 8°C, were smallest at ambient (7.3–17.4°C) and cold (3.8–14.4°C) water temperature regimens. This resulted from the temperature being below 8°C during acclimation and at the start of the experiment, although they had high FCEs at cold and ambient temperature regimens compared to the other phenotypes (Table 7). The

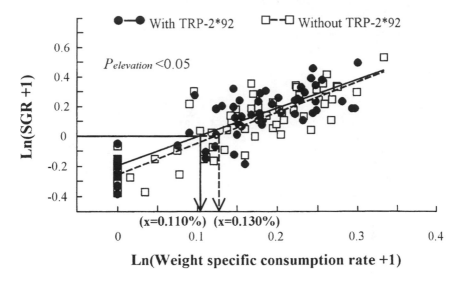

Figure 7. Relationship between weight-specific consumption rate (% body weight day⁻¹) and specific growth rate (SGR: % day⁻¹) of Atlantic salmon during starvation and restricted rations. The relationships are $Ln(y+1)=1.855 Ln(x+1) - 0.194$ ($n=59$, $r^2=0.667$, $p=0.0001$) for salmon with the variant, and $Ln(y+1)=2.053 Ln(x+1) - 0.250$ ($n=59$, $r^2=0.713$, $p=0.0001$) for salmon without the variant. The x values at y=0 indicate maintenance rations. (From Ref. 36.)

salmon (patterns 2′ and 2′+3) lacking the common isozyme TRP-2*100, which effectively functioned at temperature >8°C, had lower growth and/or FCE than the other phenotypes under the same temperature regime (Table 7). At a warm water temperature regimen of 11.9–19.9°C, the fish with different trypsin phenotypes had similar growth rates, but the TRP-2*100/92 fish carrying the trypsin variant (pattern 2) had significantly higher FCE than the other phenotypes (Table 7). The observations suggest that trypsin phenotypes affect feed utilization and growth at different temperatures. The fish possessing both common and variant trypsin isozymes will have an advantage on efficient utilization of diet at varying temperatures.

2. Free Amino Acids in Plasma and White Muscle

An effect on variations in growth rates between different trypsin phenotypes has been shown to be due to differences in protein utilization (34). Further attempts to study protein digestion were performed by using a more sensitive method through a detection of changes in tissue free amino acids (FAA) after feeding (42). Since FAA did not accumulate in the gut lumen at any time after feeding,

Table 7 Weight, Specific Growth Rate (SGR) and Feed Conversion Efficiency (FCE, wet weight gain per dry weight of feed consumed) of Scottish Atlantic Salmon Parr of Different Trypsin Phenotypes (see Fig. 1), Cultured at Different Temperature Regimens for 121 days

Water temperature	Trypsin pattern	n	Initial weight after 10 days acclimation (g)	Final weight (g)	SGR (% per day)	Weight-specific consumption rate (% body weight per day)	FCE
Warm (11.9–19.9°C)	1	70	5.68 ± 0.20	13.64 ± 0.42	0.93 ± 0.02	1.09 ± 0.03	0.81 ± 0.02
	2	22	5.29 ± 0.30	13.05 ± 0.55	0.97 ± 0.05	1.06 ± 0.07	0.91 ± 0.06*
	2'	1	3.05	7.51	0.95	1.31	0.69
	3	104	5.27 ± 0.15	12.93 ± 0.32	0.95 ± 0.02	1.13 ± 0.03	0.81 ± 0.02
	2+3	31	5.72 ± 0.24	14.23 ± 0.24	0.96 ± 0.03	1.14 ± 0.05	0.80 ± 0.02
	2'+3	6	5.11 ± 0.82	12.21 ± 1.35	0.95 ± 0.07	1.26 ± 0.03	0.69 ± 0.05
Ambient (7.3–17.4°C)	1	74	5.56 ± 0.21^{a}	10.55 ± 0.44^{a}	0.65 ± 0.02^{a}	0.90 ± 0.03	0.73 ± 0.03
	2	31	6.21 ± 0.22	12.06 ± 0.55^{b}	0.67 ± 0.04^{a}	0.89 ± 0.03	0.75 ± 0.04
	2'	10	5.84 ± 0.58	10.56 ± 1.57	0.55 ± 0.09^{a}	0.93 ± 0.07	0.56 ± 0.09^{a}
	3	86	6.40 ± 0.16^{b}	13.44 ± 0.42^{b}	0.77 ± 0.02^{b}	0.93 ± 0.02	0.82 ± 0.03^{b}
	2+3	24	6.18 ± 0.31	12.44 ± 0.76^{b}	0.72 ± 0.04	1.01 ± 0.06	0.72 ± 0.05
	2'+3	11	6.15 ± 0.53	12.77 ± 1.39^{b}	0.73 ± 0.07	0.96 ± 0.07	0.74 ± 0.08
Cold (3.8–14.4°C)	1	217	4.81 ± 0.13^{a}	9.87 ± 0.23^{a}	0.64 ± 0.01	0.72 ± 0.02	0.95 ± 0.02^{a}
	2	50	5.91 ± 0.31^{b}	10.64 ± 0.50	0.59 ± 0.03	0.68 ± 0.03	0.92 ± 0.05
	2'	21	5.64 ± 0.44^{b}	10.30 ± 0.76	0.59 ± 0.04	0.70 ± 0.04	0.90 ± 0.06
	3	69	5.48 ± 0.25^{b}	11.04 ± 0.41^{b}	0.65 ± 0.02^{a}	0.74 ± 0.03	0.97 ± 0.04^{a}
	2+3	29	7.73 ± 0.36^{c}	12.66 ± 0.37^{b}	0.54 ± 0.03^{b}	0.66 ± 0.03	0.84 ± 0.04^{b}
	2'+3	3	8.39 ± 1.84^{c}	13.04 ± 3.10	0.47 ± 0.06^{b}	0.69 ± 0.10	0.69 ± 0.02^{c}

* Within the same temperature regime, values with * and with different superscripts are significantly different ($p<0.05$).
Source: Ref. 27.

the rate of FAA increase in the plasma proved a reliable index of the rate of digestion, and incorporation of FAA into body protein may remove them almost as fast as they are absorbed (43). In Atlantic salmon, higher digestion and absorption of dietary protein were associated with the presence of trypsin variant TRP-2*92 (42). The total levels of postprandial FAA in both plasma (Fig. 8A) and white muscle (Fig. 9A) were significantly higher in salmon with than without the variant. Faster elevations of white muscle FAA within 6 h were observed in TRP-2*92 salmon, while it was 24 h in the other phenotype (Fig. 9A). Furthermore, significant increase in plasma lysine after the meal in salmon carrying the variant may also suggest a higher rate of digestion in these fish (Fig. 10), since lysine is a specific amino acid involved in peptide bonds hydrolyzed by trypsin (44, 45). Although the two phenotypes had similar growth rates (0.38 ± 0.04 % day^{-1}) during a 2 week experiment (Table 8), genetic differences in digestion, absorption, and transport of FAA were observed (Figs. 8A, 9A). Rungruangsak-Torrissen et al. (36) reported that differences in digestion, indicated by trypsin activity, affected growth 1 month later, and when the growths were different trypsin activities were similar. This should be a reason why variations in digestibility and growth may not be observed at the same time.

High absorption of amino acids in the plasma is not always due to high-quality feed. A low-quality feed could also cause a high absorption of FAA in the plasma but the transport of FAA to the muscle may not be efficient due to amino acid imbalance. It was observed by Torrissen et al. (38) that the concentration of free amino acids in the feed containing highly hydrolyzed protein resulted in low level ratio of essential to nonessential FAA with a slow assimilation of essential FAA in the white muscle (Fig. 11). It was hypothesized that imbalances in the plasma essential FAA result in increasing amino acid catabolism via TCA cycle, which in turn results in a lower proportion of absorbed essential FAA being utilized for protein synthesis (48). The TRP-2*92 fish fed partially prehydrolyzed protein had higher growth rate (0.72 ± 0.07% day^{-1}) than the fish without the variant (0.51 ± 0.06% day^{-1}) fed the same feed (Table 5) whereas there were no differences in growth rates (0.52 ± 0.06% day^{-1}) between the two phenotypes fed highly hydrolyzed protein (38). The indication of high feed utilization of high-quality feed for high growth has to be a rapid and high level of FAA, especially essential FAA, in both plasma and white muscle after feeding. The assimilation of essential FAA into protein in the white muscle must also be rapid and high. This was evidenced in the better growth of TRP-2*92 Atlantic salmon fed a better quality feed containing partially prehydrolyzed protein compared to the other phenotype and those fed the feed containing low-quality protein with high free amino acids due to high prehydrolysis (Fig. 11).

A similar experiment was performed in two strains of Arctic charr: a fast-growing Hammerfest and a slow-growing Skogseid. The levels of FAA in the two strains may not be directly comparable as in the Atlantic salmon, due to dif-

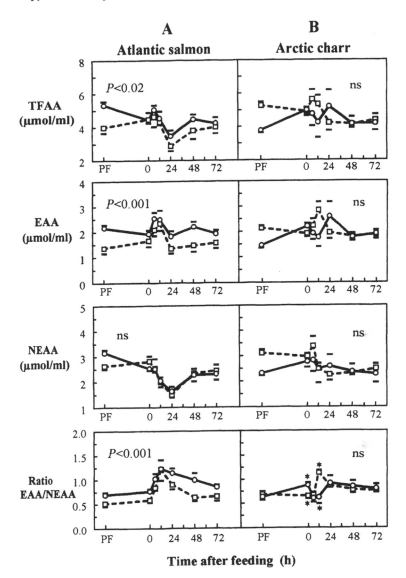

Figure 8. Postprandial total free amino acids (TFAA), essential free amino acids (EAA), and nonessential free amino acids (NEAA) in the plasma of Atlantic salmon in comparison to Arctic charr. Probability values indicate significant differences between the two phenotypes, with (—○—) and without (---□---) trypsin variant TRP-2*92, by paired analysis during the whole time course. The values with the sign * are significantly different ($p<0.05$) between the two phenotypes. PF, prefed values after 2 days starvation; ns, not significant. (Adapted from [A] Ref. 42 by permission of Academic Press and [B] Ref. 11.)

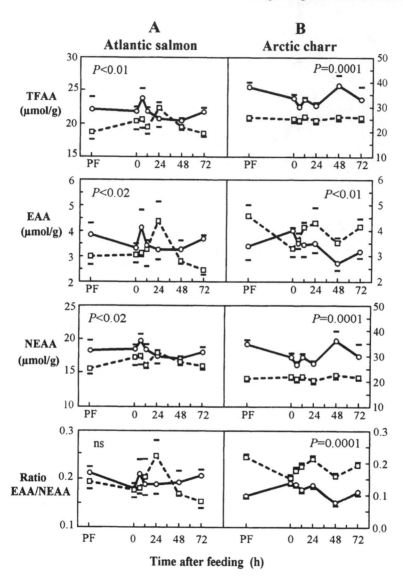

Figure 9. Postprandial total free amino acids (TFAA), essential free amino acids (EAA), and nonessential free amino acids (NEAA) in the epaxial white muscle of Atlantic salmon in comparison to Arctic charr. Probability values indicate significant differences between the two phenotypes, with (—O—) and without (---□---) trypsin variant TRP-2*92, by paired analysis during the whole time course. PF, prefed values after 2 days starvation; ns, not significant. (Adapted from [A] Ref. 42 by permission of Academic Press and [B] Ref. 11.)

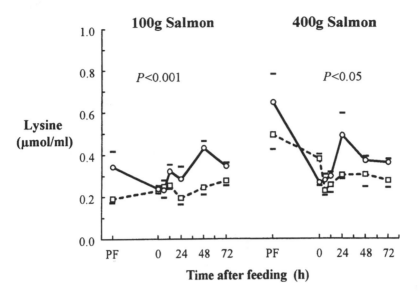

Figure 10. Postprandial lysine in the plasma of Atlantic salmon of two different sizes. Probability values indicate significant differences between the two phenotypes, with (—○—) and without (---□---) trypsin variant TRP-2*92, by paired analysis during the whole time course. PF, prefed values after 2 days starvation. (From Ref. 42.)

ferences in size at the same age (Table 8) and in the rate of transport of FAA from plasma (Fig. 8B) to muscle tissues (Fig. 9B). Higher rate of digestion and absorption of dietary protein in the fast-growing Hammerfest strain carrying the variant (see Fig. 1) may be indicated by a significantly higher elevation of total white muscle FAA after feeding, compared to the slow-growing Skogseid strain lacking the isozyme (Fig. 9B). Muscle protein synthesis was observed at 9 h after the meal in Atlantic salmon (49) and after 6 h in rainbow trout (50). The rate of muscle protein synthesis in Arctic charr may be similar to these two species. The level of essential FAA is important for protein synthesis. Therefor, decreased essential FAA after the rapid peak (0–6 h after feeding) in the white muscle in both Atlantic salmon (Fig. 9A) and Arctic charr (Fig. 9B) indicate a rapidly high assimilation of essential FAA and a more rapid rate of protein synthesis (Fig. 9).

The average concentrations of FAA in the plasma and the white muscle during the whole time course after feeding between different species at different sizes are illustrated in Table 8. The differences in total FAA in the plasma (5.78 ± 0.22 μmol ml^{-1}) and the white muscle (15.79 ± 1.22 μmol g^{-1}) after feeding for 400 g Atlantic salmon, compared to 100 g salmon (4.26 ± 0.16 μmol ml^{-1} and

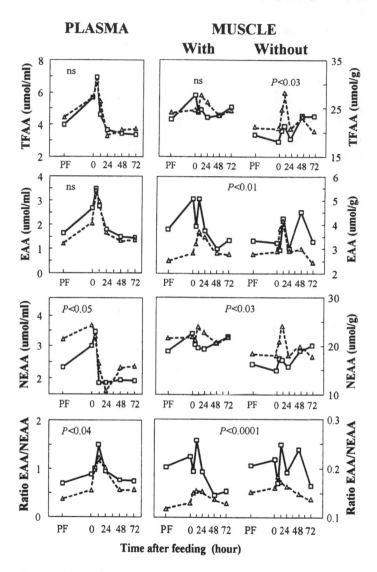

Figure 11. Postprandial total free amino acids (TFAA), essential free amino acids (EAA), nonessential free amino acids (NEAA), and the EAA to NEAA ratio in the plasma and the epaxial white muscle of Atlantic salmon with and without the trypsin variant TRP-2*92 fed feed with partially prehydrolyzed protein (—□—) or with highly prehydrolyzed protein (---Δ---). Probability values indicate significant differences between the two feed types by paired analysis during the whole time course. PF, prefed values after 2 days starvation; ns, not significant. (Adapted from Ref. 38 with permission from Prof. M.R. Jisnuson Svasti.)

20.80 ± 0.42 μmol g^{-1}, respectively, regardless of phenotypes), may be due to slower rate of transport of amino acids from plasma to the white muscle caused by slow/no growth rate. For Arctic charr, the transport seemed to be higher in 400 g Hammerfest (33.89 ± 2.27 μmol g^{-1}) than in 100 g Skogseid charr (25.48 ± 0.07 μmol g^{-1}), which had normal growth rates of 0.52 ± 0.13 and 1.12 ± 0.04% day^{-1}, respectively. Fish size may affect protein digestion, amino acid absorption, and the transport of amino acids to the white muscle, which influence growth later. Differences in trypsin phenotype affected protein utilization in both Atlantic salmon and Arctic charr. However proximate compositions were generally similar between these two species fed the same feed (Table 8).

3. Trypsin Activity

Genetic variation in trypsin isozyme pattern is proposed to be a primary factor affecting food conversion efficiency and growth under different rearing temperature (27). Trypsin activity, indicating an ability of digestion, increased in the digestive tract during early development and there were differences between Atlantic salmon families with different frequency of the variant TRP-2*92, after 80 days of the first feeding at 12°C (Fig. 6). The differences in trypsin activity between salmon possessing different trypsin phenotypes should be due to variations in the enzyme production during early development. When the studies were performed in salmon with a fully developed digestive system, a decrease in trypsin activity was detected after feeding (42, 51) and during growth (36). Elevations of total white muscle FAA after feeding also showed a negative association with trypsin activity in the pyloric cecal tissues of Atlantic salmon with different trypsin phenotypes (Table 8). The phenomenon of lower specific protease activity (μmol tyrosine h^{-1} mg protein^{-1}) in the pyloric cecal tissues of the TRP-2*92 Atlantic salmon fry with higher growth was observed, in contrast to the total protease activity (μmol tyrosine h^{-1} fry^{-1}) which was at a high level (14). This could be a consequence of variations in the secretion of the enzyme into the lumen, and therefore differences in reduced enzyme activity were observed in the region of cecal tissues sampled (51, 52).

In order to estimate secretion of enzymes in the pyloric cecal area, an experiment was set up in Atlantic salmon postsmolts, possessing or lacking the variant TRP-2*92 to examine the activities of trypsin and chymotrypsin during starvation and after feeding (R Moss and K Rungruangsak Torrissen, unpublished). The enzyme activities were determined both in the pyloric cecal tissues (empty ceca without luminal content) and in the whole cecum with luminal content from the same fish (Table 9). After 10 days starvation, both trypsin and chymotrypsin seemed to be at a basal level and associated with the cecal tissue. Enzyme activities were similarly low in the whole pyloric cecum, and the levels were lower than in the cecal tissues due to protein (mucous) in the lumen. At 5–7

Table 8 Size, Growth, Average Postprandial Free Amino Acids during the Whole Time Course, Trypsin Activity and Proximate Composition of Atlantic Salmon and Arctic Charr of Different Phenotypes, with and without Trypsin Variant TRP-2*92, in Comparison to Atlantic Salmon, Rainbow Trout and Atlantic Cod of Different Sizes

	Atlantic salmon[a]		Arctic charr[b]		Atlantic salmon[a,c]	Rainbow trout[d]	Atlantic cod[e]
	With	Without	Hammerfest (with)	Skogseid (without)			
Wet weight (g)	110±5*	92±3*	432±23*	100±8*	436±27	265±38	365±31
Specific growth rate (% day⁻¹)	0.38±0.04	0.38±0.04	0.52±0.13*	1.12±0.04*	-0.28±0.02		
Time course (h)	72	72	72	72	72	24	24
Plasma free amino acids (μmol ml⁻¹)							
Total free amino acids	4.50±0.16*	4.01±0.18*	4.47±0.56	4.84±0.36	5.78±0.22	—	6.00±0.79
Essential free amino acids (EAA)	2.15±0.09*	1.69±0.11*	1.95±0.24	2.11±0.19	2.18±0.06	—	1.89±0.35
Non-essential free amino acids (NEAA)	2.34±0.11	2.32±0.12	2.52±0.34	2.73±0.19	3.59±0.18	—	4.11±0.77
EAA/NEAA Ratio	0.95±0.05*	0.76±0.06*	0.77±0.04	0.77±0.04	0.65±0.03	—	0.56±0.18
Muscle free amino acids (μmol g⁻¹)							
Total free amino acids	21.75±0.44*	19.84±0.43*	33.89±2.27*	25.48±0.07*	15.79±1.22	32.77±0.59	15.80±0.72
Essential free amino acids (EAA)	3.56±0.18*	3.16±0.18*	3.40±0.31*	3.96±0.33*	2.80±0.13	14.24±0.74	7.49±0.57
Non-essential free amino acids (NEAA)	18.19±0.35*	16.68±0.31*	30.50±2.12*	21.53±0.87*	13.00±1.28	18.53±1.04	8.31±0.36
EAA/NEAA Ratio	0.20±0.01	0.19±0.01	0.11±0.01*	0.18±0.01*	0.22±0.03	0.79±0.09	0.91±0.07
Specific activity of trypsin (μmol p-nitroaniline h⁻¹ mg protein⁻¹)							
Intestinal content	70.14±12.9	76.89±14.5	56.64±12.01	35.15±7.33	100.3±10.3	—	—
Pyloric cecal tissues	12.29±1.23*	15.58±1.41*	13.25±1.83*	6.02±0.80*	5.17±1.02	—	—
Proximate composition[c] (% of wet weight)							
Moisture	68.5±1.0	70.1±0.3	70.2±0.5	71.9±0.6	—	—	—
Protein	16.7±0.3	17.1±0.2	15.9±0.3	15.5±0.3	—	—	—
Lipid	10.9±1.1	9.2±0.4	10.0±0.5	8.7±0.7	—	—	—
Ash	2.3±0.3	2.1±0.1	1.7±0.1	1.8±0.1	—	—	—
Ratio of protein to lipid	1.6±0.2	1.9±0.1	1.6±0.1	1.9±0.2	—	—	—

* Within the same species, significantly different ($p < 0.05$).

The average rearing temperature was about 10°C in all experiments.

Sources: [a]Calculated from Ref. 42; [b]calculated from Ref. 11; [c]K. Rungruangsak-Torrissen, E. Lied and M. Espe (unpublished); [d]calculated from Ref. 46; [e]calculated from Ref. 47.

Table 9 Enzyme Activities of Trypsin and Chymotrypsin in µmol p-Nitroaniline h^{-1} mg protein^{-1} (mean ± SEM)

Enzyme activity	Pyloric cecal tissues		Whole pyloric cecum	
	Without	With	Without	With
Trypsin (T)				
10 days starvation	31.85±5.42[a]	40.53±6.08[a]	21.93±4.32[b]	22.19±3.52[b]
5–7 h postfeeding	52.05±3.15[a]	38.10±3.77[b]	36.23±5.02[b]	41.39±4.61
Chymotrypsin (C)				
10 days starvation	18.48±1.92[a]	17.56±1.17[a]	14.08±2.47	11.12±2.08[b]
5–7 h postfeeding	28.48±2.37[a]	27.88±2.39[a]	64.41±5.37[b]	68.77±5.15[b]
Activity ratio of T/C				
10 days starvation	1.92±0.31	2.35±0.30	2.16±0.30	2.31±0.36
5–7 h postfeeding	1.95±0.15[a]	1.53±0.16[a]	0.62±0.11[b]	0.64±0.09[b]

Values are given for the pyloric cecal tissues (empty cecum without luminal content) and in the whole pyloric cecum (with luminal content), of Atlantic salmon with ($n = 19$) and without ($n = 17$) trypsin variant TRP-2*92.
The values with different superscripts are significantly different ($p<0.02$).
Source: R. Moss and K. Rungruangsak-Torrissen (unpublished).

h postfeeding, most chymotrypsin was secreted into the lumen since its activity in the whole cecum with food content (66.6 ± 5.1 µmol p-nitroaniline h^{-1} mg protein^{-1}) was higher than in the cecal tissues without food content (28.2 ± 2.2 µmol p-nitroaniline h^{-1} mg protein^{-1}), regardless of phenotypes. Trypsin appeared to be a membrane-associated enzyme in the pyloric cecal tissues, since its activity was not high in the whole cecum. Trypsin may be an apical glycocalyx-bound enzyme where secondary digestion has occurred prior to absorption, a mechanism reviewed by Ugolev and Iezuitova (53). Total enzyme activities after feeding in the whole cecum were not directly comparable to the activities in the cecal tissues due to high protein in the food content. It appeared that the total trypsin activity in the whole cecum was similar between the fish with (41.4 ± 4.6 µmol p-nitroaniline h^{-1} mg protein^{-1}) and without (36.2 ± 5.0 µmol p-nitroaniline h^{-1} mg protein^{-1}) the variant, while the remain of the activity in the cecal tissues was lower in the fish with (38.1 ± 3.8 µmol p-nitroaniline h^{-1} mg protein^{-1}) than without (52.1 ± 3.2 µmol p-nitroaniline h^{-1} mg protein^{-1}) the variant (Table 9). The results indicate higher secretion of trypsin into the lumen of the fish with than without the variant. This should be the reason why trypsin activity remained lower in the pyloric cecal tissues of the TRP-2*92 salmon. The activity ratio of trypsin to chymotrypsin was similar between the cecal tissues and the whole cecum during starvation, while it changed after feeding due to a high secretion of chymotrypsin (Table 9). There was a significant relationship between trypsin and chymotrypsin activity in the pyloric cecum during starvation (Fig. 12A), but the

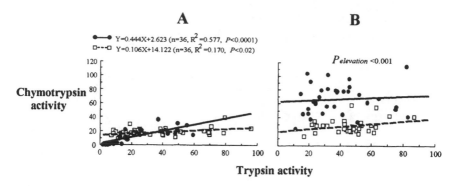

Figure 12.   Relationship between trypsin and chymotrypsin activities (μmol *p*-nitroaniline h^{-1} mg protein^{-1}) in Atlantic salmon, regardless of trypsin phenotypes, **(A)** during starvation and **(B)** during 5–7 h postfeeding, in the whole pyloric cecum with luminal content (—●—) and the pyloric cecal tissues without luminal content (---□---). (From R Moss and K Rungruangsak-Torrissen, unpublished.)

relationship disappeared after feeding (Fig. 12B). Chymotrypsin activity, in both the whole pyloric cecum and the cecal tissues, was at similar levels between different trypsin phenotypes both during starvation and after feeding, while trypsin activity in the pyloric cecal tissues after feeding was different between phenotypes (Table 9). Trypsin phenotype should be a key factor to differentiate protein digestibility in fish. In Arctic charr, the fast-growing Hammerfest strain possessing trypsin variant TRP-2*92 (see Fig. 1) has higher trypsin activity in both the pyloric cecum and the intestine than the slow-growing Skogseid charr without the variant (Table 8).

To study genetic differences, it is difficult to observe any variations in enzyme activity unless the assay is performed at around optimum temperature. Trypsin activity has been assayed at 50°C, which is close to the earlier observation of 52.5°C of trypsinlike activity by Torrissen (10), and chymotrypsin activity at 40°C (Fig. 13). The enzymes from the intestinal tissues are usually extracted with 1mM HCl, according to Rungruangsak and Utne (54). An experiment was set up to study the effect of acid and alkaline extractions on the activity of trypsin and chymotrypsin (R Moss and K Rungruangsak-Torrissen, unpublished). Pyrolic caeca were extracted with either 1mM HCl or 0.2M Tris buffer pH 8.4, and the same enzyme extracts were used for the determination of trypsin and chymotrypsin activities at different temperatures. There were no differences in trypsin activity with the substrate benzoyl-DL-arginine-p-nitroanilide (BAPNA), whether the enzyme was extracted and kept frozen in 1mM HCl or in 0.2M Tris buffer pH 8.4 (Fig. 13). Neither differences in chymotrypsin activity were observed with the substrate succinyl-Ala-Ala-Ala-Pro-Phe-p-nitroanilide,

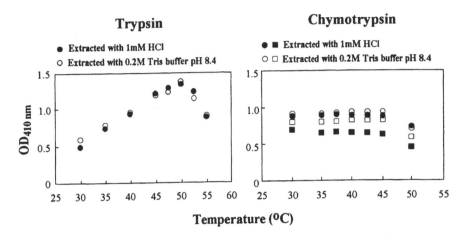

Figure 13. Enzyme activities, at different assayed temperatures, of trypsin and chymotrypsin in the pyrolic cecum extracted with either 1mM HCl or 0.2M Tris buffer pH 8.4. The substrate for trypsin was benzoyl-DL-arginine-p-nitroanilide (BAPNA). The substrates for chymotrypsin were succinyl-Ala-Ala-Ala-Pro-Phe-p-nitroanilide (● and ○) and N-acetyl-L-tyrosine-p-nitroanilide (■ and □). (From R Moss and K Rungruangsak-Torrissen, unpublished.)

but the activity could be different ($P<0.05$) between the acid and buffer extractions if the substrate was N-acetyl-L-tyrosine-p-nitroanilide (Fig. 13). It is practical to use the same enzyme extract for different assays of enzyme activities, but suitable substrates have to be chosen. Although the crude enzyme homogenate was extracted in acid solution (1mM HCl), trypsin activity was assayed in the buffer of an alkaline pH of 8.4, and the crude enzyme extract was always kept below 4°C prior to reaction. It is possible that salmon trypsin is stable in acid solution at low temperature or it is trypsinogen, which is stable in acid, that was activated during the assay as observed by Dimes et al. (55). Similar effect may also occur with chymotrypsinogen, which is probably activated by trypsin during the assay.

Torrissen (10) illustrated the effect of different assayed temperatures on the protease activities in different parts of the digestive tract in both Atlantic salmon and rainbow trout, using casein as substrate. The specific activities of alkaline proteases (μmol tyrosine h^{-1} mg protein^{-1}) were ranked in different parts of the intestinal tissues as the first half large intestine > the second half small intestine > pyloric cecum = the second half large intestine > the first half small intestine in Atlantic salmon, and the first half large intestine > pyloric cecum > the second half small intestine = the second half large intestine > the first half small intestine in rainbow trout (Fig. 14). The first half of small intestine seemed to be

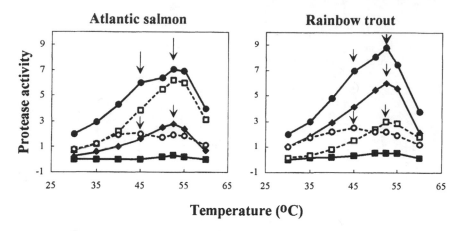

Figure 14.   Temperature profiles of alkaline protease activity (μmol tyrosine h^{-1} mg protein^{-1}), using casein as substrate, in different parts of the digestive tract: the pyloric cecum (—◆—), the first half small intestine (—■—), the second half small intestine (---□---), the first half large intestine (—●—), and the second half large intestine (---○---), in Atlantic salmon and rainbow trout. The arrows indicate two peaks at 45°C and 52.5°C. (Adapted from Ref. 10 with permission from Elsevier Science.)

mainly an absorption site, as it did not have any significant protease activities. Two peaks at temperatures of 45°C and 52.5°C were observed for alkaline proteases in the intestine (Fig. 14). Purified trypsin (56) and chymotrypsinlike (57) enzymes from the pyloric cecum of rainbow trout showed their optimum temperatures at 60°C and 55°C, respectively, in the presence of $CaCl_2$. Based on these evidences of optimum temperatures of the alkaline proteases and the results from Figures 13 and 14, it could be explained that trypsin and trypsinlike enzymes dominate in the pyloric cecum, whereas both trypsin/trypsinlike and chymotrypsin/chymotrypsinlike enzymes dominate in the small and large intestine.

It is interesting to note the work by Lemieux et al. (58) in Atlantic cod (*Gadus morhua*). Among different enzyme activities involved in digestion (acid proteases, trypsin, and chymotrypsin), absorption (alkaline phosphatase), and transport (glutamyltransferase) processes in the digestive system of Atlantic cod injected with different levels of recombinant bovine somatotropine, trypsin is the only enzyme measured that showed a significant correlation with food conversion efficiency. The work provides important support to the notion that trypsin is the key enzyme in limiting growth rate of the fish through its special role in the digestion process. Haard et al. reported that trypsin from coho salmon (*Oncorhynchus kisutch*) fingerling fed diets with heated or unheated soybean meal was less sensitive to inhibition by soybean trypsin inhibitor than fish fed no soybean meal (59). Seldal et al. demonstrated that a grazing-induced action of trypsin inhibitors in food plants was a possible cause for survival rates and the cyclic dynamics of lemming populations (60). Trypsin is influenced by external factors, such as temperature and food quality, and it is a very important enzyme involving in food utilization, growth, and survival of both animals and plants. Studies on trypsin's properties would provide information on how growth and survival are controlled through food utilization.

4. Plasma Insulin

Recent investigation aimed at explaining the mechanisms responsible for the observed differences in performance between Atlantic salmon of different trypsin phenotypes indicated that variation in trypsin isozymes also affected plasma insulin concentration (36). In addition, growth was found to be associated with a decrease in trypsin activity in the pyloric cecal tissues, due to its high secretion into the lumen (see page 237), accompanied by an increase in plasma insulin 1 month before an enhanced growth was observed. Rungruangsak-Torrissen et al. (36) reported a significant relationship between weight-specific consumption rate and plasma insulin level, and also an interesting different phenomenon of plasma insulin and growth in Atlantic salmon postsmolts of different trypsin phenotypes fed at 0.5% and 1% body weight day^{-1}. The fish possessing the variant TRP-2*92 had a greater ability to utilize the feed at a restricted ration (0.5% body weight day^{-1}). They exhibited similar plasma insulin levels to the fish fed 1% body weight day^{-1} while their growth rates were lower. This resulted in a similar growth rate 1 month later between the TRP-2*92 fish fed the two different rations (Table 10). In the fish lacking the variant fed the different rations, both plasma insulin levels and growth rates were different during the whole experimental period (Table 10). A preliminary experiment was carried out in Atlantic salmon weighing of 350–1550 g reared at a temperature of about 6°C and fed at full ration to determine general association between the levels of insulin and free amino acids (FAA) in the plasma (K Rungruangsak-Torrissen and A Sundby, unpublished). Plasma insulin levels were found to correlate with specific growth rate (Fig. 15A) and the concentrations of total FAA (Fig. 15B). The correlation with the total FAA in the plasma was due to the levels of essential FAA (Fig. 15B) that are very important precursors for protein synthesis.

Insulin is an anabolic hormone known to be involved in protein metabolism in fish (61–63), and to stimulate growth in rainbow trout (64). A significant correlation between plasma insulin levels and body weight was reported in different salmonids (36, 65). In addition, rainbow trout from fast-growing families had significantly higher plasma insulin levels than those from slow-growing families (65), and the levels of plasma insulin were reported to be higher in 3.5 years old Atlantic salmon fed to satiation than those fed half the satiation ration

Table 10 Growth and Plasma Insulin Concentration in Atlantic Salmon with and without Trypsin Variant TRP-2*92, Fed at 0.5% and 1% Body Weight Day^{-1}

Parameter	With TRP-2*92		Without TRP-2*92	
	0.5 % ration	1% ration	0.5% ration	1% ration
Initial weight (g)	88.6±1.9*		79.1±1.6*	
Growth rate at day				
164 (% day^{-1})	0.40±0.02^a	0.53±0.02^b	0.38±0.03^a	0.54±0.04^b
Plasma insulin at day				
164 (ng ml^{-1})	15.57±1.67	16.20±1.48	14.13±1.07*	17.67±1.39*
Growth rate at day				
190 (% day^{-1})	0.66±0.07	0.77±0.09	0.47±0.09*	0.79±0.09*

The fish had similar weight-specific consumption rates studied by x-radiography of individuals. The values with * or different superscripts are significantly different ($p<0.05$). *Source:* Adapted from Ref. 36

(66). Plasma insulin concentration has been observed to correlate with rates of food consumption (36), plasma FAA (Fig. 15B), and growth (36, 65). It may be possible that higher protein digestion and absorption rate of FAA in the plasma of the TRP-2*92 fish, compared to the fish lacking the variant (Fig. 8A), stimulate higher secretion of plasma insulin for anabolic stimulation of protein synthesis, which results in higher protein growth efficiency in these fish (34). The work on the relationship between protease activities, plasma FAA, and insulin in Atlantic salmon with different trypsin phenotypes support this hypothesis (K Rungruangsak-Torrissen and A. Sundby, to be published).

5. Capacity for Protein Synthesis

The capacity for protein synthesis (μg RNA mg protein^{-1}) and RNA activity (g protein synthesized g RNA^{-1} day^{-1}) has been shown to be positively correlated with rates of food consumption and growth in Atlantic salmon (40). The TRP-2*92 salmon exhibited higher capacity for protein synthesis in the white muscle, due to higher RNA concentration during starvation, than the fish lacking the variant (36). The estimate of white muscle protein synthesis in vivo by a flooding injection of [^{3}H]phenylalanine (67, 68) did not indicate any differences in the rates between the salmon with and without the variant after 2 weeks starvation (36). White muscle RNA concentrations positively correlated with growth during starvation, regardless of trypsin phenotypes. In contrast, RNA activity showed a negative correlation with RNA concentration ($r = -0.53$, $p<0.01$, Fig. 16A) (36), similar to the other observation of a significantly negative correlation between ribosomal RNA (rRNA) activity and rRNA concentration ($r =$

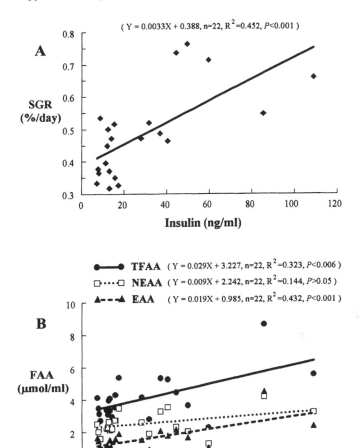

Figure 15. Relationships of plasma insulin concentration with **(A)** daily specific growth rate (SGR) and **(B)** the concentration of plasma free amino acids (FAA) in Atlantic salmon 3–9 kg fed at full ration at about 8°C. TFAA, total FAA; EAA, essential FAA; NEAA, nonessential FAA. (From K Rungruangsak-Torrissen and A Sundby, unpublished.)

-0.58, $p<0.01$, Fig. 16B) in in vitro protein synthesis studied in the white muscle of 100 g salmon fed at a restricted ration of 0.5 % body weight day^{-1} (K Rungruangsak-Torrissen, E Lied, and M Espe, unpublished). These results were suggested to be due to low rate of protein turnover in the fish with high capacity for protein synthesis (36). Carter et al. (40) attempted to explain the faster

● **With TRP-2*92**   □ **Without TRP-2*92**

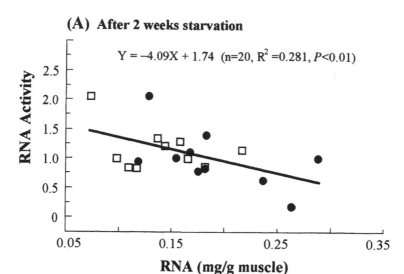

(A) After 2 weeks starvation

Y = −4.09X + 1.74 (n=20, R^2 =0.281, P<0.01)

RNA Activity

RNA (mg/g muscle)

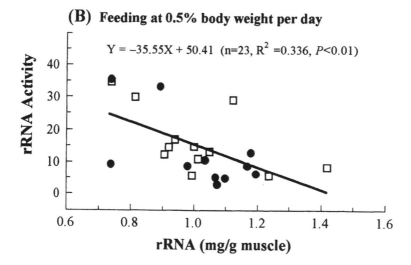

(B) Feeding at 0.5% body weight per day

Y = −35.55X + 50.41 (n=23, R^2 =0.336, P<0.01)

rRNA Activity

rRNA (mg/g muscle)

Figure 16. Relationships of **(A)** RNA activity (g protein synthesized g RNA^{-1} day^{-1}) in in vivo protein synthesis and **(B)** rRNA activity (pmol[^{14}C]phenylalanine mg rRNA^{-1} min^{-1}) in in vitro protein synthesis, with their concentrations in the white muscle of Atlantic salmon with and without the variant TRP-2*92. ([A] Data from Ref. 36 with permission from Kluwer Academic Publishers; [B] from K Rungruangsak Torrissen, E Lied and M Espe, unpublished.)

growth and higher efficiency of some salmon compared with individuals with the same food intake. They suggested that the influx of dietary protein–nitrogen caused greater anabolic stimulation of protein synthesis of which more was retained in the high efficiency fish. Thus, small differences in protein turnover that did not attain the level of statistical significance lead to the difference in protein growth efficiency between individual salmon feeding at similar rates. Therefore the TRP-2*92 salmon was defined as a high-proteingrowth-efficiency fish with low protein turnover rate (36). It was also suggested that during starvation the TRP-2*92 salmon, which have low maintenance, maintain higher RNA concentrations and are more responsive to changes in food intake. There may also be a similar difference between the two phenotypes during feeding and growth. This suggestion is supported by evidence that concentrations of white muscle rRNA in the salmon with and without the variant TRP-2*92 did not seem to be different in the white muscle unless their growth rates were different (Table 5). The rRNA activity was affected by the dietary protein quality (Table 5). In Arctic charr, the RNA in the white muscle of a fast-growing Hammerfest strain exhibited a significantly higher level than that of a slow-growing Skogseid strain (69). These results suggest that fast-growing fish have higher RNA and rRNA and a higher capacity for protein synthesis for protein growth in the white muscle.

There are several indications that the major effect of the variant TRP-2*92 is postabsorptive and related to protein metabolism. Higher postprandial elevations of alanine and glutamic acid were observed in these salmon (42). Glutamic acid is a pivotal molecule in nitrogen metabolism (70), and alanine is produced by active skeletal muscle and erythrocytes as a major raw material of gluconeogenesis (71). Gluconeogenesis plays a role in teleost white muscle as evidenced by the study of the enzyme fructose-1,6-biphosphatase (72). Almost every single FAA in the white muscle of salmon with the variant increased with the peak at 6 h postfeeding (42), which is about the time that an increase in the rate of muscle protein synthesis was observed in salmonids (49, 50). By the time the white muscle FAA concentrations were highest in the TRP-2*92 salmon, glutamine, glycine, and methionine were significantly higher than they were in salmon without the variant (42). For salmon lacking the variant, white muscle FAA peaked later at 24 h postfeeding, and while the FAA were at the highest levels no differences in FAA concentrations were observed between the two phenotypes (42). Glutamine concentration was reported to correlate with muscle protein synthesis, and it has been shown to inhibit muscle proteolysis (73). Glycine is an essential amino acid for collagen structure (74), and methionine is involved in the initiation mechanism for protein synthesis (75). These results indicate that the TRP-2*92 salmon have higher capacity for protein synthesis in the white muscle than the fish without the variant.

D. Effect on In Vitro Digestion

In vitro digestion potential, expressed as amino group liberation, of the pyloric cecal enzyme extract suggested an advantage of having diverse trypsin isozymes for efficient digestion of different-quality fish meals based on mink digestibility (35). Crude enzyme extracts from the pyloric cecum of Atlantic salmon carrying both the common isozyme TRP-2*100 and the variant TRP-2*92 (Pattern 2, see Fig. 1) had a significantly higher digestibility than the enzyme extracts from the salmon either lacking the variant (Pattern 1, see Fig. 1) or lacking the common isozyme (Pattern 2′, see Fig. 1), for both normal (90 % mink digestibility) and low- (86% mink digestibility) quality fish meals (Fig. 17). No differences in in vitro digestibility between different trypsin phenotypes were observed in high-quality fish meal of 94% mink digestibility (Fig. 17). Different trypsin pheno-

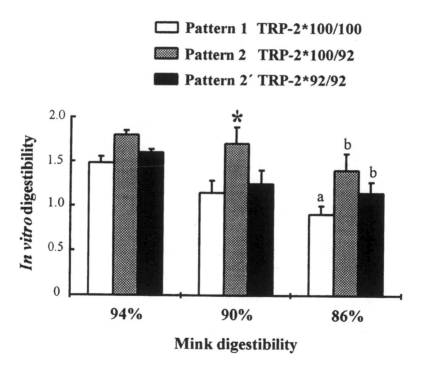

Figure 17. In vitro digestion potential (expressed as amino group liberation in 10^{-4} mole alanine equivalent) of the enzyme extracts from the pyloric cecum of Atlantic salmon possessing different trypsin phenotypes, using fish meals as substrates with different quality based on mink digestibility. Details of trypsin isozyme patterns are shown in Figure 1. The bars with the sign * or with different letters are significantly different ($p<0.05$). (Adapted from Ref. 35 with permission from Kluwer Academic Publishers.)

types also had an effect on in vivo digestion and utilization of different dietary protein qualities determined by varying the degree of prehydrolysis with pepsin (38). Salmon possessing the variant TRP-2*92 had higher growth rate and white muscle rRNA levels than those lacking the variant when they were fed a feed containing partially prehydrolyzed protein (Table 5). These results indicate an interaction between feed quality and genetic differences in feed utilization, and support an advantage of possessing variation in different isozymes for growth under varying living conditions.

An interesting aspect that should be noted is that digestibility study of amino acids in the feeds, containing different dietary proteins with variations in degree of prehydrolysis by pepsin, in Atlantic salmon of different trypsin phenotypes indicated a high degree of differences in cysteine digestibility (47–91%) (38). This may suggest differences in the content of sulfydryl group and disulfide bond in the protein molecule of different feed qualities. Lower cysteine digestibility seemed to be associated with higher degree of protein prehydrolysis (38, 76), suggesting a very low absorption of cysteine. There were evidences that disulfide bond of cystine in the protein molecules negatively affected the digestibility of the proteins by in vitro digestion (77, 78) and in vivo digestion (79) studies. It seemed unlikely that different trypsin isozymes may have different digestibility in association with the sulfydryl group and the disulfide bond. Further investigation on this aspect is required to explain whether the effect is from trypsin isozymes or other unknown covariate factor(s).

IV. OTHER POSSIBLE ASSOCIATIONS

A. Muscle Growth

Glutamic acid, glutamine, and alanine are major amino acids involved in nitrogen metabolism of active skeletal muscle (70, 71). These amino acids were higher in the white muscle of salmon with than without the variant (Table 11). In addition, the TRP-2*92 salmon also had a higher elevation of white muscle glycine and proline after feeding (Table 11). Since proline and glycine are essential for collagen (74), the presence of higher concentrations of these amino acids together with glutamic acid, glutamine, and alanine may support a higher protein metabolism for TRP-2*92 salmon in remodelling connective tissue framework, which is a rate-limiting step for muscle growth suggested by Millward (73). Another amino acid that associates with muscle is hydroxyproline. This amino acid is found in collagen, a connective tissue with a directive role in developing tissues (74). White-muscle hydroxyproline concentrations were similar in salmon with and without trypsin variant TRP-2*92 when they had similar growth rates (fed nonhydrolyzed protein), while the values were significantly different in the two phenotypes (fed prehydrolyzed protein) when they had significantly different growth

Table 11 White Muscle Free Amino Acids after Feeding in Atlantic Salmon and Arctic Charr, with and without Trypsin Isozyme TRP-2*92

White muscle free amino acids (μmol/g wet weight)	Nonhydrolyzed protein[a]		Prehydrolyzed protein[b]		Nonhydrolyzed protein[b,c]	
	Salmon with	Salmon without	Salmon with	Salmon without	Hammerfest with	Skogseid without
Total free amino acids	21.75±0.43*	19.84±0.51*	24.47±0.63*	20.70±0.81*	33.89±0.81*	25.48±0.07*
Ratio of EAA/NEAA	0.20±0.01	0.19±0.01	0.20±0.01	0.21±0.01	0.11±0.01	0.18±0.01
Glutamic acid	3.57±0.09*	3.15±0.13*	2.82±0.14*	2.44±0.42*	3.06±0.11*	4.16±0.11*
Glutamine	0.22±0.03*	0.19±0.02*	0.38±0.04*	0.26±0.04*	0.20±0.02	0.17±0.01
Alanine	2.98±0.11*	2.34±0.09*	3.45±0.13*	2.91±0.11*	3.12±0.08	2.97±0.11
Proline	0.25±0.03*	0.20±0.03*	0.31±0.03*	0.25±0.04*	0.24±0.01	0.25±0.02
Glycine	5.18±0.17*	4.94±0.23*	6.07±0.38	5.56±0.50	8.74±0.41*	6.21±0.38*
Hydroxyproline	0.51±0.04	0.58±0.04	1.00±0.05*	0.38±0.06*	0.75±0.08*	0.39±0.01*
β-Alanine	0.50±0.03	0.42±0.02	0.53±0.05*	0.34±0.03*	0.70±0.06*	1.34±0.11*
Taurine	3.53±0.16	3.45±0.09	4.27±0.19*	3.66±0.09*	4.16±0.20*	3.62±0.10*
Anserine	11.86±0.02*	11.62±0.03*	16.45±0.13*	12.86±0.43*	10.51±0.24	not detected

Results are presented as mean ± SEM. Within the same species and the same feed, values with the sign * are significantly different ($p<0.05$) between the two phenotypes, by paired analysis during the whole time course (from prefed to 72 h postfeeding). EAA, essential free amino acids; NEAA, nonessential free amino acids.

Sources: [a]Calculated from Ref. 42; [b]K. Rungruangsak-Torrissen, E. Lied and M. Espe (unpublished); [c]calculated from Ref. 11.

rates (Tables 5 and 11). Since proline is a precursor of the hydroxyproline residues in collagen whereas exogenous hydroxyproline is not, and collagen is hydrolyzed by collagenases during growth and remodeling (74), the elevation of white muscle hydroxyproline indicates hydrolysis of collagen. The fish with high growth rates have high levels of hydroxyproline in the white muscle while low levels were found in the plasma, suggesting less mobilization of collagen/muscle protein as an energy source (Fig. 18). The results suggest a higher collagen catabolism for remodeling connective tissue framework for muscle growth in the TRP-2*92 salmon with higher growth rate. A significantly lower concentration of hydroxyproline in the plasma of the TRP-2*92 salmon after 24 h postfeeding also suggests less mobilization of collagen/muscle protein as an energy source during low nutritional status in these salmon compared to the other phenotype with similar restricted growth rate (42).

Comparisons were also performed between the two strains of Arctic charr, an anadromous Hammerfest and nonanadromous Skogseid, which had different sizes at the same age. Higher levels of glutamine, alanine, and glycine were also observed in fast-growing Hammerfest strain, but not all differences were significant, compared to those of slow-growing Skogseid strain (Table 11). Glutamic acid was lower in the white muscle of Hammerfest than Skogseid, controversially to the levels found between Atlantic salmon with different trypsin pheno-

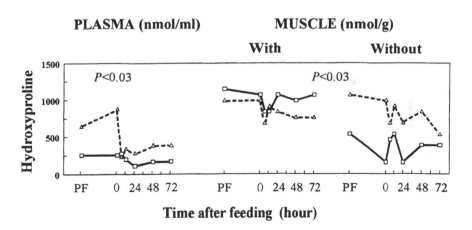

Figure 18. Postprandial hydroxyproline in the plasma and the epaxial white muscle of Atlantic salmon with and without the trypsin variant TRP-2*92 fed feed with partially pre-hydrolyzed protein (—☐—) or with highly prehydrolyzed protein (---Δ---). Probability values indicate significant differences between the two feed types by paired analysis during the whole time course. PF, prefed values after 2 days of starvation. (Adapted from Ref. 38 with permission from Prof. M.R. Jisnuson Svasti.)

types (Table 11). Proline was similar while hydroxyproline was different between the two Arctic charr (Table 11). The Hammerfest charr have higher protein metabolism indicated by higher white muscle RNA (69) and FAA in the plasma and the white muscle (Figs. 8B, 9B) than in the Skogseid strain. The different levels of glutamine, alanine, glycine, and hydroxyproline in the white muscle between the two strains of Arctic charr should also indicate differences in protein synthesis and collagen catabolism for remodeling of developing tissues for muscle protein growth, as in Atlantic salmon, and that the higher the concentration, the higher the metabolism (Table 11). High elevations of these free amino acids (Table 11) and rapid decreased levels of essential free amino acids, due to high assimilation (Figs. 9 and 11), in the white muscle could indicate a high metabolism of protein growth in the fish white muscle.

B. Muscle-Buffering Capacity

Taurine, dipeptide anserine, and β-alanine were higher after feeding in the white muscle of Atlantic salmon smolts carrying than those lacking the trypsin variant TRP-2*92 (Table 11). The differences were significant when the fish had different growth rates (Tables 5 and 11). In addition, taurine was higher after feeding in the white muscle of anadromous Hammerfest charr carrying the variant than the nonanadromous Skogseid charr lacking the variant, and anserine was high in the white muscle of Hammerfest charr while it was not detected in the Skogseid charr (Table 11). Since these nitrogen-containing compounds are believed to participate in osmoregulation and biological buffering capacity (80–82), their concentrations in the white muscle may affect smoltification for sea migration of Atlantic salmon smolts and anadromous Arctic charr. It is possible that β-alanine is the major nitrogenous osmoregulatory compound for Skogseid's muscle, since its concentration was significantly higher than in the white muscle of Hammerfest charr (Table 11).

These results indicate that the TRP-2*92 Atlantic salmon, especially of pattern 2 (TRP-2*100/92), not only grow faster in the freshwater phase but their physiological conditions are also fast developed for the next phase of sea migration and utilization of variety of feeds under different temperatures.

C. Immune Parameters

Fish health should always be questioned in the fish with high growth rate, whether or not the genetic selection of growth performance would be performed. Experiments were set up to examine some selected immune parameters and disease resistance between groups of Atlantic salmon possessing and lacking the trypsin variant TRP-2*92 (83). Genetic variation in both specific and nonspecific immune parameters were observed among salmon with different trypsin phenotypes (Table 12). The parameters that respond to vaccination, such as spontaneous complement hemolytic activity (SH50) and lysozyme activity (84), and the antibodies against *Aeromonas salmonicida,* were varied among different phenotypes. Although total serum IgM did not seem to respond to vaccination (85), the values were different between phenotypes (Fig. 15, Table 12), indicating genetic differences in immunity among the salmon with different trypsin phenotypes. The total complement hemolytic activity (CH50) (84) and the total serum protein (85), which did not respond to vaccination, were similar among different trypsin phenotypes (Table 12). The fish of patterns 1 and 2, which had lower SH50 activity, had higher lysozyme activity, compared to the fish of pattern 2′ (Table 12). This is a phenomenon similar to the observation of a negative genetic correlation between the lysozyme and SH activities in Atlantic salmon reported by Røed et al. (86). There was a negative relationship between body weight and total serum IgM, and within the same weight range of 80–170 g the fish carrying the variant had higher total serum IgM (345 ± 30 µg ml^{-1}) than the fish lacking the variant (236 ± 18 µg ml^{-1}) (Fig. 19). The concentrations of serum IgM of the fish possessing the variant TRP-2*92 were higher 4 weeks after vaccination (Fig. 19), while the values were lower at 5 months postvaccination (Table 12), compared to the salmon without the variant. The controversial

Table 12 Specific and Nonspecific Immune Parameters in Atlantic Salmon Postsmolts possessing different trypsin isozymes in the TRP-2 system (see Fig. 1), at 5 Months Postvaccination with Glucan and Oil Adjuvant Multiple Vaccine against Furunculosis, Vibriosis, Cold-water vibriosis and IPN (Intervet Norbio A/S, Norway)

Parameter	Pattern 1 TRP-2*100/100	Pattern 2 TRP-2*100/92	Pattern 2′ TRP-2*92/92
Serum protein (mg ml^{-1})	39.5 ± 1.0	38.1 ± 1.1	37.9 ± 1.6
*IgM (µg ml^{-1})	909 ± 76^a	673 ± 66^b	680 ± 71
IgM/protein (µg mg^{-1})	22.4 ± 1.7^a	16.9 ± 1.5^b	16.9 ± 1.4^b
*CH50 (U ml^{-1})	15.9 ± 0.7	16.5 ± 1.0	15.9 ± 1.0
CH50/protein (U mg^{-1})	0.40 ± 0.01	0.43 ± 0.02	0.41 ± 0.02
*SH50 (U ml^{-1})	8.9 ± 0.5	8.2 ± 0.5	10.1 ± 0.7
SH50/protein (U mg^{-1})	0.22 ± 0.01^a	0.21 ± 0.01^a	0.27 ± 0.02^b
Lysozyme (U ml^{-1})	10.6 ± 0.2^a	10.3 ± 0.2^a	9.6 ± 0.3^b
Antibody against *A. salmonicida* (OD$_{492nm}$ at dilution 1:25600)	0.68 ± 0.07	0.56 ± 0.06	0.55 ± 0.06

Within the same row, the values with different superscripts are significantly different ($p<0.05$). The parameters with the sign * correlated with protein concentration in the serum ($r^2>0.4$, $p<0.0001$), and give better comparison when calculated with the concentration of protein, as in the case of IgM and spontaneous hemolytic activity (SH50).
CH50, Total hemolytic complement activity.
Source: Adapted from Ref. 80

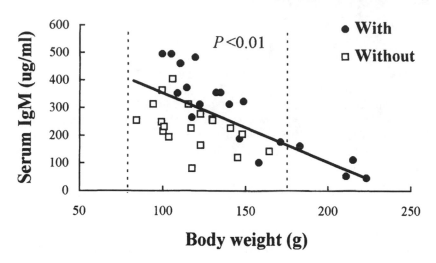

Figure 19. Relationship between body weight and total serum IgM for Atlantic salmon with and without trypsin variant TRP-2*92, 4 weeks after vaccination with a nonadjuvant vaccine against furunculosis. The relationship is described by $Y = -0.0023X + 0.5669$ ($n=39$, $r^2=0.395$, $p=0.0001$). Probability value indicates significant difference in total serum IgM between the two phenotypes at the same weight range of 80–170 g. (From Ref. 83.)

different total serum IgM was suggested to be due to a higher capacity for protein synthesis in the TRP-2*92 salmon, which may have a more rapid immune response after vaccination than the other phenotype, and kinetics of antibody production may be different among the phenotypes or the type of vaccine (83). It is not known whether unvaccinated salmon of different trypsin phenotypes exhibit similar differences in immune parameters to vaccinated fish, although they showed similar disease resistance to natural infection with furunculosis or by experimental cohabitant challenge with *A. salmonicida* ssp. *salmonicida* (83). Whether the vaccinated TRP-2*92 salmon (especially the pattern 2 fish) have higher survival rate during disease infections than the other phenotype is not clear (83). Since the rates of responding to vaccination of different immune parameters varied among the Atlantic salmon possessing different trypsin phenotypes, the immune mechanisms of the different phenotypic fish may respond differently to disease infections (83).

Based on the effects of dietary nutrients on immune system (87), and protein concentration and amino acid profile in the diets on disease resistance (88, 89), the effect of trypsin phenotypes on immune parameters observed in Atlantic salmon was suggested to be due to differences in feed utilization (83). The latter affects physiological amino acid profile and nutritional status of the fish (38, 42).

Trypsin in the mucus-secreting epithelium of Arctic charr, studied by immuno-histochemistry, was suggested to be part of the nonspecific defense mechanisms of fish (90). It is interesting to note that eight trypsinogen genes were found to constitute 4.6% of the DNA of the human β T-cell receptor locus, which has a vital role in immunity (91). This raises a question about the role of trypsinogen isozymes in nonpancreatic tissues (91). Thus trypsin phenotypes probably have an indirect effect on immune system. These issues require further careful investigations.

V. CATALYTIC ACTIVITY AND STRUCTURE OF TRYPSIN FROM FISH

Trypsin is one of the serine proteases secreted from the pancreatic acinar cells, together with chymotrypsin and elastase (2). These enzymes are structurally similar, but demonstrate different substrate specificity due to characteristics of the substrate-binding pocket and surrounding supporting structures. A wealth of structural information is available for trypsin, especially on mammalian trypsin, but also on fish trypsin. The structure of Atlantic salmon trypsin has been determined at 1.8 Å resolution (92, 93). Comparison with bovine trypsin reveals a very similar structure (93, 94). Based on the data on salmon and bovine trypsins, a model of trypsin from the Antarctic fish *Paranotothemia magellanica* has been proposed (95). The primary structure of trypsin is available for a vast number of species including at least nine species of fish. Typically, trypsin appears to be expressed as different isoforms. The most notable distinction between the enzyme species is the overall charge dividing the proteins into cationic and anionic forms. In humans, the trypsin-coding genes have been studied in detail (91). It appears that the trypsin genes are organized as 3.6 kb units within eight 10 kb tandem repeats flanking the β T-cell receptor gene cluster, five at the 5' end and three in the 3', all on chromosome 7. An additional trypsin gene together with parts of the β T-cell receptor gene cluster has been translocated to chromosome 9. Of the nine genes, only three appear to be expressed. Interestingly, both mouse and chicken appear to possess a similar organisation of trypsin and β T-cell receptor genes.

A. Primary Structure

Currently more than 400 sequences of members of the peptidase family S1, also called the trypsin family, are deposited in sequence databases. About 20% of these sequences are trypsin or trypsinlike enzymes. Analysis is made more complex by the presence of multiple isozymes within each species and because the rates of evolution apparently vary highly between both species

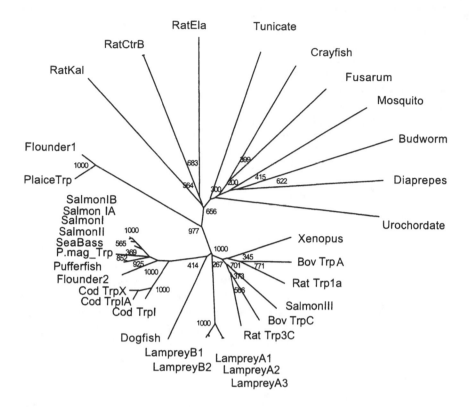

Figure 20. Phylogenetic tree of serine proteases. Unrooted phylogenetic bootstrap tree drawn by the neighbor-joining method (96) based on paired alignment of all amino acid sequences excluding gaps using the Clustal W computer program (97). The figure is drawn to scale using the Treeview program (98); the distance indicates relative divergence. Number of bootstrap trials was set to 1000. The bootstrap value are given in the figure to show confidence levels for groupings. Trypsin amino acid sequences were obtained from EMBL and Swissprot databases under the following accession numbers. Bov TrpA: Q29463, *Bos taurus,* cattle pancreas anionic pretrypsinogen; Bov TrpC: D38507, *B. taurus* trypsinogen, cationic precursor; Budworm: L04749, *Choristoneura fumiferana,* spruce budworm trypsinlike enzyme; Cod TrpI: GM47819, *Gadus morhua,* Atlantic cod trypsinogen I; Cod TrpIA: X76886, X75998, *G. morhua* mRNA for trypsinogen I; Cod TrpX: X76887, X75998, *G. morhua* trypsinogen X; Crayfish: P00765, *Astacus fluviatilis,* broad-fingered crayfish trypsin I; Diaprepes: *Diaprepes abbreviata,* sugarcane borer trypsin; Dogfish: P00764, *Squalus acanthias,* spiny dogfish; Flounder1: AF012462, *Pleuronectes americanus* trypsin I; Flounder2: AF012463, *P. americanus,* winter flounder trypsin II; Fusarium: P35049, *Fusarium oxysporum,* imperfect fungi; LampreyA1: AF011352, *Petromyzon marinus* trypsinogen a1; LampreyA2: AF011898, *P. marinus,* sea lamprey trypsinogen a2; LampreyA3: AF011899, *P. marinus* trypsinogen a3; LampreyB1: AF011900, *P. marinus* trypsinogen b1; LampreyB2: AF011901, *P. marinus* trypsinogen b2; Mosquito: P35035, *Anopheles gambiae,* African malaria mosquito trypsin I; P.mag_Trp: X82223, *Paranotothemia magellanica* trypsin; PlaiceTrp: X56744,

and isozymes. As an approach to identify amino acid residues important for enzyme function, a selection of sequences from the trypsin family was aligned and conserved structures were identified. The relationship between the sequences is presented as a phylogram (Fig. 20). In the alignment, trypsin sequences from evolutionary distant species covering fungi, urochordata, worms, insects, crustacea, fish, amphibians, and mammals were included. Certain residues are conserved over the whole range of sequences. The three residues in the catalytic triad (His57, Asp102 and Ser195) and the trypsin determinant residues (Tyr172 and Ser189) are conserved along with residues in the N-terminal. The six cysteines in three disulfide bridges and several residues situated around the catalytic center and the substrate-binding cleft are also conserved. Most of the conserved amino acids are buried residues. The conserved residues are noted in an alignment of selected sequences presented in Figure 21. Generally the fish trypsinogen genes encode a signal peptide of 13–16 amino acids, an activation peptide of 7–9 residues, and a mature trypsin molecule of most frequently 222 amino acids, but some variants with up to 228 residues. The molecular weight is typically 24 kDa. All trypsins from fish that are deposited in sequence databases have six conserved disulfide bridges. Cationic and anionic forms of trypsin are generally around 70% identical in their primary sequence and probably separated early in evolution as evident from phylogenetic analysis (15). Notably the salmon anionic and cationic trypsins (Figs. 20, 21) has identity scores similar to the overal identity between mammalian and fish trypsins. The importance of the cationic trypsin has been demonstrated by the apparent link between a mutation in this gene in humans to a heritable relative serious form of pancreatitis (113). An Arg-His substitution at residue 117 of the cationic trypsinogen makes the enzyme sensitive to autocatalytic degradation. The different anionic forms are closely related and may only vary in a few residues within a species (Fig. 22). The sequence variation among the anionic salmon trypsins are all in regions of the protein expected to be exposed to the solvent. Some variation in length

---

Pleuronectes platessa trypsinogen; Pufferfish: U25747, *Fugu rubripes;* RatCtrB: P07338, *Rattus norvegicus* chymotrypsinogen B; RatEla: P00773, *R. norvegicus* elastase 1; RatKal: P00758, *R. norvegicus* pancreatic kallikrein 1; Rat Trp1a: P00762, *R. norvegicus* trypsinogen I, anionic precursor; Rat Trp3C: P08426, *R. norvegicus* trypsinogen III, cationic precursor; SalmonI: X70075, *Salmo salar* anionic trypsin I; Salmon IA: X70071, *S. salar* anionic trypsin IA; SalmonIB: X70072, *S. salar* anionic trypsin IB; SalmonII: X70073, *S. salar* anionic trypsin II, SalmonIII: X70074, *S. salar* cationic trypsin III; SeaBass: AJ006882, *Dicentrarchus labrax* trypsin; Tunicate: AF011897, *Boltenia villosa* trypsinogen; Urochordate: X96387, anionic trypsinogen from the colonial Urochordate *Botryllus schlosseri;* Xenopus: P19799, *Xenopus laevis.*

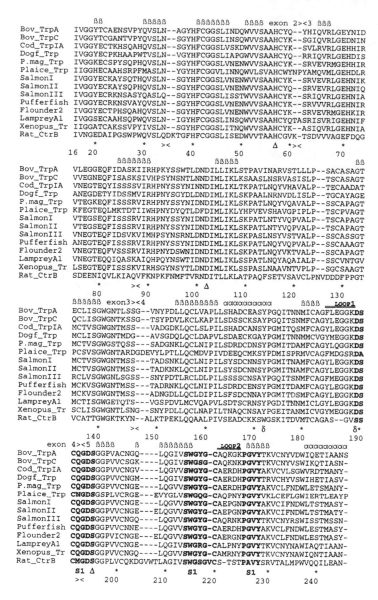

Figure 21. Alignment of amino acid sequences of serine proteases. The sequences were obtained from the Gene Bank or Swissprot databases and aligned using the Clustal W computer program (97). Bov_TrpA: bovine (*B. taurus*) anionic trypsin (99); Bov_TrpC: bovine (*B. taurus*) cationic trypsin (EMBL accession number D38507); Cod_TrpIA: cod (*G. morhua*) trypsin I (100); Dogf_Trp: dogfish trypsin (101); P.mag_Trp: *P. magellanica* trypsin (95); Plaice_TRP: plaice (*P. platessa*) trypsin (EMBL accession number X56744); SalmonI: *S. salar* anionic trypsin I (15); SalmonII: *S. salar* anionic trypsin II (15); SalmonIII: *S. salar* cationic trypsin III (15); Pufferfish: *F. rubripes*

of the trypsin sequences is evident. Most of these inserted/deleted residues are probably positioned in loop structures in the protein. The variations appear to be situated at exon/intron borders in the gene which suggest that they are gained by simple shifts in the mRNA splicing pattern.

B. Catalytic Efficiency and Protein Structure

Fish species that live in cold habitats must have enzyme systems that are adapted to low temperatures. These species must sustain temperature down to 0°C. Trypsin and other enzymes from fish have been studied from species including anchovy (114), capelin (19), cod (18, 115), salmon (116, 117), and the Antarctic fish *P. magellanica* (95). A general trend is that these enzymes display substantially higher catalytic efficiency than the mammalian counterparts over a roughly temperature range of 0–30°C (118). The increased catalytic efficiency, expressed as k_{cat}/K_m, is caused by differences in both variables (18, 115, 117). Outzen and co-workers (117) reported a K_m of salmon trypsin of approximately 10^{-1} of the bovine enzyme. The k_{cat} was increased 2–4.5-fold, depending on temperature, giving a k_{cat}/K_m near 35-fold higher for the fish trypsin than the bovine enzyme when assayed at 4°C. Interestingly, the cationic variant of trypsin from salmon does not possess a higher catalytic efficiency than the mammalian enzyme (117). Protein modeling and studies of x-ray crystallographic structures of enzymes from cold-adapted microorganisms indicate weakening in intramolecular interactions, leading to a more flexible structure, giving an enzyme with lower energy cost for catalysis (118). A model of trypsin from the Antarctic fish *P. magellanica* has been proposed (95). The authors conclusions are that the cold-adapted enzyme contains decreased number of salt bridges, lower overall hydrophobicity, and increased surface hydrophilicity, giving an overall more flexible structure compared to the bovine enzyme. Evaluation of the results from studies of the salmon enzyme has proven more complicated. The crystal structure of the major anionic form of salmon trypsin has been determined at 1.8 Å (93). Comparison

---

trypsin (102); Flounder2: Winter flounder, *P. americanus* trypsin 2 (EMBL accession number AF012463); LampreyA1: sea lamprey *P. marinus* trypsin A1 (102); Xenopus_Tr: *X. laevis* trypsin (103), Rat_CtrB: rat chymotrypsin B (104). The numbers (underneath) refers to the classic system for chymotrypsinogen (105). The residues in the catalytic triad are indicated by Δ. In accordance with Hedstrom et al. (106–108) the trypsin determinant residues 172 and 189 are marked with δ, the two surface loops (loop 1: residues 184a–188a; and loop 2: residues 221–225) are noted together with residues in the S1 binding pocket. Secondary structures in the salmon trypsin is indicated as α (α-helix) and β (β-sheet) structure according to Smalås et al. (93). Exon/intron borders are indicated as ><, in rat trypsin over the sequences (109, 110) and rat chymotrypsin B underneath (104).

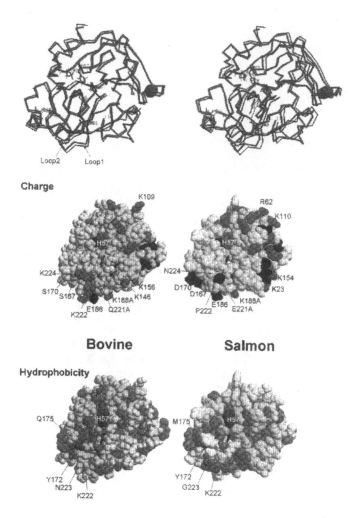

Figure 22. Structure comparison of salmon and bovine trypsin. Structure data were obtained from the Protein Data Bank (PDB) ID code 1BTY: bovine trypsin (111, 112) and 2TBS: salmon trypsin (93). The salmon trypsin is identical to SalmonI in Figures 20 and 21. Amino acids are named using the single letter code and are numbered according to the classic system for chymotrypsinogen (105). Upper panel, stereo view of salmon (thick lines) and bovine (thin lines) trypsin. Only the α carbon chain is illustrated except the residues in the catalytic triad, His57, Asp102 and Ser195 and the trypsin determinant amino acids Tyr172 and Asp 189 which are depicted. A benzamidine inhibitor molecule is situated in the substrate binding cleft. The enzyme associated Ca ion is shown as a sphere. Loop1 and Loop2 refer to structures marked in Figure 21. Middle panel depicts bovine (left) and salmon (right) trypsin as spacefill models where charged amino acids are marked. Basic residues are marked light grey, while dark grey structures refers to acidic amino acids. Lower panel shows distribution of charged amino acids in bovine and salmon trypsin. Hydrophobic residues are grey.

with the bovine counterpart gave only minor differences (93). However a difference in distribution of charged residues was noted (Fig. 22). In general, the trypsin double β-barrel secondary structure is well conserved. The two β-barrels are relatively rigid; however, the loop structures connecting the β-sheets may be more flexible and vary between bovine and salmon trypsins. The three residues in the catalytic triad (His57, Asp102, Ser195) are positioned in the junction between the two β-barrels. This makes the junction very important, but only small differences are apparent in the connections between the two domains. The intradomain hydrogen bonding between the domains seem to be conserved, but there may be some differences in hydrophobic interactions (93). The hydrophobic substrate-binding pocket is apparently structurally conserved. Similarly, comparison of temperature factors does not indicate any overall differences in the flexibility of salmon and bovine trypsin (93, 94). The main differences in the primary structure of salmon and bovine trypsins are located in loop region in the crystal structures. These loop structures are frequently solvent exposed, but the dynamic properties of solvent-exposed residues are not included in crystallographic temperature profiles. Differences in loop structures extending from the substrate-binding region and other important structures may contribute to the differences in catalytic efficiencies.

VI. CONCLUSIONS

Trypsin is a major member of the serine proteases that consists of a single peptide chain with molecular weight typically 24 kDa. It is synthesized in the pancreas and secreted as an inactive precursor, trypsinogen. Trypsin is the key enzyme for activating all other pancreatic zymogens. It may play a major role in regulating protein digestion and affect the availability of nutrients for synthesis and growth. Several isoforms of trypsin have been reported in fishes. These isozymes possess differences in their catalytic efficiency (k_{cat}/K_m) and the distribution of charged amino acids appears to be responsible for different substrate-binding preferences. The genetic presence of different trypsin isozymes is related to growth and physiological change in fishes, since trypsin appears to be the enzyme in the digestion process that is the rate-limiting step for feed utilization and growth in fishes. Studies of protein structure of trypsin and its catalytic activity are also discussed.

REFERENCES

1. L Stryer. Control of enzymatic activity. In: L Stryer, ed. Biochemistry, 3rd ed. New York: W.H. Freeman, 1988, pp 233–259.

2. WR Rypniewski, A Perrakis, CE Vorgias, KS Wilson. Evolutionary divergence and conservation of trypsin. Protein Engineering 7: 55–64, 1994.

3. N Uchida, T Obata, T Saito. Occurrence of inactive precursors of proteases in chum salmon pyloric caeca. Bull Jpn Soc Sci Fish 39: 825–828, 1973.

4. R Yoshinaka, M Sato, S Ikeda. Distribution of trypsin and chymotrypsin and their zymogens in the digestive system of catfish. Bull Jpn Soc Sci Fish 47: 1615–1618, 1981.

5. R Yoshinaka, M Sato, T Sato, S Ikeda. Distribution of trypsin and chymotrypsin, and their zymogens in the digestive organs of the eel (*Anguilla japonica*). Comp Biochem Physiol 78B: 569–573, 1984.

6. KA Walsh. Trypsinogens and trypsins of various species. In: GE Perlmann, L Lorand, eds. Proteolytic enzymes, Methods of Enzymology, Vol. 19. New York and London: Academic Press, 1970, pp 41–63.

7. M Bolognesi, G Gatti, E Menegatti, M Guarneri, M Marquart, E Pamamokos, R Huber. Three-dimensional structure of the complex between pancreatic secretory trypsin inhibitor (Kazal type) and trypsinogen at 1.8 Å resolution. Structure solution, crystallographic refinement and preliminary structural interpretation. J Mol Biol 162: 839–868, 1982.

8. W Bode, R Huber. Crystal structures of pancreatic serine endopeptidases. In: P Desnuelle, H Sjöström, O Noren, eds. Molecular and Cellular Basis of Digestion. Amsterdam: Elsevier, 1986, pp 213–234.

9. L Stryer. Mechanisms of enzyme action. In: L Stryer, ed. Biochemistry, 3rd ed. New York: W.H. Freeman, 1988, pp 201–232.

10. KR Torrissen. Characterization of proteases in the digestive tract of Atlantic salmon (*Salmo salar*) in comparison with rainbow trout (*Salmo gairdneri*). Comp Biochem Physiol 77B: 669–674, 1984.

11. K Rungruangsak Torrissen. Trypsin isozyme *TRP-2(92)*: a growth marker in Atlantic salmon (*Salmo salar* L.) and its effect on digestion and absorption of dietary protein. PhD dissertation, Institute of Marine Research and University of Bergen, Norway, 1993.

12. M Lauff, R Hofer. Proteolytic enzymes in fish development and the importance of dietary enzymes. Aquaculture 37: 335–346, 1984.

13. N Uchida, K Watanabe, H Anzai, E Nishide. Distribution of anionic and cationic trypsins in the pyloric caeca of rainbow trout, mackerel and yellowtail. Bull Coll Agric Vet Med Nihon Univ 42: 222–227, 1985.

14. KR Torrissen. Genetic variation of trypsin-like isozymes correlated to fish size of Atlantic salmon (*Salmo salar*). Aquaculture 62: 1–10, 1987.

15. R Male, JB Lorens, AO Smalås, KR Torrissen. Molecular cloning and characterization of anionic and cationic variants of trypsin from Atlantic salmon. Eur J Biochem 232: 677–685, 1995.

16. N Uchida, K Tsukayama, E Nishide. Purification and some properties of trypsins from the pyloric caeca of chum salmon. Bull Jpn Soc Sci Fish 50: 129–138, 1984.

17. KR Torrissen, TN Barnung. Genetic difference in trypsin-like isozyme pattern between two strains of Arctic charr (*Salvelinus alpinus*). Aquaculture 96: 227–231, 1991.

18. B Ásgeirsson, WJ Fox, B Bjarnason. Purification and characterization of trypsin from the poikilotherm *Gadus morhua*. Eur J Biochem 180: 85–94, 1989.

19. K Hjelmeland, J Raa. Characteristics of two trypsin type isozymes isolated from the Arctic fish capelin (*Mallotus villosus*). Comp Biochem Physiol 71B: 557–562, 1982.

20. K Murakami, M Noda. Studies on proteinases from the digestive organs of sardine. I. Purification and characterization of three alkaline proteinases from the pyloric caeca. Biochim Biophys Acta 658: 17–26, 1981.

21. R Yoshinaka, T Suzuki, M Sato. Purification and some properties of anionic trypsin from catfish pancreas. Bull Jpn Soc Sci Fish 49: 207–212, 1983.

22. TS Fletcher, M Alhadeff, CS Craik, C Largman. Isolation and characterization of cDNA encoding rat cationic trypsinogen. Biochemistry 26: 3081–3086, 1987.

23. BF Erlanger, N Kokowsky, C William. The preparation and properties of two new chromogenic substrates of trypsin. Arch Biochem Biophys 95: 271–278, 1961.

24. B Dahlmann, Kl-D Jany. A rapid sensitive detection of proteolytic enzymes after electrophoresis. J Chromatogr 110: 174–177, 1975.

25. AC Bratton, EK Marshall Jr. A new coupling component for sulfanilamide determination. J Biol Chem 128: 537–550, 1939.

26. KR Torrissen, OJ Torrissen. Protease activities and carotenoid levels during the sexual maturation of Atlantic salmon (*Salmo salar*). Aquaculture 50: 113–122, 1985.

27. K Rungruangsak-Torrissen, GM Pringle, R Moss, DF Houlihan. Effects of varying rearing temperatures on expression of different trypsin isozymes, feed conversion efficiency and growth in Atlantic salmon (*Salmo salar* L.). Fish Physiol Biochem 19: 247–255, 1998.

28. FW Allendorf, GH Thorgaard. Tetraploidy and the evolution of salmonid fishes. In: BJ Turner, ed. Evolutionary Genetics of Fishes. New York: Plenum, 1984, pp 1–53.

29. SE Hartley. The chromosomes of salmonid fishes. Biol Rev 62: 197–214, 1987.

30. KR Torrissen, R Male, G Nævdal. Trypsin isozymes in Atlantic salmon, *Salmo salar* L.: studies of heredity, egg quality and effect on growth of three different populations. Aquacult Fish Manage 24: 407–415, 1993.

31. WD Harvey, RL Nobel, WH Neill. A liver biopsy technique for electrophoretic evaluation of largemouth bass. Prog Fish-Cult 46: 87–91, 1984.

32. KR Torrissen. Genetic variation in growth rate of Atlantic salmon with different trypsin-like isozyme patterns. Aquaculture 93: 299–312, 1991.

33. KR Torrissen, OJ Torrissen. Digestive proteases of Atlantic salmon (*Salmo salar*) from different river strains: development after hatching, rearing temperature effect and effect of sex and maturation. Comp Biochem Physiol 77B: 15–20, 1984.

34. KR Torrissen, KD Shearer. Protein digestion, growth and food conversion in Atlantic salmon and Arctic charr with different trypsin-like isozyme patterns. J Fish Biol 41: 409–415, 1992.

35. M Bassompierre, TH. Ostenfeld, E McLean, K Rungruangsak Torrissen. In vitro protein digestion, and growth of Atlantic salmon with different trypsin isozymes. Aquacult Int 6: 47–56, 1998.

36. K Rungruangsak-Torrissen, CG Carter, A Sundby, A Berg, DF Houlihan. Maintenance ration, protein synthesis capacity, plasma insulin and growth of Atlantic salmon (*Salmo salar* L.) with genetically different trypsin isozymes. Fish Physiol Biochem 21:223–233, 1999.

37. GA Winans. Multivariate morphometric variability in Pacific salmon: technical demonstration. Can Fish Aquat Sci 41: 1150–1159, 1994.

38. KR Torrissen, E Lied, M Espe. Differences in utilization of dietary proteins with varying degrees of partial pre-hydrolysis in Atlantic salmon (*Salmo salar* L.) with genetically different trypsin isozymes. Proceedings of the 11th FAOBMB Symposium on Biopolymers and Bioproducts: Structure, Function and Applications, Bangkok, 1995, pp 432–442.

39. ID McCarthy, DF Houlihan, CG Carter, KA Moutou. Variation in individual food consumption rates of fish and its implications for the study of fish nutrition and physiology. Proc Nutr Soc 52: 427–436, 1993.

40. CG Carter, DF Houlihan, B Buchanan, AI Mitchell. Protein-nitrogen flux and protein growth efficiency of individual Atlantic salmon (*Salmo salar* L.). Fish Physiol Biochem 12: 305–315, 1993.

41. CG Carter, DF Houlihan, B Buchanan, AI Mitchell. Growth and feed utilization efficiencies of seawater Atlantic salmon, *Salmo salar* L., fed a diet containing supplementary enzymes. Aquacult Fish Manage 25: 37–46, 1994.

42. KR Torrissen, E Lied, M Espe. Differences in digestion and absorption of dietary protein in Atlantic salmon (*Salmo salar*) with genetically different trypsin isozymes. J Fish Biol 45: 1087–1104, 1994.

43. RA Coulson, TD Coulson, JD Herbert, MA Staton. Protein nutrition in the alligator. Comp Biochem Physiol 87A: 449–459, 1987.

44. AL Lehninger. Principles of Biochemistry. New York: Worth, 1982, pp 683–720.

45. RA Wallace, JL King, GP Sanders. Digestion and nutrition. In: RA Wallace, JL King, GP Sanders, eds. Biology the Science of Life. Illinois: Scott, Foresman, 1986, pp 731–757.

46. CG Carter, Z-Y He, DF Houlihan, ID McCarthy, I Davidson. Effect of feeding on the tissue free amino acid concentrations in rainbow trout (*Oncorhynchus mykiss* Walbaum). Fish Physiol Biochem 14: 153–164, 1995.

47. AR Lyndon, I Davidson, DF Houlihan. Changes in tissue and plasma free amino acid concentrations after feeding in Atlantic cod. Fish Physiol Biochem 10: 365–375, 1993.

48. FE Stone, RW Hardy. Plasma amino acid changes in rainbow trout (*Salmo gairdneri*) fed ensiled, liquefied, and freeze-dried fish proteins. Proceeding of Aquaculture International Congress and Exploration, Vancouver, 1988, pp 29.

49. B Fauconneau, J Breque, C Bielle. Influence of feeding on protein metabolism in Atlantic salmon (*Salmo salar*) Aquaculture 79: 29–36, 1989.

50. DN McMillan, DF Houlihan. Short-term responses of protein synthesis to refeeding in rainbow trout. Aquaculture 79: 37–46, 1989.

51. GM Pringle, DF Houlihan, KR Callanan, AI Mitchell, RS Raynard, GH Houghton. Digestive enzyme levels and histopathology of pancreas disease in farmed Atlantic salmon (*Salmo salar*). Comp Biochem Physiol 102A: 759–768, 1992.

52. T Lovell. Nutrition and Feeding of Fish. New York: Van Nostrand Reinhold, 1989, pp 73–80.

53. AM Ugolev, NN Iezuitova. Membrane digestion and modern concepts of food assimilation. World Rev Nutr Diet 40: 113–187, 1982.

54. K Rungruangsak, F Utne. Effect of different acidified wet feeds on protease activities in the digestive tract and on growth rate of rainbow trout (*Salmo gairdneri* Richardson). Aquaculture 22: 67–79, 1981.

55. LE Dimes, FL Garcia-Carreno, NF Haard. Estimation of protein digestibility—III. Studies on the digestive enzymes from the pyloric caeca of rainbow trout and salmon. Comp Biochem Physiol 109A: 349–360, 1994.

56. MM Kristjánsson. Purification and characterization of trypsin from the pyloric caeca of rainbow trout (*Oncorhynchus mykiss*). J Agric Food Chem 39: 1738–1742, 1991.

57. MM Kristjánsson, HH Nielsen. Purification and characterization of two chymotrypsin-like proteases from the pyloric caeca of rainbow trout (*Oncorhynchus mykiss*). Comp Biochem Physiol 101B: 247–253, 1992.

58. H Lemieux, PU Blier, J-D Dutil. Do digestive enzymes set physiological limit to growth rate and food conversion efficiency in Atlantic cod (*Gadus morhua*). Abstract of the VIII International Symposium on Fish Physiology. Uppsala University, Sweden, 15–18 August 1998, p 145.

59. NF Haard, LE Dimes, R Arndt, FM Dong. Estimation of protein digestibility—IV. Digestive proteinases from the pyloric caeca of Coho salmon (*Oncorhynchus kisutch*) fed diets containing soybean meal. Comp Biochem Physiol 115B: 533–540, 1996.

60. T Seldal, KJ Andersen, G Högstedt. Grazing-induced proteinase inhibitors: a possible cause for lemming population cycles. OIKOS 70: 3–11, 1994.

61. Y Inui, S Arai, M Yokote. Gluconeogenesis in the eel. VI. Effects of hepatectomy, alloxan and mammalian insulin on the behaviour of plasma amino acids. Bull Jpn Soc Sci Fish 41: 1105–1111, 1975.

62. BW Ince, A Thorpe. The effects of insulin on plasma amino acid levels in the Northern pike, *Esox lucius* L. J Fish Biol 12: 503–506, 1978.

63. CR Machado, MAR Garofalo, JEC Roselino, IC Kettelhut, RH Migliorini. Effects of starvation, refeeding, and insulin on energy-linked metabolic processes in catfish (*Rhamdia hilarii*) adapted to a carbohydrate-rich diet. Gen Comp Endocrinol 71: 429–437, 1988.

64. RF Ablett, RO Sinnhuber, RM Holmes, DP Selivonchick. The effect of prolonged administration of bovine insulin in rainbow trout (*Salmo gairdneri* R.). Gen Comp Endocrinol 43: 211–217, 1981.

65. A Sundby, K Eliassen, T Refstie, EM Plisetskaya. Plasma levels of insulin, glucagon and glucagon-like peptide in salmonids of different weights. Fish Physiol Biochem 9: 223–230, 1991.

66. A Sundby, KA Eliassen, AK Blom, T Åsgård. Plasma insulin, glucagon, glucagon-like peptide and glucose levels in response to feeding, starvation and life long restricted feed ration in salmonids. Fish Physiol Biochem 9: 253–259, 1991.

67. DF Houlihan, DN McMillan, P Laurent. Growth rates, protein synthesis, and protein degradation rates in rainbow trout: effects of body size. Physiol Zool 59: 482–493, 1986.

68. DF Houlihan, SJ Hall, C Gray, BS Nobel. Growth rates and protein turnover in cod, *Gadus morhua*. Can J Fish Aquat Sci 45: 951–964, 1988.

69. A von der Decken, M Espe, E Lied. Growth and physiological properties in white trunk muscle of two anadromous populations of Arctic charr (*Salvelinus alpinus*). Fisk Dir Skr Ser Ernæring 5: 49–57, 1992.

70. L Stryer. Biosynthesis of amino acids and heme. In: L Stryer, ed. Biochemistry. 3rd ed. New York: W.H. Freeman, 1988, pp 575–600.

71. L Stryer. Pentose phosphate pathway and gluconeogenesis. In: L Stryer, ed. Biochemistry. 3rd ed. New York: W.H. Freeman, 1988, pp 427–448.

72. RA Ferguson, KB Storey. Gluconeogenesis in trout (*Oncorhynchus mykiss*) white muscle: purification and characterization of fructose-1,6-biphosphatase activity in vitro. Fish Physiol Biochem 10: 201–212, 1992.

73. DJ Millward. The nutritional regulation of muscle growth and protein turnover. Aquaculture 79: 1–28, 1989.

74. L Stryer. Connective-tissue proteins. In: L Stryer, ed. Biochemistry. 3rd ed. New York: W.H. Freeman, 1988, pp 261–281.

75. L Stryer. Protein synthesis. In: L Stryer, ed. Biochemistry. 3rd ed. New York: W.H. Freeman, 1988, pp 733–766.

76. M Espe, E Lied, KR Torrissen. In vitro protein synthesis in muscle of Atlantic salmon (*Salmo salar*) as affected by the degree of proteolysis in feeds. J Anim Physiol Anim Nutr 69: 260–266, 1993.

77. S Boonvisut, JR Whitaker. Effect of heat, amylase, and disulfide bond cleavage on the in vitro digestibility of soybean proteins. J Agric Food Chem 24: 1130–1135, 1976.

78. M Friedman, OK Grosjean, JC Zahnley. Inactivation of soya bean trypsin inhibitor by thiols. J Sci Food Agric 33: 165–172, 1982.

79. J Opstvedt, R Miller, RW Hardy, J Spinelli. Heat-induced changes in sulfhydryl groups and disulfide bonds in fish protein and their effect on protein and amino acid digestibility in rainbow trout (*Salmo gairdneri*). J Agric Food Chem 32: 929–935, 1984.

80. GN Somero. pH–Temperature interactions on protein: principles of optimal pH and buffer system design. Mari Biol Lett 2: 163–178, 1981.

81. H Abe, PG Dobson, U Hoeger, WS Parkhouse. The role of histidine and histidine-related compounds to intracellular buffering in fish skeletal muscle. Am J Physiol 249: R449–R454, 1985.

82. A Van Waarde. Biochemistry of non-protein nitrogenous compounds in fish including the use of amino acids for anaerobic energy production. Comp Biochem Physiol 91B: 207–228, 1988.

83. K Rungruangsak-Torrissen, HI Wergeland, J Glette, R Waagbø. Disease resistance and immune parameters in Atlantic salmon (*Salmo salar* L.) with genetically different trypsin isozymes. Fish Shellfish Immunol 9:557–568, 1999.

84. AFA El-Mowafi, R Waagbø, A Maage. Effect of low dietary magnesium on im-

mune response and osmoregulation of Atlantic salmon. J Aquat Anim Health 9: 8–17, 1997.

85. GO Melingen, SO Stefansson, A Berg, HI Wergeland. Changes in serum protein and IgM concentration during smolting and early postsmolt period in vaccinated and unvaccinated Atlantic salmon (*Salmo salar* L.). Fish Shellfish Immunol 5: 211–221, 1995.

86. KH Røed, KT Fjalestad, A Strømsheim. Genetic variation in lysozyme activity and spontaneous haemolytic activity in Atlantic salmon (*Salmo salar*). Aquaculture 114: 19–31, 1993.

87. R Waagbø. The impact of nutritional factors on the immune system in Atlantic salmon, *Salmo salar* L.,: a review. Aquacult Fish Manage 25: 175–197, 1994.

88. V Kiron, H Fukuda, T Takeuchi, T Watanabe. Dietary protein related humoral immune response and disease resistance of rainbow trout, *Oncorhynchus mykiss*. In: SJ Kaushik, P Luquet, eds. Fish Nutrition in Practice. Paris: INRA Editions, 1993, pp 119–126.

89. H Neji, N Naimi, R Lallier, J De La Noüe. Relationships between feeding, hypoxia, digestibility and experimentally induced furunculosis in rainbow trout. In: SJ Kaushik, P Luquet, eds. Fish Nutrition in Practice. Paris: INRA Editions, 1993, pp 187–197.

90. SM Paulsen, K Hjelmeland. Localization of trypsin in mucus secreting epithelium of Arctic charr (*Salvelinus alpinus* L.) by immunohistochemical staining. Abstract of the 3rd International Marine Biotechnology Conference, University of Tromsø, Norway, 1994, p 112.

91. L Rowen, BF Koop, L Hood. The complete 685-kilobase DNA sequence of the human β T cell receptor locus. Science 272: 1755–1762, 1996.

89. AO Smalås, A Hordvik. Structure determination and refinement of benzamidine-inhibited trypsin from North Atlantic salmon (*Salmo salar*) at 1.82 Å resolution. Acta Cryst D49: 318–330, 1993.

90. AO Smalås, ES Heimstad, A Hordvik, NP Willassen, R Male. Cold-adaption of enzymes: Structural comparison between salmon and bovine trypsins. Proteins 20: 149–166, 1994.

91. ES Heimstad, LK Hansen, AO Smalås. Comparative molecular dynamics simulation studies of salmon and bovine trypsins in aqueous solution. Protein Engin 8: 379–388, 1995.

92. S Genicot, F Rentier-Delrue, D Edwards, J VanBeeumen, C Gerday. Trypsin and trypsinogen from an Antarctic fish: molecular basis of cold adaptation. Biochim Biophys Acta 1298: 45–57, 1996.

93. N Saitou, M Nei. The neighbor-joining method: a new method for reconstructing phylogenetic trees. Mol Biol Evol 4: 406–425, 1987.

94. JD Thompson, DG Higgins, TJ Gibson. CLUSTAL W: improving the sensitivity of progressive multiple sequence alignment through sequence weighting, positions-specific gap penalities and weight matrix choice. Nucleic Acids Res 22: 4673–4680, 1994.

95. RDM Page. TREEVIEW: An application to display phylogenetic trees on personal computers. Comput Appli Biosci 12: 357–358, 1996.

96. I Le Huerou, C Wicker, P Guilloteau, R Toullec, A Puigserver. Isolation and nucleotide sequence of cDNA clone for bovine pancreatic anionic trypsinogen. Structural identity within the trypsin family. Eur J Biochem 193: 767–773, 1990.

97. A Gudmundsdottir, E Gudmundsdottir, S Oskarsson, JB Bjarnason, AK Eakin, CS Craik. Isolation and characterization of cDNAs from Atlantic cod encoding two different forms of trypsinogen. Eur J Biochem 217: 1091–1097, 1993.

98. K Titani, LH Ericsson, H Neurath, KA Walsh. Amino acid sequence of dogfish trypsin. Biochemistry 14: 1358–1366, 1975.

99. JC Roach, K Wang, L Gan, L Hood. The molecular evolution of the vertebrate trypsinogens. J Mol Evol 45: 640–652, 1997.

100. YB Shi, DD Brown. Developmental and thyroid hormone dependent regulation of pancreatic genes in *Xenopus laevis*. Genes Dev 4: 1107–1113, 1990.

101. GI Bell, C Quinto, M Quiroga, P Valenzuela, CS Craik, WJ Rutter. Isolation and sequence of a rat chymotrypsin B gene. J Biol Chem 259: 14265–14270, 1984.

102. BS Hartley, DL Kauffman. Correction to the amino acid sequence of bovine chymotrypsinogen A. Biochem J 101: 229, 1966.

103. L Hedstrom, L Szilagyi, WJ Rutter. Converting trypsin to chymotrypsin: the role of surface loops. Science 255: 1249–1253, 1992.

104. L Hedstrom, JJ Perona, WJ Rutter. Converting trypsin to chymotrypsin: residue 172 is a substrate specificity determinant. Biochemistry 33: 8757–8763, 1994.

105. L Hedstrom, S Farr-Jones, CA Kettner, WJ Rutter. Converting trypsin to chymotrypsin: ground-state binding does not determine substrate specificity. Biochemistry 33: 8764–8769, 1994.

106. RJ MacDonald, SJ Stary, GH Swift. Two similar but nonallelic rat pancreatic trypsinogens. J Biol Chem 257: 9724–9732, 1982.

107. CS Craik, QL Choo, GH Swift, C Quinto, RJ MacDonald, WJ Rutter. Structure of two related rat pancreas trypsin genes. J Biol Chem 259: 14255–14264, 1984.

108. RM Stroud, LM Kay, RE Dickerson. The crystal and molecular structure of DIP-inhibited bovine trypsin at 2.7 Ångstroms resolution. Cold Spring Harbor Symp Quant Biol 36: 125–140, 1972.

109. BA Katz, J Finer-Moore, R Mortezaei, DH Rich, RM Stroud. Episelection: novel Ki approximately nanomolar inhibitors of serine proteases selected by binding or chemistry on an enzyme surface. Biochemistry 34: 8264–8280, 1995.

110. DC Whitcomb, MC Gorry, RA Preston, W Furey, MJ Sossenheimer, CD Ulrich, SP Martin, LK Gates Jr, ST Amann, PP Toskes, R Liddle, K McGrath, G Uomo, JC Post, GD Ehrlich. Hereditary pancreatitis is caused by a mutation in the cationic trypsinogen gene. Nature Genet 14: 141–145, 1996.

111. A Martinez, RL Olsen, JL Serra. Purification and characterization of two trypsin-like enzymes from the digestive tract of anchovy *Engraulis encracicholus*. Comp Biochem Physiol 91B: 677–684, 1988.

112. BK Simpson, NF Haard. Purification and characterization of trypsin from the Greenland cod (*Gadus ogac.*). 1. Kinetic and thermodynamic characteristics. Can J Biochem Cell Biol 62: 894–900, 1984.

113. LD Taran, IN Smovdyr. Comparative kinetic investigations of the primary specificity of bovine and salmon trypsins. Biokhimiya 57: 55–60, 1992.

114. H Outzen, GI Berglund, AO Smalås, NP Willassen. Temperature and pH sensitivity of trypsins from Atlantic salmon (*Salmo salar*) in comparison with bovine and porcine trypsin. Comp Biochem Physiol 115B: 33–45, 1996.

115. C Gerday, M Aittaleb, JL Arpigny, E Baise, J-P Chessa, G Garsoux, I Petrescu, G Feller. Psychrophilic enzymes: a thermodynamic challenge. Biochim Biophys Acta 1342: 119–131, 1997.

10
Polyphenoloxidase

Jeongmok Kim and M. R. Marshall
University of Florida, Gainesville, Florida

Cheng-i Wei
Auburn University, Auburn, Alabama

I. INTRODUCTION

Appearance, flavor, texture, and nutritional value are four attributes consumers evaluate when selecting food choices. Appearance, on which a significantly impact is made by color, is one of the first attributes used by consumers for evaluation. Color can be influenced by many compounds: naturally occurring pigments, chlorophylls, carotenoids, anthocyanins, and others; or other colors formed through enzymatic and nonenzymatic reactions. One of the most important color reactions that affect many fruits, vegetables, and seafoods, especially crustaceans, is enzymatic browning, caused by the enzyme polyphenol oxidase (1,2 benzenediol;oxygen oxidoreductase, EC1.10.3.1). This enzyme has also been labeled phenoloxidase, phenolase, monophenol and diphenol oxidase, and tyrosinase.

Enzymatic browning is one of the most studied reactions in seafoods, fruits, and vegetables. Researchers in the fields of food science, horticulture, plant and postharvest physiology, microbiology, and even insect and crustacean physiology have studied this reaction because of its impact on agriculture.

A. Problem

1. Melanosis

Phenoloxidase is responsible for a type of discoloration called melanosis in crustacean species such as lobster, shrimp, and crab. The postmortem dark

Score

0

2

4

6

8

10

Figure 1 Melanosis progression scale of shrimp. (Courtesy of WS Otwell, University of Florida.)

discoloration on crustaceans, called melanosis or blackspot, connotes spoilage, is unacceptable to consumers, and thus reduces the market value of these products. Figure 1 provides an example of a visual scale for the progression of melanosis. Table 1 shows the scale used to describe the progression of melanosis (black spot) on pink shrimp (*Penaeus dourarum*).

2. Enzyme Characteristics of Polyphenol Oxidase

Polyphenol oxidase is responsible for catalyzing two basic reactions. Using the substrates phenols and O_2, the enzyme catalyzes the hydroxylation to the *o*-position adjacent to an existing hydroxyl group. The second reaction is the oxidation of the diphenol to *o*-benzoquinones, which are then further oxidized to melanins

Table 1 Scales Used to Describe the Progression of
Melanosis (Black Spot) on Pink Shrimp

Melanosis scale	Description
0	Absent
2	Slight, noticeable on some shrimp
4	Slight, noticeable on most shrimp
6	Moderate, noticeable on most shrimp
8	Heavy, noticeable on most shrimp
10	Heavy, totally unacceptable

Source: Ref. 1.

(brown products), usually by nonenzymatic mechanisms. The monophenol oxidase reaction will be discussed followed by the diphenol oxidase mechanism. It is still not clear whether a single enzyme system is responsible for both reactions or if there are two distinct enzyme molecules. For example, the monophenol oxidase activity is not always present in plant systems. However, when both monophenol- and diphenol oxidases are present, the ratio of monophenol to diphenol oxidase is usually 1:10 or as high as 1:40 (2).

Monophenol Oxidase.   Monophenol oxidase catalyzes the hydroxylation of monophenols to *o*-diphenols. The reaction is depicted in Figure 2. In plants, this reaction is also called cresolase activity because of the ability of the enzyme to utilize the substrate cresol. In animals, the enzyme is referred to as tyrosinase activity because L-tyrosine is the major monophenolic substrate. Although many use tyrosinase activity to describe the monophenol and diphenol oxidases in plants, L-tyrosine is probably not a major substrate for plants, considering the rich abundance of other phenols. This is probably why this enzyme is commonly called polyphenol oxidase. Due to the interest in browning reactions in fruits and vegetables, cresolase activity has been overlooked because the hydroxylation reaction is dramatically slower than the diphenol oxidase reaction. In animals, tyrosinase activity has been given more attention. There is physiological significance of this enzyme with the diphenolase activity in the hardening of the cuticle for sclerotization. It has been postulated that polyphenol oxidase in plants is similar to its counterpart in animals. Only substrate specificity makes them different. There is still some discussion of whether one isoform having both activities (mono- and diphenol oxidase) is responsible for these reactions or whether two distinct isoforms exists. Enzyme multiplicity in this reaction is known, since isoforms of polyphenol oxidase have been identified.

Studies have also examined two other types of substrates, aromatic amines and *o*-aminophenols, which are very similar to mono- and diphenols (Fig. 3) (3). Kinetic data show similarities between these substrates and phenols.

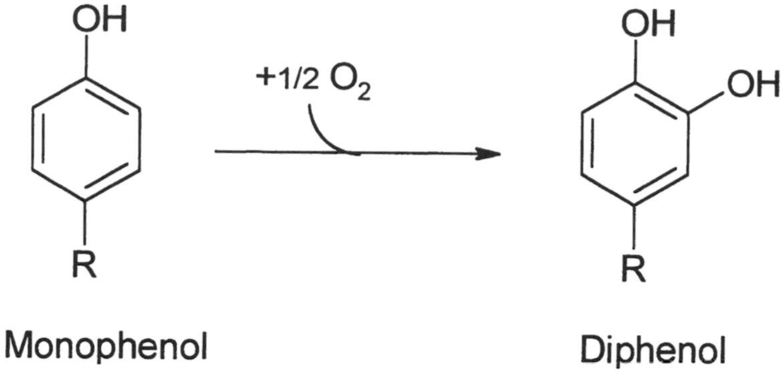

Figure 2 Monophenol oxidase pathway producing the diphenol.

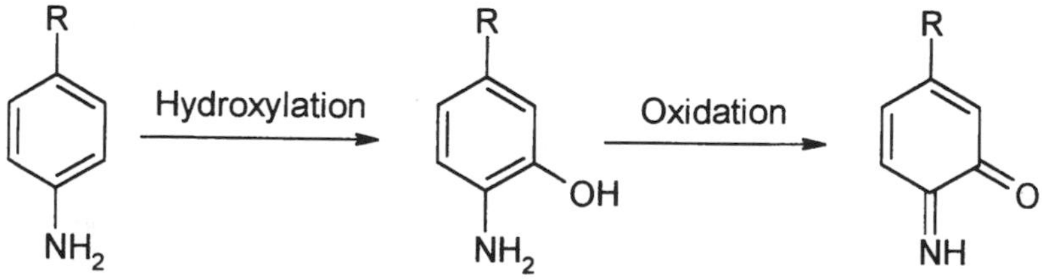

Figure 3 Polyphenol oxidase activity for aromatic amines and o-aminophenols substrates.

Diphenol Oxidase.   The second reaction, the oxidation of the diphenols to quinones, is called diphenol oxidase or catechol oxidase activity. The reaction is the oxidation of catechol to *o*-benzoquinone. Obviously, any diphenol can be oxidized to the quinones (Fig. 4). This reaction has received more attention because of its faster rate than the monophenol oxidase and its association with the formation of quinones, leading to melanin (Fig. 5).

Isozymes of polyphenol oxidase were first isolated from mushroom. It was observed that subunits of isozymes were different in chemical, physical, and kinetic properties. It was thought that these differences helped to explain the enzymes' activity toward mono- and diphenols. Polyphenol oxidase isolated from mango has two isozymes, both capable of oxidizing *o*-diphenols but not monophenols. Gross Michel banana polyphenol oxidase is capable of oxidizing *o*-diphenols but not monophenols, while other varieties of banana were shown to have both mono- and diphenol activities (4). In insects, polyphenol oxidase isolated from the cuticle appears to have only diphenol oxidase activity, while

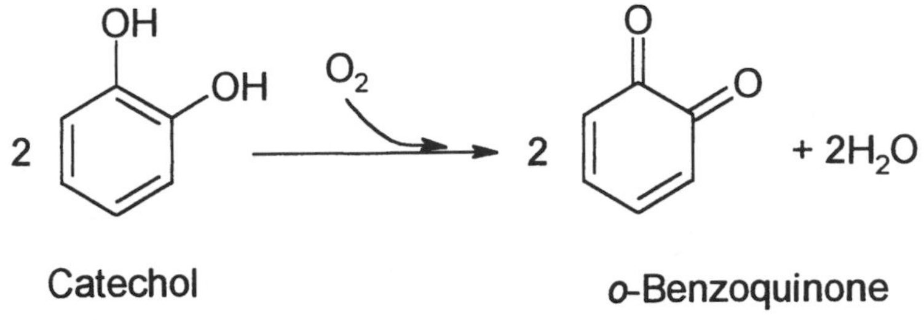

Figure 4 Diphenol oxidase pathway producing the quinones.

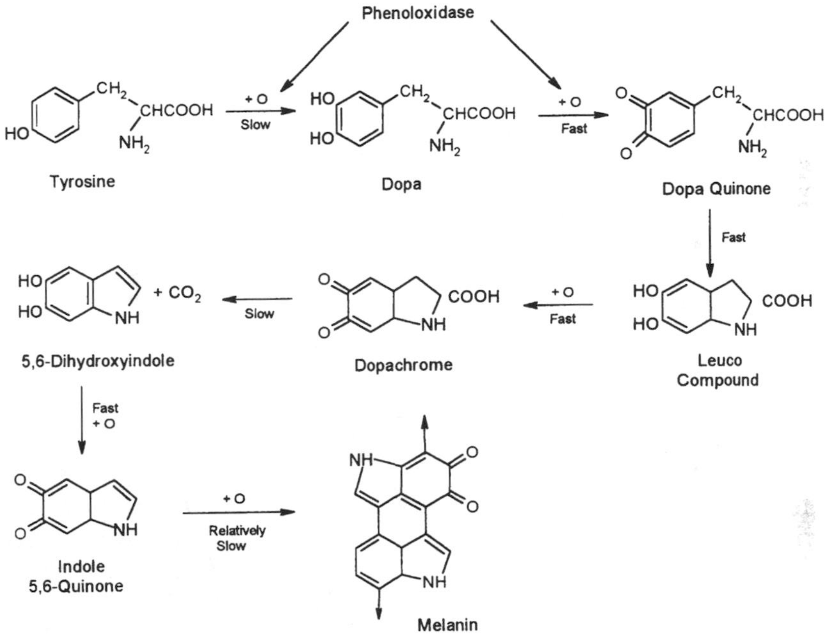

Figure 5 Formation of melanin from tyrosine. (From Ref. 5.)

polyphenol oxidase from insect hemolymph does have the monophenol oxidase activity. Polyphenol oxidase from shrimp cuticle exhibits both activities.

The diphenol oxidase reaction is classified as ordered Bi Bi because there are three substrates involved: oxygen and two diphenols (6). The diphenol oxidase activity is still somewhat unknown because no free radical intermediates are formed and oxygen seems to bind to the enzyme before the o-diphenols. The

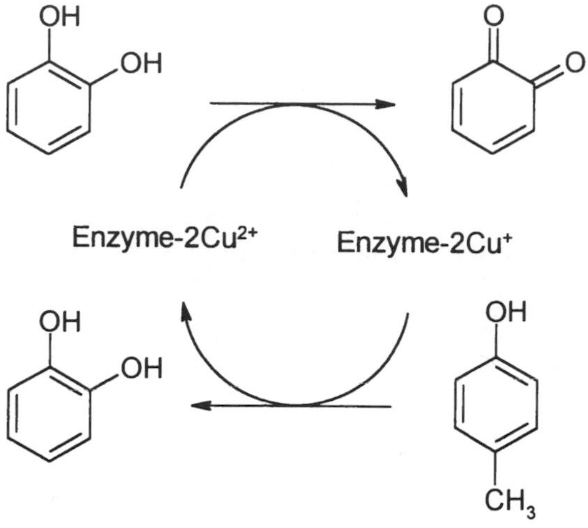

Figure 6 Simplified mechanism for the hydroxylation and oxidation by phenol oxidase(s).

mechanism for hydroxylation and oxidation by polyphenol oxidase(s) involves two copper moities with the enzyme. Figure 6 shows a simplified mechanism for the hydroxylation and oxidation of phenols by polyphenol oxidase.

Laccase.   Laccase (*p*-diphenol oxidase, E.C. 1.10.3.2) (DPO) is another type of copper-containing polyphenol oxidase. Laccases are mainly present in many phytopathogenic fungi and in certain higher plants (7). The unique ability to oxidize *p*-diphenols can be used to differentiate laccase activity from that of the catechol oxidase (Fig. 7). Laccase has a deep blue color in the pure state and is remarkably nonspecific as to its phenolic substrate. Substrate oxidation by laccase is one-electron reaction generating a free radical (8). The reduction of oxygen to water is accompanied by the oxidation of a phenolic substrate. Laccase oxidizes polyphenols, methoxy-substituted phenols, diamines, and a considerable range of other compounds, but unlike tyrosinase does not oxidize tyrosine. The *p*-DPOs or laccases are similar in many other respects to *o*-DPOs. However, another difference is that *p*-DPOs are unable to accomplish the hydroxylation of monophenols.

Phenolic Substrates.   Tyrosine has been reported to be a natural substrate for polyphenol oxidase (PPO) activity in crustaceans (9). Tyrosine is the monohydroxyl phenol. Hydroxylation of tyrosine leads to the formation of dihydroxylphenylalanine (DOPA). PPOs from shrimp and lobster are activated by trypsin or by a trypsinlike enzyme in the tissues (11). However, other proteases such as

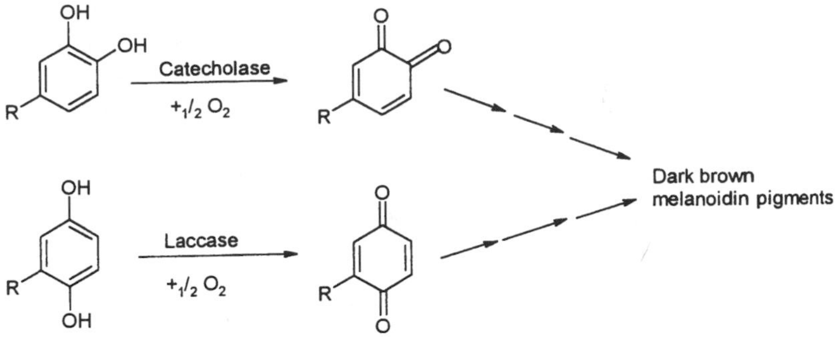

Figure 7 Comparison of reactions catalyzed by catecholase (*o*-DPO) and laccase (*p*-DPO). (From Ref. 10.)

chymotrypsin and pepsin do not activate the lobster PPO. Coralase and trypsin were also shown to activate gill pre-PPO in the golden crab (12).

B. Importance

1. Economic Losses Caused by Browning in Aquatic Foods

Crustaceans rely on polyphenol oxidases to impart important physiological functions for their development. Polyphenol oxidases are important in the sclerotization of the cuticle of insects and crustaceans such as shrimp and lobsters. Sclerotization is the hardening of the shell after molting, which is part of the growing phase for the organism. A second physiological function of polyphenol oxidase is wound healing. The mechanism of wound healing in aquatic organisms is similar to that in plants: the compounds produced from the polymerization of the quinones posses active antibacterial or antifungal activities. Unfortunately, polyphenol oxidases can cause browning of the shell postharvest, which affects the quality of these products and consumer acceptability.

Browning or melanosis in aquatic foods occurs primarily in crustaceans. These highly prized and economically valuable products are extremely vulnerable to enzymatic browning. The discoloration on the surface of the crustacean shell will eventually expand into the shell towards muscle. In lobsters, melanosis is usually more severe if the head is retained; if the head is removed, care should be taken to wash the exposed meat and tail to remove proteases that promote the browning. Although the products of melanosis are not harmful and do not influence flavor or aroma, consumers still will not select these products because they connote spoilage. Severe melanosis on these products can cause tremendous economic losses due to the high value these aquatic products command in the marketplace. To compound this problem, crustaceans differ widely in their susceptibility to melanosis. This presents the processor with the problem of deciding whether or not to treat the product to prevent melanosis. There are many examples of imported products entering the United States worth millions of dollars that were reduced markedly or lost completely because of the severity of melanosis. Many of these products came from developing countries that lack the resources (scientific and technological) and processing infrastructure to prevent this devastating reaction.

Although the economic impact of browning reactions to products imported and exported to countries all over the world can be seen, its occurrence in most food systems is considered a deteriorative process that must be controlled and/or eliminated. Controlling this reaction begins by understanding its mechanism(s) in fruits, vegetables, and seafoods, the properties of the enzyme(s), its substrates and inhibitors, and the chemical, biological, and physical factors that affect each of these. Secondary control is usually by means of extending shelf life by slowing the browning process as much as possible (sometimes eliminating it completely) for consumers to purchase and consume the products. To understand the control, one must first understand the significance of these reactions in food systems.

C. Physiological Role of Enzymatic Browning

In order for the food scientist to understand better how to prevent enzymatic browning, it is important to understand why polyphenol oxidase is present in the tissue of animals. What is the significance of the enzyme in animals? Despite knowing and hypothesizing some functions of polyphenol oxidase in these tissues, researchers are still trying to piece together the functional roles of this enzyme in animals.

1. Animals

Disease resistance is one of the functions considered for polyphenol oxidase in animals, primarily insects and crustaceans. Polyphenol oxidase is thought to be involved in immunity and self-recognition. The enzyme in animals often exists as a zymogen or propolyphenol oxidase form. It has been reported that polyphenol oxidase is released from the pro-form by microbial products such as laminarin (β-1,3-glucan), which also initiate phagocytosis. It is thought that a protease is involved in the activation and production of polymerized phenolics, which prevent bacterial spread. Crayfish propolyphenol oxidase is activated by β-1,3-glucans, fungal glycoproteins, laminaripentaose, and lipopolysaccharides (13). Some of these compounds appeared to activate a protease that serves to activate the zymogen directly. A positive correlation between disease resistance and degree of melanosis, and polyphenol oxidase activity, has been demonstrated in crustaceans.

Researchers have shown in shrimp and lobster that a latent polyphenol oxidase is activated by trypsin or an endogenous enzyme (11, 14, 15). These researchers proposed that activation in these species is the result of proteolysis that produced numerous isoenzymes. Other proteases (chymotrypsin and pepsin) do not seem to have the same activation as trypsin, although chymotrypsin was shown to activate polyphenol oxidase in insects.

The diagram in Figure 8 depicts how host-defense mechanisms activate the propolyphenol oxidase that causes the formation of melanins. These melanins and partially polymerized phenolics possess antimicrobial properties. It is believed that the specific microorganisms can induce the proteases that activate the polyphenol oxidase. The formation of secondary metabolites, such as glucans, glycoproteins, lipopolysaccharides laminarins, and others, by microorganisms may also induce the protease to activate the polyphenol oxidase. However, if the proteases are inactivated by inhibitors, some of these metabolites can still activate the polyphenol oxidase. Both of these mechanisms lead to the host–parasite defense catalyzed by these enzymes.

Although host-defense has been postulated for polyphenol oxidase activity, its major role in crustaceans and insects is most probably related to hardening of the chitin shell, which is called sclerotization. The enzyme appears to be located in the cuticle and hemolymph of insect and crustacean blood. These

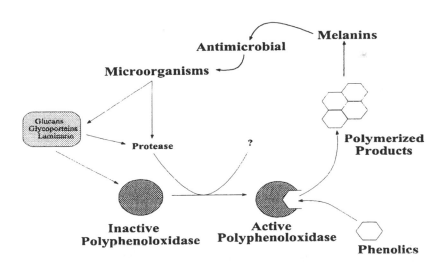

Figure 8 Schematic diagram of the activation of polyphenol oxidase by microorganisms.

enzymes are also present as zymogens (inert proenzymes) that require an activation process (Fig. 8). Since sclerotization is associated with the molting of the animal, this activation mechanism is an excellent way of controlling this growth and development stage. Polyphenol oxidases have been more widely studied in insects; however, both crustaceans and insects appear to have similar mechanisms for sclerotization.

Polyphenol oxidase oxidizes diphenols to their corresponding quinones, which then react with side groups on adjacent protein molecules, thus linking them for hardening (sclerotization). In order to increase in size, crustaceans must replace their exoskeleton, which confines them, with a larger one. Every tissue in the crustacean will be affected in some manner by molting. Each molt cycle is clearly set off by the shedding of the old exoskeleton (ecdysis). Molting includes all morphological and physiological changes in preparation for and recovery from ecdysis (16).

Four stages of molting (postmolt, intermolt, premolt, molt) have been described (17). Figure 9 demonstrates the different developmental stages of lobster cuticle formation. In stage A, there is no cuticle and only the epidermis is present. This would be the case in a lobster that has already molted or is in postmolt. The shell has already hardened and the polyphenol oxidase activity

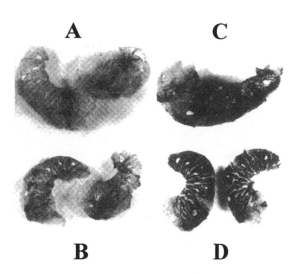

Figure 9   Developmental stages in the formation of Florida spiny lobster cuticle. A. No new cuticle, only epidermis from old cuticle; B. Beginning signs of a newly forming cuticle; C. Advanced stages of a newly forming cuticle; D. Newly formed cuticle almost completely formed as in late premolt.

is very low. In stage B, which would represent intermolt, the lobster is beginning to develop the new cuticle under the old shell. In stages C and D, the cuticle is more defined, especially in stage D. These stages probably represent
early and late premolt. During late premolt, the old shell is ready to be discarded and the new cuticle begins to harden. Stage D shows the new cuticle. At
this stage the lobster is the most vulnerable because the new cuticle will not
harden for a few days. Polyphenol oxidase levels during these stages are
shown in following Figure 10.

The polyphenol oxidase activity as affected by trypsin activation and
stages of molt demonstrated that the highest activity resulted when the lobster
was under going new cuticle formation. The results also demonstrate the zymo-
genic activity of polyphenol oxidase through the various molt stages. The apparent similar activities in late premolt suggest that the activated enzyme is
present when the new cuticle is formed and needs to begin sclerotization. Due
to the antiseptic nature of quinones, the increase in polyphenol oxidase levels
with a corresponding increase in quinones may prevent the penetration of microorganisms through the unfinished cuticle. It is functioning as part of a host-
defense mechanism.

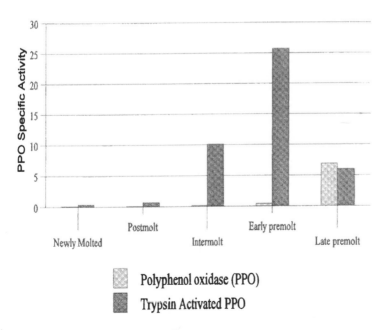

Figure 10 Polyphenol oxidase activity and trypsin-activated polyphenol oxidase activity at various stages of lobster molting.

II. BIOCHEMISTRY AND MOLECULAR BIOLOGY OF POLYPHENOLOXIDASES

PPO has been isolated from various sources. The primary structure of several PPOs from mammals, higher plants, insects, fungi, and bacteria has been characterized and their encoding genes have been cloned. However, there is little information for seafood PPOs. Table 2 summarizes PPOs for which the complete amino acid sequence is known. PPOs are copper-containing proteins. The most prominent features observed in all PPO sequences are the two Cu-binding sites, called CuA and CuB. The active site of PPO consists of a pair of copper ions, which are each bound by three conserved histidine residues (18). This copper pair is the site of interaction of PPO with both molecular oxygen and its phenolic substrates.

In crustaceans, the PPO system is localized in the hemolymph as an inactive prophenoloxidase (proPO) that has to be activated by proteases, lipids, or polysaccharides. PPO activity is abundant in the gill of crustaceans, which contains mostly hemolymph or blood. Therefore, it was suggested that PPO is derived from hemolymph. Hemocyanins are oxygen transport proteins found in the plasma fraction of hemolymph in crustaceans (19). Hemocyanins and PPOs have oxygen-binding sites that are formed by two copper ions linked directly to the protein; however, hemocyanins do not carry out catalysis. Similarity in sequence, and in chemical and spectroscopic behavior, suggests that the tertiary structure of the binuclear copper site of PPO is very similar to that found in the hemocyanins. Therefore, knowledge of the hemocyanin structure may be useful in predicting the structure of the active site of PPO.

A phenoloxidase was purified from blood cells of the crayfish, *Pacifastacus leniusculus* (20). Table 3 shows the amino acid composition of proPO from crayfish. The purified proPO had a molecular mass of 76 kDa. It was a glycoprotein with an isoelectric point around 5.4. Crayfish proPO is activated by cleavage of 177 amino acids from the N-terminus of the protein (21). The other arthropod PPOs cleave an approximately 50 amino acid peptide from the N-terminus of the protein. Thus, these PPOs differ in sequence as well as in domain structure from vertebrate, plant, and fungal PPOs. Figure 11 shows the comparisons of the deduced amino acid sequence of the Cu-binding regions for proPO cDNA with PPO or hemocyanins (21). The six histidine residues assigned to ligate the two oxygen-binding copper ions in spiny lobster hemocyanin appear to be present in a similar position in crayfish proPO. The crayfish proPO containing CuA and CuB sites shows extensive sequence homology with the corresponding site of crustacean hemocyanins. However, with the exception of the copper-binding sites, there was very little sequence similarity between proPOs and the hemocyanins or other copper-containing proteins (21).

A summary of the characteristics for crustacean PPOs is presented in Table

Table 2 Homology of Crayfish proPO and Other Proteins

Species	Number of amino acids	Copper-binding site, %		References
		A site	B site	
Pacifastacus leniusculus pro-PO (crayfish)	706			21
Limulus polyphemus (American horseshoe crab) hemocyanin II	628	69	54	22
Panulirus interruptus (spiny lobster)	~ 660			23
Hemocyanin subunit a		42	46	24
subunit b		42	45	25
subunit c		40	41	26
Drosophila melanogaster (fruit fly) Larval serum protein2 (LSP-2)	690	65	59	27
Bombyx mori (silkworm) pro-PO I	696	67	66	28
pro-PO II		67	48	28
Oryzias latipes (medaka fish) PPO	540		30	29
Mus musculus (mouse) PPO	533		33	30

Table 3 Amino Acid Composition of Prophenoloxidase Purified from Crayfish

Amino acids	*P. leniusculus*, mol %
Asp	7.2
Thr	6.7
Ser	6.4
Glu	11.1
Pro	4.9
Gly	5.6
Ala	5.1
Cys 1/2	1.9
Val	7.1
Met	1.2
Ile	7.9
Leu	12.0
Tyr	2.6
Phe	2.6
His	1.4
Lys	8.6
Arg	7.8
Trp	Not determined

Source: Ref. 20

4. PPOs from various crustaceans differ in size. The shrimp PPOs are relatively small in size, while the inactive PPO from lobster is very large. Pink shrimp PPO was most stable over a broader pH range (6.5–9.0), while white shrimp was most stable at slightly acidic to neutral pH. The lobster PPO was most stable at pH 7.5. At extremely acidic or alkaline condition, PPO activity would be reduced due to the significant conformational change of the protein. In a number of studies, the values for optimum pH of PPO activity are reported between pH 6.0 and pH 8.0.

The influence of temperature on the stability of PPO from crustaceans indicates that the enzymes appeared to have similar stability regions (20–50°C). Pink shrimp PPO was more heat-labile than white shrimp PPO. For example, about 35% of the original activity of pink shrimp PPO was destroyed after 30 min at 50°C, while the original activity of white shrimp PPO remained practically unchanged after the same treatment (31). The Florida spiny lobster PPO was stable up to 40°C, but after incubation at 60°C for 30 min the PPO was completely inactivated (34). Most PPO enzymes are heat labile; a short exposure of the enzyme to temperatures of 70–90°C causes a partial or total irreversible denaturation. The Florida spiny lobster PPO had three isoforms and Western Australian lobster PPO had two isoforms. The lobster PPO activity was enhanced by trypsin. This suggests that the lobster PPO exist in a precursor form that requires further activation by trypsin or trypsinlike enzymes (11, 14).

The stability of the PPOs was very different depending on a number of factors such as temperatures, pH, substrate used, ionic strength, buffer system, and time of incubation. Therefore, it is difficult to compare data unless the conditions are similar.

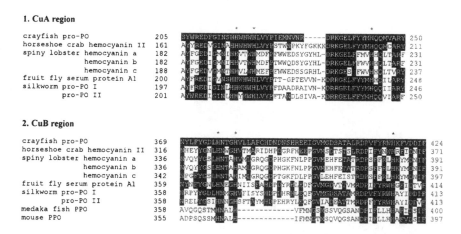

1. CuA region

crayfish pro-PO	205	250
horseshoe crab hemocyanin II	161	211
spiny lobster hemocyanin a	182	231
hemocyanin b	182	231
hemocyanin c	188	237
fruit fly serum protein A1	200	248
silkworm pro-PO I	197	246
pro-PO II	201	250

2. CuB region

crayfish pro-PO	369	424
horseshoe crab hemocyanin II	316	371
spiny lobster hemocyanin a	336	391
hemocyanin b	336	391
hemocyanin c	342	397
fruit fly serum protein A1	359	414
silkworm pro-PO I	358	413
pro-PO II	358	413
medaka fish PPO	358	400
mouse PPO	355	397

Figure 11 Comparison of putative copper-binding sequences in crayfish proPPO with sequences at CuA and CuB of other hemocyanins or protein.

Table 4 Characteristics of PPO from Crustaceans

Source	M.W. (Daltons)	Specific activity (units/mg)	Km (mM)	pH optima	pH stability	Temperature optima	Temperature stability	References
White shrimp (*Penaeus setiferus*)	30,000	59	2.83	7.5	6.0–7.5	45°C	25–50°C	31, 32
Brown shrimp (*Penaeus aztecus*)	210,000	0.26		6.0–6.5			20–40°C	33
Pink shrimp (*Penaeus duorarum*)	40,000	82.8	1.63	8.0	6.5–9.0	40°C	20–40°C	31,32
Florida spiny lobster (*Panulirus argus*)	97,000 88,000 82,000	0.36	9.85	6.5	6.5–8.0	35°C	30–40°C	15, 34
Western Australian lobster (*Panulirus cygnus*)	92,000 87,000	0.03	3.57	7.0	5.0–9.0	30°C	25–35°C	15
Lobster (*Homarus americanus*)		21.1	2.13	6.5	6.5–7.5	30°C	15–30°C	35

III. CONTROLLING BROWNING

Polyphenol oxidase (PPO) catalyzes the oxidation of phenols into *o*-quinones. The *o*-quinones are highly reactive compounds. They can polymerize spontaneously to form high-molecular-weight compounds or brown pigments (melanins), or react with amino acids and proteins that enhance the brown color produced. Preventing enzymatic browning is accomplished in many ways. The methodology is to eliminate from the reaction one or more of its essential components: oxygen, enzyme, copper, or substrate. Chemical compounds have also been used to prevent PPO action. Various techniques and mechanisms have been developed over the years to prevent PPO action in food products. This review tries to give an overview of the prevention of enzymatic browning, especially in seafood.

A. Processing

1. Heating

Heat treatment is the most frequently utilized method for stabilizing foods because of its capacity to destroy microorganisms and inactivate enzymes. Enzyme denaturation at high temperatures involves unfolding of the native structure to a random coiled structure that has lost catalytic activity. However, the thermal stability of an enzyme is related to its structure, the relative amounts of free and bound water, substrate binding, pH, and presence of salts (36).

However, heat treatment is not a practical method for treatment of fresh foods. For high-temperature treatments, the applied temperature depends on the thermostability of the enzyme to be inactivated as well as the nature of the food products. In general, pasteurization utilizes temperatures between 60 and 85°C, which do not inactivate some enzymes. Blanching techniques often involve temperature ranges of 70–105°C or higher. Thermostability of PPO and its isoenzymes varies with the source of the enzyme, but generally exposure to temperatures in the 70–90°C range results in destruction of catalytic activity (37). The time and temperature are controlled in blanching. The time taken for complete heat inactivation of PPO of various food products has been found to vary widely. White shrimp (*Penaeus setiferus*) PPO was rapidly inactivated by a 30 min incubation at 60°C or higher (38).

The disadvantages of blanching include some loss of vitamins, flavors, colors, texture, carbohydrates, and other water-soluble components when water or steam is used. There are high energy demands and waste disposal problems because of the need for large amounts of water and energy. If products after blanching are going to be frozen, a chilling step needs to be performed before transporting the product frozen. If this cooling is done with cold water, additional leaching will occur.

2. Refrigeration

Cold preservation and storage during distribution and retailing are necessary to prevent browning in food products. This is based on the idea that refrigerated temperatures are effective in reducing enzyme activity. The rate of enzyme-catalyzed reactions is controlled to a great extent by temperature. With every rise in temperature of 10°C (in biological important ranges), there is a doubling in the rate of reaction. This is known as the temperature coefficient (Q_{10}). On the other hand, every 10°C reduction in temperature gives a similar decrease in the rate of biological activity. At low temperature, the reduced kinetic energy of the reactant molecules results in a decrease in both mobility and "effective collisions" necessary to form the enzyme–substrate complex and its subsequent products.

3. Freezing

Freezing at temperatures of -18°C or below is often used for long-term preservation of food quality. There are several hypotheses to explain the mechanisms of enzyme inactivation at freezing temperatures. Many of the solutes, such as salt, sugars, and other carbohydrates, are effective inhibitors at high concentrations (39). Therefore, the increased solute concentration because of quasifrozen state would enhance substrate and product inhibition. The effects of freezing may also be attributed to changes in pH (40). Changes in buffer concentration and composition could cause changes in pH, with the effective acidity being increased by the enhanced mobility of the hydrogen ion in ice compared with water.

Like high temperature, low-temperature inactivation also has its drawbacks. Freezing tends to cause changes in texture and other freshlike characteristics. Perhaps the most best-documented example is the decompartmentalization of certain enzymes, substrates, and/or activators as a result of cell disruption. This allows for enzyme activity in the frozen state and is even further enhanced upon thawing of the food due to its release after cell disruption (41).

4. Dehydration

Water content generally exerts an enormous influence on enzyme activity by acting as a solvent in which reactants are able to dissolve, be transported, and react; and as a reactant (41).

For enzymatic reactions, the water activity (a_w) is very important. Water activity is defined as the ratio p/p_0 where p is the partial pressure of the water vapor above the sample and p_0 the partial pressure of saturated water vapor at the same temperature. The decrease of a_w causes an inhibition of enzymatic activities. In solid media, a_w can affect reactions in two ways; the lack of reactant mobility and the alteration of active conformation of substrates and enzymatic protein. Intermediate moisture food products have a moisture content of 15–40%

and a_w is controlled either by partial drying or by adding water-binding agents such as polyols, sugars, and salts in order to increase stability and inhibit microbial growth. During dehydration, there may be many different effects on enzymes because of their differing responses to the concentration of various solutes, inhibitors, or activators. Some solutes such as sugars and proteins often have a protective effect on enzymes inactivation.

Salts are used in preventing undesirable enzyme activity by modifying the thermodynamics of enzyme-catalyzed reactions. In general, polyvalent anions, such as SO_3^{2-} and PO_4^{2-}, tend to stabilize enzymes, whereas polyvalent cations, such as Ca^{2+} and Mg^{2+}, tend to be destabilizing (36). In general, high salt concentrations remove water from the reaction medium (i.e., reduce a_w) and make it unavailable for water–protein interactions, resulting in reduced activity. Sodium chloride, sucrose, and other sugars, glycerol, propylene glycol, and modified corn syrups are some of the solutes used for dehydration.

Freeze drying is the removal of moisture from foods at the frozen state by sublimation under high vacuum. The low temperature used in the process inhibits undesirable chemical and biochemical reactions and minimizes loss of volatile aromatic compounds. The dried product is lighter in weight, can be stored in airtight containers without refrigeration, and has a very attractive color, flavor, and nutritive value. The rapid-freezing process minimizes enzymatic browning and yields a freeze-dried product of excellent quality. Besides freeze drying, several other ways to achieve dehydration are spray, radiative, solar, and microwave-drying methods.

5. Irradiation

Food irradiation is increasingly recognized as a method for reducing postharvest food losses, ensuring hygienic quality, and facilitating wider trade in food products. The unique feature of food irradiation is the use of ionizing radiation to improve the shelf life or wholesomeness of the product. Microorganisms present in the food can be inactivated to secure long-term preservation. The ionizing radiation used for processing food include gamma rays (from cobalt-60 or cesium-137), x-rays, and accelerated electrons (electron beams). Ionizing energy has been approved in the United States for food preservation and processing.

Low-dose γ-irradiation of fresh shrimp reduced melanosis and resulted in a significant extension of shelf life when stored on ice (42). However, when shrimp were irradiated after the onset of melanosis with a low dose of irradiation, melanosis was accelerated. Medium-dose (5 kGy) irradiation inhibits the dispersion of erythin in chromatosomes but not the dispersion of melanins (43). This inhibition can be caused by conformational changes due to physical and chemical alterations of membranes induced by irradiation (44). Ionizing radiation at doses above 1 kGy can introduce various types of physiological disorders in food products. During ionization of foods with ionization energy, free radicals are produced. These free radicals react with various food constituents and may induce undesirable side effects, such as tissue darkening, lipid oxidation, and decreased vitamin content. Nonenzymatic browning reactions of free amino acids and proteins with reducing sugars such as glucose could also explain these discolorations. Combination treatments, using both irradiation and heat or other methods, have demonstrated a synergistic effect in preventing enzymatic browning.

6. High-Pressure and Supercritical Carbon Dioxide

Enzyme denaturation is caused by rearrangement and/or destruction of noncovalent bonds such as hydrogen bonding, hydrophobic interaction, and ionic bonding of the tertiary protein structure. Pressure can influence biochemical reactions by reducing the available molecular space and by increasing interchain reactions. In aqueous systems, the unfolding of a protein molecule causes exposure of nonpolar residues to water. The amino acid residues are not tightly packed in a folded protein because of strong constraints set by the invariance of bond angles and distances. Therefore, pressure may be a potentially viable technique for preserving food quality by inactivating enzymes. Pressures higher than 500 Mpa generally cause irreversible denaturation attributed to the weakening of the hydrophobic interactions and the breaking of intramolecular salt bridges (45, 46). The effect of high pressure treatments on enzymes can be related to reversible or irreversible changes in protein structure. However, in this case, the loss of catalytic activity can be considerably different depending on the type of enzymes, the nature of the substrates, the temperature and the length of processing (45, 47).

The high-pressure treatment is known to keep quality carriers, such as flavors, taste, and vitamins, intact (45). However, pressure application induces certain quality changes in foods, depending on the nature of the food. The proper combination of pressure and temperature (heat/cold) could be used to enhance enzyme inactivation in certain foods, and prevent off-flavors, and changes in color, nutritional value, and structure (48).

Pressure inactivation of PPO in Tris buffer required treatments for more than 30 min with 900 Mpa at 45°C (49). Inactivation of enzymes by pressure is dependent on the immersion medium, the pH, as well as the temperature and time of treatment. However, pressure treatment in real food systems may show a protective effect by other ingredients and thus require the higher pressures to inactivate PPO.

A supercritical (SC) fluid is a fluid at conditions above its critical temperature and pressure. At temperatures and pressures above its critical point a pure substance exists in a state that exhibits gaslike and liquidlike properties (50). The fluid's density could be very close to that of a liquid, the surface tension is close to zero, the diffusivity and viscosity are between that of a liquid and a gas. Carbon dioxide (CO_2) is mostly used as the SC fluid because it is virtually inert, nontoxic, nonflammable, and cheap (51). Also, its low critical temperature and pressure (52), and environmentally acceptable nature make it highly desirable as an SC fluid. SC-CO_2 can reduce the number of microorganisms and inactivate unwanted enzymes in various foods. A solution of high-pressure CO_2 in water causes carbonic acid formation while temporarily lowering the pH, and it inactivates enzymes and microorganisms. Florida spiny lobster (*Panulirus argus*) PPO exposed to CO_2 (1 atm) at 33, 38, and 43°C showed a decrease in enzyme activity with increased heating time (53). PPO inactivation by CO_2 was faster for PPO heated at 43°C than 33 or 38°C. Lobster PPO was more labile to CO_2 and heat in the range of 33–43°C than to heat alone. PPO activity was affected by temperature and pressure, resulting in a synergistic decrease in activity of a combined treatment. Purified Florida spiny lobster, brown shrimp, and potato PPOs exhibited a time-related decline in activity following treatment at 43°C with high-pressure CO_2 at 58 atm. Kinetic studies showed that crustacean PPOs were more vulnerable than potato PPO to the treatment. For instance, after subjecting the lobster, shrimp, and potato PPOs to the above treatments for 1 min, only 2, 22, and 45%, respectively, of the original activity was retained (54). PPO can be inactivated at low temperatures with CO_2 depending on the source of the enzyme. For example, spiny lobster PPO seems to be very sensitive, while potato PPO is quite resistant to CO_2 treatment (55). The circular dichroism (CD) spectra at far ultraviolet (UV) range for untreated and high-pressure CO_2-treated lobster, brown shrimp, and potato PPO were obtained. Table 5 shows CO_2-treatment caused conformational changes in the secondary structures (α-helix, β-sheet, β-turn, and random coil). Lobster and brown shrimp PPO showed the most noticeable alteration in the composition of α-helix and random coil. However, only minor alterations in secondary structure occurred in high-pressure CO_2-treated potato PPO.

The inactivation kinetics of spiny lobster PPO are given in Table 6. The determination of k values at different conditions resulted in straight lines. The D values (decimal reduction times) of CO_2 treatment are less than those for the temperature control, indicating that at the temperature ranges used, it is much easier to inactivate the enzyme by CO_2 than by heat. Also, the z value of CO_2 treatment is smaller, indicating that PPO from spiny lobster is more sensitive to an increase in temperature when it is in a CO_2 environment.

B.   Inhibitors

The use of browning inhibitors in food processing is restricted by such considerations as toxicity, wholesomeness, effect on taste, flavor, texture, and cost. Browning inhibitors have been classified according to their primary mode of ac-

Table 5 Secondary Structure Estimates of Non-CO_2-Treated and Supercritical Pressure CO_2-Treated Florida Spiny Lobster, Brown Shrimp, and Potato Polyphenol Oxidase (PPO) from Far UV CD Spectra

PPO		Percentage of secondary structure			
		α-Helix	β-Sheet	β-Turn	Random coil
Lobster	Untreated	24.4	26.2	21.4	29.9
	CO_2-treated	19.7	25.9	15.2	39.3
Brown shrimp	Untreated	20.1	22.3	15.2	42.4
	CO_2-treated	29.6	18.9	18.2	33.3
Potato	Untreated	14.8	34.6	28.4	22.2
	CO_2-treated	17.8	35.9	25.9	20.4

Source: Ref. 55

Table 6 Inactivation Kinetics for Spiny Lobster PPO: Temperature Controls and Atmospheric CO_2 Treatments

Temperature (°C)	D^a (min)	z^b (°C)	k (min^{-1})	Ea (KJ/mol)
T control				
33 °C	320		7.20×10^{-3}	
38 °C	314	69.1	7.32×10^{-3}	26.6
43 °C	229		1.0×10^{-2}	
Treatment with CO_2, 1 atm				
33 °C	17.4		1.20×10^{-1}	
38 °C	13.1	43.3	1.56×10^{-1}	39.7
43 °C	10.2		1.93×10^{-1}	

[a] Decimal reduction time: the time necessary at a given temperature to reduce the activity by 90%.
[b] Increase in temperature necessary to reduce the D value by 90%.
Source: Ref. 53.

tion. As seen in Table 7, there are six categories of inhibitors used to prevent enzymatic browning:

1. Reducing agents such as ascorbic acid and isomer erythorbic acid, and related compounds used extensively to transfer hydrogen ions
2. Chelating agents such as citric acid and EDTA
3. Acidulants
4. Enzyme inhibitors
5. Enzyme treatments
6. Complexing agents

These inhibitors of browning will be discussed in further detail.

1. Reducing Agents/Antioxidants

The major role of reducing agents or antioxidants in the prevention of browning is their ability to reduce the o-quinones to the colorless diphenols, or react irreversibly with the o-quinones to form stable colorless products. The use of reducing compounds is the most effective control method for PPO browning. The most widespread treatment used by the food industry for control of browning is the addition of sulfiting agents.

Sulfiting Agents. The most widely used inhibitors of enzymatic browning have been sulfites. Sulfiting agents include sulfur dioxide (SO_2) and several forms of inorganic sulfite that liberate SO_2 under the conditions of use. In aqueous solution, SO_2 and sulfite salts form sulfurous acid (H_2SO_3), and exist as a mixture of the ionic species bisulfite (HSO_3^-) and sulfite (SO_3^{2-}). The dibasic acid ionizes according to the following reaction scheme:

$$SO_2 \cdot H_2O \Leftrightarrow (H_2SO_3) \Leftrightarrow HSO_3^- + H^+$$
$$HSO_3^- \Leftrightarrow SO_3^{2-} + H^+$$

Table 7   Representative Inhibitors of Enzymatic Browning

Reducing agents	Sulfiting agents
	Ascorbic acid and analogues
	Cysteine
	Glutathione
Chelating agents	Phosphates
	EDTA
	Maltol
	Kojic acid
Acidulants	Citric acid
	Phosphoric acid
Enzyme inhibitors	Aromatic carboxylic acids
	Aliphatic alcohol
	Anions
	Peptides
	Substituted resorcinols
Enzyme treatments	Oxygenases
	o-Methyl transferase
	Proteases
Complexing agents	Cyclodextrins

Source: Ref. 56

with pKa values of 1.89 and 7.18 (25°C, zero ionic strength) for the first and second ionizations, respectively. The pH of the solution determines the amount of each species. At pH 4, HSO_3^- is at its highest concentration and at pH 7, both SO_3^{2-} and HSO_3^- exist approximately in the same proportion (57). At pH levels below 3.0, the dominant form is H_2SO_3 (58, 59). Sulfites added to food are present in either free or bound forms (57). Bound sulfite consists of reversibly and irreversibly bound forms. The reversibly bound sulfites are released in acidic media. Irreversibly bound sulfites form very stable addition compounds and are not affected by pH changes. In food processing, use of sulfiting agents is somewhat based on the sulfur dioxide equivalence (60). Table 8 gives a list of sulfiting agents and their theoretical yields of sulfur dioxide.

Bisulfite is a competitive inhibitor of PPO by binding a sulfhydryl group on the PPO active site (33). Bisulfite inhibition on the PPO catalyzed melanosis in lobster was accomplished by it reacting with intermediate quinones forming sulfoquinones, and by it inhibiting irreversibly PPO causing complete inactivation (14). Figure 12 shows the mechanism of action of sulfites in preventing enzymatic browning. In the case of catechol as a substrate, inhibition of browning involves nucleophilic attack by sulfite ion in position 4 of the *o*-quinone to give 4-sulfocatechol after subsequent addition of a hydrogen ion (61). Therefore, the quinone has been reduced in this reaction. It is interesting to find that 4-sulfocatechol is unreactive towards PPO.

Sulfiting agents are widely used in the seafood industry to inhibit melanosis. Sulfites are typically applied on the shrimp vessel immediately postharvest to prevent the formation of blackspot (63). Regulations allow shrimp and other crustaceans to be dipped for 1 min in a 1.25% sodium bisulfite treatment before further processing or storage to delay the formation of blackspot and help maintain quality. This treatment allows for a maximum of 100 ppm of residual sulfite on the raw, edible portion (64, 65). Residuals in excess of 10 ppm require labeling (66). The most commonly and widely used sulfiting agent in the treatment of shrimp melanosis is metasulfite salt. Metabisulfite appears to be an anhydride of bisulfite that is able to release two sulfur dioxide equivalents (60). However, when preparing treatment solutions, the sodium metabisulfite releases a higher amount of volatile sulfur dioxide than other sulfiting agents. Therefore, the use of various blends of sodium metabisulfite and sodium sulfite proved in general to be similarly effective to sodium metabisulfite alone in controlling the problem of melanosis. From the investigation on retardation of melanosis, the 1.0% concentration typically used in commercial setting gave comparably good results for the 85/15 and 60/40 sodium metabisulfite/sodium sulfite blends (Fig. 13) (67). This was seen from the melanosis rating of 4 and below (quality grading A) maintained by treated shrimp samples during 5 days of iced storage. Different levels of volatile sulfur dioxide were produced from the solutions due to the various sulfite blends. The 10% test solutions of the 60/40 sulfite blend

Table 8   Chemicals Yielding Sulfur Dioxide Currently Allowed for Use in
Food as Preservatives

Chemical	Formula	Theoretical Yield (%)	Solubility (mg/100 ml)
Sulfur dioxide	SO_2	100.0	11 at 20°C
Sodium sulfite anhydrous	Na_2SO_3	50.8	28 at 40°C
Sodium sulfite (heptahydrate)	$Na_2SO_3 \cdot 7H_2O$	25.4	24 at 25°C
Sodium hydrogen sulfite	$NaHSO_3$	61.6	300 at 20°C
Sodium metabisulfite	$Na_2S_2O_5$	67.4	54 at 20°C
Potassium metabisulfite	$K_2S_2O_5$	57.6	25 at 0°C

Source: Ref. 57.

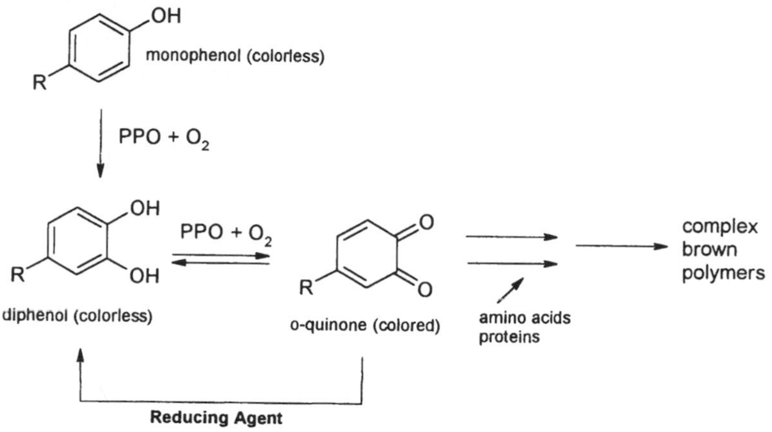

Figure 12   The primary role of reducing agents such as sulfiting agents in the inhibition of enzymatic browning is to reduce the pigment precursors (quinones) to colorless, less-reactive diphenols. (From Ref. 62.)

yielded approximately 50% less volatile sulfur dioxide (SO_2 ppm) than that from the sodium metabisulfite solution and 20% less from 85/15 sulfite blend at similar concentrations. This implies that the 60/40 blend would be an effective and safer chemical treatment, unlike the 85/15 blend solution. Sulfite blends can lessen the risk associated with residues that can result from dip treatments. The residual levels (SO_2) attained for 0.5% and 1% sulfite solutions were less than 50 ppm, including the 60/40 sulfite blend treatments. The 2.0% sulfite solution yielded residuals less than 100 ppm, which is below the Food and Drug Admin-

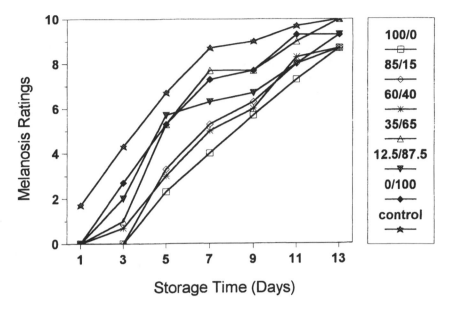

Figure 13 Mean (n=3) melanosis ratings for pink shrimp (*Penaeus duorarum*) treated with various blends of sulfites in 1.0% solutions. (From Ref. 67.)

istration's (FDA) guidelines for residuals on treated shrimp. The use of sulfite blends could provide a safer melanosis control in an industry that continues to take advantage of the benefits of sulfite treatments.

The amount of sulfites necessary for controlling enzymatic browning depends primarily on the nature of the available substrate. When only monophenols, such as tyrosine, are present, the amount of sulfite required is very low. When diphenols are present, much higher sulfite concentrations are needed. Furthermore, since sulfites do not irreversibly inhibit browning, the required concentrations also are dependent on the length of time the reaction must be inhibited (68). Therefore, sulfite treatment levels in foods vary widely, depending on the application. However, sulfites are subject to regulatory restrictions because of potential adverse effects on health. Many reports have described allergic reactions following ingestion of sulfite-treated foods by hypersensitive asthmatics. Sulfites are no longer generally regarded as safe (GRAS) chemicals for fruits and vegetables served raw and are no longer used in salad bars. Therefore, much research is being conducted to find suitable substitutes for sulfites.

L-Ascorbic Acid. *L*-Ascorbic acid and its various neutral salts and other derivatives have been the leading GRAS antioxidants for use on food products to prevent browning and other oxidative reactions (59). Ascorbic acid (vitamin C) is very water soluble and acidic in nature as well as forming neutral salts with bases. Ascorbic acid also acts as an oxygen scavenger to reduce molecular oxygen. The mechanism of ascorbic acid inhibition has generally been attributed to the reduction of enzymatically formed *o*-quinones back to their precursor diphenols, thus preventing the formation of pigments (62). However, it is irreversibly oxidized to dehydroascorbic acid during the process and browning can occur after depletion (Fig. 14). Therefore, it has led to the development of ascorbic acid derivatives such as erythorbic acid, 2- and 3-phosphate, phosphinate esters, and ascorbyl-6-fatty acid esters of ascorbic acid as more stable forms (70, 71). These esters release ascorbic acid when hydrolyzed by acid phosphatases (72). However, their relative effectiveness as inhibitors of browning is dependent on food products (69). The presence of reactive molecules with amino or thiol groups in the medium can greatly affect the reactivity of *o*-quinones. The secondary products formed may or may not be good substrates for PPO and may exhibit different reactivities with *o*-quinones. When used in shrimp to prevent melanosis, it caused a distinct yellow off color (1). Ascorbic acid is usually added together with citric acid, which tends to maintain more acidic pH levels and also acts as a chelating agent on such enzymes as copper-containing PPO (6).

Erythorbic acid. Erythorbic acid and its salt, sodium erythorbate, are strong reducing agents. They act as oxygen scavengers, thus reducing molecular oxygen. Erythorbic acid is the D-isomer of ascorbic acid but has no vitamin C activity. The use of erythorbic acid with citric acid has often been suggested as a substitute for sulfites. Most research suggests that L-ascorbic acid and erythorbic acid have about equal antioxidant properties. This combination is used

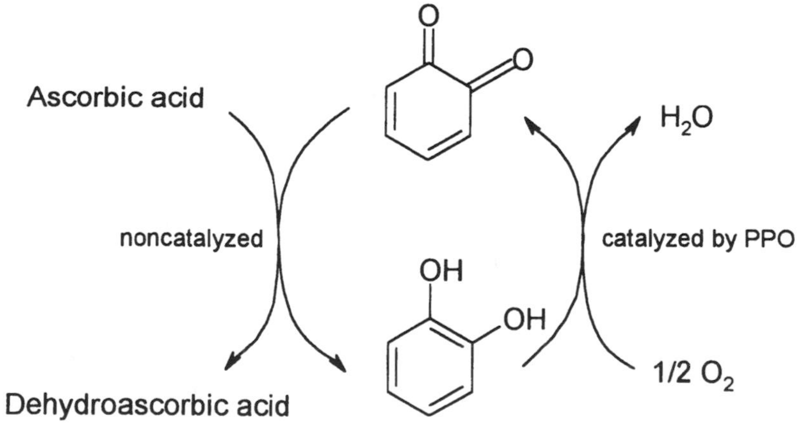

Figure 14 Mechanism of prevention of color formation by ascorbic acid.

at the retail level to inhibit oxidative rancidity and discoloration in frozen seafood.

Cysteine. Cysteine is generally assumed to inhibit melanosis by forming the colorless thiol-conjugated *o*-quinone (73), even though direct inhibition of PPO by its formation of stable complexes with copper has been proposed (74, 75). Cysteine also reduces the *o*-quinones to their phenol precursors (62, 76). The proposed mode of action for cysteine and cysteinyl addition compounds is illustrated in Figure 15. Cysteine–quinone adducts proved to be competitive inhibitors of PPO. Cysteine has been used as a commercial browning inhibitor (77). However, the concentrations of cysteine and other thiols necessary to achieve acceptable levels of browning inhibition have negative effects on taste.

Phenolic Antioxidant. Antioxidants preserve food by retarding rancidity or discoloration by interfering with oxidative processes that generate free

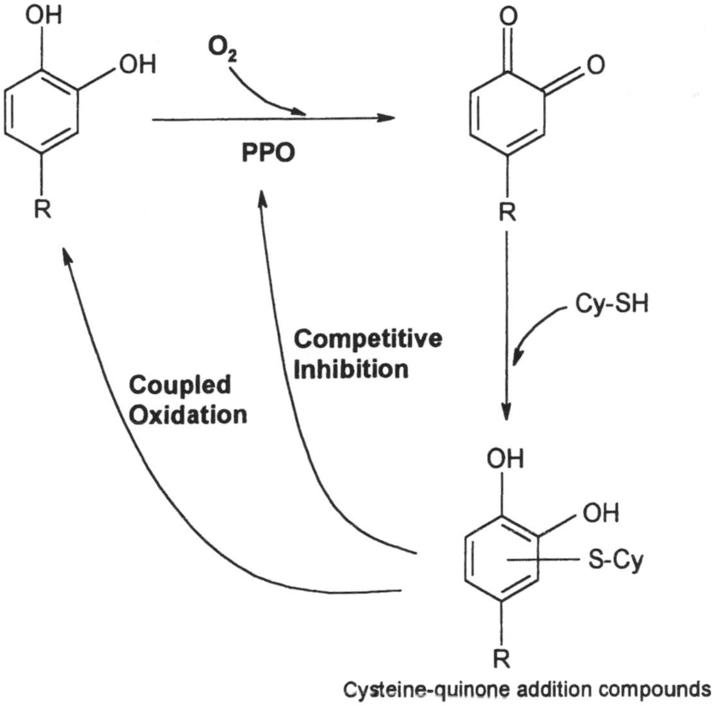

Cysteine-quinone addition compounds

Figure 15   Effect of cysteine and cysteinyl addition compounds with *o*-quinones on the enzymatic oxidation of *o*-diphenols. (From Ref. 79.)

radicals, chelating metals, and also acting as singlet oxygen scavengers. Synthetic or naturally occurring phenolic antioxidants can be used in foods. Several synthetic antioxidants such as butylated hydroxyanisole (BHA), butylated hydroxytoluene (BHT), tertiarybutylhydroxyquinone (TBHQ), and propyl gallate (PG) are permitted for use in food. The naturally occurring antioxidants are primarily plant phenolic compounds such as tocopherols, flavonoid compounds, cinnamic acid derivatives, coumarins, and others.

2. Acidulants

The pH of food can affect ionizable groups on the enzyme. These groups must be in the proper ionic form in order to maintain the conformation of the active site, bind the substrates, or catalyze the reaction (78). However, most of these changes in ionization are reversible. Irreversible denaturation occurs usually at extreme pH values. The pH may also affect the stability of the substrate. The concentration of substrates may be reduced by chemical breakdown caused by pH. These breakdown products are often enzyme inhibitors, because they share the same molecular features of the substrate (80). The stability of an enzyme is affected by pH, buffer type and concentration, the presence or absence of substrate, type of substrate, ionic strength and dielectric constant of the medium, and the effect of pH on the stability of cofactors and activators (81). This difference in pH optimum is probably due to the presence of multiple forms (isozyme) of PPO from different fractions of the source, the stage of maturity of the enzyme source, the degree of purification of the enzyme or the type of substrate (37).

The role of acidulants is to maintain the pH well below what is needed for optimum catalytic activity of PPO. Acidulants such as citric, malic, and phosphoric acids will lower the pH of the system to below 3 where PPO is inactive (82). The acidulant is often used in combination with other antibrowning agents. Citric acid is the one of the most widely used acids in the food industry with ascorbic acid. Suggested usage levels for citric acid are typically 0.1–0.3 % with the appropriate antioxidant at 100–200 ppm (83). Most of the organic acids are also very effective chelators. Citric acid exerts inhibition on PPO by reducing the pH as well as by chelating the copper at the enzyme-active site. It is used synergistically with ascorbic or erythorbic acids and their neutral salts to chelate prooxidants and inactivate PPO. Treatment with L-lactic acid (1%) in combination with 4-hexylresorcinol (0.0025 %) was effective as a melanosis inhibitor in brown shrimp (*Penaeus aztecus*) (84).

3. Chelators

Enzymes have metal ions at their active sites, so removal of these ions by chelators would render the enzymes inactive. Chelating agents complex with prooxidative agents such as copper and iron ions through an unshared pair of electrons in their molecular structure which provides the complexing or chelating action. Since PPO is a metalloprotein with copper as the prosthetic group, it can be inhibited by metal chelating agents. The best known GRAS chelating agents for use on foods are citric acid and ethylenediamine tetraacetic acid (EDTA). Other non-GRAS chelating agents demonstrating PPO inhibition were cyanide, diethyldithiocarbonate, 2-mercaptobenzothiazole, and azide, which inhibit PPO by interacting with its prosthetic group; and polyvinylpyisolidone (PVPP), which bonds the phenolic substances, preventing their conversion to quinones (85–87). Chelators used in the food industry are sorbic acid, polycarboxylic acids (citric, malic, tartaric, oxalic, and succinic acids), polyphosphates (ATP and pyrophosphates), macromolecules (porphyrins, proteins), and EDTA. These chelators have been used in various foods to inactivate enzymes (56).

EDTA. EDTA is a chelating agent permitted for use in the food industry as a chemical preservative. The major compounds approved as food additives by FDA are calcium disodium EDTA (21 CFR 172.120) and disodium EDTA (21 CFR 172.135) (88). Highly stable complexes are formed by the sequestering action of the EDTA compounds on iron, copper, and calcium. The maximum chelating efficiency occurs at the higher pH values where the carboxyl groups are dissociated (83). In the prevention of enzymatic browning, EDTA is generally used in combination with other chemical treatments to eliminate browning. Treatment of shrimp with sodium bisulfate containing 1% sodium salt of EDTA controlled melanosis in shrimp and improved the organoleptic quality (89). Lobster and shrimp PPOs are copper-dependent (35, 38). The activity of lobster PPO was stimulated by the addition of copper but inhibited by EDTA (35).

Phosphates. Polyphosphates, sodium acid pyrophosphate, and metaphosphate are chelating agents and have been used as antibrowning agents for food products in concentrations as low as 0.5–2% (final concentration in the dip solution) due to their minimal solubilities in cold water (56). Also, trisodium phosphate reduces bacterial counts in shrimp during refrigerated storage (90). Sporix, an acidic polyphosphate mixture (sodium acid pyrophosphate, citric acid, ascorbic acid, and calcium chloride), has been evaluated to inhibit oxidation and enzymatic browning (91).

Maltol. Maltol (3-hydroxy-2-methyl-4H-pyran-4-one), a γ-pyrone derivative, is a relatively weak inhibitor of pigmented products formation. Maltol does not inhibit PPO directly, but it prevents browning by its ability to conjugate with *o*-benzoquinones back to *o*-dihydroxyphenols or by irreversible inactivation of PPO (92). γ-pyrone derivatives, such as maltol and kojic acid, have an α, β-unsaturated keto-enol as a common denominator in their structure, a feature that makes them good chelators. However, it does not inhibit PPO by binding the copper at the active site of the enzyme. Maltol did not inhibit the rate of oxygen uptake but only the rate of pigmented products formation when different *o*-hydroxyphenols were acted upon by tyrosinase, this being probably due to the ability of maltol to conjugate with *o*-quinones but not to inhibit the enzyme (93).

Kojic Acid. Kojic acid (5-hydroxy-2-hydroxymethyl-4H-pyran-4-one), a γ-pyrone derivative, is a fungal metabolite produced by many species of *Aspergillus* and *Penicillium*. It is found in many fermented Asian foods (94), and possesses antibacterial and antifungal activities. A mixture of ascorbic acid and kojic acid has been patented for use as an antibrowning agent in foods (95). Kojic acid could be potentially applied to prevent melanosis in plant and seafood products. It was shown to inhibit melanosis in pink shrimp (96). A 1% kojic acid solution is comparable to the customary sulfite treatment of shrimp. It is a competitive inhibitor for the oxidation of chlorogenic acid and catechol by apple polyphenol oxidase (97). Kojic acid inhibits very effectively the rate of pigmented products formation as well as the rate of oxygen uptake when various *o*-dihydroxy- and trihydroxyphenols are oxidized by tyrosinase (92). The inhibition by kojic acid of tyrosinase is probably due to its ability to bind to copper at the active site of the enzyme, since kojic acid is a good chelator of transition metal ions such as Fe (III) and Cu (II) (98, 99). However, the toxicity of kojic acid and its production must be considered. Kojic acid shows weak mutagenic activity in a *Salmonella typhimurium* assay (100).

4. Complex Agents

Cyclodextrins. The cyclodextrins (CDs) are a class of cyclic oligosaccharides produced by the action of cyclomaltodextrin glucanotransferase on liquified starches. The major CDs produced industrially are six to eight glucose units per macrocycle, linked by α-(1, 4)-glycosidic bonds and include cyclomaltohexaose (α-CD, six units), cyclomaltoheptaose (β-CD, seven units), and cyclomaltooctaose (γ-CD, eight units). The chemical structure and a diagrammatic representation of the functional structure of β-CD are provided in Figure 16. The structure of CDs results in all of the C-6 (primary) hydroxyl groups projecting from one side of the torus, while the C-2,3 (secondary) hydroxyl groups project from the other. The central cavity is hydrophobic while the outside is hydrophilic, and is due to the location of the primary and secondary hydroxyls at the narrow and wide bases, respectively. Although CDs are very insoluble, by using the reverse action of debranching enzymes (isoamylase, pullulanase) greater solubility can be achieved (101).

The most important functional property of CDs is their ability to behave as clathratelike compounds by inclusion complex formation with a range of "guest" molecules. If the guest molecule is of suitable size and conformation to bind within the hydrophobic core, a complex will form. Greater inclusion activity can be obtained by suitable chemical modification of the CD. This applica-

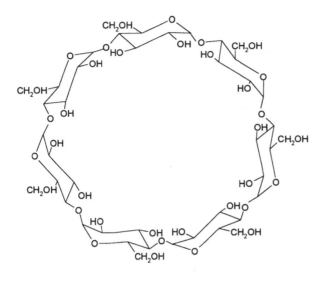

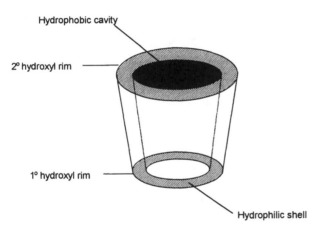

Figure 16 Schematic and chemical structure of a β-cyclodextrin molecule.

tion is of particular interest to the food industry for the molecular encapsulation of insoluble or volatile food ingredients (102). CDs have been proposed to control enzymatic browning of fruits (71, 103) and its use has been patented (104). The mechanism for browning inhibition with CD was mainly due to the binding of PPO substrates. The internal cavity of β-CD is slightly apolar and forms inclusion complexes with phenolic substrates of PPO, thereby preventing their oxidation to quinones and subsequent polymerization to brown pigments. However, cyclodextrins have not yet been approved for food use by the FDA. The adsorption of flavor or color compounds by cyclodextrins is a major drawback to their use.

Chitosan. Chitosan is a naturally abundant polymer of β-(1→4)-*N*-acetyl-D-glucosamine derived from shellfish chitin. Chitosan is nontoxic, biodegradable (105), and has antimicrobial properties (106). It is soluble in diluted organic acids. Treatment of shrimp with 2% chitosan showed consistantly lower incidence of melanosis during storage (106). The mode of action has not been known, but it is probably a consequence of the ability of the positively charged polymer to adsorb suspended PPO, its substrates, or products. Chitosan can form a semipermeable film. Its coating might be expected to inhibit browning by reducing the amount of oxygen for enzymatic oxidation of phenolics.

5. Enzyme Inhibitors

4-Hexylresorcinol. The substituted resorcinols, which are also structurally related to phenolic substrates, are recognized as PPO inhibitors (56, 107). Resorcinol derivatives are *m*-diphenols compounds that may inhibit browning reactions by competitive inhibition of PPO due to structural resemblance to phenolic substrates. Hydrophobic substitution in the 4-position in the aromatic resorcinol ring, such as hexyl, dodecyl, and cyclohexyl, increases the effectiveness of PPO inhibition (56). The 4-cyclohexyl has the lowest inhibitor concentration that yields 50 % inhibition of PPO activity (I_{50}) (Fig. 17). Both the monophenolase and diphenolase activities of tyrosinase are inhibited by 4-hexylresorcinol (4-HR).

4-HR has several advantages over sulfites in food including its specific mode of inhibitory action, its lower use level required for effectiveness, its inability to bleach preformed pigments, and its chemical stability. To show an ef-

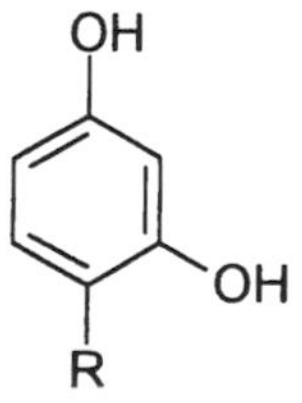

R	I_{50} μM
H	2700
Hexanoyl	750
Carboxyl	150
Ethyl	0.8
Hexyl	0.5
Dodecyl	0.3
Cyclohexyl	0.2

Figure 17 I_{50} values for synthetic 4-substituted resorcinols. (From Ref. 56.)

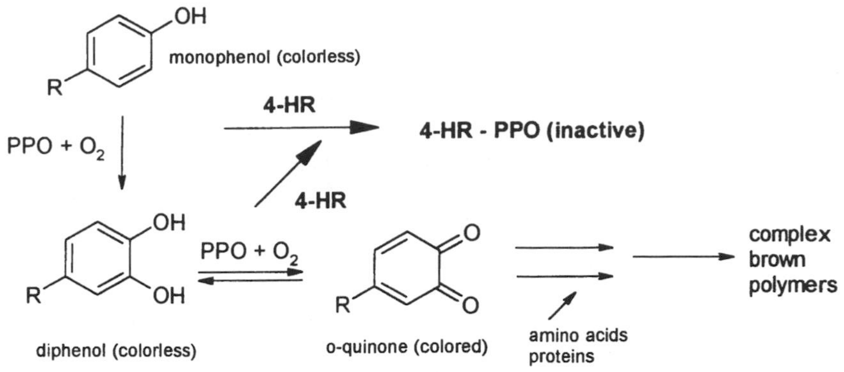

Figure 18 The inhibitory effect of 4-hexylresorcinol on PPO. (From Ref. 108.)

fective synergistic antibrowning effect, 4-HR is used in combination with ascorbic acid. Ascorbic acid reduces quinones generated by polyphenoloxidase to retard browning, whereas 4-HR is a specific inhibitor of PPO. 4-HR interacts with PPO and renders it incapable of catalyzing the enzymatic reaction (Fig. 18) (108).

4-HR acts as an enzyme-competitive inhibitor. Several studies have shown the effectiveness of 4-HR in controlling enzymatic browning in shrimp (107, 109). A 1 min dip into a 50 ppm 4-HR solution in sea water with subsequent storage on crushed ice inhibits black spot formation for up to 14 days (110). Ever-Fresh is a patented product (patent # 5,049,438) composed of 4-HR as the active ingredient and sodium chloride as the carrier agent. It was used for controlling enzymatic browning, or black spot, in crustaceans as an alternative to sulfites (108). The headless shrimp dipped in 4-HR for 1 min controlled black spot for a longer period of time than fresh water (control) or 1.25% sodium metasulfite. After 7 days of storage at 2°C, raw head-off brown shrimp treated with water showed 54% black spot and the sulfite-treated shrimp had 11% black spot, whereas 4-HR-treated shrimp had only 3.6% black spot. At day 14, the black spots of control and sulfite-treated shrimp were increased to 75% and 25%, respectively. However, the 4-HR treated shrimp did not show an increase in black spots.

4-HR is a chemically stable and water-soluble compound. It has a long history of use in both human and veterinary medicine. A review of toxicological, mutagenic, carcinogenic, and allergenic studies showed that 4-HR presents no risk at the levels used to treat shrimp (111). Currently in the United States, 4-HR has been rated as GRAS (65) and was accepted by the FDA for use on shrimp (112). The use of 4-HR for the inhibition of shrimp melanosis has no impact on taste, texture, or color when the 4-HR residuals on shrimp meat are less than 1 ppm (113, 114).

Halide Salts.   Inorganic halides are well-known inhibitors of PPO (37). NaF was the most potent inhibitor on PPO followed by NaCl, NaBr, and NaI (115). The inhibition of enzymatic browning by halides decreases as the pH is increased. Sodium and calcium chloride at concentrations of 2–4 % (w/v) are most commonly used in the food industry for the inhibition of browning (116). The activity of PPO decreased as the concentration of NaCl increased. Sodium zinc chloride has been shown to be a highly effective browning inhibitor when used in combination with calcium chloride, ascorbic acid, and citric acid (117).

Honey.   Honey has been shown to inhibit enzymatic browning. It should be of great consumer interest as a natural browning inhibitor. Purification of the honey by Sephadex G-15 column chromatography showed that the compound responsible for the inhibition of PPO appeared to be a small peptide with a molecular weight about 600. Since proteins, peptides, and amino acids can affect PPO activity by chelating the essential copper at the active site of PPO and forming stable complexes with Cu^{2+} (75), this may be the mechanism of action of the honey peptide.

Honey has been shown to contain antioxidants: tocopherols, alkaloids, ascorbic acid, flavonoids, and phenolics. The antioxidant content and the efficacy of honeys in inhibiting PPO activity depend on the various types of honey (118). Commercial browning inhibitors (ascorbate and sodium metasulfite) and various honeys were compared in a study on PPO activity and browning. Antioxidant content was positively correlated with honey color.

Amino Acids, Peptides, and Proteins.   Amino acids, peptides, or proteins can affect PPO-catalyzed browning by direct inhibition of the enzyme or by reaction with the quinone products of PPO catalysis (56). They can also affect PPO activity by chelating the essential copper at the active site of PPO for various phenolic substrates. Proteins, peptides, and α-amino acids can form stable complexes with Cu^{2+} (119). Histidine and cysteine have particularly high affinities for Cu^{2+} since histidine has an imidazole ring and cysteine a thiol group, both of which can bind metals (120).

Sulfhydryl (thiol) compounds, such as cysteine, *N*-acetylcysteine (NAC), tripeptide reduced glutathione (GSH), mercaptoethanol, dithiothreitol, and thiourea, were good inhibitors of browning in a variety of foods (121, 122). Another possibility is that the SH-containing compounds reduce carbonyl groups and double bonds in brown products to form colorless materials. The quinone intermediate may be preferentially trapped as the sulfur adduct, thus preventing enzymatic browning (123). The oxidation of a dihydroxybenzene by PPO to *o*-quinone, which then can participate in nucleophilic addition reactions with

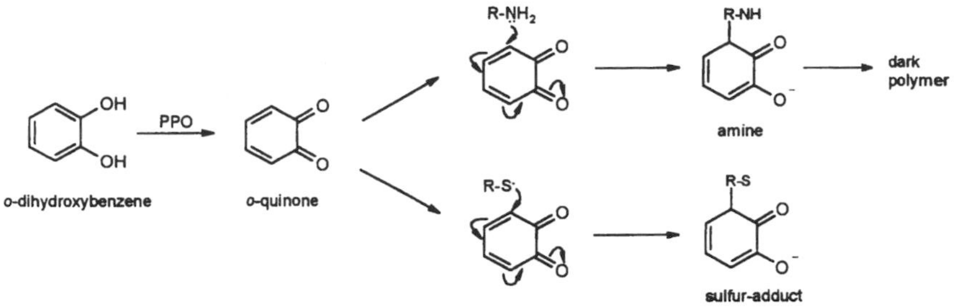

Figure 19 Postulated mechanism for the inhibition of PPO-induced enzymatic browning by thiols. (From Ref. 123.)

amino groups, will lead to the formation of dark browning products. However thiol inhibitors form sulfur adducts with the *o*-quinone, thus blocking polymer formation and preventing browning (Fig. 19). Aliphatic amino groups (secondary amines in amino acids), *N*-terminal primary amino groups, and thiol-containing amino acids react with *o*-benzoquinones and 4-methyl-*o*-benzoquinone, while thiol-containing compounds and aromatic amines react with oxidation products of dehydroxyphenylalanine (DOPA) (124).

Aromatic Carboxylic Acids.   Aromatic carboxylic acids of benzoic acid and cinnamic series are inhibitors of PPO due to their structural similarities to its phenolic substrates (125). Cinnamic acid and its analogues, *p*-coumaric, ferulic, and sinapic acids, were found to be potent inhibitors of plant PPO (126–128). Cinnamic acid at levels of 0.01 % or less was reported to be the most effective antibrowning agent providing long-term inhibition. The main effect of the combination was to increase the lag time for the onset of browning. The degree of inhibition by carboxylic acids is also pH dependent, increasing as the pH decreases. The undissociated form of the acid is responsible for inhibition by forming a complex with copper at the active site of PPO (115, 129). When a carboxyl group was present, either directly bound to the benzene ring (benzoic series) or the conjugated double bonds (cinnamic series), it could form a complex with the copper at the active site.

6.   Others

Killer Enzyme.   Enzyme action can be exploited to prevent other undesirable enzyme activities in three ways:

1.   Substrate and/or product modification by enzymes other than the target enzymes
2.   Direct inactivation of the target enzyme by other enzymes
3.   Inactivation by secondary reactions of highly reactive products

The activities of "killer enzymes" or "antienzyme enzymes," which inactivate other enzymes via direct proteolytic activity, have been demontrated. The use of enzymes to control other enzyme-related processes also has been reported in the control of enzymatic/nonenzymatic browning.

Kiwi fruit is known to contain a highly active protease (actinidine) (130). Three plant proteases (ficin from figs, papain from papaya, and bromelain from pineapple) proved to be effective browning preventors. All three proteases are sulfhydryl enzymes of broad specificity (130). The ficin, a sulfhydryl protease extracted from the genus *Ficus,* has been used as a possible sulfite substitute for inhibiting shrimp melanosis as well as enzymatic browning of apple and potato slices (131). The ficin was effective in preventing black spot formation under re-frigerated storage for more than 1 week. The mechanism of PPO inhibition by ficin is presumed to be the inactivation of PPO by binding or hydrolysis at spe-cific sites necessary for the activity of PPO. Bromelain, organic acids, sulfhydryl compounds, and certain metals in pineapple juice may be responsible for brown-ing inhibition. Glucose oxidase/catalase has been reported as a treatment to con-trol microbial growth in flounder (132), cod fillets (133), and shrimp (134, 135). Also, shrimp treated in glucose oxidase/catalase solution inhibited the enzymatic reaction, thus preventing discoloration (135). The use of catechol transferase (EC 2.1.1.6) has been proposed for the prevention of browning (136). Unfortu-nately, these enzyme treatments are too expensive to be of commercial use. The use of combinations of antibrowning agents gives more effective inhibition of browning than a single agent. A typical combination may include a chemical re-ducing agent (ascorbic acid), an acidulant (citric acid), and a chelating agent (EDTA).

Edible Coating. Use of edible coatings to minimize undesirable changes due to minimal processing has been reported for several commodities (137). Coatings are also useful as carriers of antioxidants and preservatives (138). Edi-ble coatings have the potential to retard water loss, to form a barrier to oxygen, and to hold antioxidants as well as preservatives on the surface of cut tissue to control discoloration. A polysaccharide/lipid bilayer formulation reduced respi-ration of fruits through modification of the gas exchange between the processed tissue and the external environment (139). The application of an edible coating in the glaze of frozen seafoods may be possible; however, little research in this area has occurred with seafoods.

IV. CONCLUSION

Enzymatic browning can be controlled in many different ways. In addition to physical treatment, a wide range of chemicals inhibit PPO activity, but only a limited number of them are considered potential alternatives to sulfites. New approaches to control enzymatic browning are under study at universities, government, research laboratories, and industry. These alternatives must be evaluated on the basis of efficacy, cost, method of application, safety, and regulatory status. Also, antibrowning agents should not affect the flavor, taste, texture, or color of food products.

REFERENCES

1. WS Otwell, MR Marshall. Studies on the use of sulfites to control shrimp melanosis (blackspot): Screen alternatives to sulfiting agents to control shrimp melanosis. Florida seagrant technical paper No. 46:1–10, 1986.
2. JJ Nicolas, FC Richard-Forget, PM Goupy, MJ Amoit, SY Aubert. Enzymatic browning reactions in apple and apple products. Crit Reviews Food Sci Nutr 32(2):109–157, 1994.
3. O Toussaint, K Lerch. Catalytic oxidation of 2-aminophenols and ortho hydroxylation of aromatic amines by tyrosinase. Biochemistry 26:8567–8571, 1987.
4. JK Palmer, Banana polyphenoloxidase, preparation and properties. Plant Physiol 38:508–513, 1963.
5. AB Lerner. Metabolism of phenylalanine and tyrosine. Adv Enzymol 14:73–128, 1953.
6. JR Whitaker, Polyphenol oxidase. In: JR Whitaker, ed. Principles of Enzymology for the Food Sciences. New York: Marcel Dekker, 1972, pp 571–582.
7. AM Mayer, E Harel. Phenoloxidase in plants. Phytochemistry 18:193–215, 1979.
8. B Reinhammar, BG Malstrom. "Blue" copper-containing oxidases. In: TG Spiro, ed. Copper Proteins. New York: John Wiley and Sons, 1981, pp 109–149.
9. JA Koburger, ML Miller, WS Otwell. Thawed lobster tails: some quality changes. Proc Tenth Annual Trop Subtrop Fish Conf Americas 10:201–206, 1985.
10. JRL Walker. Enzymatic browning in fruits: Its biochemistry and control. In: CY Lee, JR Whitaker, eds. Enzymatic Browning and Its Prevention. Washington, DC: American Chemical Society, 1995, pp 8–22.
11. KA Savagaon, A Sreenivasan. Activation mechanism of pre-phenoloxidase in lobster and shrimp. Fish Technol 15:49–55, 1978.
12. MR Marshall, WS Otwell, B Walker. Preliminary study on the isolation and activation of polyphenoloxidase from deep sea crab (*Geryon* sp.). Proc Ninth Annual Trop Subtrop Fish Technol Conf Americas 9:118–125, 1984.
13. K Söderhäll, T Unestam. Activation of serum prophenoloxidase in arthropod immunity. The specific cell wall glucan activation and activation by purified fungal glycoproteins of crayfish phenoloxidase. Can J Microbiol 25:406–414, 1979.
14. OJ Ferrer, JA Koburger, WS Otwell, RA Gleeson, BK Simpson, MR Marshall. Phenoloxidase from the cuticle of Florida spiny lobster (*Panulirus argus*): mode of activation and characterization. J Food Sci 54:63–67, 1989.

15. JS Chen, RS Rolle, MR Marshall, CI Wei. Comparison of phenoloxidase activity
     from Florida spiny lobster and Western Australian lobster. J Food Sci 56:154–157,
     1991.
16. DE Aiken. Molting and growth. In: JS Cobb BF Phillips, eds. The Biology and
     Management of Lobsters. New York:Academic Press, 1980, pp 91–163.
17. DF Travis. The molting cycle of the spiny lobster, *Panulirus argus* Latreille. I.
     Molting and growth in laboratory maintained individuals. Biol Bull 107:433–450,
     1954.
18. MP Jackman, A Hajnal, K Lerch. Albino mutants of *Streptomyces glaucescens* ty-
     rosinase. Biochem J 274:707–713, 1991.
19. WH Lang, KE Van Holde. Cloning and sequencing of *Octopus dofleini* hemo-
     cyanin cDNA: Derived sequences of functional units Ode and Odf. Proc Natl Acad
     Sci USA 88:244–248, 1991.
20. A Aspán, K Söderhäll. Purification of prophenoloxidase from crayfish blood cells,
     and its activation by an endogenous serine proteinase. Insect Biochem
     21:363–373, 1991.
21. A Aspán, TS Huang, L Cerenius, K Söderhäll. cDNA cloning of prophenoloxidase
     from the freshwater crayfish *Pacifastacus leniusculus* and its activation. Proc Natl
     Acad Sci USA 92:939–943, 1995.
22. H Nakashima, PQ Behrens, MD Moore, E Yokota, AF Riggs. Structure of hemo-
     cyanin II from the horseshoe crab, *Limulus polyphemus*. J Biol Chem
     261:10526–10553, 1986.
23. WPJ Gaykema, WGJ Hol, JM Vereijken, NM Soeter, HJ Bak, and JJ Beintema.
     3.2Å structure of the copper-containing, oxygen-carrying protein *Panulirus inter-
     ruptus* hemocyanin. Nature 309: May 3, 23–29, 1984.
24. HJ Bak, B Neuteboom, PA Jekel, NM Soeter, JM Vereijken, JJ Beintema. Structure
     of arthropod hemocyanin. FEBS Lett 204:141–144, 1986.
25. PA Jekel, HJ Bak, NM Soeter, JM Vereijken, JJ Beintema. *Panulirus interruptus*
     hemocyanin. The amino acid sequence of subunit b and anomalous behaviour of
     subunits a and b on polyacrylamide gel electrophoresis in the presence of SDS. Eur
     J Biochem 178:403–412, 1988.
26. B Neuteboom, PA Jekel, JJ Beintema. Primary structure of hemocyanin subunit c
     from *Panulirus interruptus*. Eur J Biochem 206:243–249, 1992.
27. K Fujimoto, N Okino, SI Kawabata, S Iwanaga, E Ohnishi. Nucleotide sequence
     of the cDNA encoding the proenzyme of phenol oxidase A sub(1) of *Drosophila
     melanogaster*. Proc Natl Acad Sci USA 92:7769–7773, 1995.
28. T Kawabata, Y Yasuhara, M Ochiai, S Matsura, M Ashida. Molecular cloning of
     insect pro-phenol oxidase: a copper-containing protein homologous to arthropod
     hemocyanin. Proc Natl Acad Sci USA 92:7774–7778, 1995.
29. H Inagaki, Y Bessho, A Koga, H Hori. Expression of the tyrosinase-encoding gene
     in a colorless melanophore mutant of the medaka fish, *Oryzias latipes*. Gene
     150:319–324, 1994.
30. BS Kwon, M Wakulchik, AK Haq, R Halaban, D Kestler. Sequence analysis of
     mouse tyrosinase cDNA and the effect of melanotropin on its gene expression.
     Biochem Biophys Res Commun 153:1301–1309, 1988.

31. BK Simpson, MR Marshall, WS Otwell. Phenoloxidases from pink and white shrimp: kinetic and other properties. J Food Biochem 12:205–217, 1988.

32. MR Marshall, BK Simpson, O Ferrer, WS Otwell. Purification and charactrization of phenoloxidases in crustaceans: shrimp and lobster. Proc Twelfth Annual Trop Subtrop Fish Technol Conf Americas. 12:1–12, 1988.

33. CF Madero, G Finne. Properties of polyphenoloxidase isolated from gulf shrimp. Proc Seventh Annual Trop Subtrop Fish Technol Conf Americas 7:328–339, 1982.

34. MT Ali, MR Marshall, CI Wei, RA Gleeson. Monophenol oxidase from the cuticle of Florida spiny lobster (*Panulirus argus*). J Agric Food Chem 42:53–58, 1994.

35. A Opoku-Gyamfua, BKSimpson, EJ Squires. Comparative studies on the polyphenol oxidase fraction from lobster and tyrosinase. J Agric Food Chem 40:772–775, 1992.

36. JB Adams. Review: Enzyme inactivation during heat processing of food-stuffs. Int J Food Sci Technol 26:1–20, 1991.

37. L Vámos-Vigyázó. Polyphenoloxidase and peroxidase in fruits and vegetables. Crit Rev Fd Sci Nutri 15:49–127, 1981.

38. BK Simpson, MR Marshall, WS Otwell. Phenoloxidase from shrimp (*Paneus setiferus*): Purification and some properties. J Agric Food Chem 35:918–921, 1987.

39. AL Tappel. Effects of low temperatures and freezing on enzymes and enzyme systems. In: HT Meryman, ed. Cryobiology. New York: Academic Press, 1966.

40. OR Fennema. Activity of enzymes in partially frozen aqueous systems. In: RB Duckworth, ed. Water Relations of Foods. New York: Academic Press, 1975, pp 397–413.

41. INA Ashie, BK Simpson. Application of high hydrostatic pressure to control enzyme related fresh seafood texture deterioration. Food Res Int 29:569–575, 1996.

42. AF Novak, RM Grodner, MR Ramachandrarao. Radiation pasteurization of fish and shell fish. In: RF Gould, ed. Radiation Preservation of Foods. Washington, DC: American Chemical Society, 1967, pp 142–151.

43. JS Yang, FS Perng, SE Liou, JJ Wu. Effects of gamma irradiation on chromatophores and volatile components of grass shrimp muscle. Radiat Phys Chem 42:319–322, 1993.

44. R Voisine, LP Vezina, C. Willemot. Induction of senescence-like deterioration of microsomal membranes from cauliflower by free radicals generated during gamma irradiation. Plant Physiol 97:545–550, 1991.

45. JC Cheftel. Effects of high hydrostatic pressure on food constituents: overview. In: C Balny, R Hayashi, K Heremans, P Masson, eds. High Pressure and Biotechnology. London: John Libbey Eurotext, 1992, pp 195–209.

46. P Masson. Pressure denaturation of proteins. In: C Balny, R Hayashi, K Heremans, P Masson, eds. High Pressure Technology. London: John Libbey Eurotext, 1992, pp 89–99.

47. S Kunugi. Effect of pressure on activity and specificity of some hydrolytic enzymes. In: C Balny, R Hayashi, K Heremans, P Masson, eds. High Pressure Technology. London: John Libbey Eurotext, 1992, pp. 129–137.

48. DF Farkas. Novel processes-ultra high pressure processing. In: CW Felix, ed. Food Protection Technology, Chelsea. MI: Lewis Publishers, 1987, pp 393–396.

49. I Seyderhelm, S Boguslawski, G Michaels, D Knorr. Pressure induced inactivation of selected food enzymes. J Food Sci 61:308–310, 1996.

50. E Kiran, W Zhuang. Miscibility and phase separation of polymers in near- and supercritical fluids. In: MA Abraham, AK Sunol, eds. Supercritical Fluids. Washington, DC: American Chemical Society, 1997, pp. 2–36.

51. I Hardardottir, JE Kinsella. Extraction of lipid and cholesterol from fish muscle with supercritical fluids. J. Food Sci 1988, 53:1656–1658.

52. SS Rizvi, HAL Benado, JA Zollweg, JA Daniels. Supercritical fluid extraction fluid extraction: fundamental principles and model methods. Food Technol 40 (6):55–65, 1986.

53. JS Chen, MO Balaban, CI Wei, RA Gleeson, MR Marshall. Effect of carbon dioxide on the inactivation of Florida spiny lobster polyphenol oxidase. J Sci Food Agric 61:253–259, 1993.

54. JS Chen, MO Balaban, CI Wei, MR Marshall, WY Hsu. Inactivation of polyphenol oxidase by high-pressure carbon dioxide. J Agric Food Chem 40:2345–2349, 1992.

55. MO Balanban, S Pekyardimci, JS Chen, A Arreola, G Zemel, MR Marshall. Enzyme inactivation by pressurized carbon dioxide. In: M Yalpani, ed. Science for the Food Industry of the 21st Century. Biotechnology, Supercritical Fluids, Membranes, and other Advanced Technologies for Low Calorie, Healthy Food Alternatives. Mount Prospect, IL:ATL Press, 1993, pp 235–252.

56. AJ McEvily, R Iyengar, WS Otwell. Inhibition of enzymatic browning in foods and beverages. Crit Rev Food Sci Nutr 32:253–273, 1992.

57. LF Green. Sulphur dioxide and food preservation—a review. Food Chem 1:103–124, 1976.

58. LA Sayavedra-Soto, MW Montgomery. Inhibition of polyphenoloxidase by sulfite. J Food Sci 51:1531–1536, 1986.

59. RC Lindsay. Food additives. In: OR Fennema ed. Food Chemistry. New York: Marcel Dekker, 1985, pp 629–687.

60. JP Modderman. Technological aspects of use of sulfiting agents in food. J Assoc Off Anal Chem 69:1–3, 1986.

61. BL Wedzicha, I Bellion, SJ Goddard. Inhibition of browning by sulfites. Adv Exp Med Biol 189:217–236, 1991.

62. JRL Walker. Enzymatic browning in foods. Its chemistry and control. Food Technol NZ 12:19–25, 1977.

63. MB Faulkner, BM Watts, HJ Humm. Enzymatic darkening of shrimp. Food Res 19:302–310, 1954.

64. Federal Register. Notice to shippers, distributors, packers, and importers of shrimp containing sulfites. 1985, 50:2957–2958.

65. Federal Register. Sulfiting agents in standardized foods; labeling requirements. 1988, 53:51062–51065.

66. Federal Register. Sulfiting agents; revocations of GRAS status for use on fruits and vegetables intended to be served or sold raw to consumers. 1986, 51:25021–25026.

67. IC Forbes. Sodium metabisulfite and sodium sulfite blends to control shrimp melanosis and reduce sulfur dioxide. MS thesis, University of Florida, Gainesville, FL, 1996.

68. SL Taylor, NA Higley, RK Bush. Sulfites in foods: uses, analytical methods, residues, fate, exposure assessment, metabolism, toxicity, and hypersensitivity. Adv Food Res 30:1–76, 1986.

69. JC Bauernfeind, DM Pinkert. Food processing with added ascorbic acid. Adv Food Res 18:219–315, 1970.

70. ML Liao, PA Seib. A stable form of vitamin C: L-ascorbate 2-triphosphate. Synthesis, isolation and properties. J Agric Food Chem 38:355–366, 1990.

71. GM Sapers, KB Hicks. Inhibition of enzymatic browning in fruits and vegetables. In: JJ Jen, ed. Quality Factors of Fruits and Vegetables: Chemistry and Technology, Washington, DC : American Chemical Society, 1989, pp 29–43.

72. ML Liao, PA Seib. Chemistry of L-ascorbic acid related to foods. Food Chem 30:289–312, 1988.

73. WS Pierpoint, The enzymatic oxidation of chlorogenic acid and some reactions of the quinone produced. Biochem J 98:567–580, 1966.

74. E Valero, R Varón, F García-Carmona. A kinetic study of irreversible enzyme inhibition by an inhibitor that is rendered unstable by enzymatic catalysis. The inhibition of polyphenol oxidase by L-cysteine. Biochem J 277:869–874, 1991.

75. V Kahn. Effects of proteins, protein hydrolyzates, and amino acids on o-dihydroxyphenolase activity of polyphenol oxidase of mushroom, avocado and banana. J Food Sci 50:111–115, 1985.

76. JJL Cilliers, VL Singleton. Caffeic acid autooxidation and the effects of thiols. J Agric Food Chem 38:1789–1796, 1990.

77. J Cherry, SS Singh. Discoloration preventing food preservative and method. 1990, U.S. patent 4,937,085.

78. IH Segel. Biochemical Calculations, 2nd ed. New York: John Wiley and Sons, 1976, pp 273–277.

79. FM Richard-Forget, PM Goupy, JJ Nicolas. Cysteine as an inhibitor of enzymatic browning. II. Kinetic studies. J Agric Food Chem 40:2108–2113, 1992.

80. KF Tipton, HBF Dixon. Effect of pH on enzymes. In: DL Purich, ed. Contemporary Enzyme Kinetics and Mechanism. New York: Academic Press, 1983, pp 97–148.

81. OR Fennema. Effect of pH on rates of enzyme-catalyzed reactions. In: JR Whitaker, ed. Principles of Enzymology for the Food Sciences. New York: Marcel Dekker, 1972, pp 287–317.

82. T Richardson, DB Hyslop. Enzymes. In: OR Fennema, ed. Food Chemistry, New York: Marcel Dekker, 1985, pp 371–476.

83. JD Dziezak. Presertative systems in foods, antioxidants and antimicrobial agents. Food Technol 40(9):94–136, 1986.

84. RA Benner, R Migit, G Finne, GR Acuff. Lactic acid/Melanosis inhibitors to improve shelf life of brown shrimp (*Penaeus aztecus*). J Food Sci 59:242–245, 250, 1994.

85. GM Sapers, KB Hicks, JG Philips, L Garzarella, DL Pondish, RM Matulaitis, TJ

McCormack, SM Sondey, PA Seib, YS. El-Ataway. Control of enzymatic browning in apple with ascorbic acid derivatives, polyphenol oxidase inhibitors, and complexing agents. J Food Sci 54:997–1002, 1989.

86. JP Van Buren. Causes and prevention of turbidity in apple juice. In: DD Downing, ed. Processed Apple Products. New York: AVI-Van Nostrand Reinhold, 1989, pp 97–120.

87. D Osuga, A van der Schaaf, JR Whitaker. Control of polyphenoloxidase activity using a catalytic mechanism. In: RY Yada, RL Jackman, JL Smith, eds. Protein Structure–Function Relationships in Foods. New York: Blackie Academic & Professional, 1994, pp 62–88.

88. Code of Federal Regulations, Food and Drugs 21 Parts 170.199, 1992. Washington, DC: Office of the Federal Register National Archives and Records Administration.

89. DW Cook, RE Bowman. Retardation of shrimp spoilage with ethylenediaminetetraacetic acid. J Miss Acad Sci 17:38–43, 1973.

90. D Mu, YW Huang, KW Gates, WH Wu. Effect of trisodium phosphate on Listeria monocytogenes attached to rainbow trout (*Oncorhynchus mykiss*) and shrimp (*Penaeus spp.*) during refrigerated storage. J Food Safety 17:37–46, 1997.

91. J Gardner, S Manohar, WS Borisenok Method and composition for preserving fresh peeled fruits and vegetables. 1991, U.S. patent 4,988,523.

92. V Kahn. Multiple effects of maltol and kojic acid on enzymatic browning. In: CY Lee, JR Whitaker, eds. Enzymatic Browning and Its Prevention. Washington, DC: American Chemical Society, 1995, pp 277–294.

93. V Kahn, F Schved, P Lindner. Effect of maltol on the oxidation of *o*-dihydroxyphenols by mushroom tyrosinase and by sodium periodate. J Food Biochem 17:217–233, 1993.

94. R Kinosita, T Ishiko, S Sugiyama., T Seto, S Igarasi, IE Goetz. Mycotoxins in fermented foods. Cancer Res 28:2296–2311, 1968.

95. R Fukusawa, H Wakabayashi, T Natori. Inhibitor of tyrosinases in foods. 1982, Japanese patent 57-40875.

96. L Applewhite, WS Otwell, MR Marshall. Effect of kojic acid on pink shrimp phenoloxidase. Proc Fifteenth Annual Trop Subtrop Fish Technol Conf Americas 15:141–146, 1990.

97. JS Chen, CI Wei, RS Rolle, WS Otwell, MO Balaban, MR Marshall. Inhibitory effect of kojic acid on some plant and crustacean polyphenol oxidase. J Agric Food Chem 39:1396–1401, 1991.

98. A Beélik. Kojic acid. Adv Carb Chem 11:145–183, 1956.

99. JW Wiley, GN Tyson, JS Steller. The configuration of complex kojates formed with some transition elements as determined by magnetic susceptibility measurements. Am Chem Soc J 64:963–964, 1942.

100. CI Wei, SY Fernando, TS Huang. Mutagenicity studies of kojic acid. Proc Fifteenth Annual Trop Subtrop Fish Technol Conf Americas 15:464–470, 1990.

101. Y Okada, Y Kubota, K Koizumi, S Hizukuri, T Ohfuji, K Ogata. Some properties and the inclusion behavior of branched cyclodextrins. Chem Pharm Bull 36:2176–2185, 1988.

102. JS Pagington. β-Cyclodextrin and its uses in the flavor industry. In: GG Birch, MG Lindley. Developments in Food Flavours. New York: Elsevier Applied Science Publishers, 1986, pp 131–150.

103. C Billaud, E Regaudie, N Fayad, F Richard-Forget, J Nicolas. Effect of cyclodextrins on polyphenol oxidation catalyzed by apple polyphenol oxidase. In: CY Lee, JR Whitaker, eds. Enzymatic Browning and Its Prevention. Washington DC: American Chemical Society, 1995, pp 295–312.

104. KB Hicks, GM Sapers, PA Seib. Process for preserving raw fruit and vegetable juices using cyclodextrins and compositions thereof. 1990, U.S. patent 4,975,293.

105. S Hirano, C Itakura, H Seino, Y Akiyama, I Nonaka, N Kanbara, T Kawakami. Chitosan as an ingredient for domestic animal feeds. J Agric Food Chem 38:1214–1217,1990.

106. BK Simpson, N Gagne, INA Ashie, E Noroozi. Utilization of chitosan for preservation of raw shrimp (Pandalus borealis). Food Biotech 11:25–44, 1997.

107. AJ McEvily, R Iyengar, A Gross. Compositions and methods for inhibiting browning in foods using resorcinol derivatives. 1991, U.S. patent 5,059,438.

108. HS Lambrecht. Sulfite substitutes for the prevention of enzymatic browning in foods. In: CY Lee, JR Whitaker, eds. Enzymatic Browning and Its Prevention. Washington, DC: American Chemical Society, 1995, pp 313–323.

109. WS Otwell, R Iyengar, A McEvily. Inhibition of shrimp melanosis by 4-hexylresorcinol. J Aquat Food Prod Technol 1:53–65, 1992.

110. AJ McEvily, R Iyengar, WS Otwell. A new processing aid for the inhibition of shrimp melanosis. Proc Fifteenth Annual Trop Subtrop Fish Technol Conf Americas 15:147–153, 1990.

111. VH Frankos, DF Schmitt, LC Haws, AJ McEvily, R Iyengar, SA Miller, IC Munro, FM Clydesdale, AL Forbes, RM Sauer. Generally recognized as safe (GRAS) evaluation of 4-hexylresorcinol for use as a processing aid for prevention of melanosis in shrimp. Regul Toxicol Pharmacol 14:202–212, 1991.

112. Federal Register Filing of petition for affirmation of GRAS status. 1992, 57:8461.

113. R Iyengar, CW Bohmont, AJ McEvily. 4-Hexylresorcinol and prevention of shrimp melanosis: residual analyses. J Food Comp Anal 4:148–157, 1991.

114. JM King, AJ McEvily, R Iyengar. Liquid chromatographic determination of the processing aid 4-hexylresorcinol in shrimp. J Assoc Off Anal Chem 74:1003–1005, 1991.

115. AH Janovitz-Klapp, FC Richard, PM Goupy, JJ Nicolas. Inhibition studies on apple polyphenol oxidase. J Agric Food Chem 38:926–931, 1990.

116. F Steiner, TE Rieth. Preservative method and preserved fruit or vegetable, using citric acid, sodium and calcium chloride containing preservative composition. 1989, U.S. patent 4,818,549.

117. HR Bolin, CC Huxsoll. Storage stability of minimally processed fruit. J Food Proc Preserv 13:281–292, 1989.

118. L Chen, A Mehta, M Berenbaum, N Engeseth. The potential use of honey as an inhibitor of enzymatic browning. Institute of Food Technologists, June 20–24, Atlanta, GA, 1998, abstract no. 20B-29.

119. WJ O'Sullivan. Stability constants of metal complexes. In: RMC Dawson, DC Elliot, WH Elliot, KM Jones, eds. Data for Biochemical Research. Oxford University Press, England, 1969, pp 423–435.
120. CF Bell. Principles and applications of metal chelation. Oxford: Clarendon Press, 1977.
121. I Molnar-Perl, M Friedman. Inhibition of browning by sulfur amino acids. 2. Fruit juices and protein-containing foods. J Agric Food Chem 38:1648–1651, 1990.
122. I Molnar-Perl, M Friedman. Inhibition of browning by sulfur amino acids. 3. Apples and potatoes. J Agric Food Chem, 38:1652–1656, 1990.
123. M Friedman, I Molnar-Perl. Inhibition of browning by sulfur amino acids. 1. Heated amino acid-glucose systems. J Agric Food Chem 38:1642–1647, 1990.
124. HS Mason, EW Peterson. Melanoproteins. I. Reactions between enzyme-generated quinones and amino acids. Biochim Biophys Acta 111:134–146. 1965.
125. RC Krueger. The inhibition of tyrosinase. Arch Biochem Biophys 57:52–60, 1955.
126. AR Macrae, RG Duggleby. Substrates and inhibitors of potato tubers phenolase. Phytochemistry 7:855–861, 1968.
127. PG Pifferi, L Baldassari, R Cultrera. Inhibition by carboxylic acids of an o-diphenol oxidase from *Prunus avium* fruit. J Sci Food Agric 25:263–270, 1974.
128. JRL Walker, EL Wilson. Studies on the enzymatic browning of apples. Inhibition of apple o-diphenol oxidase by phenolic acids. J Sci Food Agric 26:1825–1831, 1975.
129. DA Robb, T Swain, LW Mapson. Substrates and inhibitors of the activated tyrosinase of broad bean (*Vicia faba L.*). Phytochemistry 5:665–675, 1966.
130. MF Miller, GW Davis, SC Seideman, CB Ramsey, TL Rolan. Effects of papain, ficin and spleen enzymes on textural, visual, cooking and sensory properties of beef bullock restructured steaks. J Food Qual 11:321–330, 1989.
131. PS Taoukis, TP Labuza, JH Lillemo, SW Lin. Inhibition of shrimp melanosis (black spot) by ficin. Lebensm Wiss Technol 23:52–54, 1990.
132. CE Field, LF Pivarnik, SM Barnett, AG Rand. Utilization of glucose oxidase for extending the shelf-life of fish. J Food Sci 51:66–70, 1986.
133. SJ Shaw, EG Bligh, AD Woyewoda. Spoilage pattern of Atlantic cod fillets treated with glucose oxidase/gluconic acid. J Can Inst Food Sci Technol 19:3–6, 1986.
134. M Dondero, W Egaña, W Tarky, A Cifuentes, and JA Torres. Effectiveness of glucose oxidase/catalase in the preservation of shrimp (*Heterocarpus reedi.*). Institute of Food Technologists, June 16–20, Anaheim, CA, 1990, paper no. 362.
135. CA Kantt, J Bouzas, M Dondero, JA Torres. Glucose oxidase/catalase solution for on-board control of shrimp microbial spoilage: model studies. J Food Sci 58:104–107, 1993.
136. BJ Finkle, RF Nelson. Enzyme reactions with phenolic compounds: effect of o-methyl-transferase on a natural substrate of fruit polyphenolxoidase. Nature 197:902–903, 1963.
137. EA Baldwin, MO Nisperos-Carriedo, RA Baker. Edible coatings for lightly processed fruits and vegetables. Hort Sci 30:35–38, 1995.

138. SL Cuppett. Edible coatings as carriers of food additives, fungicides and natural antagonists. In: JM Krochta, EA Baldwin, MO Nisperos-Carriedo, eds. Edible Coatings and Films to Improve Food Quality. Lancaster/Basel: Technomic Publishing, 1994, pp. 121–137.

139. WS Wong, SJ Tillin, JS Hudson, AE Pavlath. Gas exchange in cut apples with bi-layer coatings. J Agric Food Chem 42:2278–2285, 1994.

11
Lipoxygenases

Bonnie Sun Pan
National Taiwan Ocean University, Keelung, Taiwan, Republic of China

Jen-Min Kuo
Chia-Nan College of Pharmacy, Tainan, Taiwan, Republic of China

I. INTRODUCTION

Lipoxygenase (LOX) (E.C. 1.13.11.12) catalyzes the incorporation of oxygen into the moiety of *cis, cis*-1,4 pentadiene in polyunsaturated fatty acids such as arachidonic acid to form the hydroperoxide derivatives at a specific position (Fig. 1). It is widely distributed in the plant kingdom (1), such as in peas (2), soybean (3), broad bean (4), in the germs of corn (5), rice (6), and in fruits such as tomato (7) and cucumber (8). The LOX isozymes in plants play important roles in wound healing, disease resistance (1), and flavor biogenesis (7).

LOXs are also found in animals' sanguineous tissues, such as platelet (9, 10), leukocyte (11), reticulocyte (12), basophil (13), and neutrophil (14). They participate in the metabolism of eicosanoids (Fig. 1): formation of leukotrienes, a key mediator of allergy and inflammation (15); syntheses of lipoxins (16); and in the insulin secretory response to glucose (11). LOXs are also present in aquatic animals and algae (Table 1).

II. OCCURRENCE IN AQUATIC ORGANISMS

A. Aquatic Animals

In aquatic animals, LOX activity was identified in coral (17), in skin and gill of trout (18), ayu (19, 20), and sardine (21, 22), in eggs of starfish (23, 24), sea

317

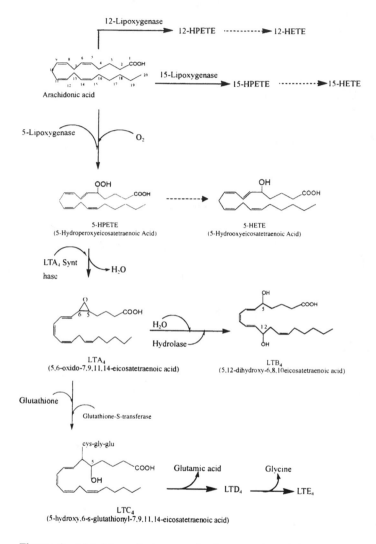

Figure 1   Main biosynthetic routes for the generation of eicosanoids via site-lipoxygenase activities. Solid lines, enzymatic steps; dotted lines, nonenzymatic reactions; LT, leukotriene.

urchin (25), and grey mullet (26). LOX in eggs was considered important in oocyte maturation (23, 24, 26) or regulating the membrane permeability (27). LOX of fish skin and gill was proposed to be responsible for flavor formation of fresh and processed fish (28–31).

LOX activity was found in the hemolymph (32, 33) but negligible in the muscle, branchiae, midgut gland (hepatopancreas) of live tiger shrimp, eye, brain, or

Table 1   Lipoxygenase Activity Observed in Aquatic Organisms

Lipoxygenase[a]	Origin	Tissue	Reference
	Teleost		
12-	Rainbow trout	Skin	18
12-	Ayu	Skin and gill	19, 20
13(S)-[b]	Sardine	Skin	21, 22
5-, 12-, 15-	Grey mullet	Gill	34
5-, 12-, 15-	Grey mullet	Platelet	35, 36
	Grey mullet	Blood plasma	35
	Grey mullet	Erythrocyte	35
	Grey mullet	Ovary	26
12-, 15-	Rainbow trout	Gill	37, 38, 39, 40
12-	Rainbow trout	Skin	40
	Rainbow trout	Brain	40
	Lake herring	Muscle	41
12-	Tilapia	Gill	42
lipoxygenase	Menhaden	Gill	42a
	Crustacean		
12-	Shrimp	Hemolymph	32, 33
12-	Crab	Gill	42
	Bivalve		
5-, 12-	mussel	gill	43
	Anthozoa		
8-	Gorgonian coral	Branches	17
	Echinoderm		
8-, 12-, 15-	Starfish	Eggs	23, 24
11(R)-, 12(R)-	Sea urchin	Eggs	25
	Alga		
	Green alga		
9-, 13-[b]	*Oscillatoria sp.*		44
9-, 13-[b]	*Enteromorpha intestinalis*		46, 47
12-, 15-			
9-, 13-[b]	*Ulva lactuca*		47
12-, 15-			
9-, 13-[b]	*Ulva conglobata*		48
12-, 15-			
12-	Red alga		49

[a] Assayed with arachidonate, C20:4 as substrate.

[b] Assayed with linoleate, C18:2.

Table 2 Lipoxygenase Activity of Erythrocytes, Platelet, and Blood Plasma of Cultured Grey Mullet (*Mugil cephalus*) and Tiger Shrimp (*Penaeus japonicus* Bate)

Tissue	5-HETE	12-HETE	15-HETE
Mullet			
erythrocyte	1	4	0.3
platelet	139	2817	95
blood plasma	2	28	1
Shrimp			
hemolymph	—[a]	31	—
muscle	—	—	—
branchiae	—	—	—
midgut gland	—	—	—
eye	—	—	—
brain & spinal cord	—	—	—

Activity was assayed on 100μM arachidonic acid as substrate in 0.2M phosphate buffer (pH 6.5) at 25°C for 30 min. Products were analyzed with HPLC. Values are given as pmole/min/mg protein.
[a]—, Negligible.

spinal cord. Neither was LOX found significant in muscle or liver in comparison to that in gill, skin, and platelet of grey mullet (Table 2). However, in lake herring LOX activity was reported highest in light muscle, lowest in skin, and intermediate in dark muscle (41).

Three types of arachidonate LOX (5-LOX, 12-LOX, and 15-LOX) are present in the sanguineous tissues of shrimp (32, 33) and grey mullet (34, 35) (Table 2). The 12-LOX has the highest activity, being present at more than 20-fold greater concentration than that of 5-LOX or 15-LOX. In fish blood tissues, LOX activity in platelets is higher than that found in plasma or erythrocytes. Gill has even higher 12-LOX activities than the platelet does (Table 3). 12-LOX is widely distributed in different organs / tissues of rainbow trout (40). Gill and skin have the highest activity followed by brain, ovary, muscle, eye, liver, spleen, and heart having 12-LOX activities less than 10% of those found in gills and skin. The physiological significance of having high 12-LOX activity in fish, aside from generating eicosanoids, has not been discussed in the literature.

B. Algae

LOXs are present in fresh water green algae, *Oscillatoria sp.* (44) marine macro green algae, *Enteromorpha intestinalis,* (45, 46), *Ulva lactuca* (47) and *Ulva conglobata* (48), marine red algae (49), and micro green algae, chlorella (50). They are important in the production of antimicrobial substances (45), and formation of seaweed and fresh fish flavor (45–48).

Table 3 12-lipoxygenase (LOX) Activity Identified in 1-Year Grey Mullet (Sampled in October)

tissue	12-HETE (pmole/min/mg protein)	LOX activity (%)	opt. pH
platelet	227.0	100.0	6.5
gill	329.5	145.0	8.0
ovary	40.4	17.8	6.5
muscle	—	—	
liver	—	—	

Lipoxygenase activity was assayed on 100 μM arachidonic acid substrate in 0.2M phosphate buffer (pH 6.5) at 25°C for 30 min. Products were analyzed with HPLC.

III. CHEMICAL STRUCTURE AND PROPERTIES

A. Catalytic Site

1. Iron

A non-heme Fe center is the catalytic site of LOX. The ferrous site consists of oxygen and nitrogen ligands in a roughly octahedral field of symmetry (51). The iron is at high-spin Fe(II) state and is activated by oxidation of the Fe(II) to Fe(III) by 1 mole of the hydroperoxy product.

Soybean LOX-1 and 2 contain one atom of Fe per mole (51). Mammalian LOXs contain 1 atom per mol of rabbit reticulocyte 15-lipoxygenase, 0.74 atom of iron per mol of porcine leukocyte 12-lipoxygenase, and 1.1 or 0.5 atom of iron per mol of recombinant 5-lipoxygenase of human origin (52). The iron content in LOX of aquatic organisms has not been investigated.

2. Histidines

Sequences of LOX from plants and mammalian species display a motif of 38 amino acid residues including five conserved histidines and a sixth histidine about 160 residues downstream (53). These residues are at positions 494, 499, 504, 522, 531, and 630 in soybean LOX-1. The six conserved histidine residues are possible metal ligands in LOX and contribute to the catalytic effect of LOX. The number of imidazole ligands is estimated at 4 ± 1, while the remaining ligands are proposed to be carboxylate oxygens (54).

For human 5 LOX, His-367, -372, and -551 are crucial, while deletion of Gln-558 and the six C-terminal amino acids also results in loss of activity (55). For porcine leukocyte arachidonate 12-LOX, His-361, -366, and -541 play important roles in iron binding (56).

B. Three-Dimensional Structure

The structure of LOX from aquatic organisms has not been reported. Soybean lipoxygenases structures have been studied in greatest detail, followed by mammalian LOXs (57–59).

Soybean LOX-1 (5-LOX) consists of 838 amino acids (60), while LOX-2 (12-LOX) has 859 residues (61). Both LOX have a single chain in two domains. Domain I is an eight-stranded antiparallel β-barrel with an hydrophobic interior contributed from some of the side chains of the 146- N terminal residues. Domain I makes loose contact with the C-terminal of domain (II).

The mammalian LOXs do not have domain I. Domain II consists of the C-terminal residues: 663 residues in rat pineal gland 12-LOX (62). It has 23 helices and 2 antiparallel β-sheets (57).

The overall structure of LOX has two internal cavities (57, 58). One cavity is shaped like a channel and faces three conserved histidines, presenting a path for the molecular oxygen moving from outside to the iron center. The second cavity is comparatively long and narrow, which faces the C-terminal and two of the histidine ligands (His-499, -690). This cavity has two bends, one of which is adjacent to the iron center and close to the end of the cavity. The reacting pentadiene system is situated opposite His-690. Most of the residues lining this cavity are either hydrophobic or neutral for the reacting fatty acid substrate to fit into this channel. When the O_2 and the fatty acid are coordinated to the iron center simultaneously, the iron has six ligands and catalyzes the dioxygenation of the reacting fatty acid.

C. Reaction Characteristics

1. Positional and Stereo Specificity

The LOX-catalyzed reaction is characterized by the position-specific and stereospecific elimination of the methylene hydrogen atom from the *cis, cis* 1,4-pentadiene structure of the PUFA. The oxygenation followed is also stereospecific, producing a hydroperoxide of either R- or S- configuration. Some of the LOX is not 100% stereoselective that produces a minor anomer (52). The positional specificity is not absolute. It is affected by substrate structure (63) and reaction condition such as temperature and pH (1).

Dual positional specificity also characterizes some LOXs. For example, leukocyte 12-LOX or erythroid 15-LOX oxygenates arachidonate at C-12 or C-15 as the major catalytic site but also oxygenates at an alternative site (64, 65). Our purified mullet gill 15-LOX after storage in liquid nitrogen for a period of time also produces minor 12-HETE when assayed on arachidonate (66).

Dihydroperoxidation or additional dioxygenation can be catalyzed by the same LOX. For example, 15-HPETE is further oxygenated at either C_8 or C_{14} by 15-LOX (67). The dihydroperoxidation is promoted by reaction pH lower than the optimum for the monohydroperoxidation (1).

2. Substrate Specificity

Polyunsaturated Fatty Acids. LOX from different aquatic origanisms consists of different isozymes and shows a different preference for polyunsaturated fatty acids. Shrimp hemolymph LOX increases in reactivity with PUFA substrate of increased unsaturation (Fig. 2) (32), while grey mullet platelet shows higher LOX activity on $C_{20:5}$ than on $C_{22:6}$ (Fig. 3) (36). The 12-LOX from trout gill and ayu skin show similar preference for $C_{20:5}$, $C_{22:6}$, and $C_{20:4}$, and significantly lower reactivity on $C_{18:3}$ and $C_{18:2}$ (20, 38). On the other hand, partially purified sardine skin LOX has higher reaction velocity with $C_{18:2}$ than with $C_{20:4}$, or $C_{20:5}$ (21). This order of substrate reactivity is different from that observed with most of the animal LOXs or autoxidation.

Both soybean and mammalian 15-LOX show a broad substrate speci-

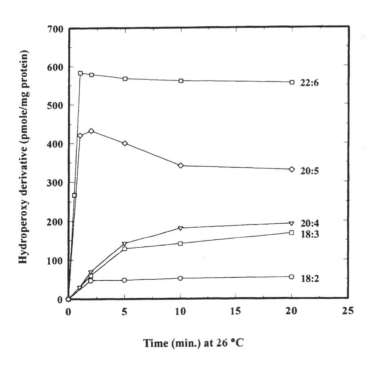

Figure 2 Time-course of hydroperoxide formed from 100 μM PUFA reacted with shrimp hemolymph lipoxygenase at 26 °C.

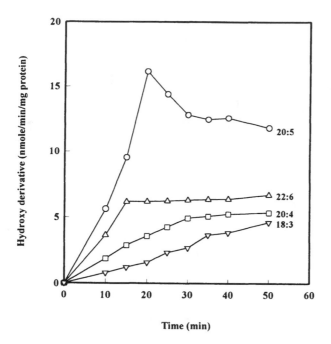

Figure 3   Time-course of hydroxy dervatives formed from 100 μM PUFA reacted with mullet platelets extract at 25 °C.

ficity with reference to carbon chain length, while mammalian 5-LOX has narrow specificity for $C_{20:4}$. Mammalian platelet 12-LOX prefers $C_{20:4}$ to C_{18} fatty acids. On the other hand, leukocyte 12-LOX has similar affinity toward C_{18}-C_{22} PUFA (52). The epithelial 12-LOX has markedly higher activity for linoleic acid than does the corresponding enzyme from leukocyte (68). The distinct substrate specificity of the 3 types of 12-LOX may relate to their functions in different cell types.

Esterified Polyunsaturated Fatty Acids.   Esterified linoleic acids (i.e., trilinolein and methyl linoleate) are more favorably oxidized than the free linoleic acid by the sardine skin linoleate 13(s)-LOX (22). Mammalian 5-LOX does not catalyze phospholipid peroxidation until phospholipase A_2 splits off the PUFA (i.e., arachidonic acid) from the membrane phospholipids (68). However, the reticulocyte 15-LOX oxygenates the esterified membranes in vivo (63).

Lipoproteins.   Mammalian reticulocyte 15-LOX and leukocyte 12-LOX are capable of oxygenating very-low-density lipoprotein (VLDL), low-density lipoproteins (LDL), and high-density lipoproteins (HDL) without the preceding action of a phospholipase and/or cholesterol esterase. The reactivity of mammalian LOX with lipoproteins is much lower than with free polyenoic fatty acids. Among the lipoprotein classes LDL is the best substrate for the reticulocyte 15-LOX while HDL is preferred by the leukocyte 12-LOX (69). The oxidative modification of LDL by 15-LOX plays an important role in inflammation and atherogenesis (69–71).

3. Temperature and pH Optima

The mullet platelet 12-lipoxygenase showed optimal pH at 6.5, optimal temperature at 38°C by measuring 12-HETE with high-performance liquid chromatography (HPLC). However, a pH optimum of 7.0 and a temperature optimum of 20°C was obtained based on oxygen consumption (36). Since 12-HETE is a secondary product nonenzymatically reduced from 12-HPETE (Fig. 1), it is possible that the temperature optimum for formation of 12-HPETE by lipoxygenase is 20°C, while the optimum for 12-HETE formation from HPETE is 38°C. LOX is known to be "suicidal" (72, 73). Therefore, it is also possible that the temperature coefficients for catalysis and inactivation are different.

The pH optimum of rainbow trout 12-LOX is 7.5 (38), ayu skin 12-LOX is 7.4 (20), sardine skin 13(s)-LOX is 7.0 (21). Two types of 12-LOX isozymes in grey mullet differed in optimum pH, showing values of 6.5 and 8.0 (Table 3). Generally, the optimal pH of LOX from fish ranges around neutral pH. However, menhaden LOX has optimal pH around 9~10 (42a).

D. Self-Inactivation

When mullet gill or platelet extract was stored at 4°C, the LOX activities decreased rapidly (34, 36). Shrimp hemolymph LOX showed a similar pattern (32). The oxidation of C20:4 with mullet platelet LOX was inhibited by high concentrations of substrate (i.e., C20:4 and C22:6) (36). The suicidal effect is a result of the hydroperoxy derivatives produced rapidly from the PUFA by the LOX activity (13, 73). Nearly stoichiometric amounts of hydroperoxylinoleate were sufficient to inactivate the reticulocyte LOX, which resulted from oxidation of a single methionine to methionine sulfoxide in the active center of the enzyme (72). Soybean 15-LOX was also inactivated after a methionine residue was oxidized with acetylenic fatty acids (69).

Glutathione was added to facilitate the reduction of hydroperoxides to their stable hydroxy forms to enhance the stability of LOX (74). However, glutathione at a concentration higher than 1 mM inhibited LOX activity (36). Other protective compounds (i.e., imidazole or histidine) are added to improve the stability of LOX during purification (34).

Since many metal ion-binding proteins are susceptible to H_2O_2 inactivation, LOX, being an iron-binding enzyme, becomes very unstable after purification. Small amounts of glutathione peroxidase and superoxide dismutase are able to stabilize purified human 5-LOX, while 2-mercaptoethanol is the most efficient hydrogen donor substrate for glutathione peroxidase in the protection of 5-LOX (75).

In the preparation of LOX sanguineous tissue, contamination with hemoproteins (i.e., hemoglobin) results in inactivation of LOX. Hydroxyapatite chromatography effectively removes the hemeprotein from LOX and improve the stability of LOX (34).

IV. PHYSIOLOGICAL FUNCTIONS

A. Role in Ovarian Development

The LOX pathway is likely involved in the oocyte maturation process. Both 12- and 15-hydroxyeicosatetraenoic acids (HETE) as well as eicosatetraenoic ($C_{20:4}$) and eicosapentaenoic acids ($C_{20:5}$) were found to induce oocyte maturation in starfish and sea urchin (21–23).

The platelet 12-LOX activity of cultured grey mullet changes during the months of ovarian development (September to December). At the time when the size of ovary is maximal (i.e., November for the cultured and December for the wild grey mullet), the platelets have the lowest 12-LOX activity (Fig. 4). Dietary estradiol 17-increases the 12-LOX activity in grey mullet platelet, while increased dietary intake of vitamin E decreases the LOX activity (76). In rat, platelet 12-LOX was also significantly increased by dietary estradiol in a dose-dependent manner (77). The stimulatory effect of estradiol on platelet 12-LOX was inhibited by the antiestrogen agent nafoxidine (77). These findings point to the fact that 12-LOX in platelet has an effect on ovary development.

Three types of LOX are present in the ovary of grey mullet: 5-, 12-, 15-LOX (Fig. 5). During the early ovarian development, 15-LOX was the dominant form. As the ovary develops, the activity of 15-LOX reduces. When the ovary reaches maximum size (i.e., the gonadosomatic index is maximal), the three LOXs are at their minimal activities. When the ovary undergoes atresia, both 5-LOX and 12-LOX increase rapidly, the 5-LOX is about four times the activity of the 12-LOX, while the 15-LOX remains the least active. In vitro culture of grey mullet oocyte of 0.7 mm diameter showed that 15-LOX accelerated the growth of oocytes (26). These observations seem to imply that 15-LOX is probably involved in egg formation while 5-LOX may play a role in atresia.

The LOX activity varies not only with the species and tissues but also with seasons and maturation.

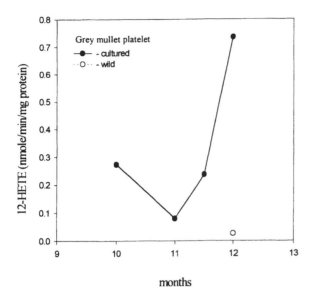

Figure 4 Change in 12-lipoxygenase of blood platelets isolated from cultured grey mullet during the season of ovary development.

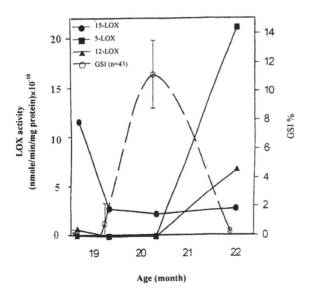

Figure 5 Changes in gonadosomatic index (GSI) and ovarian lipoxygenase (LOX) activities of cultured grey mullet (*Mugil cephalus*) during ovarian development in a sample population of 43 fish.

B. Immune Regulation

LOX activities in gill tissues of both normal and diseased mullet were activated by Ca^{2+}, ATP, or ionophore A23187. The diseased cultured grey mullet showed exophthalmia, hemorrhaged eye, pink-discolored gills having dioxygenation velocity fivefold greater than that of the normal gill. 12-LOX activity was the highest among the three LOXs. No 15-LOX activity was found. The 5-LOX of diseased fish was 1.5-fold of that in the normal fish, but 12-LOX levels were not significantly different (78, 79).

An interesting phenomenon was observed: in the gill tissue of diseased grey mullet, content of 15-HETE was higher than 12-HETE, while no 5-HETE was detected in both normal and diseased gill. However, the enzyme extract of gill showed activities of 5-LOX as well as 12- and 15-LOX (79). It seemed that the product of 5-LOX, 5-HPETE, was immediately metabolized into other metabolites, possibly leukotrienes (68). Therefore, it is possible that in mullet gill both activities of 5-LOX and leukotriene (LT) A_4 synthase reside on a single protein similar to the protein possessing both activities characterized from rat basophilic leukemia (RBL-1) cells (80, 81).

Eicosanoid generated by leukocytes and thrombocytes and LOX activities were observed in rainbow trout, carp, tilapia, rudd, catfish (82), and plaice (83), suggesting that eicosanoids play a role in immune regulation in fish in a similar way to that reported in mammals (84).

C. Defense Against Pathogens

Our preliminary survey of green macroalgae from Keelung waters showed that LOX activities differed by 20 times from month to month in *Ulva* spp of similar size harvested at the same location. Therefore, we propose that the stress developed in the intertidal zone may be the cause of the increases in the defense mechanisms reflected by the synthesis of LOX and accumulation of LOX-catalyzed products, similar to the role of LOX played in land plants. For example, the LOX activity in soybean cotyledon on phospholipid increases after treatment with fungal elicitor (85); LOX is induced in tomato leaves infected with powdery mildew; LOX shows reaction specificity in tomato fruits (86, 87); and the resistance of maize genotypes to aflatoxin production is influenced by LOX (88). These observations indicate that LOX-catalyzed peroxidation of polyenoic acids may play a role in the defense against pathogens.

In animals, inflammation may be regarded as a defense mechanism against endothelial injury in the early stages of plaque development. The intracellular action of LOX may be involved in a rather beneficial process in early stages of the plaque formation. In later stages, the intracellular LOX may be released from the foamy macrophages and may oxidize extracellular LDL into its atherogenic form or may produce free radicals, causing other types of pathogenesis (69).

V. SIGNIFICANCE IN FLAVOR FORMATION

A. Undesirable Flavor

LOX present in fish skin tissue (18, 31) generates hydroperoxy arachidonic acid (12-HPETE), and 14-hydroperoxydocosahexaenoic acid (14-HPDHE). The alpha-cleavage at either side of the carbon atoms bearing the hydroperoxy groups can result in the volatile compounds responsible for the typical oxidative fishy odor, such as 2-nonenal from 12-HPETE and 3,6-nonadienal from 14-HPDHE (29).

B. Seafood Flavor

Characteristic aromas for freshly harvested fish are derived from PUFA reactions catalyzed by the endogenous LOX yielding volatile alcohols and aldehydes of 6, 8-, and 9-carbons (29, 89). Fresh shrimp flavor, characterized by total volatiles and 1-octen-3-ol, increases with increases in LOX activities from either endogenous or exogenous sources when added to shrimp homogenate (Fig. 6)(90–92). Alcohols (i.e., 1-octen-3-ol) are likely formed from the hydroperoxy PUFA (31).

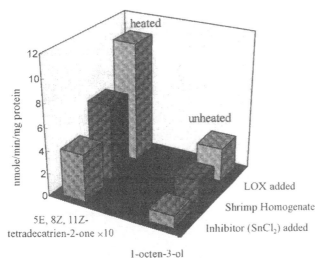

Figure 6 Amount of key volatile compounds formed in shrimp with added lipoxygenase or its inhibitor, $SnCl_2$.

Table 4   Odor Modification by Lipoxygenase Activity of
Shrimp Hemolymph (SH-LOX) and Mullet Gill (MG-LOX)

Fatty acid	Odor		
	Control	+SH-LOX	+MG-LOX
18:2	Rancid	Dried grass, rancid	Green
18:3	Rancid	Green	Green
20:4	Oily	Egg-drop souplike	Green, fishy
20:5	Fish-liver oil	Fish-liver oil	Fresh fish
22:6	Fish-liver oil	Katsuobushilike	Fresh fish

Source: Kuo, Tsai, and Pan, unpublished.

The (9-carbon unsaturated carbonyl and alcohols that characterize the aroma of
ayu and smelt are likewise reduced in amount when endogenous lipoxygenase
activity is inhibited (19, 20, 93).

    5,8,11-Tetradecatrien-2-one, a keynote compound of cooked shellfish (94),
also increases with increased LOX activity in shrimp homogenate (Fig 6) (91).

    Different fatty acids treated with LOX extracts from shrimp hemolymph
or mullet gill result in different fresh seafood aroma (Table 4). Algal LOX
was also able to modify commercial fish oil, yielding more desirable flavor
notes at the expense of about 1% PUFA (48). The role of lipoxygenase in fish
flavor (Chap. 13) and seaweed flavor (Chap. 14) is discussed in more detail
elsewhere.

VI. CONCLUSION

Despite of the wealth of information on the structure and function of plant and
mammalian LOXs, much less is known about the LOXs in tissues of aquatic organisms. One possible reason for this paucity of literature on fish LOXs is that
only partial purification has been achieved with fish LOXs. The self-inactivation
or suicidal effect seems to be more detrimental during the preparation of fish
LOXs than mammalian LOXs. Fish tissues often consist of much higher concentrations of EPA and DHA than their mammalian counterparts in cell membranes
coexisting with LOX in cells. These highly unsaturated fatty acids find their way
into LOX extracts and form hydroperoxy derivatives, which can accelerate the
self-inactivation of LOX. In order to purify and carry on further studies on fish
LOX, effective protective means need to be developed for fish LOXs.

    Another unique phenomenon of fish LOXs is the high variation in activity
found in a given tissue, resulting into a relatively low reproducibility (see also
Chap. 1). Since LOX can be induced by stress and fish have more direct contact

with the environment, the differences in LOX activities may come from various environmental impacts. It is our current interest to examine the relationship between fish gill LOX activities and the environmental conditions, especially in culturing conditions. We suspect that LOX in gills may serve as an overall stress index before plaque development or mass mortality outbreak.

The ovarian LOXs also change during the reproductive cycle. The timing of harvest may cause differences in LOX activities, leading to changes in flavor and color development of roe products. Since LOXs activities are involved in both undesirable and desirable seafood flavor formation, the adjustment of handling and processing parameters may result in pronounced flavor differences in similar products.

Algal LOXs seem to be an underutilized resource. Studies on algal LOXs have been reported only in the last decade. Algal LOX is more stable than fish LOX. Its structure may be closer to plant LOX than fish or mammalian LOXs. It can form persistent algal and fresh seafood aroma.

Treatment of fish oil with algal LOX results in more desirable flavor notes than the original undesirable fishy odor. If a continuous process can be developed during the culturing and and extracting of algal LOX, it may just be possible to use it to modify fish oil to have a fresh seafood aroma.

REFERENCES

1. HW Gardner. Recent investigations into the lipoxygenase pathway of plants. Biophys Acta 1084:221–239, 1991.
2. AO Chen, JR Whitaker. Purification and characterization of a lipoxygenase from immature English peas. J. Agric Food Chem 34:203–211, 1986.
3. YB Man, LS Wei, AI Nelson. Acid inactivation of soybean lipoxygenase with retention of protein solubility. J Food Sci 54(4):963–967, 1989.
4. HM Alobaidy, AM Siddiqi. Properties of broad bean lipoxygenase. J Food Sci 46:622–625, 1981.
5. HW Gardner. Sequential enzyme of linoleic acid oxidation in corn germ lipoxygenase and linoleate hydroperoxide isomerase. J Lipid Res 11:311–320, 1970.
6. A Yamamoto, Y Fujii, K Yasumoto, H. Mitsuda. Partial purification and study of some properties of rice germ lipoxygenase. Agric Biol Chem 44(2):443–445, 1980.
7. T Galliard, JA Matthew. Lipoxygenase-mediated cleavage of fatty acids to carbonyl fragments in tomato fruits. Phytochem 16:339–343, 1977.
8. DR Phillips, T. Galliard, Flavour biogenesis: Partial purification and properties of a fatty and hydroperoxide cleaving enzyme from fruits of cucumber. Phytochem 17:355–358, 1978.
9. DP Wallach, VR Brown. A novel preparation of human platelet lipoxygenase. Characteristics and inhibition by a variety of phenyl hydrazones and comparisons with other lipoxygenases. Biochem Biophys Acta 663:361–372, 1981.

10. M Croset, M Lagarde. Enhancement of eicosapentaenoic acid lipoxygenation in human platelets by 12-hydroperoxy derivative arachidonic acid. Lipids 20:743, 1985.

11. AA Spector, JA Gordon, SA Moore. Hydroxyeicosatetraenioc acid (HETES). Prog Lipid Res 27:271–323, 1988.

12. H Kuhn, R Brash. Occurrence of lipoxygenase products in membranes of rabbit reticulocytes. J Biol Chem 265, 1454–1458, 1990.

13. EMM Donk J van der Verhagen, GA Veldink, JFG Vliegenthart. 12-Lipoxygenase from rat basophilic leukemia cells: seperation from 5-lipoxygenase and temperature-dependent inactivation by hydroperoxy fatty acid. Biochem Biophys Acta 1081:135–140, 1991.

14. AJ Marcus, LB Safier, HL Ullman, N Isiam, MJ Broekmen, CV Schacky. Studies on the mechanism of w- hydroxylation of platelet 12-hydroxyeicosatetraenoic acid (12-HETE) by unstimulated neutrophils. J Clin Invest 79:179–187, 1987.

15. I Honda, M Noguchi, M Furuno, T Matsumoto, M Shibagaki, M Noma, K Yoneyama. Inhibition of human 5-lipoxygenase by 3-nitro-2, 4, 6,-trihydroxybenzamide derivatives. Agric Biol Chem 55:833–837, 1991.

16. AW Ford-Hutchinson. Arachidonate 15-lipoxygenase characteristics and potential biological significance. Eicosanoids 4:65–74, 1991.

17. GL Bundy. Discovery of an arachidonic acid C-8 lipoxygenase in the Gorgonian coral *Pseudoplexaura porosa*. J Biol Chem 261(2):747–751, 1986.

18. JB German, JE Kinsella. Lipid oxidation in fish tissue. Enzymatic initiation via lipoxygenase. J Agric Food Chem 33:680–683, 1985.

19. CH Zhang, T Hirano, T Suzuki, T Shirai. Enzymatically generated specific volatile compounds in ayu tissues. Nippon Suisan Gakk 58:559–565, 1992.

20. CH Zhang, T Shirai, T Suzuki, T Hiramo. Lipoxygenase-like activity and formation of characteristic aroma compounds from wild and cultured *ayu*. Nippon Suisan Gakk 58:959–964, 1992.

21. S Mohri, SY Cho, Y Endo, K Fujimoto. Lipoxygenase activity in sardine skin. Agric Biol Chem 54:1889–1991, 1990.

22. S Mohri SY Cho, Y Endo, K Fujimoto. Linoleate 13(s)-lipoxygenase in sardine skin. J Agric Food Chem 573–576, 1992.

23. L Meijer, P Guerrier, J Maclouf. Arachidonic acid, 12- and 15-hydroxyeicosatraenoic acid, eicosapentaenoic acid, and phospholipase A2 induce starfish oocyte maturation. Dev Biol 106:368–378, 1984.

24. AR Meijer, AR Brash, RW Bryant, K Ng, J Maclouf, H Sprecher. Stereospecific induction of starfish oocyte maturation by (8R)-hydroxyeicosatetraenoic acid. J Biol Chem 261(36):17040–17047, 1986.

25. DJ Hawkins, AR Brash. Eggs of the sea urchin, *Strongylcentrotus purpuratus,* contain a prominent (11R) and (12R) lipoxygenase activity. J Biol Chem 262(16):7629–7634, 1987.

26. MC Liao, BS Pan. Changes in ovarian lipoxygenase activities of cultured grey mullet during ovary development. Food Science Dep't, National Taiwan Ocean University. Keelung, Taiwan (unpublished), NSC 87-2312-B-019-048

27. G Perry, D Epel. Fertilization stimulates lipid peroxidation in sea urchin egg. Dev Biol 107:58–65, 1985.

28. DB Josephson, RC Lindsay, DA Stuiber. Variations in the occurrences of enzymically derived volatile aroma compounds in salt and freshwater fish. J Agric Food Chem 32:1344–1347, 1984.

29. DB Josephson, RC Lindsay. Enzymatic generation of volatile aroma compounds from fresh fish. In: TH Parliment, R Croteau, eds. Biogeneration of Aromas. Washington, D.C.: American Chem. Soc., 1986, pp 201–221.

30. DB Josephson, RC Lindsay, DA Stuiber. Influence of processing on the volatile compounds characterizing the flavor of pickled fish. J Food Sci 52:10–14, 1987.

31. RJ Hsieh, JE Kinsella. Lipoxygenase generation of specific volatile flavor carbonyl compounds in fish tissues. J Agric Food Chem 37(2):280–286, 1989.

32. JM Kuo, BS Pan. Occurrence and properties of 12- lipoxygenase in the hemolymph of shrimp (*Penaeus japonicus* Bate) J Chin Biochem Soc 21, 9–16, 1992.

33. JM Kuo, BS Pan, H Zhang, JB German. Identification of 12-lipoxygenase in the hemolymph of tiger shrimp (*Penaeus japonicus* Bate). J Agric Food Chem 42:1620–1623, 1994.

34. HH Hsu, BS Pan. Effects of protector and hydroxyapatite partial purification on stability of lipoxygenase from grey mullet gill. J Agric Food Chem 44(3):741–745, 1996.

35. BS Pan, HH Hsu, SF Chen, HM Chen. Effect of α-tocopherol on lipoxygenase-catalyzed oxidation of highly unsaturated fatty acids. In: JR Whitaker, NF Haard, RP Singh, C. Shoemaker, eds. Food for Health in the Pacific Rim. Food & Nutrition Press, 1999, pp. 76–85.

36. NF Haard, SF Chen, BS Pan. Identification and stabilization of lipoxygenase from platelet of grey mullet (*Mugil cephalus*). NSC 82-0409-B-019-013, 1994.

37. RJ Hsieh, JB German, JE Kinsella. Relative inhibitory potencies of flavonoids on 12-lipoxygenase of fish gill. Lipids 23(4):322–326, 1988.

38. RJ Hsieh, JB German, JE Kinsella. Lipoxygenase in fish tissue: some properties of the 12-lipoxygenase from trout gill. J Food Chem 36:680–685, 1988.

39. JB German, RK Creveling. Indentification and characterization of a 15-lipoxygenase from fish giill. J Agric Food Chem 38:2144–2147, 1990.

40. J Knight, JW Holland, LA Bowden, K Hallidary, AF Rowley. Eicosanoid generating capacities of different tissues from rainbow trout, *Oncorhynchus mykiss*. Lipids 30:451–458, 1995.

41. YJ Wang LA Miller, PB Addis. Effect of heat inactivation of lipoxygenase on lipid oxidation in lake herring *Coregonus artedii*. JAOCS 68(10):752–757, 1991.

42. JM Kuo, BS Pan. Preliminary survey on lipoxygenases from fish and shellfish, unpublished.

42a. IU Grun, WE Bareau. Lipoxygenase activity in menhaden gill tissue and its effect on odor of n-3 fatty acid ester concentrates. J. Food Biochem 18:199–212, 1995.

43. AF Hagar, DH Hwang, TH Dietz. Lipoxygenase activity in the gills of freshwater mussel, *Ligumia subroserata*. Biochem Biophys Acta 1005:162–169, 1989.

44. JL Beneytout, RH Andrianarison, Z Rakotoarisoa, M Tixier. Properties of a lipoxygenase in green algae (*Oscillatoria sp.*) Plant Physiol 91:367–372, 1989.

45. JM Kuo, A Hwang, HH Hsu, BS Pan. Preliminary identification of lipoxygenase in algae (*Enteromorpha intestinalis*) for aroma formation. J Agric Food Chem 44(8):2073–2077, 1996.

46. JM Kuo, A Hwang, HH Hsu, BS Pan. Identification of lipoxygenase isozymes in marine green algaes for aroma formation. Abstract of IFT Annual Meeting, Section 80D-18, 1996.

47. JM Kuo, A Hwang, DB Yeh. Purification, substrate specificity and products of a Ca^{2+}-stimulating lipoxygenase from sea algae (*Ulva lactuca*). J Agric Food Chem 45(6):2055–2060, 1997.

48. S-P Hu, BS Pan. Aroma modification of fish oil using macro algae lipoxygenase. J Am Oil Chem Soc, 1999, (in press).

49. MF Moghaddan, WH Gerwick. 12-Lipoxygenase activity in the red marine alga *Gracilariopsis lemaneiformis*, Phytochem 29(8):2457–2458, 1990.

50. DC Zimmerman, BA Vick. Lipoxygenase in *Chlorella pyrenoidosa*. Lipids 8:264–266, 1973.

51. WR Dunham, RT Carroll, JF Thompson, RH Sands, MO Funk. The initial characterization of the iron environment in lipoxygenase by Mossbauer spectroscopy. Eur J Biochem 190:611–617, 1990.

52. S Yamamoto. Mammalian lipoxygenases: molecular structures and functions. Biochem Biophys Acta 1128:117–131, 1992.

53. J Steczko, GP Donoho, JC Clemens, JE Dixon, B Axelrod. Conserved histidine residues in soybean lipoxygenase: functional consequences of their replacement. Biochem 31:4053–4057, 1992.

54. S Navaratnam, MC Feiters, M Al-Hakim, JC Allen, GA Veldink, JFG Vliegentharn. Iron environment in soybean lipoxygenase-1. Biochem Biophys Acta 956:70–76, 1988.

55. YY Zhang, O Radmark, B Samuelsson. Mutagenesis of some conserved residues in human 5-lipoxygenase: effects on enzyme activity. Proc Natl Acad Sci USA 89:485–489, 1992.

56. H Suzuki, K Kishimoto, T Yoshimoto, S Yamamoto, F Kanai, Y Ebina, A Miyatake, T Tanabe. Site-directed mutagenesis studies on the iron-binding domain and the determinant for the substrate oxygenation site of porcine leukocyte arachidonate 12-lipoxygenase. Biochem Biophys Acta 1210:308–316, 1994.

57. JC Boyington BJ Gaffeny , LM Amzel. Structure of soybean lipoxygenase-1. Biochem Soc Trans 21:744–748, 1993.

58. JC Boyington, BJ Gaffeny, LM Amzel. The three-dimensional structure of an arachidonic acid 15-lipoxygenase. Science 260:1482–1486, 1993.

59. EI Solomon, J Zhou, F Neese, EG Pavel. New insights from spectroscopy into the structure/function relationships of lipoxygenases. Chem Biol 4:795–808, 1997.

60. D Shibata, J Steczko, JE Dixon, M Hermodson, R Yazdanparast, B Axelrod. Primary structure of soybean lipoxygenase-1. J Biol Chem 262(21):10080–10085, 1987.

61. D Shibata, J Steczko, JE Dixon, M Hermodson, R Yazdanparast, B Axelrod. Primary structure of soybean lipoxygenase L-2. J. Biol Chem 263(14):6816–6821, 1988.

62. T Hada H Hagiya, H Suzuki, T Arakawa, M Nakamura, S Matsuda, T Yoshimoto, S Yamamoto, T Azekawa, Y Morita, K Ishimura, H-Y Kim. Arachidonate 12-

lipoxygenase of rat pineal glands: catalytic properties and primary structure deduced from its cDNA. Biochem Biophys Acta 1211:221–228, 1994.

63. H Kuhn, J Belkner, R Wiesner, AR Brash. Oxygenation of biological membranes by the pure reticulocyte lipoxygenase. J Biol Chem 265:18351–18361, 1990.

64. T Schewe, SM Rapoport, H Kuhn. Enzymology and physiology of reticulocyte lipoxygenase: comparison with other lipoxygenases. Adv Enzymol Rel Areas Mol Biol 58:141–172, 1986.

65. JR Hansbrough, Y Takahashi, N Ueda, S Yamamoto, MJ Holtzman. Identification of a novel arachidonate 12-lipoxygenase in bovine tracheal epithelial cells distinct from leukocyte and platelet forms of the enzyme. J. Biol. Chem. 265:1771–1776, 1990.

66. H-M Chen, BS Pan. Characteristic and purification of lipoxygenase of grey mullet gill. NSC 87-2811-B-019-0005, 1998.

67. RW Bryant, T Schewe, SM Rapoport, JM Bailey. Leukotriene formation by a purified reticulocyte lipoxygenase enzyme. Conversion of arachidomic acid and 15-hydroperoxyeicosatetraenoic acid to 14,15-leukotriene A_4. J Biol Chem 260:3548–3555, 1985.

68. WR Henderson, Jr. The role of leukotrienes in inflammation. Ann Intern Med 121:684–697, 1994.

69. H Kuhn, J Belkner, H Suzuki, S Yamanoto. Oxidative modification of human lipoproteins by lipoxygenases of different positional specificities. J Lipid Res 35:1749–1759, 1994.

70. S Yla-Herttuala. Macrophages and oxidized low density lipoproteins in the pathogenesis of atherosclerosis. Ann Med 23:561–567, 1991.

71. E Sigal, CW Laughton, MA Mulkins. Oxidation, lipoxygenase, and atherogenesis. Ann NY Acad Sci 714:211–224, 1994.

72. SM Rapoport, B Hartel, G Hausdorf. Methionine sulfoxide formation: the cause of self-inactivation of reticulocyte lipoxygenase. Eur J Biochem 139:573–576, 1984.

73. B Hartel, P Ludwig, T Schewe, SM Rapoport. Self-inactivation by 13-hydroperoxylinoleic acid and lipohydroperoxidase activity of the reticulocyte lipoxygenase. Eur J Biochem 126:353–357, 1982.

74. JB German, JE Kinsella. Hydroperoxide metabolism in trout gill tissue: Effect of glutathione on lipoxygenase products generated from arachidonic acid and docosahexaenoic acid. Biochem Biophys Acta 879:378, 1986.

75. YY Zhang, M Hamberg, O Radmark, B Samuelsson. Stabilization of purified human 5-lipoxygenase with glutathione peroxidase and superoxide dismutase. Anal Biochem 220:28–35, 1994.

76. SF Chen. Effects of dietary vitamin E and fish oil on blood lipoxygenase characteristics and viscosity of cultured grey mullet (*Mugil cephalus*). Masters thesis, National Taiwan Ocean Unversity, 1993.

77. W-C Chang, J Nakao, H Orimo, H-H Tai, S-I Murota. Stimulation of 12-lipoxygenase activity in rat platelets by 17β-estradiol. Biochem Pharmacol 31(16):2633–2638, 1982.

78. HM Chiou. The characteristics of gill lipoxygenase in diseased cultured grey mullet (*Mugil cephalus*). Masters thesis, National Taiwan Ocean University, 1997.

79. HM Chiou AJ Chen, BS Pan. The characteristics of lipoxygenase in diseased cultured grey mullet (*Mugil cephalus*) gill. Pacific Fisheries Technologists Conference, Astoria Oregon, April, 20–23, 1997.

80. GK Hogaboom, M Cook, A Newton Varrichio, RGL Shorr, HM Sarau, ST Cooke. Purification, characterization and structure properties of a single protein from rat basophilic leukencia(RBL-1) cells possessing 5-lipoxygenase and leukotriene A_4 synthatase activities, Mol Pharmacol 30:510–519, 1986.

81. AL Maycock, SS Pong, JF Evans, DK Miller. Biochemistry of the lipoxygenase pathways. In: J Rokach, eds. Leukotrienes and Lipoxygenases. New York: Elsevier Science Publishers, 1989, pp 146–195.

82. AF Rowley. Lipoxin formation in fish leukocytes. Biochem Biophys Acta 1084:303–306, 1991.

83. DR Tocher, JR Sargent. The effect of calcium ionophore A23187 on the metabolism of arachidonic and eicosapentaenoic acids in neutrophils from a marine teleost rich in n-3 polyunsaturated fatty acids. Comp Biochem Physiol 87B:733–739, 1987.

84. AF Rowley, J Knight, P Hoyd-Evans, JW Holland, PJ Vickers. Eicosanoids and their role in immune modulation in fish—a brief overview. Fish Shellfish Immunol 5(8):549–567, 1995.

85. Y Kondo, Y Kawai, T Hayashi, M Ohnishi, T Miyazawa, S Itoh, J Mizutan. Lipoxygenase in soybean seedlings catalyzes the oxygenation of phospholipid and such activity changes after treatment with fungal elicitor. Biochemi Biophys Acta 1170:301–306, 1993.

86. D Regdel, H Kuhn, T Schewe. On the reaction specificity of the lipoxygenase from tomato fruits. Biochem Biophys Acta 1210:297–302, 1994.

87. T Kato, Y Maeda, T Hirukawa, T Namai, N Yoshioka. Lipoxygenase activity increment in infected tomato leaves and oxidation product of linolenic acid by its in vitro enzyme reaction. Biosci Biotech Biochem 56(3):373, 1992.

88. HJ Zeringue, RL Brown, JN Neucere, TE Cleveland. Relationships between C_6-C_{12} alkanal and alkenal volatile contents and resistance of maize genotypes to *Aspergillus flavus* and aflatoxin production. J Agric Food Chem 44:403–407, 1996.

89. RC Lindsay. Fish flavors. Food Rev Int 6(4):437–455, 1990.

90. BS Pan, JR Tsai, LM Chen, CM Wu. Lipoxygenase and sulfur-containing amino acid in seafood flavor formation. In: F Shahidi, KR Cadwallader, eds. Flavor and Lipid Chemistry of Seafoods. Washington DC: Amer. Chem. Soc. Symp. Series 674:64–75, 1997.

91. JM Kuo, BS Pan. Effect of lipoxygenase on formation of cooked shrimp flavor compound- 5, 8, 11-tetradecatrien-2-one. Agric Biol Chem 55:827–848, 1991.

92. BS Pan, JM Kuo. Flavor of shellfish and kamaboko flavorings. In: F Shahidi, JR Botta, eds. Seafoods: Chemistry, Processing Technology and Quality. New York: Elsevier USA, 1994, pp 85–110.

93. T Hirano, CH Zhang, A Morishita, T Suzuki, T Shirai. Identification of volatile compounds in ayu fish and its feeds. Nippon Suisan Gakk 58:547–557, 1992.

94. A Kobayashi, K Kubota, M Iwamoto, H Tamura. Syntheses and sensory characterization of 5,8,11-tetradecatrien-2-one isomers. J Agric Food Chem 37:151, 1989.

12
Enzymes and Enzyme Products as Quality Indices

K. Gopakumar
Indian Council of Agricultural Research, New Delhi, India

I. INTRODUCTION

Enzyme activity has been used extensively as an indicator for quality changes in fish and fishery products. Most enzyme indicators are endogenous (i.e., inherently present in the fish). Postmortem changes, preprocessing and processing methods, particularly freezing, icing, and thawing, can all lead to tissue damage and disruption of such cellular organelles as mitochondria and lysosomes, releasing several enzymes bound to these structures into the cellular fluid. Thus, cell disruption will increase or decrease the activity of enzymes. The measurement of the rate of change of this enzyme activity is usually done by the estimation of the reaction product(s). Of the various enzymes, ATPase (EC 36.1.8, ATP pyrophospho hydrolase) and lactic dehydrogenase (LDH, L-lactate: NAD oxydo-reductase, EC 1.1.1.27) of the glycolytic pathway and associated ATP breakdown and rigor mortis have been extensively investigated (1).

Watabe (2) reported that in sardine and mackerel the progress of rigor mortis in association with ATP depletion and lactate accumulation is dependent on temperature. Nambuthiri and Gopakumar (1) used ATPase and LDH assay as an index of assessing freshness of freshwater and brackishwater fish, and found that these indices correlate well with other biochemical parameters and sensory evaluation. The activities of the lysosomal enzymes beta-hydroxy-acyl-CoA-dehydrogenase (HADH), alpha-glucosidase, and acid phosphatase have also been successfully used to differentiate the quality of extremely fresh fish and fish held on ice for only 1 day (3).

II. ENZYMES AS QUALITY INDICES

A. Adenosine Triphosphatase (EC 3.6.1.3)

ATPase activity is associated with prerigor change and the onset of rigor mortis. The biochemistry of myosin ATPase is discussed in detail in Chapter 3. One of the important factors determining the onset of spoilage in a freshly caught fish is the stiffening of the body called rigor mortis. Usually this occurs within 1–7 h after death. Rigor mortis sets in and passes quickly in active fish but rather slowly in passive and/or anesthetized fish. As long as the fish is alive, its circulatory system functions even after capture. With the stoppage of the circulatory system, there is cessation of oxygen supply to the muscle and this disrupts normal biological functions. However, the oxygenated glycogen rich muscle can remain metabolically active for hours in the prerigor condition. In the living animal, aerobic respiration is the normal reaction that takes place, in which every molecule of glucose oxidized results in the synthesis of 36 molecules of ATP. Soon after death, anaerobic metabolism of glucose takes place. These reactions are discussed in more detail in Chapter 2. Glycolytic rate is mediated by ATP concentration and hence by the enzyme(s) ATPase that catalyses the hydrolysis of ATP to ADP as given below:

$$\text{ATP} \xrightarrow{\text{ATPase}} \text{ADP} + \text{Pi} \dots \tag{1}$$

Due to this reaction a rapid decrease in the level of ATP is observed after death, suggesting that the measurement of ATPase activity could serve as an index of postmortem changes in fish muscle and hence its quality.

ATP content in *prerigor* fish is variable (e.g., 0.69–5.19 mg % P, averaging 2.4 mg % P in redfish fillets). In *postrigor* fish fillets only small amounts of ATP are found (e.g., 0.7–0.80 mg % P, averaging 0.35 mg % P). This reflects a high rate of ATPase activity during the *prerigor* phase. In *postrigor* fillets 19.8 % less myosin ATPase was extracted than in the *prerigor* processed fillets (4). This was attributed to the myosin ATPase forming an actomyosin complex that becomes insoluble. Curran et al. (5), in their study on handling of tropical fish, observed that the postmortem metabolism of tilapia, which were immediately iced following death, was accelerated and ATP degradation and lactic acid accumulation were faster than in tilapia held at ambient temperatures. However, Fraser et al. (6) found that ATP breakdown and IMP dephosphorylation occurred much earlier in Nova Scotia mackerel stored at 13 to 20°C for 7h before icing than in fish immediately iced on capture. All these observations evidently indicate that measurement of ATPase activity has certain limitations; however, it could serve as an index for assessing the freshness of specific fish and fishery products.

Activities of some fish muscle enzymes, including malic enzyme, aldolase, or ATPase, show characteristic changes during ice storage. Several workers (2, 7,

8) have suggested that determination of ATPase activity might be a more accurate index of change in quality of chill stored fish than measurement of metabolites. Kielley and Mayerhof (9) showed that Mg^{2+} ATPase could be successfully extracted from rat muscle by an aqueous medium. Employing this aqueous extraction, Yamanka and Mackie (10) isolated ATPase from cod (*Gadus morhua*) caught from North Sea waters and showed that this enzyme was activated by Mg^{2+} ion and inhibited by Ca^{2+}ion. They also showed that when cod was stored in ice the activity was lost over 2 weeks. Activity was lost in only 1 week at -15°C. The authors concluded that loss of ATPase activity is a potential index for evaluating quality of ice-stored but not frozen-stored cod. Seki and Narita (11), while measuring the ATPase activities of minced and block meats of carp, showed that EDTA–ATPase activities of the myofibrils from minced fish held under ice storage for up to 16 days decreased rapidly, while changes in Ca^{2+} and Mg^{2+} ATPase activities were gradual. This suggests that ATPase activity varies from fish to fish. However, the same authors noted few changes when fish was held under ice storage as blocks, evidently suggesting more release of enzyme by cellular damage during the mincing operation. They found that the ATP-sensitivity of the myosin B served as a good quality marker for carp muscle during ice storage; there was up to 45% reduction of this sensitivity after 16 days. Most ATPases are sensitive to KCl concentrations and optimum activity was seen for 0.1 M KCl at 25 C.

1. Assay of ATPase Activity

An aqueous extract of fish muscle is prepared by homogenizing 1.0 g muscle in 80 ml distilled water at 0°C in a high-speed blender at 20,000 rpm. The extract is centrifuged for 20 min. in a refrigerated centrifuge at 0°C and the supernatant is taken for assay.

An aliquot (2 ml) of this aqueous extract is added to an assay system containing the following components:

    1.0 ml 0.1 m histidine pH 6.9
    0.1 ml 0.15 M magnesium chloride
    0.5 ml 0.003 M adenosine triphosphate pH 7.5
    0.4 ml distilled water

The mixture is then incubated at 25°C in a water bath, aliquots of assay solution are taken at different intervals of time, such as 2, 4, 6, 8 min. Reaction is stopped by adding 4.0 ml of 5 % trichloroacetic acid solution and the mixture is centrifuged. The supernatant is taken and the inorganic phosphate released (Pi) is measured using a spectrophotometer. The ATPase activity is usually expressed as micromoles of phosphorus released per minute per milligram of protein.

The estimation of ATPase activity is now a well-standardized technique

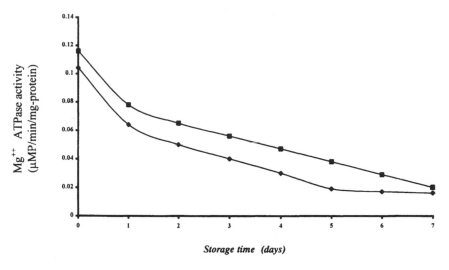

Figure 1   Changes in Mg⁺⁺ ATPase activity of aqueous extracts of cod flesh during iced storage. Fish no. 1; ◆——◆; fish no. 2, ■——■. (From Ref. 10.)

and assay kits are available commercially, making the whole procedure easy and accurate.

2.   ATPase Activity During Ice Storage

Yamanka and Mackie (10) have shown that Mg^{2+} ATPase activity of the aqueous extract of cod muscle falls steadily during ice storage from 0.10 Pi/min/mg protein initially to approximately 0.002 Pi/min/mg protein after 14 days of storage (Fig. 1).

3.   ATPase Activity During Frozen Storage

It was observed that Mg^{2+} ATPase activity of cod falls very rapidly during the first few days of storage and approaches asymptotic value after 4 days (10) (Fig. 2). Nambudiri and Gopakumar (12) observed that specific activities of ATPase showed a gradual decline for five species of Indian fishes when held under frozen storage at -20°C for 180 days. Measurement of ATPase activity can therefore be used an index of the quality of these fish during prolonged frozen storage. Exposure to low pH and high salt concentration appears to destroy ATPase activity and such changes in the frozen muscle may be responsible for the variations in ATPase activity during frozen storage of different types of fish.

Denaturation of Ca^{2+} ATPase has been found to be accelerated by freezing

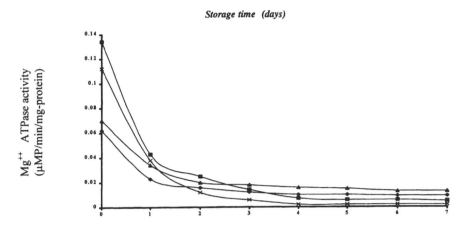

Figure 2 Changes in Mg⁺⁺ ATPase activity of aqueous extracts of cod flesh during frozen storage at -15°C. Fish no. 1, ◆——◆; fish no. 2, ■——■; fish no. 3, ▲——▲; fish no. 4, X——X. (From Ref. 10).

and some sugar compounds such as sorbitol and sucrose have a protective effect on actomyosin Ca^{2+} ATPase denaturation (13).

4. ATPase Activity as a Quality Index of Surimi

Surimi is a washed fish mince containing chiefly myofibrillar proteins. ATPase activity has been used to evaluate the quality of frozen surimi from Alaska pollack (13, 14). High Ca^{2+} ATPase activity was noted for surimi of high quality (CA) and this agrees well with another excellent quality criterion for surimi: gel strength. The myofibrillar Ca^{2+} ATPase total activities from surimi grades were C 102 ± 15 (gel strength 250 ± 50), A 136 ± 18 (gel strength 1200 ± 100), CA 175 ± 11 (gel strength 1400 ± 200) and 423 ± 14, 670 ± 40, and 721 ± 10, respectively, for Mg^{2+} ATPase. This evidently suggests that Mg^{2+} ATPase activity is an excellent index for assessing quality of surimi as well as for estimating the freshness of raw material used for the preparation of surimi.

It should be noted that it is always advisable to assess the quality of fish and fishery products with another quality index, such as sensory tests, gel strength, or others along with ATPase activity measurement to arrive at an objective evaluation of the product's edibility.

B. Lactate Dehydrogenase (LDH, L-Lactate: NAD Oxido-Reductase EC 1.1.127)

Lactate dehydrogenase, widely distributed in all animal tissues, is a key enzyme in the glycolytic pathway. It is a bisubstrate enzyme acting on both lactate and pyruvate. In glycolysis, this enzyme catalyzes the reduction of pyruvate to lactate in the presence of NADH according to the reaction:

$$CH_3\text{-}C\text{=}O\text{-}COOH + NADH + H^+ \Leftrightarrow CH_3\text{-}CHOH\text{-} COOH + NAD^+ \ldots (2)$$

High LDH activity is normally observed in many species of fish and in such species accumulation of lactic acid is faster and the amount of lactic acid formed is a measure of the enzyme activity. The latter could be used to assess the freshness of fish.

In postmortem fish muscle, in the absence of oxygen and citric acid cycle activity, the normal pathway of glycogen degradation involves a step in which pyruvate is reduced by LDH to lactic acid. The LDH molecule has polypeptide subunits and normally exists in five different permutations. The distribution of these isoforms is dependent on the tissue of origin (e.g., in heart muscle the H4 isomer predominates and in skeletal muscle the M4 isomer predominates). In tuna and flounder muscle, LDH isoforms are quite similar. However, the heart isoforms are different in the two species. Cahn et al. (15) reported the differences between LDH isoforms seen in the heart and skeletal muscle of fishes and reported that flat fish (sole and halibut) are the only group of vertebrates for which identical types of LDH are found in heart and muscle in the adult stage. Curran et al. (5) showed that in the tropical fish tilapia (*Oreochromis mossambicus*), held under ice storage, lactic acid almost reached double its initial concentration 50 h after death and reached a very low value after 240 h. Rossman and Lijas (16) reported high LDH specific activity (770 mole/min/mg protein) in fish muscle. Janicke and Knopf (17) reported higher specific activity in the skeletal muscle than the cardiac muscle of chicken. Significant loss of enzyme activity was reported (18) for lobster muscle when held at 37°C. ATP was degraded more rapidly at 22°C than at 0°C in carp, whereas in tilapia the rate of ATP depletion was not significantly different at the two different temperatures. In general, fish held at higher temperature had greater ATPase activity and rate of depletion of ATP, ADP, AMP, and IMP. This observation is supported by the results of several workers.

Table 1 Specific LDH Activity Values

Fish/Shrimp	LDH activity NADH µ mole/min.mg.protein
Mrigal	545.0
Mullet	384.7
Pearl spot	312.1
Tilapia	513.0
Milk fish	933.3
P. indicus	21.5

Source: Ref. 19.

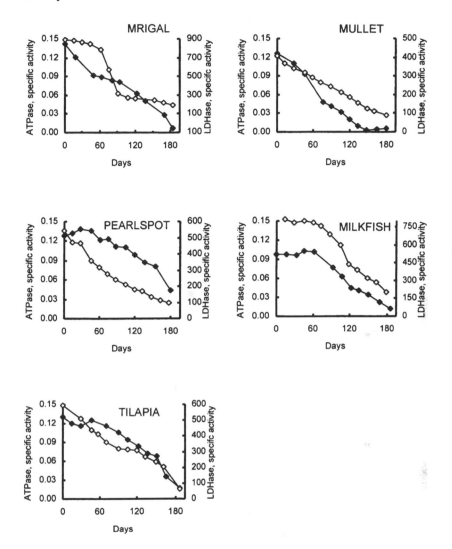

Figure 3 Adenosine triphosphatase (◆) and lactate dehydrogenase (□) activities in various fish during frozen storage. (Redrawn with permission from Ref. 1.)

Nambudiri (19) and Nambudiri and Gopakumar (12) studied LDH activities of five species of tropical fish—mrigal (*Cirrhinus mrigala*), tilapia (*Oreochromis mossmbicus*), pearl spot (*Etroplus suratensis*), milk fish (*Chanos chanos*), and mullet *Liza parsia*)—held under both iced and frozen storage. There was significant loss in LDH activity under both storage conditions (Fig. 3). Nambudiri (18), after calculating specific activities, observed that these values agree well with sensory scores and chemical indices of quality including TMA, PV, FFA, and others. He recommended critical specific LDH activity values for five species of fish and one species of shrimp (Table 1). When enzyme activity falls below the indicated level, these species can be considered inedible as supported by sensory scores (Fig. 3).

The author also noted that LDH activity in shrimp is very low compared to fish and it disappeared within 15 days of ice storage. Disappearance of LDH activity during frozen storage is attributed to protein denaturation caused by accumulation of free fatty acids from lipid hydrolysis. Fatty acids cause aggregation of myofibrilar proteins and associated decreased enzyme activity. Fish LDH is also sensitive to pH and temperature.

Measurement of LDH activity has been standardized and enzyme assay kits are now available commercially to ensure reliability and accuracy. However, since the values of specific activities of LDH and ATPase are found to vary for different species of fish, the procedure has to be standardized for each specie.

C. Lysosomal and Mitochondrial Enzymes

Rapid freezing and thawing of animal tissues result in the irreversible damage of mitochondria, releasing a number of enzymes into the sarcoplasm. Several workers (20–26) have shown that an increase in the lysosomal and mitochondrial enzyme activities can differentiate fresh (unfrozen) fish tissues from frozen thawed tissues. Some of these enzymes play a significant role in fish muscle damage, particularly alpha glucosidases (EC 3.2.1.20), beta-N-glucosaminidase (EC 3.2.1.30), and acid phosphatase (EC 3.1.3.2). Ham and Kormandy (20) found that in bovine and porcine muscle freezing and thawing cause a remarkable increase in mitochondrial glutamic oxalacetic (GOT) activity in the muscle press juice. Ham and Masic (21) measured the release on freezing and thawing of the mitochondrial form of glutamate oxaloacetate transaminase in carp fillets, with the aim of developing an analytical method of differentiating fresh fish fillets from frozen and thawed fillets. However, not much success was achieved since the mitochondria were extensively destroyed, with concomitant release of GOT isoenzyme, during normal storage of carp fillets in ice. Gould (22) found that changes in malate dehydrogenase are an excellent index for determining whether fish have been frozen using cod and haddock. The freezing and thawing process also causes a 2.5-fold increase in the activity of cytochrome oxidase in trout muscle (24).

The specific activity of the lysosomal enzyme glucose oxidase in press juice of trout was found to increase significantly on freezing and thawing, while no difference was noted in the activity of β- N acetyl glycosaminidase (27). Fresh and thawed frozen meat was identified from the ratio of β-hydroxy acyl CoA dehydrogenase activity of a partially disintegrated sample of the total ex-

Table 2 Activity of Lipoamide Reductase and 5′ AMP Deaminase

	Specific Activity μ mole Pi/min/mg protein					
	Mrigal	Mullet	Pearl Spot	Milk fish	Tilapia	P. Indicus
Lipoamide reductase						
Initial	0.040	0.044	0.026	0.019	0.050	0.051
Final	0.062	0.050	0.033	0.078	0.070	0.078
% increase	55.0	11.1	8.3	310.5	40.0	52.9
5′ AMP deaminase						
Initial	0.50	0.13	3.50	0.20	2.70	0.75
Final	0.72	0.18	6.50	0.40	3.50	1.00
% increase	44.0	44.0	85.7	100.0	29.6	33.3

Source: Ref. 19

tracted enzyme activity. Hamm and Badawi (27) observed no significant changes in the activities of the mitochondrial acotinase, fumarate, and glutamate dehydrogenase in the bovine muscle press juice at -20°C. However, study on changes in mitochondrial enzymes in fish muscle appears more encouraging and these changes have been used to differentiate fresh fish from frozen and thawed fish (12, 19).

Numbuthiri (17) reported that freezing and thawing of tropical farmed fish and shellfish cause substantial increase in the activity of lipoamide reductase and 5′ AMP deaminase (Table 2) and these enzyme activities can be used as indices of determining whether the fish is fresh or frozen thawed. Disruption of mitochondria due to freezing and subsequent thawing appears to be a possible reason for the release of more enzymes and increased enzymatic activity during frozen storage.

Despite all these observations, the measurement of lysosomal enzyme activities can seldom be adopted as a routine test for measurement of fish quality because the methods are very cumbersome and results are difficult to reproduce. However, they form an excellent tool for interpreting the cellular damage and its impact on fish muscle spoilage during frozen storage.

III. ENZYME PRODUCTS AS QUALITY INDICES

A. Creatine

Another important enzyme associated with fish spoilage is creatine kinase. This enzyme acts on creatine phosphate with the formation of ATP as:

$$CRP + ADP \Leftrightarrow CR + ATP. \ldots \tag{3}$$

where CRP is creatine phosphate and CR is creatine. The creatine reserve in the fish is finite and is soon depleted. The small rate of ATP synthesis by Reaction (3) can no longer compete with the Reaction (1) and the ATP concentration begins to fall to near zero. The level of free creatine is often estimated to determine the rate of spoilage of fresh fish, especially when held under ice storage.

Oehlenschlager (28) measured the creatine content of five species of cold-water fish (haddock, cod, plaice, red fish, and whiting) and observed that in all these fish creatine content declined with storage period. In haddock, creatine changes were significant: from an initial level of 500 mg % fresh flesh to 400 mg % at 15 days and 300 mg % at 18 days. Changes in creatine values were in good correlation with other usual indices of freshness, particularly trimethyl-amine (TMA), trimethylamine oxide (TMAO), total volatile bases (TVB), and sensory score. The author recommended analysis of a pooled sample of at least three fishes, since there is individual variation in the creatine content from fish to fish. Quantitative measurement of creatine content of fish, its changes during storage conditions, as well as correlation with other indices of quality assessment are very few and further studies are needed in this subject.

B. Consequences of Glycolysis

The concentration of lactic acid formed in the muscle due to the breakdown of glucose via glycolysis is found to depend on the glycogen reserves of the muscle prior to death. In wild cod it is seen that lactic acid concentration reaches almost double its initial level within 24 h after capture and thereafter its level remains stationary up to 72 h. This accumulation of lactate is related to levels of glycogen and pH decline. The glycogen reserve of the muscle at harvest is directly related to the struggle the fish has undergone prior to capture and death. The greater the struggle the fish has experienced, the lower its glycogen reserves and the more rapid the onset of rigor mortis.

One of the consequences of the accumulation of lactic acid in the muscle is the lowering of the pH from near neutral to acid range. This fall in muscle pH affects the quality of fish tissues: in some species as pH declines the tissue becomes firmer and tendency to drip is enhanced.

Glycogen is the reservoir of energy in live fish for its various physiological functions. In the live fish respiratory cataboism of glucose is aerobic and results in the synthesis of adenosine triphosphate and pyruvic acid. Soon after death, in the absence of oxygen in fish tissues, anaerobic catabolism of glucose takes place via the glycolytic pathway. In the anaerobic breakdown of glucose, pyruvate is converted to lactate. Accumulation of lactic acid takes place in the fish muscle with resultant lowering of muscle pH. The pH of a fresh fish muscle is normally in the range of 6.4. The pH is reported to be a poor indicator of freshness since there is wide variation between individual specimens. However, the pH of postrigor muscle rises during spoilage and when it reaches 7.0 the fish are usually spoiled.

During the early *postmortem* stage, the most important change is the breakdown of ATP since it is accompanied by a number of major changes in fish muscle. With the loss of ATP, membranes may undergo changes leading to loss in function. ATP is the energy source for both membrane repair after damage and pumping mechanisms. Calcium ions play a major role in muscle cell physiology (28). Maintenance of ionic concentration is critical in all living cells. During the *postmortem* phase, damage to cell membranes causes calcium ions leak into the sarcoplasm, activating the contractile processes in the muscle by binding to troponin. In addition to this, calcium ions activate many tissue enzyme including lipases, phospholipases, and proteases. The net result is the hydrolysis of proteins and lipids and making the fish muscle tender.

C. Nucleotide Degradation Products

1. Hypoxanthine Phosphoribosyl Transferase (EC 2.4.2.8) and Xanthine Oxidase (EC 1.2.3.2.)

The breakdown of the ATP to the end product urea and the role of various enzymes involved in the process are documented in Figure 4. Some of the enzymes involved in this pathway may also be of microbial origin (see Chapter 2 for details). Although several enzymes are involved in the catabolism of ATP to ADP, AMP, IMP, and finally to hypoxanthine, two enzyme activities monitored frequently are the hypoxanthine phosphoribosyl transferase and xanthine oxidase levels. Hypoxanthine (Hx) content of the fish muscle increases during storage and measurement of its value is an excellent index of measuring spoilage of fish (29).

Jones et al. (30) proposed a freshness estimation method using Hx. Many researchers of that period used Hx concentration as an index of fish quality. Since the Hx formation is a result of both autolytic and bacterial activities, it has an advantage over TMA assay, which measures a microbial product.

The rate of Hx formation during chill storage of fish varies with the treatment: gutted fish stored in chilled sea water (CSW) have a fast rate of HX formation than whole fish in ice. Within each chill storage treatment, Hx content increased linearly with storage time and its concentration appears to be a useful indicator of storage time. Jacober and Rand (31) observed a remarkable increase of Hx with chill storage in winter flounder and suggested that Hx can serve as a good method for following deteriorative changes. These authors suggested standards of Hx and diamines along with analysis for spoilage compounds as a measure of grading the quality of fish.

Crawford and Finch (32) followed the quality changes of albacore tuna in

Breakdown of ATP in post-mortem fish muscle

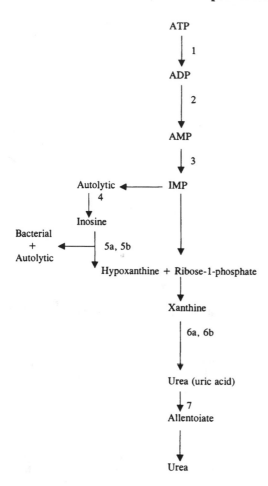

Figure 4 Breakdown of ATP in postmortem fish muscle. **1.** Adsnosine triphosphatase; **2.** myokinase (nucleoside monophosphate kinase); **3.** AMP deaminase; **4.** IMP phosphohydrolase; **5a.** muscle phosphorylase; **5b.** inosine nuclcosidase; **6a.** xanthine osidase; **6b.** xanthine dehydrogenase; **7.** allentoinase.

ice and refrigerated seawater (RSW). They observed that a reasonable chemical assessment of albacore quality could be determined by measuring both Hx (autolysis) and TMA (bacterial activity).

Hypoxanthine is a major catabolite of ATP in rock cod, yellow fin, flat head, and Dover sole. The gradual accumulation of Hx in rock cod and yellow fin and soles made this catabolite a useful freshness indicator for these species. The nucleoside phosphorylase activities of Hx-accumulating species were generally high. Nakano et al. (33) observed that some species of plaice, squid, and octopus showed relatively higher activities and hence Hx can form a very good objective quality index. Nunes et al. (34) found that Hx values and Torry meter readings are good indicators for the quality of sardines *(Sardina pilchardus)* during iced storage. The rate of nucleotide catabolism was sufficiently rapid even in sterile chill stored and frozen bastard trumpeter, *(Lactridopsis forsteri)*. The formation of Hx from inosine at 4°C was also rapid, with Hx accounting for 33 % and 63 % of the total nucleotide pool after 7 and 21 days, respectively. Hx production was found to be linear with time in iced albacore. The Hx content increased from 0.52 to 2.93 µmol/g during 29 days in ice. The quality changes in sea frozen whole and filleted rock cod. *Epinephelus sp.* was studied by following sensory, biochemical, and microbiological changes in frozen material. The Hx content in these samples showed a steady increase with time and the values remained below 2 µmol/g up to 36 months. Thereafter it increased considerably to exceed 2 µmol/g. The Hx content showed significant negative correlation with sensory score.

However, Hx cannot be considered an index of quality for all types of fish species since the rate of formation of Hx and inosine differs between species (35). It should be noted that a poor correlation between Hx concentration and quality changes has been observed in a number of fish species.

Formation or accumulation of hypoxanthine is initially mediated by tissue enzymes and later by bacterial enzymes. The enzymes responsible for the breakdown of ATP and its catabolites to Hx do not operate at the same rate through out the year (37). Xanthine oxidase activity is considered critical to the levels of Hx. The value of Hx progressively increases from near zero in extremely fresh fish to levels as high as 8 µmoles/g when the fish is considered spoiled.

Hence, the formation of Hx can be considered as a measure of both autolytic deterioration and bacterial spoilage. The values of Hx and freshness of fish correlate well with flavor changes during storage. Therefore, it is now generally considered as a useful index of freshness of fish, provided the limits of acceptability are known for a particular species of fish. Hx accumulation during chill storage of the fish has been studied for over 150 species of fish so far and it was reported that it correlates well with spoilage index for about 120 species (35). There are, however, certain abnormalities with regard to Hx values in certain species when assessment of freshness is measured based on other indices such as TMA, TVN, TVB, and sensory score. The main reason indicated for this abnormality is that in these species hypoxanthine is immediately broken down to uric acid due to a high degree of native xanthine oxidase activity. In certain other species, the breakdown of IMP to hypoxanthine is slow due to reduced activity of the enzyme hypoxanthine phosphoribosyl transferase. Fish with low xanthine oxidase activity can also cause accumulation of high concentrations of Hx in muscle but the fish could be rated as fresh by sensory analysis.

Despite all these limitations, Hx value is one of the best indices so far available for measurement of fish spoilage.

2. K-Value

Since the measurement of Hx does not take into account all nucleotide breakdown products, Saito et al. (38) proposed a new concept called K-value as an index of spoilage.

K-value is calculated from the values of Hx, inosine (HxR) and total nucleotide levels in fish at the point of measurement. It is defined as:

$$K\% = \frac{[HxR + Hx]}{[ATP + ADP + AMP + IMP + HxR + Hx]} \times 100$$

where:

ATP = adenosine 5 - triphosphate; ADP = adenosine - 5 - diphosphate; AMP = adenosine monophosphate; IMP = inosine monophosphate; HxR = inosine; and, Hx = hypoxanthine.

The K-value measurement takes into account the role of most enzymes in the breakdown of ATP. Hence, it is a more accurate measure of loss of fish freshness. Saito et al. (38) also observed that in freshly caught fish, K-value could be as low as zero, 10–20 for moderate-quality fish, and could go up to 90 when the fish is found to be spoiled. K-values were found to have excellent agreement with sensory data (Fig. 3) when measured for a number of species of fish. Studies conducted by the Lakshmanan and Gopakumar (39) have shown that ex-

Table 3 Correlation between K Values and Sensory Score

Fish species	K value % Initial	K value % Final	Storage time	Sensory value
Silver pomfret	3.5	> 50	14 days	6-5
Rainbow sardine	-	> 50	14 days	6-5
Jew fish	-	> 50	16 days	6-5
Ribbon fish	-	> 50	16 days	6-5
Squid	-	> 50	16 days	6-5
M. dobsoni	-	> 50	8–10 days	6-5
P. stylifera	-	> 50	8–10 days	6-5
Pearl spot	2.8	55	14 days	5
Mullet (Liza corsula)	4.2	51	8 days	5

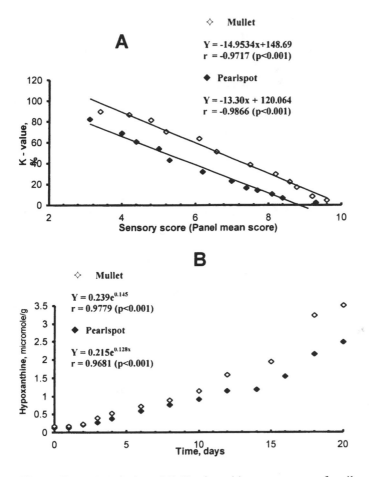

Figure 5 A. Correlation of % K-value with sensory score of mullet and pearlspot. B. Correlation of hypoxanthine concentration with storage time on ice for mullet and pearlspot. (*Redrawn with permission from Food Control* 7(6): 282, 1996.)

tremely fresh fish, when sampled onboard, have K (%) values ranging from 3 to 5 (Table 3).

K-value increases to more than 50 after 10 days of iced storage in most cases, when score (hedonic scale, initial value 10) reached 6 (good), showing excellent correlation.

K-value is now considered one of the most appropriate indicators of freshness of fish. Analyses of several species of coldwater and tropical fish confirmed that there is an excellent correlation between freshness of fish and K-values. The mean K-value, if prime quality, should be around 20 %. Based on nucleotide breakdown analysis, fish species can be classified into three groups: HxR-forming, Hx-forming, and intermediate species. Freshness of HxR-forming species cannot be measured by estimating Hx alone. Using K-value, freshness of both HxR- and Hx-forming species can be effectively measured.

Marine mollusks and crustaceans degrade ATP in a different way from finfish. Marine invertebrates have less or are devoid of AMP deaminase activity and the major pathway of ATP degradation in them is via the adenosine pathway. This mechanism does not influence the results of measuring K-value in mollusks or crustaceans.

Extensive studies conducted on measurement of quality in ice storage of brackishwater fish mullet *(Liza corsula)* and pearl spot *(Etroplus suratensis)* showed that K value is a useful assessment of quality because it correlated well with sensory tests (39). It is seen that ATP, ADP, and AMP levels decreased sharply in both species but comparatively more quickly in mullet than pearl spot. IMP is the prominent nucleotide present in both species in early stages of iced storage and reached a maximum within 1 day postmortem (Fig. 5). It was also seen that HxR does not split easily to Hx and ribose in mullet as it does in pearl spot, showing the difference in nucleotide catabolism. However, a definite uniform pattern of accumulation of Hx was seen in both species and it correlates well with iced storage shelflife and sensory values (39).

3. Enzyme-Sensor Method for Determination of K-Value

An area of rapid development recently is the use of an enzyme-sensor technique to monitor quality of fish electronically by K-value measurement. This technique involves the use of a multienzyme sensor system to measure the freshness indicator K-value.

This equipment developed by Watanabe et al. (40) and others basically consists of a multielectrode enzyme sensor, relay controller, A/D converter, microcomputer, floppy disk, monitor, and printer. The freshness indicator K_1 is defined as:

$$K_1 = \frac{[HxR + Hx]}{[IMP + HxR + Hx]} \times 100$$

where: HxR = inosine; Hx = hypoxanthine; and IMP = inosine - 5 - monophosphate.

In order to avoid the tedious extraction procedure for preparation of fish muscle exudate, Watanabe et al. (40–42) suggested heating the fish muscle at 80°C for 5–10 min prior to obtaining the exudates. The K_1 values of the exudates were calculated by measurement of the output current of the sensor by computer and then converted to digital signals. Freshness (K_1) determined for five species of fish was: mackerel = 8.1; skipjack = 6.6; saurel = 4.1; sardine = 36.1; and bluefin tuna = 31.1. Based on these studies, the authors expressed freshness of the fish in the following scale: very fresh, K_1- 10; fresh, K_1- 40; not fresh, K_1->40. However, variations were noted for fatty fishes such as sardine. Therefore, the authors suggested that while determining freshness of each species of fish heating conditions should be examined. However, it is better to standardize the K-values for each species after comparing with sensory values.

Development of bioenzyme sensors has now emerged as an expanding field. Several reports (43–45) have appeared for specific determination of biogenic amines such as putrescine, Hx, HxR, and others using sensor techniques. The reader is referred to Chapter 22 for more details on biosensors.

D. Products of Decarboxylases

There are several decarboxylases, of endogenous and exogenous origin (mainly bacterial origin), responsible for fish spoilage. They produce biogenic amines including histamine, putrecine, cadaverine, and also a number volatile bases such as ammonia, TMA, DMA, and others (46,47). Most mesophilic bacteria seen in tropical fish contain the enzyme histidine decarboxylase, particularly *Pseudomonas, Bacillus, Clostridium,* and *Aeromonas* groups of bacteria. Visceral proteinase or other proteolytic enzymes are reported to accelerate the quantity of histamine production by releasing more histidine into the pool of free amino acids (48).

Although decarboxylases can release a variety of amines from the free amino acid pool seen in fish muscle, decarboxylase activity is usually measured by estimating the amount of histamine released from histidine. Histidine carboxylase activity is usually expressed as µmole histamine/mg protein.

1. Histamine

Among the decarboxylases, the most important one responsible for fish quality control is histidine decarboxylase, producing histamine from the parent amino acid histidine. A number of bacteria isolated from tropical fish contain this enzyme. Histidine is a constituent of muscle proteins. Most pelagic fish and scombroid fish contain an appreciable amount of histidine in free state as well as with proteins. In postmortem fish, the free histidine is converted by the bacterial enzyme histidine decarboxylase into free histamine. Histamine develops in freshly caught fish 40–50 h after death when fish is not chilled properly. In tuna, skipjack, and mackerel fisheries, care is taken to chill and freeze the fish as quickly as possible to minimize bacterial growth and avoid histamine formation. Histamine causes a form of food poisoning known as scombroid poisoning. The name arose because it is linked with eating tuna, mackerel, and other species of the scombroid family. Since then, histamine formation has also been demonstrated with other pelagic fish species. The usual symptoms of scombroid poisoning are facial flushing, rashes, headache, and gastrointestinal disorders.

Histamine alone is not responsible for scombroid poisoning, since toxicity seems strongly influenced by other related amines such as cadaverine and putrescine (49, 50). However, histamine is rather easy to detect and estimate. Hence, it is the histamine content that is specified in the standard regulations and is used as an index to assess the danger of scombroid poisoning. The U.S. standard for canned tuna specifies that a consignment shall be rejected if out of 24 samples one contains over 50 mg % histamine or two contain over 20 mg %.

The Food and Drug Administration (FDA) of the United States has identified detection and prevention of histamine contamination in fishery products as a Critical Control Point in the Hazards Analysis (HACCP). Recently, a rapid test based on the enzymes linked immunosorbent assay (ELISA) test has been devised. The test, which utilizes a quick water extraction, will provide results in 30 min. The test will determine if a sample contains more or less than 50 ppm histamine, which is the current FDA action level. The test has several advantages over existing test protocols as it can be evaluated visually. It is also insensitive to salt content, which is a problem with other tests.

2. Total Volatile Base

There are a number of decaboxylases of endogenous and exogenous origin responsible for fish spoilage. Several bacterial decarboxylases are now known to produce biogenic amines. Another volatile base is ammonia, which can be formed by the decomposition of urea by enzyme urease. The total volatile bases are expressed in terms of nitrogen content and values reflects both TMA nitrogen and other metabolites that result from either bacterial or enzymatic breakdown of fish flesh. The total volatile base (TVB) value cannot be used as a freshness indicator. In the initial stages of ice storage, up to 7–8 days, the TVB nitrogen values remain fairly constant or slowly decrease (57) and thereafter start increasing (44), with the onset of spoilage quickened by the rapid growth of microorganisms.

According to The Food and Agriculture Organization (FAO) (32), a TVB nitrogen level of 28–35 mg % is considered the limit of acceptability of a good-quality fish. For very-good-quality cod, 35–40 mg % TVB nitrogen is regarded as the limit beyond which whole or round chilled fish can be considered too spoiled for most purposes. In case of different fish and fishery products, varying levels of TVB nitrogen values are followed. For frozen tuna and swordfish not greater than 30 mg %, salted and dried fish 20 mg %, and fish meant for canning 20 mg % are the prescribed limits of TVB nitrogen values (44).

E. Products of TMAO Degradation

Trimethylamine oxide (TMAO) is a compound universally found in most species of marine fish but not in freshwater fish. During postmortem changes and associated biochemical and microbial reactions on fish muscle, TMAO is degraded into trimethylamine (TMA), dimethylamine (DMA), and formaldehyde (FA). Two of the enzymes responsible are an oxidoreductase and demethylase. These enzymes are discussed in detail in Chapter 7. While TMA is associated with changes in odor of fish, both DMA and FA are reported to cause denaturation of proteins under frozen storage.

1. TMAO Oxidoreductase

Many species of marine fish contain an odorless compound called trimethylamine oxide (TMAO). During fish spoilage this compound is reduced to trimethylamine (TMA), a volatile base. TMA is associated with fatty substance and is alleged to be responsible for the fishy smell of spoiled fish. TMAO is reduced partly by intrinsic and mainly by bacterial enzymes of the group "reductases." Most spoilage organisms such as *Pseudomonas* contain this enzyme and are responsible for this reaction. The production of TMA is exponential: slow initially and increasing rapidly after few days of chilled storage. During storage, changes in odor go more or less hand in hand with values of TMA.

2. TMAO Demethylase

Another spoilage index of freshness evaluation for fish is TMAO demethylase activity. This enzyme is mainly found in gadoid fish and is responsible for the textural deterioration in fish during frozen storage (53). In the absence of oxygen, TMAO is degraded rapidly by an endogenous enzyme TMAO demethylase to dimethylamine (DMA) and formaldehyde. Formaldehyde production is accompanied by an increase in the firmness of texture of fish muscle held under frozen storage. TMAO demethylase activity has been demonstrated in a number of species of fish and the amount of formaldehyde or DMA produced can be related to the freshness of fish (45,54–56). It has been shown that within the gadoid family, species with higher rates of production of DMA and FA undergo more rapid deterioration in texture during frozen storage. For example, red hake, with high levels of production of DMA and FA, has a considerably shorter storage life than cod. TMAO-degrading enzymes are most abundant in kidneys and other tissues of viscera (53). Although there is no definite generalization regarding the involvement of TMAO degrading enzymes such as TMAO demethylase in textural hardening of frozen fish, there are reports indicating that at least in gadoid species this mechanism dominates. In gadoids, the FA released from TMAO by these enzymes has been postulated to interact with myofibrillar proteins, forming methylene crosslinks and noncovalent bonds, which reduce the water-holding capacity of proteins, and thus produce toughening and other deleterious textural changes during frozen storage. Some fish species do not appear to contain TMAO demethylase. Redfish is one such species for which DMA does not change, remaining constant at a level of 0.6 mg % throughout the frozen storage time (57).

The most extensively used method of estimation of TMA is the Conway microdiffusion procedure (44) and another colorimetric method using picric acid with which it forms a colored complex. TMA may also be estimated by gas chromatography and high-pressure liquid chromatography (HPLC). Recently Wong and Gill (58) developed an enzyme assay for determination of TMA from the aqueous extracts of fish muscle. This method employs trimethylamine dehydrogenases (TMA-DH) isolated from the cultures of *HyphomicrobiumX*. Wong et al. (59) also developed a paper strip method for the rapid routine assay of TMA using immobilized trimethylamine dehydrogenase. Trimethylamine-specific gas sensor technology is also now available for rapid assay (60).

The TMA values do not increase noticeably in the very early stages of fish spoilage. For a good-quality fish a TMA nitrogen level of 1.25–2 mg % is recommended. TMA nitrogen levels of of 10–15 mg % can be considered the safety limit beyond which most chilled fish become spoiled (44). It has been suggested that for species such as plaice, herring, and, possibly, some others, TMA values cannot be considered as an index of quality since these species do not form sufficient amount of this compound.

F. Lipoxygenases

Lipoxygenase is discussed in detail in chapter 11. Lipoxygenase catalyzes the addition of oxygen in the 5, 12, or 15 position of various eicosanoic acids. Fish oil is an ideal substrate for lipoxygenase as it contains a large quantity of these polyunsaturated fatty acids. Lipoxygenase is also described as a pro-oxidant initiating lipid oxidation, which follows a free radical mechanism (29, 61). The mechanism of lipoxygenase reaction is a simple dioxygenase reaction with the formation of a *cis–trans*-conjugated diene. The hydroperoxide so formed is subsequently reduced to an alcohol of the parent acid. Hydroperoxides are capable of producing a number of carbonyl compounds responsible for the rancid flavor in fish muscle. However, there are a number of compounds, including tocopherols, ubiquinones, and others, and enzymes such as catalase and glutathione peroxidase with remarkable antioxidant activity capable of inhibiting or controlling lipoxygenase activity in fish muscle. The role of lipoxygenase in the degradation of quality and shelf life of fish by promoting oxidation of polyunsaturated fatty acids is now well documented (62–65).

The role of lipoxygenases in the degradation of quality of fish by promoting oxidative rancidity and development of rancid flavor in fish muscle is well established. Quality and shelf life of fish are generally influenced by the development of rancidity brought on by oxidation of highly polyunsaturated acids present in fish. Presence of lipoxygenases has been shown in rainbow trout, mackerel, sardine, and wild and cultured ayu (62). Fish muscle is reported to contain a glutathione peroxidase (GSH-Px) activity that is presumed to protect muscle from oxidative deterioration. Glutathione peroxidase is found in both Japanese jack mackerel and skipjack tuna and several other fish species (66, 67).

The lipoxygenase-mediated degradation of polyunsaturated fatty acids results in the formation of several alcohols and carbonyls. These were detected in highest concentrations by Josephson et al. (63) in skin and slime fractions of fish. The preferential oxidation of lipids in skin of both lean and fatty fish is also well established (68).

It is extremely difficult to define an index of fish spoilage based on estimation of lipoxygenase activity. Fish with lipoxygenase and other means of enzyme-mediated lipid oxidation can, however, be considered susceptible to rapid onset of rancidity. Measurement of the products of lipid oxidation is now an accepted way of assessing fish quality. Two common indices used for this are the peroxide value (PV) and thiobarbituric acid number (TBA).

1. Peroxide Value

The peroxide value is a measure of the early stage of oxidation of lipids. The measurement of PV is based on the release of iodine from potassium iodide by hydroperoxide in the lipid. The liberated iodine is titrated against standard 0.002 N sodium thiosulfate using starch as indicator. The values of PV are expressed as milliliters of 0.002 N sodium thiosulphate required to titrate 1 g extracted fat in fish. The value will be equivalent to the number of μmoles of hydroperoxide. Hence, peroxide value is also often referred to as PV number. At a peroxide number above 20, most fish smell rancid.

2. Thiobarbituric Acid Number

Oxidative rancidity, especially in fatty species of fish, is a very complex phenomenon. Unsaturated fatty acid in the lipids reacts with oxygen-forming hydroperoxides that then break down to a number of substances called carbonyls, which give the rancid flavor to fish meat. Peroxide value is the measurement of the first stage of rancidity and TBA is a measure of the second stage.

Malonaldehyde is one of the principal end products of lipid oxidation. In alcoholic solution the reaction between 2-thiobarbituric acid and malonaldelyde forms a pink complex having absorption maxima at 532 nm and this is utilized to estimate the extent of rancidity in food products. Oxidative rancidity is frequently expressed in terms of TBA number or mg malonaldehyde per kilogram material or μmoles malonaldehyde per 1 g fat extracted from sample. Fish develop a rancid smell when the TBA number is above 1–2 (μmoles/g of oil). Neither peroxide value nor thiobarbituric acid value often correlates well with sensory evaluation of rancidity, which is perhaps the best quality index for consumer acceptability.

G. Other Products

1. Phenol Oxidase (EC 1.14.18.1; Monophenol Monoxygenase)

Tyrosinase is a very important enzyme in shrimp and lobsters because it is responsible for black spot formation (69–72; see also Chap. 10). In animals phenol oxidase is found in the organelles known as melanosomes, which are present in pigment-producing melanocytes. Melanocytes are located in the epidermis and certain other tissues. One of the important routes of tyrosine metabolism in crustaceans is that leading to the synthesis of the black pigment called melanin. It is a two-stage attack on tyrosine by the enzyme tyrosinase, first forming dihydroxyphenylalanine (dopa). The dopa is again oxidized by tyrosinase to yield the 3,4-quinone. The quinone is unstable and undergoes a series of spontaneous reactions, with the formation of an intermediate, 5–6 dihydroxyindole and leading ultimately to the synthesis of melanin.

Phenol oxidase present in shrimp and lobsters is responsible for melanosis or blackening. Melanosis that occurs in shrimp and lobster indicates spoilage and consumers do not like to purchase products that are discolored. In the muscle of lobster and shrimp the amino acid tyrosine is present in free form and is converted enzymatically to the black pigment melanin.

It is reported that phenol oxidase is present in shrimp in a latent form and is activated by tyrosine and an endogenous enzyme with tryptic activity (70). One of the best ways to control blackening in shrimp is to slow down or stop phenol oxidase activity. Since oxygen is essential for this reaction, immersion in chilled water is commonly practiced onboard trawlers. Phenol oxidase is reported to be present in high concentrations in shrimp heads (71). Hence, beheading and use of effective enzyme inhibitors or reducing agents also can control the blackening. The most commonly used chemicals are ascorbic acid, sodium bisulfite, and certain aluminium compounds.

Sodium bisulfite is a permitted preservative for shrimp processing and is commercially employed onboard fishing trawlers and in industrial shrimp processing to prevent blackening.

2. Tryptophanase (EC 5.3.1.4)

This enzyme plays an important role in the quality assessment of shrimp. Tryptophanase converts free tryptophan present in fish muscle to indole, which imparts a typical foul smell to shrimp. Formation of excess indole in shrimp is hazardous to human health and limits are often prescribed by importing countries.

Tryptophanase is not an endogenous enzyme in shrimp. It is of bacterial origin, chiefly from *E. coli* and some *Pseudomonas sp.* In fact, indole cannot be considered as an index of spoilage since the bacteria producing indole are not generally regarded as fish spoilers. At best, estimation of indole can be considered only as a measure of the tryptophanase activity. The maximum permissible level of indole in processed shrimp is 25 mg %. Frozen shrimp containing indole above this level are likely to be rejected by importing countries. It is interesting to observe than frozen shrimp containing this concentration of indole are normally unacceptable by chemical and sensory methods.

IV. CONCLUSIONS

Enzyme activity has been used extensively as an indicator for quality changes in fish and fishery products. Some enzyme indicators are endogenous (i.e., inherently present in the fish) while others are of microbial origin. Postmortem changes and processing methods, particularly freezing, icing, and thawing, can all lead to tissue damage and disruption of cellular organelles including mitochondria and lysosomes, releasing several enzymes bound to these structures into the cellular fluid. Thus, cell disruption will increase or decrease the activity of enzymes. However, one of the problems associated with using enzyme assays as quality indicators is that they are time consuming and should be carried out by an expert. This makes it difficult for fish technologists to use them routinely as indices of quality control. The measurement of the rate of change of enzyme activity is most often done by the estimation of the reaction product(s).

REFERENCES

1. DD Nambudiri, K Gopakumar. ATPase and lactate dehydrogenase activities in frozen stored fish muscle as indices of cold storage deterioration. J Food Sci 57(1): 72–76, 1992.
2. S Watabe, H Ushio, M Iwamoto, M Kamal, H Ioka, K Hashimoto. Rigor mortis progress of sardine and mackerel in association with ATP degradation and lactate accumulation. Bull Jpn Soc Sci Fish 55: 1833, 1989.
3. K Nilsson, B Ekstrand. The effect of storage on ice and various freezing treatments on enzymatic leakage in muscle tissues of rainbow trout *(Oncorhynchus mykiss)*. Z Lebensm Unters Forsch 197(1):3–7, 1993.
4. DJ Tilgner, W Rogos. Influence of time lag prior to freezing and quality of red fish. In: R Kreuzer, ed. Freezing and Irradiation of Fish. London: Fishing News (Books) Ltd., 1964, pp 76–79.

5. CA Curran, RE Poulter, A Brueton, NR Jones, NSD Jones. Effect of handling treatment on fillet yields and quality of tropical fish. J Food Technol 21: 301–310, 1986.

6. CA Curran, RG Poulter, A Brueton, NSD Jones. Cold shock reactions in iced tropical fish. J Food Technol 21: 289–299, 1986.

7. T Sato, K Arai. Studies on the organic phosphates in the muscle of aquatic animals III. Effects of storing temperature upon adenosine polyphosphate content of carp muscle. Bull Jpn Soc Sci Fish 122:569–574, 1957.

8. E Noguchi, J Yamamoto, Studies on the arai phenomenon (the muscle contraction caused by perfusing water) IV. On the rigor mortis of white meat fish and red meat fish. Bull Jpn Soc Sci Fish 20: 1023–1026, 1955.

9. WW Kielley, O Mayerhof, Studies on adenosinetriphosphatase of muscle. II A new magnesium-activated adenosinetriphosphatase. J Biol Chem 176: 591–601, 1948.

10. A Yamanka, IM Mackie. Changes in the activity of a sarcoplasmic adenosine triphosphatase during iced-storage and frozen storage of cod. Bull Jpn Soc Sci Fish 37(11): 1105–1109, 1971.

11. N Seki, N Narita. Changes in ATPase activities and other properties of carp myofibrillar proteins during ice storage. Bull Jpn Soc Sci Fish 46(2): 207–213, 1980.

12. DO Nambudiri, K Gopakumar. Effect of freezing and Thawing on press juice and enzyme activity in the muscle of farmed fish and shellfish I.I.F. - I.I.R. Aberdeen, U.K.: Commission, 1990, pp 183–187.

13. N Katoh. N Nozaki, K Komatsu, K Arai. A new method for evaluation of quality of frozen surimi from Alaska pollack. Relationship between myofibrillar ATPase activity and Kamaboko forming ability of frozen surimi. Bull Jpn Soc Sci Fish 48(8): 1027–1032, 1979.

14. T Kawashima, K Arai, T Saito. Studies on muscular proteins of fish-X. The amount of actomyosin in frozen surimi from Alaska-pollack. Bull Jpn Soc Sci Fish 39(5): 525–532, 1973.

15. RD Cahn, NO Kaplan, L Levine, E Zwilling. Nature and development of lactic dehydrogenase. Science 136: 962, 1962.

16. MG Rossman, A Liljas. X-ray studies of protein interactions. In: EE Snell, ed. Annual Review of Biochemistry. California: Annual Review Inc., 1971, 43: 501.

17. R Janicke, S Knopf Molecular weight and quarternany structure of lactic dehydrogenase. Eur J Biochem 4:157, 1968.

18. HD Kaloustian, FE Stolzenbach, J Everse, MO Kaplau. Lactate dehydrogenase of lobster (*Homarus americanus*) tail muscle. J Biol Chem 244: 2891, 1969.

19. DD Nambudiri. Enzyme reactions as index of freshness of fish and shellfish. Ph.D. thesis, Cochin University of Science and Technology, Cochin, India, 1987.

20. R Hamm, L Kormandi. Transaminases of skeletal muscle. 3. Influence of freezing and thawing on the sub cellular distribution of GOT in bovine and poreine muscle. J Food Sci 34: 452–455, 1969.

21. SK Chhatbar, NK Velankar. A biochemical test for the distinction of fresh fish from frozen and thawed fish. Fish Technol 14: 131–136, 1977.

22. R Hamm, D Masic. Influence of freezing and thawing of carp on the sub-cellular distribution of asparate aminotransferase in the skeletal muscle. Arch Fisch 22: 121–124, 1971.

23. E Gould. An objective test for determining whether fresh fish have been frozen and thawed. In: R Kreuzer, ed. Fish Inspection and Quality Control. London: Fishing News (Books), 1971, pp 72–74.

24. C Barbagli, GS Crescenzi. Influence of freezing and thawing on the release of cytochrome oxidase from chicken liver and from beef and trout muscle. 3. J. Food Sci 46: 491–493, 1981.

25. W Demmer, K Werkmeister. Distinguish between fresh and thawed pork. Arch Labensmitt Hyg 36: 15–20, 1985.

26. K Yoshioka. Differentiation of freeze thawed fish from fresh fish by the examination of blood. Bull Jpn Soc Sci Fish 51: 1331–1336, 1985.

27. R Hamm, AA Badawi. Activity and subcellular distribution of mitochondrial enzyme Lisomiali in trote dopo scongelamento. Arch Vet Ital 33: 96–97, 1982.

28. HO Hultin. Biochemical deterioration of fish muscle. In: HH Huss et al, eds. Quality Assurance in the Fish Industry. Proceedings of an International Conference, Copenhagen, Denmark 26–30 August New York: Elsevier, 1991, pp 125–138.

29. T Tomiyama, K Kabayashi, K Kitahava, E Shiraishi, N Ohba. A study on the change in nucleotide and freshness of carp muscle during the chill-storage Bull Jpn Soc Sci Fish 32: 262–266, 1966.

30. NR Jones, J. Murray, EI Livingstone, CK Murray. Rapid estimation of hypoxanthine concentrations as indices of the freshness of chilled stored fish. J Food Sci 30: 791–794, 1964.

31. LF Jacober, AG Rand. Biochemical evaluation of seafoods. In: RE Martine, GJ Flick, CE Hebard, DR Ward, eds. Chemistry and Biochemistry of Foods. AVI Publishing Company, 1982, pp 347–365.

32. L Crawford, R Finch. Quality changes in albacore tuna during storage in ice and in refrigerated seawater. Food Technol 22: 87, 1968.

33. T Nakano, E Ito, T Nakagawa, F Nagayama. Properties of inosine-decomposing enzymes from aquatic animal muscle. Bull Jpn Soc Sci Fish 56: 633–639, 1990.

34. ML Nunes, I Batista, R Morao. Physical chemical sensory analysis of sardine *(Sardinella pilehardus)* stored in ice. J Sci Food Agric 59: 37–43, 1992.

35. JR Burt. Hypoxanthine as biochemical index of fish quality. Proc Biochem 11: 10, 1976

36. A Hoffman, JG Disney, A Finegar, JD Cameron. The preservation of some East African freshwater fish African J Trop Hydrobiol Fish 3(1): 1–3, 1974.

37. GD Stroud, JC Early, GL Smith. Chemical and sensory changes in iced *Nephrops norvegicus* as induces of spoilage. J Food Technol 17: 541–551, 1982.

38. T Saito, K Arai, H Matsuyoshi. A new method for estimating the freshness of fish. Bull Jpn Soc Sci Fish 24: 749–750, 1959.

39. PT Lakshmanan, PD Antony, K Gopakumar. Nucleotide degradation and quality changes in mullet *(Liza corsula)* and pearlspot *(Etroplus suratensis)* in ice and at ambient temperatures. Food Control 7(6): 277–283, 1996.

40. E Watanabe, H Endo, N Takeuchi, T Hayashi, K Toyama. Determination of freshness with a multi-electrode enzyme sensor system. Bull Jpn Soc Sci Fish 52(2): 489–495, 1986.

41. E Watanabe, S Tokimatsu, K Toyama. Simultaneous determination of hypoxanthine,

inosine, inosine-5-phosphate and adenosine-5-phosphate with a multi-electrode sensor. Anal Chim Acta 164: 139–146, 1984.

42.   E Watanabe, K Ando, I Karube, H Matsuoka, S Suzuki. Determination of hypoxanthine in fish with an enzyme sensor. J Food Sci 48(2): 496–500, 1983

43.   GC Chemnitius, M Suzuki, K Isobe, J Kimura, I Karube, RD Schmid. Thin-film polyamine biosensor: substrate specificity and application to fish freshness determination. Anal Chim Acta 263 (1–2): 93–100, 1992.

44.   JJ Connell. Control of Fish Quality. London: Fishing News (Book) Ltd., 1975.

45.   H Rehbein, W Schreiber. TMAO ase activity in tissues of fish species from the Northeast Atlantic. Comp Biochem Physiol 7B(3): 447–452, 1984.

46.   J Liston. Bacterial spoilage of seafood. In: HH Huss, ed. Quality Assurance in The Fish Industry. New York: Elsevier, 1992, pp 93–105.

47.   BS Pan, D James (eds). Histamine in Marine Products: Production by Bacteria Measurement and Prediction Formation. Rome, Italy: FAO Fisheries Technical Paper, 252, 1982.

48.   V Venugopal. Extracellular proteases of contaminant bacteria in fish spoilage; a review. J Food Protect 5: 341–350, 1990.

49.   SL Taylor. Histamine food poisoning: toxicology and clinical aspects. CRC Crit Rev Toxicol 17: 19–128, 1986.

50.   SL Taylor, JE Stratton, JA Nordlee. Histamine poisoning: an allergy like intoxication. Clin Toxicol 27: 225–240, 1989

51.   T. Rustad. Muscle chemistry and quality of wild and farmed cod In: HH Huss, ed. Quality Assurance in the Fish Industry. New York: Elsevier, 1991, pp 19–27.

52.   CAM Lima dos Santos. Guidelines for chilled fish storage experiments. Rome, Italy: FAO Fisheries Technical Paper 210, 1981, p 7.

53.   WM Laird, IM Mackie. Deterioration of frozen cod and haddock minces. Torry Memoirs No. 640°, 395–400, 1981

54.   H Rehbein. Formaldehyde and TMAO ase activity in fillets of the rough-tail from South Atlantic. Inf Fischwirtsch 38(4): 136–143, 1991

55.   P Reece. The role of oxygen in the production of formaldehyde in frozen minced cod muscle. J Sci Food Agric 34(10): 1108–1112, 1983.

56.   RC Lundstrom, PF Correia, KA Wilhelm. Dimethylamine production in fresh hake *(Urophycis chuss):* the effect of packaging material oxygen permeability and cellular damage. J Food Biochem 6(4): 229–241, 1982.

57.   J. Ochlenschlager. Evaluation of some well established indices for the determinations of freshness and/or spoilage of ice stored wet fish. In: HH Huss et al, eds. Quality Assurance in The Fish Industry. Proceedings of an International Conference, Copenhagen, Denmark, 26–30 August 1991. New York: Elsevier, 1991 pp 339–350.

58.   K Wong, TA Gill. Enzymatic determination of trimethylamine and its relationship to fish quality. J Food Sci 52: 1–3, 1987.

59.   K Wong, F Bartlett, TA Gill. A diagnostic test. Strip for the semi quantitative determination of trimethylamine in fish. J Food Sci 53: 1653–1655, 1988.

60.   RM Storey, HK Davis, D Owen, L Moore. Rapid approximate estimation of volatile amines in fish. J Food Technol 19: 1–10, 1984.

61.   A Khayat, D Schwall. Lipid oxidation in seafood. Food Technol 130–240, 1983.

62. C Zhang, T Shirai, T Suzuki, T Hirano. Lipoxygenase-like activity and formation of characteristic aroma compounds from wild and cultured *Ayu.* Bull Jpn Soc Sci Fish 58(5): 959–964, 1992.

63. DB Josephson, RC Lindsay, DA Stuiber, Enzymatic hydroperoxide initiated effects in fresh fish. J Food Sci 52(3): 596–600, 1987.

64. RJ Hsieh, JE Kinsella. Lipoxygenase-catalysed oxidation of n-6 and n-3 polyunsaturated fatty acids: relevance to and activity in fish tissue. J Food Sci 51(4): 940–945, 1986.

65. V Massey. Lipoyl dehydrogenase from pig heart. Methods Enzymol 9:272–276, 1966.

66. F Watanabe. M Goto, K Abe, Y Nakano. Glutathione peroxidase activity during storage of fish muscle. J Food Sci 61(4): 734–735, 1996.

67. T Nakano. M Sato, M Takeuchi. Glutathione peroxidase of fish. J Food Sci 57(5): 1116–1119, 1992.

69. H Wu, Y Lou. Studies on the mechanism of prevention of prawns from blackening 1. Extraction of the pheyoloxidase from *(P. hardwickii),* purification and analysis. J Zhejang Coll Fish Ghejiant Shuichan Xueyuan Xuebao 10(2): 85–91, 1991.

70. OJ Ferrer, MR Marshall. Isolation and characterisation of phenoloxidase from white shrimp *(Penaeus setiferus)* and spiny lobster *(Panulirus argus)* Rev Fac Agron Univ Cent Venez 8(2): 107–121, 1991.

71. KA Savagon, A Sreenivasan, Activation mechanism of pre-phenolase in lobster and shrimp. Fish Technol 15(1): 49–55, 1978.

72. T Nakagawa, Y Makinodan, M Hujita, Distribution and properties of phenoloxidase from shrimp *Penaeus* spp. Mem Fac Agric Kinki Univ Kinkidai No Kiyo 25: 27–32, 1992.

13
Enzymes and Flavor Biogenesis in Fish

Keith R. Cadwallader
Mississippi State University, Mississippi State, Mississippi

I. INTRODUCTION

The aroma of fish is an important quality indicator and is often the sole basis for product acceptance or rejection (1). For many people, the term "fishy" carries a negative connotation and is generally associated with the strong and objectionable aroma of stale or rancid fish. This term also has been used in sensory analysis to describe an element of the overall odor associated with stale, stored, or rancid lipid-containing foods (2, 3). Use of the term "fishy," however, to describe or differentiate fish aromas should be avoided since fresh fish and stored fish can have distinctly different aroma attributes. That is, the green, planty, and melonlike aroma of fresh fish is different than that of frozen-stored fish, which may develop stale and rancid notes.

For the most part, the aroma of fresh or stored fish can be attributed to lipid oxidation that is mediated via hydroperoxides formed by either enzymatic or nonenzymatic pathways. The primary difference in the two types of reaction mechanisms is that enzymes such as lipoxygenases are able to produce position-specific hydroperoxides (4–10), whereas the nonenzymatic processes are less specific in the type of peroxides generated (7, 8). Karahadien and Lindsay (8) reviewed the role of both types of oxidative processes in the generation of flavor compounds in fresh and stored fish.

Both enzymatic and autoxidative mechanisms can lead to the formation of short- and intermediate-chain aldehydes and ketones (1, 7, 8). Furthermore, both mechanisms, but especially enzymatic processes, cause the formation of compounds that impart distinct green, planty, and melonlike aroma notes reminiscent of fresh fish (1, 7–10). However, in addition to these so-called desirable aroma compounds, oxidative processes also may give rise to components responsible for the stale and rancid notes associated with stored fish (1, 8, 9). It is not clear as to what extent each of the above mechanisms contributes to the aroma of fresh fish, since there is considerable overlap and interaction between the enzymatic and chemical pathways. It has been suggested that lipoxygenase action may be most important in the generation of fresh fish flavor because of the site specificity in hydroperoxide formation, which leads to specific compounds providing characteristic fresh fish aromas (1, 8, 9, 11).

There have been several reviews written on the subject of fish flavor. Fish flavor is composed of nonvolatile taste-active compounds and volatile aroma-active compounds. Konosu and Yamaguchi (12) reviewed the occurrence of extractive (taste) compounds in fish flavor and concluded that these compounds play only minor roles in fish flavor. Others focused on various aspects of fish flavors from an aroma standpoint and discussed the effects of environment, processing, storage, cooking, and specifically on the enzyme biogenesis of fish aroma (1, 6, 8, 9, 11, 13). The role of lipoxygenase in aroma formation continues to be of interest, both from a quality aspect (5, 6) and because of the potential use of this enzyme in commercially viable processes (6, 14–16). This chapter focuses on the biogenesis of fish aroma via lipoxygenase and related enzymes and considers the roles that biogenerated compounds play in the overall aroma of fresh, stored, processed, and cooked fish.

II. AROMA OF FISH

Fish flavor is influenced by many factors such as species type, environment, storage and handling conditions, processing, and method of cooking. The flavor characteristics of fresh fish may be quite different from those of stored fish. Furthermore, thermal reactions that occur during cooking modify fish flavor due to Maillard reactions and from further breakdown of lipid-derived components. The following sections provide an overview of fish aroma as it is affected by these processes and puts into perspective the role lipoxygenase-catalyzed reactions play in the overall flavor of fish and fish products.

A. Volatile Components of Fresh Fish

Live and very fresh fish from saltwater generally have mild or faint odors (11, 17). Several components are known to contribute to the aroma of saltwater fish. It has been reported that lipid-derived carbonyls and alcohols play only a minor role in the aromas of very fresh saltwater fish (17). In addition, trimethylamine and dimethylsulfide also may have an impact on the mild aroma of fresh saltwater fish and these compounds may be especially important in the aromas of less than fresh saltwater fish (11). Trimethylamine has a distinctive fishy, aminelike odor and is derived from trimethylamine oxide, which is abundant in saltwater fish (18). Although trimethylamine is found at only very low levels in very fresh fish, it may make a substantial contribution to the overall aroma profile due to its low odor detection threshold (19) and because of a general absence of other aroma-active compounds that might serve to mask its odor. Dimethylsulfide contributes a canned-corn-, shellfish-, and seashore-like aroma note to the aroma profile of very fresh saltwater fish. Dimethylsulfide is derived by hydrolysis of dimethyl-β-propiothetin (20). Some microalgae, plankton, and seaweeds are known to contain high levels of dimethyl-β-propiothetin, which then is accumulated in fish via the food chain (20). Presence of high levels of dimethylsulfide can cause an undesirable odor in fish (21).

Other environmentally derived compounds have been reported in saltwater fish. The presence of iodoformlike off flavors have been reported in saltwater fish caught off the eastern coast of Australia (22). The compounds responsible for this defect were identified as 2- and 6-bromophenol, 2,4- and 2,6-dibromophenol, and 2,4,6-tribromophenol. These compounds are accumulated via consumption of algae and bryozoa, which contain appreciable levels of the bromophenols. The bromophenols may not necessarily be detrimental to fish flavor and at low concentrations may impart desirable marine-associated odors (1, 23, 24)

In contrast with saltwater fish, fish from freshwater environments do not contain appreciable levels of trimethylamine and dimethylsulfide in their aroma profiles. Furthermore, the aroma profiles of freshwater fish are affected to a greater extent by carbonyls and alcohols formed by the action of lipoxygenases and related enzymes on endogenous polyunsaturated fatty acids of the fish (6, 17). These lipid-derived compounds are found at even greater levels in euryhaline fish, which are able to tolerate both salt and fresh water environments and include trout, salmon, smelt, and many other species. Live and very fresh euryhaline fish typically have intense green, planty, and melonlike aromas due to the predominance of numerous unsaturated six (C_6), eight (C_8), and nine (C_9) carbon carbonyls and alcohols in their volatile profiles (17).

In addition to the lipid-derived compounds, the aromas of freshwater fish are more likely than those of saltwater fish to be affected by the occurrence of the earthy and musty off-flavor compounds geosmin and 2-methylisoborneol (11). Both geosmin and 2-methylisoborneol are of environmental origin and have odor detection thresholds in the low parts per trillion range (25). These off-flavor compounds can be produced by *Actinomycetes* (geosmin or 2-methylisoborneol) and blue-green algae (geosmin) (26–28). Occurrence of these off flavors in aquacultured fish, such as freshwater catfish, has been a significant economic problem for the industry (28). In other species such as Atlantic salmon, both 2-methylisobor-neol and geosmin were found to cause earthy off flavors in river caught fish but were not detected in the farmed or sea-caught fish (29).

Earthy off flavors in freshwater fish also can be caused by presence of high levels of amines, such as piperidine in postspawn salmon and piperidine and pyrrolidine in carp (11). Biogenesis of piperidine and pyrrolidine occurs by cyclization of 1, 5-diaminopentane (cadaverine) and 1, 4-diaminobutane (putrescine), respectively (11).

B. Volatile Components in Fish as Affected by Storage, Processing, and Cooking

The aromas of stored, processed, and cooked fish and fish products depend on many factors. The aromas of freshly harvested fish are highly perishable and are replaced during storage by stale and rancid notes, and sometimes by ammoniacal and putrid odors in improperly stored fish. The effects that processing and cooking have on the aroma of fish are beyond the scope of the present review. There are several thorough reviews of this topic (1, 8, 9, 11, 13). The following discussion will focus on defining the relative impact of lipoxygenase-derived volatiles on the flavor of fish during storage, processing, and cooking.

In stored fish, aroma compounds arise primarily from enzymatic action and oxidative processes. From a product quality standpoint, most aroma changes that occur during storage have a negative impact. Off flavors are often encountered in refrigerated- and frozen-stored saltwater fish due to the development of trimethylamine and dimethylamine, which are produced from trimethylamine oxide via microbial action and endogenous enzyme activity, respectively (18). Of the two compounds, trimethylamine exhibits more intense fishy and ammoniacal notes and this compound has been used as an indicator of fish flavor quality (1, 18). Increases in taste-active nucleotide degradation products in stored fish could potentially contribute to the flavor of fish (12), but these increases are generally considered to be indicative of a decline in product freshness (30).

Lipid oxidation plays an equally important role in the development of aromas in stored and cooked fish. The polyunsaturated fatty acids of fish are extremely prone to autoxidation and the hydroperoxides formed can be readily broken down during cooking. The lipoxygenase pathway also may participate as an initiator in the development of aroma in stored and cooked fish. That is, the peroxides formed by the action of lipoxygenase on polyunsaturated fatty acids can serve as a source of free radicals and accelerate the lipid oxidation cascade (7, 31). The lipid-derived compounds, whether from enzyme or autoxidative pathways, affect fish flavor to varying degrees during processing and cooking.

During storage, fish or fish oils develop strong oxidized odors that have been described as rancid, stale, fishy, and cod liver oil-like (8, 32). The development and chemistry of off-flavors in stored fish and fish oils have been reviewed (32, 33). The presence of a complex mixture of mostly unsaturated carbonyl compounds is primarily responsible for the oxidized off flavors of fish products. The importance of (Z)-4-heptenal to the off odor of cold-stored cod was established by McGill (34). Josephson and Lindsay (35) demonstrated that (Z)-4-heptenal can be generated from (E,Z)-2,6-nonadienal, which primarily arises due to lipoxygenase action in fresh fish but also can, to a lesser extent, be derived through lipid autoxidation (6). Most volatile compounds contributing oxidized-fishy or cod liver oil-like odors in stored fish and fish oils are generated via autoxidation of the polyunsaturated fatty acids (8, 32). These volatiles include hexanal and 2,4-heptadienal, 3,5-octadien-2-one 2,4-decadienal, and 2,4,7-decatrienal (8, 9, 32, 33, 36). Although these compounds are strictly lipid autoxidation products, lipoxygenase may acts as an initiator/promoter of lipid autoxidation by its initial formation of fatty acid hydroperoxides.

Lipid-derived volatile carbonyls have been reported to be the most important contributors to the flavor of dried and salted fish (37). These compounds are less important in the flavors of fermented fish, fish sauce, and fish hydrolysates. In these products the predominant volatiles are volatile fatty acids, esters, sulfur-containing compounds, and Maillard and Strecker degradation reaction products (38–42). In pickled (brined and acidified) fish, only modest levels of volatile carbonyls and alcohols are found due to their extraction into the pickling solution (43). It was speculated that the mild aroma of pickled fish was primarily due a general decline in the volatile constituents initially present in the fresh fish (43).

Lipoxygenase-derived carbonyls have been found to affect the aromas of cooked fish, but lipid autoxidation products seem to be of greater importance. Occurrence of (Z)-3-hexenal, (Z,Z)-3,6-nonadienal in the aroma profiles of boiled trout (44) and salmon and cod (45) indicates the initial involvement of lipoxygenase in formation of the site-specific hydroperoxides. However, most aroma-active components of these cooked fish were lipid autoxidation products and these were found to increase during storage of the raw material (44, 45). The predominance of lipid-derived compounds also was reported in the aroma profiles of cooked channel catfish (46) and sardines (47). Despite the importance of lipid-derived volatiles in the above cooked fish, there are several types of cooked fish in which lipid oxidation products were found to make only a minor contribution. For example, the character-impact aroma component of canned tuna is 2-methyl-3-furanthiol (48), which can be thermally derived from reaction of ribose and cysteine (49).

III. AROMA BIOGENESIS IN FISH VIA LIPOXYGENASE AND ASSOCIATED ENZYMES

Lipoxygenases catalyze the oxidation of polyunsaturated fatty acids to produce hydroperoxides that can undergo further breakdown by other enzymes, such hydroperoxide lyases, Z,E-enal isomerases, and alcohol dehydrogenases to produce the characteristic aroma components of fresh fish. As mention earlier, the fatty acid hydroperoxides also may undergo chemical breakdown in fish during storage, processing, or cooking to form unpleasant stale and rancid odors. The chemical breakdown of fatty acid hydroperoxides and their role as initiators of lipid autoxidation was described previously. The following sections deal with the occurrence of lipoxygenases in fish, the possible physiological roles these enzymes play in living fish, and the lipoxygenase-initiated biogeneration of aroma components of freshly harvested fish.

A. Lipoxygenase in Fish

Lipoxygenase (EC 1.13.11.12, linoleate; oxygen oxidoreductase) is an iron-containing enzyme that catalyzes the oxidation of polyunsaturated fatty acids containing Z,Z-1,4-pentadiene units to produce conjugated unsaturated fatty acid hydroperoxides (50). The existence of lipoxygenases in plants has been known for some time and was first discovered in soybeans over 65 years ago (51). A great deal was known about lipoxygenases in plants prior to its discovery and documentation in animals. In fact, it has only been in the last two decades that significant strides have been made in our understanding of lipoxygenases in animals (4, 5). Their occurrence in fish has been extensively studied in the past 15 years (5, 6). Josephson, Lindsay, and co-workers at the University of Wisconsin (7, 17, 52) and German and Kinsella at Cornell University (31) were among the first to demonstrate the existence of lipoxygenase in fish. In these early studies, the existence of lipoxygenase in fish was demonstrated in studies in which the oxidative potential of homogenized live fish (7, 53) or crude enzyme preparations from trout skin tissue (31) were tested in the absence or presence of known lipoxygenase inhibitors. These inhibitors were found to suppress the formation of aroma compounds (7, 53) or decrease the enzyme activity (31) and it was thus concluded that lipoxygenases were indeed present and active in fish. The lipid peroxidation potential of trout skin tissue was tested against docosahexaenoic acid and arachidonic acids by German and Kinsella (31). The presence of 12-hydroxyeicosatetraenoic acid as the major monohydroxy derivative suggested the presence of 12-lipoxygenase activity in the skin tissue. A mechanism and the potential consequences of this reaction were proposed (Fig. 1) (31). These researchers realized the importance of lipoxygenase as a potential initiator and accelerator of lipid autoxidation and the impact this would have on the development of oxidative rancidity in fish (6, 7, 8, 31).

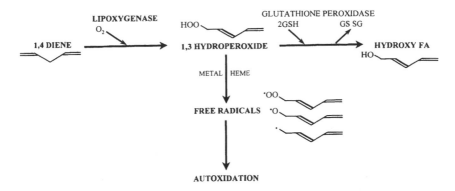

Figure 1 Proposed mechanism for initiation of tissue lipid peroxidation via lipoxygenase activity. Normal physiologically controlled pathway is shown in the horizontal direction and the proposed consequences of tissue destruction and lipoxygenase product proliferation are suggested in the vertical direction. (from Ref. 31 with permission.)

Since these early investigations, lipoxygenase activity has been reported in other species of fish (54–57). The presence of a 15-lipoxygenase had been postulated by several researchers based on the presence of specific volatile compounds formed in freshly harvested fish (1, 6). The presence of this lipoxygenase isozyme was demonstrated in fish gills (58, 59). Later, in sardine skin the presence of a 13-lipoxygenase with high activity toward linoleic acid was reported (60). Hsu and Pan (61) isolated three lipoxygenase isozymes from the gill of cultured grey mullet (*Mugil cephalus*). The major isozyme was 12-lipoxygenase followed by 15-lipoxygenase, which had much higher activities than that of 5-lipoxygenase.

B. Physiological Role of Lipoxygenase

Information on the physiological role of lipoxygenase in fish is limited; however, several researchers have focused on the effects of lipoxygenase products in other animal tissues (62). The possible physiological roles of some animal lipoxygenase-derived products have been discussed (5). Josephson and Lindsay (6) speculated that, in fish, lipoxygenases may produce physiologically active (slime secretion regulation) or osmoregulatory (cellular) compounds (e.g., leukotrienes and hydroxy fatty acids) from polyunsaturated fatty acids and that these compounds could be inactivated by hydroperoxide lyases. Additional evidence of the physiological role of lipoxygenase is that the rate of formation of volatile carbonyls and alcohols appears to be stress related and is apparently higher in the gills, outer skin, and mucus layer of the fish than in the muscle (6, 17, 52). The biochemistry and physiological role of lipoxygenases are discussed in more detail in Chapter 11.

C. Mode of Action

Lipoxygenases act on polyunsaturated fatty acids with Z,Z-1,4-pentadiene groups that contain an activated methylene group between the two double bonds. The number and position of pentadiene groups in the polyunsaturated fatty acid molecule are important since particular lipoxygenase isozymes can have high substrate (position) and conformational specificity (5, 50, 63). Lipoxygenase isozymes with different position specificity can attack the same polyunsaturated fatty acid molecule to produce polyhydroxy derivatives (64). It has been proposed that lipoxygenase-catalyzed peroxidation can occur by either aerobic or anaerobic pathways (50). Detailed discussions of the occurrence and mode of action of lipoxygenases can be found elsewhere (50, 51, 65) (see also Chap. 11).

The primary products of lipoxygenase-action are fatty acid hydroperoxides. The hydroperoxides may undergo chemical or enzymatic breakdown to form lower-molecular-weight secondary products. Chemical breakdown can lead to generation of free radicals that can further increase the rate of lipid oxidation. One the other hand, enzyme-mediated breakdown of fatty acid hydroperoxides leads to generation of specific end products. Hydroperoxide lyase catalyzes the hydrolysis of the hydroperoxide moiety to form C_6 and C_9 Z-aldehydes, which may undergo isomerization to form E-aldehydes, possibly by involvement of a Z,E-enal isomerase (6). Reduction of the aldehydes to their corresponding alcohols also may occur by action of nonspecific alcohol dehydrogenases (6).

D. Substrate Specificity

The substrate specificity of a lipoxygenase is generally defined in terms of its ability to dioxygenate a specific position on polyunsaturated fatty acids possessing 1,4-Z,Z-pentadiene moieties. Some naturally occurring n-3 and n-6 fatty acids that contain the required 1,4-Z,Z-pentadiene structure and also serve as common substrates for lipoxygenases include linoleic, linolenic (α and γ isomers), arachidonic, eicosapentaenoic, and docosahexaenoic acids. The major polyunsaturated fatty acids occurring in fish belong to the n-3 (or omega-3) group of fatty acids and include 22-carbon eicosapentaenoic and docosahexanenoic acids. The common feature of the n-3 fatty acids is that the first double bond exists at carbon number three from the methyl terminus of the molecule. Fish also contain appreciable levels of arachidonic acid, which is a 20-carbon n-6 fatty acid. In contrast to the n-3 fatty acids, the n-6 fatty acids possess the first

EICOSAPENTAENOIC ACID (C20:5 n-3)

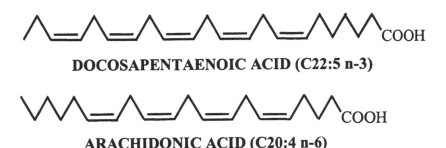

DOCOSAPENTAENOIC ACID (C22:5 n-3)

ARACHIDONIC ACID (C20:4 n-6)

Figure 2 Predominant polyunsaturated fatty acid precursors of enyzmatically derived aroma components of freshly harvested fish.

double bond at carbon number six from the methyl terminus of the molecule. The three principal polyunsaturated fatty acids involved in aroma development in fish are shown in Figure 2. All have been shown to be substrates of trout gill lipoxygenase (66).

In addition to fatty acids, lipoxygenases can act on fatty acid methyl esters and fatty acid constituents of phospholipids. In general, fish lipoxygenases have low activity toward linoleic and linolenic acids. Lipoxygenase substrate specificity may be influenced by environmental factors such as pH, presence of calcium, and others (5, 50). One important factor that clearly differentiates lipoxygenase-initiated lipid peroxidation from that of autoxidation is the high stereospecificity of the enzyme-mediated oxygen addition. For example, the 12-lipoxygenase from teleost fish gills catalyzes the formation of only the 12(S)-hydroxyeicosatetraenoic acid isomer, whereas with autoxidation this compound would be generated as a racemic mixture (63).

E. Lipoxygenases and Aroma Formation in Fresh Fish

The pleasant green, planty, and melonlike aromas of freshly harvested fish are caused by the presence of C_6, C_8, and C_9 carbonyls and alcohols (Table 1). Collectively, the C_8 compounds contribute heavy plantlike notes to fresh fish aromas, while the C_9 alcohols and aldehydes are responsible for fresh melon and

Table 1 Enzymatically Derived Aroma Components of Freshly Harvested Fish

Compounds	Concentration Range (µg/kg)*	Aroma quality†	Odor threshold‡
Alcohols			
1-Penten-3-ol	0.5–49	Grassy, green	400[a]
(Z)-3-Hexen-1-ol	0–0.3	Green, cut-leaf	70[a]
1-Octen-3-ol	18.6–110	Mushroomlike	1[b]
(Z)-1,5-Octadien-3-ol	24.8–94.8	Earthy, mushroom	10[c]
(Z)-2-Octen-1-ol	6.3–18.8	Fatty, beany	40[d]
(E,Z)-2,5-Octadien-1-ol	3.8–24.3	Earthy, mushroom	10[d]
(Z)-6-Nonen-1-ol	trace–4.8	Green, melon	n.a.
(Z,Z)-3,6-Nonadien-1-ol	trace–30.1	Watermelon, fatty	10[e]
Aldehydes			
Hexanal	8.3–418	Green, cut-grass	4.5[a]
(E)-2-Hexenal	0.5–6.3	Green apple	17[a]
(E)-2-Octenal	0–5	Tallowy, nutty	3[b]
(E)-2-Nonenal	Trace–5.8	Tallowy, stale	0.08[b]
(E,Z)-2,6-Nonadienal	0.3–17.2	Cucumber	0.01[e]
Ketones			
1-Penten-3-one	12.2–20.1	Pungent, green	1.25[a]
1-Octen-3-one	0.1–10	Earthy, mushroom	0.005[f]
(Z)-1,5-Octadien-3-one	0.1–5	Geraniumlike, metallic	0.001[g]

*Compiled from Refs. 7,43,52,53,67.
†Author's personal description.
‡Orthonasal odor detection threshold in water (µg/l): [a]68, [b]69, [c]70, [d]71, [e]72, [f]73, [g]74; n.a., not available.

cucumberlike attributes (6, 9). The C_6 volatiles impart intense green, cut-leaf notes to fresh fish aroma. These compounds are derived from long-chain polyunsaturated fatty acids primarily through the action of endogenous lipoxygenases; however, minor amounts of aroma compounds are derived by auoxidation processes (6). As previously mentioned, the oxidation-derived compounds may contribute rancid, stale, and cod liver oil-like notes (8). Different species of fish contain different types and concentrations of lipoxygenases, which accounts for different aromas associated with various freshwater and saltwater fish.

The main enzymes involved in biogeneration of aroma in fresh fish are the 12- and 15-lipoxygenases and hydroperoxide lyase (6). The 12-lipoxygenases act on specific polyunsaturated fatty acids to produce n-9 hydroperoxides. Hydrolysis of the 9-hydroperoxide of eicosapentaenoic acid by specific hydroperoxide lyases leads to the formation of mainly (Z,Z)-3,6-nonadienal, which can undergo spontaneous or enzyme-catalyzed isomerization to (E,Z)-2,6-

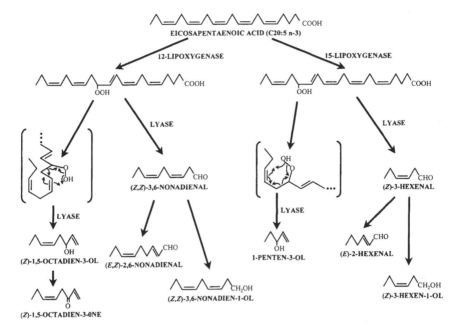

Figure 3 Proposed mechanism for the biogenesis of selected aroma components of freshly harvested fish. (From Ref. 6 with permission.)

nonadienal. These aldehydes may undergo reduction to their corresponding alcohols. A general scheme of these reactions was proposed by Josephson and Lindsay (6) (Fig. 3). The conversion of the aldehydes to their corresponding alcohols is a significant step since it leads to a general decline in aroma intensity due to alcohols having somewhat higher odor detection thresholds than the aldehydes (Table 1).

Some of the more important aroma components of freshly harvested fish are the C_8 ketones and alcohols (6). These compounds appear to be present in all species of fish and include 1-octen-3-ol, 1-octen-3-one, 1,(Z)-1,5-octadien-3-ol, and 1,(Z)-1,5-octadien-3-one (6). They do not have typical fresh fish aroma attributes and generally impart mushroom, metallic, and crushed geranium leaf aroma notes. Josephson and Lindsay (6) postulated that the formation of these compounds occurred by action of 12-lipoxygenase on a polyunsaturated fatty acid, such as eicosapentaenoic acid, to form a 12-hydroperoxide that would undergo hydrolysis by a hydroperoxide lyase to form a C_8 vinyl alcohol. Formation of the ketone could then occur by action of a nonspecific dehydrogenase on the vinyl alcohol. The proposed pathway for the formation of (Z)-1,5-octadien-3-ol and (Z)-1,5-octadien-3-one from eicosapentaenoic acid is shown in Figure 3.

This pathway is supported by the results of other researchers (10). The formation of 1-octen-3-ol and 1-octen-3-one could occur via an analogous system from lipoxygenase action on n-6 polyunsaturated fatty acids such as arachidonic acid. The conversion of the alcohols to ketones is of great significance to fresh fish aroma, since the ketones have much lower odor detection thresholds (Table 1).

The action of 15-lipoxygenases in conjunction with hydroperoxide lyases on the available polyunsaturated fatty acids gives rise to mostly C_6 aldehydes such as hexanal from n-6 polyunsaturated fatty acids and (Z)-3-hexenal from n-3 polyunsaturated fatty acids (Fig. 3). (Z)-3-hexenal can undergo isomeration by enzyme-catalyzed or chemical means to (E)-2-hexenal. The occurrence of enzyme-derived unsaturated C_6 aroma compounds has been documented in freshly harvested freshwater fish, but these compounds have not been detected in freshly harvested saltwater fish (6, 17). Only low levels of hexanal were detected in moderately fresh saltwater fish, but it was postulated that its formation was by lipid autoxidation rather than enzyme biogenesis (6).

Site specificity of the endogenous lipoxygenases is an important determinant in the aroma components formed in fresh fish (4, 6). Differences in the relative abundances of lipoxygenase isozymes in different species are responsible for the distinctive species-specific aroma profiles (4). Trout were found to have initially high 12-isozyme activity, while the 15-isozyme was barely detectable. Likewise in carp the 12-isozyme was predominant, but the 15-isozyme also was present at relatively high activity. In sturgeon, the 15-isozyme was found to be the predominant lipoxygenase. It was further pointed out that the aroma profiles of fish could change or shift in quality during refrigerated storage since lipoxygenases are highly unstable and the isozymes differ from one another in regard to their stability (4). Futhermore, the physiological state of the living fish can have an effect on lipoxygenase activity. It was observed that lipoxygenase-derived aroma compounds were in higher abundance in salmon during spawning than in the prime (saltwater) fish (67). During spawning salmon move from salt to fresh-water environments and undergo physiological changes due to both osmoregulation and spawning. Both Pacific Ocean and Great Lakes salmon contained C_8 ketones and alcohols regardless of their stage of maturity. Spawning salmon were found to have high levels of C_9 aldehydes and alcohols, which were absent from prespawn salmon from either saltwater (Pacific Ocean) or freshwater (Great Lakes) environments. The overall effect was an increase in the formation of carbonyls and alcohols in freshly harvested spawning fish over their prime counterparts. The saltwater salmon were considered to be of higher flavor quality since they had mild and fresh plantlike aromas, whereas the freshwater salmon (either prime or spawning condition) had more pronounced and heavier plantlike aromas. In later studies with fresh-baked and canned salmon, Josephson and Lindsay (68) reported the occurrence of an alkyl furanoid-type compound wth a distinctive salmon loaf-like odor. They speculated that this compound could be produced by co-oxidation of salmon pigments by the action of lipoxygenase.

Therefore, environmental conditions in which the fish live as well as storage and aging of fish tissue affect aroma profiles due to shifts in the overall lipoxygenase activity and changes in lipoxygenase isozyme ratios. Biogenesis of aromas in freshly harvested fish is also affected by the self-inhibition or suicidal nature of enzymes, the nature and abundance of available substrates (i.e., fatty acid profile), and the inhibitory effects of reaction products (5, 6).

F. Lipoxygenase in the Bioprocessing of Foods

Lipoxygenases are not only important in the endogenous biogeneration of aromas of plant and animal tissues but also may be intentionally added to foods. This might be for the purpose of bleaching of pigments and/or for the generation of aroma components. In addition, the enzymes could be used alone or with other enzymes for the production of useful chemicals (51). Soybean flour that contained an active lipoxygenase system was used to bleach pigments and improve dough properties of bread by acting upon available free fatty acids such as linoleic acid (76, 77). Active lipoxygenases and hydroperoxide lyases present in the soybean flour caused an increase in levels of several low-molecular-weight aldehydes and alcohols including hexanal, hexanol, 1-penten-3-ol, 1-pentanol, and 2-heptanone.

Developments in the use of fish lipoxygenases in such processes have been slow due to several inherent problems with fish lipoxygenase, such as the suicidal (self-inactivation) nature of the enzyme (5, 6). Lipoxygenase isolated from shrimp hemolymph and gray mullet gill was reported to produce seafood aroma from marine polyunsaturated fatty acids (14). Because of its high instability, the use of fish lipoxygenase as an industrial biocatalyst is impractical at the present time. However, there has been some recent success in the use of protector compounds such as imidazole to increase the stability of fish gill lipoxygenase (61).

Josephson and Lindsay (6) proposed the use of lipoxygenases from plants to restore fresh aroma in stored fish. These researchers rationalized that since the C_8 carbonyls and alcohols appear to play predominant roles in fish flavor, the analogous enzyme system of mushrooms or geranium leaves should generate a similar aroma profile from the polyunsaturated fatty acids of the fish. Using sensory screening techniques, they found that typical marine, green, and, in some cases, fresh fish aroma profiles, could be generated by application of the plant enzymes to the surface of the fish. Hsieh and Kinsella (78) found that soybean lipoxygenase catalyzed the oxidation of both n-3 and n-6 fatty acids, including arachidonic, eicosapentaenoic, and docosahexaenoic acids. Therefore, it appears that the use of plant lipoxygenase to form fresh fish-like aromas from marine lipid substrates has some potential.

Instead of the direct formation of aroma compounds from polyunsaturated fatty acids, it might be possible to use lipoxygenases (either plant- or fish-derived) in the indirect degradation of carotenoid pigments to produce cooked salmonlike aroma compounds (6). This has some merit since the role of lipoxygenase in the co-oxidation of carotenoids in plants to produce key aroma components has been firmly established (79, 80).

IV. CONCLUSIONS

Biogenesis of aroma in freshly harvested fish is initiated by the position-specific peroxidation of polyunsaturated fatty acids by 12- and 15-lipoxygenases. The resulting hydroperoxides subsequently undergo further breakdown to form characteristic aroma components of fresh fish. Many other factors affect fish flavor, but enzyme-derived C_6, C_8, and C_9 carbonyls and alcohols predominate in freshly harvested fish. Species type and environment can have a profound effect on the balance and overall composition of the aroma profiles of fresh fish. The aromas of fresh fish are mild and are immediately affected by conditions of handling, storage, processing and cooking. Such processes may lead to development of undesirable flavor caused by lipid autoxidation during storage. Processing and cooking can also positively affect fish flavor. Researchers have attempted to utilize fish-derived lipoxygenase for generation of fish-like aroma, but this approach is complicated because of the suicidal (self-inactivation) nature of this enzyme. Additional basic research on fish lipoxygenases is needed before practical use of this enzyme in a flavor biogeneration system will be realized.

REFERENCES

1. RC Lindsay. Fish flavors. Food Rev Int 6:437–455, 1990.
2. PB Johnson, GV Civille. A standardized lexicon of meat WOF descriptors. J Sensory Stud 1:99–104, 1986.
3. EG Pont, DA Forss, EA Dunstone. Fishy flavour in diary products. I. General studies on fishy butterfat. J Dairy Sci 27:205–209, 1960.
4. JB German, H Zhang, R Berger. Role of lipoxygenases in lipid oxidation in foods. In: AJ St. Angelo, ed. Lipid Oxidation in Foods. ACS Symposium Series 500. Washington DC: American Chemical Society, 1992, pp 74–92.
5. RJ Hsieh. Contribution of lipoxygenase pathway to food flavors. In: C-T Ho, TG Hartman, eds. Lipids in Food Flavors. ACS Symposium Series 558. Washington DC: American Chemical Society, 1994, pp 30–48.
6. DB Josephson, RC Lindsay. Enzymic generation of volatile aroma compounds from fresh fish. In: TH Parliment, R Croteau, eds. Biogeneration of Aromas. ACS Sym-

posium Series 317. Washington DC: American Chemical Society, 1986, pp 201–219.

7. DB Josephson, RC Lindsay, DA Stuiber. Enzymic hydroperoxide initiated effects in fresh fish. J Food Sci 52:596–600, 1987.

8. C Karahadian, RC Lindsay. Role of oxidative processes in the formation and stability of fish flavors. In: R Teranishi, RG Buttery, F Shahidi, eds. Flavor Chemistry: Trends and Developments. ACS Symposium Series 388. Washington DC: American Chemical Society, 1989, pp 60–75.

9. DB Josephson. Seafood. In: H Maarse, ed. Volatile Compounds in Foods and Beverages. New York: Marcel Dekker, 1991, pp 179–202.

10. RJ Hsieh, JE Kinsella. Lipoxygenase generation of specific volatile flavour carbonyl compounds in fish tissues. J Agric Food Chem 37:279–286, 1989.

11. T Kawai. Fish flavor. Crit Rev Food Sci Nutr 36:257–298, 1996.

12. S Konosu, K. Yamaguchi. The flavor components in fish and shellfish. In: R Martin, GJ Flick, CE Hebard, DR Ward, eds. Chemistry and Biochemistry of Marine Food Products. Westport, CT: AVI Publishing Company, 1982, pp 367–404.

13. E Durnford, F Shahidi. Flavour of fish meat. In: F Shahidi, ed. Flavor of Meat, Meat Products, and Seafoods, 2nd ed. New York: Blackie Academic & Professional (Chapman & Hall), 1998, pp 131–158.

14. BS Pan, JM Kuo. Flavor of shellfish and kamaboko flavourants. In: F Shahidi, JR Botta, eds. Seafoods: Chemistry, Processing Technology and Quality. Glasgow, UK: Blackie Academic & Professional (Chapman & Hall), 1994, pp 85–114.

15. BS Pan, JR Tsai, LM Chen, CM Wu. Lipoxygenase and sulfur-containing amino acids in seafood flavor formation. In: F. Shahadi, KR Cadwallader, eds. Flavor and Lipid Chemistry of Seafoods. ACS Symposium Series 674. Washington DC: American Chemical Society, 1997, pp. 64–75.

16. J Tsai, S Hu, BS Pan. Odour modification of fish oil using lipoxygenase from underutilized aquatic sources. 43rd Atlantic Fisheries Technology Conference, St. John's, Newfoundland, July 25–29, 1998, p 26.

17. DB Josephson, RC Lindsay, DA Stuiber. Variations in the occurrences of enzymatically derived volatile aroma compounds in salt- and freshwater fish. J Agric Food Chem 32:1344–1347, 1984.

18. JM Regenstein, MA Schlosser, A Samson, M Fey. Chemical changes of trimethylamine oxide during fresh and frozen storage of fish. In: RE Martin, GJ Flick, CE Heberd, DR Ward, eds. Chemistry and Biochemistry of Marine Food Products. Westport, CT: AVI Publishing Co., 1982, pp 137–149.

19. T Kikuchi, S Wada, H Suzuki. Significance of volatile bases and volatile acids in the development of off-flavor of fish meat. Eiyo Shokuryo 29:147–152, 1976.

20. H Iida. Studies on the accumulation of dimethyl-β-propiothetin and the formation of dimethyl sulfide in aquatic organisms. Bull Takai Res Fish Res Lab 124:35–110, 1988.

21. JC Sipos, RG Ackman. Association of dimethyl sulfide with the 'blackberry' problem in cod from the Labrador area. J Fish Res Bd Can 21:423–425, 1964.

22. FB Whitfield, F. Helidoniotis, D. Svoronos, KJ Shaw, GL Ford. The source of bromophenols in some species of Australian ocean fish. Water Sci Tech 31(11):113–120, 1995.

23. JL Boyle, RC Lindsay, DA Stuiber. Bromophenol distribution in salmon and selected seafoods of fresh- and saltwater origin. J Food Sci 57:918–922, 1992.

24. JL Boyle, RC Lindsay, DA Stuiber. Contributions of bromophenols to marine-associated flavors of fish and seafoods. J Aquatic Food Product Technol 1:43–63, 1992.

25. JA Maga. Musty/earthy aromas. Food Rev Int 3:269–284, 1987.

26. K Sivonen. Factors influencing odour production by Actinomycetes. Hydrobiologia 86:165–170, 1982.

27. N Sugiura, O Yagi, R Sudo. Musty odor from blue-green algae, *Phormidium tenue* in Lake Kasumigavra. Environ Technol Lett 7:77–86, 1986.

28. CP Dionigi. Physiological strategies to control geosmin synthesis in channel catfish aquaculture systems. In: AM Spanier, H. Okai, M. Tamura, eds. Food Flavor and Safety, Molecular Analysis and Design. ACS Symposium Series 628. Washington DC: American Chemical Society, 1993, pp 322–337.

29. LJ Farmer, JM McConnell, TDJ Hagan, DB Harper. Flavour and off-flavour in wild and farmed Atlantic salmon from locations around northern Ireland. Water Sci Tech 31:259–264, 1995.

30. DH Greene, EJ Bernatt-Byrne. Adenosine triphosphate catabolites as flavor compounds and freshness indicators in Pacific cod (*Gadus macrocephalus*) and pollock (*Theragra chalcogramma*). J Food Sci 55:257–258, 1990.

31. JB German, JE Kinsella. Lipid oxidation in fish tissue. Enzymatic initiation via lipoxygenase. J Agric Food Chem 33:680–683, 1985.

32. C Karahadian, RC Lindsay. Evaluation of compounds contributing characterizing fishy flavors in fish oils. JAOCS 66:953–959, 1989.

33. CF Lin. Flavor chemistry of fish oil. In: C-T Ho, TG Hartman, eds. Lipids in Food Flavors. ACS Symposium Series 558. Washington DC: American Chemical Society, 1994, pp 208–232.

34. AS McGill, R Hardy, JR Burt, FD Gunstone. Hept-*cis*-4-enal and its contibution to the off-flavour in cold stored cod. J Sci Food Agric 25:1477–1489.

35. DB Josephson, RC Lindsay. Retro-aldol degradations of unsaturated aldehydes: role in the formation of c4-heptenal from t2, c6-nonadienal in fish, oyster and other flavors. JAOCS 64:132–138.

36. PW Meijboom, JBA Stroink. 2-trans, 4-cis, 7-cis-Decatrienal, the fishy off-flavor occurring in strongly autoxidized oils containing linolenic acid or ω 3,6,9, etc., fatty acids. JAOCS 49: 555–558,1972.

37. H. Sakakibara, M. Hosokawa, I Yajima, K Hayashi. Flavor constituents of dried bonito (katsuobushi). Food Rev Int 6:553–572, 1990.

38. YJ Cha, KR Cadwallader. Aroma-active compounds in skipjack tuna sauce. J Agric Food Chem 46:1123–1128, 1998.

39. YJ Cha, GH Lee, KR Cadwallader. Aroma-active compounds in salt-fermented anchovy. In: F Shahidi, KR Cadwallader, eds. Flavor and Lipid Chemistry of Seafoods. ACS Symposium Series 674. Washington, DC: American Chemical Society, 1997, pp 131–147.

40. RC McIver, RI Brooks, GA Reineccius. Flavor of fermented fish sauce. J Agric Food Chem 30:1017–1020, 1982.

41. K Sugiyama, H Onzuka, K Takada, K Oba. Volatile components of fish protein hydrolysates. Nippon Nogeikagaku Kaishi 64:1461–1465, 1990.
42. R Trique, H Guth. Determination of potent odorants in ripened anchovy (*Engraulis encrasicholus* L.) by gas chromatography-olfactometry of headspace samples. In: F. Shahidi, KR Cadwallader, eds. Flavor and Lipid Chemistry of Seafoods. ACS Symposium Series 674. Washington, DC: American Chemical Society, 1997, pp 31–38.
43. DB Josephson, RC Lindsay, DA Stuiber, Influence of processing time on the volatile compounds characterizing the flavor of pickled fish. J Food Sci 52:10–14, 1987.
44. C. Milo, W. Grosch. Changes in the odorants of boiled trout (*Salmo fario*) as affected by the storage of the raw material. J Agric Food Chem 41:2076–2081, 1993.
45. C Milo, W Grosch. Changes in the odorants of boiled salmon and cod as affected by the storage of the raw material. J Agric Food Chem. 44:2366–2371, 1996.
46. OE Mills, CA Kelly, PB Johnsen. The volatile flavor of raw and cooked channel catfish. In: G Charalambous, ed. Food Flavors, Ingredients and Composition. New York: Elsevier Science Publishers BV, 1993, pp 221–232.
47. K Nakamura, H Iida, T Tokunaga, K Miwa. Volatile flavor compounds of grilled red-fleshed fish volatile carbonyl compounds and volatile fatty acids. Bull Jpn Soc Sci Fish 46:221–224, 1980.
48. DA Withylcombe, CJ Mussinan. Identification of 2-methyl-3-furanthiol in the steam distillate from canned tuna fish. J Food Sci 53:658,659, 1988.
49. DS Mottram. The chemistry of meat flavour. In: F Shahidi, ed. Flavor of Meat, Meat Products, and Seafoods, 2nd ed. New York: Blackie Academic & Professional (Chapman & Hall), 1998, pp 5–26.
50. DS Robinson, Z Wu, C Domoney, R Casey. Lipoxygenases and the quality of foods. Food Chem 54:33–43, 1995.
51. HW Gardner. Recent investigation into the lipoxygenase pathway of plants. Biochem Biophys Acta 1084:221–239, 1991.
52. DB Josephson, RC Lindsay, DA Stuiber. Identification of compounds characterizing the aroma of fresh whitefish (*Coregonus clupeaformis*). J Agric Food Chem 31:326–330, 1983.
53. DB Josephson, RC Lindsay, DA Stuiber. Biogenesis of lipid-derived volatile aroma compounds in the emerald shiner (*Notropis atherinoides*). J Agric Food Chem 32:1347–1352, 1984.
54. S Mohri, SY Cho, Y. Endo, K Fujimoto. Lipoxygenase activity in sardine skin. Agric Biol Chem 54:1889–1891, 1990.
55. M Winkler, G Pilhofer, JB German. Stereochemical specificity on the n-9 lipoxygenase of fish gill. J Food Biochem 15:437–448, 1991.
56. YJ Wang, LA Miller, PB Addis. Effect of heat inactivation of lipoxygenase on lipid oxidation in lake herring. JAOCS 68:752–757, 1991.
57. CH Zhang, T. Shirai, T Suzuki, T Hirano. Lipoxygenase-like activity and formation of characteristic aroma compounds from wild and cultured ayu. Nippon Suisan Gakki 58:959–964, 1992.
58. JB German, RK Creveling. Identification and characterization of a 15-lipoxygenase from fish gills. J Agric Food Chem 38:2144–2147, 1990.

59. JB German, R Berger. Formation of 8, 15-dihydroxy eicosatetraenoic acid via 15- and 12-lipoxygenases in fish gill. Lipids 25:849–853, 1990.

60. S Mohri, SY Cho, Y. Endo, K Fujimoto. Linoleate 13(S)-lipoxygenase in sardine skin. J. Agric Food Chem 40:573–576, 1992.

61. HH Hsu, BS Pan. Effects of protector and hydroxyapatite partial purification on stability of lipoxygenase from gray mullet gill. J Agric Food Chem 44:741–745, 1996.

62. JY Vanderhoek, NW Schoene, P-PT Pham. Inhibitory potencies of fish oil hydroxy fatty acids on cellular lipoxygenases and platelet aggregation. Biochem Pharmacol 42:959–962, 1991.

63. M Winkler, G Pilhofer, JB German. Stereochemical specificity of the n-9 lipoxygenase of fish gill. J Food Biochem 15:437–448, 1991.

64. JB German, JE Kinsella. Lipoxygenase activity in trout gill tissue; production of trihydroxy fatty acids. Biochim Biophys Acta 877:290–298.

65. JR Whitaker. Lipoxygenases. In: DS Robinson, NAM Eskin, ed. Oxidative Enzymes in Foods. London, UK: Elsevier Applied Science, 1991, pp 175–215.

66. JB German, G. Bruckner, JE Kinsella. Lipoxygenase in trout gill tissue affecting on arachidonic, eicosapentaenoic, and docosahexaenoic acids. Biochem Acta 875:12–19, 1985.

67. DB Josephson, RC Lindsay, DA Stuiber. Influence of maturity on the volatile aroma compounds from fresh Pacific and Great Lakes salmon. J Food Sci 56:1576–1579, 1991.

68. RG Buttery, RM Seifert, DG Guadagni, LC Ling. Characterization of additional components of tomato. J Agric Food Chem 19:524–529, 1971.

69. RG Buttery, JG Turnbaugh, LC Ling. Contribution of volatiles to rice aroma. J Agric Food Chem 36:1006–1009, 1988.

70. FB Whitfield, DJ Freeman. Off-flavors in crustaceans caught in Australian coastal waters. Water Sci Technol 15:85–95, 1983.

71. H Pyysalo, M Suihko. Odour characterization and threshold values of some volatile compounds in fresh mushrooms. Lebensm Wiss Technol 9:371–373, 1976.

72. RG Buttery. Vegetable and fruit flavors. In: R Teranshi, RA Flaith, H Sugisawa, eds. Flavor Research: Recent Advances. New York: Marcel Dekker, 1981, pp 180–184.

73. RG Buttery, R Teranishi, RA Fliath, LC Ling. Identification of tomato paste volatiles. J Agric Food Chem 38:792–795.

74. PAT Swoboda, KE Peers. Volatile odorous compounds responsible for metallic, fish taint formed in butterfat by selective oxidation. J Sci Food Agric 28:1010–1018, 1977.

75. DB Josephson, RC Lindsay, DA Stuiber. Volatile carotenoid-related oxidation compounds contributing to cooked salmon flavor. Lebenm Wiss Technol 24:424–432, 1991.

76. PA Luning, JP Roozen, RAF Moest, MA Posthumus. Volatile composition of white bread using enzyme active soya flour as improver. Food Chem 41:81–91, 1991.

77. K Addo, D Burton, MR Stuart, HR Burton, DF Hildebrand. Soybean flour lipoxygenase isozyme mutant effects on bread dough volatiles. J. Food Sci 58:583–585,608, 1993.

78. RJ Hsieh, JE Kinsella. Lipoxygenase-catalyzed oxidation of n-6 and n-3 polyunsaturated fatty acids: relevance to and activity in fish tissue. J Food Sci 51:940–945, 1986.

79. CL Allen, JW Gramshaw. The cooxidation of carotenoids by lipoxygenase in tomatoes. In: AJ Taylor, DS Mottram, eds. Flavour Science: Recent Developments. The Proceedings of the Eighth Weurman Flavour Research Symposium. Cambridge, UK: The Royal Society of Chemistry, 1996, pp 32–37.

80. P Winterhalter. Carotenoid-derived aroma compounds: biogenetic and biotechnological aspects. In: GR Takeoka, R Teranishi, PJ Williams, A Kobayashi, eds. Biotechnology for Improved Foods and Flavors. ACS Symposium Series 637, Washington DC: American Chemical Society, 1996, pp 295–308.

14
Enzymes and Seaweed Flavor

Taichiro Fujimura and Tetsuo Kawai
Shiono Koryo Kaisha, Ltd., Osaka, Japan

I. INTRODUCTION

The production of edible seaweed in Japan in 1993 was nearly 700,000 metric tons (MT). Kinds of seaweed include a red seaweed called nori (mainly *Porphyra tenera* and *P. yezoensis*) at 400,000 MT; two brown seaweeds, kombu (*Laminaria japonica* and its analogues) at 194,000 MT and wakame (*Undaria pinnatifida*) at 93,000 MT (1). The total of green seaweeds amounts to less than 1000 MT. These figures indicate the order of significance of red (Rhodophyta), brown (Phaeophyta), and green seaweeds (Chlorophyta) in Japan, where only green seaweed is consumed in small quantity. This consumption trend is also common elsewhere in the world (2). As will be seen, study of both seaweed flavor and seaweed enzymes has recently proceeded with green seaweeds. Therefore, there remains a gap between commercially important species and scientific knowledge.

Almost all the commonly occurring seaweeds are considered edible, according to the literature summarizing edible seaweeds throughout the world (3–6). Seaweeds with good flavor and texture have been selected for consumption according to regional customs. Their healthy nutritive properties have been reassessed. The Food and Drug Administration (FDA) has affirmed 16 species of the brown and 8 species of the red to be generally recognized as safe (7).

Natural products of seaweed, which have been identified in the literature, are compiled in detail by Faulkner (8–14). The volatile compounds of seaweeds are reviewed by Noda (15) and Kajiwara (16). Seaweed enzymes were more recently reviewed by Kajiwara et al. (17).

Most seaweeds do not have the obvious aromas characteristic of terrestrial plants, especially those of flowers and fruits, probably due to smaller amounts of flavor compounds. However, similar to flowers, odoriferous ova secreting odor compounds that serve as attractants for pollination are mainly found from some genera of the brown Dictyotales (18, 19).

Due to the faint odors, some characteristic flavor compounds of edible seaweeds have been researched using capillary column gas chromatography (GC) and GC-mass spectrophotometry (MS) by many workers: Kajiwara et al. (20–23), Flament and Ohloff (24), Sugisawa et al. (25), Sakagami et al. (26), Gally et al. (27), and Tamura et al. (28).

In 1950s and 1960s, before the development of capillary column GC, identification of compounds of dried seaweed was eagerly carried out using a short packed column by Katayama (29). The principal volatile compounds, in addition to scattered compounds published in the literature, were summarized by Tsuchiya (30) as follows: fatty acids (CnCOOH; n = 0, 5, 7, 13, 15), methylamine, trimethylamine, 3-octenol, heptanol, octanol, p-cresol, propanal, 2-methylpropanal, pentanal, heptenal, benzaldehyde, furfural, 5-methylfurfural, α-pinene, d-limonene, carvone, α-terpineol, terpinolene, cadinene, 1,8-cineol, linalool, ocimene, geraniol, myrcene, cadinol, hydrogen sulfide, methanethiol, and dimethyl sulfide. All the compounds, except bromophenol and iodine, have also been found in terrestrial plant volatiles.

Bromophenol found in the red, and iodine in the brown and the red, are characteristic compounds for the seaweeds. The former has a disinfectant-like odor, and the latter an odor of sea air (31). However, all volatile halogen-containing compounds are strictly prohibited for use as flavoring ingredients, even though they are natural products or nature identical compounds. This is due to their carcinogenic or toxic properties (32). Practically none of the volatile halogen-containing compounds are flavor compounds.

On the other hand, kelps contain abundant iodine (e.g., 70–2300 ppm in the fresh fronds) (30), and a quantity of iodine is often used as an additive in table salt consumed in inland areas to prevent thyroid gland disease.

This chapter reviews characteristic seaweed flavor compounds and the enzymes responsible for their biogenesis, and will discuss the influence of environmental conditions on the flavor.

II. GREEN SEAWEED FLAVOR

A. Dimethyl Sulfide

The odor profile of green seaweed is mainly characterized by a large quantity of dimethyl sulfide (DMS), which gives the impression of a fresh seashore product. DMS is a common flavor component among the green, the brown, and the red, as

Table 1 Content (μ g/g) of DMS and DMPT in
Seaweeds

Species	DMS	DMPT
Chlorophyta		
Ulva pertusa	1.15	339.00
Enteromorpha sp.	5.80	450.00
Monostroma nitidum	3.20	286.00
Cladophora wrightiana	0.04	1.36
Phaeophyta		
Undaria pinnatifida (A)	tr.	0.25
Undaria pinnatifida (B)	0.12	1.80
Ishige okamurai	tr.	0.38
I. sinicola	tr.	0.59
Sargassum ringgoldianum	tr.	0.38
S. serratifolium	0.01	0.47
S. horneri	0.01	0.55
S. patens	tr.	1.48
S. fulvellum (A)	0.06	2.88
S. fulvellum (B)	0.10	3.00
S. sagamianum	0.02	3.98
S. micracanthum	0.38	6.93
S. thunbergii (A)	0.50	12.80
S. thunbergii (B)	0.14	2.96
S. tortile	0.13	23.30
S. piluliferum	0.38	29.40
Eisenia bicyclis (A)	0.01	0.64
Eisenia bicyclis (B)	0.16	19.00
Dictyota dichotoma	0.02	0.81
Ecklonia cava	tr.	1.36
Hydroclathrus clathratus	0.09	1.70
Padina arborescens	0.05	3.14
Hizikia fusiformis	0.01	3.77
Scytosiphon lomentarius	1.25	11.02
Rhodophyta		
Gelidium amansii	0.03	3.40
Plocamium telfairiae	0.01	1.20
Gracilaria textorii	0.48	46.60
G. bursa-pastoris	0.84	3.80
Pachymeniopsis elliptica	0.05	3.40
Gloiopeltis complanata	0.18	3.40

tr., trace (<0.01 μ g/g).
Source: Ref. 33.

shown in Table 1 (33). In particular, DMS is abundant in green seaweeds, with the exception of *Cladophora wrightiana.* The DMS content ranges from 1 to 6 ppm (fresh frond), showing considerably higher concentrations than that of other seaweeds. These concentrations are much higher than the odor threshold value, which was reported as 0.3, 1, or 10 ppb in water by Fors (34). Therefore, DMS definitely contributes to green seaweed flavor as exemplified by *Enteromorpha* sp. containing 5.8 ppm DMS.

The browns also contained DMS, ranging from less than 10 ppb to 1 ppm (Table 1). Brown species of the same genus may differ in DMS content. One group has DMS in small amounts (*Sargassum ringgoldianum, S. serratifolium,* and *S. horneri*) and another group has DMS in large amounts (*S. tortile* and *S. piluliferum*). Also, there is a discrepancy between the amount of DMS and its precursor, dimethyl-β-propiothetin (DMPT). For example, there are very small amounts of DMS compared with DMPT in some browns (*Hizikia fusiformis, S. sagamianum,* and *Padina arborescens*). However, the concentration of DMS is nearly proportional to DMPT in other species. Four pairs of the browns in Table 1 illustrate seasonal and regional variation in DMS and DMPT between the fronds collected on March (A) and those in June (B) at two locations only 100 km apart. Large differences in DMS and DMPT concentration between the same species suggest that environmental factors can be more important than morphological similarity based on taxonomy (35). The environmental influence may also cause a wide variation in other components as well. Some examples will be discussed below. The quantitative instability, along with the qualitative variability, seems to be one of the most characteristic properties of seaweed flavor.

Among the greens, *U. pertusa* grows most rapidly in areas with water pollution, and often takes on a seawater taint. In addition, because *U. pertusa* has a harder texture than *Entermorpha,* it is mainly used for chicken feed or limited amounts of condiments. On the other hand, air dried *Entermorpha* has been favored due to its softness and high DMS content after thermal processing. The increase in DMS is caused by thermal decomposition of DMPT. This increase was also observed during the drying of a red seaweed *Porphyra* sp. (36).

B. Dimethyl-β-Propiothetin - Dethiomethylase

In addition to the environmental effects, difference in DMS content between the species is likely caused by an enzyme-catalyzed reaction converting DMPT into DMS. The enzymatic formation of DMS from DMPT in seaweed has been characterized (37–39). The difference in enzyme activity among several species was reported by Iida et al. (33). The authors determined the change in DMS content of an extract of each seaweed homogenate during incubation.

Table 2 Formation of DMS (µ g/g) in Extract of
Seaweeds during Incubation at 25°C for 2 h

Species	Before incubation	After incubation
Chlorophyta		
Ulva pertusa	1.15	82.50
Enteromorpha sp.	5.80	90.00
Monostroma nitidum	3.20	14.00
Cladophora wrightiana	0.04	0.24
Phaeophyta		
Undaria pinnatifida	tr.	tr.
Sargassum fulvellum	0.06	0.60
S. micracanthum	0.38	0.92
S. thunbergii	0.50	3.00
S. piluliferum	0.38	0.92
Dictyota dichotoma	0.02	0.50
Ecklonia cava	tr.	0.02
Hydroclathrus clathratus	0.09	0.18
Padina arborescens	0.05	0.27
Hizikia fusiformis	0.01	0.02
Scytosiphon lomentarius	1.25	3.60
Rhodophyta		
Gelidium amansii	0.03	0.16
Plocamium telfairiae	0.01	0.03
Gracilaria textorii	0.48	0.74
G. bursa-pastoris	0.84	1.38
Pachymeniopsis elliptica	0.06	0.08
Gloiopeltis complanata	0.18	0.30

tr., trace (< 0.01 µg/g).
Source: Ref. 33.

As Table 2 shows, all greens except the *Cladophora* sp. exhibited a greater
DMS increase than the browns and the reds, indicating greater enzyme-catalyzed formation of DMS in green than in the other two seaweed groups. An
enzyme was obtained from *Ulva conglobata* and partially purified. The optimum pH was 7.0–7.5 and the optimum temperature was 40°C (35). Two enzymatic reaction products were the acids, 2-propenoic acid (acrylic acid) and
possible hydrohalogenic acid as seen in Scheme 1. The enzymatic activity is
probably inhibited by the acid products. DMPT is decomposed with ease by
three processes: enzymatic, alkaline, and thermal decomposition (35). It has
not been ascertained whether hydrohalogenic acid produces halogen-containing volatiles.

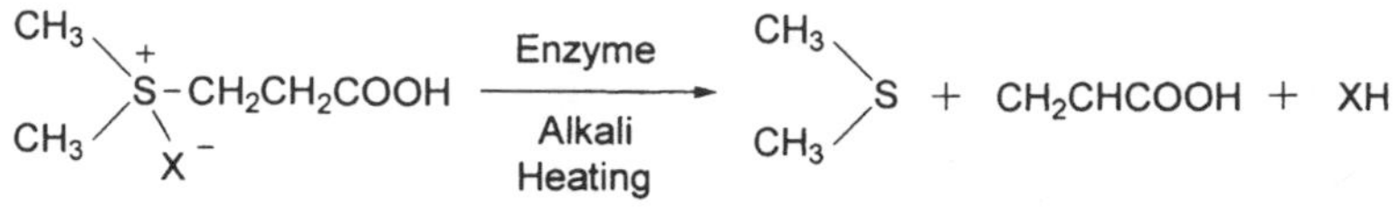

$$CH_3 \diagdown \atop CH_3 \diagup \overset{+}{\underset{X^-}{S}} - CH_2CH_2COOH \quad \xrightarrow[\substack{Alkali \\ Heating}]{Enzyme} \quad CH_3 \diagdown \atop CH_3 \diagup S \; + \; CH_2CHCOOH \; + \; XH$$

DMPT   DMS

Scheme 1   Decomposition of DMPT to DMS.

C. Unsaturated Fatty Aldehydes

Besides DMS, characteristic flavor components of the greens are unsaturated fatty aldehydes. Two clusters of C_6–C_9 and C_{15}–C_{17} unsaturated aldehydes found in *Ulva pertusa* were shown to be responsible for the "green" aroma. The former group has a fresh green or green-fatty odor and the latter a seaweed odor (25). The flavor of fresh *U. pertusa* consists mainly of the following compounds with accompanying odor profiles and odor threshold values: DMS, hexanal (grassy [25], green, aldehyde [40], 4.5 ppb [40]); 7-heptadecene (water off flavor [25, 28], weak, seaweed); (2*E*)-octenal (seaweed [25]); (2*E*)-nonenal (cucumber [25], cucumber, cardboard [40], 0.08 ppb [40]); (2*Z*,6*E*)-nonadienal (cucumber [25]); (2*E*,4*E*)-decadienal (seaweed [25]); (8*Z*, 11*Z*)-heptadecadienal (seaweed [20, 25], fresh cucumber [41]); (8*Z*, 11*Z*, 14*Z*)-heptadecatrienal (Scheme 2) (seaweed [20, 25]); and (8*Z*)-heptadecenal (seaweed [20, 25]). Tamura et al. (28) evaluated additional contributors to *Ulva pertusa* flavor: (2*Z*,4*Z*)-decadienal (seaweed [28]); (2*E*,4*E*)-heptadienal (seaweed [28]), (2*E*,6*Z*)-nonadienal (cucumber [28, 42], cucumber peel [40], powerful cucumber and violet aroma: 0.01 ppb [43]); (2*E*,4*E*)-nonadienal (seaweed [28]); and β-ionone (sweet powdery [25], nori [28]). Among the compounds, 7-heptadecene was reported to be the major component (4 ppm) reaching nearly 50 % of the essential oil obtained from *Ulva pertusa* in April (28). Its amount was nearly consistent with the estimated value (8 ppm) on the same month from the report by Okano and Aratani (44). On the other hand, Kajiwara et al. (20) reported that (8*Z*, 11*Z*, 14*Z*)-heptadecatrienal occupied 35% of the essential oil obtained from *Ulva pertusa* growing in the northern clean sea, while (7*Z*)-heptadecene was 12 % of the oil. Because unsaturated fatty aldehdes may give stronger off odors than the corresponding structural alcohols (43) and the hydrocarbons have weak odors, an odor of *Ulva pertusa* is likely to be attributable to (8*Z*, 11*Z*, 14*Z*)-heptadecatrienal. The synthesized aldehyde has a pleasant odor associated with fresh cucumber-like seaweed, similar to that of (8*Z*, 11*Z*)-heptadecadienal (41). Authentic C_{16}-unsaturated aldehydes have a chicken-fatty odor with citrus note (45).

The large difference in the volatile constitution of *U. pertusa* (20, 25, 28)

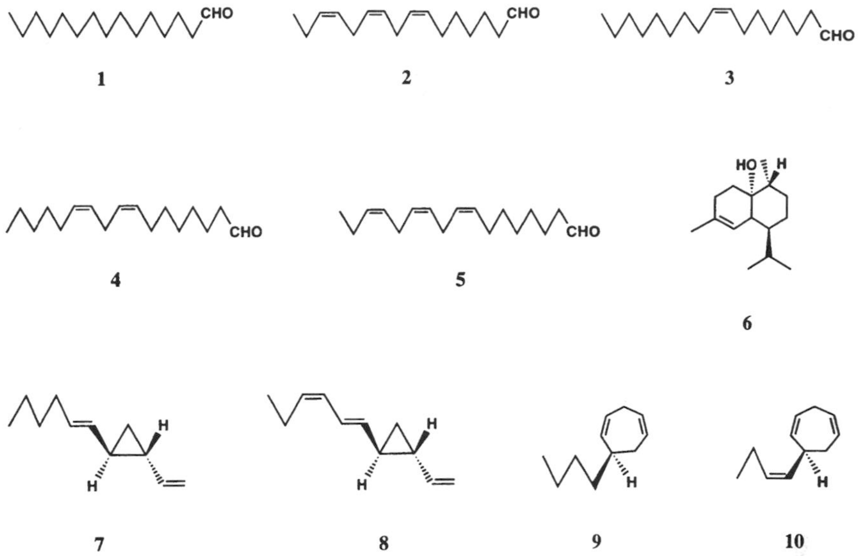

Scheme 2 Chemical structures of seaweed flavors. 1. Pentadecanal; 2. (7Z, 10Z, 13Z)-hexadecatrienal; 3. (8Z)-heptadecenal; 4. (8Z, 11Z)-heptadecadienal; 5. (8Z, 11Z, 14Z)-heptadecatrienal; 6. cubenol; 7. dictyopterene A; 8. dictyopterene B; 9. dictyopterene C′; 10. dictyopterene D′.

is likely responsible for its aroma being affected by seasonal, regional, and other environmental conditions.

D. Long-Chain Aldehyde-Forming Enzyme

Each long chain (un)saturated C_{17} aldehyde found in seaweeds, such as pentadecanal, (8Z)-heptadecenal, (8Z, 11Z)-heptadecadienal, and (8Z, 11Z, 14Z)-heptadecatrienal, increased during incubation of a homogenized mixture of *Ulva pertusa* with the corresponding (un)saturated fatty acid having one additional C atom (20). These aldehydes ($C_{17}CHO$) were elucidated to be enzymatically formed from the fatty acids ($C_{18}COOH$); stearic acid (octadecanoic acid), oleic acid ([9Z]-octadecenoic acid), linoleic acid ([9Z, 12Z]-octadecadienoic acid), and α-linolenic acid ([9Z, 12Z, 15Z]-octadecatrienoic acid), respectively (Scheme 3) (17, 46). Thus long-chain aldehyde-forming enzyme (LAFE) catalyzes a biotransformation of aldehydes (C_nCHO) from fatty acids ($C_{n+1}COOH$). The LAFE isolated from *U. pertusa* has an optimum pH of 8.5–9.0. The LAFE is a specific seaweed enzyme playing an important role in the formation of unsaturated fatty aldehydes, because, in land plants, these

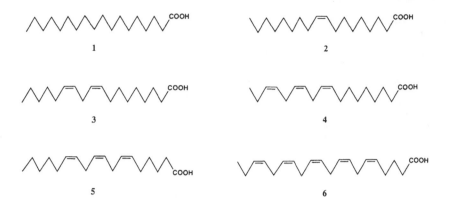

Scheme 3 Chemical structures of long-chain (un)saturated fatty acids. 1. Stearic acid; 2. oleic acid; 3. linoleic acid; 4. α-linolenic acid; 5. γ-linolenic acid; 6. (5Z, 8Z, 11Z, 14Z, 17Z)-eicosapentanoic acid (EPA).

Table 3 Long-Chain Aldehydes in Essential Oils from Green, Brown, and Red Algae

Aldehydes	Green alga		Brown alga		Red alga	
	UP	ES	HF	LJ	CC	GA
(8Z, 11Z, 14Z)-Heptadecatrienal	+ + +	+ +	+	-	-	-
(8Z, 11Z)-Heptadecadienal	+ +	+ +	+	-	+	-
(8Z)-Heptadecenal	+ +	+ +	-	-	+	-
Pentadecanal	+ + +	+ + +	+	+	+	+
Tetradecanal	+	+	-	-	-	-
Tridecanal	+	+ +	-	-	+	-

UP, *Ulva pertusa;* ES, *Enteromorpha* sp; HF, *Hizikia fusiformis;* LJ, *Laminaria japonica;* CC, *Chondria crassicaulis;* GA, *Gracilaria asiatica.*
-, Not detected; +, detected; + + +, major compound.
Source: Ref. 46.

aldehydes are trace compounds considered to be unstable intermediates in a pathway of α-oxidation of fatty acids:

$$C_{n+1}COOH \rightarrow C_n COOH \rightarrow C_{n-1}COOH$$

As Tables 3 and 4 show, large quantities of the long chain aldehydes are found in seaweeds, especially from the greens (46), that have a large amount of LAFE activity (17, 46). The biosynthesis of the aldehydes from fatty acids is shown in Figure 1 (17), where the LAFE exhibits its characteristic property of

Table 4 Distribution of Long-Chain Aldehyde-Forming Enzyme Activity in Marine Algae

	Alga	Activity	Alga	Activity
Green algae				
Ulvales	*Ulva pertusa*	+++	*Enteromorpha* sp.	+++
Siphonales	*Codium fragile*	-		
Brown algae				
Dictyotaceae	*Dictyota dichotoma*	++	*Dilophus okamurae*	+
	Dictyopteris divaricata	++	*Dictyopteris prolifera*	+
	Pachydictyon coriaceum	+	*Padina arborescens*	+
Scytosiphonaceae	*Scytosiphon lomentaria*	+	*Endarachne binghamiae*	+
	Colpomenia bullosa	+		
Asperococcaceae	*Myelophycus simplex*	-		
Leathesiaceae	*Leathesia difformis*	+		
Ralfsiaceae	*Analipus japonicus*	+		
Ishigeaceae	*Ishige okamurae*	+		
Laminariaceae	*Laminaria japonica*	-	*Laminaria angustata*	-
Alariaceae	*Alaria crassifolia*	+	*Undaria pinnatifida*	-
Fucaceae	*Pelvetia wrightii*	+		
Cystoseiraceae	*Cystoseira hakodatensis*	-		
Sargassaceae	*Hizikia fusiformis*	+	*Sargassum horneri*	+
	Sargassum thunbergii	+	*Sargassum miyabei*	+
	Sargassum nigrifolium	+	*Sargassum confusum*	+
Red algae				
Gracilariaceae	*Gracilaria asiatica*	++		
Gigartinaceae	*Rhodoglossum japonica*	-	*Chondrus yendoi*	+
	Chondrus ocellatus	+		
Petrocelidaceae	*Mastocarpus pacificus*	+		
Nemastomaceae	*Schizymenia dubyi*	-		
Halymeniaceae	*Grateloupia filicina*	+		
Tichocarpaceae	*Tichocarpus crinitus*	-		
Dumontiaceae	*Neodilsea yendoana*	+		
Rhodomelaceae	*Neorhodomela larix*	-	*Rhodomela teres*	-
	Odonthalia corymbifera	-	*Chondria crassicaulis*	++
Gelidiaceae	*Gelidium elegans*	+		
Bonnemaisoniaceae	*Delisea japonica*	+		

-, Not detected; +, detected; ++, strong; +++, very strong.
Source: Refs. 17, 46.

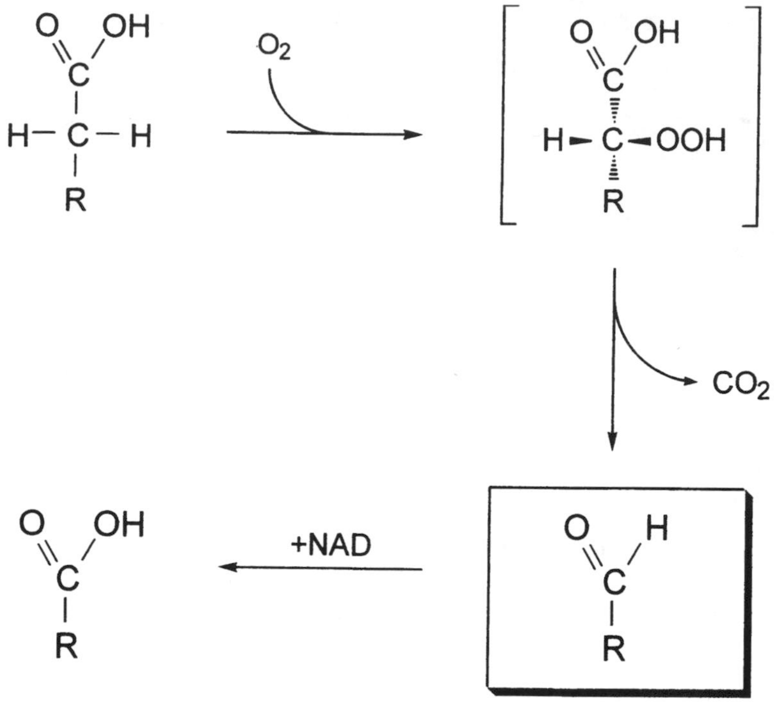

Figure 1 A possible biogeneration mechanism for long-chain aldehydes from fatty acids in seaweeds. (Modified from Ref. 17.)

a chiral selectivity for an intermediate, (2R)-hydroperoxy-fatty acid. The chiral selectivity is quite the opposite of lipoxygenase (LOX) giving (S)-hydroperoxy-fatty acid. Almost all the LOX, obtained from terrestrial plant and animal and also fish tissues, produced (S)-formed intermediates, as reviewed by Hsieh (47). Another characteristic property of LAFE is to alter the acid moiety of (2R)-hydroperoxy-fatty acid into the aldehyde moiety with decarboxylation. Only alteration occurs, with no cleavage.

E. Lipoxygenase and Hydroperoxide Lyase

In addition to LAFE, LOX also forms relatively low fatty aldehydes, such as (3Z)-nonenal, (2E)-nonenal, and hexanal, in seaweeds. LOX is a dioxygenase that oxidizes polyunsaturated fatty acids (PA) with two oxygen atoms and then converts PA into hydroperoxy-fatty acids called oxylipins (see Chap. 11). A LOX

Table 5 Summary of Bioactive Oxylipins Discovered in Red and Brown Seaweeds

Species	Bioactive oxylipins
Rhodophyta	
Gracilariopsis lemaneiformis	(12*S*)-HETE, (12*S*)-HEPE, (12*R*, 13*S*)-diHETE, (12*R*, 13*S*)-diHEPE
Gracilaria lichenoides	Prostaglandins PGE_2, $PGF_{2\alpha}$
G. verrucosa	Prostaglandin E_2
Rhodymenia pertusa	(6*E*)-Leukotriene B_4, (5*R*, 6*S*)-diHETE, (5*R*, 6*S*)-diHEPE
Cottoniella filamentosa	Hepoxilin B_3
Polyneura latissima	9-HETE
Phaeophyta	
Ecklonia stolonifera	Ecklonialactones A & B (prostanoid)
Laminaria setchellii	(15*S*)-HETE, (15*S*)-HEPE, (13*S*)-HOTE, (13*S*)-HODTA
L. sinclairii	(15*S*)-HETE, (15*S*)-HEPE, (13*S*)-HOTE, (13*S*)-HODTA
L. saccharina	(15*S*)-HETE, (15*S*)-HEPE, (13*S*)-HOTE, (13*S*)-HODTA
Cymathere triplicata	Cymathere ether A (prostanoid)

HETE, hydroxyeicosatetraenoic acid; HEPE, hydroxyeicosapentaenoic acid; HOTE, hydroxyoctade-
catrienoic acid; HODTA, hydroxyoctadecatetraenoic acid.
Source: Ref. 52.

that specifically oxidizes the 12-unsaturated position of PA (12-LOX) was recently found in the red seaweed, *Gracilariopsis lemaneiformis* (48), although it
is recognized to be widely distributed among fish and terrestrial organisms (47).
Novel (5*Z*, 8*Z*, 10*E*, 12*E*, 14*Z*)-eicosapentaenoic acid was found as a major enzymatic product of a mixture of arachidonic acid ([5*Z*, 8*Z*, 11*Z*, 14*Z*]-eicosate-
traenoic acid) in the red seaweed, *Bossiella orbigniana* (49, 50). These findings
suggested that LOX was rare or inactive in seaweed so that PA was little decomposed. Also, the 12- and 15-LOXs active with archidonic acid have recently been
found in *Enteromorpha intestinalis* (51). Oxylipins discovered thus far are listed
in Table 5 (52), where oxylipins have S-formed structure. The variety of oxylipins was much less than that found in terrestrial organisms. Incidentally, oxylipins
having (13*S*)- and (15*S*)-forms are mainly derived from linoleic and linolenic
acids, respectively (52).

As Figure 2 shows, hydroperoxidase lyase (HOL) following LOX produced only (2*E*)-nonenal on the pathway from 12-hydroperoxyeicosatetraenoic
acid (12-HPETE). Small quantities of the aldehyde were found in *Ulva pertusa*
(25) and five kelps (21).

As described above, inactivity of both LOX and HOL maintains PA almost
intact. This property should be of benefit to human health, although it probably
gives seaweeds a monotonous tone of flavor.

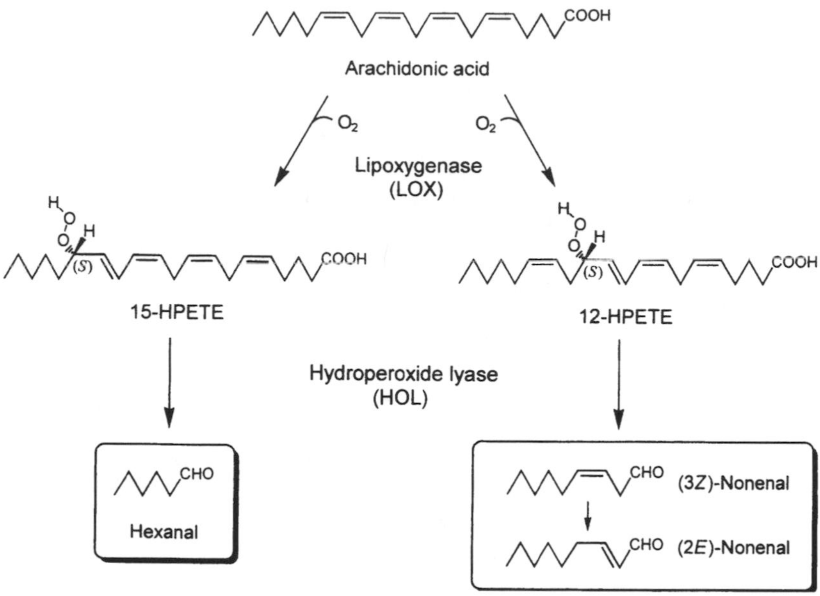

Figure 2 A possible mechanism for biogenesis of C_6 - and C_9 - aldehydes in seaweeds. (From Ref. 16.)

F. Bioflavor Produced by Seaweed Cell Cultures

The genuine flavor of intact seaweed, which is not contaminated by the products of microorganisms adhering to the thalli, was studied using the protoplasts of *U. pertusa* (53). The greenish and spherical protoplasts (Fig. 3) were prepared by digestion with combinations of polysaccharide lyases. During cultivation in a large mass for 3 months, the bacteria-free thalli were regenerated from the protoplasts, showing themselves in two different morphologies: tubular and normal thalli (54). The form of the former was quite different from that of the field fronds. The essential oil of both types of thalli was analyzed by GC-MS. Four aldehydes with (7Z, 10Z, 13Z)-hexadecatrienal were identified as the main components (Table 6). There was no large difference in volatile constitution between the thalli and the field fronds. Also, LAFE activity on (un)saturated fatty acids showed no difference between the cultured thalli and the field fronds, except for oleic acid (Table 7). The reason for the different results obtained with oleic acid has not been resolved. As Tables 6 and 7 show, these aldehydes were confirmed to be derived from the LAFE of the seaweed, not from the activity of the associated microorganisms (53).

The flavor of live-cultured thalli from protoplasts and field-cultured fronds

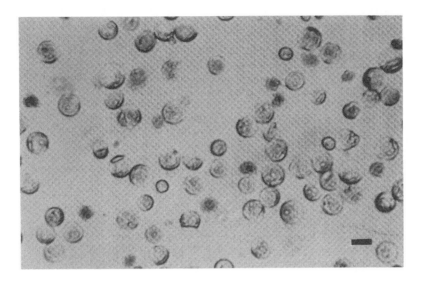

Figure 3   Protoplasts isolated from *Ulva pertusa.* Scale bar = 25 μ m.

was determined using a closed looping head space device shown in Figure 4 (55). Both gave almost the same data as shown in Table 6. Also, volatile compounds released from the protoplasts of the brown *Dictyopteris prolifera* were similar to those from the field fronds (56, 57). These findings indicated that adhesive microorganisms have little influence on the flavors.

For high-yield production of bioflavors, LAFE activity was enhanced by the immobilization of cultured cells of *U. pertusa* (55). Figure 5 shows the immobilized cells in polymer matrices of calcium alginate beads (58). The long-chain aldehydes, secreted from the alginate-embedded cells into the liquid medium, increased more rapidly than those from free living cells during immobilization for 50 days (Table 8). Thus, the polymer-embedding cells functioned to stimulate the cells to produce the aldehydes as an elicitor (59).

III. BROWN SEAWEED FLAVOR

Almost all the fresh brown seaweeds, except for *Dictyotaceae,* have a moderate and common seaweedlike or kelplike odor. The organoleptic similarity seems to be based on the small quantities of flavor components. As a characteristic compound for the browns, especially for kelps, iodine is found in large amounts (see Sec. I). Iodine (38 ppm/frond), arising in the red *Dilsea edulis* at low tide in the

Table 6 Volatile Compounds Identified in the Essential Oils from
Ocean Culture of *Ulva pertusa*

	Peak area (%)	
Compounds	Ocean culture	Field fronds
Aldehydes		
(8Z, 11Z, 14Z)-Heptadecatrienal	27.52	24.11
(8Z, 11Z)-Heptadecadienal	2.16	2.16
(8Z)-Heptadecenal	5.27	3.33
(7Z, 10Z, 13Z)-Hexadecatrienal	2.06	0.83
(6Z, 9Z, 12Z)-Pentadecatrienal	—	1.44
Pentadecanal	5.60	11.46
Tetradecanal	—	1.55
Tridecanal	—	1.67
(2E)-Nonenal	—	0.15
(2E)-Octenal	—	0.29
Alcohols		
Phytol	1.55	2.55
Octadecanol	6.44	—
α-Cadinol	2.46	—
Cubenol	1.46	6.93
α-Terpineol	1.66	—
2-Ethylhexanol	6.62	—
Ketones		
6,10,14-Trimethylpentadecan-2-one	0.67	0.62
α-Ionone	1.39	0.64
β-Ionone	—	0.35
Lactones		
Dihydroactinidiolide	1.61	0.25
2,3-Dimethyl-2-nonen-4-olide	—	1.00
Carboxylic acids		
Palmitic acid	2.14	0.60
Myristic acid	1.07	—
Ester		
(S)-3-Acetoxy-1,5-undecadiene	—	0.24
Hydrocarbons		
(7Z)-Heptadecene	10.74	26.79
Tetradecane	—	0.10
Limonene	0.89	—
Sulfur compounds		
Benzothiazole	0.49	0.61
Dimethyl trisulfide	1.20	—
Halide		
Chlorobenzene	0.91	—

Source: Ref. 53.

Table 7 Substrate specificity for Long-Chain
Aldehyde-Forming Activity from *Ulva pertusa*

Substrate	Relative activity (%)	
	Ocean culture	Field fronds
Stearic acid	0	0
Oleic acid	139	6.5
Linoleic acid	100	100
α-Linolenic acid	81.4	89.0
γ-Linolenic acid	11.3	15.5

Source: Ref. 53.

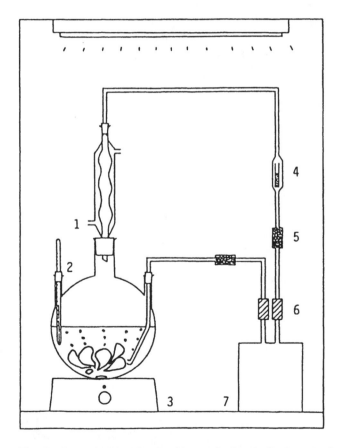

Figure 4 Closed looping head space device. 1. Cool water; 2. thermometer; 3. magnetic stirrer; 4. charcoal; 5. cotton plug; 6. sterilized filter; 7. pump. (From Ref. 55.)

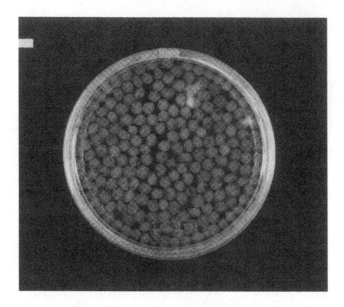

Figure 5 Immobilized cells regenerated from protoplasts of *Ulva pertusa*. Scale bar = 0.6 cm. (From Ref. 58.)

Table 8 High Production of Long-Chain Aldehydes by Immobilized Cells from *Ulva pertusa*

Cells	Products (nmol / g fr. wt.)		
	HD	HDD	HDT
Cultured cells	0.24	13.6	0.51
Immobilized cells	7.31	27.4	9.23

HD, (8Z)-heptadecenal; HDD, (8Z, 11Z)-heptadecadienal; HDT, (8Z, 11Z, 14Z)-heptadecatrienal.
Source: Ref. 55.

sun, emits an odor of sea air (31). Therefore, many kelps, on dying, probably give off a similar odor of iodine (30).

The volatile components of fresh kelps, including wakame and kombu, were principally characterized by cubenol and β-ionone, together with some unsaturated carbonyls that were are common in the greens (21): (2*E*)-nonenal, (2*E*,6*Z*)-nonadienal, (2*E*)-octenal, (2*E*)-octenol, and (2*E*,4*E*)-decadienal. A sesquiterpene, cubenol (Scheme 2), was reported to have a moderate odor of kelp with hay note and its threshold value is in the range of 100–250 ppm (21). It is remarkable that only two compounds were found in wakame: cubenol reaching to 88% of the essential oil and β-ionone reaching 2%. Despite the high odor threshold, the sniffing tests evaluated that no compounds other than cubenol showed a characteristic kelplike odor (21). The weak odor based on the simple volatile constituents supports the fact that only wakame has been greatly eaten raw in Japan, probably because of its mild flavor. Other kelps are usually eaten after being dried. The recent home consumption of dried wakame products follows a trend towards low aroma in these products.

Cubenol is also widely found in other fresh browns. For example, *Dictyopteris divaricata* contained this compound as over 90% of the essential oil, and also *Fucus evanescens*, 82%; *Hydrolathus alathratus*, 50%; *Scytosiphon lamentarius*, 25%; *Pelvetia wrightii*, 19%; *Dictyopteris undalata* 17%; *Padina arborescens*, 16%; *Diloplus okamurai*, 15%; and five Sargassaceae species 6–13% (22, 60). In the Mediterranean, the volatiles of brown *Cystoseira stricta* growing abundantly in the Aegean Sea consist mainly of cubenol (31%) followed by hexanol (12%) and octanol (9%) (27). Furthermore, cubenol was found in several reds and greens in amounts less than 10% (60). Thus, it is distributed widely; however, its physiological roles along with the mechanism of biosynthesis or accumulation are not well understood.

Compounds giving an "ocean smell" associated with a clean and rocky seashore were isolated from some genera of *Dictyotacea*, especially from *Dictyopteris*, and investigated in detail (16, 18, 19, 61–65). It is interesting that the characteristic flavors of *Dictyopteris plagiogramma* and *D. australis* are highly favored in Hawaii, and the seaweeds are used for a spice instead of pepper and sage (5, 61). The flavor compounds were identified to be dictyopterenes A, B, C', and D' (Scheme 2), of which cyclopropane- or cycloheptane-rings in the odd C_{11}-numbered chemical structures have been rare compared to many even-numbered compounds in terrestrial plants. Dictyopterenes A and B have more characteristic flavors associated with seashore than dictyopterenes C' and D' (66). Those compounds were also found as major volatile components of *D. prolifera* growing along the coast of Japan (22). Because of the oval sexual attractants, dictyopterenes may possibly be used for pollination or fertilizing sea creatures. The possible enzymatic formation of dictyopterene A from linoleic acid through (3Z,6Z,9Z)-dodecatrienoic acid by α- and β-oxidations was proposed by Kajiwara et al. (19).

IV. RED SEAWEED FLAVOR

Some essential oils obtained from fresh red seaweeds have some offensive notes and show clear differences from both kelplike and green vegetable-like odors. Little information concerning red seaweed flavor has been published. However, the flavor features can allow us to classify the reds into four types: nori-type having DMS in large quantity; *Asparagopsis*-type having a variety of halogen-containing compounds in abundance; *Chondria*-type having several heterocyclic sulfur-containing compounds; and miscellaneous.

A. Nori

The main nori product is a dried sheet used in Japan for wrapping food. This product is convenient and edible with an appetizing flavor. Fresh laver has three principal sulfur compounds: i.e., DMS, H_2S, and CH_3SH (0.4, 0.1, and 0.01 ppm, respectively). The compounds increase during mechanical drying (3.1, 13.8, and 0.01 ppm) and subsequent toasting (7.5, 131.4, and 3.9 ppm) during manufacture of nori sheets (36). These values are far above the corresponding odor threshold values in water: DMS: 0.3, 1, or 10 ppb (34); H_2S: 10 ppb (34); CH_3SH: 0.02 ppb (34). A stepwise increase in H_2S occurs, while absolute DMS content does not change during the process. In addition to DMS, H_2S as well as CH_3SH was shown to be a characteristic flavor compound for nori (36). The precursor of H_2S in nori has not been ascertained. Consumers in Japan prefer a toasted and seasoned sheet over the unseasoned dried sheet (36).

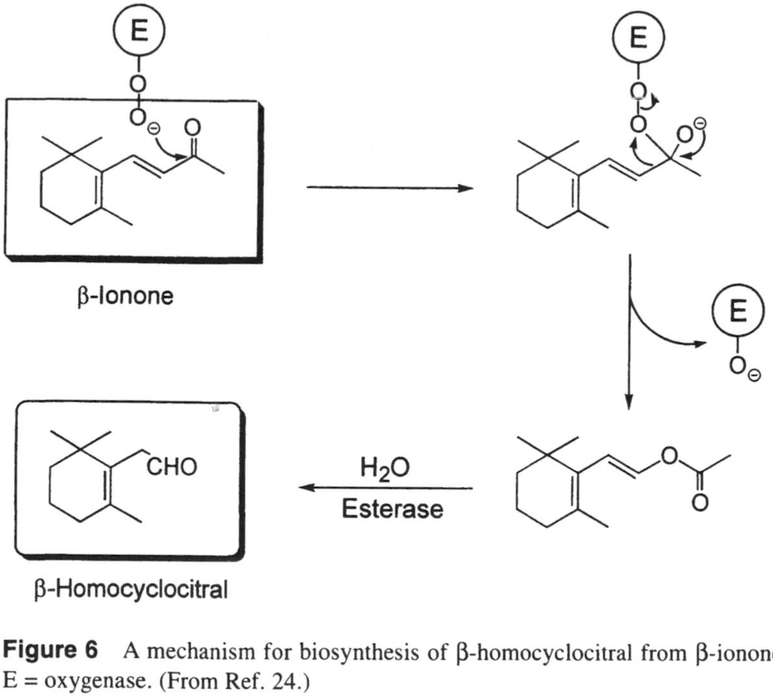

Figure 6   A mechanism for biosynthesis of β-homocyclocitral from β-ionone in nori. E = oxygenase. (From Ref. 24.)

Nor-carotenoid degradation products, such as 2,2,6-trimethylcyclohexanone, β-cyclocitral, β-ionone, and dihydroactinidiolide, have been identified as flavor components of nori (24). The formation of β-homocyclocitral was proposed in the biotransformation of β-ionone degradation by enzymatic actions (Fig. 6). The oxygenase-type enzyme similar to LOX is likely responsible for the degradation of the ionone side chain. The intermediates were probably the peroxide and its enol ester. The ester was hydrolyzed by an esterase and then formed β-homocyclocitral. The enzymatic oxygen insertion is found in many natural products (67).

Long-chain unsaturated aldehydes, such as (8Z,11Z)-heptadecadienal and (8Z)-heptadecenal, both having cucumber-like (41) or seaweed-like odors (25), were detected in nori in smaller amounts than found in the greens and the browns (19). In addition to DMS and the nor-carotenoid products, small amounts of the long-chain aldehydes possibly contribute to the nori flavor.

B. *Asparagopsis* and Analogues

More than 400 halogen-containing compounds found in the reds were summarized by Faulkner (8–14). In his reviews, the reds contained the greatest number of halogen-containing compounds among the seaweeds.

In Hawaii, *Asparagopsis taxiformis* has been highly prized for its aroma and flavor (5). From this red, Burreson et al. (32) identified 42 halogen-containing compounds (halogen = Cl, Br, and I). Among them, tribromomethane ($CHBr_3$) was the major volatile component, representing 80% of the essential oil, followed by dibromoiodomethane ($CHBr_2I$) 5%, and others less than 2%. The compound was most likely formed nonenzymatically at pH 8 from 1,1,1-tribromoacetone (CH_3COCBr_3). The latter is readily produced from bromoacetone (CH_3COCH_2Br) when treated with bromine in an aqueous $NaHCO_3$. However, in an acidic medium, different components, such as $BrCH_2COCH_2Br$ and $BrCH_2COCHBr_2$, are produced. These findings indicate that pH greatly influences volatile formation. Seaweeds having large quantities of halogen-containing compounds may be poisonous to eat; however, there has never been a single case of illness to researchers' knowledge, as described by Burreson et al. (32). The preponderant occurrence of halogen-containing compounds in red seaweed suggests that the plants have a specific ability to concentrate the haloforms in the cells.

Some bromoperoxidases that brominate various compounds in the presence of Br^- and H_2O_2 have been found from coralline algae by Yamada et al. (68). The enzymes are specific for I^- and Br^-, and do not act on Cl^- and F^-. Of 42 seaweeds tested, including *Ulva pertusa*, *Enteromorpha intestinalis*, wakame, kombu, and nori, bromoperoxidase activities were found in half of the 21 reds, none of 15 browns, and 1 of 6 greens. Among the 12 positive seaweeds, 3 edible seaweeds, *Gelidium amansii*, *Hypnea chararoides*, and *Codium fragile*, were found to have slight activities.

C. *Chondria* and Analogues

Several heterocyclic sulfur-containing compounds were found in *Chondria califor-nica* growing in the Gulf of California by Wratten and Faulkner (69). A strong sulfur odor of the red seaweed was caused by seven sulfur-heterocyclics: 1,2,4-trithiolane, 1,2,4,6-tetrathiepane, 1,2,3,5,6-pentathiepane, and some cyclic sulfoxides. The an-tibiotic activity of the red was confirmed with a mixture of the compounds. These compounds and the antibiotic activity also occurred in *C. californica* specimens from another location, but not in two analogous *Chondria* species (69).

The red seaweeds contain a large amount of sulfur in the form of sulfated polysaccharides, such as carrageenan for Gigartinales, porphyran for nori, furo-ran for *Gloiopeltis,* furcellarin for Danish agar (6), and fucoidin for the browns, especially for kelps. Also, the sulfur occurs in the forms of sulfur-containing amino acids such as cysteine, and aminosulfuric acids such as taurine. A sulfury-lase, sulfate-adenylyltransferase, which is the initial enzyme in the inorganic sul-fate assimilation pathway, is widely distributed in seaweeds. The enzyme activity was greater in the reds (*Porphyra, Chondria, Chondurus,* and *Gloiopeltis*) than the greens, browns and a land plant, spinach (70). The reaction is irreversible, resulting in accumulation of only sulfate.

Large amounts of sulfuric acid were released from the brown *Desmarestia* after death, changing the seaweed color to blue. Simultaneously, an offensive odor was emitted (71). These phenomena seem to be caused by enzymatic reac-tions; however, little information is available on the enzyme(s). Almost all the reds contain large quantities of sulfated polysaccharides; therefore, a small re-lease of sulfuric acid is also possible. The acid formed may also complicate the flavor, because flavor is influenced by pH (72).

D. Miscellaneous

Many volatile compounds of seaweed, especially the reds, have yet to be identi-fied. This is probably due to incomplete MS data on many compounds. In our previous analyses of some red seaweed essential oils, resultant data were very poor although many compounds appeared as peaks in GC charts. For example, only 3-hydroxy-1,(5Z)-octadiene was identified in *Chondrus occellatus;* only α- and β-ionones *Gelidium amansii,* while several compounds were identified in *Chondria crassicaulis:* (2E,6Z)-nonadienal, (7Z)-heptadecene, β-ionone, pen-tadecane, cubenol, and C_{14}- and C_{16} acids.

V. CONCLUSIONS

The major flavor compound common in all three groups of seaweed is DMS. Green seaweed flavor is mostly due to DMS and a group of unsaturated fatty aldehydes: (8Z, 11Z, 14Z)-heptadecatrienal. Both are enzymatically formed; the former is derived from DMPT by DMPT-dethiomethylase, and the latter from PA by LAFE. Green seaweed flavor is most similar to terrestrial plant flavor.

Brown seaweed flavor is not due as much to DMS and aldehydes as in the greens. β-Ionone and cubenol probably contribute to kelp flavor. Dictyopterenes, which are sexual attractant compounds, characterize the flavor of Dictyotales.

Red seaweed flavor is least influenced by aldehydes among the seaweeds. Nori flavor is mainly due to DMS and enzymatic nor-carotenoid products and the aldehydes. The flavor of many reds remains unknown since enzymatic products of PA have not been found.

There are some differences in the quantity and quality of flavor constituents in the published data. The multiplicity of flavors possibly results from the properties of seaweed: rapid growth in the sea, short life of 1 or 2 years, seasonal and regional differences, postmortem changes, and so on. Multiplicity is found not only in the flavor but also in the enzyme activities (e.g., for DMPT degradation). The differences reflect the highly adaptable nature of seaweed to environmental conditions. Another property of seaweed is to retain PA in the cells: seaweeds allow LAFE to alter PA to long-chain unsaturated fatty aldehydes, as can be seen in the greens. However, LAFE or LOX does not decompose PA. This behavior seems to be basically different from that of terrestrial plants. The high adaptability to the environment and the lack of adaptability to PA decomposition are considered as characteristic properties of seaweed.

REFERENCES

1. A Honma. The production of seaweed. In: M Abe, A Honma, Y Yamamoto, eds. Modern Encyclopedia of Fish. Tokyo: NTS, 1997, pp 38–40 (in Japanese).
2. H Tokuda. Seaweed resources and production in the world. In: H Tokuda, M Ohno, H Ogawa, eds. The Resources and Cultivation of Seaweeds. Tokyo: Midori Shoboh, 1987, pp 69–98 (in Japanese).
3. S Ueda. Edible seaweed. In: S Ueda, K Iwamoto, A Miura, eds. Fishery Botany. Tokyo: Koseisha-koseikaku, 1963, pp 475–497 (in Japanese).
4. S Arasaki. Utilization of seaweed. In: S Arasaki ed. How to Know the Seaweeds of Japan and Its Vicinity. Tokyo: Hokuryukan, 1981, pp 187–192 (in Japanese).
5. IA Abbott. Limu. An Ethnobotanical Study of Some Hawaiian Seaweeds, 3rd ed. Hawaii: Pacific Tropical Botanical Garden, 1984, pp 1–35.
6. H Tokuda. Utilization of seaweed. In: H Tokuda, M Ohno, H Ogawa, eds. The Resources and Cultivation of Seaweeds. Tokyo: Midori Shoboh, 1987, pp 37–66 (in Japanese).
7. United States Food and Drug Administration. GRAS status of certain red and brown algae and their extractives. Fed Reg 47:47373–47376, 1982.

8. DJ Faulkner. Marine natural products: metabolites of marine algae and herbivorous marine molluscs. Nat Prod Rep 1:251–280, 1984.
9. DJ Faulkner. Marine natural products. Nat Prod Rep 3:1–33, 1986.
10. DJ Faulkner. Marine natural products. Nat Prod Rep 4:539–576, 1987.
11. DJ Faulkner. Marine natural products. Nat Prod Rep 5:613–663, 1988.
12. DJ Faulkner. Marine natural products. Nat Prod Rep 7:269–309, 1990.
13. DJ Faulkner. Marine natural products. Nat Prod Rep 8:97–147, 1991.
14. DJ Faulkner. Marine natural products. Nat Prod Rep 9:323–364, 1992.
15. H Noda. Seaweed products. In: C Koizumi, ed. Odor of Marine Products. Tokyo: Koseisha-koseikaku, 1989, pp 83–92 (in Japanese).
16. T Kajiwara. Dynamic studies on bioflavor of seaweeds. Koryo No. 196: 61–70, 1997 (in Japanese with English summary).
17. T Kajiwara, K Matsui, Y Akakabe. Biogeneration of volatile compounds via oxylipins in edible seaweeds. In: GR Takeoka, R Teranishi, PJ Williams, A Kobayashi, eds. Biotechnology for Improved Foods and Flavors. ACS Symposium Series 637. Washington, DC: American Chemical Society, 1996, pp 146–166.
18. RE Moore. Volatile compounds from marine algae. Acc Chem Res 10:40–47, 1977.
19. T Kajiwara, K Kodama, A Hatanaka, K Matsui. Volatile compounds from Japanese marine brown algae. In: R Teranishi, RG Buttery, H Sugisawa, eds. Bioactive Volatile Compounds from Plants. ACS Symposium Series 525. Washington, DC: American Chemical Society, 1993, pp 103–120.
20. T Kajiwara, A Hatanaka, T Kawai, M Ishihara, T Tuneya. Long chain aldehydes in the green marine algae Ulvaceae. Nippon Suisan Gakk 53:1901, 1987.
21. T Kajiwara, A Hatanaka, T Kawai, M Ishihara, T Tsuneya. Study of flavor compounds of essential oil extracts from edible Japanese kelps. J Food Sci 53:960–962, 1988.
22. T Kajiwara, A Hatanaka, Y Tanaka, T Kawai, M Ishihara, T Tsuneya, T Fujimura. Volatile constituents from marine brown algae of Japanese *Dictyopteris*. Phytochemistry 28:636–639, 1989.
23. T Kajiwara, A Hatanaka, K Kodama, S Ochi, T Fujimura. Dictyopterenes from three Japanese brown algae. Phytochemistry 30:1805–1807, 1991.
24. I Flament, G Ohloff. Volatile constituents of algae. Odoriferous constituents of seaweeds and structure of nor-terpenoids identified in Asakusa-nori flavour. In: J Adda, ed. Progress in Flavour Research 1984. Amsterdam: Elsevier Science Publishers BV, 1985, pp 281–300.
25. H Sugisawa, K Nakamura, H Tamura. The aroma profile of the volatiles in marine green algae (*Ulva pertusa*). Food Rev Int 6:573–589, 1990.
26. H Sakagami, J Iseda, T Fujimori, Y Hara, M Chihara. Volatile constituents in marine brown and red algae. Nippon Suisan Gakk 56:973–983, 1990 (in Japanese with English summary).
27. A Gally, N Yiannovits, C Poulos. The aroma volatiles from *Cystoseira stricta* var. *amentacea*. J Essent Oil Res 5:27–32, 1993.
28. H Tamura, H Nakamoto, R-H Yang, H Sugisawa. Characteristic aroma compounds in green algae (*Ulva pertusa*) volatiles. Nippon Shokuhin Kagaku Kogaku Kaishi 42:887–891, 1995 (in Japanese with English abstract).

29. T Katayama. Volatile constituents. In: RA Lewin ed. Physiology and Biochemistry of Algae. New York: Academic Press, 1962, pp 467–473.

30. Y Tsuchiya. Extractive and volatile components of seaweed. In: Y Tsuchiya ed. Fishery Chemistry. Tokyo: Koseisha-koseikaku, 1965, pp 389–399 (in Japanese).

31. Y Tsuchiya, H Baba. On the irritative odor emitted by *Dilsea edulis*. Tohoku J Agric Res 1:103–105, 1950.

32. BJ Burreson, RE Moore, PP Roller. Volatile halogen compounds in the alga *Asparagopsis taxiformis* (Rhodophyta). J Agric Food Chem 24:856–861, 1976.

33. H Iida, K Nakamura, T Tokunaga. Dimethyl sulfide and dimethyl–β–propiothetin in sea algae. Bull Jpn Soc Sci Fish 51:1145–1150, 1985 (in Japanese with English summary).

34. S Fors. Sensory properties of volatile Maillard reaction products and related compounds: a literature review. In: GR Waller, MS Feather, eds. The Maillard Reaction in Foods and Nutrition. ACS Symposium Series 215. Washington, DC: American Chemical Society, 1983, pp 185–286.

35. H Iida. Studies on the accumulation of dimethyl–β–propiothetin and the formation of dimethyl sulfide in aquatic organisms. Bull Tokai Reg Fish Res Lab No.124:35–111, 1988 (in Japanese with English abstract).

36. Y Osumi, K Harada, N Fukuda, H Amano, H Noda. Changes of volatile sulfur compounds of 'Nori' products, *Porphyra* spp. during storage. Nippon Suisan Gakk 56:599–605, 1990.

37. F Challenger, MI Simpson. Studies on biological methylation. Part XII. A precursor of the dimethyl sulphide evolved by *Polysiphonia fastigiata*. Dimethyl-2-carboxyethylsulphonium hydroxide and its salts. J Chem Soc 1591–1597, 1948.

38. GL Cantoni, DG Anderson. Enzymatic cleavage of dimethylpropiothetin by *Polysiphonia lanosa*. J Biol Chem 222:171–177, 1956.

39. RC Greene. Biosynthesis of dimethyl–β–propiothetin. J Biol Chem 237:2251–2254, 1962.

40. DB Josephson. Seafood. In: H Maarse ed. Volatile Compounds in Foods and Beverages. Food Science and Technology Series 44. New York: Marcel Dekker, Inc, 1991, pp 179–202.

41. T Kawai, T Kajiwara, M Ishihara, K Yonetani, H Horie, M Irie. Flavor composition. Japan Kokai Tokkyo Koho (Japan Patent) 63-233914, 1988 (Sept 29), pp 83–89 (in Japanese).

42. RC Lindsay. Fish flavors. Food Rev Int 6:437–455, 1990.

43. DB Josephson, RC Lindsay. Enzymic generation of volatile aroma compounds from fresh fish. In: TH Parliment, R Croteau, eds. Biogeneration of Aromas. ACS Symposium Series 317. Washington, DC: American Chemical Society, 1986, pp 201–219.

44. M Okano, T Aratani. Constituents in marine algae—I. Seasonal variation of sterol, hydrocarbon, fatty acid, and phytol fractions in *Ulva pertusa*. Bull Jpn Soc Sci Fish 45:389–393, 1979.

45. T Kawai, T Kajiwara, M Ishihara, K Yonetani, H Horie, M Irie. Flavor composition. Japan Kokai Tokkyo Koho (Japan patent) 63-233913, 1988 (Sept 29), pp 75–81 (in Japanese).

46. T Kajiwara, K Matsui, A Hatanaka, T Tomoi, T Fujimura, T Kawai. Distribution of an enzyme system producing seaweed flavor: conversion of fatty acids to long-chain aldehydes in seaweeds. J Appl Phycol 5:225–230, 1993.

47. RJ Hsieh. Contribution of lipoxygenase pathway to food flavors. In: C-T Ho, TG Hartman, eds. Lipids in Food Flavors. ACS Symposium Series 558. Washington, DC: American Chemical Society, 1994, pp 30–48.

48. MF Moghaddam, WH Gerwick. 12-Lipoxygenase activity in the red marine alga *Gracilariopsis lemaneiformis*. Phytochemistry 29:2457–2459, 1990.

49. JR Burgess, RI de la Rosa, RS Jacobs, A Butler. A new eicosapentaenoic acid formed from arachidonic acid in the coralline red algae *Bossiella orbigniana*. Lipids 26:162–165, 1991.

50. J Burgess. Arachidonic acid metabolism in calcifying red algae: a model for studying the role of eicosanoids in the biomineralization process in mammals. In: RS Jacobs, M de Carvalho, eds. Marine Pharmacology: Prospects for the 1990s. La Jolla, CA: The California Sea Grant College, University of California, 1991, pp 54–55.

51. J-M Kuo, A Hwang, HH Hsu, BS Pan. Preliminary identification of lipoxygenase in algae (*Enteromorpha intestinalis*) for aroma formation. J Agric Food Chem 44:2073–2077, 1996.

52. GL Rorrer, J Modrell, C Zhi, H-D Yoo, DN Nagle, WH Gerwick. Bioreactor seaweed cell culture for production of bioactive oxylipins. J Appl Phycol 7:187–198, 1995.

53. T Fujimura, T Kawai, M. Shiga, T. Kajiwara, A Hatanaka. Long-chain aldehyde production in thalli culture of the marine green alga *Ulva pertusa*. Phytochemistry 29:745–747, 1990.

54. T Fujimura, T Kawai, M. Shiga, T. Kajiwara, A Hatanaka. Regeneration of protoplasts into complete thalli in the marine green alga *Ulva pertusa*. Nippon Suisan Gakk 55:1353–1359, 1989.

55. T Fujimura, T. Kajiwara. Production of bioflavor by regeneration from protoplasts of *Ulva pertusa* (Ulvales, Chlorophyta). Hydrobiologia 204/205:143–149, 1990.

56. T Fujimura, T Kawai, T. Kajiwara, Y Ishida. Volatile components in protoplasts isolated from the marine brown alga *Dictyopteris prolifera* (Dictyotales). Plant Tissue Culture Lett 11:34–39, 1994.

57. T Fujimura, T Kawai, T. Kajiwara, Y Ishida. Protoplast isolation in the marine brown alga *Dictyopteris prolifera* (Dictyotales). Plant Cell Rep 14:571–574, 1995.

58. T Fujimura, T Kawai, M. Shiga, T. Kajiwara, A Hatanaka. Preparation of immobilized living cells from a marine green alga *Ulva pertusa*. Nippon Suisan Gakk 55:2211, 1989.

59. AG Darvill, P Albersheim. Phytoalexins and their elicitors—a defense against microbial infection in plants. Ann Rev Plant Physiol 35:243–275, 1984.

60. T Kajiwara, A Hatanaka, M Ishihara, T Kawai, T Tsuneya. Distribution of cubenol in seaweeds. Abstracts, Meeting of the Japanese Society of Fisheries Science, Tokyo, 1986, pp 199.

61. RE Moore. Chemotaxis and the odor of seaweed. Lloydia 39:181–191, 1976.

62. K Yamada, H Tan, H Tatematsu. Isolation and structure of dictyoprolene, a possible precursor of various undecanes in brown algae from *Dictyopteris prolifera*. J C S Chem Commun 572–573, 1979.

63. W Boland, K Mertes. Synthetic routes to trace constituents of algal pheromone bouquets and other information-imitating substances. Helv Chim Acta 67:616–624, 1984.
64. W Boland, DG Müller. On the odor of the Mediterranean seaweed *Dictyopteris membranacea;* new C_{11} hydrocarbons from marine brown algae—III. Tetrahedron Lett 28:307–310, 1987.
65. T Kajiwara, K Kodama, A Hatanaka. Male-attracting substance in marine brown algae the genus *Dictyopteris*. Bull Jpn Soc Sci Fish 46:771–775, 1980.
66. T Kajiwara. Property of marine perfumes and possibility of application to cosmetics and toiletries. Fragrance J (Tokyo) No. 73:106–111, 1985 (in Japanese).
67. V Krasnobajew, D Helmlinger. Fermentation of fragrances: biotransformation of β–ionone by *Lasiodiplodia theobromae*. Helv Chim Acta 65:1590–1601, 1982.
68. H Yamada, N Itoh, S Murakami, Y Izumi. New bromoperoxidase from coralline algae that brominates phenol compounds. Agric Biol Chem 49:2961–2967, 1985.
69. SJ Wratten, DJ Faulkner. Cyclic polysulfides from the red alga *Chondria californica*. J Org Chem 41:2465–2467, 1976.
70. N Kanno, M Sato, Y Sato. 3′-Phosphoadenosine 5′-phosphosulfate synthesis by cell-free enzyme system of marine alga *Porphyra yezoensis*. Nippon Suisan Gakk 54:1063–1066, 1988.
71. M Chihara. Common Seaweeds of Japan in Color, 11th ed. Osaka: Hoikusha Publishing Co, Ltd, 1982, pp 35–36 (in Japanese).
72. T Kawai. Fish flavor. Crit Rev Food Sci Nutr 36:257–298, 1996.

15

Enzymes and Their Effects on Seafood Texture

Shann-Tzong Jiang
National Taiwan Ocean University, Keelung, Taiwan, Republic of China

I. INTRODUCTION

Thousands of enzymes have been found in animal muscles and organs. However, this chapter will only focus on those affecting the texture of fish muscle and processed seafood. Some proteinases, found to be involved in the texture changes of postmortem muscle and surimi-based product, will be reviewed and discussed. Transglutaminase (TGase), which is detected in fish muscles and surimi wash water, is also discussed because of its role in cooked surimi texture. Transglutaminases are discussed in further detail in Chapter 6.

II. TYPES AND DISTRIBUTION OF ENDOGENOUS SEAFOOD ENZYMES

Proteinases exist in the muscle fiber and cytoplasm, and in the extracellular matrix of the connective tissue surrounding the cells of fish muscle (1–4). Most of the proteinases are lysosomal and cytosolic enzymes, but some of them are in the sarcoplasm and associated with myofibrils or external to the cell in macrophages. After disintegration of tissues, the enzymes are found mainly in the sarcoplasmic protein fraction. However, the tissue enzymes may also originate from other sources, such as those in the muscle of Pacific hake infected by myxosporidian parasites (5).

There are two main proteinase groups that play important roles in the texture change of postmortem fish muscle. The lysosomal cathepsins and calpains are involved in the postmortem tenderization process and softening of fish gels. According to the type of active site, the endogenous proteinases of seafood can also be classified as serine, cysteine, aspartic, and metalloproteinase. Their activities are controlled by specific and endogenous inhibitors, activators, pH, and temperature of the environment. However, the proteolytic activity varies greatly among species (6) and with the harvesting season, gender maturation, spawning, and other variables. (7) (see also Chap. 1).

A. Cathepsins

Lysosomes are known to harbor about 13 cathepsins which play key roles in the protein turnover in vivo and in the postmortem rheological changes of fish muscle (8). Among these lysosomal enzymes, cathepsins B, D, H, L, L-like and X have been purified and characterized from fish and shellfish muscles (9–23). The properties of these proteinases will be discussed in the following sections. Recovery of catheptic enzymes from surimi wash water is discussed in chapter 23.

B. Calpains

A neutral calcium-activated proteinase, requiring cysteine for activity, was first identified in terrestrial animal tissues (24, 25). Since then, similar proteinases have been reported to be involved in degradation of myofibrillar and neurofilament proteins, activation of phosphorylase b kinase and protein kinase C, and transformation of steroid hormone-binding proteins. Names given to these proteinases include calcium-activated factor (CAF), kinase-activating factor (KAF), calcium-activated neutral proteinase (CANP) and calpain. More recently, calpain has been separated from its endogenous inhibitor. It has been found widely in mammalian, avian, and fish muscles (26–31). Calpain and calpastatin constitute an intracellular regulatory system responsible for the protein turnover (32).

C. Transglutaminase

Transglutaminase (TGase), which catalyzes the intermolecular ε-(γ-glutamyl)lysine crosslinkage, is widely distributed in various tissues and body fluids (33). It has been purified from various animal tissues or organs, such as liver (34) and hair follicle (35) of guinea pig, pig plasma (36), human epidermis (37), erythrocyte (38), plasma (39), and placenta (40), and also isolated from muscle and surimi of Alaska pollack (41). This enzyme is known to contribute to the gelation of fish muscle proteins (see Chap. 6).

III. CHARACTERISTICS OF CATHEPSINS B, L, L-LIKE, AND CYSTATIN

During the last decade, our understanding of cellular processes involving protein degradation has reached the level at which we can explain with more certainty the regulation of proteinase activities (16, 42–48). Among the four classes of proteinases, aspartic, serine, metallo, and cysteine proteinases, more recent attention has focused on the cysteine proteinases. Currently it is known that the lysosomal cathepsins B, H, L, L-like, S, and X are cysteine proteinases and papainlike enzymes (14, 15, 18–20). They are well-characterized proteins with known primary structures. The structure of rat cathepsin B has been determined by computer modeling (49) and x-ray crystallography (50). The overall folding pattern of this enzyme and the arrangement of the active-site residues are similar to the other cysteine proteinases such as papain and actinidin. During the last 10 years, it had also become evident that inhibitors of cysteine proteinases, named cystatins, are involved in the regulation of these enzymes (51–52). These inhibitors might protect the cells from inappropriate endogenous or external proteolysis and might be involved in the control mechanism responsible for intracellular or extracellular protein turnover. The cystatins are tight and reversible binding inhibitors of the papainlike cysteine proteinases (53). Proteinase inhibitors are further discuseed in Chap. 18.

A. Cathepsin B

Cathepsin B, a cysteine carboxypeptidase, was purified from mackerel dorsal muscle and characterized by Jiang et al. (14, 15). It had a molecular weight (MW) of 30 kDa and activity optimum at pH 6.0 for the hydrolysis of Z-Phe-Arg-MCA. It hydrolyzed various peptides and released tripeptides and dipeptides (16, 17). It can also attack myofibrillar proteins, degrade myosin heavy chain into fragments of 150 kDa and 170 kDa, and slightly hydrolyze actin and troponin T at pH 6.0. However, cathepsin B purified from chum salmon, with a mass of 28 kDa, pI of 4.9, and optimal pH at 5.7, was shown to hydrolyze myosin, connectin, and nebulin (54).

B. Cathepsin L

The cathepsin L from rat liver, with MW of 30 kDa and pI of 5.2, is an endopeptidase with a preference for hydrophobic amino acid residues Phe-Val, Leu-Val, Leu-Tyr, and Phe-Phe (55). The optimal pH of cathepsin L from salmon white muscle was 5.6 for the hydrolysis of Z-Phe-Arg-MCA (54, 56–61). The cathepsin L and cathepsin L-like enzyme purified from mackerel dorsal muscle are thiol-dependent proteinases and are inhibited by cystatin. Their molecular weights are 30 and 55 kDa and pH optima for the hydrolysis of Z-Phe-Arg-MCA are 5.0 and 5.5, respectively (14–15). Both proteinases lost activity rapidly at pH above the optimum value. The optimum temperatures are 50°C and 40°C, respectively, but they are inactivated at 65–70°C.

C. Cysteine Proteinase Inhibitors

1. Cystatins

Whitaker and his co-workers isolated and partially characterized a protein from chicken egg-white that could inhibit ficin and papain (62). Keilova and Tomasek (63, 64) further reported that this protein could also inhibit cathepsins B and C. Barrett (65) named this protein "cystatin," because of its distinctive inhibitory activity on cysteine proteinases.

Based on partial amino acid sequence homology, four members of the superfamily have thus far been characterized, including kininogens, stefins, cystatins, and fetuins, from vertebrate, insects, plants, and microorganisms (66–68). Cystatins are a family of cysteine proteinase inhibitors (65) and contain approximately 115 amino acids. They are mostly acidic and contain four conserved residues noted as the "QVVAG" sequence (69–70). Cystatins may be glycosylated and/or phosphorylated.

2. Kininogens

Proteinaceous kininogens are inhibitory against many cysteine proteinases (68, 71). There are three types of kininogens, designated as high-molecular-weight kininogen (H-kininogen) with a mass of 120,000; low-molecular-weight kininogen (L-kininogen) with molecular weight of 68,000; and T-kininogen (also called thiostatin) with mass of 68,000, in mammalian blood plasma (72). Among these kininogens, T-kininogen has only been detected in rat plasma. Kininogen consists of a heavy chain and a light chain. Both L- and H-kininogens are strong inhibitors of cathepsins B and L (73). To explain the inhibitory sites in the heavy chain of the kininogen molecule, L-kininogen was subjected to limited proteolysis with trypsin and divided into three functional domains (74). Among these fragments, only domain 1 has no inhibitory activity on cysteine proteinases; domain 2 had high inhibitory activity against m-calpain, papain, and cathepsin L; whereas domain 3 did not inhibit m-calpain, but inhibited papain and cathepsin L strongly (70).

IV. CHARACTERISTICS OF CALPAIN AND CALPASTATIN

A. Common Properties

The common properties of calpains from various sources are endopeptidase activity catalyzing limited proteolysis; existing principally in cytosol; requiring calcium and SH-reducing agent for activity; sensitivity to leupeptin and E-64; pH optimum ranging from 7.0 to 8.5 for the hydrolysis of casein; inability to hydrolyze various synthetic substrates; having a mass of approximately 110 kDa, with 72–82K and 25–30K subunits; and specifically inhibited by calpastatin.

A calpain, which was activated by 10^{-5}M calcium, was first found in canine cardiac muscle and subsequently in various tissues of vertebrate organisms. Calpains usually require 10^{-3}M calcium for full activity, which is much higher than that in the cell (75, 76). Cells contain two groups of calpains, μ- and m-calpains, which differ in calcium requirement. No evidence, thus far, is available to indicate whether μ- or m-calpains are interconvertible in vivo, although autolyzed chicken skeletal muscle m-calpain is activated at as low calcium concentrations as μ-calpain.

B. Calcium Requirement

Calpains (intracellular cysteine proteinases) are Ca-dependent. They interact with calpastatin in vivo and are easily autolyzed in the presence of calcium. So far, only m-calpain has been found in fish or shellfish. In the presence of calcium, but absence of substrate, the catalytic subunit (80 kDa) and regulatory subunit (30 kDa) are dissociated and further autolyzed into 78 and 18 kDa fragments, respectively. The requirement of calcium for proteolytic activity and membrane-binding ability decreases gradually during autolysis (77, 78). Although autolysis lowers the calcium requirement for both μ- and m-calpains (77, 78), the calcium required for autolysis is much higher than the calcium concentration in vivo (79). According to the membrane/activation theory (80, 81), the unautolyzed calpains are considered to be proenzymes. According to the study by Coolican and Hathaway (82), the addition of phosphatidylinositol decreases the calcium required for the half-maximal rate of autolysis. Limited proteolysis could activate proenzymes of calpains. The autolyzed calpains still remain associated with cell membranes and/or are released into the cytosol. Even in the presence of phosphatidylinositol, the calcium requirement for autolysis is at least 5–10 times higher than the physiological free calcium concentration. However, Cong et al. (79) indicated that unautolyzed m-calpain was an active protease, not a proenzyme. Kinetic studies further suggested that unautolyzed m-calpain was also an active protease, and easily autolyzed in the presence of calcium (83). Calpain autolysis could not be completely inhibited by excess calpastatin. Inhibitor studies showed that calpain could degrade calpastatin into 15 kDa fragments that retain inhibitory activity (84). Although calpains are considered to be one of the major proteolytic systems in protein degradation of postmortem muscle, the regulation of their activity in vivo is still unclear.

C. Calpain Autolysis

Leupeptin (1–20 mM), a specific inhibitor for calpain, substantially inhibited the autolysis of tilapia m-calpain in the presence of 1 mM calcium. However, the 80 kDa subunit was degraded into a minor 78 kDa fragment in the presence of leupeptin (24, 25, 27, 85–88). According to native polyacrylamide gel electrophoresis (PAGE) analysis, calpain is autolyzed into nine fragments after 15 min incubation at 20°C in the presence of 1.0 mM calcium, while it is autolyzed into three components in the presence of both 1.0 mM calcium and 20 mM leupeptin (27). Thus, with or without leupeptin, calpain might be dissociated and further aggregated in the presence of calcium.

The rate of tilapia calpain autolysis decreased when temperature was lowered at pHs ranging from 5.5 to 7.5 (27). However, autolysis was still observed even at 0°C. According to previous studies (29, 31), the activities of calcium-autolyzed and intact m-calpain of tilapia muscle still had 20% and 16% of their maximal activities, respectively, even at 0°C. The optimal pH for tilapia calpain was broad, and the maximal activity occurred at pH 7.5 (31). The effect of pH on the autolysis of calpain was very similar to that of pork calpain (87, 88).

D. Calpain and Postmortem Tenderization

Calpains are believed to participate in postmortem muscle tenderization (42, 89–91). The activity of these enzymes in the cell is controlled by calcium, phospholipids, calpastatin, and activators (80, 81, 92). Only m-calpain is contained in carp muscle, eggs, and erythrocytes (93–94), tilapia muscle (10, 26), lobster claw, and abdominal muscles. Four calpains with a mass of 310, 195, 125, and 59 kDa were purified from lobster.

Calpains from crustacean muscle completely hydrolyzes myofibrillar proteins, but those from vertebrate muscle only selectively cleave myofibrillar proteins into fragments (95, 96). Furthermore, calpains obtained from crustacean muscle appeared to be more stable than those from vertebrate muscles in the presence of calcium (24, 25, 81).

E. Calpastatin

Calpastatin is a collective name given to a family of calpain-specific, endogenous inhibitor proteins that is as widely distributed among mammalian and avian cells as calpain. It was first identified in 1978, from the soluble fractions of bovine cardiac muscle and rat liver homogenates. Until now, several calpastatin proteins from different sources have been purified. All of these calpastatins are extremely heat-stable, maintaining their inhibitory activity at 100°C for 5–10 min. Calpastatin is strictly calpain-specific and completely ineffective on trypsin, chymotrypsin, and papain. However, a calpastatin isolated from one tissue can inhibit calpain from other tissue of the same or even different animal species. It is a specific endogenous competitive inhibitor for calpain and coexists with calpains in the same cellular area. It possesses three or four repetitive inhibitory domains. Each domain may bind one calpain in the presence of calcium, but inhibitory activities of the repetitive domains are not identical. The autolysis of calpains cannot be completely inhibited by an excess amount of calpastatin. Only unautolyzed m-calpain requires more calcium for half-maximal binding ability to calpastatin than for half-maximal activity (97). According to Nakamara et al. (84), calpain could degrade calpastatin into 15 kDa fragments that still have inhibitory activity. However, the inhibitory mechanism of calpain by calpastatin is still unclear.

The inhibitory activity of calpastatin can only be observed in the presence of calcium, which is also essential for the assay of calpain (97). Calpain and calpastatin, which coexist in tissue homogenates, are readily separated from one another by either ion-exchange or gel chromatography, in the presence of a Ca-chelating agent such as EDTA. When a solution containing equivalent concentrations of calpain and calpastatin is gel-filtered in the presence of calcium, the complex no longer shows either enzymatic or inhibitory activities.

V. CHARACTERISTICS OF TRANSGLUTAMINASE

A. Animal Sources

Transglutaminase is widely distributed in various tissues and body fluids (33). It has been purified from liver (34) and hair follicle (35) of guinea pig, rabbit liver (98), epidermal tissue (37), and erythrocytes (38) of human, walleye pollack liver (99), Japanese oyster (100), and also isolated from muscle and surimi of Alaska pollack (41). The properties of the enzyme have been extensively examined (33, 34, 98–105). Typically, endogenous TGase is calcium dependent and has a mass ranging from 77 to 90 kDa. The optimal temperature and pH are normally 35–50°C and 7.5–9.0, respectively. However, the mass of microbial TGase ranges from 30 to 45 kDa and it is calcium independent. In general, the mass of TGase from animals is much higher than those from microbial sources, such as guinea pig liver TGase, 76.6 kDa (106); human erythrocyte TGase, 82 kDa (38); human placenta TGase, 76 kDa (40); and rabbit liver TGase, 80 kDa (98). The pI of animal-origin TGase has been found to be lower than 7.0, such as, guinea pig liver TGase, 4.5 (101); rabbit liver TGase, 5.4 (98); and human placenta TGase, 5.1(40). Optimal temperature and pH are normally 35–50°C and 5.5–7.0, respectively. The pI of animal-origin TGase has been found to be lower than 7.0: guinea pig liver TGase:4.5 (101); rabbit liver TGase:5.4 (98); and human placenta TGase:5.1(40). Factor XIII (protransglutaminase) was purified from human plasma (112) and platelet (113). The catalytic properties of factor XIII were found to be different from those of transglutaminase obtained from tissues (35) and microorganisms (107–109, 114, 115). However, because of the limited amounts and difficulties of isolation and purification from animals, more economical sources need to be found for the commercial utilization of TGase.

B. Microbial Sources

Few works have been published on TGase from microbial sources. Extracellular TGases were purified from cultural filtrate of *Streptoverticillium mobarense* (116, 117), from *Streptoverticillium sp.* (108) and from *Streptoverticillium ladakanum* (107, 109, 114, 115). Intracellular TGases were also found in *Bacillus subtilis* (118) and in the spherules of *Physarum polycephalum* (119). The effects of the MTGase (microbial TGase) from *Streptoverticillium mobarense* and *Streptoverticillium sp.* on protein gels have been extensively studied (117, 120–124).

The mass of TGase from *S. ladakanum* was similar to that from *Streptoverticillium mobarense* (40 kDa), while the pI (pI:7.9) (107) was lower than that from *S. mobarense* (pI:8.9) (108). The MTGase was stable at pH 5.0–7.0 after a 30 min incubation at 25°C, in which more than 90% activity was retained (107, 109). This suggested that the MTGase was quite resistant to variations of pH, compared with the narrow working pH of TGase from animal sources (102, 110, 111). According to dynamic study on MTGase from *S. ladakanum,* the rate constants for thermal inactivation at 50 and 55 C were 6.2×10^{-4}, and 2.1×10^{-3} sec^{-1}, respectively (107), which were higher than those from pig plasma (36). The ΔH^* and Ea at 55 C of the MTGase were 33.6 and 34.3 kcal/mol, respectively (107), which were lower than those from pig plasma (60.3 and 47.2 kcal/mol) (36). These results suggested that the MTGase was less thermally stable and less dependent on change in temperature than Factor XIIIa from pig plasma.

VI. EFFECTS OF CATHEPSINS AND CALPAINS ON THE TEXTURE OF SEAFOOD

Tenderization—one of the most important quality changes of red meat (125) and seafood (26)—is mainly caused by enzymatic breakdown of the contractile proteins. Postmortem time–temperature relationships (rigor mortis development) have profound effects on the myofibrillar toughness of muscles (126). The enzymes involved in the tenderization process are assumed to be endogenous muscle proteinases that are active at the postmortem pH of muscles (127).

Calpain, cathepins B, L, and L-like have proteolytic activity for a variety of protein substrates (16, 21, 27, 60, 61), and are considered to participate in postmortem muscle tenderization (16, 18, 21, 27, 128–131) since they can rapidly hydrolyze muscle proteins at postmortem pH. Cathepsins B and L are present in macrophage-like phagocytes between muscle fibers of mature salmon during spawning migration and are known to cause muscle softening (56–58, 60, 61). Based on the optimal pH for hydrolysis of myofibrillar and cytoskeletal proteins, cathepsins B, L, and calpain are considered to be more critical than cathepsin D, since the pH optima for cathepsins B and L (pH 5.5~6.5) (14, 18) and for calpain (pH 7.0–7.5) are much closer to that of postmortem muscle than cathepsin D (pH 3.5–4.5). However, the question is whether these proteinases are involved in gel softening (*modori* in Japanese). Studies on their effects on softening of surimi gels thus far are still limited. Although heat-stable alkaline proteinases (HAP) have been considered to be the "modori-inducing" or "gel-softening" factor in some fish (132–137), gel softening in threadfin-bream surimi was not caused by HAP (138). Some modori-inducing proteinases (MIP) have also been found in sarcoplasmic fraction of fish muscle (2, 4). Accordingly, a leaching (washing) process has long been employed to improve the quality of surimi (139). However, although cathepins B and H in Pacific whiting could be removed by leaching, cathepsin L could not (140). In addition, more than 80% residual activities of calpains, cathepsins B, L, and L-like remained in frozen mackerel surimi or surimi gel after the grinding process (16). These phenomena suggest that these proteinases are very difficult to remove during the leaching process. The presence of these residual proteinases were also found to affect the texture of surimi-based products (5, 141).

Several fish species, such as Pacific whiting (142, 143), arrowtooth flounder (142), menhaden (144–145), white croaker (146), and mackerel (16), are known to have soft texture associated with proteolytic activity. Although full utilization of these species for conventional seafood products has been hampered, some have been successfully processed into surimi with the aid of protein additives. Since the textural characteristics developed during gelation are critical to surimi quality and price, this section emphasizes the effects of cathepsins B, L, L-like, and calpains on the texture of gels or gel softening (degradation of surimi gels).

A. Effects on Texture

1. Development of Rigor Mortis

Rigor mortis is the first step in the conversion of muscle into meat. The onset and extent of rigor mortis are biochemically characterized by the content of energy-rich compounds, including ATP, creatine phosphate (Cp), and glycogen, together with the activities of ATPase, kinase, and glycolytic enzymes in the muscle. In living muscles or early postmortem, ATP, which is continuously hydrolyzed by intracellular ATPase and especially by myosin, is resynthesized through either rephosphorylation of liberated ADP by creatine kinase or glycolytic degradation of glucose (147, 148). In the early postmortem stage of muscle, the ATP content consequently remains at almost a constant level as long as the Cp concentration is high enough; it then decreases (148). The consequences of ATP catabolism in postmortem muscle are discussed further in Chapter 2. In resting muscles, an ATP molecule binds to each myosin head (so-called charged myosin), and further reactions between actin and myosin in thin and thick filaments are prevented by the intrusion of tropomyosin and troponin. Contraction in living muscle is initiated by the release of calcium from the sarcoplasmic reticulum. During muscle contraction, ATP is degraded into ADP with a release of energy while the charged myosin remains tightly attached to actin and swivels to cause filament movement. The myosin heads, which are still attached to their sites on the actin, can only detach themselves if new ATP is available for binding. When muscle is converted to meat, myosin heads are still locked to actin and even passive filament sliding is impossible (149). When the pH of muscle declines to around 6.0, which is the optimal pH for myosin ATPase activity, the ATP decreases rapidly to a level insufficient to keep the principal contractile proteins (actin and myosin) apart; they irreversibly form inextensible actomyosin and are manifested as the stiffness of rigor mortis (150).

Due to the accumulation of lactic acid and protons formed during ATP hydrolysis, the pH of meat decreases with postmortem time. The rate of pH decrease is highly related to the contractile and metabolic type of muscle and to meat texture. The rate and extent of pH decline depend primarily on the levels of glycogen and energy-rich phosphate compounds at death, the rate of ATP turnover, and the buffering capacity of muscle tissue, and thus reflect muscle metabolism (i.e., oxidative capacity and glycolytic potential together with muscle contractile properties) (151, 152).

2. Tenderness of Meat

Sometime after death an opposing process called tenderization begins and continues after the onset of rigor. To maximize the benefits of postmortem storage on meat tenderness, beef should be stored for 10–14 days, lamb for 7–10 days, and pork for 5 days at 15°C. Unlike the toughening phase, tenderization does not occur equally in all animals. In fact, it is well documented that there is a large variation in the rate and extent of postmortem tenderization (87, 88, 153–157). This variability in the tenderization process results in inconsistency in meat tenderness at the consumer level. The reasons for the variability in the rate and extent of postmortem tenderization must be identified so that the tenderization process can be manipulated and equalized between carcasses; otherwise technology must be developed to identify those carcasses that will not respond to post-mortem tenderization. Without this information, inconsistency in meat tenderness will continue to exist at the consumer level.

Meat tenderization during storage of carcasses at refrigerated temperatures has been studied in many laboratories (8, 87, 88, 153–156, 158–165). Current evidence suggests that proteolysis of key myofibrillar and associated proteins is the cause of meat tenderization. These proteins are involved in either inter- (e.g., desmin and vinculin) and intramyofibrils (e.g., titin, nebulin, and possibly, troponin-T) linkages or in the linking myofibrils to the sarcolemma by costameres (e.g., vinculin, dystrophin) as well as in the attachment of muscle cells to the basal lamina (e.g., laminin, fibronectin, and the newly described 550 kDa protein) (166). The function of these proteins is to maintain the structural integrity of myofibrils (167). Degradation of these proteins will, therefore, cause weakening of myofibrils and, thus, tenderization. Although the list of these proteins may change over the years, the principle will stand the test of time: proteolysis of key myofibrillars and associated proteins is responsible for postmortem tenderization.

Although tenderization of meat has been extensively studied in beef, pork, and chicken, the process appears, from histological studies, to be similar in different species. Dissolution or breaks in the I-band at the position of the Z-line have been observed in conditioned meat from different species. These phenomena occur earlier in chicken (168) than in beef (169). There also appears to be a difference between the fiber types since Z-line fractures have been found to occur earlier in white than in red fibers of beef (170), pork (171), and chicken (168). Many biochemical changes during postmortem tenderization have generally been assumed to arise from the release of endogenous muscle proteases that are active at the postmortem pH of muscles (127). Many studies have been conducted to clarify the mechanism of muscle tenderization. It has been variously postulated that muscle tenderization results from the disappearance of Z-disks, dissociation of actomyosin complex, destruction of connectin, or denaturation of collagen (90, 91, 127, 148, 172–176).

3. Muscle Proteinase and Meat Tenderization

The mechanisms involved in the meat tenderizing process are considered to be enzymatic and physicochemical reactions (154). In living muscles, intracellular protein degradation is controlled, at least partly, by a number of different endogenous proteolytic systems (160, 177). Since most of the postmortem changes occurring in the process of meat tenderization are considered to be the result of proteolysis, proteinases located inside muscle cells or cytosol can be potential contributors to meat tenderness. It should be emphasized that the regulation of the enzymes involved in postmortem tenderization process is presently unclear.

Although several proteolytic systems related to the tenderization process are described in the literature (i.e., calpains-calpastatin, cathepsins-cystatin, and proteasome or macropain), only two of them have attracted much attention from meat scientists. They are μ- and m-calpains, and lysosomal proteinases, namely, cathepsins D, B, H, and L (160). Four phenomena reveal the involvement of calpain in postmortem muscle tenderness (45, 178). First, the ultrastructural degradation of postmortem myofibrils is quite similar to that of myofibrils treated with calpain. Second, postmortem myofibrillar proteins, untreated or treated with calpain, have similar electrophoretical degradation patterns. Third, the Z-disk, where calpain is localized, is extremely susceptible to calpain-catalyzed hydrolysis. Fourth, the higher the level of calpain in muscle, the faster the rate of postmortem tenderization. However, in these studies the concentration of calcium and the existence of the specific endogenous inhibitor for calpain, calpastatin, were ignored. Since the endogenous protein inhibitors may constitute a powerful regulatory system for muscle proteinases, interest in their identification and characterization has recently increased markedly. Many studies have accordingly examined the existence in skeletal muscle of protein inhibitors inactivating specifically cysteine proteinases, serine proteinases, and, in some cases, both types of proteinases (179–181).

There has been considerable debate about the specific protease responsible for postmortem changes. A protease must meet certain criteria to be considered a possible candidate for involvement in postmortem tenderization (163). Goll et al. (8) provided the logic for the first criteria on: that the protease must be endogenous to skeletal muscle cells. Second, the protease must have the ability to reproduce postmortem changes in myofibrils in vitro. Finally, the protease must have access to myofibrils in tissue. If a protease does not have these characteristics, it cannot be considered as a candidate in the postmortem tenderization process. Likewise, if a protease meets these criteria, it is impossible to exclude its possible involvement in the tenderization process. Of all the potential candidates, calpains are the proteases that meet all of the above requirements. Based on the results of numerous experiments reported by different laboratories, it can be concluded that proteolysis of key myofibrillar proteins by calpains is the underlying mechanism of meat tenderization that occurs during storage of meat at refrigerated temperatures (8, 87, 88, 153, 159, 182). Much evidence supports the idea that proteolysis of Z-lines by calpains causes tenderization (8, 10, 26, 29–31, 87, 153–157, 159, 161, 163, 164, 182).

According to Ouali and Talmant (183) and Valin et al. (151), the meat-aging rate is related to the enzyme/inhibitor ratio but not to calpain activity (151, 183). However, the activity of calpastatin in beef decreased while the activity of calpains remained unchanged at a high level at -70°C (163). Although some investigations have indicated that μ-calpain and calpastatin of beef muscle lose their activities more rapidly than does *m*-calpain at 4°C (174, 184), the thermal stability of μ-calpain and calpastatin was found to be higher than that of *m*-calpain in vitro (32, 185, 186). The resistance of myofibrillar proteins to calpains and the regulation and stability of calpains in muscles might also be major factors affecting tenderness. According to Valin et al. (151), no direct relationship was found between muscle proteinase content and the rate of tenderization, but a positive relationship was found in the activity ratio of proteinase to inhibitor.

Differing from the calpains, which are considered to specifically attack certain proteins of the Z-line, such as desmin, filamin, nebulin, and, to a lesser extent, connectin (187–190), cathepsins preferentially attack myosin and actin (11, 12, 16, 191). Furthermore, they can attack contractile proteins at various strategic points (150). Cathepsin B can rapidly degrade myosin heavy chains (14, 16) while cathepsin L degrades the troponins T and I, and C-protein rapidly, and degrades myosin, actin, tropomyosin, nebulin, titin, and α-actinin slowly. Although calpains have an optimum pH range for activity near neutrality, cathepsins, especially cathepsin B and L, have pH optima more closely associated with the pH range (5.5–6.5) found in many postmortem skeletal muscles (14, 18, 175, 192). Nevertheless, Koohmaraie (163) did not consider the cathepsins to be responsible for the tenderization of meats. The reason is that lysosomal proteases might still be located in lysosomes and have no access to myofibrils or cytosol. However, the fall in pH during postmortem glycolysis weakens the walls of organelles such as lysosomes (192), which consequently causes the release of lysosomal proteinases, such as cathepsins B, H, and L, that have pH optima at around 5.5–6.5 (14, 18, 192, 193). Furthermore, when aged muscle is extended, fractures mainly appear close to the Z-lines and, although less frequently, at the junctions of A-bands and I-bands (194–195). Increase fragility of these regions in stored meat can perhaps be ascribed to the action of lysosomal proteinases (154). A number of in vitro studies have clearly demonstrated the high susceptibility to proteolysis of numerous myofibrillar proteins by calpains and lysosomal proteinases. Both calpains and lysosomal proteinases degrade troponin T, troponin I, tropomyosin, C-protein, desmin, titin, and nebulin while myosin heavy chains, myosin light chains, α-actinin, troponin C, and actin appear to be sensitive to the action of lysosomal proteinases, especially to cathepsins D, B, and L (8, 11, 12, 16, 128, 129, 190, 196–199). The general findings summarized above lead to the conclusion that almost all the modifications so far identified at the myofibrillar structure level in aged meat can be explained by the proteolytic action of muscle proteinases. These findings also substantiate the necessary synergistic contribution of calpains and lysosomal proteinases to the postmortem tenderization of meats.

Myofibrils are a poor substrate for the multicatalytic protease complex (known as proteasome or macropain; MCP) and MCP does not degrade the same proteins degraded postmortem (15, 163, 165). The role of MCP in postmortem tenderization needs to be clarified further.

On the other hand, fast-contracting muscles that tenderize more quickly exhibit the highest osmotic pressure values. Osmotic pressure values so far obtained correspond to an ionic strength equivalent to those of NaCl concentrations ranging from 0.2 M to 0.35 M. These salt concentrations are high enough to cause solubilization of myosin and partial dissociation of the filament structures of contractile muscles. All the data so far available support the concept that the synergistic action of at least lysosomal and calcium-dependent proteinases explains various observed postmortem structural and biochemical changes. Neither proteolytic system alone is able to cause all the postmortem physical and biochemical changes so far described. In view of these contradictory results presented in the literature, more detailed investigation is needed to clarify which of the parameters so far discussed, such as the pH-temperature-induced release of cytosol calpains and/or membrane-bound lysosomal enzymes, or other CDP or proteasomes, has the largest impact on meat tenderness.

According to the literature reviewed here, in spite of recent advances in both in vivo and postmortem muscle biochemistry, there are still many points, such as the changes responsible for postmortem improvement in meat tenderness, that remain to be clarified. Although most scientists recognize that the weakening of the Z-disk by Ca-dependent proteinases contributes to meat tenderness, the increased fragility of myofibrils at the junctions of A-bands and I-bands or in myosin domains due to lysosomal proteinases, which can lead to total transversal disruption of the structure, might also be an important factor. Even though myofibril rupturing is not complete during the tenderization process, it seems likely that the weakening of the structures involved is sufficient to cause a significant decrease in the mechanical resistance of myofibrils. However, further studies are needed to determine how these regions of the sarcomere are structurally organized and the nature of the proteins involved.

Most of the myofibrillar ultrastructural and biochemical changes identified so far in stored meat can perhaps be explained by the synergistic action of calpains and lysosomal proteinases. Nevertheless, unexpected contradictory results have been revealed by comparing the proteolytic enzyme levels found in different muscles. Therefore, some basic studies on the in vivo and postmortem mechanisms by means of which the activities of these proteinases are regulated, and on their real potential activities in meats, are needed.

Besides the calpains and lysosomal proteinases mentioned above, muscle also contains a number of other proteinases that have been more or less well characterized. Among them, the multicatalytic proteinases are also important and need to be investigated in the future. In conclusion, meat tenderization is a very complex multifactorial process controlled by a number of various endogenous proteinases and, as yet, poorly understood biological parameters. More detailed investigation is indeed needed to clarify which of the parameters so far discussed, including shortening, pH-temperature-induced release of cytosol calpains and/or membrane-bound lysosomal enzymes, electrical stimulation (ES) induced release of proteases, filament disrupture, and other factors have the largest impact on the tenderness of meats at temperatures that cause minimal cold shortening.

B. Effects on Processed Seafood

1. Gelation (Suwari) of Surimi

Surimi gels are composed of a three-dimensional protein network formed mainly by actomyosin. For elucidating the heat-induced gelation mechanism, actomyosin, myosin, and myosin subfragments have been studied for protein–protein interactions occurring in surimi gels (200–205). This approach allows for the study of the contribution of individual myofibrillar protein components to surimi gelation without interference from endogenous enzymes. According to these studies, myosin is the most important component in the formation of gel products. Sano et al. (203) found that the gel strength and elasticity of myosin gels were higher than those of natural actomyosin gels and the gelation of carp actomyosin took place in two stages: at 30–41°C and 51–80°C as observed by dynamic viscoelastic behavior and differential scanning calorimetric (DSC) analysis. Differential shear modulus for natural actomyosin from Argentinean hake (*Merluccius hubbsi*) also produced two transitional temperatures ranges at 36–38°C and 48°C (202). Sano et al. (203) thus proposed that the development of gel elasticity at the first stage is due to the interactions between the tails and the second stage due to the interactions among the heads of myosins. Ziegler and Foegeding (206) further summarized that gelation of myosin proceeds by losing its noncovalently stabilized α-helical structure due to heating and then by increasing turbidity due to intermolecular association, which leads to the formation of a rigid protein network structure that is stabilized by covalent disulfide bonds and various noncovalent interactions.

Gelation characteristics of surimi such as firmness, cohesiveness, and water-holding capacity can be increased by incubating surimi paste below 40°C (also referred to as "setting" or "suwari") (207). Suwari can be achieved within a short period of time (2–4h) near 40°C (high-temperature setting) or within an extended period (12–24 h) below 40°C (low-temperature setting) (200, 201). According to Wu et al. (201–202), low temperature setting is related to transglutaminase activity, while high temperature setting is related to the transition in rheological properties of actomyosin observed at 36–38 C. α-Helical structure of myosin, prevalent in the tail portion, unfolded at 30–40° C, corresponding to low temperature setting (204). Lanier (208) also observed a great increase in gel strength and, to a smaller extent, the elasticity of Atlantic croaker and sand trout surimi gels preincubated at 40°C. Furthermore, of the 14 fish species studied, the gel strength was highly correlated with the decrease in α-helicity (r=0.85), therefore, Ogawa et al. (205) proposed that the setting of surimi is initiated by the unfolding of α-helix.

Gel strength of myofibrillar proteins can be influenced by factors affecting myosin structure. The gelation properties of myosin are highly related to the length of the double stranded α-helical tail. Myosin rod (140 nm long) showed higher rigidity than light meromyosin (80 nm long) at all salt concentrations studied (96). Accordingly, proteolysis of myosin was shown to lower surimi gel strength (209). The native conformation of myosin is of primary importance for proper gelation. Maximum gel strength cannot be obtained if myosin is denatured or degraded prior to initiation of gelation (210). Myofibrillar proteins from fish arc more susceptible to thermal denaturation than those from terrestrial animals (204). Since the hydrophobic amino acid residues of actomyosin exposed during freezing and subsequent storage were mainly from myosin and, to a much lesser extent, from actin, repeated freezing and thawing of surimi denatured myosin, resulting in a substantial decrease in gel strength (211).

2. Softening (Modori) of Surimi-Based Products

Temperature plays an important role in surimi gelation. In addition to its effects on the conformation of myofibrillar proteins, temperature can activate endogenous enzymes (Fig. 1). Fish muscles from various species revealed similar responses to temperature (212): a structure-setting reaction for gelation below 40°C (suwari) and a structure disintegration reaction at 50–70°C (modori). Low temperature setting is associated with transglutaminase activity (207, 213), while modori is induced by endogenous thermal stable proteinases that can degrade myosin rapidly (16–17, 140, 214).

Postmortem fish muscle is susceptible to proteolysis by endogenous proteinases, resulting in a texture often described as soft or mushy. Proteolytic disintegration of surimi gels is characterized by high activity at temperatures above 50°C and by the rapid and severe degradation of myofibrillar proteins, particularly myosin (215, 216). This disintegration has detrimental effects on surimi quality, which substantially lowers the gel strength and elasticity (144, 209). Among numerous proteinases presented in muscle, endoproteolytic cysteine proteinases have the most serious effect on texture due to their thermostability and the ability to cleave the internal peptide bonds, while exopeptidase hydrolyzes terminal peptide bonds (48). The most active proteinases in fish muscle that can soften the surimi gels vary with species, but are generally categorized into two major groups: cathepsins (5, 16–17, 54, 56–61, 217) and heat-stable alkaline proteinases (144, 146, 215, 216, 218).

High levels of cysteine proteinase activity caused by cathepsins B, H, L,

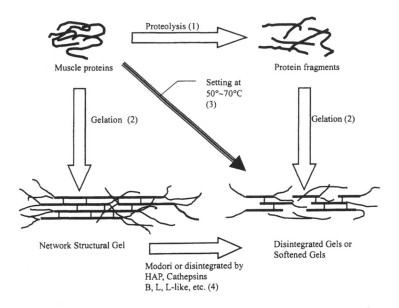

Figure 1. Proposed model of the gelation and disintegration of surimi gels. (1) Proteolysis by calpains, cathepsins, and other proteinases during storage; (2) gelation; (3) proteolysis by HAP, cathepsins B, L, and L-like during setting at 50–70°C; (4) modori or disintegration by HAP, cathepsins, calpains, and other proteinases if the formed network structure is not fixed by heating at 85–100°C.

and L-like have been observed in Pacific whiting and arrowtooth flounder (140, 215, 216), chum salmon during spawning migration (56–59), and mackerel (18). Arrowtooth flounder softening is due to a cysteine proteinase with maximum proteolytic activity at 50–60°C (219). When the Pacific whiting muscle was incubated at 60°C for 30 min prior to cooking at 90°C, most myosin heavy chain (MHC) was degraded, and surimi did not form a gel with a measurable gel strength (209). The proteolysis of muscle proteins and connective tissue of fish is also induced by infection with *Myxosporea,* which releases proteinases into the fish muscle tissue. In Pacific whiting, these parasites have been identified as *Kudoa paniformis* and *K. thyrsitis* (220), and in arrowtooth, *K. thyrsitis* (219). The degree and stage of the infection have been linked to the variation in proteolytic activity between individual fish, which consequently results in the variation in gel strength of surimi (5, 143). The majority of proteolytic activity in Pacific whiting muscle is due to cathepsin L (140, 217), while that in mackerel is due to cathepsins B, L, and L-like (16, 17). Cathepsin L has a high affinity for myosin and is not completely

removed by leaching during surimi processing (140). Purified cathepsin L
from Pacific whiting and mackerel consists of a single peptide with mass of
28,800 and 30,000 Da and a temperature optimum at 55°C and 50°C, and pH
optimum at 5.25–5.5 and 5.0, respectively (18, 141, 142, 217). According to
Lee et al. (18), the mass, optimal temperature, and pH for cathepsin L-like enzyme are 58,000, 40°C and 5.5, respectively.

Cathepsin L activity was also shown to be the major contributing factor
to the softening of chum salmon muscle during spawning migration but not
feeding migration (56–59). The degradation of myofibrillar proteins by purified cathepsin L observed with sodium dodecyl sulfate (SDS)-PAGE was identical to that of muscle stored at 4°C for 7 days (56–59). In the extensively
softened muscle of chum salmon, substantial degradation of myosin heavy
chain into fragments with a mass of 160,000 Da, and degradation of troponin
T, C, I, and myosin light chains were observed (61). Purified cathepsin L from
chum salmon or mackerel could hydrolyze myosin, desmin, collagen, and
gelatin, and was capable of initiating the hydrolysis of numerous other proteins, indicating its possible role in severe proteolytic degradation of muscle
(16, 18, 60).

Heat-stable alkaline proteinase (HAP) has often been reported as responsible for textural degradation of surimi gels. Its activity has been found in muscles
from a large number of fish including rainbow trout, sardine, white croaker, carp,
common mackerel, cod, herring, and Atlantic salmon (218, 221, 222). White
croaker meat paste formed a poor elastic gel when heated around 60°C, while
purified actomyosin from croaker muscle did not (146). Although cathepsin D,
neutral proteinase, and calpain were also found in white croaker muscle, Makinodan et al. (146) concluded that only alkaline proteinase could act at the pH of
meat paste and 60°C. Alkaline proteinase purified from white croaker was a
heat-stable cysteine proteinase with temperature optimum at 60°C and pH opti-
mum at 8.0 (223, 224). It is composed of four different subunits ($\alpha\beta\gamma2\delta4$) with
molecular weights ranging from 45 to 57 KDa. These complex subunits con-
tribute to the thermostability of the enzyme (223, 224). A HAP was also purified
from Atlantic menhaden, which showed similar characteristics to white croaker,
with an optimum activity at 60°C and pH 7.5–8.0 (144). HAP is characterized by
its narrow range of pH and temperature optima. It is usually not detected below
50°C (225–226).

In spite of the extensive studies cited above, the following questions related to surimi must be answered: How much of these proteinases are left in
the surimi? What is the frozen stability of these proteinases? Are the cryoprotectants such as polyphosphate, sorbitol, and sucrose usually used in surimi
manufacturing affecting these proteinases? To address these questions, research has been conducted and some of the results are summarized in the following sections.

C. Effects of Endogenous Enzymes on Surimi

1. Changes in Cathepsins B, L, L-Like, and Calpain Activities during Surimi Processing

The activity of cathepsins B+L+L-like remaining in mackerel surimi after mincing, leaching, and NaCl-grinding processes, which was assayed with the synthetic peptide Z-Phe-Arg-MCA, was 6.02, 5.23 and 4.07 units/g of muscle (Table 1) (17). Chang-Lee et al. (227) reported that the protease activity of Pacific whiting muscle decreased gradually during surimi processing. However, as much as 87% of cathepsins B+L+L-like activity was left in mackerel surimi after washing treatment (Table 1), and there was no significant decrease in calpain activity after surimi processing and frozen storage (unpublished data). The results indicate that cathepsins B, L, L-like, and calpain are very difficult to remove during surimi processing. According to Kinoshita et al. (139), some softening-inducing proteinases in mackerel muscle were not easily removed by washing. Based on our data, cathepsins B, L, L-like, and calpain are hypothesized to be the active proteases in mackerel surimi. However, in the case of Pacific whiting, cathepsin B activity sharply decreased after surimi process (140). Therefore, in surimi processing, removal of cathepsins B, L, and L-like seemed to be species-dependent.

Cathepsin L had a higher proteolytic activity for hydrolysis of myofibrillar proteins than cathepsin B (54, 60–61). It was regarded as an important proteinase in muscle softening of chum salmon (54) or in surimi degradation of Pacific whiting (140). Since cathepsins B, L, and L-like are lysosomal proteinases, the pH, temperature, and ionic strength around lysosomes could affect the release of these proteinases into sarcoplasmic fluids. Conditions of postmortem fish and surimi processing would, therefore, highly affect the removal of these proteinases from surimi. Accordingly, useful approach to inhibit the gel softening may be to produce the natural inhibitor cystatin economically or to rupture the

Table 1 Cathepsin B and L Activity Remaining in Mackerel Meat After Each Processing Step

Process	Enzyme activity (units/g)	Relative activity (%)
Mincing	6.02±0.11[a]	100.00[a]
Leaching	5.23±0.12[b]	86.88[b]
NaCl-grinding	4.07±0.09[c]	67.61[c]

[a–c] Means of three determinations, expressed as mean ± standard deviation. Values in same column with different letters differ significantly (p<0.05).

membrane of lysosomes without deteriorating the myofibrillar proteins for easy removal of cathepsins B, L, and L-like before or during washing treatment.

After grinding with 2.5% NaCl, about 68% cathepsins B+L+L-like activity remained in the ground surimi (Table 1). According to Kinoshita et al. (228), the proteolytic activity of sarcoplasmic-50°C-MIP (modori-inducing proteinases) from threadfin bream against actomyosin was not affected by treatment with 10% NaCl. Furthermore, carp cathepsin B activity increased in the presence of 0.1–0.5 M NaCl (96). The loss of cathepsins B, L, and L-like activities (about 32%) after grinding with 2.5 % NaCl (16) might be due to the denaturation of enzymes during grinding.

2. Frozen Stability of Cathepsins B, L, and L-Like in Surimi

Although cathepsins B, L, and L-like activities in mackerel surimi decreased progressively during frozen storage at -40°C, there was still 82% activity left after 8 weeks storage (Table 2) (16). This might be because cathepsins B and L still existed in lysosomes, which consequently protected them from denaturation during frozen storage. Cryoprotectants, such as sucrose, sorbitol, and polyphosphates, have been used to prevent protein denaturation in frozen surimi (229). These reagents showed little effect on cathepsin B and L, which retained at least 83% of their activities after 8 weeks of storage (Table 3) (16). Although a decrease in cathepsin L-like activity was much faster than that in cathepsins B and L during 8 weeks storage at -40°C, there was still about 40% activity left. The activity of cathepsins B and L in mackerel surimi without cryoprotectants only slightly decreased during frozen storage. This result further supports the hypothesis that these proteinases are important in the gel softening of surimi.

3. Degradation of Surimi Proteins by Purified Cathepsin B, L, L-Like, and Calpain

Degradation of MHC in mackerel surimi with purified cathepsin B, L, and L-like during 5 h incubation at 40 or 55°C was observed on SDS-PAGE. The result was compared with those containing E-64 and cathepsins B, L, and L-like inhibitor.

Table 2 Cathepsin B and L Activity in Mackerel Surimi During Storage at -40°C

		Storage time (weeks)				
	Unfrozen	0	2	4	6	8
Activity (units/g)	6.02±0.08[a] (100.0)	5.93±0.11[a] (98.5)	5.73±0.12[a] (95.2)	5.60±0.09[a,b] (93.0)	5.25±0.11[b] (87.2)	4.92±0.09[b] (81.7)

[a,b] Means of three determinations, expressed as mean ± standard deviation. Values in same row with different letters differ significantly (p<0.05).

Table 3 Effects of Sucrose, Sorbitol, and Polyphosphate on Purified Cathepsin B Activity During Frozen Storage

Storage time (weeks at −40°C)	Cathepsin B Activity (unit/ml)			
	Control	3% Sucrose	3% Sorbitol	0.2% PP
0	13.66±0.25[a,A]	12.57±0.26[b,A]	12.76±0.15[b,A]	12.64±0.18[b,A]
2	12.39±0.22[a,B]	12.69±0.21[a,A]	12.76±0.15[a,A]	13.02±0.20[a,A]
4	11.58±0.16[b,B,C]	12.31±0.17[b,A]	12.00±0.19[b,A,B]	13.80±0.09[a,A]
6	11.33±0.24[a,C]	11.97±0.13[a,A,B]	11.53±0.21[a,B]	11.77±0.21[a,B]
8	11.27±0.19[a,C]	11.69±0.16[a,B]	11.33±0.25[a,B]	11.60±0.16[a,B]

PP, mixture of 50% sodium tripolyphosphate and 50% potassium pyrophosphate.
[a,b,A,B] Means of three determinations, expressed as mean ± standard deviation. Values the same row with different lower-case superscripts differ significantly (p<0.05); values in same column with different upper-case superscripts differ significantly (p<0.05).

Proteolysis was faster in pH 6.5 samples than in pH 7.0 samples, and faster in samples incubated at 55°C than that at 40°C (16). Degradation of MHC by calpain was also observed on SDS-PAGE of mackerel surimi with purified calpain during 3 h incubation at 40 and 55°C when compared to those incubated with EDTA. Samples incubated at pH 7.5 had more extensive degradation than samples at pH 7.0 (unpublished data). According to Jiang et al. (16), cathepsins B, L, and L-like not only degraded MHC but also partially degraded actin at pH 6.5 or 7.0 after 5 h incubation at 55°C. These results suggested that the thermally unfolded globular actin might be hydrolyzed by cathepsins B, L, and L-like. However, degradation of surimi proteins by cathepsins B, L, and L-like could be effectively inhibited by E-64 (16). Some protease inhibitors from beef plasma (145, 209, 230), fish plasma (3), potato and egg white (209, 231–232), and legume seed extracts (233) were also effective in preventing degradation of fish myofibrillar proteins. Akazawa et al. (234) indicated that bovine plasma powder was an effective inhibitor in degradation of Pacific whiting surimi proteins. Plasma factor XIIIa (an active transglutaminase) purified from pig plasma also had very strong crosslinking abilities on MHC of mackerel surimi and this crosslinked protein gel was strongly resistant to proteolysis (36).

Results suggested that cathepsins B, L, and L-like participated in the thermal degradation of mackerel surimi. Cathepsins B, L, and L-like are active proteinases in many fish, such as Pacific whiting (140), chum salmon (56–58), and mackerel (18). Muscle softening (54, 60, 61) and surimi gel softening (140) have been attributed mostly to cathepsin L. However, in our study, cathepsins B, L, and L-like could not be completely washed out from the surimi (Table 1) and revealed myosin-degrading activity on samples incubated at 40 and 50°C, pH 6.5 and 7.0 (16). Therefore, cathepsin B, L, and L-like might also be important in the gel softening of mackerel surimi. Refrigerated (0–5°C), ambient (25–30°C), and warm temperatures (50–65°C) are usually used for setting of surimi gels. No significant changes in MHC of surimi gels with purified cathepsin B, L, or L-like were observed at 25°C (16). It might be suitable for mackerel surimi gel to set at temperatures below 25°C.

4. Gel Softening by Purified Cathepsin B, L, and L-Like

The breaking force of surimi gels ground with purified cathepsins B (5 unit/g of meat) decreased from 426 (pH 7.0) and 477 (pH 6.5) g to 345 and 343 g, respectively. Those ground with purified cathepsin L (5 unit/g of meat) decreased to 450 (pH 7.0) and 417 (pH 6.5) g, and that with purified cathepsin L-like (5 unit/g of meat) decreased to 449 (pH 7.0) and 430 (pH 6.5) g ($p<0.05$) after 2 h incubation at 55°C. In addition, the deformation of surimi gels with purified cathepsins B, L, and L-like decreased ($P<0.05$) after 2 h incubation at 55°C, compared to that of control (Table 4) (16). Our results provided direct evidence to demonstrate that cathepsins B, L, and L-like could induce softening in fish gels. There were about 5.2 units of cathepsins B, L, and L-like contained in 1 g leached mackerel surimi (Table 1). Consequently, when extra cathepsins B, L, or L-like (5 units/g surimi) were added, the strength of protease-added gels (2 h setting at 55°C followed by 90°C for 20 min) was reduced by about 50% as compared to the control gel (without setting at 55°C) or by 6–28% as compared to samples without enzymes after 2 h setting (Table 4) (16). However, the gel strength of surimi with added calpain did not change significantly (unpublished data) compared to the control gel (without protease addition) (Table 4). This might be due to high endogenous calpastatin activity (specific inhibitor of calpain) that was present in surimi. Heat-stable alkaline proteinases (HAP) also deteriorate surimi gels (135, 136). The strength of HAP-added gel decreased to about half that of the control gel (223, 224). However, Toyohara et al. (2, 4) found that HAP isolated from threadfin bream had no effect on disintegration of surimi gels. According to these reports, cathepsins B, L, and L-like seemed to contribute more to the softening of surimi gels.

The gel strength of surimi with addition of purified cathepsins B, L, and L-like plus E-64 were 504, 540, 505 (pH 7.0), and 469, 490, 533 (pH 6.5) g × cm, respectively (Table 4) (16). Compared with the control, the strength of inhibitor-added gels was slightly lower. E-64 could not completely inhibit the softening of mackerel protein gels. This suggested that, in addition to the cysteine proteinases, some other factors causing gel softening existed in the mackerel surimi. According to Iwata et al. (235), some softening-inducing proteins without proteolytic activity also existed in fish muscle. The mechanism of protein-induced softening is still not clear. On the other hand, Sp-50-MIP activity was not affected even in the presence of 10% NaCl (228). HAP proteolytic activity against

Table 4 Gel Strength of Minced Mackerel after 2 h Incubation with Cathepsins B and L and E-64[1] at 55 C, PH 7.0 and 6.5

pH 7.0

Treatments	H at 55°C	Breaking force (g)	Deformation (mm)	Gel strength (g.cm)
None	0	630±15[a]	10.6±0.8[a]	668[a]
None	2	426±18[c]	8.2±0.4[b]	349[c]
Cathespin B	2	399±11[d]	7.9±0.4[b]	315[d]
Cathepsin B and E-64	2	537±21[b]	9.4±0.5[a]	504[b]
Cathespins B+L	2	345±12[e]	7.5±0.3[b]	259[e]
Cathepsins B+L+E-64	2	557±16[b]	9.5±0.5[a]	529[b]
pH 6.5				
None	0	613±15[a]	9.8±0.4[a]	601[a]
None	2	477±11[c]	8.3±0.5[b]	396[c]
Cathespin B	2	373±13[d]	8.4±0.4[b]	312[d]
Cathepsin B and E-64	2	515±18[b]	9.1±0.7[a,b]	469[b]
Cathespins B+L	2	343±12[e]	7.4±0.3[c]	254[e]
Cathepsins B+L+E-64	2	535±18[b]	9.2±0.8[a,b]	492[b]

E-64, L-trans-epoxysuccinyl-leucylamido-4-(guanidino)butane.
[a-e] Means of 10 determinations, expressed as mean ± standard deviation. Values in same column with different letters differ significantly ($p < 0.05$).

carp actomyosin gradually decreased with increasing concentration of NaCl (0.1–0.5M) (134, 236). Our data indicated that cathepsins B, L, and L-like could hydrolyze surimi protein in the presence of 0.6 M NaCl and result in gel softening of mackerel surimi (16). If the muscle proteins are not properly stored, they will be hydrolyzed into fragments by various proteinases depending on the species, storage temperature, and pH of postmortem muscle. However, both intact muscle proteins and their proteolytic fragments form two different types of gels: firm network or loose (disintegrated) gel structures. The degree of softening varies with the extent of proteolysis. The intact or native muscle proteins can form firm network structure gels. However, if the well cross-linked network structural sols is not fixed at optimal time by heating at 85–100°C, or the setting process is over the optimal time, the network structural gels will be disintegrated into softened gels. This phenomenon is usually called modori (in Japanese) or disintegration. As mentioned previously, HAP or cathepsin B, L, and L-like have high activity at 50–70°C and a substantial amount of activity is retained in frozen surim even after NaCl-grinding. Therefore, when the surimi with native muscle proteins is set at 50–70°C, it will easily form a soft gel due to the action of these proteinases.

In summary, results obtained thus far suggest that cathepsins B, L, and L-like play an important role in the disintegration of surimi gels (gel softening) by the hydrolysis of myosin heavy chain, light chains, actin, and troponins. Disintegration could be substantially inhibited by adding cysteine proteinase inhibitors such as cystatin or E-64. Development of molecular biological technique for producing economical natural inhibitor cystatin would be an ideal way to prevent gel softening of surimi.

VII. EFFECTS OF TRANSGLUTAMINASE ON THE TEXTURE OF FOOD MYOSYSTEMS

Transglutaminase (TGase) can catalyze the formation of ε-(γ-glutamyl) lysyl crosslinks and improve functionality of food proteins, such as casein, β-lactoglobulin, and soybean proteins (117, 237–242; see also Chap. 6). TGase can also catalyze the incorporation of essential amino acids for fortifying food proteins (243, 244). The formation of crosslinks by TGase between different food proteins, such as casein and soybean globulin (245), casein and myosin (246), myosin and soybean protein (239), whey proteins and caseins (247), and soybean proteins and meat (111) shows promising potential in the development of new protein foods. Protein solutions with high concentration can be firmly crosslinked by TGase (248). This technique is applicable to the preparation of edible films (249) and surimi-based products (36).

TGase can also form isopeptide bonds among fish myosin molecules, which is considered to be related to the gel strength of minced fish (250–257). Seki et al. (41) isolated TGase from Alaska pollack and found that it could induce the gelation of minced fish. Tsukamasa and Shimizu (213) further reported that the strong gel-forming ability of sardine was due to the formation of the nondisulfide bond, which later was shown to be due to the action of TGase (258). Jiang et al. (11, 12) also reported that the gel strength of mackerel surimi was greatly improved by the addition of pig plasma TGase. Addition of calcium to activate the endogenous TGase or incubation at higher than 35°C to inactivate endogenous TGase during setting process indicated that this enzyme contributes greatly to the gelation of pollack paste (259, 260). Recently, the Ca-independent microbial TGase (MTGase) from *Streptoverticillium mobarense* (116, 117, 261) or from *Streptoverticillium ladakanum* (105, 107, 114, 115) has shown high potential to increase the gel strength of fish surimi as well.

Besides TGase catalysis of covalent crosslinking, the gel strength of fish paste can also be improved by ultraviolet (UV) irradiation (115, 262–266). Although the effect of UV irradiation on protein structures is unknown in detail, there is evidence that it can cause the oxidative fragmentation (267–270) or polymerization of proteins (266, 271–273), and thereby change protein functionality. Taguchi et al. (263) proved that UV irradiation could activate Mg-ATPase, which is responsible for a stronger binding of myosin to actin. Ishizaki et al. (266) demonstrated that UV irradiation fragmented flying fish myosin, and caused an increase in surface hydrophobicity and the polymerization of myosin heavy chains. Combination use of MTGase and UV irradiation substantially accelerated the gelation of mackerel muscle proteins (115)

VIII. CONCLUSIONS

Numerous enzymes influence the physical properties of seafood. Among the most important enzymes that can influence seafood texture are those involved with energy metabolism and the onset of rigor mortis, endogenous and exogenous proteolytic enzymes, and transglutaminase.

REFERENCES

1. Y Makinodan, M Yamamoto, W Simidu. Protease in fish. Nippon Suisan Gakk 29:776–780, 1963.
2. H Toyohara, M Kinoshita, Y Shimizu. Proteolytic degradation of threadfin bream meat gel. J Food Sci 55:259–260, 1990.
3. H Toyohara, T Sakata, K Yamashita, M Kinoshita, Y Shimizu. Degradation of oval-filefish meat gel caused by myofibrillar proteinase(s). J Food Sci 55:364–368, 1990.
4. H Toyohara, K Sasaki, M Kinoshita, Y Shimizu. Effect of bleeding on the modori-phenomenon and possible existence of some modori-inhibitor(s) in serum. Nippon Suisan Gakk 56:1245–1249, 1990.
5. H Toyohara, M Kinoshita, I Kimura, M Satake, M Sakaguchi. Cathepsin L-like protease in Pacific hake muscle infected by myxosporidian parasites. Nippon Suisan Gakk 59:1101–1107, 1993.
6. MB Wojtowicz, PH Odense. Comparative study of the muscle catheptic activity of some marine species. J Fish Res Bd Can 29:85–90, 1972.
7. H Toyohara, K Ito, M Ando, M Kinoshita, Y Shimizu, M Sakaguchi. Effect of maturation on activities of various proteases and protease inhibitors in the muscle of ayu (*Plecoglossus altivelis*). Comp Biochem Physiol 99B:419–424, 1991.
8. DE Goll, Y Otsuka, PA Nagainis, JD Shannon, AK Sathe, M Mururuma. Role of muscle proteinases in maintenance of muscle integrity and mass. J Food Biochem 7:137–177, 1983.
9. ST Jiang, YT Wang, BS Gau, CS Chen. Role of pepstatin-sensitive proteases on the postmortem changes of tilapia (*Tilapia nilotica X T. aurea*) muscle myofibrils. J Agric Food Chem 38:1464–1468, 1990.
10. ST Jiang, YT Wang, CS Chen. Purification and characterization of a protease identified as cathepsin D from tilapia muscle (*Tilapia nilotica X T. aurea*). J Agric Food Chem 39:1597–1601, 1991.

11. ST Jiang, YT Wang, CS Chen. Lysosomal enzyme effect on the postmortem changes in tilapia (*Tilapia nilotica X T. aurea*). J Food Sci 57(2):277–279, 1992.
12. ST Jiang, FP Nei, HC Chen, JH Wang. Comparative study on the cathepsin D from banded shrimp (*Penaeus japonicus*) and grass shrimp (*Penaeus monodon*). J Agric Food Chem 40:961–966, 1992.
13. ST Jiang, YH Her, JJ Lee, JH Wang. Comparison of the cathepsin D from mackerel (*Scomber australasicus*) and milkfish (*Chanos chanos*) muscle. Biosci Biotech Biochem 57:571–577, 1993.
14. ST Jiang, JJ Lee, HC Chen. Purification and characterization of cathepsin B from ordinary muscle of mackerel (*Scomber australasicus*). J Agric Food Chem 42(5):1073–1079, 1994.
15. ST Jiang, JJ Lee, HC Chen. Purification and characterization of a novel cysteine proteinase from mackerel (*Scomber australasicus*). J Agric Food Chem 42:1639–1646, 1994.
16. ST Jiang, JJ Lee, HC Chen. Proteolysis of actomyosin by cathepsins B, L, L-like and X from mackerel (*Scomber australasicus*). J Agric Food Chem 44:769–773, 1996.
17. ST Jiang, BL Lee, CY Tsao, JJ Lee. Mackerel cathepsin B and L effects on thermal degradation of surimi. J Food Sci 62:1–6, 1997.
18. JJ Lee, HC Chen, ST Jiang. Purification and characterization of proteinases identified as cathepsin L and L-like (58 kDa) from mackerel (*Scomber australasicus*). Biosci Biotech Biochem 57:1470–1476, 1993.
19. JJ Lee, HC Chen, ST Jiang, Comparison of the kinetics of cathepsins B, L, L-like and X from the dorsal muscle of mackerel on the hydrolysis of methylcoumaryla- mind substrates. J Agric Food Chem 44:774–778, 1996.
20. AJ Barrett, H Kirschke. Cathepsin B, cathepsin H, and cathepsin L. Methods En- zymol 80:535–561, 1981.
21. A Okitani, U Matsukura, H Kato, M Fujimaki. Purification and some properties of a myofibrillar protein degrading protease, cathepsin L, from rabbit skeletal muscle. J Biochem 87:1133–1143,1980.
22. A Okitani, M Matsuishi, T Matsumoto, E Kamoshida, M Sato, U Matsukura, M Watanabe, H Kato, M Fujinaki. Purification and some properties of cathepsin B from rabbit skeletal muscle. Eur J Biochem 171:377–381, 1988.
23. WN Schwartz, WC Bird. Degradation of myofibrillar proteins by cathepsins B and D. Biochem J 167:811–820, 1977.
24. K Suzuki, S Tsuji, S Ishiura. Autolysis of calcium-activated neutral protease of chicken skeletal muscle. FEBS Lett 136:119–122, 1981.
25. K Suzuki, S Tsuji, S Ishiura, Y Kimura, S Kubota, K Imahori. Autolysis of cal- cium-activated neutral protease of chicken skeletal muscle. J Biochem 90:1787–1790, 1981.
26. ST Jiang, JH Wang, CS Chen. Purification and some properties of calpain II from tilapia muscle (*Tilapia nilotica X T. aurea*). J Agric Food Chem 39:237–241, 1991.
27. ST Jiang, JH Wang, CS Chen, JC Su. Substrate and calcium effects on the autoly- sis of tilapia muscle *m*-calpain. Biosci Biotech Biochem 59(1):119–120, 1995.
28. T Murachi, M Hatanaka, Y Yasumoto, N Nakayama, K Tanaka. A quantitative dis-

tribution study on calpain and calpastatin in rat tissues and cells. Biochem Int 2:651–656, 1981.

29. JH Wang, ST Jiang. The properties of calpain II from tilapia muscle (*Tilapia nilotica X T. aurea*). Agric Biol Chem 55:339–345, 1991.

30. JH Wang, ST Jiang. Stability of calcium-autolyzed calpain II from tilapia muscle (*Tilapia nilotica X T. aurea*). J Agric Food Chem 40:535–539, 1992.

31. JH Wang, WC Ma, JC Su, CS Chen, ST Jiang. Comparison of the properties of *m*-calpain from tilapia and grass shrimp muscles. J Agric Food Chem 41:1379–1384, 1993.

32. T Murachi. Calpain and calpastatin. Trends Biochem Sci 8:167–169, 1983.

33. JE Folk, JS Finlayson. The ε-(γ-glutamyl) lysine crosslinking and the catalytic role of transglutaminases. Adv Protein Chem 31:1–133, 1977.

34. JM Connellan, SI Chung, NK Whetzel, LM Bradley, JE Folk. Structural properties of guinea pig liver transglutaminase. J Biol Chem 246:1093–1095, 1971.

35. SI Chung, JE Folk. Kinetic studies with transglutaminase: the human blood enzymes (activates coagulation factor XIII) and the guinea pig hair follicle enzyme. J Biol Chem 247:2798–2807, 1972.

36. ST Jiang, JJ Lee. Purification, characterization, and utilization of pig plasma factor XIIIa. J Agric Food Chem 40:1101–1107, 1992.

37. H Hanigan, LA Goldsmith. Endogenous substrates for epidermal transglutaminase. Biochim Biophys Acta 522:589–601, 1978.

38. S Brenner, F Wold. Human erythrocyte transglutaminase purification and properties. Biochim Biophys Acta 522:74–83, 1978.

39. T Takagi, RF Doolittle. Amino acid sequence studies on factor XIII and the peptide released during its activation by thrombin. Biochemistry 13:750–756, 1974.

40. BRC De, F Traore, JC Meunier. Purification and properties of factor XIII from human placenta. Biochem 24:91–97, 1992.

41. N Seki, H Uno, NH Lee, I Kmura, K Toyoda, T Fujita, K Arai. Transglutaminase activity in Alaska pollack muscle and surimi, and its reaction with myosin B. Nippon Suisan Gakk 56:125–132, 1990.

42. A Asghar, AR Bhatti. Endogenous proteolytic enzymes in skeletal muscle: their significance in muscle physiology and during postmortem aging events in carcasses. Adv Food Res 343–451, 1987.

43. Y Bando, E Kominami, N Katunuma. Purification and tissue distribution of rat cathepsin L. J Biochem 100:35–42, 1986.

44. WR Dayton, JV Schollmeyer. Localization of a Ca^{2+}-activated neutral protease in skeletal muscle. J Cell Biol 87:267–273, 1980.

45. WR Dayton, JF Schollmeyer, RA Lepley. A calcium-activated protease possibly involved in myofibrillar protein turnover. Isolation of a low-calcium-requiring form of the protease. Bichim Biophys Acta 659:48–61, 1981.

46. GN DeMartino. Calcium-dependent proteolytic activity in rat liver: Identification of two proteases with different calcium requirement. Arch Biochem Biophys 211:253–257, 1981.

47. S Ishiura, H Sugita, I Nonaka, K Imahori. Calcium-activated neutral protease. Its localization in the myofibril, especially at the Z-band. J Biochem 86:579–581, 1980.

48. H Kirschke, AJ Barrett. Chemistry of lysosomal proteases. In: H Glaumann, FJ Ballard, ed. Lysosomes: Their Role in Protein Breakdown. London: Academic Press, 1987, pp 193–238.

49. P Lindahl, D Ripoll, M Abrahamson, JS Mort, AC Storer. Evidence for the interaction of valine-10 in cystatin C with the S_2 subsite of cathepsin B. Biochemistry 33:4384–4392, 1994.

50. D Musil, D Zucic, D Turk, RA Engh, I Mayr, R Huber, T Popovic, V Turk, T Towatari, N Katunuma, W Bode. The refined x-ray crystal structure of human liver cathepsin B: the structural basis for its specificity. EMBO J 10(9):2321–2330, 1991.

51. A Anastasi, MA Brown, AA Kembhavi, MJH Nicklin, CA Sayers, DC Sunter, AJ Barrett. Cystatin, a protein inhibitor of cysteine proteinases. Biochem J 211:129–138, 1983.

52. R Colella, Y Sakaguchi, H Nagase, JWC Bird. Chicken egg white cystatin. J Biol Chem 264(29):17164–17169, 1989.

53. I Bjork, E Pol, E Raub-Segall, M Abrahamson, AD Rowan, JS Mort. Differential changes in the association and dissociation rate constants for binding of cystatins to target proteinases occurring on N-terminal truncation of the inhibitors indicate that the interaction mechanism varies with different enzymes. Biochem J 299:219–225, 1994.

54. M Yamashita, S Konagaya. Immunochemical localization of cathepsins B and L in the white muscle of chum salmon (Oncorhynchus keta) in spawning migration probable participation of phagocytes rich in cathepsins in extensive muscle softening of the mature salmon. J Agric Food Chem 39:1402–1405, 1991.

55. T Towatari, N Katunuma. Selective cleavage of peptide bonds by cathepsins L and B from rat liver. J Biochem 93(4):1119–1128, 1983.

56. M Yamashita, S Konagaya. High activities of cathepsins B, D, H and L in the muscle of chum salmon in spawning migration. Comp Biochem Physiol 95B:149–152, 1990.

57. M Yamashita, S Konagaya. Purification and characterization of cathepsin L from the white muscle of chum salmon, *Oncorhynchus keta*. Comp Biochem Physiol 96B:247–252, 1990.

58. M Yamashita, S Konagaya. Purification and characterization of cathepsin B from the white muscle of chum salmon *Oncorhynchus keta*. Comp Biochem Physiol 96B:733–737, 1990.

59. M Yamashita, S Konagaya. Participation of cathepsin L into extensive softening of the muscle of chum salmon caught during spawning migration. Nippon Suisan Gakk 56:1271–1277, 1990.

60. M Yamashita, S Konagaya. Hydrolytic action of salmon cathepsins B and L to muscle structural proteins in respect of muscle softening. Nippon Suisan Gakk 57:1917–1922, 1991.

61. M Yamashita, S Konagaya. Proteolysis of muscle proteins in the extensively softened muscle of chum salmon caught during spawning migration. Nippon Suisan Gakk 57:2163–2169, 1991.

62. LC Sen, JR Whitaker. Some properties of a ficin-papain inhibitor from avian egg white. Arch Biochem Biophys 158:623–632, 1973.

63. H Keilova, V Tomasek. Effect of papain inhibitor from chicken egg white on cathepsin B1. Biochem Biophys Acta 334:179–186, 1974.

64. H Keilova, V Tomasek. Inhibition of cathepsin C by papain inhibitor from chicken egg white and by complex of this inhibitor with cathepsin B1. Coll Czech Chem Commun 40:218–224, 1975.

65. AJ Barrett. Cystatin, the egg white inhibitor of cysteine proteinases. Methods Enzymol 80:771–778, 1981.

66. AJ Barrett, ND Rawlings, ME Davies, W Machleidt, G Salvesen, V Turk. Cysteine protease inhibitors of the cystatin superfamily. In: J J Barrett, G Salvesen, eds. Proteinase Inhibitors Amsterdam: Elsevier 1986, pp 515–569.

67. B Lenarcic, A Ritonja, A Sali, M Kotnik, B Turk, W Machleidt. In: V Turk, ed. Cysteine Proteinases and Their Inhibitors. Berlin: Walter de Gruyter, 1986, pp 473–487.

68. I Ohkubo, K Kurachi, T Takasawa, H Shiokawa, M Sasaki. Isolation of a human cDNA for a thiol proteinase inhibitor and its identity with low molecular weight kininogen. Biochemistry 23:5691–5697, 1984.

69. A Grubb, H Lofberg, AJ Barrett. The disulphide bridges of human cystatin C (g-trace) and chicken cystatin. FEBS Lett 170:370–374, 1984.

70. V Turk, W Bode. The cystatins: protein inhibitors of cysteine proteinases. FEBS Lett 285:213–219, 1991.

71. T Sueyoshi, K Enjyoji, T Shimada, H Kato, S Iwanaga, Y Bando, E Kominami, N Katunuma. A new function of kininogens as thio-proteinase inhibitors: inhibition of papain and cathepsins B, H and L by bovine, rat and human plasma kininogens. FEBS Lett 182:193–195, 1985.

72. W Muller-Esterl, M Vohle-Timmermann, B Boos, M Dittman. Purification and properties of human low molecular weight kininogen. Biochim Biophys Acta 706:145–152, 1982.

73. S Higashiama, I Ohkubo, H Ishiguro, M Kunimatsu, K Sawaki, M Sasaki. Human high molecular weight kininogen as a thio proteinase inhibitor: presence of the entire inhibition capacity in the native form of heavy chain. Biochemistry 25:1669–1675, 1986.

74. G Salvesen, C Parkes, M Abrahamson, A Grubb, AJ Barrett. Human low Mr-kininogen contains three copies of a cystatin sequence that are divergent in structure and in inhibitory activity for cysteine proteinases. Biochem J 234:429–434, 1987.

75. A Kishimoto, N Kajikawa, H Tabuchi, M Shiota, Y Nishizuka. Calcium-dependent neutral proteases, widespread occurrence of a species of a protease active at lower concentration of calcium. J Biochem (Tokyo) 90:889–892, 1981.

76. T Murakami, M Hatanaka, T Murachi. The cytosol of human erythrocytes contains a highly Ca-sensitive thiol protease (calpain I) and its specific inhibitor protein (calpastatin). J Biochem 90:1809–1816, 1981.

77. WR Dayton. Comparison of low- and high-calcium-requiring forms of the calcium-activated protease with their autocatalytic breakdown products. Biochim Biophys Acta 709:166–173, 1982.

78. PA Nagainis, SK Sathe, DE Goll, T Edmunds. Autolysis of high-Ca^{2+} and low-Ca^{2+}

forms of the Ca^{2+}-dependent proteinase from bovine skeletal muscle. Fed Proc 42:1780–1786, 1983.

79. J Cong, DE Goll, AM Peterson, HP Kapprell. The role of autolysis in activity of the calcium dependent proteinases (mu-calpain and m-calpain). J Biol Chem 264(17):10096–10103, 1989.

80. RL Mellgren. Calcium-dependent protease: an enzyme system active at cellular membranes? FASEB J 1:110–115, 1987.

81. K Suzuki, S Imajoh, Y Emori, H Kawasaki, Y Minami, S Ohno. Calcium-activated neutral protease and its endogenous inhibitor. FEBS Lett 220–271, 1987.

82. SA Coolican, DR Hathaway. Effect of L-a-phosphatidylinositol on a vascular smooth muscle Ca^{2+}-dependent protease. J Biol Chem 261:11627–11631, 1986.

83. DE Goll, WC Kleese, A Okitani, T Kumamoto, J Cong, H-P Kapprell. Historical background and current status of the Ca^{2+}-dependent proteinase system. In: R L Mellgren, T Murachi, eds. Intracellular Calcium-Dependent Proteolysis. Boca Raton, FL: CRC Press, 1990, pp 13–15.

84. M Nakamura, M Inomata, S Imajoh, K Suzuki, S Kawashima. Fragmentation of an endogenous inhibitor upon complex formation with high and low-Ca^{2+}-requiring forms of calcium-activated neutral protease. Biochemistry 28:449–455, 1989.

85. RL Mellgren, A Repetti, TC Muck, J Easly. Rabbit skeletal muscle calcium-dependent protease requiring millimolar Ca^{2+}. J Biol Chem 257:7203–7209, 1982.

86. UJP Zimmerman, WW Schlaepfer. Two-stage autolysis of the catalytic subunit initiated activation of calpain I. Biochim Biophys Acta 192:1078–1082, 1991.

87. M Koohmaraie. The role of calcium-dependent proteases calpains in post-mortem proteolysis and meat tenderness. Biochimie (Paris) 74:239–245, 1992.

88. M Koohmaraie. Effect of pH, temperature, and inhibitors on autolysis and catalytic activity of bovine skeletal muscle m-calpain. J Anim Sci 70:3071–3080, 1992.

89. A Asghar, RL Henrickson. Postmortem stimulation of carcasses: effects on biochemistry, biophysics, microbiology and quality of meat. Crit Rev Food Sci Nutr 18:1–58, 1982.

90. M Koohmaraie, AS Babiker, RA Merkel, TR Dutson. Role of Ca-dependent proteases and lysosomal enzymes in post-mortem changes in bovine skeletal muscle. J Food Sci 53:1253–1257, 1988.

91. M Koohmaraie, AS Babiker, AL Schoreder, RA Merkel, TR Dutson. Acceleration of postmortem tenderization in bovine carcasses through activation of Ca-dependent protease. J Food Sci 53:1638–1641, 1988.

92. T Murachi. Intracellular Ca protease and its inhibitor protein: calpain and calpastatin. In: WY Cheung ed. Calcium and Cell Function, vol. IV. Academic Press, 1983, pp 377–410.

93. T Taneda, T Watanabe, N Seki. Purification and some properties of a calpain from carp muscle. Bull Jpn Soc Sci Fish 49:219–228, 1983.

94. H Toyohara, Y Makinodan, K Tanaka, S Ikeda. Purification and properties of carp (*Cyprinus carpio*) muscle calpain II (high-Ca^{2+}-requiring form of calpain). Comp Biochem Physiol 81B:573–578, 1985.

95. M Ishioroshi, K Samejima, T Yasui. Further studies on the roles of the head and

tail regions of the myosin molecule in heat-induced gelation. J Food Sci 47:114–1120, 1982.

96. K Hara, A Suzumatsu, T Ishihara. Purification and characterization of cathepsin B from carp ordinary muscle. Nippon Suisan Gakk 54:1243–1252, 1988.
97. HP Kapprell, DE Goll. Effect of Ca²⁺ on binding of the calpains to calpastatin. J Biol Chem 264(30):17888–17896, 1989.
98. T Abe, SI Chung, RP DiAugustine, JE Folk. Rabbit liver transglutaminase: physical, chemical, and catalytic properties. Biochemistry 16:5495–5501, 1977.
99. Y Kumazawa, K Nakanishi, H Yasueda, M Motoki. Purification and characterization of transglutaminase from walleye pollack liver. Fisheries Sci 62:959–964, 1996.
100. Y Kumazawa, K Nakanishi, H Yasueda, M Motoki. Purification and characterization of transglutaminase from Japanese oyster (*Crassostrea gigas*). J Agric Food Chem 45:604–610, 1997.
101. JE Folk, PW Cole. Identification of a functional cysteine essential for the activity of guinea pig liver transglutaminase. J Biol Chem 41:3238–3240, 1966.
102. JE Folk, PW Cole. Transglutaminase: mechanistic features of the active site as determined by kinetic and inhibitor studies. Biochim Biophys Acta 122:244–264, 1966.
103. JE Folk, SI Chung. Molecular and catalytic properties of transglutaminases. Adv Enzymol 38:109–191, 1973.
104. JE Folk. Transglutaminases. Annu Rev Biochem 49:517–531, 1980.
105. L Lorand, SM Conrad. Transglutaminases. Mol Cell Biol Chem 58:9–35, 1984.
106. K Ikura, T Nasu, H Yoshikawa, Y Tsuchiya, R Sasaki, H Chiba. Amino acid sequence of guinea pig liver transglutaminase from its cDNA sequence. Biochemistry 27:2898–2905, 1988.
107. GJ Tsai, SM Lin, ST Jiang. Transglutaminase from *Streptoverticillium ladakanum* and application to minced fish product. J Food Sci 61:1234–1238, 1996.
108. H Ando, M Adachi, K Umeda, A Matsuura, M Nonaka, R Uchio, H Tanaka, M Motoki. Purification and characteristics of a novel transglutaminase derived from microorganisms. Agric Biol Chem 53:2613–2617, 1989.
109. GJ Tsai, JW Wu, ST Jiang. Screening the microorganism and some factors for the production of transglutaminase. J Chinese Agric Chem Soc 34:228–240, 1996.
110. WSD Wong, C Batt, JE Kinsella. Purification and characterization of rat liver transglutaminase. Int J Biochem 22:53–59, 1990.
111. C De Backer-Royer, JC Meunier. Effect of temperature and pH on factor XIIIa from human placenta. Biochem 24:637–642, 1992.
112. AG Loewy, K Dunathan, R Kriel, HL Wolfinger. Fibrinase. I. Purification of substrate and enzyme. J Biol Chem 236:2625–2633, 1961.
113. ML Schwartz, SV Pizzo, RL Hill, PA Mckee. Human factor XIII from plasma and platelets. J Biol Chem 248:1395–1407, 1973.
114. GJ Tsai, FC Tsai, ST Jiang. Cross-linking activity of mackerel muscle proteins by microbial transglutaminase. J Fish Soc (Taiwan) 22:125–135, 1995.
115. ST Jiang, SZ Leu, GJ Tsai. Crosslinking of mackerel surimi actomyosin by microbial transglutaminase and UV irradiation. J Agric Food Chem 46:5278–5282, 1998.

116. U Gerber, U Jucknischke, S Putzien, HL Fuchsbauer. A rapid and simple method for the purification of transglutaminase from *Streptoverticillium mobaraense*. Biochem J 299:825–829, 1994.

117. M Nonaka, H Tanaka, A Okiyama, M Motoki, H Ando, K Umeda, A Matsuura. Polymerization of several proteins by Ca^{2+}-independent transglutaminase derived from microorganisms. Agric Biol Chem 53:2619–2623, 1989.

118. MV Ramanujam, JH Hageman. Intracellular transglutaminase (EC 2.3.2. 13) in a procaryote evidence from vegetative and sporulating cells of *Bacillus subtilis* 168. FASEB J 4:A2321, 1990.

119. JD Klein, E Gozman, GD Koehn. Purification and partial characterization of transglutaminase from *Physarum polycephalum*. J Bacteriol 174:2599–2605, 1992.

120. M Nonaka, S Toiguchi, H Sakamoto, H Kawajiri, T Soeda, M Motoki. Changes caused by microbial transglutaminase on physical properties of thermally induced soy protein gels. Food Hydrocolloids 8:1–8, 1994.

121. H Sakamoto, Y Kumazawa, M Motoki. Strength of protein gels prepared with microbial transglutaminase as related to reaction conditions. J Food Sci 59:866–871, 1994.

122. H Sakamoto, Y Kumazawa, S Toiguchi, K Seguro, T Soeda, M Motoki. Gel strength enhancement by addition of microbial transglutaminase. J Food Sci 53:924–928, 1988.

123. K Seguro, Y Kumazawa, T Ohtsuka, S Toiguchi, M Motoki. Microbial transglutaminase and ε-(γ-glutamyl) lysine crosslink effects on elastic properties of kamaboko gels. J Food Sci 60:305–311, 1995.

124. H Tanaka, M Nonaka, M Motoki. Polymerization and gelation of microbial transglutaminase. Nippon Suisan Gakk 56: 1341–1348, 1990.

125. RA Lawrie. Meat Science, 4th ed. Oxford: Pergamon Press, 1985, pp 401–459.

126. BB Marsh. Electrical stimulation. In: Advances in Meat Research, vol. 1. Westport, CT: AVI Publishing Co., Ltd., 1985, p 227.

127. FM Robbins, JE Walker, SH Cohen, S Charterjee. Action of proteolytic enzymes on bovine myofibrils. J Food Sci 44:1672–1677, 1979.

128. M Mikami, AH Whiting, MAJ Taylor, RA Maciewicz, DJ Etherington. Degradation of myofibrils from rabbit, chicken and beef by cathepsin L and lysosomal lysates. Meat Sci 21:81–97, 1987.

129. A Ouali, N Garrel, A Obled, C Deval, C Valin. Comparative action of cathepsins D, B, H, L and of a new lysosomal cysteine proteinase on rabbit myofibrils. Meat Sci 19:83–88, 1987.

130. DJ Etherington, MA Taylor, E Dransfield. Conditioning of meat from different species-relationship between tenderising and the levels of cathepsin B, cathepsin L, calpain I, calpain II and beta-glucuronidase. Meat Sci 20(1):1–18, 1987.

131. I Kolodziejska, ZE Sikorski, M Sadowska. Texture of cooked mantle of squid Illex argentinus as influenced by specimen characteristics and treatments. J Food Sci 52:932–935, 1987.

132. Y Makinodan, S Ikeda. Studies on fish muscle protease. IV. Relation between Himodori of kamaboko and muscle proteinase. Bull Jpn Soc Sci Fish 37:518–523, 1971.

133. Y Makinodan, S Ikeda. Relation between himodori of kamaboko and muscle proteinase. Nippon Suisan Gakk 37:518–523, 1971.

134. K Iwata, K Kobashi, J Hase. Some enzymatic properties of carp muscular alkaline protease. Nippon Suisan Gakk 40:189–200, 1974.

135. H Su, TS Lin, TC Lanier. Contribution of retained organ tissues to the alkaline protease content of mechanically separated Atlantic croaker (*Micropogon undulatus*). J Food Sci 46:1650–1653, 1981.

136. H Su, TS Lin, TC Lanier. Investigation into potential sources of heat-stable alkaline protease in mechanically separated Atlantic croaker (*Micropogon undulatus*). J. Food Sci 46:1654–1656, 1981.

137. L Busconi, EJ Folco, RE Martone, JJ Sanchez. Identification of two alkaline proteases and trypsin inhibitor from muscle of white croaker (*Micropogon opercularis*). FEBS Lett 176:211–216, 1984.

138. H Toyohara, Y Shimizu. Relation between the modori phenomenon and myosin heavy chain breakdown in threadfin bream gel. Agric Biol Chem 52:255–257, 1988.

139. M Kinoshita, H Toyohara, Y Shimizu. Diverse distribution of four distinct types of modori (gel degradation)-inducing proteinases among fish species. Nippon Suisan Gakk 56:1485–1492, 1990.

140. H An, TA Seymour, JW Wu, MT Morrissey. Assay systems and characterization of Pacific whiting (*Merluccius productus*) protease. J Food Sci 59:277–281, 1994.

141. T Masaki, M Shimomukai, Y Miyauchi, S Ono, T Tuchiya, T Mastuda, H Akazawa, M Soejima. Isolation and characterization of the protease responsible for jellification of pacific hake muscle. Nippon Suisan Gakk 59:683–690, 1993.

142. RW Porter, B Koury, G Kudo. Inhibition of protease activity in muscle extracts and surimi from pacific whiting, *Merluccius productus,* and arrowtooth flounder, *Atheresthes stomias.* Marine Fish Rev 55(3):10–15, 1993.

143. MT Morrissey, PS Hartley, H An. Proteolytic activity in pacific whiting and effect of surimi processing. J Aquat Food Prod Technol 4(4):5–18, 1995.

144. S Boye, TC Lanier. Effects of heat-stable alkaline protease activity of Atlantic menhaden (*Brevoorti tyrannus*) on surimi gels. J Food Sci 53:1340–1346, 1988.

145. DD Hamann, PM Amato, MC Wu, ER Jones. Inhibition of modori (gel-weakening) in surimi by plasma hydrolysate and egg white. J Food Sci 55:665–669, 1990.

146. Y Makinodan, H Toyohara, E Niwa. Implication of muscle alkaline proteinase in the texture degradation of fish meat gel. J Food Sci 50:1351–1355, 1985.

147. JR Bendall. Chapter 5, Postmortem changes in muscle. In: The Structure and Function of Muscle, vol. 2, 2nd ed. New York: Academic Press, 1973, pp 243.

148. M Suyama, A Konosu. Chapter 5, Changes in postmortem fish. In: Suisan Shokuhin Gaku (Marine Food Science) Tokyo: Koseisha, Koseikaku, 1987, pp 93–107.

149. HJ Swatland. The conversion of muscles to meat. In: Structure and Development of Meat Animals and Poultry. Lancaster, PA: Technomic Pub. Co. Inc., 1994, p 495.

150. Lawrie, R.A. Conversion of muscle into meat: Biochemistry. In: DE Johnston, MK Knight, DA Ledward, ed. The Chemistry of Muscle-Based Foods. Cambridge: The Royal Society of Chemistry, 1992, p 43.

151. C Valin, C Tassy, G Farias, G Geesink, A Ouali. International Workshop on Proteolysis and Meat Quality, 24–28, May, 1993, Clermont-Ferrand, France.

152. G Monin, A Ouali. In: R Lowrie, ed. Development in Meat Science 5. London: Elsevier Appl. Sci. Pub, 1991, p 99.

153. M Koohmaraie. Muscle proteinases and meat aging. Meat Sci 36(1/2):93–104, 1994.

154. A Ouali. Meat tenderization: possible causes and mechanisms. A review. J Muscle Foods 1:129–165, 1990.

155. A Ouali. Proteolytic and physicochemical mechanisms involved in meat texture development. Biochimie 74:251–265, 1992.

156. IF Penny. Chapter 5, the enzymology of conditioning. In: R Lawrie, ed. Development in Meat Science—I. Englewood, WJ: Applied Science Publishers Inc., 1980, p 115.

157. RG Taylor, GH Geesink, VF Thompson, M Koohmaraie, DE Goll. Is Z-disk degradation responsible for postmortem tenderization. J Anim Sci 73:1351–1367, 1995.

158. CL Davey. Post-mortem chemical changes in muscle-meat aging. Proc Recip Meat Conf 108–115, 1983.

159. DE Goll. Role of proteinases and protein turnover in muscle growth and meat quality. Recip Meat Conf (American Meat Science Association) 44:25–36, 1991.

160. DE Goll, WC Kleese, A Szpacenko. Chapter 8, Skeletal muscle proteases and protein turnover. In: DR Campion, GJ Hausman, RJ Martin, ed. Animal Growth Regulation. New York: Plenum Press, 1989, p 141.

161. DE Goll, VF Thompson, RG Taylor, K Christiansen. Role of the calpain system in muscle growth. Biochimie 74:225–237, 1992.

162. Greaser, M.L. Poultry muscle as food. In: PJ Bechtel, ed. Muscle as Food. New York: Academic Press, 1986, p 395.

163. M Koohmaraie. Biochemical factors regulating the toughening and tenderization processes of meat. Meat Sci 43(S):S193–S201, 1996.

164. M Koohmaraie, SD Shackelford, TL Wheeler, SM Lonergan, ME Doumit. A muscle hypertrophy condition in lamb (callipyge): characterization of effects on muscle growth and meat quality traits. J Anim Sci 73:3596–3607, 1995.

165. RG Taylor, C Tassy, M Briand, N Robert, Y Briand, A Ouali. Proteolytic activity of proteasome on myofibrillar structures. Mol Biol Rep 21:71–73, 1995.

166. A Hattori, T Ishii, R Tatsumi, K Takahasi. Changes in the molecular types of connectin and nebulin during development of chicken skeletal muscle. Biochim Biophys Acta 1244:179–184, 1995.

167. MG Price. In: SKJ Melhorta ed. Advances in Structural Biology, vol. I. Greenwich, CT: AI Press, 1991, p 175.

168. JD Hay, RW Currie, FH Wolfe. Effects of postmortem aging on chicken muscle lipids. J Food Sci 38(4):696–699, 1973.

169. CL Davey, KV Gilbert. Studies in meat tenderness. VII. Changes in the fine structure of meat during ageing. J Food Sci 34:69–74, 1969.

170. GL Gann, RA Merkel. Ultrastructural changes in bovine longissimus muscle during postmortem ageing. Meat Sci 2(2):129–144, 1978.

171. MT Abbott, AM Pearson, JF Price, GR Hooper. Ultrastructural changes during autolysis of red and white porcine muscle. J Food Sci 42:1185–1188, 1977.

172. A Hattori. Aging of meat. Chem Biol 24:789–791, 1986.

173. M Koohmaraie, JE Schollmeyer, TR Dutson. Effect of low-calcium-requiring cal-

cium activated factor on myofibrils under varying pH and temperature conditions. J Food Sci 51:28–32, 1986.

174. M Koohmaraie, SC Seideman, JE Schollmeyer, TR Dutson, JD Crouse. Effect of post-mortem storage on Ca^{2+}-dependent proteases, their inhibitor and myofibril fragmentation. Meat Sci 19(3):187–196, 1987.

175. M Koohmaraie, JD Crouse, HJ Mersmann. Acceleration of postmortem tenderization in ovine carcasses through infusion of calcium chloride:effect of concentration and ionic strength. J Anim Sci 67:934–942, 1989.

176. M Koohmaraie, G Whipple, JD Crouse. Acceleration of postmortem tenderization in lamb and Brahman-cross beef carcasses through infusion of calcium chloride. J Anim Sci 68:1278–1283, 1990.

177. JS Bond, PE Butler. Intracellular proteases. Annu Rev Biochem 56:333–364, 1987.

178. WR Dayton, WJ Reville, DE Golf, MH Stromer. A Ca^{2+}-activated protease possibly involved in myofibrillar protein turnover. Partial characterization of the purified enzyme. Biochemistry 15:2159–2167, 1976).

179. M Matsuishi, A Okitani, Y Hayakawa, H Kato. Cysteine proteinase inhibitors from rabbit skeletal muscle. Int J Biochem 20(3):259–264, 1988.

180. A Ouali, L Bige, A Obled, A Lacourt, C Valin. In: V Turk, ed. Cysteine Proteinases and Their Inhibitors. Berlin: Walter de Gruyter, 1986, p 545.

181. TL Wheeler, M Koohmaraie. A modified procedure for simultaneous extraction and subsequent assay of calcium-dependent and lysosomal protease systems from a skeletal muscle biopsy. J Anim Sci 69:1559–1565, 1991.

182. DE Goll, GH Geesink, RG Taylor VF Thompson. Does proteolysis cause all post-mortem tenderization, or are changes in the actin/myosin interaction involved? Proc Int Cong Meat Sci Technol 41:537, 1995.

183. A Ouali, A Talmant. Calpains and calpastatin distribution in bovine, porcine and ovine skeletal muscles. Meat Sci 28:331–348, 1990.

184. A Ducastaing, C Valin, J Schollmeyer, R Cross. Effect of electrical stimulation on post-mortem changes in the activities of two Ca dependent neutral proteinases and their inhibitor in beef muscle. Meat Sci 15:193–202, 1985.

185. E Dransfield. Modelling post-mortem tenderization IV. Role of calpains and calpastatin in conditioning. Meat Sci 34:217–234, 1993.

186. E Dransfield. Modelling post-mortem tenderization V. Inactivation of calpains. Meat Sci 37(3):391–398, 1994.

187. IF Penny, DJ Etherington, JL Reeves, MAJ Taylor. Proceedings, 30th European Meeting, Meat Research Workers, Bristol, 1984, p 133.

188. PJA Davies, D Wallach, MC Willingham, I Paston, M Yamaguchi, RM Robson. Filamin-actin interaction. Dissociation of binding from gelation by Ca^{2+}-activated proteolysis [phosphoproteins, chickens]. J Biol Chem 253(11):4036–4042, 1978.

189. ML Lusby, JF Ridpath, Jr FC Parrish, RM Robson. Effect of postmortem storage on degradation of the myofibrillar protein titin in bovine longissimus muscle. J Food Sci 48:1787–1790, 1983.

190. NL King. Breakdown of connectin during cooking of meat. Meat Sci 11:27–43, 1984.

191. NFS Gault. In: DE Johnston, MK Knight, DA Ledward, eds. The Chemistry of Muscle-Based Foods. Cambridge: The Royal Society of Chemistry, 1992, p 98.

192. DJ Etherington. The contribution of proteolytic enzymes to postmortem changes in muscle. J Anim Sci 59(6):1644–1650, 1984.

193. IF Penny, E Dransfield. Relationship between toughness and troponin T in conditioned beef. Meat Sci 3:135–141, 1979.

194. CL Davey, MA Dickson. Studies in meat tenderness VIII. Ultrastructural changes in meat during aging. J Food Sci 35:56–64, 1970.

195. RH Locker, DJC Wild. Myofibrils of cooked meat are a continuum of gap filaments. Meat Sci 7:189–195, 1982.

196. DG Olson, FC Parrish, WR Dayton, DE Goll. Effect of postmortem storage and calcium activated factor on the myofibrillar proteins of bovine skeletal muscle. J Food Sci 42:117–120, 1977.

197. MG Zeece, RM Robson, ML Lusby, FC Parrish. Effect of calcium activated protease on bovine myofibrils under different conditions of pH and temperature. J Food Sci 51:79–84, 1986.

198. MG Zeece, K Katoh, RM Robson, FC Parrish. Effect of cathepsin D on bovine myofibrils under different conditions of pH and temperature. J Food Sci 51:769–772, 1986.

199. A Ouali, A Obled, C Deval, N Garrel, C Valin. Proteolytic action of lysosomal proteinases on the myofibrillar structure, Comparison with CaANP effects and postmortem changes. Proceeding, of 30th European Meeting Meat Research Workers, 1984, pp 126–138.

200. MC Wu, T Akahane, TC Lanier, DD Hamann. Thermal transitions of actomyosin and surimi prepared from Atlantic croaker as studied by differential scanning calorimetry. J Food Sci 50:10–13, 1985.

201. MC Wu, TC Lanier, DD Hamann. Rigidity and viscosity changes of croaker actomyosin during thermal gelation. J Food Sci 50:14–19&25, 1985.

202. V Beas, M Crupkin, RE Trucco. Gelling properties of actomyosin from pre- and post-spawning hake (*Merluccius hubbsi*). J Food Sci 53:1322–1326, 1988.

203. T Sano, SF Noguchi, JJ Marsumoto, T Tsuchiya. Thermal gelation characteristics of myosin subfragments. J Food Sci 55:55–58&70, 1990.

204. M Ogawa, T Ehara, T Tamiya, T Tsuchiya. Thermal stability of fish myosin. Comp Biochem Physiol 106B:517–521, 1993.

205. M Ogawa, J Kanamaru, H Miyashita, T Tamiya, T Tsuchiya. Alpha-helical structure of fish actomyosin: Changes during setting. J Food Sci 60:297–299, 1995.

206. GR Ziegler, EA Foegeding. The gelation of proteins. Adv Food Nutr Res 34:203–297, 1993.

207. I Kimura, M Sugimoto, K Toyoda, N Seki, K Arai, T Fujita. A Study on the cross-linking reaction of myosin in kamaboko "Suwari" gels. Nippon Suisan Gakk 57:1389–1396, 1991.

208. TC Lanier. Functional properties of surimi. Food Technol 40(3):107–114, 1986.

209. MT Morrissey, JW Wu, DD Lin, H An. Effect of food grade protease inhibitor on autolysis and gel strength of surimi. J Food Sci 58:1050–1054, 1993.

210. E Niwa. Measurement of surimi composition and functional properties. In: TC Lanier, CM Lee, eds. Surimi Technology. New York: Marcel Dekker, Inc., 1992, pp 389–427.

211. BY Kim, DD Hamann, TC Lanier, MC Wu. Effects of freeze–thaw abuse on the viscosity and gel-forming properties of surimi from two species. J Food Sci 51:951–956&1004, 1986.

212. Y Shimizu, R Machida, S Takenami. Species variations in the gel-forming characteristics of fish meat paste. Nippon Suisan Gakk 47:95–104, 1981.

213. Y Tsukamasa, Y Shimizu. Setting property of sardine and pacific mackerel meat. Nippon Suisan Gakk 56:1105–1112, 1990.

214. J Yongsawatdigul, JW Park, YA Dagga, E Kolbe. Ohmic heating maximizes gel functionality of Pacific whiting surimi. J Food Sci 60:10–14, 1995.

215. DH Wasson, KD Reppond, JK Babbitt, JS French. Effects of additives on proteolytic and functional properties of arrowtooth flounder surimi. J Aquat Food Prod Technol 1(3/4):147–165, 1992.

216. D Wasson, JK Babbitt, JS French. Characterization of a heat stable protease from arrowtooth flounder: Atheresthes stomias. J Aquat Food Prod Technol 1(3/4):167–182, 1992.

217. TA Seymour, MT Morrissey, MY Gustin, H An. Purification and characterization of pacific whiting proteases. J Agric Food Chem 42:2421–2427, 1994.

218. Y Makinodan, H Toyohara, S Ikeda. Comparison of muscle proteinase activity among fish species. Comp Biochem Physiol 79B:129–134, 1984.

219. DH Greene, J Babbitt. Control of muscle softening and protease-parasite interactions in arrowtooth flounder (*Atheresthes stomias*). J Food Sci 55:579–580, 1990.

220. JF Morado, A Sparks. Observations on the host-parasite relations of the pacific whiting, *Merluccius productus* (Ayres), and two Myxosporean parasites, *Kudoa thyrsitis* (Gilchrist, 1924) and K. paniformis Kabata and Whitaker. J Fish Dis 9:445–455, 1986.

221. I Stoknes, R Rustad, V Mohr. Comparative studies of the proteolytic activity of tissue extracts from cod (*Gadus morhua*) and herring (*Clupea harengus*). Comp Biochem Physiol 106B:613–619, 1993.

222. I Stoknes, T Rustad. Proteolytic activity in muscle from Atlatic salmon (*Salmo salar*). J Food Sci 60:711–714, 1995.

223. Y Makinodan, Y Yokoyama, M Kinoshita, H Toyohara. Characterization of an alkaline proteinase of fish muscle. Comp Biochem Physol 87B:1041–1046, 1987.

224. Y Makinodan, T Kitagawa, H Toyohara, Y Shimizu. Participation of muscle alkaline protease in himodori of kamaboko. Nippon Suisan Gakk 53:99–101, 1987.

225. Y Makinodan, S Ikeda. Alkaline proteinase of carp muscle: effects of some protein denaturing agents on the activity. J Food Sci 42:1026–1028&1033, 1977.

226. EJ Folco, L Busconi, CB Martone, RE Trucco, JJ Sanchez. Activation of an alkaline proteinase from fish skeletal muscle by fatty acids and sodium dodecyl sulfate. Comp Biochem Physiol 91B:473–476, 1988.

227. MV Chang-Lee, R Pacheco-Aguilar, DL Crawford, LE Lampila. Proteolytic activity of surimi from Pacific whiting (*Merluccius productus*) and heat-set gel texture. J Food Sci 54:1116–1119, 1989.

228. M Kinoshita, H Toyohara, Y Shimizu, M Sakaguchi. Modori-inducing proteinase active at 50°C in threadfin bream muscle. Nippon Suisan Gakk 58:715–720, 1992.

229. S Noguchi. The control of denaturation of fish muscle proteins during frozen storage. 1974, Doctoral Dissertation, Sophia University, Tokyo, Japan.

230. JW Park, J Yongsawatdigul, TM Lin. Rheological behavior and potential crosslinking of Pacific whiting surimi gel. J Food Sci 59:773–776, 1994.

231. K Piyachomkwan, MH Penner. Inhibition of pacific whiting surimi-associated protease by whey protein concentrate. J Food Biochem 18:341–353, 1995.

232. KD Reppond, JK Babbitt, S Berntsen, M Tsuruta. Gel properties of surimi from Pacific herring. J Food Sci 60:707–710, 1995.

233. FI Garrcia-Carreno, MA Navarrete Del Toro, M Diaz-Lopez, MP Hernandez-Cortes, JM Ezquerra. Proteinase inhibition of fish muscle enzymes using legume seed extracts. J Food Protect 59:312–318, 1996.

234. H Akazawa, Y Miyauchi, K Sakurada, DH Wasson, KD Reppond. Evaluation of protease inhibitors in Pacific whiting surimi. J Aquatic Food Product Technol 2(3):79–95, 1993.

235. K Iwata, K Kobashi, J Hase. Studies on muscle alkaline protease-VI. Purification of proteins which induce the modori phenomenon during kamaboko production and of cathepsin A from carp muscle. Nippon Suisan Gakk 43:181–193, 1977.

236. K Iwata, K Kobashi, J Hase. Effect of carp musclar alkaline protease from carp muscle. Nippon Suisan Gakk 40:1325–1337, 1974.

237. K Ikura, T Kometani, R Sasaki, H Chiba. Crosslink ing of soybean 7S and 11S proteins by transglutaminase. Agric Biol Chem 44:2979–2984, 1980.

238. K Ikura, T Kometani, M Yoshikawa, R Sasaki, H Chiba. Crosslinking of casein components by transglutaminase. Agric Biol Chem 44:1567–1573, 1980.

239. L Kurth, PJ Rogers. Transglutaminase catalyzed cross-linking of myosin to soya protein, casein and gluten. J Food Sci 49:573–576, 1984.

240. M Motoki, N Nio, K Takinami. Functional properties of food proteins polymerized by transglutaminase. Agric Biol Chem 48:1257–1261, 1984.

241. G Matheis, JR Whitaker. A review: enzymatic cross-linking of proteins applicable to foods. J Food Biochem 11:309–327, 1987.

242. SY Tanimoto, JE Kinsella. Enzymatic modification of proteins effects of transglutaminase cross-linking on some physical properties of β-lactoglobulin. J Agric Food Chem 36:281–285, 1988.

243. K Ikura, M Yoshikawa, R Sasaki, H Chiba. Incorporation of amino acids into food proteins by transglutaminase. Agric Biol Chem 45:2587–2592, 1981.

244. K Ikura, K Okumura, M Yoshikawa, R Sasaki, H Chiba. Incorporation of lysyldipeptides into food protein by transgluta minase. Agric Biol Chem 49:1877–1878, 1985.

245. M Motoki, H Aso, K Seguro, N Nio. α-S$_1$-casein film prepared using transglutaminase. Agric Biol Chem 51:993–996, 1987.

246. L Kurth. Crosslinking of myosin and casein by the enzyme transglutaminase. Food Technol Aust 35:420–423, 1983.

247. F Traore, JC Meunier. Cross-linking activity of placental factor XIIIa on whey proteins and caseins. J Agric Food Chem 40:399–402, 1992.

248. N Nio, M Motoki, K Takinami. Gelation of casein and soybean globulin by transglutaminase. Agric Biol Chem 49:2283–2286, 1985.

249. M Motoki, N Nio, K Takinami. Functional properties of heterologous polymer prepared by transglutaminase between milk casein and soybean globulin. Agric Biol Chem 51:237–239, 1987.

250. D Joseph, TC Lanier, DD Hamann. Temperature and pH affect transglutaminase-catalyzed "setting" of crude fish actomyosin. J Food Sci 59:1018–1023, 1994.

251. HG Lee, TC Lanier, DD Hamann, JA Knopp. Transglutaminase effects on low temperature gelation of fish protein sols. J Food Sci 62:20–24, 1997.

252. A AKM Nowsad, S Kanoh, E Niwa. Electrophoretic behavior of cross-linked myosin heavy chain in suwari gel. Nippon Suisan Gakk 59:667–671, 1993.

253. A AKM Nowsad, S Kanoh, E Niwa. Effects of amine salts on the elasticity of suwari gel from Alaska pollack. Nippon Suisan Gakk 59:1017–1021, 1993.

254. A AKM Nowsad, S Kanoh, E Niwa. Setting of surimi paste in which transglutaminase is inactivated by p-chloromercuribenzoate. Fisheries Sci 60:185–188, 1994.

255. A AKM Nowsad, S Kanoh, E Niwa. Setting of surimi paste in which transglutaminase is inactivated by N-ethylmaleimide. Fisheries Sci 60:189–191, 1994.

256. A AKM Nowsad, S Kanoh, E Niwa. Setting of transglutaminase-free actomyosin paste prepared from Alaska pollack surimi. Fish Sci 60:295–297, 1994.

257. A AKM Nowsad, S Kanoh, E Niwa. Contribution of transglutaminase on the setting of various actomyosin pastes. Fish Sci 61:79–81, 1995.

258. Y Tsukamasa, K Sato, Y Shimizu, C Imai, M Sugiyama, Y Minegishi, M Kawabata. ε-(γ-Glutamyl) lysine crosslink formation in sardine myofibril sol during setting at 25°C. J Food Sci 58:785–787, 1993.

259. H Takeda, N Seki. Enzyme-catalyzed cross-linking and degradation of myosin heavy chain in walleye pollack surimi paste during setting. Fish Sci 62:462–467, 1996.

260. J Wan, I Kimura, M Satake, N Seki. Effect of calcium ion concentration on the gelling properties and transglutaminase activity of walleye pollack surimi paste. Fish Sci 60:107–113, 1994.

261. YP Huang, K Seguro, M Motoki, K Tawada. Cross-linking of contractile proteins from skeletal muscle by treatment with microbial transglutaminase. J Biochem 112:229–234, 1992.

262. T Taguchi, S Ishizaki, M Tanaka, Y Nagashima, K Amano. Effect of ultraviolet irradiation on the thermal gelation of fish meat pastes. Nippon Shokulin Kogyo Gakk 35:452–455, 1988.

263. T Taguchi, S Ishizaki, M Tanaka, Y Nagashima, K Amano. Effect of ultraviolet irradiation on thermal gelation of muscle pastes. J Food Sci 54:1438–1440, 1989.

264. S Ishizaki, M Hamada, N Iso, T Taguchi. Effect of ultra-violet irradiation on rheological properties of thermal gels from sardine and pork meat pastes. Nippon Suisan Gakk 59:1219–1224, 1993.

265. S Ishizaki, M Hamada, M Tanaka, T. Taguchi. Conformational changes in actomyosins from fish and pork muscles by ultraviolet irradiation. Nippon Suisan Gakk 59.2071–2077, 1993.

266. S Ishizaki, M Ogasawara, M Tanaka, T Taguchi. Ultraviolet denaturation of flying fish myosin and its fragments. Fish Sci 60:603–606, 1994.

267. SF Curran, MA Ammoruso, BD Goldstein, RA Berg. Degradation of soluble collagen by ozone or hydroxyl radicals. FEBS Lett 176:155–160, 1984.

268. K Uchida, Y Kato, S Kawakishi. A novel mechanism for oxidative cleavage of propyl peptides induced by the hydroxyl radical. Biochem Biophys Res Commun 169:265–271, 1990.

269. K Uchida, Y Kato, S Kawakishi. Metal-catalyzed oxidative degradation of collagen. J Agric Food Chem 40:9-121992.

270. Y Kato, K Uchida, S Kawakishi. Oxidative degradation of collagen and its model peptide by ultraviolet irradiation. J Agric Food Chem 40:373–379, 1992.

271. E Fujimori. Changes induced by ozone and ultraviolet light in type I collagen. Eur J Biochem 152:299–306, 1985.

272. E Fujimori. Cross-linking of CNBr peptide by ozone or UV light. FEBS Lett 235:98–102, 1988.

273. Y Kano, Y Sakano, D Fujimoto. Cross-linking of collagen by ascorbate-copper ion system. J Biochem 102:839–842, 1987.

16

Endogenous Enzyme Activity and Seafood Quality: Influence of Chilling, Freezing, and Other Environmental Factors

Zdzisław E. Sikorski
Technical University of Gdańsk, Gdańsk, Poland

Edward Kołakowski
Agricultural University of Szczecin, Szczecin, Poland

I. INTRODUCTION

One of the most important factors in seafood quality is freshness. It determines, to a large extent, the market price of fish, especially of those eaten raw or processed to delicatessen products. Second-quality fish of highly valued species that is still edible and suitable for smoking, marinading, or canning may bring only about 10% of the price paid for the same product of prime freshness. Live fish are regarded highest in freshness and are best to be consumed raw just after killing. Commercially valuable seafood that is supplied to the market in a pre-rigor state may receive the same price as live fish. Contrary to beef meat that normally undergoes natural, enzymatic tenderization in refrigerated storage to increase its eating quality, fish is generally tender enough that tenderization is not required. On the other hand, the enzymatic, postmortem reactions lead as a rule to rapid loss of prime freshness.

The normal way of decreasing the very high rate of quality deterioration in seafood due to the activity of endogenous enzymes and, later, bacterial spoilage is fast chilling to about 0°C, in ice or cold water. If the catch has to be kept longer than about 2 weeks, the storage method of choice is freezing. Various techniques are used experimentally or in commercial scale to extend the market life of chilled fish: irradiation, modification of the atmosphere during storage, and addition of chemical preservatives. Generally, these treatments do not decrease the rate of prime quality loss of seafood, since this depends not on the microbial activity but rather on changes brought about by endogenous enzymes (1, 2). Refrigeration and freezing, however, do not completely arrest the endogenous and bacterial enzyme activity in fish. In a few cases very rapid cooling may even increase the rate of some undesirable, enzyme-catalyzed reactions.

II. THE MAJOR ENZYMATIC REACTIONS IN CHILLED SEAFOOD

A. Introduction

The initial loss of the attributes of prime freshness in seafoods results primarily from catabolic changes in nucleotides and carbohydrates, which are rapidly followed by degradative reactions of nitrogenous compounds as well as hydrolysis and peroxidation of lipids. These reactions are catalyzed mainly by endogenous enzymes. During further chill storage of the catch bacterial activity contributes to the quality deterioration. The rate of these processes is controlled mainly by endogenous factors related to species, condition, infestation with parasites, and initial bacterial contamination of the fish, as well as by handling of the catch and preservative treatments.

B. Nucleotide Catabolism

The most important factors responsible for the loss of prime freshness in seafood are nucleotide catabolism and glycolysis (3). Some very early breakdown products of nucleotides may improve the sensory properties of seafoods. The final catabolites, however, have a deteriorative effect on quality. The flavor intensity and acceptability of fish meat is related to the concentration of inosine monophosphate (IMP) in the muscle. Thus the high-quality life of fish lasts as long as the pool of IMP is not depleted. Fish species known to degrade IMP slowly are regarded as inosine producers, while those that breakdown IMP rapidly are hypoxanthine producers (4).

Depletion of ATP in muscle normally leads to rigor mortis. Rigor state is generally a good sign of prime fish freshness. The onset and resolution of rigor are affected by the biochemical properties characteristic for different fish species. In sardine the fast, (i.e., white or ordinary) muscles are mainly responsible for rigor tension of the whole body during postmortem changes, while the slow, red muscles do not participate significantly (5). There is a general view that the muscles of fish in rigor are tough and become tender in the postrigor state. However, Ando et al. (6) found, that in plaice, parrot bass, yellowtail, carp, red sea-bream, striped grunt, and rainbow trout the breaking strength of the muscles decreased at 5°C sharply within 24 h after death, although the maximal rigor continued for at least 72 h.

Since the changes of nucleotides in fish muscle postmortem in the first period after catch are catalyzed mainly by endogenous enzymes, the concentration of major adenine nucleotides and their degradation products has been long used as a measure of fish freshness in the early phase of storage. Saito et al. (7) proposed to use the k value (i.e., the sum of the contents of inosine and hypoxanthine over that of the total adenosine triphosphate [ATP] and its degradation products), while Jones (8) described the suitability of hypoxanthine as a freshness index. The rates and pathways of degradation of nucleotides, however, differ in fish of various species and in muscles of different types. In the flesh of Japanese baking scallop, Yesso scallop, abalone, kuruma prawns, plaice, and red sea bream the rate of ATP degradation has been found by various investigators to be higher at about 0°C than at higher temperature (9). The content of nicotinamide adenine dinucleotide in fish muscle decreases very rapidly during storage in ice (10).

In the muscles of shrimp, ammonia is produced in the early stage postmortem due to deamination of adenosine monophosphate and adenosine by endogenous enzymes. The deaminases of shrimp muscles are not stable during iced storage of the crustacean and lose their activity within about 10 days (11). Hypoxanthine produced as a result of enzymatic degradation of inosine accumulates to a peak concentration and than gradually levels off at a much lower value. The appearance of hypoxanthine in the muscles may be regarded as the first sign of quality decrease (12). The products of the muscular nucleotide metabolism are further degraded by microbial enzymes in the later stages postmortem. The enzymes involved with nucleotide catabolism are discussed in detail in Chapter 2.

C. Carbohydrate Catabolism

In the anoxic conditions of postmortem fish flesh, glycogen and glucose continue to be degraded in reactions catalyzed by endogenous enzymes, leading to accumulation of mainly lactate. These reactions proceed concurrently with the catabolism of ATP and creatine phosphate. The correlation coefficient for ATP content versus L-lactate in the muscle of sardine and mackerel in the temperature range 0–10°C is about –0.8 (13).

The postmortem degradation of glycogen in fish muscles proceeds by phosphorolysis and by the route catalyzed by glycogenolytic hydrolases (3, 14, 15):

$$\text{Glycogen} \Rightarrow \text{maltose} \Rightarrow \text{glucose}$$
$$\Uparrow \qquad \searrow$$
$$\text{UDP-glucose} + \text{phosphate} \Rightarrow \text{glucose-1-phosphate}$$
$$\Downarrow$$
$$\text{glucose-6-phosphate}$$
$$\Downarrow$$
$$\text{L-lactate}$$

The changes in the concentration of sugars and sugar phosphates in muscle (Fig. 1) contribute to the formation and gradual loss of the sweet, meaty flavor of very fresh fish. The free sugars and sugar phosphates produced due to glycogen and ATP catabolism participate in browning that decreases the quality of frozen, dried, salted, and canned fish and marine invertebrates. One of such quality defects due to reactions of reducing sugars with amino groups is the orange discoloration in canned skipjack and albacore tuna. Delay in canning, leading to

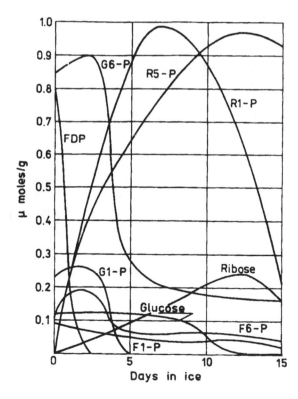

Figure 1   Sugars and sugar phosphates in the muscle of cod during ice storage. Glucose × 0.1, ribose and ribose phosphates × 1000. (From Ref. 16.)

accumulation of reaction substrates, increases the darkening of the product (17). The discolorations are especially significant in the flesh of marine invertebrates rich in free amino acids.

In the muscles of mollusks, anaerobic glycolysis leads to generation of D-lactate and four opines: octopine, strombine, alanopine, and tauropine. The accumulation of opines results from reactions of pyruvate with arginine, glycine, alanine, and taurine, catalyzed by opine dehydrogenases. Tauropine and D-lactate have been proposed for monitoring the exposure of live abalone to anaerobic stress during culture, transport, and storage (18).

The catabolism of ATP and glycogen leads to a decline of pH in the meat. The lowest level of pH reached due to accumulation of the breakdown products is called the "ultimate pH." This value normally depends on the amount of the substrate (mainly glycogen) prior to death of the animal and on the buffering capacity of the meat. In the white muscles of fish and marine invertebrates the ultimate pH is normally in the range of 6.2–6.6. This is significantly higher than in the meat of terrestrial animals, because of the high content of nonprotein nitrogenous compounds in fish flesh that contributes to the buffering capacity (19).

D. Changes in Proteins and Other Nitrogenous Compounds

1. Protein Hydrolysis

Changes in proteins and nonprotein nitrogenous compounds other than nucleotides affect various aspects of fish quality, including color and flavor as well as the rheological and functional properties. In the early stage after catch many of these reactions are catalyzed by endogenous enzymes; later they proceed mainly due to the activity of microflora.

The hydrolysis of proteins affects principally the rheological properties of seafood. The weakening of the fish muscle in early hours postmortem at abuse temperature results from enzymatic breakdown of some cytoskeleton proteins and of connective tissues responsible for the integrity of the muscles. The collagenous structures hold together the main components of the fish muscle by forming a fibrous network surrounding each muscle fiber and linking it to the myocommata. Enzymatic hydrolysis of these connective tissue structures results in breakdown of the muscle integrity and in rheological changes that decrease the suitability of the fish for different forms of utilization. As stated by Bremner (20): "Post mortem changes occur within the muscle cell in the elements of the cytoskeleton, in interactions between the proteins in the cell and externally in the links between the cell and its envelope and in the myotendinous junction. The relative importance of these phenomena varies with species and circumstance."

The collagen structures may be degraded in a concerted action of different enzymes. Initially undenatured molecules can be attacked by collagenases. The collagen thus fragmented in specific sites can be further hydrolyzed by other proteinases. The hydrolytic changes of collagen and of other extracellular matrix proteins are probably to some extent catalyzed by the heat-stable metalloproteinases, identified in Pacific rockfish muscles (21). Degradation at the interface between the connective tissue of the myocommata and the muscle cell precedes any significant structural alterations within the muscle fiber itself (20, 22–24). Scanning electron microscopy of ice-chilled whole freshwater prawn has shown a gradual disintegration of collagenous structures: perimysium and endomysium leading to separation of the muscle fibers (25). Mizuta et al. (26) suggested that enzymatic degradation of nonhelical domains in collagen of the perimysium and endomysium may contribute to the decrease in the penetration resistance of the muscles of kumara prawn stored for 24 h at 5°C.

The participation of various proteinases in autolytic processes of ice-stored fish depends on the following: location of the enzymes in the muscle and impact of factors affecting tissue compartmentization; seasonal changes in enzyme concentration; the synergistic effect of various enzymes in proteolysis; the presence of activators and inhibitors in significant concentration; and the susceptibility of the proteins responsible for muscle integrity to degradation in situ by the respective enzymes in the pH range of about 5.4–7.1 at the given temperature (27). Endogenous proteinases are also responsible for the extremely rapid quality deterioration of whole krill and loss in yield of the peeled meat of Antarctic krill, although the proteolytic activity may be advantageous for preparing protein precipitates (28, 29).

According to Watson et al. (30), the development of soft and pale, muddy brown, or turbid meat with a stringent aftertaste in fresh flesh of struggling tuna (burned tuna) is caused by endogenous, calcium-activated neutral proteinases. The activity of these enzymes is enhanced by high levels of catecholamines in the blood of struggling fish.

2. Changes in Other Nitrogenous Compounds

In fish muscle in the rigor state, no significant changes occur in total free amino acids (31). Similarly in Antarctic krill, autolysis leads mainly to polypeptides and dipeptides, while the amount of free amino acids is comparatively small (32). In the flesh of mollusks and many crustaceans, free amino acids are the most important taste compounds. They may constitute the majority of water-soluble nitrogenous components. About 80% of the total free amino acid pool in Japanese baking scallop is made up of glycine, taurine, and arginine (9). Changes in the concentration of the free amino acids during refrigerated storage due to the activity of endogenous and bacterial enzymes in the meat, however, do not follow a strict pattern.

In the prerigor muscles of some rays and sharks, a strong ammoniacal odor develops early after catch. This is the result of hydrolysis of urea to ammonia and carbon dioxide by endogenous and bacterial urease. Such meat may have a bitter taste. Degradation of trimethylamine oxide (TMAO) catalyzed by endogenous and bacterial enzymes leads to accumulation of volatile amines, characteristic for the odor of fish that has lost prime freshness. The occurrence of TMAO and its derivatives in fish and marine invertebrates, as well as their significance for the quality of the products and use as freshness indicators, has been reviewed by Hebard et al. (33). Enzymes involved with TMAO metabolism in fish are discussed in Chapter 7.

Some fish and marine invertebrates contain thiaminase, an enzyme classified as a transferase. It catalyses the degradation of thiamine in a reaction with amines:

$$\text{Pyr-CH}_2\text{-thiaz}^+ + \text{R-NH}_2 \Rightarrow \text{Pyr-CH}_2\text{-NH-R} + \text{Thiaz} + \text{H}^+$$

The enzyme has been found in fresh-water fishes (e.g., carp, chub, smelt, and minnow) and in sea fishes (e.g., capelin, herring, menhaden, garfish, and whiting) (34). In carp, the thiaminase is located mainly in the viscera and in the gills. Some of the thiaminase in fish may be of bacterial origin (35). Seafood rich in thiaminase contain very low amounts of thiamine. Furthermore, raw fish offal fed to fur animals may destroy the enzyme contained in the feed and induce thiamine deficiency. Cooking of the fish offal destroys the enzyme activity.

E. Lipid Changes

1. Hydrolysis

Hydrolysis of lipids and oxidative instability of the polyenoic fatty acids in fish lipids are also important factors in seafood quality deterioration. Fish muscle lipolysis was thoroughly reviewed more than a decade ago (36). Most of the published papers at this time dealt with frozen fish. Phospholipases and lipases are also discussed in Chapters 4 and 5, respectively.

Lipid hydrolysis can be catalyzed by endogenous lipases and phospholipases as well as by bacterial enzymes. The majority of the free fatty acids accumulating in the flesh of refrigerated fish originate, however, from phospholipids. In minces of Baltic cod and herring kept at 4°C, the highest rate of lipid hydrolysis is at pH 6.65–6.85 (37). In gutted herring containing 4–9% fat, stored 3 weeks on ice, the free fatty acids make up about 0.3% of the muscle wet weight. All the free fatty acids appear to be derived almost entirely from phospholipids. The phospholipid content in herring is about 0.8% of the wet muscle (38). In cod stored in ice, phosphatidylcholine and phosphatidylethanolamine is hydrolyzed slowly during the first 4 days of storage and then rapidly until 20 days. After 30 days, the content of both phospholipids decreases by about 70% and that of free fatty acids increases to about 175 mg/100 g flesh (39). The rapid increase in the rate of hydrolysis during the latter stages of storage may be caused by bacterial enzymes. In salted fishery products that can be stored over several weeks at refrigerated temperature, significant enzymatic lipid hydrolysis can occur. In rainbow trout roe preserved with 5% salt, stored for 52 weeks at 2°C, about 15% of total lipids are hydrolyzed. The free fatty acids are released both from the neutral lipids and from phospholipids. The phospholipids constitute about 50% of the fish roe (40). In Antarctic krill stored 3 days at 3°C, about 20% of the phospholipids are hydrolyzed, with no changes in neutral lipids. The content of free fatty acids increased to about 6% of the total lipids. The phospholipids constitute about 80% of total krill lipids (41). Enzymes that hydrolyze lipids are discussed in detail in Chapters 4 and 5.

2. Oxidation

Lipid oxidation in fish muscles may be caused by nonenzymatic processes such as autoxidation and photosensitized oxidation, as well as by catalysis with lipoxygenase of fish gill and skin (42), by the peroxidase of fish blood, and the microsomal NADH peroxidase of fish muscle (43). The lipid peroxides may be decomposed by homolytic cleavage and beta-scission, producing various short-chain products: alcohols, aldehydes, ketones, and hydrocarbons. Some of these compounds are associated with the desirable, fresh aroma of seafood (44, 45) (see also Chaps. 11, 13, and 14). Conversion of several carbonyl compounds to alcohols has been found to cause the loss of the initial flavor of seafood (46). Some decomposition products are known to contribute to the unpleasant taste and smell of stale fish. In a model system, involving rainbow trout gill 12-lipoxygenase plus arachidonic and eicosapentaenoic acids, the following major volatile compounds were formed: 1-octen-3-ol, 2-octenal, 2-nonenal, 2-nonadienal, 1,5-octadien–3-ol, and 2,5-octadien-3-ol. The experiment was conducted at 25°C during 2 h (47).

Endogenous lipoxygenase-type enzymes have long been known to degrade fish carotenoids. Astaxanthin, tunaxanthin, and β-carotene can be degraded in the dark to colorless, carbonyl compounds (48). Thus the fading of the pink or red color of skin or flesh of various fish at chilling temperature has also been associated with lipoxygenase activity (17). The discoloring activity of the enzyme in the fish skin has been found to decrease with the deterioration of fish freshness. In experiments with rainbow trout, the stability of astaxanthin was higher in samples that deteriorated faster than in fish that was better protected from loss of freshness. The higher color stability was interpreted as possibly resulting from lower activity of the enzyme in faster-spoiling fish (49).

Fish muscles contain various endogenous antioxidants that may be lipid-soluble, cytosolic, or enzymatic in nature. Among the cytosolic compounds that have antioxidant activity in muscle systems, glutathione is important. Glutathione may inhibit lipid oxidation by supplying electrons needed by the endogenous glutathione peroxidase to decompose the hydrogen and lipid peroxides. A significant role in retarding lipid oxidation in muscle systems may be also played by the cytosolic superoxide dismutase and catalase (50). The relative concentrations of pro- and antioxidants present in various tissues depend on the type of muscle and species and on the condition of the fish (51) (see Chap. 1). Light muscle is comparatively low in pro-oxidants, whereas dark muscle and skin are rich in microsomal enzymes, hemoproteins, and low-molecular-weight metal compounds.

One of the possible controlling factors in lipid oxidation and its subsequent effects in fish muscle may be hydrolysis of phospholipids by endogenous phospholipase A. In a model system involving flounder muscle microsomes, at 20°C, addition of phospholipase A inhibited subsequent oxidation in membrane phospholipids (36). In earlier experiments conducted by other groups, similar effects were observed upon addition of phospholipase A to tissue homogenates or as a result of increase in the concentration of free fatty acids in situ.

In fish of various species two types of lipid peroxidation protection factors were found by different groups. One of the factors was nondialyzable and heat-labile, which might perhaps be superoxide dismutase or catalase. The other factor was dialyzable and heat-resistant, which could perhaps be an iron chelating reagent (52). The lipoxygenase-initiated oxidation can be effectively retarded by butyl hydroxyanisole and butylhydroxytoluene, but especially by several flavonoids, which have been found to be very potent inhibitors of 12-lipoxygenase in fish gill (53). Several natural plant flavonoids are effective in retarding lipid oxidation, as measured by accumulation of thiobarbituric acid-reactive substances, in ground muscle of *Scomberomorus commersoni* during storage for 2 weeks at 4°C (54).

The susceptibility of lipids to oxidation, even in fish of the same species, depends on the biological state of the specimens and on the conditions of postmortem processing and storage. In minces of Baltic cod and Baltic herring kept at 4°C, the lowest rate of peroxide formation was at pH 6.20–6.60 and 6.25–6.30, respectively (37). Lipid oxidation has been found to play a significant role in the quality loss of iced, whole, prespawning and spawning Baltic herring. After 7 days the peroxide value in these fish exceeded that regarded as rejection limit in frozen product (55). When the proteolytic activity is inhibited, lipid oxidation may be the major cause of quality deterioration. Seafood lipoxygenases are discussed in detail in Chapter 11.

F. Other Major Enzymatic Reactions

In shellfish, the freshness deterioration is additionally related to enzymatic discolorations known as blackspot development. The polyphenol oxidase present in the tissues of crustaceans participates in sclerotization of the exoskeleton after molting by catalyzing crosslinking reactions between adjacent protein chains. After catch, the enzyme is responsible for the formation of melanins causing darkening of the meat and shell. The susceptibility of various crustaceans to melanosis is different. This difference may result from variations in enzyme activity (56). Furthermore, the polyphenol oxidases are present in some crustaceans in latent, proenzyme forms. These may be activated by some proteolytic enzymes or other factors, including fatty acids, or lipopolysaccharides that activate the respective proteases. Refrigerated storage was found to activate the polyphenol oxidase in different crustaceans (57, 58). Black spots occur on raw and undercooked prawns. Their formation is accelerated by increase in pH and temperature in the tissues as well as by oxygen, copper, and light. In Norway lobster raw handling during and after catch increases the activity of polyphenol oxidase (59).

The participation of enzymes in color change of seafood was extensively treated in the review by Haard (17). Polyphenol oxidases are discussed in detail in Chapter 10.

III. ENZYMATIC REACTIONS IN FROZEN SEAFOOD

A. Introduction

Frozen fish produced according to the requirements of good manufacturing practice retain their high quality for several months when stored below –30°C, depending on the raw material, packaging, and storage temperature. Later, a perceptible decrease in the quality occurs, mainly in flavor, color, and texture of the tissues, gradually making the product unacceptable. The rate and the contribution of these different quality defects in total quality loss depend again on the species and form of processing of the seafood, packaging, and storage temperature.

Each aspect of quality deterioration is caused at least in part by enzyme-catalyzed reactions, similar to those occurring in chilled seafood. Freezing and cold storage may decrease or increase the activity of different enzymes. The loss of ATPase activity in frozen meat homogenates and in protein solutions may reach 50–80% (60). In solutions of milkfish actomyosin, the Ca-ATPase activity decreased in 8 weeks at –20°C and –30°C by about 80% and 35%, respectively (61). The specific Ca-ATPase activity in frozen amberfish, carp, mackerel, and mullet stored for 12 weeks at –20°C was about 30% of that in unfrozen fish samples (62). The ATPase and lactate dehydrogenase activities in the muscles of fresh water and brackish water fish have been found to decrease by about 75% during 6 months' storage at –20°C (63). The specific activity of lactate dehydrogenase extracted from the thawed muscles of carp that had been stored 1 week at –20°C reached only about 28% of the activity determined in extracts of fresh muscle (64). This decrease in enzyme activity may be used as a quality indicator in frozen fish (see Chapter 12).

However, freezing and thawing may cause lysis of mitochondria and lysosomes and alter the distribution of enzymes and factors affecting the rate of enzyme reactions in the tissues, e.g., of lysosomal cathepsin D and mitochondrial cytochrome C oxidase in Baltic herring muscle (65). Chawla et al. (66) found that the activity of phospholipase in the microsomal fraction isolated from myotomal tissue of frozen Atlantic cod was significantly higher after 4 and 8 weeks at −30°C than before storage, but after 12 weeks was lower than the initial values. On the other hand, the activity of 5′ nucleotidase, acid phosphatase, and succinic dehydrogenase was negatively correlated with the duration of frozen storage. Delocalization of several mitochondrial enzymes due to freezing and thawing of mammalian skeletal muscles has been described by Hamm (67). Freezing and thawing increases by 2.5 times the activity of cytochrome oxidase in trout muscle (68). The increase in enzyme activity in extracts of thawed muscle has been proposed as a test to differentiate between fresh and frozen/thawed fish (69) (see also Chapter 12). Furthermore, low temperature and changes in the microenvironment induced in the muscles due to ice formation affect the reaction rate.

The decrease in the rate of deteriorative changes in fish caused by lowering the temperature can be roughly described by the known Arrhenius equation. However, the reaction rate is affected not only by temperature. Other important factors resulting from ice crystallization are

Denaturation of proteins, including enzymes
Release of enzymes, especially of ATPase, phospholipases, and cathepsins, and of substrates due to freeze damage of cellular structures and compartmentization
Selective increase in the concentration of different solutes in the unfrozen fraction of the tissue fluids, including enzymes, substrates, and inhibitors
A slight acidification and decrease in water activity

The listed factors are especially important in modifying the reaction rates of frozen fish held between −1 and −6°C. Under these conditions, the temperature effect is still comparatively low, while the concentration change is dramatic. At −6°C, about 80% of the tissue water in the muscle of cod turns to ice (70). Thus the rate of several enzymatic reactions in muscle foods is greater in the critical temperature zone −1 to −6°C than above the freezing point. Reay (71) found in 1933 that denaturation of proteins in frozen haddock muscle proceeded at a maximum rate between −2°C and −4°C.

The biochemical reactions in frozen fish have been thoroughly reviewed by Haard (72).

B. Changes in Carbohydrates and Nucleotides

In frozen fish muscle, the glycolytic reactions are to a large extent retarded at very low temperature, although at −29°C in cod muscle glycolysis still proceeds to a significant extent (73). On the other hand, in the critical temperature zone there is an increase in the reaction rate. In haddock muscle glycolysis proceeds at −2°C at least twice as fast as at 0°C and the maximum rate is at −3.2 to −3.7°C (74). Sharp noted (75) that in muscle tissue with low initial glycogen, degradation to lactic acid was continuous. At high initial reserves of glycogen, however, the reaction was stopped at an acid concentration of about 0.45%. A significant increase was observed in the degradation rate of glycogen and organic phosphates in cod muscle in the range of temperatures -1 to −5°C over that in unfrozen fish (76). At lower temperatures the reactions gradually become very slow. NAD in the white muscle of trout and carp was degraded after 1 month at −8°C by about 95% and 70%, respectively. At -18°C, about 60% of the nucleotide was split in trout, while in carp muscle the initial concentration remained unchanged (77). In the muscles of immediately killed, eviscerated, and frozen mullet the K value increased after 18 weeks at −20°C from the initial 14–18% (78). Dyer and Hiltz (79) found that in swordfish muscle about 10 times less hypoxanthine accumulated at −26°C than at -18°C.

C. Changes in Nitrogenous Compounds

The quality of frozen fish after several months of storage, even at temperature as low as 30°C, decreases due to loss of desirable, rheological, and functional properties, caused by freeze denaturation of proteins. The changes in proteins are brought about mainly by interactions with formaldehyde, free fatty acids and oxidized lipids, and by -S-S- -SH exchange reactions, depending on the characteristic species properties of the fish and the conditions of frozen storage. These lead to the formation of new hydrogen bonds, hydrophobic interactions, and crosslinking via covalent bonds (80–84). Enzymatic reactions are involved in these protein changes by generation of endogenous formaldehyde from trimethylamine oxide (85, 86), release of free fatty acids, oxidation of lipids (84), and by transglutaminase-catalyzed crosslinking.

Enzymatic reactions resulting in extensive protein hydrolysis and development of various low-molecular weight flavor compounds are mainly responsible for the rapid decline in the sensory-determined quality of iced fish. These reactions are of much lower significance at temperatures below freezing, and negligible at about −30°C. In the muscles of frozen fish, stored for 12 weeks at −20°C, the content of free amino nitrogen was higher in mackerel, amberfish, mullet, and carp by 90 %, 115 %, 125 %, and 170 %, respectively, than in unfrozen samples (62). In the meat of frozen mackerel stored up to 6 months at

−25°C, the content of nonprotein nitrogenous compounds was significantly higher than in unfrozen samples (87). During storage of prerigor grass shrimp at −20°C, the cathepsin D-like activity of intact lysosomes was negligible after 4 weeks (88). Recently Ben-Gigirey et al. (89) reported on a significant, several-fold increase in putrescine in the white muscle of albacore after 9 months at −25°C. They suggested that bacterial decarboxylating enzymes released prior to freezing were responsible for accumulation of biogenic amines in the frozen muscle.

D. Changes in Lipids

1. Hydrolysis

The endogenous fish lipases and phospholipases retain much of their activity in frozen fish and some may even be activated (87). Slow freezing and fluctuations of storage temperature from -12°C to −35°C caused in rainbow trout muscles significant release of lipase from the lysosomes. During frozen storage the release of acid lipase was maximal during the first month. The rate of release of the enzyme was higher at higher frozen storage temperature (90). In frozen fish of lean species, accumulation of free fatty acids is mainly due to hydrolysis of phospholipids (91) and the activity of phospholipase may be increased by freezing and frozen storage (92). Storage of Atlantic cod muscles at -12°C leads to extensive membrane condensation associated with sarcoplasmic reticulum that may facilitate enzymatic hydrolysis of phospholipids. After 4 months in uneviscerated, frozen Baltic cod, glazed and packed in polyethylene, stored at -18°C, free fatty acids constituted about 58% of total lipids. After 8 months about 95% of the liberated fatty acids originated from hydrolysis of phospholipids. In Baltic herring and sprat held in the same conditions, free fatty acids made up about 14% and 16% of total lipids and originated in 85% and 70%, respectively, from triacylglycerols (93). According to Hanaoka and Toyomizu (94), the rate of hydrolysis of phosphatidylcholine has a maximum at −5°C. In chum salmon and cod at −20°C the contents of phospholipids decreased to 20–40% of the initial value after 1 year. The phospholipases had similar activity in dark and white muscles (95). In frozen hake mince the rate of hydrolysis of phospholipids decreased with decreasing temperature significantly faster than that of neutral lipids (96). Differences in the activity of phospholipase and lysophospholipase lead to accumulation of various products of hydrolysis in the muscles of different fish (97). In Antarctic krill, stored for 6 months at −22°C, hydrolysis of phospholipids, triacylglycerols, diacylglycerols, and cholesterol esters occurred, although the phospholipid hydrolysis was dominating, leaving only 30 % of its original content (98). Prefreezing storage at 3°C increased the lipolytic activity in frozen krill (99).

2. Oxidation

Lipid oxidation in frozen seafood is affected by the lipid content and composition, presence of pro- and antioxidants, activity of enzymes, integrity of the flesh, pH, and temperature. Among the important factors is also lipolysis in the fish flesh. According to Han and Liston (100), who worked with frozen stored muscle of rainbow trout, the extent of lipid peroxidation correlated with that of phospholipid hydrolysis. Some experimental results indicate that not only the presence of free fatty acids but also the pretreatment of the fish, i.e., filleting, skinning, and application of antioxidants, must be considered (101). Other results do not support the notion that hydrolysis facilitates oxidation. For example, in Pacific mackerel at –5°C the release of fatty acids from phospholipids did not lead to inhibition of lipid oxidation (102). It seems that hydrolysis of triacylglycerols stimulates lipid oxidation, while phospholipid hydrolysis has a retarding effect. This subject was reviewed by Shewfelt (103).

Since there are so many factors involved, it is expected, that in different parts of one fish and in various batches of fish of the same species, stored under identical conditions, the degree of oxidation may differ significantly. Specimens belonging to one species, of the same lipid content, may differ in the degree of oxidation by a factor of three after 3 months' frozen storage. In frozen cod, lipid oxidation is extremely slow and occurs primarily in the phospholipid fraction according to Hardy et al. (104). Nevertheless, during frozen storage sufficient amounts of volatile oxidation products are formed to decrease the sensory properties of the product. Mince prepared from mackerel dark muscle is more susceptible to oxidation than mince from light muscle (105). In frozen herring and mackerel the largest rate of lipid oxidation is in the muscles of emaciated fish. The skin on fillets may retard lipid oxidation by restricting the access of oxygen to the flesh, although the lipoxygenase may have an catalytic effect, especially in the presence of skin fragments in fish minces (106). Generally, the susceptibility to lipid oxidation increases in the following order: ordinary muscle, dark muscle, skin (105, 107). In frozen herring fillets packed in air-tight containers and stored at -18°C for up to 18 weeks, the content of pro-oxidants has a greater effect than the composition of the lipids on the degree of oxidation. These conclusions were based on the concentration of hydroperoxides and oxidized fatty acids containing conjugated diene and triene systems (108).

Freeze texturization may decrease the oxidation of lipids in fish minces. In minced Baltic cod and bream meat, freeze-textured, the peroxide value after 6–8 months at –20°C was four to nine times lower and the docosahexaenoic acid level was significantly higher than in conventionally frozen minces (109).

Carotenoids are known to have an antioxidative effect by quenching radicals generated by lipoxygenase-catalyzed oxidation of polyenoic fatty acids in fish muscle. However, in farmed rainbow trout fed a diet supplemented with astaxanthin, no effect of the carotenoid on the oxidation of lipids in the meat during frozen storage was seen (110).

IV. BIOLOGICAL AND ENVIRONMENTAL FACTORS AFFECTING ENZYME ACTIVITY

A. Introduction

The suitability of fish as raw material for the preparation of different products depends on properties related to the species and stock as well as the season of catch and fishing area. This principle has been utilized for centuries, for example, in traditional European herring fishery, which supplies the raw material for producing different assortments of salted herring. Later came the recognition that the quality of the salted products depends on the activity of proteolytic and lipolytic enzymes, which are controlled by the life cycle of the fish. The environmental and biological factors must be considered also in evaluating the technological suitability of various species of fish for making other products (e.g., fresh fillets or surimi). Recently artificial neural network systems have been used for evaluating the effects of these factors on the technological value of raw materials. The role of intraspecific factors on enzymes is discussed in detail in Chapter 1.

B. Maturation and Spawning

The autolytic activity of fish muscles at pH 4 is markedly increased during the spawning season. In Baltic herring high autolytic activity has been found in the prespawning and spawning period (111), and in typical anodromous fish such as salmon (112, 113) during the spawning migration and in the mature fish.

It is widely accepted that cathepsins are responsible for many of the physiological changes associated with the sexual maturation of fish. The results of several experiments point to the dominating role of cathepsin D in the intracellular breakdown of tissue proteins during the spawning period (112, 114). However, Yamashita (115) concluded, that cathepsin L is the enzyme most probably responsible for drastic muscle autolysis and it causes extensive softening of mature salmon. This enzyme was found in the phagocytes near the muscle fibers. The degree of muscle softening during storage was closely correlated to the autolytic activity, as well as that of cathepsin B and cathepsin L. The activities of cathepsins B, D, E, and L in mature chum salmon muscle were shown to be higher than in immature salmon. A significant increase in the activity of neutral and alkaline proteinases occurs in ordinary muscle of common mackerel, when the gonad–somatic index (gonad mass × 100/body mass) reaches the maximum (116).

The decrease in protein-bound hydroxyproline and increase in free hydroxyproline in the muscle of spawning ayu suggested proteolytic breakdown of collagen (117). The rate of collagen hydrolysis appears, however, to be smaller than that of other muscle proteins in some fish. According to Hughes (118), the relative concentration of collagen in the muscles of spawning fish is higher than during the prespawning period. The transport of collagen breakdown products from the muscle to blood begins earlier in females than in the male fish in the spawning stage (119). Thus it appears that the mobilization of flesh proteins for building the gonads during maturation is due to a concerted action of different groups of proteolytic enzymes. In Baltic herring, the first proteins depleted are the sarcoplasmic, followed by the myofibrillar, proteins. The regeneration of the protein pool takes place in a reversed order (111). The spawning effort leads to depletion of about 60% of lipids and saccharides and about 15% in proteins in fish muscle (120).

The biological condition of the fish related to the spawning season also affects the functional properties of proteins. During spawning the gel-forming ability of myofibrillar proteins is very low because of the low ratio of myosin to actin concentration (121, 122) and the reduced viscosity of actomyosin in solution (123).

In migratory salmonid fish that use large amounts of energy for sexual maturation and also for spawning migrations, significant softening of the flesh often occurs. Histological investigations of such pastelike meat have revealed that the muscle fibers are broken down to irregular fragments and amorphous granules that appeared to be lysed (124). Such softening is ascribed to hydrolysis of the nonhelical region in the collagen molecule as well as degradation of myofibrillar proteins by cathepsin L (125). Decrease in the shear rate of raw and cooked Baltic herring meat due to maturation of fish gonads was found by Lachowicz and Kołakowski (126).

C. Feeding Habits and Intensity

The feeding intensity of fish is affected mainly by their biological cycle. Prespawning and spawning fish feed only sparsely or not at all. Their belly cavity is mainly filled by the maturing gonads. The metabolites and energy required for building the gonads are taken from breakdown of the fish flesh. After spawning the fish feed heavily to rebuild the muscle tissue. Thus the technological value and the biochemical properties of the catch in different periods of the biological cycle vary significantly.

Heavily feeding small fishes, just after spawning, are very susceptible to autolysis that quickly leads to breakdown and solubilization of the belly tissue. This makes them unsuitable for refrigerated storage and for marinading, canning, and other processing. The autolytic degradation of small, heavily feeding fish is usually called "belly burst" because of the bursting belly early after catch. The belly burst is initiated by enzymes leaking from the digestive tract into the surrounding tissues. The feeding fish usually contain significantly more digestive proteases than the fish before spawning. This has been well documented for capelin (127, 128), Baltic herring (129), other small seafish, and fresh-water fish as well (130). Siebert (131) found a severalfold increase in the activity of trypsin in the belly muscles of ungutted cod after a few hours postmortem.

The proteases from prey, stomach, intestine, muscle, and bacteria contribute in a concerted action to the total proteolytic activity. The digestive proteases participating in anchovy tissue degradation have optimum pH in the alkaline range. At pH 6.5 and higher, which prevails in the fish tissues postmortem, the trypsinlike enzymes play an important role in the degradation of myofibrillar proteins and, to a limited degree, also in the solubilization of connective tissue. In fish, as well as in mammals, trypsin and chymotrypsin in the intestinal contents originate from the pancreas or other secretory tissues. Their zymogens are stored in the pancreas to provide a ready supply of the enzymes to digest ingested protein even during the starved condition (132). For the degradation of connective tissue other proteases are mainly responsible. Collagenolytic activity, detected in various digestive organs of several teleosts, could be correlated to the kind of the usual diet of the fish in nature (133). Animal components in the diet stimulate proteolytic activity, while large amounts of detritus depress it. Thus, the proteolytic activity of the alimentary tract is much higher in predatory fish than in those belonging to omnivorous species (134).

The digestive enzymes of fish are strongly temperature dependent. The rate of belly tissue degradation in anchovy during storage can be effectively reduced when the fish is chilled by refrigerated sea water or iced sea water to about 0°C (135).

D. Environmental Temperature

The waters inhabited by fish vary in temperature from slightly below zero to about 50°C. Acclimation to warm water takes place in about a day, while complete adjustment to cold may take as long as 20 days (136). The environmental temperature is known to affect the metabolic rate, enzyme structure and functions, properties of the contractile machinery, and the fluidity of biological membranes (137). The muscle of fish adapted to lower temperature contains a greater number of mitochondria and other bodies rich in enzymes than the muscles of fish living in warmer waters (136). The activity of enzymes of the tricarboxylic acids cycle and of cytochrome oxidase was shown to increase in the flesh during cold acclimation (138). The acclimation temperature has a considerable effect on both the period required for reaching the peak of rigor and on maximal rigor strength. Rigor mortis proceeds more quickly with increasing difference between the acclimation and storage temperature of carp. Also the rate of ATP and glycogen catabolism and of lactate accumulation is larger in the flesh of warm-acclimated fish than in the cold-acclimated fish. The rate of ATP degradation may be attributed to the activity of the myofibrillar ATPase (139). The activity of Mg^{2+}-ATPase from cold-acclimated carp was about twice as high as that from warm-acclimated fish (140). The enthalpy and entropy of activation of Ca^{2+}-ATPase and Mg^{2+}-ATPase from cold-water fish were significantly lower than for the enzyme from warm-water fish (141, 142).

E. Effects of Parasites

The most important foodborne parasites transmitted to humans from ingested fish and other aquatic animals have been reviewed by Higashi (143), Olson (144), Rodrick and Cheng (145), and Grabda (146). There are about 85 different parasitic organisms that may be transmitted to humans by way of food and drink. However, only a few of them account for the majority of food parasitic illnesses in the human population. In fresh fish the larvae of *Anisakis, Phocanema, Terranova, Phocascaris, Diphyllobothrium, Cryptocotyle,* and *Clonorchis* are mainly responsible for health hazards. There are, however, several groups of parasites regarded as harmless to humans, but their presence in the fish meat or on the skin and gills makes the product unfit for human consumption on esthetic grounds. It may also decrease the technological value of utilizing the catch.

   The products of parasite metabolism include more than waste material that is secreted and excreted at the surface of the organism. Cytostatic effects of anisakid products on mammalian cells have been described by Raybourne et al. (147). Many helminths and protozoa have externalized digestive enzymes. The excreted enzymes of *Kudoa* (*Myxobolidae*) destroy the structure of the infested muscles. Breakdown of the structure begins at the infected muscle fiber, extends progressively to the surrounding, uninfected fibers, and develops into visible, disrupted areas as milky pockets. Jellification of such muscle proceeds very rapidly above 20°C, with a maximum rate at about 55°C (148). Frozen products thawed at room temperature also exhibit tissue breakdown due to proteolysis. In Pacific whiting infested with *Kudoa paniformis,* the flesh becomes mushy on slow cooking, as the proteolytic enzymes of the parasite have a temperature optimum above 60°C. In highly parasitized muscle of Pacific whiting, the protein hydrolysis determined in a test at 45°C may be about 20 times higher than in nonparasitized samples (149). Rapid deep frying or treatment with oxidants, including hydrogen peroxide and potassium bromide, are effective in inactivating the proteinases.

   Ichthyophonus hoferi is a potent pathogen for several important, commercial fish: herring, plaice, haddock, and mackerel (150). Herring fillets infected with *I. hoferi* are very soft and slimy and have strong off odors (151). Although the parasite is a fungus, its effect on the fish tissue is similar to that in muscles infested with *Kudoa*. The major problem in the factory is sorting of herring attacked by *I. hoferi*, since macroscopic changes brought about by the fungus are negligible (152). *I. hoferi* grows within the temperature range 0–25°C; however, at 0–5°C the growth rate is very slow.

Accelerated muscle breakdown in whiting infested by myxosporidian parasites has been shown to be caused by cathepsins B and C. The activity of these enzymes in the attacked muscles of whiting is much higher than in unparasitized cod (153). According to Masaki et al. (154) and Toyohara et al.(155), the enzyme responsible for jellification of fresh Pacific hake muscle is a cathepsin L-like protease. The myxosporidian salmon parasite *Henneguya salminiloca* contains a soluble, thermostable protease that has maximum activity with muscle proteins at pH 4.5. Although the temperature maximum for this enzyme was 30–40°C, some activity was displayed also at refrigerated temperature. Freezing arrested the activity; however, long-term storage at –28°C inactivated the enzyme only slightly (156).

Among the parasites present in fish, those most resistant to freezing, cooking, salting, smoking, and marinading are the nematodes. Thus it is generally assumed that measures taken for killing the nematodes are sufficiently effective also against other parasites (146). Nematodes in herring are inactivated by freezing of the fish within 2–3 h to a core temperature of -18°C and additional cold storage for at least 24 h at that temperature. The larvae can survive plate freezing without additional cold storage as well as a slow freezing process to –20°C in 20 kg containers (157).

F. Other Factors

The activity of myofibrillar ATPase declines with the length of the fish. It is a size rather than an age effect (158). This corresponds to the decrease in maximum contraction rate in skeletal muscle with length of the fish (159).

V. EFFECT OF FISHING TECHNIQUES AND HANDLING ON ENZYMATIC PROCESSES

A. Introduction

Capture, handling on board the vessel and ashore, as well as transportation of the fish are traumatic procedures. They may cause various physiological reactions in the animals. The general adaptation syndrome causes a release of adrenalin and cortisol. This is followed by secondary changes such as increased muscle activity, use of energy reserves in the muscle and liver, and change in the acid–base balance. There may also be an increase in blood plasma ion concentration and reduced tissue water content in the fish (160, 161). These reactions, by changing the conditions in the tissues, affect the enzyme activity in postmortem fish and the rate of freshness degradation. Improvement in postharvest fish muscle quality and a delay in onset of rigor mortis are associated mainly with low-stress and rapid killing methods.

B. Fishing Gear and Conditions

The circumstances of fish capture by the harvest gear have a great effect on the initial quality of the catch and rate of quality degradation during subsequent storage. They can either lead to anaerobic metabolism and acidity in the muscles of the struggling fish, if the exercise is very strenuous, or else allow for metabolic transformation of the lactic acid at longer capture times at low activity. In large tuna, caught by techniques that haul the fish on board within minutes after hooking, the strenuous struggling and restricted oxygen supply may bring about acidosis in the muscle. Since it is difficult to chill a large fish very rapidly, symptoms of burned tuna can develop in the meat. Under anaerobic conditions, at low pH and relatively high temperature, myoglobin can be oxidized to brown metmyoglobin and protein denaturation may occur. Exhaustive exercise, by increasing the production of lactic acid in the muscle, may favor rapid proteolysis, because lactate is a potent activator of lysosomal proteolytic enzymes.

The importance of reducing preharvest exercise in the production of high-quality fish is illustrated with the examples of chinook salmon and salmon (162, 163). Handling stress led to rapid exhaustion of phosphocreatine and ATP in the muscles of salmon. This decreased the prerigor period severalfold and softened the fillets. However, no detectable effects of stress were found on flavor and color of the meat (163).

C. Slaughtering, Bleeding, and Handling

In large-scale commercial processing of cultured fish, the technique of slaughtering must take into consideration not only the efficiency of the process flow and ethical aspects but also the effect on the quality of the fish. Various slaughtering techniques, mainly asphyxiation, percussive stunning, CO_2 narcotization, electrocution, tail bleeding, and hypothermia, influence the extent of antemortem stress brought upon the fish and might have different effects on the rate of biochemical processes and possibly on the quality of the product. Cultured and wild plaice, when killed while struggling, entered into rigor mortis at 10°C about three times faster than the fish killed without struggle (164). Stressing salmon to exhaustion during handling before killing shortened the prerigor period and increased the rate of degradation of muscle high-energy phosphates during storage in ice (165). Onset of rigor and degradation of ATP were delayed in unstressed, iced snapper compared with the muscles of line-captured and exercised fish (166). A small effect on the rate of some biochemical postmortem reactions has been shown in experiments with 223 rainbow trout bred in captivity (167). Asphyxiation of the fish 15 min in tanks without water left only 1.9 µmol glycogen in 1g muscle. It caused a large accumulation of lactic acid (48.7 µmol/g) and short duration of rigor mortis of the fish in crushed ice. Percussive stunning, (i.e., blow with a wooden club on the cephalic region), taking 30–90 s, preserved the highest glycogen reserve (12.3 µmol/g fresh muscle) and led to low pH (6.56) and extended rigor mortis. In CO_2 narcotized fish, by immersing it into water at about 15°C, to which CO_2 was percolated until the concentration reached 250–300 ppm, only a small amount of glycogen remained (2.8 µmol/g) and the pH was 6.83. Electrocution led to a loss in glycogen and accumulation of lactic acid, similar to those caused by asphyxiation and narcotization. Earlier exeriments (168) have shown that electrocution and CO_2 narcotization induced greater initial production of lactic acid. However, no significant differences were found in the sensory and functional properties of the rainbow trout stored ungutted in ice for up to 15 days.

In slaughtering of cultured fish a wire can be pushed into the spinal cord. It is believed that such spiking maintains the freshness of the fish longer. It has been shown, that destruction of the spinal cord decreased the rate of ATP catabolism and delayed by 6–12 h the appearance of full postmortem rigor in yellowtail and sea bream, while it had no effect in plaice (169).

Enzymatic protein degradation and rough handling of fish in rigor may bring about undesirable texture changes in the muscles. The flesh softens and gapes, the yield of fillets decreases, and the fillets fall apart during skinning.

D. Rate of Chilling of the Catch

In fish from cold waters the rate of biochemical, postmortem processes decreases with lowering the temperature down to 0°C (74). Thus, in order to extend the state of fish freshness, it has been recommended to chill the catch as rapidly as possible in melting ice (170). However, in sardine and mackerel spiked in the brain, the rate of rigor mortis development is lower in samples stored at 10°C than in those stored at 0°C (13). In fish of several tropical species an acceleration of postmortem stiffening at 0°C, known as cold shock, was reported. Curran et al. (171) found that tilapia stiffened within minutes at 0°C while at 22°C this occurred only after 7 h. However, no acceleration of ATP and glycogen degradation was evident. In later experiments (172), ATP degradation and lactic acid accumulation were faster in immediately chilled fish than in those aged before chilling. Rapid chilling of bighead in ice-water slurry immediately after killing evoked strong cold shock. The degree of stiffening was similar when the fish were chilled to 0. 2, 5, and 11°C, but the cold shock was not observed at 15°C.

Aging bighead at ambient temperature before icing led to less stiffening in ice. The longer the aging, the less rigid were the iced fish during *rigor* (173). In yellowtail, bartailed flathead, and Japanese striped knifejaw spiked in the brain, rigor mortis proceeded roughly twice as fast and the rate of ATP degradation was two to three times greater at 0°C than at 10°C (174). In wild and cultured plaice spiked in brain, stiffness started at 0°C after 3 h while at 10°C it occurred after 6 h and developed fully after 21 and 32 h, respectively. The ATP degradation and lactate accumulation proceeded essentially in parallel to rigor (164). Cold shock has been observed also in fish of several other warm-water species. During onboard handling of such fish, the effect of cold shock on the filleting yield, drip loss, gaping of the fillets, and shelf life in ice must be considered. According to Curran et al. (172) "Aging fish at ambient conditions for a few hours before icing could produce higher yields in some tropical fish intended for filleting."

VI. EFFECT OF PROCESSING ON ENZYMATIC REACTIONS

A. Introduction

To avoid undesirable enzymatic changes in the catch the processor can remove parts of the fish that harbor most of the enzymes or parasites, inactivate the enzymes by heating or chemical treatment (see Chaps. 17 and 19), or else otherwise apply conditions unfavorable for enzyme activity. The effectiveness of processing depends on the properties of the fish and on the existing conditions. Not all treatments effective in the laboratory bring the expected results when applied in a fishing vessel or in the fish plant and may not be applicable under existing regulations.

B. Removing of Undesirable Enzymes

In processing fish for short-term storage, i.e., in ice or otherwise chilled as fillets on trays without ice, it must be decided at which stage postmortem the gutting and filleting should take place. Handling and processing of fish during rigor mortis can result in loss of quality due to gaping and decrease in fillet yield (175). A point of considerable interest is whether or not the fish should be gutted and filleted as soon after catch as possible. Shaw et al. (176) have shown that storage of cod as iced, gutted fish rather than in the form of fillets is an effective way to prolong the freshness. Factors that affect the rate of quality decrease of the catch in any commercial operation include the delay in chilling caused by gutting and filleting as well as the hazard of bacterial contamination of the cut surfaces of the fish, related to the standards of hygiene in processing.

In order to eliminate rapid autolytic changes in Antarctic krill, manifested by development of a flabby texture and excessive drip, it is necessary to separate the parts of the crustacean that constitute the principal source of enzymes or to inactivate the enzymes by cooking as fast as possible after harvest. The hepatopancreas and stomach of krill are very rich in a variety of hydrolytic enzymes, including proteases, cellulases, chitinases, and lipases (see Chap. 18). The proteolytic activity of the gut content is at least five times higher than that of the whole krill. According to practical experience, the catch should not be kept on board before processing longer than 1 h at 10°C or 3–4 h at 0–7°C. The method of choice in processing krill for high yield and meat quality is mechanical peeling on board early after catch (28).

C. Inactivating of Enzymes

Holding rock lobsters in slush ice for 18 h before tailing does not affect the yield of cooked tail flesh or its eating quality (177). The formation of black spots on crustacean during transportation from the aquatic farm to the processing plant can be avoided by keeping prawns alive in ice-cold water supplied with oxygen gas (178). A widely accepted treatment prior to chilling, retarding the reactions leading to black-spot formation in prawns and shrimp during refrigerated storage, is dipping the crustaceans in a solution of sodium metabisulfite. US regulations allow a 1 min dip of shrimp into a 1.25% sodium metabisulfite solution and an residual content of 100 ppm SO_2 in the shrimp meat (179, 180). In Australia dipping prawns for 30 s in 1% sodium metabisulfite solution is recommended and the maximum permitted concentration of SO_2 in the muscle is below 30 ppm for the domestic market. A similar result can be achieved using a 15 min dip in a 3% solution of a mixture of sodium bisulfite, sodium chloride, and dextrose (HQ Bacterol F) approved by the National Food Authority (181). Equally effective treatment, and less hazardous in application on commercial vessels than the use of sulfiting agents, may be a 1 min dip in 0.005% solution of the potent polyphenol oxidase inhibitor 4-hexyresorcinol. This treatment leaves a residue of less than 1 ppm of the agent in the shrimp meat (182). Methods to prevent enzyme browning in seafood are discussed in more detail in Chapter 10.

D. Other Treatments

Freeze texturization of fish mince may be accelerated by limited proteolysis via trypsin treatment. According to Kołakowski et al. (183), the proteolytic changes induced by trypsin favor the transglutaminase-mediated crosslinking of proteins in the fish mince. Freeze texturization may decrease the oxidation of lipids in fish minces. In minced Baltic cod and bream meat, freeze-textured, the peroxide value after 6–8 months at −20°C was four to nine times lower and the docosahexaenoic acid content was significantly higher than in conventionally frozen minces (109). The stabilizing effect of freeze-texturization may be caused by limited oxygen penetration into the fibrous–laminar texture of the mince (184). Storage stability of frozen fish mince can be increased by admixing shrimp muscle. This is probably the effect of a natural antioxidant present in the shrimp tissue, possibly a polyhydroxylated derivative of an aromatic amino acid (185).

Prefreezing storage at 3°C decreases the susceptibility of lipids, including carotenoids, to oxidation in frozen krill (99).

The development of ammoniacal odor of fresh meat of rays and sharks may be prevented by very early and possibly complete bleeding of the fish by severing the tail vein, followed by gutting, filleting, and washing, to remove the urea. Leaching in a solution containing 2–2.5% NaCl and 0.02–0.04% acetic acid is particularly effective. The leached meat of the great blue shark has very acceptable sensory properties after cooking (186).

Removing of belly flaps, practiced on automatic lines for filleting large cod, guards against loss in quality caused by the parasites that tend to accumulate in the belly parts of the fish. Careful skinning removes the skin lipoxygenase.

Mincing fish and crustacean meat generally increases the rate of deteriorative enzyme reactions by destroying the compartmentization of the tissues. To avoid the undesirable changes many special treatments are used, mainly washing of the mince to remove enzymes, substrates, and factors promoting enzyme reactions, as well as addition of protecting agents (187, 188).

VII. ENZYME TREATMENTS IN CHILLED SEAFOOD

Endogenous enzymes and added enzyme preparations play an important role in producing different assortments of salted and marinaded fish, being responsible for development of the desirable flavor and texture of such delicatessen-type products (189). On the other hand, only very limited information exists on possibilities of using enzymes for improving the quality of chilled seafood. The enzymatic system glucose oxidase/catalase/glucose, known for its activity as an preservative agent, has been proposed in form of a dip for extending the shelf life of fish fillets, especially in hypobaric storage (190). The system was also effective in retarding lipid oxidation in minced mackerel as well as in increasing the storage life of iced shrimp and inhibiting black-spot formation. The use of transglutaminase for inducing covalent crosslinking of proteins has been a major research topic in the last decade and commercial preparations of the enzyme are available in the market for the food industry (see Chapter 6). In fish processing, transglutaminase preparations may be used to improve the texture of surimi and fish sausages by inducing crosslinking of the myosin heavy chains (191–194) and for binding of fish meats in various products. The various methods of employing enzyme catalyzed reactions in fish are reviewed in Chapters 21 and 22.

VIII. CONCLUSIONS

The quality deterioration in chilled and frozen fishery products results from physical, chemical, and enzymatic processes. These reactions proceed in the tissues or minces simultaneously, often in a concerted action. In specific situations, it is impossible to decide on the extent to which the process is chemical or enzymatic, e.g., phospholipid hydrolysis in cooked, minced carp during frozen storage [195] or lipid oxidation in fish tissues. In many cases, it is equally difficult to know whether endogenous or bacterial enzymes play a primary role in the process. It is, however evident, that even slight enzymatic changes in the early postmortem period may initiate significant quality deterioration during subsequent storage by altering the integrity and compartmentization of the muscle tissue, activating enzymes, and affecting the balance of compounds promoting and inhibiting enzymatic processes. Better insight into the mechanisms of these reactions may help to evaluate the technological suitability of the catch and increase the effectiveness of different processing techniques. In order to control effectively the deteriorative, enzymatic reactions in chilled and frozen seafood it is generally recommended to:

Handle the catch as carefully as possible.

Slaughter and bleed the fish under conditions minimizing additional stress.

Minimize the time elapsing between hauling and chilling the catch down to about 0°C, avoiding, however, cold shock reactions in warm-water fish.

Remove the entrails as soon as possible, especially from feeding fish, in hygienic conditions, avoiding bacterial contamination of the cut surfaces.

Fillet the fish in appropriate time after catch.

In preparing fish minces, remove carefully all bloody parts and skin as well as all components involved in the generation of formaldehyde.

Add to minces, if feasible, agents that are effective in inhibiting endogenous enzymes known to participate in lipid oxidation and crosslinking of proteins.

Use approved chemical additives, possibly in form of dips, to prevent black-spot formation in crustaceans.

Apply conditions favoring phospholipid hydrolysis to inhibit lipid oxidation in frozen products.

Freeze and thaw the product in conditions preventing significant histological changes in the tissues and avoid exposure to temperature in the range −1°C to −5° C,

Keep the temperature of frozen fish sufficiently low, depending on the required storage time.

REFERENCES

1. NF Haard. Technological aspects of extending prime quality of seafood: a review. J Aquat Food Prod Technol 1(3/4).9–27, 1992.

2. ZE Sikorski BS Pan. Preservation of seafood quality. In: F Shahidi, JR Botta, eds. Seafood: Chemistry, Processing Technology and Quality. London, Blackie Academic and Professional, 1994, pp 168–195.

3. HLA Tarr. Post-mortem changes in glycogen, nucleotides, sugar phosphates, and sugars in fish muscle—a review. J Food Sci 31:846–854, 1966.

4. HA Bremner, J Olley, JA Statham, AMA Vail. Nucleotide catabolism: influence on the storage life of tropical species of fish from the north west shelf of Australia. J Food Sci 53:6–11, 1988.

5. S Watabe, M Kamal, K Hashimoto. Postmortem changes in ATP, creatine phosphate, and lactate in sardine muscle. J Food Sci 56:151–153, 1991.

6. M Ando, H Toyohara, Y Shimizu, M Sakaguchi. Post-mortem tenderization of fish muscle proceeds independently of resolution of rigor mortis. Nippon Suisan Gakk 57:1165–1169, 1991.

7. T Saito, K Arai, M Matsuyoshi. A new method for estimating the freshness of fish. Bull Jpn Soc Sci Fish 24:749–751, 1959.

8. NR Jones. Hypoxanthine and other purine containing fractions in fish muscle as indices of freshness. In: R Kreuzer, ed. The Technology of Fish Utilization. London:Fishing News Books, 1965, p 179.

9. S Wongso, H Yamanaka. Extractive components of the adductor muscle of Japanese baking scallop and changes during refrigerated storage. J Food Sci 63:772–776, 1998.

10. M Murata, M Sakaguchi. Storage if yellowtail (Seriola quinqueradiata) white and dark muscle in ice: changes in content of adenine nucleotides and related compounds. J Food Sci 51:321, 1986.

11. G Finne. Enzymatic ammonia production in penaied shrimp held on ice. In: RE Martin, GE Flick, CE Hebard, DW Ward, eds. Chemistry and Biochemistry of Marine Food Products. Westport, CT: AVI, 1982, pp 323–331.

12. LF Jacober, AG Rand. Biochemical evaluation of seafood. In: RE Martin, GE Flick, CE Hebard, DW Ward, eds. Chemistry and Biochemistry of Marine Food Products. Westport, CT: AVI, 1982, pp 347–363.

13. S Watabe, H Ushio, M Iwamoto, M Kamal, H Ioka, K Hashimoto. Rigor mortis progress in sardine and mackerel in association with ATP degradation and lactate accumulation. Nippon Suisan Gakk 55:1833–1839, 1989.

14. JR Burt. Glycogenolytic enzymes of cod (Gadus callarias) muscle. J Fish Res Bd Can 23:527, 1966.

15. JR Burt, GD Stroud. The metabolism of sugar phosphates in cod muscle. Bull Jpn Soc Sci Fish 32:204, 1966.

16. NR Jones. Observations on the relations of flavor and texture to mononucleotide breakdown and glycolysis in fish muscle. In: DJ Tilgner and A Borys, eds. Proceedings, 2nd International Congress of Food Science and Technology. Warsaw: Wydawnictwo Przemyslu Lekkiego i Spożywczego (in Polish), 1967, pp 109–119.

17. NF Haard. Biochemistry and chemistry of color and color change in seafoods. In:

GJ Flick, RE Martin, eds. Advances in Seafood Biochemistry, Composition and Quality. Lancaster-Basel:Technomic Publishing Co, 1992, pp 305–360.

18. J Baldwin, RMG Welles, M Lowe, JM Ryder. Tauropine and D-lactate as metabolic stress indicators during transport and storage of live paua, (New Zealand abalone) (*Haliotis iris*). J Food Sci 57:280–282, 1992.

19. NF Haard. Foods as cellular systems: impact on quality and preservation. A review. J Food Biochem 19:191–238.

20. HA Bremner. Fish flesh structure and the role of collagen—its post mortem aspects and implications for fish processing. In: HH Huss, M Jacobsen, J Liston, eds. Quality Assurance in the Fish Industry. London: Elsevier, 1992, pp 39–62.

21. GE Bracho, NF Haard. Identification of two matrix metalloproteinases in the skeletal muscle of Pacific rockfish (*Sebastes* sp.). J Food Biochem 19:299–319, 1995.

22. M Ando, H Toyohara, Y Shimizu, M Sakaguchi. Postmortem tenderization of fish muscle due to weakening of pericellular connective tissue. Bull Jpn Soc Sci Fish 59:1073–1076, 1993.

23. HA Bremner. Post mortem breakdown of the myotendinous junction in fish. PhD dissertation, University of Tasmania, Hobart, Tasmania, 1992.

24. K Sato, A Koike, R Yoshinaka, M Sato, Y Shimizu. Postmortem changes in type I and V collagens in myocommatal and endomysial fractions of rainbow trout (*Oncorhynchus mykiss*) muscle. J Aquat Food Prod Technol 3(2):5–11, 1994.

25. WK Nip, JH Moy. Microstructural changes of ice-chilled and cooked freshwater prawn, *Macrobrachium rosenbergii*. J Food Sci 53:319–322, 1988.

26. S Mizuta, R Yoshinaka, M Sato, M Sakaguchi. Histological and biochemical changes of muscle collagen during chilled storage of the kuruma prawn *Penaeus japonicus*. Fish Scie 63:784–795, 1997.

27. I Kołodziejska, ZE Sikorski. The properties and utilization of proteases of marine fish and invertebrates. Pol J Food Nutr Sci, 4/45:5–12, 1995.

28. E Budzinski, P Bykowski, D Dutkiewicz. Possibilities of processing and marketing of products made of Antarctic krill. FAO Fish Technical Paper 268:46, 1985.

29. E Kołakowski, L Gajowiecki, Z Szybowicz, T Chodorska. Application of partial autoproteolysis to extraction of protein from Antarctic krill (*Euphausia superba*). Part 1. Effect of pH on protein extraction intensity. *Nahrung* 24 (6):127–134, 1980.

30. C Watson, RE Bourke, RW Brill. A comprehensive theory on the etiology of burned tuna. Fish Bull 86:367–372, 1988.

31. CY Shiau, YJ Pond, TK Chiou, YY Tin. Biochemical changes in milkfish during postmortem rigor-mortis. J Chinese Agric Chem Soc 34:355–363, 1996.

32. E Kołakowski. Changes of non-protein nitrogen fractions in Antarctic krill (*Euphausia superba* Dana) during storage at 3°C and 20°C. Z Lebensm Unters Forsch 183:421–425, 1986.

33. CE Hebard, GJ Flick, RE Martin. Occurrence and significance of trimethylamine oxide and its derivatives in fish and shellfish. In: RE Martin, GE Flick, CE Hebard, DW Ward, eds. Chemistry and Biochemistry of Marine Food Products. Westport, CT: AVI, 1982, pp 149–304.

34. T Richardson, DB Hyslop. Enzymes. In: OR Fennema, ed. Food Chemistry. New York and Basel:Marcel Dekker, 1985, pp 371–476.

35. R Jacquot. Organic constituents of fish and other aquatic animal foods. In: G Borgstrom, ed. Fish as Food, Vol. 1. New York and London: Academic Press, 1961, pp 145–209.

36. RL Shewfeld, RE McDonald, HO Hultin. Effect of phospholipid hydrolysis on lipid oxidation in flounder muscle microsomes. J Food Sci 46:1297–1301, 1981.

37. L Stodolnik. The effect of pH on hydrolysis and oxidation of lipids in the muscle tissue of Baltic cod and herring (in Polish). Przemysl Spożywczego 42:295–297, 1988.

38. J Olley, H Watson. Phospholipase activity in herring muscle. In: JM Lectch, ed. Proceedings of the International Congress of Food Science and Technology, Vol 1. Chemical and Physical Aspects of Foods. New York:Gordon and Break, 1962.

39. T Oshima, S Wada, C Koizumi. Enzymatic hydrolysis of phospholipids in cod flesh during storage in ice. Bull Jpn Soc Sci Fish 50:107–114, 1984.

40. JK Kaitaranta 1982. Hydrolytic changes in the lipids of fish roe products during storage. J Fd Technol 17:87–98, 1982.

41. A Kołakowska. Changes in lipids during the storage of krill (*Euphausia superba* Dana) at 3°C. Z Lebensm Unters Forsch 186:519–523, 1988.

42. RJ Hsieh, JB German, JE Kinsella. Lipoxygenase in fish tissue: some properties of the 12-lipoxygenase from trout gill. J Agric Food Chem 36:680–685, 1988.

43. RE McDonald, SD Kelleher, HO Hultin. Membrane lipid oxidation in a microsomal fraction of red hake muscle. J Food Biochem 3:125–134, 1979.

44. RC Lindsey. Flavor of fish. Paper read at the 8th World Congress of Food Science and Technology. 29 September–4 October, 1991, Toronto, Canada.

45. DB Josephson, RC Lindsay, DA Stuiber. Identification of compounds characterizing the aroma of fresh whitefish (*Coregonus clupeaformis*). J Agric Food Chem 31:326–330, 1983.

46. RC Lindsey. Chemical basis of the quality of seafood flavors and aromas. Marine Technol Soc J 25:16–22, 1991.

47. RJ Hsieh, JE Kinsella. 1989. Lipoxygenase generation of specific volatile flavor carbonyl compounds in fish tissues. J Agric Food Chem 37:279–286, 1989.

48. KL Simpson. Carotenoid pigments in seafood. In: RE Martin, GE Flick, CE Hebard, DW Ward, eds. Chemistry and Biochemistry of Marine Food Products. Westport, CT: AVI, 1982, pp 115–136.

49. HM Chen, SP Meyers, RW Hardy, SL Biede. Color stability of astaxanthin pigmented rainbow trout under various packaging conditions. J Food Sci 49:1337–1340, 1984.

50. EA Decker, Z Xu. Minimizing rancidity in muscle foods. Food Technol 52(10):54–59, 1998.

51. HO Hultin. Lipid oxidation in fish muscle. In: GJ Flick, RE Martin, eds. Advances in Seafood Biochemistry, Composition and Quality. LancasterBasel: Technomic Publishing Co, 1992, pp 99–122.

52. TJ Han, J Liston. 1989. Lipid peroxidation protection factors in rainbow trout (*Salmo gairdneri*) muscle cytosol. J Food Sci 54:809–813, 1989.

53. RJ Hsieh, JB German, JE Kinsella. 1988a. Relative inhibitory potencies of flavonoids on 12-lipoxygenase of fish gill. Lipids 23:322–326, 1988.

54. L Ramanathan, NP Das. Studies on the control of lipid oxidation in ground fish by some polyphenolic natural products. *J Agric Food Chem* 40:17–21, 1992.

55. A Kołakowska, B Czerniejewska-Surma, L Gajowiecki, K Lachowicz, L Zienkowicz. Effect of fishing season on shelf life of iced Baltic herring. In: HH Huss, ed. Quality Assurance in the Fish Industry. Amsterdam:Elsevier Science Publishers, 1992, pp 81–91.

56. JS Chen, RS Rolle, MR Marshall, CI Wei. Comparison of phenoloxidase from Florida spiny lobster and Western Australian lobster. J Food Sci 56:154–157, 1991.

57. RS Rolle, MR Marshall, CI Wei, JS Chen. Phenoloxidase forms of the Florida spiny lobster: immunochemical and spectropolarimetric characterization. Comp Biochem Physiol 97B:483–489, 1990.

58. X Yan, KDA Taylor. Studies of the mechanism of phenolase activation in Norway lobster (*Nephrops norvegicus*). Food Chem 41:11–21, 1991.

59. I Bartolo, EO Birk. Some factors affecting Norway lobster (*Nephrops norvegicus*) cuticle polyphenol oxidase activity and blackspot development. Int J Food Sci Technol 33:329–336, 1998.

60. H Buttkus. Accelerated denaturation of myosin in frozen solution. *J Food Sci* 35:558–562, 1970.

61. ST Jiang, DC Hwang, CS Chen. Effect of storage temperatures on the formation of disulfides and denaturation of milkfish actomyosin (*Chanos chanos*). J Food Sci 53:1333–1335, 1988.

62. ST Jiang, TC Lee. Changes in free amino acids and protein denaturation in fish muscle during frozen storage. J Agric Food Chem 33:839–844, 1985.

63. DD Nambudiri, K Gopakumar. ATPase and lactate dehydrogenase activities in frozen stored fish muscles as indices of cold storage deterioration. J Food Sci 57:72–76, 1992.

64. S Nakajima, T. Tamiya, T Akahane, T. Tsuchiya, JJ Matsumoto. Inactivation of lactate dehydrogenase derived from carp muscle during frozen storage and its prevention. Nippon Suisan Gakk 55:1119, 1989.

65. VP Karvinen, DH Bamford, B Granroth. Changes in muscle subcellular fractions of Baltic herring (*Clupea harengus membras*) during cold and frozen storage. J Sci Food Agric 33:763–770, 1982.

66. P Chawla, B MacKeigan, SP Gould, RF Ablett. Influence of frozen storage on microsomal phospholipase activity in myotomal tissue of Atlantic cod (*Gadus morhua*). Can Inst Food Sci Technol J 21:399–402, 1988.

67. R Hamm. Delocalization of mitochondrial enzymes during freezing and thawing of skeletal muscle. In: O Fennema, ed. Proteins at Low Temperature. Washington DC:American Chemical Society, Advances in Chemistry Series 180, 1979, pp. 192–204.

68. C Barbagli, GS Crescenzi. 1981. Influence of freezing and thawing on the release of cytochrome oxidase from chicken's liver and from beef and trout. J Food Sci 46:491–493, 1981.

69. L Hoz, C Yustes, JM Camara, MA Ramos, GDG Fernando. 1992. Beta-hydroxya-

cyl-CoA-dehydrogenase (HADH) differentiates unfrozen from froze-thawed crawfish (*Procambarus clarkii*) and trout (*Aalmo gairdneri*) meat. Int J Food Sci Technol 27:133–136, 1992.

70. L Riedel. Kalorimetrische Untersuchunges über das Gefrieren von Seefischen. *Kältetechnik* 8:374–377, 1956.

71. GA Reay. The influence of freezing temperatures on haddock's muscle. Part I. J Soc Chem Ind 52(34):265T–270T, 1933.

72. NF Haard. Biochemical reactions in fish muscle during frozen storage. In: EG Bligh, ed. Seafood Science and Technology. London:Fishing News Books, 1992, pp 176–209.

73. JR Burt. Changes in sugar phosphate and lactate concentration in trawled cod (*Gadus callarias*) muscle during frozen storage. J Sci Food Agric 22:536–539, 1971.

74. JG Sharp. Post mortem breakdown of glycogen and accumulation of lactic acid in fish muscle at low temperatures. *Biochem J* 24:850–853, 1935.

75. JG Sharp. Post mortem breakdown of glycogen and accumulation of lactic acid in fish muscle. I. Proc R Soc B 114:506–512, 1934.

76. SS Nowlan, WJ Dyer. Glycolytic and nucleotide changes in the critical freezing zone –0.8 to –5°C. J Fish Res Bd Can 26:2625, 1969.

77. W Partmann. Investigation on the breakdown of nicotinamide adenine dinucleotide (NAD) in fish muscle during frozen storage. (in German). *Lebensm Wiss Technol* 6:155–157, 1973.

78. ST Jiang, BS Hwang, CY Tsao. Protein denaturation and changes in nucleotides of fish muscle during frozen storage. J Agric Food Chem 35:22–27, 1987.

79. WJ Dyer, DI Hiltz. Nucleotide degradation in frozen swordfish muscle. J Fish Res Bd Can 26:159, 1969

80. ZE Sikorski, J Olley, S Kostuch. Protein changes in frozen fish. Crit Rev Food Sci Nutr 8:97–129, 1976.

81. JJ Matsumoto. Denaturation of fish muscle proteins during frozen storage. In: O Fennema, ed. Proteins at Low Temperatures, Advances in Chemistry Series 180, Washington DC:American Chemical Society, 1979, pp. 205–224.

82. SV Shenouda. Theories of protein denaturation during frozen storage of fish flesh. In: CO Chichester, EM Mrack, GF Stewart, eds. Advances in Food Research, vol. 26. New York:Academic Press, 1980, pp 275–311.

83. ZE Sikorski, A Kołakowska. Changes in proteins in frozen stored fish. In: ZE Sikorski, BS Pan, F Shahidi, ed. Seafood Proteins. New York-London:Chapman & Hall, 1994, pp 99–112.

84. NF Haard. Biochemical reactions in fish muscle during frozen storage. In: EG Bligh, ed. Seafood Science and Technology. London:Fishing News Books, 1992, pp. 176–209.

85. ZE Sikorski, S Kostuch. Trimethylamine N-oxide demethylase. Its occurrence, properties, and role in technological changes in frozen fish. *Food Chem* 9:213–222, 1982.

86. HO Hultin. Trimethylamine-N-oxide (TMAO) demethylation and protein denaturation in fish muscle. In: GJ Flick, RE Martin, eds. Advances in Seafood Biochem-

istry, Composition and Quality. Lancaster-Basel: Technomic Publishing Co, 1992, pp 25–42.

87. TK Levskaya, NG Dolgikh, GM Dubinskaya, BC Tuchkov. Ultrastructural and biochemical changes of the muscle tissue of mackerel during long term frozen storage (in Russian). Khol Tekh (11):32–35, 1984.

88. BS Pan, WT Yeh. Biochemical and morphological changes in grass shrimp (*Penaeus monodon*) muscle following freezing by air blast and liquid nitrogen. J Food Biochem 17:147–160, 1993.

89. B Ben-Gigirey, JMVB De Sousa, TG Villa, J Barros-Velazquez. Changes in biogenic amines and microbiological analysis in albacore (*Thunnus alalunga*) muscle during frozen storage. J Food Prot 61:608–615, 1998.

90. EJ Geromel, MW Montgomery. Lipase release from lysosomes of rainbow trout (*Salmo gairdneri*) muscle subjected to low temperatures. J Food Sci 45:412–415, 1980.

91. J Olley, R Pirie, H Watson. Lipase and phospholipase activity in fish skeletal muscle and its relationship to protein denaturation. J Sci Food Agric 13:501, 1962.

92. J Olley, JA Lovern. Phospholipid hydrolysis in cod flesh stored at various temperature. J Sci Food Agric 11:644–652, 1960.

93. L Stodolnik, Z Podeszewski. Lipids of industrial Baltic fish. IV. Changes of lipid fractions during storage of frozen fish. (in Polish). Bromat Chem Toksykol 8:9–21, 1975.

94. K Hanaoka, M Toyomizu. Acceleration of phospholipid decomposition in fish muscle by freezing. Bull Jpn Soc Sci Fish 45:465–468, 1979.

95. N Tsukuda. Changes in the lipids of frozen fish. II. Changes in the lipids of chum salmon and cod muscle during frozen storage. Bull Tokai Reg Fish Res Lab 87:1, 1976.

96. AJ De Koning, TH Mol. Rates of free fatty acid formation from phospholipids and neutral lipids in frozen cape hake (*Merluccius* spp) mince at various temperatures. J Sci Food Agric 50:391–398, 1990.

97. T Oshima, S Wada, C Koizumi. 1985. Accumulation of lyso-form phospholipids in several species of fish flesh during storage at –5°C. Bull Jpn Soc Sci Fish 52:965–971,1985.

98. A Kołakowska. Lipid composition of fresh and frozen stored krill. Z Lebensm Unters Forsch 182:475–478, 1986.

99. A Kołakowska. Krill lipids after frozen storage of about one year in relation to storage time before freezing. Nahrung 33:241–244, 1989.

100. TJ Han, J Liston. 1987. Lipid peroxidation and phospholipid hydrolysis in fish muscle microsomes and frozen fish. J Food Sci 52:294–296&299, 1989.

101. JC Deng. Effect of iced storage on free fatty acid production and lipid oxidation in mullet muscle. J Food Sci 43:337–340, 1978.

102. M Toyomizu, K Hanaoka, K Yamaguchi. Effect of release of free fatty acids by enzymatic hydrolysis of phospholipids on lipid oxidation during storage of fish muscle at –5°C. Bull Jpn Soc Sci Fish 47:615–620, 1981.

103. RL Shewfelt. Fish muscle lipolysis—a review. J Food Biochem 5:79–100, 1981.

104. R Hardy, AS McGill, FD Gunstone. Lipid and autoxidative changes in cold stored cod. J Sci Food Agric 30:999–1006, 1979.

105. SD Kelleher, LA Silva, HO Hultin, KA Wilhelm. Inhibition of lipid oxidation during processing of washed, minced Atlantic mackerel. J Food Sci 57:1103–1108, 1992.

106. A Kołakowska. Effect of initial processing of fat rancidity dynamics during frozen storage of fish. Proceedings of 15th International Congress of Refrigeration. Bull Int Inst Refrig 4:1184, C2-54, 1978.

107. M Toyomizu, K Hanaoka. Lipid oxidation of the minced ordinary muscle of fish during storage at –5°C and susceptibility to lipid oxidation. Bull Jpn Soc Sci Fish 46:1007–1010, 1980.

108. J Undeland, B Ekstrand, H Lingnert. Lipid oxidation in herring (*Clupea harengus*) light muscle, dark muscle, and skin, stored separately or as intact fillets. J. Am. Oil Chem. Soc. 75:581–590, 1998.

109. A Kołakowska, M Szczygielski. Stabilization of lipids in minced fish by freeze texturization. J Food Sci 59:88–90, 1994.

110. T Ingemansson, A Pettersson, P Kaufman. Lipid hydrolysis and oxidation related to astaxanthin content in light and dark muscle of frozen rainbow trout (*Oncorhynchus mykiss*). J Food Sci 58(3):1–6, 1993.

111. E Kołakowski, G Bortnowska. Seasonal variations in technological properties of Baltic herring. II. Extractability of skeletal muscle protein. *Acta Ichthyol Piscat* (in press).

112. H Nomata, H Toyohara, Ch Koizumi. The existence of proteinases in chum salmon muscle and their activities in the spawning stage. Bull Jpn Soc Sci Fish 51:1799–1804, 1985.

113. S Ando, M Hatano. Biochemical characteristics of chum salmon muscle during spawning migration. Bull Jpn Soc Sci Fish 52:1229–1235, 1986.

114. MC Gómez-Guillen, I Batista. Seasonal changes and preliminary characterization of cathepsin D-like activity in sardine (*Sardina pilchardus*) muscle. Int J Food Sci Technol 32:255–260.

115. M Yamashita. Studies on cathepsins in the muscles of chum salmon. Bull Natl Res Inst Fish Sci 5:9–116, 1993.

116. M Matsumiya, A Mochizuki, S Otaka. Seasonal variation of alkaline proteinase activity in the ordinary muscle of common mackerel. Bull Jpn Soc Sci Fish 57:1411, 1991.

117. T Hirano, M Suyama. Quality of life and cultured ayu—III. Seasonal variation of nitrogenous constituents in the extracts. Bull Jpn Soc Sci Fish 46:215–219, 1980.

118. RB Hughes. Chemical studies on herring (*Clupea harengus*). VII. Collagen and cohesiveness in heat-processed herring, and observations on a seasonal variation in collagen content. J Sci Food Agric 14:432–441, 1963.

119. H Toyohara, K Ito, K Touhata, M Kinoshita, S Kubota, K Sato, K Ohtsuki, M Sakaguchi. Effect of maturation on content of free and bound forms of hydroxyproline in ayu muscle. Fish Sci 63:843–844, 1997.

120. E Kołakowski, R Jarosz, M Wianecki. Seasonal variation in technological properties of Baltic herring. I. Proximate composition. Acta Ichthyol Piscat (in press).

121. M Crupkin, CL Montecchia, RE Trucco. Seasonal variations in gonadosomatic index, liver-somatic index and myosin/actin ratio in actomyosin of mature hake (*Merluccius hubbsi*). Comp Biochem Physiol 89A:7–10, 1988.

122. Y Shimizu, N Wendakoon. Effects of maturation and spawning on the gel-forming ability of lizardfish (*Saura elongata*) muscle tissues. J Sci Food Agric 52:331–338, 1990.

123. SI Roura, C Montecchia, L Goldenberg, RE Trucco, M Crupkin. Biochemical and physicochemical properties of actomyosin from pre- and post-spawned hake (*Merluccius hubbsi*) stored on ice. J Food Sci 55:688–692, 1990.

124. S Konagaya. Enhanced protease activity in muscle of chum salmon *Oncorhynchus keta* during spawning migration. Bull Jpn Soc Sci Fish 48:1504, 1982.

125. M Yamashita, S Konagaya. Proteolysis of muscle proteins in the extensively softened muscle of chum salmon caught during spawning migration. Bull Jpn Soc Sci Fish 57:2163, 1991.

126. K Lachowicz, E Kołakowski. Seasonal variation in technological properties of Baltic herring. III. Texture. Acta Parasithol Piscat (in press).

127. A Gildberg. Proteolytic activity and the frequency of burst bellies in capelin. J Food Technol 13:409–416, 1978.

128. A Gildberg, J Raa. Solubility and enzymic solubilization of muscle and skin of capelin (*Mallotus villosus*) at different pH and temperature. Comp Biochem Physiol 63B:309–314, 1979.

129. E Kołakowski, A Kołakowska, K Lachowicz, L Gajowiecki, A Protasowicka, B Czerniejewska-Surma, M Wianecki, G Bortnowska, R Jarosz. Technological value of Baltic herring. Part 2 (in Polish). Research Report CPBR 10–16. Agricultural University in Szczecin, 1989.

130. AV Ananichev. Comparative biochemical data for some freshwater invertebrates and fish. Biochemistry 26:16–26, 1961.

131. G Siebert. Enzymes of marine fish muscles and their role in fish spoilage. In: E Heen, R Kreuzer, eds. Fish in Nutrition. London:Fishing News, 1961.

132. R Yoshinaka, M Sato, S Ikeda. Distribution of trypsin and chymotrypsin and their zymogens in digestive system of catfish. Bull Jpn Soc Sci Fish 47:1615–1618, 1981.

133. R Yoshinaka, M Sato, S Ikeda. Distribution of collagenase in the digestive organs system of some teleosts. Bull Jpn Soc Sci Fish 44:263–267, 1977.

134. R Hofer. The adaptation of digestive enzymes to temperature, season and diet in roach (*Rutilus rutilus*), and rudd (*Scardinius erythrophthalmus*) proteases. J Fish Biol 15:373–379, 1979.

135. A Martinez, A Gildberg. Autolytic degradation of belly tissue in anchovy (*Engraulis encrasicholus*). Int J Food Sci Technol 23:185–194, 1988.

136. RM Love. The Chemical Biology of Fishes. London and New York:Academic Press, 1970.

137. PW Hochachka, GN Somero. Biochemical Adaptation. Princeton, NJ: Princeton University Press, 1984, pp 355–449.

138. J Freed. Changes in activity of cytochrome oxidase during adaptation of goldfish to different temperatures. Comp Biochem Physiol 14:651–659, 1965.

139. H Abe, E Okuma. Rigor mortis in carp acclimated to different water temperatures. Bull Jpn Soc Sci Fish 57:2095–2100, 1991.

140. XF Guo, S Watabe. ATPase activity and thermostability of actomyosin from thermally acclimated carp. Bull Jpn Soc Sci Fish 59:363–369, 1993.

141. A Hashimoto, A Kobahashi, K Arai. Thermostability of fish myofibrillar Ca-ATPase and adaptation of environmental temperature. Bull Jpn Soc Sci Fish 48:671–684, 1982.

142. A Hashimoto, K Arai. Temperature dependence on Mg-ATPase activity and its Ca-sensitivity of fish myofibrils. Bull Jpn Soc Sci Fish 50:853–864, 1984.

143. GI Higashi. Foodborne parasites transmitted to man from fish and other aquatic foods. Food Technol 39(3) 69–74, 1985.

144. RE Olson. Marine fish parasites of public health importance. In: DE Kramer, J Liston, ed. Seafood Quality Determination. Proceedings of International Symposium on Seafood Quality Determination, Anchorage, Alaska, 1986, pp 339–355.

145. GE Rodrick, TC Cheng. Parasites: occurrence and significance in marine animals. Food Technol. 43(11):98–102, 1989.

146. J Grabda. Outline of Marine Fish Parasitology. Warsaw:PWN, Polish Sci Publ., 1991.

147. RB Raybourne, TL Deardorff, JW Bier. *Anisakis simplex:* larval excretory secretory protein production and cytostatic action in mammalian cell cultures. Exp Parasitol 62:92–95, 1986.

148. S Konagaya. Studies on the jellied meat of fish, with special reference to that of yellowfin tuna. Bull Tokai Reg Fish Res Lab 114:1–101, 1984.

149. V Stout, G Carter. Ames test for mutagenicity on Pacific whiting treated with hydrogen peroxide. J Food Sci 48:492–495, 1983.

150. GA Neish, GC Hughes. Fungal diseases of fishes. In: SF Snieszko, HR Axelrod, ed. Diseases of Fishes; vol. 6. Neptune, NJ: TFH Publications, 1980, pp 61–100.

151. B Spanggaard, HH Huss. Growth of the fish parasite *Ichthyophonus hoferi* under food relevant conditions. Int J Food Sci Technol 31:427–432, 1996.

152. JA Machado-Cruz, JC Eiras, D Marques. 1982. Two new hosts (*Mugil auratus* and *Blennius pholis*) of *Ichthyosporidium* and diagnosis in asymptomatic carriers. Publ do Instituto de Zoologia "Dr Augusto Nobre", Facultade de Ciencias do Porto, 167:5–11, 1982.

153. MC Erickson, DT Gordon, AF Anglemier. 1983. Proteolytic activity in the sarcoplasmic fluids of parasited Pacific whiting (*Merluccius productus*) and unparasited true cod (*Gadus macrocephalus*). J Food Sci 48:1315–1319, 1983.

154. T Masaki, M Shimomukai, Y Miyauchi, S Ono, T Tuchiya, T Matsuda, H Akazawa, M Soejima. Isolation and characterization of the protease responsible for jellification of Pacific hake muscle. Bull Jpn Soc Sci Fish 59:683–690, 1993.

155. H Toyohara, M Kinoshita, I Kimura, M Satake, M. Sakaguchi. Cathepsin L-like protease in Pacific hake muscle infected by myxosporidian parasites. Bull Jpn Soc Sci Fish 59:1101–1107, 1993.

156. E Biliński, NP Boyce, REE Jonas, MD Peters. Characterization of protease from myxosporen salmon parasite, *Henneguya salminicola*. Can J Fish Aquatic Sci 41:371–376, 1984.

157. H Karl, M Leinemann. Überlebensfähigkeit von Nematodenlarven (*Anisakis* sp.) in gefrosteten Heringen. Arch Lebensmittelhyg 40:14–16, 1989.
158. PR Witthams, MG Walker. The activity of myofibrillar and actomyosin ATPase in the skeletal muscle of some marine teleosts in relation to their length and age. J Fish Biol 20:471–477, 1982.
159. CS Wardle. Limit of swimming speeds in fish. Nature 255:725–727, 1975.
160. MM Mazeaud, F Mazeaud. Adrenergic responses to stress in fish. In: AD Pickering, ed. Stress and Fish. London:Academic Press, 1981, pp 49–75.
161. CM Wood. Acid-base and ion balance, metabolism, and their interaction, after exhaustive exercise in fish. J Exp Biol 160:285–308, 1991.
162. AR Jerrett, J Stevens, AJ Holland. Tensile properties of white muscle in rested and exhausted chinook salmon (*Oncorhynchus tshawutscha*). J Food Sci 61:527–532, 1996.
163. T Sigholt, U Erikson, T Rustad, S Johansen, TS Nordtveldt, A Seland. Handling stress and storage temperature affect meat quality of farmed-raised Atlantic salmon (*Salmo salar*). J Food Sci 62:898–905, 1997.
164. M Iwamoto, H Yamanaka, S Watabe, K Hashimoto. Comparison of rigor-mortis progress between wild and cultured plaices. Nippon Suisan Gakk 56:101–104, 1990.
165. U Erickson, AR Beyer, T Sigholt. Muscle high-energy phosphates and stress affect K-values during ice storage of Atlantic salmon (*Salmo salar*). J Food Sci 62:43–47, 1997.
166. TE Lowe, JM Ryder, JF Carragher, RMG Wells. Flesh quality in snapper, *Pagrus auratus*, affected by capture stress. J Food Sci 58:770–773, 1993.
167. P Sebastio, F Ambroggi, G Baldrati. Influence of slaughtering method on rainbow trout bred in captivity. I. Biochemical considerations. Ind Conserve 71:37–49, 1996.
168. K Azam, IM Mackie, J Smith. The effect of slaughter method on the quality of rainbow trout (*Salmo gairdneri*) during storage on ice. Int J Food Sci Technol 24:69–79, 1989.
169. M Ando, A Banno A, M Hitani, H Hirai, T Nakagawa, Y Makinodan. Influence on post-mortem rigor of fish body and muscular ATP consumption by the destruction of spinal cord in several fishes. *Fish Sci* 62:796–799, 1996.
170. FAO Reference manual to codes of practice for fish and fishery products. FAO Fisheries Circular (750), Rome:FAO, 1982.
171. CA Curran, RG Poulter, A Brueton, NSD Jones. Cold shock reactions in iced tropical fish. J Food Technol 21:289–299, 1986.
172. CA Curran, RG Poulter, A Brueton, NR Jones, NSD Jones. Effect of handling treatment on fillet yields and quality of tropical fish. J Food Technol 21:301–310,1986.
173. RWH Parry, MV Alcasid, EB Panggat. Cold shock in fish: its characteristics in bighead. Int J Food Sci Technol 22:637–642, 1987.
174. M Iwamoto, H Yamanaka, H Abe, S Watabe, K Hashimoto. Rigor-mortis progress and its temperature-dependency in several marine fishes. Nippon Suisan Gakk 56:93–99, 1990.

175. J Lavety. Gaping in farmed salmon and trout. Torry Advisory Note 90. Aberdeen:Torry Research Station.

176. SJ Shaw, EG Bligh, AD Woyewoda. Effects of delayed filleting on quality of cod flesh. J Food Sci 49:979–980, 1984.

177. HA Bremner, G Veith. Effects on quality attributes of holding rock lobsters in slush ice before tailing. J Food Sci 45:657–660, 1980.

178. HC Chen, MW Moody, ST Jiang. 1990. Changes in biochemical and bacteriological quality of grass prawn during transportation by icing and oxygenation. J Food Sci 55:670–673, 1990.

179. Sulfiting agents: Proposed affirmation of GRAS status with specific limitations. Fed Reg 41(15):29956, 1982.

180. Notice to shippers, distributors, packers and importers of shrimp containing sulfites. Fed Reg 50(15):2957, 1985.

181. S Slattery, A Bremner, D Williams. An alternative to meta? Austral Fish 51(11):30–31, 1992.

182. AJ McEvily, R Iyengar, S Otwell. Sulfite alternative prevents shrimp melanosis. Food Technol 45(9), 80–86, 1991.

183. E Kołakowski, M Wianecki, G Bortnowska, R Jarosz. Trypsin treatment to improve freeze texturization of minced bream. J Food Sci 62:737–743 + 752, 1997.

184. A Kołakowska, E Kołakowski, M Szczygielski. Effect of unidirectional freezing on lipid changes during storage of minced bream. Proceedings of the 19th International Congress of Refrigeration, Vol II, 196–201, 1995.

185. LJ De Rosenzweig Pasquel, JK Babbitt. Isolation and partial characterization of a natural antioxidant from shrimp (*Pandalus jordani*). J Food Sci 56, 143–145, 1991.

186. A Koczot, M Szewczuk, J Zalewski, M Iadkowska, I Barska, M Brzeski, W Korzewa, M Prada. The technological suitability of elasmobranch and teleostid fish of the open ocean. (in Polish). Research Report No. T-133. Sea Fisheries Institute, Gdynia, 1981.

187. JJ Matsumoto, SF Noguchi. Cryostabilization of protein in surimi. In: TC Lanier, CM Lee, eds. Surimi Technology. New York: Marcel Dekker, 1992, pp 357–388.

188. TC Lanier. Functional food protein ingredients from fish. In: ZE Sikorski, BS Pan, F. Shahidi, eds. Seafood Proteins. New York London: Chapman & Hall, 1994, pp 127–159.

189. ZE Sikorski, A Ruiter. Changes in proteins and nonprotein nitrogen compounds in cured, fermented, and dried seafoods. In: ZE Sikorski, BS Pan, F Shahidi, eds. Seafood Proteins. New York London:Chapman & Hall, 1994, pp 113–126.

190. AG Rand, LF Pivarnik. Enzyme preservation of fresh seafoods. In: GJ Flick, RE Martin, ed. Advances in Seafood Biochemistry, Composition and Quality. Lancaster Basel:Technomic Publishing, 1992, pp 135–150.

191. I Kimura, M Sugimoto, K Toyoda, N Seki, K Arai, T Fujita. A study on the crosslinking reaction of myosin in kamaboko "suwari" gels. Bull Jpn Soc Sci Fish 57:1389–1396, 1991.

192. GG Kamath, TC Lanier, EA Foegeding, DD Hamann. Non-disulfide covalent crosslinking of myosin heavy chain in "setting" of Alaska pollock and Atlantic croaker surimi. J Food Biochem 26:151–172, 1992.

193. J Wan, J Miura, N. Seki. Effect of monovalent cations on cross-linking of myosin heavy chain in surimi gel from walleye pollock. Bull Jpn Soc Sci Fish 58:583–590, 1992.

194. HG Lee, TC Lanier, DD Hamann, JA Knopp. Transglutaminase effects on low temperature gelation of fish protein sols. J Food Sci 62:20–24, 1997.

195. J Mai, JE Kinsella. Changes in lipid composition of cooked minced carp (*Cyprinus carpio*) during frozen storage. *J Food Sci* 44:1619–1624, 1997.

17
Chemical Treatments to Control Enzymes in Fishery Products

Marilyn Erickson
University of Georgia, Griffin, Georgia

I. INTRODUCTION

Enzymes are macromolecular proteins whose unique primary and three-dimensional structure leads to specific catalytic activity in food and other biological systems. Disruption of that structure invariably leads to loss of activity and would be desirable in the case of enzymes whose activities are associated with deterioration in seafood quality. Several studies, for example, concluded that chemical/biochemical reactions as opposed to microbial growth were responsible for the deterioration in initial (prime) quality of unprocessed fish. In the case of snapper (1), yellow-eyed mullet (2), and trumpeter (3), the shelf lives of sterile and nonsterile fish were the same during initial stages of storage. Hence, maintenance of the prime quality of fish would require effective strategies directed at controlling chemical/biochemical reactions (4). When fish have been exposed to processes that lead to release of degradative enzymes from storage compartments and interaction of substrate with these enzymes is accelerated, the need to control chemical/biochemical reactions becomes even greater. In other cases, enzyme action is desirable and maximization of that activity requires maintenance of that structure. While heat and pressure (see Chap. 20) are commonly used for enzyme denaturation, control of enzyme structure and hence activity may also be achieved through modification of the enzyme's chemical environment. The latter process is the focus of this chapter. To address this issue, the chapter is organized into two parts. In the first part, the fundamental chemical mechanisms for control of enzyme catalyzed reactions will be explored. In the second part, specific examples will be given on the response of enzymic systems in fishery products to chemical treatments.

II. FUNDAMENTAL MECHANISMS FOR CHEMICAL CONTROL OF ENZYME REACTIONS

A. Removal of Cofactor

To function, many enzymes require the presence of small amounts of chemical agents other than the reactants of the catalyzed reaction. These compounds are called cofactors and may be simply metal ions, more complex organic molecules, or even other proteins. The purpose of these cofactors is usually to supply a specific chemical function that is not possible with the enzyme alone. Thus a metal cofactor may undergo a cyclic oxidation–reduction process or act as a chelating agent for the substrate. In many of these cases, the cofactor can be separated by dialysis from the enzyme that catalyzes the specific reaction. In other cases, the enzyme cofactors are more tightly bound to the enzyme molecule and are called prosthetic groups. Examples of prosthetic groups are the hemes, flavins, and biotin.

Any substance that complexes with and/or removes the cofactor from interaction with the enzyme will inhibit that enzyme. Some examples of these types of inhibitors include cyanide, sulfide, azide, carbon monoxide, ethylenediaminetetraacetic acid (EDTA), 1,10-phenanthroline, imidazole, mercaptoethanol, and diethyldithiocarbamate. In complexing with iron, copper, zinc, or molybdenum, cyanide can inhibit enzyme systems such as cytochrome oxidase, catalase, and ascorbic acid oxidase. EDTA is also capable of forming strong complexes with divalent and higher oxidation state cations (Fe^{2+}, Zn^{2+}, Cu^{2+}, Co^{2+}, Mn^{2+}, Mg^{2+}, Ca^{2+}, Fe^{3+}, and Mo^{6+}) and consequently could inhibit enzymes requiring those cations. One must bear in mind that the ability of a compound to inhibit will depend on many other environmental factors including pH, temperature, source and concentration of enzyme, substrate and cofactor concentrations, and concentration of inhibitor. It is also important to note that the toxic nature of many of these inhibitors prevents their application to foods.

B. Substrate Modification or Competition

To exhibit enzyme activity, the substrate must form a complex with the enzyme. Consequently, modification of the substrate may alter enzyme activity by altering its binding affinity with the enzyme. The efficacy of this approach ultimately depends on the specificity of the enzyme. For example, the lipase from *Candida rugosa* shows preference for the C18:1 *cis*-9 fatty acid whereas the microbial lipase from *Corynebacterium acnes* is nonspecific (5). Oxidation of esterified oleic acid would therefore make it unavailable to the lipase from *C. rugosa.*

Another means to make the substrate unavailable to the enzyme is to apply an inert substance having similar characteristics as the substrate such that it effectively replaces the substrate in the vicinity of the enzyme. For example, enzymes requiring oxygen or carbon dioxide as one of the substrates may be inhibited by replacement of the necessary gaseous substrate with an inert gas.

A much more common method for making a substrate unavailable to an enzyme is to expose the enzyme to a substrate analogue. Substrate analogues are broadly defined as any compound structurally related to the substrate such that it binds to the active site of an enzyme. Enantiomeric analogues, anomeric analogues, positional isomers, geometrical isomers, and products of the reaction are examples of substrate analogues. While the substrate analogue may participate in the catalytic process to a greater or lesser extent than the normal substrate, the substrate analogue is also capable of completely interfering with the catalytic process by preventing binding of the substrate. In most cases, however, the presence of a substrate analogue leads to competitive inhibition that can be reversed by increasing substrate concentration to saturating levels.

C. Modification of Enzyme Structure

To maintain conformation of the active site or to hold a multisubunit enzyme together, specific groups on the enzyme are involved. Modification of those groups through adjustments in pH, adjustments in ionic composition, or through direct reaction of the group will invariably affect the enzyme activity.

1. pH

An enzyme has maximum activity only within a narrow pH range. Factors responsible for the effect of pH on enzyme-catalyzed reactions can be classified into two categories: those that influence the stability of the enzyme, and those that influence binding and catalysis. Enzymes are ampholytes: they have dissociation constants for both their acidic and alkaline groups. Changes in pH will therefore affect the ionization of these groups and in turn the enzyme stability or the ability of the enzyme to react with the substrate. Bear in mind that adjustments of pH to muscle tissues, fruits, or vegetables may have deleterious side effects. In any case, a more detailed explanation on effect of pH on rates of enzyme-catalyzed reactions may be found in work by Whitaker (6).

2. Ionic Environment

Although ions may be required as components of the active site, they also have nonspecific effects on the enzyme, such as changing the surface charge on the enzyme protein, changing the equilibrium constant of the enzyme reaction, displacing an ineffective metal ion from the active site or from the substrate, or shifting the equilibrium of a less active conformer to a more active conformer. Ionic effects vary from one enzyme to another, such that an ion could be an activator to one enzyme yet be an inhibitor to another. Furthermore, the concentration of the ion may have an effect on the response in which it could activate the enzyme at one concentration and inhibit at another.

3. Direct Chemical Reaction

A number of groups on amino acid residues may be essential for enzyme activity and include sulfhydryl groups; the carboxyl groups of glutamic and aspartic acids; the α-carboxyl group at the end of the polypeptide chain; the imidazole group of histidine; the aliphatic hydroxyl groups of serine and threonine; the ϵ-amino group of lysine and the terminal α-amino group; the phenolic hydroxyl group of tyrosine; the guanidino group of arginine; and the indole nucleus of tryptophan. When these groups are essential, chemical modification will lead to enzyme inhibition. Some of the reagents used in chemical modification of enzymes include sulfiting agents, alkylating agents, oxidizing agents, reducing agents, mercaptide-forming agents, and acetylating agents. In most cases, however, the modifications are applied to purified enzymes to understand the nature of those enzymes. Because the modifying agents are often nonselective, modifying groups on proteins other than the targeted enzyme, application to complex food systems is often not taken.

III. RESPONSE OF ENZYMIC SYSTEMS IN FISHERY PRODUCTS TO CHEMICAL TREATMENTS

A. Oxidoreductase Enzymes

Enzymes that oxidize or reduce substrates by transfer of hydrogen or electrons are designated as oxidoreductase enzymes. Examples of oxidoreductase enzymes in seafood products in which chemical control is desirable include polyphenoloxidase and lipoxygenase. Another oxidoreductase enzyme applied to seafood products whose action has beneficial consequences on seafood quality is glucose oxidase.

1. Polyphenoloxidase (Tyrosinase)

Melanosis in shellfish, commonly referred to as "blackspot," is a surface color defect caused by polyphenoloxidase (PPO) activity. More specifically, PPO catalyzes the oxidation of mono (tyrosine) and orthodiphenols to quinones (see Chap. 10). Subsequently, a nonenzymatic polymerization of the quinones occurs, leading to the formation of dark, high-molecular-weight pigments. In the case of shrimp, the black spots begin to form on the head, then proceed down the shrimp, forming black lines just under the shell that outlines the sections of the tail. PPO enzyme systems remain active during iced or refrigerated storage to the extent that raw, untreated pink shrimp developed melanosis in less than 3 days and within 7 days progressed to a severe product defect (7). These defects decrease the perceived quality of the product by consumers and reduce its commercial value (8).

To thwart PPO's activity, several chemicals have been tested as potential inhibitors against the purified enzyme (9). Phosphate compounds (sodium acid pyrophosphate, sodium hexametaphosphate, sodium tripolyphosphate), tryptophan, and the sulfhydryl-inhibiting compounds, N-ethyl-maleimide and Ellman's reagent, showed no inhibitory activity while phenylthiourea, sodium bisulfite, ascorbic acid, cysteine, and methyl mercury were found to be effective.

When tested against endogenous PPO in seafood, sulfites have proved to be one of the most effective agents used. Immersion of prawns in a brine containing 0.7% sodium bisulfite prevented blackening (10) whereas frozen raw crab meat required a treatment in a solution containing ≥ 1% sodium bisulfite (11). The primary mechanisms responsible for bisulfite inhibition of lobster shell melanosis have been ascribed to irreversible reaction with PPO possibly through disruption of disulfide bonds present in PPO; and bisulfite interaction with intermediate quinones and their inability to form brown pigments. In the latter case, the enzyme per se has not been inhibited; rather the products of the enzyme reaction have been diverted to non-melanoidin-generating compounds. Sulfites, through their reducing properties, may likewise cycle quinones back to the original phenols and hence prevent formation of brown pigment. With both the latter mechanisms, the amount of sulfites necessary for controlling the enzymatic browning would be dependent on the concentration of the tyrosine substrate present in the product.

Allergic reactions among sulfite-sensitive individuals and asthmatics have led to increasing concern regarding sulfite as a melanosis inhibitor (12). Residual concentrations of the inhibitor are therefore of concern. While residual SO_2 concentrations in crab tissues have been shown to be dependent on the sodium bisulfite concentrations in the dipping solutions, considerable fluctuations exist among the tissues, with higher residuals in the carapace than in the muscle tissues (10, 11). Size of sample is also a factor in levels of residual SO_2, with smaller samples containing higher concentrations than larger samples (10). Reduction in residual SO_2 may occur by rinsing in brine immediately after sodium bisulfite treatment with little decline in effectiveness. Despite these attempts to optimize the sulfite treatment, alternative chemical methods for controlling polyphenoloxidase activity have been explored.

Kato et al. (13) recently reported on the capacity of 0.5% and 1.0% sericin, a silk protein, to inhibit mushroom tyrosinase in an in vitro system. Another chemical treatment explored for prevention of melanosis was a dip treatment employing a glucose oxidase/catalase enzyme system in combination with 0.5% D-glucose (14). Unfortunately, this treatment had only a marginal effect on prevention of black spot development in iced shrimp. In contrast, dipping of shrimp in dilute solutions of the sulfhydryl protease ficin appeared to be an effective treatment against the development of melanosis (15). Binding or hydrolysis at specific sites necessary for the activity of PPO was presumed to be the mechanism of inactivation but subsequent evaluation of the ficin dip solutions identified the presence of another PPO inhibitor, 4-hexylresorcinol. At concentrations of 0.025%, 4-hexylresorcinol was proven to be effective as an inhibitor of shrimp melanosis (16) and in 1991, Enzytech, Inc. of Cambridge, Massachusetts, was awarded a patent for use of this chemical on shrimp (17). When used in the field as a dip solution, however, it was noted that a heavy bacterial load developed with time in the dip solution. To prevent the potential for microbial contamination of shrimp, incorporation of lactic acid (1.0%) in the dip solution was recommended (16). Addition of lactic acid may also contribute to the inhibition of PPO through its ability to alter the pH environment of the enzyme. Exposing krill to acidic baths decreased tyrosinase activity (18) and a similar situation could occur with shellfish PPO, which has an optimum pH of 6.2 (9).

2. Lipoxygenase

Three detrimental effects have been identified in response to the activity of the enzyme, lipoxygenase: destruction of essential fatty acids, linoleic, linolenic and arachidonic acids; damage to vitamins and proteins by the radicals that are produced as a byproduct; and development of off flavors and odors. The specific activity of the enzyme is to catalyze the addition of molecular oxygen to a *cis-cis*-4-pentadiene-containing unsaturated fatty acid, releasing a stereospecific conjugated diene hydroperoxy fatty acid product (see Chap. 11 for details). Although sometimes referred to as an initiator of lipid oxidation, strictly speaking that is not the case since preformed lipid hydroperoxides are required for lipoxygenase activity. Once the enzyme is activated, their products may further break down and contribute to the latter two detrimental effects listed above.

Lipoxygenase activity has been reported in fish tissues from several different species (19–26); therefore, its activity is of concern. Reported inhibitors of lipoxygenase include the antioxidants nordihydroguaiacetic acid, propyl gallate and α-tocopherol (6). The antioxidant 1,3-dihydro-7-methyl-4,5,6-trihydroxy-isobenzofuran, isolated from the fungus, *Aspergillus terreus,* has also been shown to inhibit the oxidation of linoleic acid by sardine flesh lipoxygenase (27). When the relative efficacy of flavonoid compounds was evaluated on 12-lipoxygenase from fish gill, the inhibition was determined to be noncompetitive with fisetin and quercetin the most potent. In all these cases, the antioxidants' mode of action is presumably through their ability to break the subsequent chain of free radical events rather than by direct inhibition of the enzyme. Saturated monohydric alcohols, on the other hand, may inhibit by hydrophobically binding to the site of interaction of the $CH_3(CH_2)_4$ group of the substrate with the enzyme (6).

3. Glucose Oxidase

An oxidoreductase enzyme reported to have a beneficial effect on quality of seafood is glucose oxidase. When this enzyme was applied together with glucose to the surface of fish, the treatment was found effective in extending the period of sensory acceptance of flounder fillets at 2–4°C by about 6 days (28). The gluconic acid generated by the applied enzyme was thought to reduce the pH and ultimately lower the growth of the microorganisms. In contrast, Shaw et al. (29) found that a glucose oxidase enzyme treatment on cod fillets showed insignificant inhibition of the microbial indicators and little if any benefit to the chemical, raw quality, or sensory assessment of the fillets. Differences in the chemical environment during application of the enzyme treatment could account for these contrasting findings. For example, the level of enzyme and substrate applied may be adjusted to ensure maximum benefit by the enzyme treatment application. In the case of shrimp, optimal levels of glucose oxidase and glucose were shown to be 1 unit and 4%, respectively (30). The inclusion of catalase in the glucose oxidase solution (31) would also be expected to be beneficial. Catalase's activity serves to remove H_2O_2, the second product generated in the glucose oxidase reaction, which otherwise could eventually inactivate the glucose oxidase enzyme.

B. Hydrolytic Enzymes

Hydrolases or hydrolytic enzymes are catabolic enzymes that catalyze the transfer of groups to water for degradation and turnover of cellular macromolecules. Examples of hydrolytic enzymes important in seafoods include lipolytic and proteolytic enzymes. While they are important physiologically, their activity in stored seafood products often leads to sensorially unacceptable product.

1. Lipolytic Enzymes

The two primary categories of lipolytic enzymes are lipases and phospholipases that produce free fatty acids following the hydrolysis of ester linkages of triacylglycerols and phospholipids, respectively. These enzymes are discussed in more detail in Chapters 4 and 5. In seafood products, the free fatty acids produced are of concern for two reasons: they may interact with proteins to form insoluble protein–lipid complexes that modify texture; or they may be more susceptible to lipid oxidation, as is the case for free fatty acids arising from triacylglycerols.

Endogenous lipolytic enzymes are relatively resistant to low temperatures and retain much of their activity in the frozen state (32–34), hence, inactivation would be desirable. A variety of chemical substances have been shown to inhibit lipase activity and include anionic surfactants, certain proteins, metal ions, boronic acids, phosphorus-containing compounds such as diethyl-p-nitrophenyl phosphate, phenylmethyl sulfonylfluoride, certain carbamates, β-lactones, and diisopropylfluorophosphate (35). Unfortunately, these chemicals are not applied commercially to seafood products due to limitations of toxicity and/or inability to get the chemical to the enzyme in the product. Other chemicals, applied for purposes other than inhibition of lipolysis, however, have been shown to modify lipolytic activity. For example, both Damodaran Nambudiry (36) and Takiguchi (37) found that high levels of NaCl inhibited lipid hydrolysis in sardines. Immersion of fish samples in $MgSO_4$ or sodium hypochlorite has also led to a significant inhibition in free fatty acid formation (38, 39). In the former case, magnesium may compete with endogenous calcium, which is known to stimulate lipase-catalyzed hydrolytic activity (5). In the latter case, hypochlorite may directly oxidize groups on the lipolytic enzymes and render them inactive.

2. Proteolytic Enzymes

Proteases catalyze the hydrolysis of proteins to short-chain peptides and amino acids. While their initial activity is considered beneficial, providing tenderization to muscle in rigor mortis, excessive proteolysis can lead to liquefaction or mushiness of the product. Depending on the desired final state of the product, this excessive activity may or may not be desired. In the case of fish protein concentrate (FPC), or fish sauce, breakdown of the protein is required whereas breakdown in commercially sold fillets or mince, such as occurs with the Pacific whiting species, is considered unacceptable.

To maximize the production of FPC, enzyme activity should be optimized. Ibrahim and Nyns (40) found that papain or pepsin was effective in FPC solubilization if a pretreatment with NaOH (pH 10, 80–85°C for 20 min) was used. Since the majority of endogenous proteases and proteases added to the system are most effective at acidic pHs, adjustment of pH is often applied. Fik et al. (41) found that the reagent used for adjusting pH (HCl vs. acetic acid) had an important effect on rate of hydrolysis, with HCl the most effective. Salt concentration of the system has also been shown to have a major effect on the proteolytic activity, with alkaline and acid proteinases inhibited by > 15% NaCl in an in vitro system but minimally inhibited when sardine muscle proteins were used as the substrate (42). Susceptibility to inactivation by NaCl was greater with endogenous herring proteolytic enzymes than with papain or a *Bacillus subtilis* protease added to minced herring samples (43).

Chemical inhibitors to control autolysis of seafood muscle samples include a number of proteins and protein sources including α_2-macroglobulin (MAC), beef blood plasma, egg white and potato extract. For example, a MAC-rich extract from beef blood plasma inhibited gel weakening in surimi of New Zealand hoki and Alaska pollack (44). The MAC molecule is believed to undergo limited proteolysis, leading to a conformational change in the protein that traps the protease within it and prevents its interaction with large protein substrates (45). Application of egg white and potato extract to surimi formulations has also been successful in improving the texture of gels (46–49). More information on these types of inhibitors may be found in Chapter 19.

Other chemicals, besides NaCl and proteins, have been shown to inhibit proteases. These include spices and potassium bromate (50). In Pacific whiting surimi sols, a bromate level of 0.075% was found to inactivate 89.9% of the protease activity. Lower proteolytic activity in fish sausages, compared to corresponding minced meat samples, was attributed to the effects of the spices in the sausages (51). In a subsequent study, however, pepper spices added to mackerel stuffing, showed more autoproteolysis than the control stuffing (52). These contradictory results point out the necessity for additional controlled studies on the effect of spices and their constituents on protease activity.

C. Transferase Enzymes

Enzymes that remove groups (excluding H) from substrates and transfer them to acceptor molecules (excluding water) are called transferase enzymes. An example of a transferase enzyme in gadoid fish in which activity should be minimized is trimethylamine oxide demethylase (see Chap. 7) and an example of a transferase enzyme in which activity should be optimized is transglutaminase (see Chap. 6).

1. Trimethylamine Oxide Demethylase

Breakdown of trimethylamine N-oxide (TMAO) by the enzyme TMAO demethylase in gadoid fish species generates dimethylamine (DMA) and formaldehyde (FA). The FA, in turn, is extremely reactive, and crosslinks with the myofibrillar proteins, causing a toughening of the texture and a loss of water holding capacity.

Numerous chemical additives have been tested for their ability to inhibit freeze-toughening of fish associated with TMAO demethylase activity. In general, oxidizing agents (oxygen, H_2O_2) have been found to inhibit the enzyme, while reducing agents (NADH) activate the enzyme (53–56). To determine the effectiveness of different oxidizing agents (H_2O_2, NaOCl, and $KBrO_3$), Raciciot et al. (53) applied the chemicals to minced red hake samples at four different levels. The most effective treatment was with H_2O_2 based on quantities of DMA, FA, and TMAO and on sensory panel analysis.

Sodium citrate has also been shown to reduce the amount of toughening that developed during storage of gadoid species (55,56). Although sodium citrate appeared to inhibit the amount of DMA produced, the reduced toughening in pollack fillets was instead attributed to their improvement in water-holding capacity, a condition that would have lessened the interaction of substrate with enzyme or formaldehyde with protein.

Varied success has been obtained with anionic hydrocolloids in the diminishment of production of DMA and FA during storage of whiting and cod minced fillets. (57–60). Whereas Kelcosol, Xanthan, Iota carrageenan, and Kappa carrageenan were all capable of inhibiting TMAO demethylase activity, only xanthan was capable of inhibiting the enzyme reaction in a subsequent study. Again, the hydrocolloids may act through their ability to increase water-holding capacity and dilute the interaction of enzyme with substrate.

2. Transglutaminase

To aid in gelation and binding of proteins in surimi and restructured tissue products, transglutaminase activity is beneficial. More specifically, transglutaminase generates ε-(γ-glutamyl) lysine crosslinking by catalyzing an acyl transferase reaction between the γ-carboxamide groups of peptide-bound glutaminyl residues and a variety of primary amines or lysyl residues of proteins (61). The enzyme is calcium dependent, therefore modification of the calcium concentration has proven to affect gelation. In the case of walleye pollack surimi, maximum gelling occurred in the range of 2–5 mM Ca^{2+} (62). Addition of EDTA, which sequesters calcium, on the other hand, only affected endogenous transglutaminase activity in fish protein gels and not microbial transglutaminase added to the protein sols (63).

D. Lyase Enzymes

Enzymes that remove groups from their substrates (not by hydrolysis) to leave a double bond, or that conversely add groups to double bonds, are designated as lyases. One lyase enzyme found in seafood products that could benefit from chemical control is histidine decarboxylase.

1. Histidine Decarboxylase

The major route of histidine breakdown in fish muscle is decarboxylation to histamine by enzymes of contaminating bacteria (64). Since histamine is a capillary dilator responsible for allergic responses in some humans and is often implicated as a causative agent in scombroid poisoning, inhibition of histidine decarboxylase is desirable.

Several food additives (glucose, glycine, sucrose, sorbic acid, sodium chloride, citric acid, malic acid, and succinic acid) have been tested by a Korean group of researchers for their effect on histamine formation and histidine decarboxylase activity (65, 66). With the exception of glucose and sucrose, all these additives were found to inhibit the enzymic activity and hence generation of histamine. Concentration of the additive was a factor in the degree of inhibition by salt and sorbic acid, with higher levels of inhibition being observed at higher concentrations.

Spices have also been evaluated for their inhibitory capacity on histidine decarboxylase activity (67). Demonstrating a significant inhibitory effect on this enzyme from the bacteria *Morganella morganii* were the spices cinnamon and cloves. In a similar system, cardamom and turmeric had a moderate effect while pepper exhibited no effect. A similar pattern of inhibition in the formation of histamines and biogenic amines was observed in response to application of these spices (3%) to whole Indian mackerel and storage of the product at 30°C for up to 24 h.

IV. CONCLUSIONS

A variety of chemical agents have been shown to modify enzymatic activity in seafood products either directly or indirectly. Of these chemical agents, the number that have potential for application in a commercial food processing setting is limited. Chemical toxicity, limited dispersion within the product, and/or interaction with constituents other than the desired enzyme may all contribute to the inappropriateness of a chemical agent being applied commercially to the seafood product. Additional research on the inhibitory capacity of novel natural compounds to enzymatic activity could expand our arsenal of useful agents. Compounds capable of specifically inhibiting lipolytic enzymes would be particularly beneficial in light of the fact that no chemical treatments exist today for that purpose.

REFERENCES

1. GC Fletcher, JA Hodgson. Shelf-life of sterile snapper (*Chrysophrys auratus*). J Food Sci 53:1327–1332, 1988.
2. GC Fletcher, JA Statham. Shelf-life of sterile yellow-eyed mullet (*Aldrichetta forsteri*) at 4°C. J Food Sci 53:1030–1035, 1988.
3. GC Fletcher, JA Statham. Deterioration of sterile chill stored and frozen trumpeter fish (*Latridopsis forsteri*). J Food Sci 53:1336–1339, 1988.
4. NF Haard. Technological aspects of extending prime quality of seafood: a review. J Aquat Food Prod Technol 1:9–27, 1992.

5. JD Weete. Microbial lipases. In: CC Akoh, DB Min, eds. Food Lipids. Chemistry, Nutrition, and Biotechnology. New York: Marcel Dekker, 1998, pp. 641–664.

6. JR Whitaker. Principles of Enzymology for the Food Sciences, 2nd ed. New York: Marcel Dekker 1994.

7. WS Otwell, M. Marshall. Screening alternatives to sulfiting agents to control shrimp melanosis. Florida Cooperative Extension Service, Sea Grant Extension Program, 1986, Technical Paper No. 26.

8. PL Cooper, MR Marshall, JF Gregory III, WS Otwell. Ion chromatography for determining residual sulfite on shrimp. J Food Sci 51:924–928, 1986.

9. CF Madero Farias. Purification and characterization of phenoloxidase from brown shrimp (*Penaeus aztecus*). Ph.D. dissertation, Texas A & M University, College Station, Texas, 1982.

10. N Tsukuda, K Amano. Effects of sodium bisulphite on prevention of blackening of prawns, and residues of SO_2 in treated prawns. Bull Tokai Region Fish Res Lab No. 72:9–19, 1972.

11. T Kinumaki, T Watanabe. Studies on prevention of blue discoloration of raw frozen wary crab meat. Bull Tokai Region Fish Res Lab No. 92:65–77, 1977.

12. SL Taylor, RK Bush. Sulfites as food ingredients. Food Technol 40(6):47–52, 1986.

13. N Kato, S Sato, A Yamanaka, H Yamado, N Fuwa, M Nomura. Silk protein, sericin, inhibits lipid peroxidation and tyrosinase activity. Biosci Biotechnol Biochem 62:145–147, 1998.

14. MSH Al-Jassir. Polyphenol oxidase inhibition by glucose oxidase in pink shrimp (*Pandalus borealis*). Ph.D. Dissertation, University of Rhode Island, Kingston, RI, 1989.

15. PS Taoukis, TP Labuza, JH Lillemo, SW Lin. Inhibition of shrimp melanosis (black spot) by ficin. Lebensm Wiss Technol 23:52–54, 1990.

16. RA Benner, R Miget, G Finne, GR Acuff. Lactic acid/melanosis inhibitors to improve shelf life of brown shrimp (*Penaeus aztecus*). J Food Sci 59:242–245, 250, 1994.

17. AJ McEvily, R Iyengar, A Gross. Compositions and methods for inhibiting browning in foods using resorcinol derivatives. U.S. Patent 5,059,438. 1991.

18. T Onishi, T Watanabe, M Suzuki. Tyrosinase of Antarctic krill and prevention of its blackening by dipping in acidic solutions. Bull Tokai Region Fish Res Lab No. 96:1–9, 1978.

19. JB German, JE Kinsella. Lipid oxidation in fish tissue. Enzymatic initiation via lipoxygenase. J Agric Food Chem 33:680–683, 1985.

20. RJ Hsieh, JB German, JE Kinsella. Lipoxygenase in fish tissue. Some properties of the 12-lipoxygenase from trout gill. J Agric Food Chem 36:680–685, 1988.

21. RJ Hsieh, JE Kinsella. Lipoxygenase generation of specific volatile flavor carbonyl compounds in fish tissues. J Agric Food Chem 37:279–286, 1989.

22. JB German, RK Creveling. Identification and characterization of a 15-lipoxygenase from fish gills. J Agric Food Chem 38:2144–2147, 1990.

23. S-Y Mohri, S Cho, Y Endo, K Fujimoto. Lipoxygenase activity in sardine skin. Agric Biol Chem 54:1889–1891, 1990.

24. YJ Wang, LA Miller, PB Addis. Effect of heat inactivation of lipoxygenase on lipid

oxidation in lake herring (*Coregonus artedii*). J Am Oil Chem Soc 68:752–757, 1991.

25. CH Zhang, T Shirai, T Suzuki, T Hirano. Studies on the odour of fishes. IV. Lipoxygenase-like activity and formation of characteristic aroma compounds from wild and cultured ayu. Nippon Suisan Gakk 58:959–964, 1992.

26. P Harris, J Tall. Substrate specificity of mackerel flesh lipoxygenase. J Food Sci 59:504–506, 1994.

27. Y Ishikawa, T Kato, K Hsiung Lee. Inhibition of sardine flesh lipoxygenase by a new antioxidant from *Aspergillus terreus*. J Jpn Oil Chem Soc 45:1321–1325, 1996.

28. CE Field, LF Pivarnik, SM Barnett, AG Rand Jr. Utilization of glucose oxidase for extending the shelf-life of fish. J Food Sci 51:66–70, 1986.

29. SJ Shaw, EG Bligh, AD Woyewoda. Spoilage pattern of Atlantic cod fillets treated with glucose oxidase/gluconic acid. Can Inst Food Sci Technol 19:3–6, 1986.

30. K Jukokusumo, AG Rand, Jr. Glucose oxidase treatment to preserve fresh shrimp. In: M Kroger, R Shapiro, eds. Changing Food Technology. Lancaster, PA: Technomic Publishing Co., 1987, p. 164.

31. P Wesley. Glucose oxidase treatment prolongs shelf life of fresh seafood. Food Dev 16:36–38, 1982.

32. AJ de Koning, S Milkovitch, TH Mol. The origin of free fatty acids formed in frozen cape hake mince (*Merluccius capensis,* Castelnau) during cold storage at -18°C. J Sci Food Agric 39:79–84 1987.

33. AJ de Koning, TH Mol. Rates of free fatty acid formation from phospholipids and neutral lipids in frozen cape hake (*Merluccius* spp.) mince at various temperatures. J Sci Food Agric 50:391–398, 1990.

34. Y Jeong, T Ohshima, C Koizumi, Y Kanou. Lipid deterioration and its inhibition of Japanese oyster *Crassostrea gigas* during frozen storage. Nippon Suisan Gakk 56:2083–2091, 1990.

35. S Patkar, F Bjorkling. Lipase Inhibitors. In: P Woolley, SB Petersen, eds. Lipases. Cambridge, MA: Cambridge University Press, 1994, p. 207.

36. D Damodaran Nambudiry. Lipid oxidation in fatty fish: The effect of salt content in the meat. J Food Sci Technol India 17:176–178, 1980.

37. A Takiguchi. Effect of NaCl on the oxidation and hydrolysis of lipids in salted sardine fillets during storage. Nippon Suisan Gakk 55:1649–1654, 1989.

38. REE Jonas. Effect of magnesium sulphate and cryoprotective agents on free fatty acid formation in cold-stored fish muscle. J Fish Res Bd Can 26:2237–2240, 1969.

39. PJ Ke, RG Ackman. Inhibition of formation of free fatty acids in cod liver with sodium hypochlorite solution (Javex). J Fish Res Bd Can 32:297–300, 1975.

40. AA Ibrahim, E Nyns. Solubilization of fish protein concentrate. Rev Ferment Ind Aliment 34:14–20, 1979.

41. M Fik, E Kolakowski, QAV Mecola. Effect of pH, temperature and incubation time on proteolysis rate of sardine meat. Przemysl Spozywczy 35:195–197, 1981.

42. M Nada, TV Van, I Kusakabe, K Murakami. Substrate specificity and salt inhibition of five proteinases isolated from the pyloric caeca and stomach of sardine. Agr Biol Chem 46:1565–1569, 1982.

43. M Fik, L Sypniewska, N Pham Thi. Effect of selected physico-chemical factors on

Baltic herring meat hydrolysis rate and proteolytic enzyme activity. Acta Aliment Pol 11:305–312, 1985.

44. DD Hamann, PM Amato, MC Wu, EA Foegeding. Inhibition of modori (gel weakening) in surimi by plasma hydrolysate and egg white. J Food Sci 55:665–669, 795, 1990.

45. AJ Barrett, PM Starkey. The interaction of α_2-macroglobulin with proteinases. Characteristics and specificity of the reaction, and a hypothesis concerning its molecular mechanism. Biochem J 133:709–724, 1973.

46. MW Chang-Lee, R Pacheco-Aguilar, DL Crawford, LE Lampila. Proteolytic activity of surimi from Pacific whiting (*Merluccius productus*) and heat-set gel texture. J Food Sci 54:1116–1119, 1124, 1989.

47. TC Lanier, TS Lin, DD Hamann, FB Thomas. Effects of alkaline protease in minced fish on texture of heat processed gels. J Food Sci 46:1643–1645, 1981.

48. Y Makinodan, H Toyohara, E Niwa. Implication of muscle alkaline protease in the textural degradation of fish meat gel. J Food Sci 50:1351–1355, 1985.

49. DH Wasson. Fish muscle proteases and heat-induced myofibrillar degradation: a review. J Aquat Food Prod Technol 1:23, 1992.

50. R Pacheco-Aguilar, DL Crawford. Potassium bromate effects on gel-forming ability of Pacific whiting surimi. J Food Sci 59:786–791, 1994.

51. E Kolakowski, M Fik, L Gajowiecki. Comparison of proteolytic processes in frozen minced meat and sausages from Baltic herring. Nahrung 18:503–509, 1974.

52. M Fik, E Kolakowski, I Strzelec. Effect of natural spices on intensity of autoproteolysis of frozen mackerel stuffing. Przemysl Spozywczy 31:68–69, 1977.

53. LD Raciciot, RC Lundstrom, KA Wilhelm, EM Ravesi, JJ Licciardello. Effect of oxidizing and reducing agents on trimethylamine N-oxide demethylase activity in red hake muscle. J Agric Food Chem 32:459–464, 1984.

54. P Reece. The role of oxygen in the production of formaldehyde in frozen minced cod muscle. J Sci Food Agric 34:1108–1112, 1983.

55. KL Parkin, HO Hultin. Some factors influencing the production of dimethylamine and formaldehyde in minced and intact red hake muscle. J Food Proc Preserv 6:73–97, 1982.

56. DJ Krueger, OR Fennema. Effect of chemical additives on toughening of fillets of frozen Alaska pollack (*Theragra chalcogramma*). J Food Sci 54:1101–1106, 1989.

57. DJB da Ponte, JP Roozen, W Pilnik. Effects of additions on the stability of frozen stored minced fillets of whiting. I. Various anionic hydrocolloids. J Food Qual 8:51–69, 1985.

58. DJB da Ponte, JP Roozen, W Pilnik. Effects of additions on the stability of frozen stored minced fillets of whiting. II. Various anionic and neutral hydrocolloids. J Food Qual 8:175–190, 1985.

59. DJB da Ponte, JP Roozen, W Pilnik. Effects of different types of carrageenans and carboxymethyl cellulose on the stability of frozen stored minced fillets of cod. J Food Technol 20:587–598, 1985.

60. DJB da Ponte, JP Roozen, W Pilnik. Effects of additions, irradiation and heating on the activity of trimethylamine oxide-ase in frozen stored minced fillets of whiting. J Food Technol 21:33–43, 1986.

61. JE Fold. Mechanisms and basis for specificity of transglutaminase catalyzed ε-(γ-glutamyl) lysine bond formation. In: A Meister, ed. Advances in Enzymology and Related Areas in Molecular Biology. New York: Wiley, 1983, p. 1.

62. W Jianrong, I Kimura, M Satake, N Seki. Effect of calcium ion concentration on the gelling properties and transglutaminase activity of walleye pollack surimi paste. Fish Sci 60:107–113, 1994.

63. HG Lee, TC Lanier, DD Hamann. Covalent cross-linking effects on thermo-rheological profiles of fish protein gels. J Food Sci 62:25–28, 32, 1997.

64. JM Regenstein, CE Regenstein. Introduction to Fish Technology. New York: Van Nostrand Reinhold, 1991.

65. JH Kang, YH Park. Effect of salt, acidulants and sweetenings on histamine formation during processing and storage of mackerel. Bull Korean Fish Soc 17:383–390, 1984.

66. JH Kang, YH Park. Effect of glucose, sucrose and sorbic acid on histamine formation during processing and storage of mackerel. Bull Korean Fish Soc 17:485–491, 1984.

67. R Jeya Shakila, TS Vasundhara, D Vijaya Rao. Inhibitory effect of spices on in vitro histamine production and histidine decarboxylase activity of *Morganella morganii* and on the biogenic amine formation in mackerel stored at 30°C. Z Lebensm Unters Forsch 203:71–76, 1996.

18

Endogenous Enzymes in Antarctic Krill: Control of Activity During Storage and Utilization

Edward Kołakowski
Agricultural University of Szczecin, Szczecin, Poland

Zdzisław E. Sikorski
Technical University of Gdańsk, Gdańsk, Poland

I. INTRODUCTION

The small crustacean Antarctic krill (*Euphausia superba* Dana) forms a unique marine resource that plays a crucial role in the Antarctic ecosystem. Krill is the most important food for whales, seals, penguins, squid, and fish of many Antarctic species. It also constitutes a very large reservoir of animal protein suitable for being exploited for food purposes and for farm animal feeding. The actual landings of krill are still rather small: about 4×10^5 metric tons (mt) annually (1). The potential sustainable annual catch that would not adversely affect the Antarctic ecosystem has been estimated as high as 10^8 mt, (i.e., almost equal to the landings of marine fish).

In the 1960s and 1970s an enormous international research effort was put into understanding the biology, biochemistry, nutrition, fishing techniques, preservation on board, and processing of krill to edible products and as animal fodder (2, 3). The seasonal fluctuation in krill concentration affects the feasibility of commercial fishing in different parts of the Antarctic waters. Efficient krill fishing techniques have been developed and methods of processing the little crustacean to peeled meat and various food products have been established.

Also, the nutritional value of different krill products has been determined. A very comprehensive chapter on the utilization of krill for human consumption, based mainly on Japanese investigations and industrial experience, has been published by Suzuki (4). Also, ways of utilization of chitin, the main component of the krill shell, have been proposed (5, 6).

It is now recognized the digestive enzymes of the crustacean play a crucial role in its quality, deterioration, and processing. About 30 different krill hydrolases have been described, some of them fully purified (7). The temperature of krill's Antartic water habitat Antarctic waters averages about 0°C, declining in winter to -2°C. Thus the enzyme systems of krill are adapted to very low temperature (see Chap. 1). After harvest, at higher ambient temperatures on board the vessel, the activity of krill digestive enzymes brings about very rapid auytolysis and deteriorative processes. Tissue enzymes also contribute to quality changes. Later during storage microbial enzymes play the major role in spoiling the crustacean.

II. DEGRADATION PROCESSES IN KRILL AFTER CATCH

A. Glycolysis, ATP Degradation, and pH

The rate of biochemical processes in krill stored on board the fishing vessel is more than one order higher than that for fin fish. The content of glycogen in the muscle of krill decreases fourfold in 1 h at 2°C and ATP content decreases sevenfold in the same time (8). The ratio of AMP+IMP to inosine+hypoxanthine +adenosine+IMP+AMP decreased from the initial 82% to 68%, and the ratio hypoxanthine to adenosine increases from the initial 18% to 32% after 48 h at 5°C (9). The ATPase activity in krill muscles decreases rapidly during cold storage of the harvested crustacean. The first-order rate constant (K_D) of inactivation of the myofibrillar Ca-ATPase of krill at 5°C and 10°C was 1.2×10^{-5} sec^{-1} and 4.4×10^{-5} sec^{-1}. The loss of activity of the enzyme in fish muscle is very much lower. At 20°C, K_D of inactivation of the krill ATPase is 9.5 times and 34 times higher than that for the enzyme in grenadier and Alaska pollack muscle, respectively (10). A rapid decrease in myofibrillar Ca-ATPase and Mg-ATPase activities in krill stored at 2°C was shown also by Shibata et al. (11). Most of the loss in activity occurred in 1–3 days of storage. The krill myofibrillar Ca-ATPase has maximal activity at about 17°C, while the enzyme is fish muscle has a temperature optimum in the range 30–36°C (10; see also Chap. 3).

Shibata and Nakamura (12) monitored guanosine triphosphate breakdown products in the tail meat of krill and in whole krill. The ratio of the total amount of these compounds to that of ATP breakdown products was 2:5 and 4:5, respectively, in the tail meat and whole krill.

Rigor mortis in krill, marked by the characteristic folding of the abdomen of the crustacean due to muscle contraction, starts after about 1 h and lasts usu-

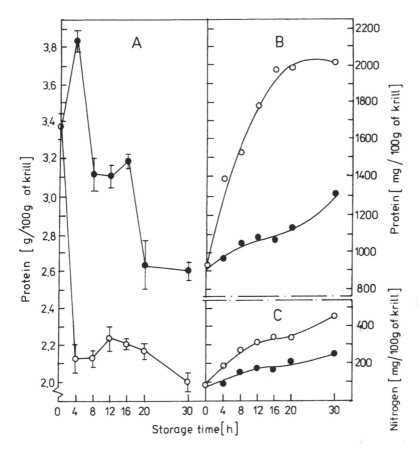

Figure 1 Changes in hot-water-soluble polypeptides (A) and polypeptides soluble in 4% TCA (B) as determined by the biuret method, and in amino-acid nitrogen (C) content during storage of krill at 3°C (o-o) and 20°C (•-•). (From Ref. 14.)

ally 2 h at 3°C. Although the degradation of glycogen proceeds at a high rate, the pH in krill muscles does not decrease, as is characteristic of other food myosystems, but increases up to 8.0 in the first 2–4 h after catch due to rapidly proceeding proteolysis and other enzyme-catalyzed reactions (13).

B. Autolytic Changes

Degradation of proteins by endoproteinases and exoproteinases to polypeptides, olygopeptides, and amino acids occurs rapidly in postharvest krill. During the first 20–24 h after catch, there is a very rapid decrease in the concentration of muscle

WHOLE KRILL   ABDOMEN

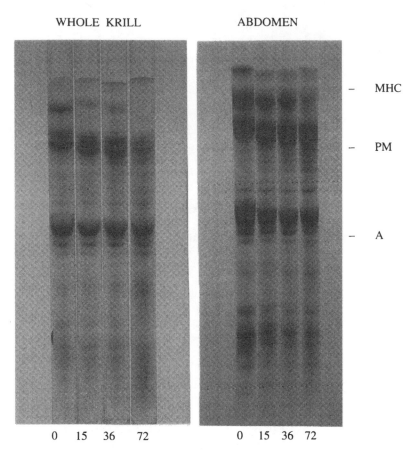

Figure 2   Changes in SDS–polyacrylamide electrophoretic fractions of krill myofibrillar proteins during storage at 3°C. MHC, myosin heavy chain; PM, paramyosin; A, actin.
(From Ref. 15.)

protein. The accumulation of polypeptides proceeds at a lower rate, while that of
amino acid occurs even more slowly, especially at higher temperature (Fig. 1).

Electrophoretic separation of hydrolysis products revealed that the most
susceptible protein to degradation was the myosin heavy chain, while actin and
paramyosin were comparatively resistant to proteolysis by endogenous enzymes
of krill (Fig. 2).

The rate of proteolysis is significantly higher in whole krill than in the isolated abdominal muscle. For isolated krill muscle, protein extractability and viscosity decrease slowly only after 8–9 days of refrigerated storage, except for a
slight loss during the *rigor mortis* period (11, 13).

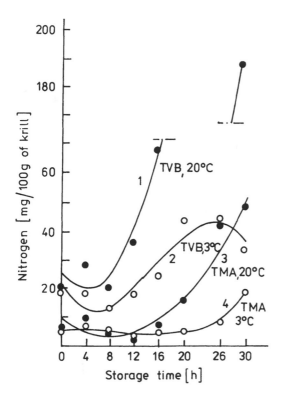

Figure 3 Changes in total volatile bases (TVB) and trimethylamine (TMA) content in Antarctic krill during storage at 3°C (o-o) and 20°C (•-•). (From Ref. 14.)

Severe proteolysis results in formation of drip containing protein and products of protein degradation. The drip originates mainly from the cephalothorax, since this part of the crustacean is especially rich in digestive enzymes and is most succeptible to mechanical damage. The amount of drip and the rate of its formation depend mainly on the feeding intensity of the krill and on the damage done to the animal during fishing operations and handling on board, as well as on the temperature of storage of the catch. When krill is stored at room temperature, the drip after 10 h may reach as much as 25% of the total mass of the crustacean (16).

C. Microbiological Processes

In krill stored at 3°C (i. e., in conditions similar to those prevailing on board fishing vessels), the concentration of total volatile bases and of trimethylamine starts to increase after 8 and 24 h, respectively. At 20°C the rate of accumulation of these nonprotein nitrogenous compounds is much higher (Fig. 3).

These results indicate that microbiological changes do not contribute significantly to the degradation processes during the first hours of storage of the catch on board. Ellingsen and Mohr (17) demonstrated that microbial activity in krill postmortem was negligible even during the first week of storage at 0°C. This, however, does not correspond to other published results indicating that rapid growth of microflora in krill at 0–2°C starts after 2–3 days and makes the crustacean unfit for human consumption (18, 19). In iced, small marine fish early bacterial spoilage is noticeable usually 6–8 days after catch.

Much more rapid deterioration of the sensory quality of krill is due to dark discoloration, similar to that in shrimps and other crustacea (see Chap. 10). In krill stored at 3°C, enzyme browning, measured by absorbance at 450 nm in the trichoracetic acid extracts, was noticeable after 8 h, while at 20°C browning is observed after only 3–4 h (14). The darkening involves first a "reddening" caused probably by enzymatic hydrolysis of carotenoprotein complexes, followed by darkening of the cephalothorax and later also of the abdomen (16).

III. FREEZING CHANGES IN KRILL PROTEINS

Prolonged frozen storage brings about severe loss in quality of krill or krill meat, similar to that known for frozen fish (20). The quality changes are caused mainly by denaturation and aggregation of proteins and are manifested by loss in ATPase activity and deterioration of several functional properties, mainly protein extractability, water-holding capacity, and gel-forming ability. Losses in ATPase activity and protein extractability are often used as indicators of protein denaturation in frozen seafood (see also Chap. 12). The krill proteins are especially liable in this respect (21–23). Nishimura et al. (24) found that in frozen krill stored 3 months at -20°C, the ATPase activity was completely lost and this coincided with the disruption of the myofibrillar structure. About 50% of the total amount of myofibrillar protein in peeled, frozen krill was denatured within 3 months at -20°C or 9 months at -27°C. No loss in myofibrillar protein extractability was evidenced in samples stored 9 months at -40°C (4, 23). Thus the recommendable storage temperature for frozen, peeled krill meat is below -40°C. Cooking of krill prior to freezing prevents quality deterioration of the products during frozen storage (4, 25).

Changes in protein of frozen fish can be effectively retarded by using cryoprotectants (26). Cryoprotectant are also effective in preventing quality deterioration of frozen krill. Ca-ATPase in krill can be protected against inactivation by using monosodium glutamate, sorbitol, lactitol, sucrose, and glucose (27, 28).

IV. KRILL PROTEASES

Krill proteases have been investigated by the comparative biochemist as well as the seafood technologist because of their significant role in krill processing. Some digestive enzymes have potential applications in the cosmetic and tanning industries, in household chemistry, and in medicine. In most cases, enzyme activity was determined in extracts from whole krill that were frozen and stored under different conditions. In some experiments liquid digesta obtained by centrifugation from live specimens with different feeding status was used as the source of enzymes (7). Thus in discussing the characteristics of different enzymes, the origin and preparation of the samples should be considered.

Among the proteases displaying high activity in the neutral and slightly alkaline environment are trypsin and trypsinlike enzymes, aminopeptidase, carboxypeptidase A and B, and an enzyme of unknown specificity (Table 1 A). Proteinases active at low pH are mainly cathepsins A, L, B, H, and D among others (Table 1 B). The presently known proteases found in different parts of the body of krill should not be regarded as incomplete at this time. The proteases are mainly concentrated in the hepatopancreas and stomach of krill located in the cephalothorax, while their activity in the abdomen and the muscles is poor (Fig. 4). Thus in experiments aimed at understanding the role of proteases in krill processing it is advisable to determine separately the enzyme activity in the cephalothorax and the abdomen.

A. Trypsinlike Enzymes

In the cephalothorax, the most abundant proteinases are the trypsinlike enzymes. This group appears to consist of two or three glycoproteins. They are stable in the pH range 5.5–9.0 and their pH optimum is 8.2. Unlike mammalian trypsins, they are relatively unstable at low pH. At pH 4.5 they are inactivated in less than 1 h, whereas at pH 7.5 they do not lose significant activity in 17 days at 0°C (34). The acid lability of krill trypsinlike enzymes results from the strong anionic nature of these enzymes, since they contain very many residues of aspartic and glutamic acids. In this respect they are similar to trypsins of other marine crustaceans and unlike mammalian trypsins (see Chapter 9). Krill trypsinlike enzymes are serine proteinases and are totally inhibited by soybean tripsin inhibitor or phenylmethanesulfonyl fluoride. They are unaffected by the chymotrypsin inhibitor tosyl phenyl chloromethyl ketone (34), although other authors suggest that this group of enzymes includes also chymotrypsin-like enzymes. The krill trypsinlike enzymes are mainly endopeptidases, in that they lack the ability to split amino acids from casein, although one of the isoform also exhibits some exopeptidase activity (34). With respect to resistance to thermal inactivation, the

Table 1 Characteristics of Proteinases of Antarctic Krill (*Euphasia superba* Dana).

Enzyme Code	Molecular weight, KDa	Stable pH range	Optimum pH	Optimum temperature (°C)	Classification	Reference
A. Acting Optimally at Neutral and Weakly Alkaline pH						
A	—	6.0–10.0	8.0	60	Unknown	29
B	—	6.0–10.0	8.0	40	Trypsinlike	
C	—	6.0–10.0	8.0	45	Trypsinlike	
A_1	24	—	9.0	48	Carboxypeptidase A	30
A_2	—	—.	—	—	Trypsinlike	
B	24	—	8.0	50	Trypsinlike	
C	28	—	7.5	55–60	Trypsinlike	
D	27	—	7.5–8.0	55	Trypsinlike	
E-I	100	8.0[a]	—	—	Carboxypeptidase A type	31
E-II	38	8.0[a]	—	—	Carboxypeptidase A type	
E-III	35	8.0[a]	—	—	Carboxypeptidase A type	
E-IV	32	8.0[a]	—	—	Carboxypeptidase B type	
E-V	33	8.0[a]	—	—	Trypsin type	
A, B, C	28–30	5.9–9.5	8.0	55	Anionic trypsyn-like	32
A_2	29	5.5–9.0	8.0	45	New type of anionic Chymotrypsin	33
I	30	4.5–9.0	8.2	45	Anionic serine-type peptide	34, 35
II	31	4.5–9.0	8.2	50	Hydrolase	
III	31	4.5–9.0	8.2	50	Anionic serine-type peptide Hydrolase Anionic serine-type peptide Hydrolase	
CPB 1						36
	31	5.5–9.0	~7.0	~40	Carboxypeptidase B	
CPB 2						36, 37
AP	~60	7.0–10.0[a]	8.7	~50	Aminopeptidase	
P II	30.5	5.0–8.0	6.0–6.4 8.0[d] 8.7–8.9[e]	~40	Anionic serine proteinase	7
B. Acting Optimally at Acid, and Weakly Acid pH						
P	26K	5.0–7.0	6.0	40	?	(38)
A	—	—	2.5–3.0	—	—	(39)
B	78K	—	2.5–3.0	—	—	
C	36K	—	2.5–3.0	—	—	
A	45K	2.5–6.0	2.5–3.0	40	Acid proteinase inhibited by pepstatin;	(40)
B	64K	2.5–6.0	2.5–3.0	40	Acid proteinase less sensitive to pepstatin	

Table 1 Continued

Enzyme Code	Molecular weight, KDa	Stable PH range	Optimum pH	Optimum temperature (°C)	Classification	Reference
E-VII	60K	—	~5.0[a]	—	Cathepsin A type	(31)
E-VIII	25K	—	~5.0[a]	—	Carboxypeptidase A type	
E-IX	130K	—	~5.0[a]	—	Aminopeptidase type	
E-X	—	—	~5.0[a]	—	Several types	
I	80K	3.0–8.0	4,0[b]	45	Cathepsin L type	(41)
II	52K	5.0–8.5	6,0[c]	45	Chymotrypsin-like (?)	
III	58K	3.0–7.0	3,0[b]	45	Typical acid proteinase pepstatin sensitive	
CPA 1						(36)
	27K	4.0–9.0	5.5–6.5	~60	Carboxypeptidase A	
CPA 2						

[a] Investigated only at this pH.
[b] Hemoglobin substrate.
[c] Casein substrate.
[d] Against BzTyrOE.
[e] Against BzArgOE.

krill trpsinlike enzymes are similar to trypsins. All three isoforms of krill trypsin-like enzyme were inactivated at 50°C after 3–6 min (34).

B. Serine Collagenase

Turkiewicz (7) isolated and purified to homogeneity (from whole krill or liquid digesta) a serine proteinase with the specificity of tissue collagenases: capable of splitting soluble collagen into products of three-quarters and one-quarter of the initial length. The enzyme also hydrolyzed hemoglobin, casein, myosin, and fibrinogen. The proteinase also exhibited strong chymotrypsinlike and weak trypsinlike activities. The yield of the enzyme was severalfold higher from heavilly feeding krill than from animals with a low food intake. The enzyme is a 30.5 kDa glycoprotein, the saccharide moiety makes 5–10% of the molecular weight. The isoelectric point is below 2.9. Complete inactivation of the enzyme occurs at 50°C in 10 min. The collagenase is well adapted to catalysis at low temperature (0–12°C).

C. Digestive Peptidases

The group of proteases present in the krill cephalothorax, with optimum activity around neutral pH, includes carboxypeptidases and aminopeptidases (Table 1).

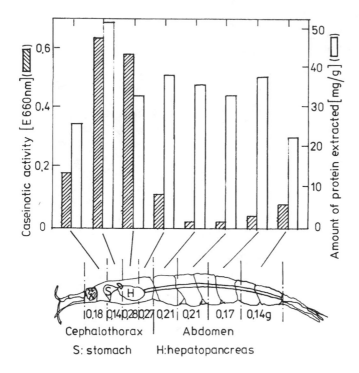

Figure 4   Distribution of caseinolytic activity in krill body. (From Ref. 42.)

Carboxypeptidase B enzymes is most active at pH close to neutrality, carboxypeptidase A at slightly acidic pH, and aminopeptidase(s) at pH 8.7. The enzymes do not exhibit any pronounced activity towards casein; however, in mixtures with trypsinlike enzymes a significant synergistic effect in liberating amino acids is evident (36). Such synergistic effects of concerted action of exopeptidases and endopeptidases are characteristic of other digestive proteases in other animals (43).

D.   Cathepsins

In the abdomen muscle of Antarctic krill, the dominant acid proteinases that hydrolyze hemoglobin optimally at pH 3.0–4.0 and azocasein at pH 5.0 are cathepsins. Their substrate specificity and sensitivity to inhibitors indicate the presence of cathepsins A, B, L, and probably also H (31, 41, 44). Kimoto et al. (41) found in krill abdomen muscle a typical acid proteinase inhibited by pepstatin and thus classified as cathepsin D. This enzyme, however, was not found in krill homogenate by Nishimura et al. (31).

E. Calpains

Calpains, the calcium activated neutral proteinases, known to play a significant part in proteolytic changes in the muscles of other animals (see chapter 15), were not detected in krill abdomen muscle (41).

V. PROTEOLYSIS IN KRILL

A. Effect of pH and Methods of Inactivation

The autoproteolysis in Antarctic krill is very rapid and extensive. Even in the intact abdomen, excised from live crustacean and stored at 3°C, degradation of myosin can be noticed after several hours. In the cephalothorax, the changes are much more rapid and pronounced (Fig. 2). The myosin heavy chain is the component most sensitive to autolysis.

Autoproteolysis proceeds at a broad range of pH (i.e., 2–12) (Table 2), with the highest rate between pH 7 and 10. About 75% of the total proteinases in the whole krill are active at pH 7–10, whereas in the acidic range the activity is about 2.5 times lower than in neutral and slightly alkaline environment (51). This difference in activity of alkaline and acidic proteinases increases with krill that had been intensely feeding prior to harvest (52, 53).

There are at least five autoproteolysis optima, at pH ranges 2.0–3.0, 5.0–6.0, 7.5–8.0, 9.0–10, and 11.0–12.0. In the range 2.0–3.0, the peak of proteolysis is probably caused mainly by the pepstatin-sensitive acid proteinases including possibly pepsin. The peak at pH 5.0–6.0 results from the activity of cathepsins A, L, and B that at pH 7.5–8.0 is caused by the trypsinlike and chymotrypsinlike enzymes, and the enzyme responsible for the maximum at pH 9.0–10.0 is probably an aminopeptidase. In discussing the activity peaks at pH above 9.0, however, the possibility of accumulation of products of hydrolysis of nucleic acids should be taken into consideration (44).

For industrial utilization of Antarctic krill most important is the autoproteolytic activity displayed in the neutral and slightly alkaline pH, which is natural for the crustacean.

The rate of autoproteolysis increases severalfold after disintegration and mixing of krill (54) due to more uniform distribution of the proteases located in the digestive organs hepatopancreas and stomach as well as due to increased synergism (55). Likewise, after thawing of krill, autolysis starts with higher intensity irrespective of the preceding frozen storage temperature (56).

Boiling of krill immediately after harvest seems to be the only way to stop the autolysis completely (4, 57). Thermal inactivation of krill proteases is greater at lower pH values. Inactivation at 70°C and pH 7.0 is almost equal to that at 60°C and pH 4.0 or 50°C and pH 2.8. Thus the autolysis of frozen krill

Table 2 Conditions of Autoproteolysis in Whole Krill

Reference	Optimum pH	pH range of proteolytic activity	Optimum temperature (°C)	Temperature range (°C)	Effective inhibitors
45	5.0–6.0	Wide	a	a	at pH 3–4: pepstatin, DEP at neutral and alkaline pH: SBI, leupeptin, DEP
46	7.0	2.2–7.0	45–50	4–70	—
47	4.2–6.2, 7.7–8.5 and 9.2–9.7	3.5–11.5	45–55	8–60	—
48,49	5.5–6.7 and 8.2–9.0	4.0–10.5	45–55	25–60	—
42	3.0 and 7.0	2.0–10.5	45–55	10–70	at pH 6.2: leupeptin, DEP, EDTA
50	6.2	4.0–8.5	40–45	20–70	EDTA, pCMB
44	5.0–8.0	2.0–10.0	a	a	at pH <5.0: pepstatin, leupeptin, IAA at pH 6–8: STI, DEP
37	6.0–8.0 and 3.0–4.0	2.0–10.5	50 (for neutral pH) 35–40 (for acid pH)	0–70	—

a Investigated only at 20°C.

pCMB, p-chloromercuribenzoate; DEP, diisopropyl fluorophosphate; EDTA, ethylenediaminetetraacetic acid disodium salt; IAA, iodoacetic acid; SBI, soybean trypsin inhibitor.

homogenate may be effectively inhibited by acidification to pH 2.8 and heating 10 min above 55°C or acidifying to pH 2.2 and heating at 45°C (46).

The use of chemical protease inhibitors is impractical because of their antinutritional effects (58, 59; see also Chap. 19). On the other hand, addition of natural sources of protease inhibitors (e.g., soybean powder or egg white) would have to be economically justified (4).

B. Seasonal Variation in Autoproteolytic Activity

Experiments with live krill have shown that the autoproteolytic activity over the year follows a sinusoid curve with a maximum during the summer in the southern hemisphere (January) and a minimum during the winter (July–August) (52). The maximum initial reaction rate is about 10 times that of the minimum (Fig. 5). The highly significant statistical correlation ($\alpha<0.001$) between the degree of filling of the alimentary tract and the proteolytic activity of whole krill (53) points to the importance of feeding intensity for the suitability of the crustacean for processing to food products. The seasonal changes in the autoproteolytic activity of krill also coincide markedly with the annual cycle of Antarctic phytoplankton primary production, which directly affects the growth rate of the crustacean (60, 61). The proteinase activity of whole krill against denatured hemoglobin at pH 8.3 and 4.0 during the spring–summer season increases in time according to the exponential curve in subsequent days of the year (53). This indicates, that physiological processes in krill involve both the trypsinlike enzymes and acid proteases. The proteolytic activity at pH 8.3. is, however, five to seven

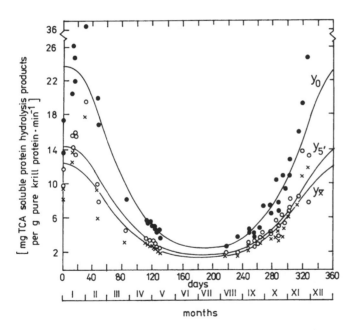

Figure 5 Relationship between the first-order reaction rate of krill autoproteolysis and the consecutive day of the year, as described by the sinusoidal equation

$$y = \exp \{ a + b \sin [\varpi/x - 1/\varphi] \}$$

where: y = reaction rate (mg TCA-soluble protein hydrolysis products per 1 g pure krill protein per min) after 0 time (y_0), 5 min (y_5), and the mean reaction rate ($y_{\bar{x}}$) after 0, 5, 10, 15 and 20 min incubation of the sample at 40±0,2 C; a = harmonic mean of the results over the period of study (365,25 days); b = maximum amplitude; $\varpi = 2 / 365.25$; x = number of the day in the calendar; φ = phase shift. (From Ref. 52.)

times higher in October–November than the activity at pH 4.0. This confirms the opinion that in krill autoproteolysis a major role is played by the enzymes acting at neutral and slightly alkaline pH and that the domination of these enzymes increases with the intensity of feeding of the crustacean. Low activity of the acid proteinases in krill caught in early spring does not support the suggestion of Ikeda and Dixon (62) that during Antarctic winter krill takes energy mainly from autohydrolysis of its own body proteins.

C. Effect of Autoproteolysis on the Solubility of Krill Protein

Using different procedures for extraction of proteins from krill, there is substantial protein hydrolysis (13). Nevertheless, water extracts of krill contain considerable amounts of high-molecular-weight proteins (63). Very careful extraction of proteins from live krill indicates that the ratio of sarcoplasmic to myofibrillar proteins in this crustacean is higher than in fish muscles (16). The salt-soluble protein fraction of krill, containing the myofibrillar proteins, showed no peak of actomyosin in the ultracentrifugal sedimentation patterns (64). This indicates that the molecular properties of the myofibrillar proteins in the extracts from krill are different from those of extracts from other sources.

Partial autoproteolysis has been used for increasing the extraction of protein from Antarctic krill in the process of manufacturing protein precipitate (65, 66). The highest solubility of protein from frozen krill occurs at pH 5.8 (4.5–6.2) at 6–10°C (47, 48) and at 20–30°C from fresh krill (67). The largest yield of thermally precipitated protein was obtained after extraction of the frozen krill at 10°C and the fresh material at 20–30°C (67).

VI. USE OF ENDOGENOUS PROTEASES IN KRILL PROCESSING

Kołakowski et al. (47–49) employed directed, partial autoproteolysis of krill to obtain the highest possible yield and quality of a food-grade product (precipitate) manufactured under industrial conditions.

Fresh krill is mixed gently (68 rpm) with water (3:1 w/v) for 15–90 min at 12–55°C, in order to decrease the viscosity of the mixture to below 0.8 Ns/m². With these conditions, the mixture can be separated in a decanter into the shell-free solution: hydrolysate and the chitin waste. The dominating factor increasing the efficiency of protein extraction is autohydrolysis, which leads under appropriately low temperature to the formation of products that are soluble in water but still susceptible to thermal precipitation (67). The rate of autoproteolysis is highest within a mean integrated temperature interval (T) of about 20–30°C. At T > 34°C and 60 min the reaction rate drops and the yield of the overhydrolyzed product decreases significantly. The second important factor is avoiding shell crushing during mixing, otherwise the needlelike fragments are difficult to separate from the liquid phase (68). A partial digestion of the junction of the shell with the abdomen facilitates the transfer of unhydrolyzed parts of the muscle into the liquid portion during decanting. The highest separation yield can be obtained in mixtures containing more than 625 mg Folin-Ciocalteau-positive compounds per 100 g autoproteolysate (66). Green, heavily feeding krill, containing very active proteolytic enzymes, should be centrifuged prior to processing in order to remove partially the semiliquid fraction of the cephalothorax, containing mainly the hepatopancreas and stomach (69).

Thermal coagulation of the protein proceeds during 3–5 min at 92–93°C, although other procedures of protein separation may be used. The separated protein gel is chilled, strained, homogenized, and frozen in blocks. In industrial conditions, the yield is about 400 kg of precipitate from 1000 kg of fresh krill (68). The protein recovery is about 80% of that present in the whole krill. The precipitate contains 71–75% water, 18–22% protein, 3–7% crude fat, 0.0005–0.04% shell material, and 7–29 mg F^r/kg (66). Owing to its high carotenoid content, about 4.5 mg/100g (mainly astaxanthin), the product has a desirable, pinkish–red color (70). The precipitate produced on board a Polish fishing vessel has received toxicological clearance to be used in the food industry.

Ellingson and Mohr (71) developed a process for large-scale production of free amino acids from krill. The ground raw material is stirred 12–20 h in a reactor at 40–50°C and strained to remove the undissolved matter, mainly the chitinous exoskeleton. The liquid phase is clarified by centrifuging and concentrated to a dry matter content of 30–50% under conditions involving thermal pasteurization of the product. The yield of dry amino acid mixture is 75–90 kg from 1000 kg krill.

Experiments were also carried out to prepare a new type of sauce (72–74) and of seafood extracts (75) from Antarctic krill, using autolysis, fermentation, and commercial proteolytic enzymes.

VII. CATECHOL OXIDASE AND DARKENING OF KRILL

According to many sources (76–78), the enzyme catalyzing the darkening of krill is tyrosinase. However, Nagayama et al. (79) and Ohshima and Nagayama (80) have shown that the bifunctional tyrosinase (EC 1.14.18.1) does not occur in krill, but rather a catechol oxidase (EC 1.10.3.1), similar to that found in microorganisms, plants, and other animals, especially in crustaceans, is responsible for the darkening process (81; see also Chap. 10).

The krill catechol oxidase, a metalloprotein, is a glycoprotein capable of oxidizing only a limited number of o-diphenols, such as catechol and trihydroxyphenols such as pyrogallol. Its optimum pH for catechol oxidation is 6.5 and the isoelectric point of the enzyme is 4.5. Reducing agents, such as potassium sulfite and ascorbic acid, prevent krill discoloration (80). According to Rasulova (78), the enzyme has optimum activity at 6.5–7.5. At pH below 5.5. and above 8.0 it undergoes complete inactivation. At pH 7.0, 10 min heating at 50°C destroys 85% of the enzyme activity, while at 70°C complete inactivation occurs in 4 min (77). In producing precooked frozen krill the crustacean is heated 5 min at 90°C prior to freezing to prevent darkening of the product (3).

Blackening of krill after thawing can be retarded by washing in running water (82), dipping in an acid solution at pH below 4.5 (76, 83), or in a mixture of vitamin E, citric acid, and gelatine hydrolysate (84). Dipping of krill immediately after catch in 0.3% sodium hydrosulfite solution prevents darkening for 5 days at 0°C (85).

VIII. LIPASES/PHOSPHOLIPASES AND LIPOLYSIS

Galas et al. (86) demonstrated the presence of lipase in the alimentary tract of two samples of frozen krill, and Nagayama et al. (79) showed that the lipase extracted from frozen krill did not differ much in the substrate specificity from mammalian lipase. The enzyme hydrolyzed tributyrate at a rate of 182 µmol of free fatty acids $h^{-1}g^{-1}$ at pH 8.6 and 30°C. The thermal stability of the krill lipase is relatively poor. Thermal inactivation of the crude enzyme extract proceeded slowly at 50°C but rapidly at 55°C, while above 60°C the enzyme is completely inactivated.

Indirect evidence for the presence of lipolytic enzymes in Antarctic krill includes the rapid change in the composition of the lipids of frozen stored crustacean. The concentration of free fatty acids in frozen krill is 11–32% of total lipids. Kołakowski (87) found that in krill sampled live, the contents of free fatty acids ranged from 1.6 to 7.4% of the total lipids and was affected by the fishing season. After 6 months at -20°C, the content of free fatty acids increases to 39–53% of total lipids and that of phospholipids decreases from the initial 74–87% to 22–32% and was also affected by the season. The decrease in the content of triacylglycerols was lower than that of phospholipids. This is indirect evidence for higher activity of phospholipases than of other lipolytic enzymes.

Ellingsen and More (17) reported that each lipid fraction, and particularly the triacylglycerols and phospholipids, changed notably during *postmortem* storage of krill at 0°C. The fatty acids 14:0, 16:1, 18:0, 18:1, and 20:1 originated mainly from hydrolysis of triacylglycerols, whereas other fatty acids were liberated from triacylglycerols and/or phospholipids. The more extensive hydrolysis of phospholipids than triacylglycerols in Antarctic krill may be due to a better contact of the enzyme with the more hydrophilic phospholipids. According to Seather et al. (88) in North Atlantic krill (*Meganyctiphanes norvegica*) the activity of lipase was of the same order as that of phospholipase.

In krill of various species, lipolysis proceeds without any distinct, initial lag-phase (17, 88, 89). The rapid accumulation of free fatty acids in frozen krill is not reflected in sensory changes of the frozen product. This might suggest that the free fatty acids interact with proteins in frozen stored krill. The role of lipases and phospholipases in seafood quality is further discussed in Chapters 4 and 5.

Despite the high contents of polyunsaturated fatty acids (90, 91), krill lipids are relatively resistant to autoxidation. One of the proposed explanations for the high oxidation stability is the interaction of phospholipids with hydroperoxides (89, 92).

IX. KRILL CARBOHYDRASES

Antarctic krill feed mainly on phytoplankton (93) and have an enzyme system capable of hydrolyzing different polysaccharides. The *Euphausia superba* carbohydrases prefer to hydrolyze (1-3)-β-D-glycosidic bonds and are not resistant to hydrolysis by the endogenous proteinases after catch. The (1,3)-β-D-glucanase activity in the just-harvested krill digestive tract is 35–40% greater than it is 2–3 h after harvest (7). A very informative study on krill carbohydrases has been published by Turkiewicz (7).

A. Laminarinase

The viscera of krill contains at least three enzymes that hydrolyze laminarin. They are capable of depolymerizing (1,3)-β-D-glucans contained in the diet: an exo-(1,3)-β-D-D-glucanase, an endo-(1,3)-β-D-glucanase, and a β-D-glucosidase (94, 95). More than 70% of the glucanase activity is localized in the hepatopancreas. The glucanases have a pH optimum of 5.4, a temperature optimum of 50°C and are optimally extracted at pH 7.2. A laminarinase isolated and purified by Suzuki et al. (96) had a molecular weight of 65–70 kDa and optimum pH of 4.3–5.0. The enzyme retained 80% of its maximum activity on heating for 10 min at 55°C and pH 5.0. It hydrolyzed laminarin to glucose and laminarioligosaccharides but did not split off laminariobiose. Thiol groups and tryptophan residues were found essential for the activity of the enzyme. The laminarinase isolated and purified by Turkiewicz (7) has a molecular weight of 55 kDa, isoelectric point 5.2, and a saccharide moiety that constitutes about 9% of the enzyme's weight. The optimal conditions for hydrolysis of laminarin in vitro are pH 5.0 and temperature 60–65°C. At 0–4°C the enzyme has 20% of its maximal activity. At pH 5.0, total loss of enzyme activity occurs at 70°C after 2 min, while in the presence of bovine serum albumin, 10 min is required for inactivation. Under optimal conditions the product of 10 min hydrolysis of laminarin is a mixture containing 40% trisaccharides, 18% disaccharides, and 11% glucose.

B. Other Carbohydrases

α-Amylase and endo-(1,4)-β-D-glucanase activity are also present in krill extracts. High activity of β-D-glucosidase, β-D-galactosidase, β-D-fructosidase, amylase, carboxymethyl cellulase, and galactomannoglycanase, as well as lower activity of agarase, mannanase, carrageenanase, and pectinase were found by Chen and Gau (30). Chitinolytic activity in krill extracts was described by Hornung and Ross (97) and Gallas et al. (98). Other saccharides hydrolyzed by extracts from krill include yeast mannan, β-1,4-xylan, pullulan, dextran, β-xylosides, β-glucuronides, and hyaluronic acid. Several polysaccharidases have been purified and characterized in detail (7). Two β-1,4-xylanases isolated by Turkiewicz (7) have pH optimum between 5.8 and 5.9, temperature optimum 35–37°C and completely lose activity after 5 min at 45°C.

The digestive activity of krill against several acidic polysaccharides, including the mucilage polysaccharides of the ice diatom *Staurones amphioxys,* is minimal and is not induced when the polysaccharide is present in the diet of *Euphausia superba.* These results indicate that some, but not all, components of algae in the krill diet can be hydrolyzed and assimilated (95).

The correlation between the feeding intensity and the activity of digestive enzymes in Antarctic krill (7, 99) coincides with the seasonal changes in the contents of saccharides in the crustacean (100).

X. CONCLUSIONS

The Antarctic krill has a very active system of digestive enzymes, capable of hydrolyzing proteins in a concerted action to polypeptides, oligopeptides, and amino acids, as well as a large variety of polysaccharidases. Most of the enzymes are located in the cephalothorax. The activity of these enzymes is positively correlated with the feeding intensity of the crustacean and activities are comparatively high at low temperatures. Some of the digestive enzymes retain high activity after prolonged frozen storage of krill.

The sensory quality of krill after catch is affected mainly by reactions catalyzed by proteases and catechol oxidase. The oxidative stability of krill lipids is higher than might be expected from the high content of polyenoic fatty acids. In order to control the reactions leading to rapid loss in quality, the krill should be frozen or otherwise processed in less than 4 h after catch.

On the other hand, the endogenous proteases have been used for producing food-grade krill protein precipitates in a process based on pH- and temperature-controlled autoproteolysis. The cephalothorax and the liquid digesta separated from krill by centrifugation can be used as a rich source of enzymes for different technical and medical applications. Further biochemical characterization of these enzymes may lead to improved technologies of processing of krill for food purposes and to better utilization of the enzyme preparations.

REFERENCES

1. FAO Yearbook of fishery statistics. Captures 82: 1–678, 1996.
2. ZE Sikorski, P Bykowski, J Knyszewski: The utilization of krill for food. In: P. Linko, ed. Food Process Engineering, vol. 1. London: Applied Science, 1980, pp 845–855.
3. E Budzialski, P Bykowski, D Dutkiewicz. Possibilities of processing and marketing of products made of Antarctic krill. FAO Fish Tech Pap 268, 1985, p 46.
4. T Suzuki. Fish and Krill Protein: Processing Technology. London: Applied Science, 1981, pp 193–251.
5. ZS Karnicki, A Wojtasz-Pajalk, MM Brzeski, PJ Bykowski, eds. Chitin World. Bremerhaven Wirtschaftsverlag, 1994.
6. J Synowiecki, ZE Sikorski, M Naczk. The activity of immobilized enzymes on different krill chitin preparations. Biotechnol Bioeng 23: 2211–2215, 1981.
7. M Turkiewicz. Characterisation of some digestive enzymes from *Euphausia superba* Dana. In: S, Rakusa-Suszczewski SP, Donachie eds. Microbiology of Antarctic Marine Environment and Krill Intestine, its Decomposition and Digestive Enzymes. Warsaw: Polish Academy of Sciences, 1995.
8. N Shibata. Food biochemical study on utilization of muscular proteins in Antarctic krill for human consumption. Bull Tokai Reg Fish Lab 111: 63–141, 1983.
9. N Shibata, K Nakamura. Post-mortem degradation of ATP in Antarctic krill. Bull Jpn Soc Sci Fish 47: 1341–1345, 1981.
10. A Hashimoto, K Arai. Thermo-stability of myofibrillar Ca-ATPase of Antarctic krill. Bull Jpn Soc Sci Fish 45: 1453–1460, 1979.
11. N Shibata, H Ozaki, Y Fujii. Changes in physicochemical and biochemical properties of muscular proteins from Antarcic krill muscle stored at low temperature. Bull Jpn Soc Sci Fish 49: 1721–1729, 1983.
12. N Shibata, K Nakamura. Guanosine triphosphate (GTP) breakdown products in Antarctic krill *Euphausia superba*. Bull Jpn Soc Sci Fish 47 (11): 1527, 1981.
13. N Shibata. Relation between freshness of Antarctic krill and extractability of its muscular protein. Bull Jpn Soc Sci Fish 49: 1089–1096, 1983.
14. E Kołakowski. Changes of non-protein nitrogens fractions in Antarctic krill (*Eu-*

phausia superba Dana) during storage at 3°C and 20°C. Z Lebensm Unters Forsch 183: 421–425, 1986.

15. E Kołakowski. The dynamics of autoproteolysis in fresh (unfrozen) Antarctic krill (in Polish). XXX Sesja Naukowa Komitetu Technologii i Chemii żywności PAN, Kraków, Sept. 14–15, 1999, p 150.

16. E Kołakowski. The characteristics of krill (*Euphausia superba*) as a raw material and the possibilities of its processing into food. Przem Spoż (Poland) 36(3): 88–92, 1982.

17. T Ellingsen, V Mohr. Antarktisk krill biokjemi og teknologi. Rapport II. Lipider. Seksjon for fiskerikjemi. Institutt for teknisk biokjemi. Trontheim: Norges tekniske hØgskole, 1981, pp 1–109.

18. GA Locati, ME Espeche, ER Fraile. Changes in the bacteriological, chemical and organoleptic characteristics of Antarctic krill (*Euphausia superba*) during storage at 0–2°C. Rev Arg Microbiol 12(2): 44–51, 1980.

19. SE Fevolden, G Eidsa. Bacteriological characteristics of Antarctic krill (Crustacea, Euphausiacea). Saria 66(1): 77–82, 1981.

20. NF Haard. Biochemical reactions in fish muscle during frozen storage. In: EG Bligh, ed. *Seafood Science and Technology.* Oxford:Fishing New Books, 1992, pp 170–209.

21. T Suzuki, JJ Hashimoto. Proteins of frozen stored Antarctic *Euphausia superba* muscle. In: Food Availability and Quality through Technology and Science, Fifth Interantional Congress of Food Science & Technology. Abstracts, Kyoto, Japan, September 17–22, 1978, p 69.

22. T Suzuki, K Kanna. Denaturation of muscle proteins in krill and shrimp during frozen storage. Bull Tokai Reg Fish Lab No. 91, October, 1977, pp 67–72.

23. S Konagaya, T Watanabe, K Kanna. Some properties of myofibrillar proteins of Antarctic krill (Euphausia superba) and their freezing denaturation. Refrigeration [Reito] 56(639): 15–19, 1981.

24. K Nishimura, Y Kawamura, Y Matoba, D Yonezawa. Deterioration of Antarctic krill muscle during freeze storage. Agric Biol Chem 47: 2881–2888, 1983b.

25. TT Kozima, DTN Loan. Frozen storage of Antarctic krill. J Tokyo Univ Fish 68(1–2): 68–85, 1982.

26. JJ Matsumoto, SF Noguchi. Cryostabilization of protein in surimi. In: TC Lanier, CM Lee, eds. Surimi Technology. New York: Marcel Dekker, 1992, pp 357–388.

27. K Nishita, Y Takeda and K Arai. Biochemical characteristics of actomyosin from Antarctic krill. Bull Jpn Soc Sci Fish 47: 1237–1244, 1981.

28. T Ooizumi, M Nakamura, A Hashimoto, K Arai. Thermal denaturation of myofibrillar protein of Antarctic krill and protective effect of sugar. Bull Jpn Soc Sci Fish 49: 967–974, 1983.

29. N Seki, H Sakaya, T Onozawa. Studies on proteases from Antarctic krill. Bull Jpn Soc Sci Fish 43: 955–962, 1977.

30. C Chen, S Gau. Polysaccharidase and glycosidase activities of Antarctic krill *Euphausia superba.* J Food Biochem 5: 63–68, 1981.

31. K Nishimura, Y Kawamura, Y Matoba, D Yonezawa. Clasification of proteinases in Antarctic krill. Agric Biol Chem 47: 2577–2583, 1983.

32. K Kimoto, S Kusama, K Murakami. Purification and characterization of serine proteinases from *Euphausia superba*. Agric Biol Chem 47: 529–534, 1983.

33. K Kimoto, T Yokoi, K Murakami. Purification and characterization of chymotrypsin-like proteinase from *Euphausia superba*. Agric Biol Chem 49:1599–1603, 1985.

34. KK Osnes and V Mohr. On the purification and characterization of three anionic, serine-type peptide hydrolases from Antarctic krill, *Euphausia superba*. Comp Biochem Physiol 82B: 607–619, 1985b.

35. KK Osnes, T E Ellingsen and V Mohr. Hydrolysis of proteins by peptide hydrolases of Antarctic krill, *Euphausia superba*. Comp Biochem Physiol 83B: 801–805, 1986.

36. KK Osnes and V Mohr. On the purification and characterization of exopeptidases of Antarctic krill, *Euphausia superba*. Comp Biochem Physiol 83B: 445–458, 1986.

37. KK Osnes and V Mohr. Peptide hydrolases of Antarctic krill, *Euphausia superba*. Comp Biochem Physiol 82B: 599–606, 1985.

38. A Noguchi, M Yanagimoto, S Kimura. Purification and some properties of protease of *Euphausia superba*. Agric Chem Soc Jpn 50:415–421, 1976.

39. K Murakami, K Kimoto. Acid proteases of the Antarctic krill, *Euphausia superba:* purification and some properties. In: Food Availability and Quality Through Technology and Science, Fifth International Congress of Food Science & Technology. Abstracts, Kyoto, Japan, September 17–22, 1978, p 234.

40. K Kimoto, VV Thanh, K Murakami. Acid proteinase from Antarctic krill, *Euphausia superba:* partial purification and some properties. J Food Sci 46:1881–1884, 1981.

41. K Kimoto, A Fukamizu, K Murakami. Partial purification and characterization of proteinases from abdomen part muscle of Antarctic krill. Bull Jpn Soc Sci Fish 52: 745–749, 1986.

42. S Konagaya. Protease activity and autolysis of Antarctic krill. Bull Jpn Soc Sci Fish 46: 175–183, 1980.

43. W Goettlich-Riemann, JO Young, AL Tappel. Cathepsins D, A and B and the effect of pH in the pathway of protein hydrolysis. Biochim Biophys Acta 243: 137–146, 1971.

44. Y Kawamura, K Nishimura, T Matoba, D Yonezawa. Effects of protease inhibitors on the autolysis and protease activities of Antarctic krill. Agric Biol Chem 48: 923–930, 1984.

45. E Doi, Y Kawamura, S Igarashi, D Yonezawa. Autolysis of Antarctic krill *Euphausia superba*). In: Food Availability and Quality through Technology and Science, Fifth International Congress of Food Science & Technology. Abstracts, Kyoto, Japan, September 17–22, 1978, p 68.

46. Kubota M., K Sakai. Autolysis of Antarctic krill protein and its inactivity by combined effects of temperature and pH. Trans Tokyo Univ Fish No. 2, March, 53–63, 1978.

47. E Kołakowski, L Gajowiecki, T Chodorowska, Z Szybowicz. The establishment of optimum conditions of protein extraction by the cold autoproteolysis method in or-

der to obtain chitin-free protein concentrates from krill. Przem Spoż 43(7): 263–265, 1980.

48. E Kołakowski, L Gajowiecki, Z Szybowicz, T Chodorska. Application of partial autoproteolysis to extraction of protein from Antarctic krill (*Euphausia superba*). Part 1. Effect of pH on protein extraction intensity. Nahrung 24, 499–506, 1980.

49. E Kołakowski, L Gajowiecki, Z Szybowicz, T Chodorska. Application of partial autoproteolysis to extraction of protein from Antarctic krill (*Euphausia superba*). Part 1. Influence of temperature on protein extraction intensity. Nahrung 24: 507–512, 1980.

50. M Fik. Partial purification and some properties of protease from Antarctic krill. Z Lebensm Unters Forsch 179, 296–300, 1984.

51. LG Mitskevich, VV Mosolov. The study of proteolytic enzymes of krill. In: VP Bykov, LR Kopylenko, FM Rzhavskaya, eds. Krill Processing Technology, Collected Papers (in Russian with English summary). Moscow: All-Union Research Institute of Marine Fisheries and Oceanography (VNIRO), 1981, pp 21–24.

52. E Kołakowski. Seasonal variations of autoproteolytic activity in the Antarctic krill (*Euphausia superba* Dana). Pol Polar Res 7(3): 275–282, 1986.

53. E Kołakowski. Proteolytic activity of Anatrctic krill in relation to its feeding intensity in spring and summer. Pol Polar Res 10(2): 141–150, 1980.

54. E Kołakowski. The dynamic of autoproteolysis in whole and minced krill (in Polish with English summary). In: XV Sympozjum Polarne. Stan obecny i wybrane problemy polskich badańl polarnych. Wrocław: Uniwersytet Wrocławski, 19–21. V. 1988, 345–350.

55. N Shibata, H Ozaki. Effect of contamination of various tissues on the deterioration of fresh muscle from Antarctic krill. Bull Jpn Soc Sci Fish 49:1097–1101, 1983.

56. O Christians, M Leinemann. Die Eiweissautolyse im Muskelfleisch von gefriergelagertem antarktischen Krill (*Euphausia superba* Dana) in Abhanggigkeit von der Lagertemperatur nach 12-monatiger Lagerung. Arch FischWiss 34(1): 97–102, 1983.

57. K Kuwano, Y Osawa, N Sekiyama, A Tukui, T Mitamura. Boiling process for the prevention of the loss and autolysis of protein from Antarctic krill (*Euphausia superba*). J Jpn Soc Food Nutr. 28: 191–194, 1975.

58. J Yamamoto, Y Kaji, K Ishida, E Naune (Suisan Kaisha Ltd., Kyowa Hakko Kogyo Co., Ltd). Krill preservation. Jpn Tokkyo Koho 79 00, 986 (Cl. A23L1/325), 18 Jan 1979, Appl. 75/14,888, 06Feb 1975; 3pp.

59. ST Jiang. Effects of pretreatments on improving the quality of frozen Antarctic krill *Euphausia superba*. J Chinese Agric Chem Soc 18(1–2): 69–76, 1980.

60. R Ligowski. Net phytoplankton of the Admiralty Bay (King George Island, South Shetland Islands) in 1983. Pol Polar Res 7: 127–154, 1986.

61. R Tokarczyk. Annual cycle of chlorophyll A in Admiralty Bay 1981–1982 (King George, South Shetlands). Pol Arch Hydrobiol 33(2): 177–188, 1986.

62. T Ikeda, P Dixon. Body shrinkage as a possible overwintering mechanism of the Antarctic krill, *Euphausia superba* Dana. J Exp Mar Biol Ecol 62, 143–151, 1982.

63. E Kołakowski. Comparison of krill and Antarctic fish with regard to protein solubility. Z Lebensm Unters Forsch 188: 419–425, 1989.

64. S Konagaya, T Watanabe, K Kanna. Some properties of myofibrillar proteins of Antarctic krill (*Euphausia superba*) and their freezing denaturation. Refrigeration [Reito] 56(639): 15–19, 1981.

65. E Kołakowski, L Gajowiecki. Sposób wytwarzania precypitatu białkowego z kryla. Patent PL Nr 135 379, Int. Cl A23L 1/33, Zgloszono 83 08 01 (P. 243253), Akademia Rolnicza w Szczecinie, 1987.

66. E Kołakowski, L Gajowiecki. Optimization of autoproteolysis to obtain an edible product "precipitate" from Antarctic krill (*Euphausia superba* Dana). In: Seafood Science and Technology. Proceeding of the International Conference Seafood 2000. Canadian Institute of Fisheries Technology of the Technical University of Nova Scotia, 13–16 May. Halifax, Canada: Fishing News Books, 1992, pp 331–336.

67. E Kołakowski, K Lachowicz. Application of partial autoproteolysis to extraction of protein from Antarctic krill (*Euphausia superba*). Part 3. Changes in and yield of nitrogen substances during autoproteolysis of fresh and frozen krill. Nahrung 26: 933–939, 1982.

68. E Kołakowski, L Gajowiecki, B Nowak, M Wianecki, B Czerniejewska-Surma. Technology of obtaining a protein precipitate from krill *Euphausia superba* in the light of industrial investigation on a B-414 vessel. Bull Sea Fish Inst Gdynia 1–2: 47–58, 1984.

69. L Gajowiecki. Effects of technological processes on volatile componds and sensory properties of proten concentrate obtained from Antarctic krill (*Euphausia superba* Dana). Wydawnictwo Akademii Rolniczej w Szczecinie, Rozprawy 179: 1–79, 1997 (in Polish with English summary).

70. A Protasowicka. Carotenoids in frozen Antarctic krill and in some of its semifinished products. Acta Alim Pol 9(2): 235–245, 1985.

71. T Ellingsen and V Mohr. A new process for the utilization of Antarctic krill. Proc Biochem 14(10), 14–19, 1979.

72. K Abe, K Suzuki, K Hashimoto. Utilization of krill as a fish sauce material. Bull Jpn Soc Sci Fish 45: 1013–1017, 1979.

73. H Nakamura, Y Mohri, I Muraoka, K Ito. Studies on brewing food containing Antarctic krill. I. Production of soysauce with cryo-ground Antarctic krill, *Euphausia superba*. Bull Jpn Soc Sci Fish 45: 1389–1393, 1979.

74. EH Lee, SY Cho, HS Park, CS Kwon. Studies on processing krill sauce. J Korean Soc Food Nutr 13(1): 97–106, 1984.

75. H Ochi. Production and applications of natural seafood extracts. Food Technol 34(11): 51–53, 68, 1980.

76. T Onishi. Prevention of blackening of Antarctic krill during thawing. Bull Tokai Reg Fish Res Lab 100: 13–16, 1979.

77. A Noguchi, K Sugawara, K Umeda, S Kimura. Some properties of partially purified tyrosinase of *Euphausia superba*. National Food Res Institute Tokyo Report 35:231–236, 1979.

78. TA Rasulova. Activity of poliphenoloxsidase-tyrosinase in fresh and frozen krill

(in Russian). Kaliningrad: Spravocznik Naucznych Trudov AtlantNIRO, 1986, pp 17–24.

79. F Nagayama, T Yasuike, K Ikeru, Ch Kawamura. Lipase, carboxylesterase and catechol oxidase of the Antarctic krill. Tokyo Univ Fish Transact 3: 153–159, 1979.

80. T Ohshima, F Nagayama. Purification and properties of catechol oxidase from the Antarctic krill. Jpn Soc Sci Fish Bull 46(8): 1035–1042, 1980.

81. T Nakagawa and F Nagayama. Distribution of catechol oxidase in crustaceans. Bull Jpn Soc Sci Fish 47: 1645, 1981.

82. M Tanaka, A Maamoen, T Taguchi, K Suzuki. Prevention of browing in canned Antarctic krill. J Tokyo Univ Fish 66(1): 1–7, 1979.

83. T Onishi, T Watanabe, M Suzuki. Tyrosinase of Antarctic krill and prevention of its blackening by dipping in acidic solutions. Bull Tokai Reg Fish Res Lab 96: 1–9, 1978.

84. M Ono, K Nozaki, S Fukumoto. Prevention of discoloration of krill. Jpn Kokai Tokkyo Koho 79 59,362 (Cl. A23L1/31), 12 May 1979, Appl 77/126, 535, 20 Oct 1977 (Tanabe Seiyaku Co., Ltd).

85. T Kinumaki. Reito 53(611): 821–835, 1978 (in Japanese).

86. E Galas, Z Libudzisz, M Turkiewicz. General characteristics of oxidases from krill (in Polish with English summary). In: Kryl antarktyczny przetwórstwo i wykorzystanie. Gdynia: Studia i Materialy MIR, Seria S, Nr 1, 1979, pp 359–372.

87. E Kołakowski. Lipid composition of fresh and frozen-stored krill. Z Lebensm Unters Forsch 182: 475–478, 1986.

88. O Saether, TE Ellingsen, V Mohr. Lipolysis post mortem in North Atlantic krill. Comp Biochem Physiol 83B: 51–55, 1986.

89. A Kołakowski. Changes in lipids during the storage of krill (Euphausia superba Dana) at 3°C. Z Lebensm Unters Forsch 186: 519–523, 1988.

90. NR Bottino. Lipid composition of two species of Antarctic krill: Euphausia superba and E. crystallorophias. Comp Biochem 50B: 479–484, 1975.

91. N Shibata. Effect of fishing season on lipid content and composition of Antarctic krill. Bull Jpn Soc Sci Fish 49: 259–264, 1983.

92. J-H Lee, K Fujimoto, T Kaneda. Antioxygenic and peroxide decomposition properties of Anatractic krill lipids. Bull Jpn Soc Sci Fish 47: 881–888, 1981.

93. S Rakusa-Suszczewski eds. The Maritime Antarctic Coastal Ecosystem of Admiralty Bay. Warszaw: Department of Antarctic Biology, Polish Academy of Sciences, 1993, pp 1–215.

94. T E Nelson, J V Scarletti, F Smith, S Kirkwood. The use of enzymes in structural studies of polysaccharides. I. The mode of attack of an exo-β-$(1 \rightarrow 3)$-glucanase on laminarin. Can J Chem 41: 1671–1678, 1963.

95. MJ McConville, T Ikeda, A Bacic, AE Clarke. Digestive carbohydrases from the hepatopancreas of two Antarctic euphausiid species (Euphausia superba and E. crystallorophias). Marine Biol 90: 371–378, 1986.

96. M Suzuki, T Horii, R Kikuchi, T Ohnishi. Purification of laminarinase from Antarctic krill Euphausia superba. Bull Jpn Soc Sci Fish 53: 311–317, 1987.

97. HE Hornung, SJ Ross. Changes in the rate of chitin synthesis during the crayfish molting cycle. Comp Biochem Physiol 40(2):, 1971

98. E Galas, M Turkiewicz, M Kalulewska, T Antczak, H Bratkowska, B Kldziora. Initial studies on krill enzymatic system (in Polish with English summary). In: Kryl antarktyczny przetwórstwo i wykorzystanie. Gdynia: Studia i Materialy MIR, Seria S, Nr 1: 336–338, 1979.

99. T Ikeda, B Bruce. Metabolic activity and elemental composition of krill and other zooplankton from Prydz Bay, Antarctica, during early summer (November–December). Marine Biol 92: 545–555, 1986.

100. E Kołakowski, L Szyper-Machowska. Seasonal changes of carbohydrate content in Antarctic krill (*Euphausia superba* Dana). Pol Polar Res 10(2): 133 139, 1989.

19

Use of Protease Inhibitors in Seafood Products

Fernando Luis García-Carreño and Patricia Hernández-Cortés
CIBNOR, La Paz, BCS, Mexico

I. INTRODUCTION

Protease is the generic name given to those enzymes hydrolyzing the peptide bond in proteins and some synthetic substrates and coded as the EC 3.4.11-99.X. subgroup in the enzyme classification. Proteases, including peptidases and proteinases, are polyfunctional enzymes catalyzing the hydrolytic degradation of proteins. They are involved in several physiological functions, both extracellular and intracellular, such as proenzyme activation in coagulation, complement activation, food digestion, endocrine and neural communication, and developmental and differentiating processes. Intracellular protein degradation of functional proteins and turnover of amino acids is achieved by cathepsins and proteasomes. Proteases are involved in some of these functions by exerting control over developmental and cellular maintenance processes. Proteases are also involved in the response of organisms to infections and pathologies. In turn, protease activity is controlled by several mechanisms, including enzyme inhibition, by naturally occurring protease inhibitors. Most protease inhibitors controlling physiological processes affect the enzyme activity reversibly. In food technology, proteolysis is used to modify the functional and nutritional properties of food proteins. However, in some examples in food processing, proteolysis by endogenous proteases causes deterioration of functional proteins (autoproteolysis). For more detail on the latter the reader is referred to Chapter 15.

    This chapter focuses on the control of unwanted proteolysis in seafood processing by inhibition of the responsible enzymes. A functional definition of protease inhibitors will be provided to differentiate them from other agents that otherwise serve to decrease the rate of enzyme activity. Sources of protease inhibitors from natural sources will be emphasized. Preparation of inhibitors, properties and modes of action of protease inhibitors, some examples of modes of action, molecular aspects of inhibitors, inhibition as a problem, antinutritional factors, fish softening protection, and potential applications of protease inhibitors in seafood processing and products are discussed.

II. ENDOGENOUS INHIBITORS

A. Protease Inhibitors: A Functional Definition

An enzyme inhibitor is any substance that reduces the measured rate of an enzyme-catalyzed reaction (1). When competing substrates are incubated in an enzymatic reaction, one of them can decrease the apparent rate of hydrolysis of the other and therefore appear to inhibit the enzyme (2). This may mislead one into classifying substrates as inhibitors. Inhibitors are also set apart from other substances affecting the rate of an enzyme reaction by less specific mechanisms (e.g., by protein denaturation). These latter substances are called inactivators. Chelators that remove cations from metal-dependent enzymes and denaturing agents that alter the catalytic sites are some examples of enzyme inactivators (3).

Enzyme inhibition can be either reversible or irreversible. In irreversible inhibition, the enzyme activity cannot be regained by physical means, whereas in reversible inhibition the enzyme activity is regenerated by displacement of the inhibitory molecule. Irreversible inhibition is usually quantified in terms of the rate of inhibition whereas reversible inhibition follows kinetic mechanisms and is expressed in terms of an equilibrium constant, K_I, for the enzyme and the inhibitor.

Protease inhibitors may be either specific or nonspecific for some enzyme or class of enzymes. Specific inhibitors are active-site-directed substances and combine with the catalytic or substrate-binding sites of the enzyme to form a stable complex. For kinetic details see the comprehensive chapter by Salvesen and Nagase (2). Specific protease inhibitors of natural origin are either proteins or peptides with all the typical characteristics of the substrate, but lacking the scissile peptide bond.

Nonspecific inhibitors are rare in nature, and the only one known is the plasma protein(s) α_2-macroglobulin. This is a family of high-molecular-weight proteins that bind and inhibit most proteases. Synthetic nonspecific inhibitors belong to peptide aldehydes and peptide chloromethyl ketone derivatives. The former are reversible protease inhibitors and the latter are irreversible protease inhibitors.

An operational classification of inhibitors is into natural and synthetic types. Synthetic inhibitors are tailor-made to mimic natural ones, which is useful in classifying proteases. They are used in studies of mechanisms of catalysis. Recently, a number of inhibitors have reported inhibiting the human immunovirus (HIV) protease and are being designed to restrict the assembly of the virus.

B. Sources of Protease Inhibitors: A Natural Approach

Protease inhibitors function in one of the most important biological phenomena: the control of protease activities. Protease functions are biologically important because they control protein and polypeptide functions. In turn, protease activities are controlled by several means, including protease inhibitors. Mammal plasma comprises hundreds of proteins and a large number of zymogens. Homeostasis depends on a perfect control of activities, and protease inhibitors are involved. The importance of these compounds is demonstrated by the observation that protease inhibitors represent more than 10% of the total protein in human plasma (3). Besides their function in normal cellular processes, enzyme inhibitors function by affecting protease from competing and predatory organisms. Plant storage tissues usually possess so-called antinutritional substances that affect the digestive enzymes of predators. Legume seeds are known to synthesize several inhibiors, including lectins, phenols, and several protease and other enzyme inhibitors. Plants protect themselves against predation by several mechanisms of protection including predator digestive protease inhibitors to impair predator digestive physiology. Insects belonging to the orders Lepidoptera and Coleoptera have digestive serine and cysteine proteases. Plants have evolved to produce specific protease inhibitors to specific predator digestive systems. Biotechnology has determined the agronomic characteristics of crops by generating transgenic plants containing genes encoding protease inhibitors (4).

Some plants possess wound-induced small multigene families that encode for protease inhibitors (5). Another protein storage material protected by protease inhibitors is egg white. Table 1 shows a list of inhibitors in egg white.

Where, then, should the natural product chemist look for protease inhibitors in the vast immensity and biological diversity of the ocean? A way to start is to look for tissues containing significant amounts of protease, and in which protease activities are under physiological control. Nature seems to accumulate inhibitors in tissues containing storage proteins.

Not long ago, sources of natural products with desired activities were reduced to those tissues containing an abundant quantity. However, with the advent of genetic engineering techniques, a suitable protein gene can now be

Table 1   Protease Inhibitors in Egg White

Inhibitor	Specificity
Cystatin	Cysteine proteases: several cathepsins, ficin
Ovoinhibitor	Serine proteases: elastase and chymotrypsin
Ovomacroglobulin	Aspartic proteases: pepsin and renin
Ovomucoid	Trypsin

Table 2   Selected Sources of Protease Inhibitors

Organism	Inhibitor	Characteristic	Affected enzyme
Microorganisms			
Actinomycetes	Leupeptine	Tripeptide	Calpain, cathepsins B, H, and L, and chymotrypsin
Streptomyces testaceous	Pepstatin	Pentapeptide	Aspartic proteases: pepsin, cathepsins and HIV-1
Invertebrates: salivary gland of leech			
Hirudo medicinalis	hirudin	65 aa peptide	Thrombin
Anemonia sulcata	NA	Protein	Elastase
Plants			
Curcubita maxima	NA	Protein	trypsin
Soybean	NA	Protein	trypsin
Lima bean	NA	Protein	trypsin
Vertebrate organs and tissues			
plasma	α_2-macroglobulin	Protein	All classes
Pancreas	Aprotinin	Protein	Trypsin
Egg white	Ovomucoid	Protein	Serine proteases

expressed in another expression system, such as an microorganism, to overproduce the protein.

C.   Preparation of Inhibitors

Inhibitors are obtained from a variety of organisms, including microorganisms, plants, invertebrate, and vertebrate organs. Table 2 gives a list of organisms producing protease inhibitors.

The first step in preparing protease inhibitors is to develop a technique of quantification of the amount of the inhibitor. This is usually done by inhibition of the activity of a suitable test enzyme. Whatever the method of protease inhibition chosen, it is important to be sure that the inhibition test is done under controlled experimental conditions. The inhibitory activity may be quantitated as residual activity of the enzyme or percentage of inhibition, and inhibitor units may be defined. A methodology to quantitate protease inhibitors is reported by García-Carreño (6) and a methodology to determine the number and the molecular weight of inhibitors in a complex mixture by substrate sodium dodecyl sulfate–polyacrylamide gel electrophoresis (SDS-PAGE) is reported by Garía-Carreño et al. (7).

The second step is to make an inhibitor preparation and to try a purification procedure that will depend on the molecular and catalytic properties of the inhibitor. How far to go with the purification will depend on the intended use of the inhibitor.

D. Some Modes of Action of Protease Inhibitors

Protease inhibitors are ubiquitous regulatory molecules in all tissue cells thus far studied. The mechanisms of action are diverse. Table 3 shows a list of selected inhibitors, some properties, and mode of action.

Table 3 Modes of Action of Selected Protease Inhibitors[a]

Inhibitors	Source	Characteristics
Inhibitor of factor Xa	Human	Specific, competitive, tight-binding; two regions (residues 1–10 and 41–54) are responsible for the union to the factor Xa
Serine protease phosphonyl inhibitor[b]	Synthetic	Irreversible, forms a stable enzyme derivative with the active site Ser-195
Serpins[c]	Ubiquitous	44–100 kDa, undergoes conformational changes upon complex formation with enzyme that involves partial insertion of the reactive center loop into the -sheet of the inhibitor
Ovalbumin	Egg white	Although a serpin in nature, it is not a serine protease inhibitor because a charged residue, Arg, prevents the loop insertion into the -sheet of enzyme
Aspartyl protease inhibitors	Synthetic	Transition state analogue -hydroxylphosphonate group in a P3-P1 framework increased the inhibitory activity ($I_{50} = 10$ nM)

[a] Data taken from oral lectures given at the Keystone Symposia on Structural and Molecular Biology of protease function and inhibition. Santa Fe, New Mexico, USA. 1994.
[b] Alpha-aminoalkylphosphonate diphenyl esters.
[c] Serine proteinase inhibitors

III.   PROTEASES AND PROTEASE INHIBITORS IN FOOD TECHNOLOGY

Because proteins have both nutritive and functional roles, they are important ingredients in food products. Functional properties of proteins can sometimes be improved by limited proteolysis. Common functional properties of food proteins include solubility, gelation, emulsification, and foaming. However, these properties are impaired near their pI. Enzyme modification of food proteins can enhance their functional properties over a wide pH range (8). An example of nutritional enhancement of food protein is achieved by the plastein reaction that improves the amino acid composition by protein resynthesis reactions (9).

A highly developed strategy in protein modification by proteases is the production of protein hydrolysates with particular interesting functional and organoleptic properties. More information on protein hydrolysates in seafood products is provided in Chapter 21. The concept of degree of hydrolysis (DH; the percentage of peptide bonds cleaved) can be used to quantify the hydrolysis of a food protein (10) The DH concept allows the prediction of four out of five variables in a given enzyme–substrate system. DH is used to predict some functional properties of a protein hydrolysate.

The processing and modification of raw marine materials by enzyme technology are achieved both by using endogenous enzymes in autolytic processes and by added enzymes. At present, the production of fish protein hydrolysates is one of the most important food processes using exogenous enzymes. The process is mostly carried out with microbial enzymes. However, enzymatic methods for accelerated fish ripening and selective tissue degradation are currently in development (11).

Proteases are used in a number of food processes. Sometimes proteases are pervasive, perplexing, persistant, and pernicious for proteins, but with proper precaution, preventable (the 9Ps rule). In these cases, a reduction of protease activities is desirable (3), and protease inhibitors are deliberately used as aids in food processing.

A.   Potential Applications of Protease Inhibitors in Seafood Processing and Products

Surimi production is the best known example of food processing with the aid of protease inhibitors. Inhibitors are normally added to surimi before freezing the product. Potato, bovine plasma, and egg white are used because they contain several inhibitors for fish cathepsins (3). Some limitations of these inhibitor products are the off flavor and off color imparted to the final product. These are the reasons why alternative inhibitor sources are currently studied. Inhibitors may be used as processing aids in all those food operation units affected by un-

Table 4 Potential Uses of Inhibitors in Food Industry

Seafood	Target Enzyme or Process
Underused marine resources	
Squid (*Illex illecebrosus* and *Loligo bleekeri*)	Serine protease and cathepsin C reducing the gel-forming ability of muscle
Caught or aquafarmed crustaceans	
Krill (*Euphasia superva*)	Digestive enzymes-autolysis of tail
Crayfish (*Procambarus clarkii*)	muscle
Langostilla (*Pleuroncodes planipes*)	
Shrimp (*Penaeus sp.*)	
Fish processing	
Sardine (*Episthonema sp.*)	Cathepsins; reducing the gel-forming
Haddock (*Melanogrammus aeglefinus*)	ability of muscle in surimi production
several species	Limited collagen hydrolysis in descaling
Egg cells in roe and caviar production	Limited adhering connective tissue hydrolysis
Several inhibitor sources	Production of specialty marine enzymes; purification by affinity chromatography

wanted proteolysis, provided the inhibitor can be inactivated or be inactive against the digestive enzymes of the consumer of the processed food. Table 4 gives a list of potential uses of protease inhibitors in food processing.

B. Fish Softening Protection

Functional properties of food ingredients are their individual ability to contribute to the taste, texture, and appearance of a formulated food (12). Fish meat treated with salt and heat turns into a viscoelastic gel, whose rheological characteristic is critical to the use of surimi (13). Surimi is a Japanese term for mechanically deboned fish flesh that has been washed with water and mixed with cryoprotectants (14). The gel of surimi has firm and cohesive properties because of the quantity and quality of myosin from fish meat. Fish muscle softening by protease enzyme is one problem of the surimi industry. Myofibrillar protein degradation by proteases reduces the gel strength and elasticity of surimi. The protein degradation is caused by endogenous proteases of fish muscle. Storage conditions may reduce enzyme activity, but chilling and freezing only retard the proteolytic activity. Once the product has been thawed and the material returned to more favorable temperature, the proteolytic activity can act on the fish muscle. The enzyme-catalyzed damage of muscle proteins is irreversible and the use of inhibitors can prevent this degradation.

Several proteinase inhibitor sources have been used to avoid muscle softening: egg white, mammalian blood plasma, and potato inhibitor. Egg white and potato contain competitive inhibitors that act on the active site on the enzyme molecule. Mammalian blood has a macrogobulin protein that inactivates the enzymes by a trapping mechanism. The gel properties of surimi—strength, torsion, and punch—are improved when substances containing protein proteinase inhibitors are added to the product. Only 2% of these inhibitors can give better physical properties to Alaska pollock (*Theragra chalcogramma*) and arrowtooth flounder (*Atheresthes stomias*) surimi. Whey protein concentrate was also used as an inhibitor of autolysis of Pacific whiting surimi during processing (15). The residual proteinase activity in surimi containing this concentrate was 14.2%.

Proteases that contribute to postmortem softening of fish muscle have been identified. Cathepsin B is the most active protease in fish fillet, and cathepsin L in surimi (16). Cathepsin is a term referring to proteinases that are stored in the lysosome of the cell. Cathepsin L and B are cysteine proteinases. The functional group responsible for the catalytic activity of this proteinase family is the triad Cys-25, His-159, and Asp-158. The name of this class of proteinase is based on the nature of the most prominent amino acid residue at the active site; the number is the position in the amino acid sequence. These enzymes are a major causal factor in the degradation of surimi from some species of fish. Cathepsin L is highly active at 55°C, the temperature at which surimi is conventionally held in the suwari process (see Chap. 6). The inhibitors used to control fish muscle autolysis must be generally regarded as safe. However, the cost of the inhibitor may also be a limiting factor. Cysteine inhibitors (cystatins) from plants such as rice (oryzacystatin), legume seed meals, and tomato leaf extracts are under study (17).

Even though cystatins are useful to protect the gel strength of fish muscle during cooking, cystatin C can be inactivated by hydrolysis of hydrophobic sites with cathepsin D, an aspartyl proteinase (18). A recombinant, glycosylated cystatin with a polymannosyl chain was shown to protect surimi from gel weakening efficiently (19). The recombinant cystatin is more thermostable and the gel strength of cooked herring surimi was greater than surimi gels prepared with the unglycosylated control. The susceptibility of the glycosylated cystatin to hydrolysis by cathepsin D was lower than the wild type. Thus, the glycosylation of cystatin appears to help in the conformational stability of the inhibitor against heating and proteolysis.

Calpains are cysteine proteinases that are calcium-dependent (EC 3.4.22.17). A detailed discussion of role of calpains in impacting fish texture is provided in Chapter 15. There are two form of calpains in carp muscle. Calpain I is heat stable at 50–60°C and could contribute to fish softening (20). However, there are also calpastatins (proteinase inhibitors of calpains) found both in mammals and fish (21) that could help to reduce gel weakness.

IV. MOLECULAR ASPECTS OF INHIBITORS

With the exception of α-macrogobulin, which inhibits proteinases of all classes by trapping them, individual protein inhibitors influence only proteinases of the same class. The classification of proteinaceous inhibitors is based on family relationships because some inhibitors can have a broad specificity. In other cases, a single amino acid substitution can alter specificity. A general characteristic of most protein inhibitors is that they are cysteine-rich proteins. The disulfide bridges are important in the stability and active conformation structure of many of these proteins. Many inhibitors are multiheaded and some of them bifunctional, having two heterologous domains that inhibit two different enzymes. Other inhibitors form multimers of homologous domains either posttranslationally or by tandemly repeated gene structure.

A. Serine Proteinases

In any inhibitor there is at least a reactive site (P1) that combines with the enzyme in a substrate-like manner. The reactive site residue generally corresponds to the target enzyme. An inhibitor with a P1 Lys or Arg tends to inhibit trypsins, and those with Ala or Ser inhibit elastase. Most inhibitors have several homologous reactive sites on the same polypeptide chain that arise from gene elongation by repeated duplication. This characteristic explains why the same inhibitor can inhibit two or more enzymes. If the reactive site is replaced, which normally can happen in the course of evolution, the inhibitor retains its inhibitory activity. The change of the P1 residue(s) leads to a change of specificity. This variability of the residue contrasts with the evolution of other proteins, where active sites are strongly conserved and substitution at active sites leads to the loss of activity.

Proteinaceous inhibitors of proteinases are usually named according to the biological source and the inhibited enzyme. Both criteria are confusing. A single source could be present in more than one inhibitor family and the same inhibitor could act on several enzymes. The use of the last name of the discoverer of the inhibitor has been used as a distinguishing characteristic. However, there could be more than one inhibitor in the same family and the designation then includes letters and numbers. Laskowski and Kato (22) suggested a system for inhibitor nomenclature involving the inhibitor family, the number of domains (if more than one), and the species from which it is isolated. The continuing sequencing of inhibitors is helping the assignment to families and therefore a systematic nomenclature.

Serine proteinase inhibitors are grouped into homologous families in which similarities arose by convergent evolution. This classification is based on sequencing and x-ray crystallographic studies of the inhibitors (23). There

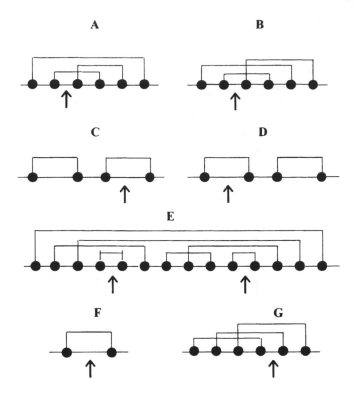

Figure 1 Common structures of serine proteinase inhibitors. A solid circle indicates a half cystein residue and an arrow indicates a reactive site. (A) family I (Kuniz); (B) family II (Kazal); (C) family III; (D) family IV (Kuniz); (E) family V (Bowman-Birk); (F) family VI; (G) family VII. Some members of family I (Kuniz) and family II (Kazal) have more than one reactive site in repeated homologous structures but the positions of all intrachain disulfide bridges and the reactive site are completely conserved (Modified from Ref. 22.)

are 10 ten families from which at least one of the member of the family has been sequenced, the reactive site has been assigned, and the standard mechanism of inhibition is obeyed. The first characteristic used to identify inhibitor families is the relation between the disulfide bridges and the location of the reactive site (Fig. 1). With the exception of potato II and *Ascaris* trypsin inhibitor, the positions of all intrachain disulfide bridges are completely conserved in each family.

The main families of protein inhibitors that inhibit serine proteinases (most of them named after the last name of the discoverer) are as follows:

```
            1        2        3
Rat I       CPKQIMGCPRIYDPVCG
Human       CYNELNGCTKIYDPVCG
Porcine I   CTSEVSGCOKIYNPVCG
Bovine      CTNEVNGCPRIYNPVCG
Ovine       CTNEVNGCPRIYNPVCG
Canine      CNLKVNGCNKIYNPICG
```

Figure 2   Amino acid sequence of the reactive site region (residues 9–25) of pancreatic secretory trypsin inhibitor (Kazal family). The P1 residue is in bold and corresponds to arginine or lysine, from which specificity to trypsin arose. The numbers above the sequence show three of the six cysteine residues present in these inhibitors.

I. Bovine pancreatic trypsin inhibitor (Kunitz) family. Present in bovine organs, snake venoms, and red sea turtle egg white (chelonianin).

II. Pancreatic secretory trypsin inhibitor (Kazal) family. This inhibitor has been found in several vertebrates and is stored in zymogen granules and secreted with the zymogens in the pancreatic juice (Fig. 2). Ovomucoids have a Kazal-type domain in which the P1 residue of the reactive site varies widely.

III. *Streptomyces* subtilisin inhibitor family.

IV. Soybean trypsin inhibitor (Kuniz) family. This inhibitor can also inhibit trypsin and chymotrypsin enzymes and is confused with Bowman-Birk inhibitor though they differ in their structure. The polypeptide chain in single-headed Kunitz soybean inhibitor has 181 residues and only two disulfide bridges while Bowman-Birk is double headed, has seven disulfides, and only 70 residues.

V. Soybean proteinase inhibitor (Bowman-Birk) family. These inhibitors are found in the seeds of leguminous plants. The sequence consists of two homologous regions linked by interhomologous region disulfide bonds. In most Bowman-Birk inhibitors the P1 residue in the first homologous region is Lys and in the second region is a Leu. This inhibitor acts on trypsin and chymotrypsin enzymes.

VI. Potato I inhibitor family. The inhibitor is a noncovalent tetramer of a single-chain inhibitor with only a single intrachain disulfide bridge.

VII. Potato II inhibitor family.

VIII. *Ascaris* trypsin inhibitor family.

IX. Other families

There are several kinds of protein inhibitors in blood plasma. Serpins refer to a superfamily of serine proteinase inhibitors. Their structure is a single-chain protein with a conserved domain of 370–390 residues. Serpins are involved in the regulation of blood coagulation, fibrinolysis, complement activation, and inflammatory responses in mammals (24).

B. Cysteine Proteinases

In contrast to serine–proteinase inhibitors, inhibitors for cysteine proteinases appear to have evolved from a common ancestor, based on their molecular structure. These inhibitors act on cathepsins, ficin, papain, and calpain enzymes (EC 3.4.22.X). Three families make up the cystatin superfamily (25). The superfamily supercystatin conserves the motif Gln-Val-Val-Ala-Gly or its homologues (Gln-X-Val-X-Gly) that are involved in the inhibition of the cysteine proteinases. Family I, known also as the stefin family, is identified by the lack of disulfide bonds. Human cystatin A and B belong to this family (26). The members of family I also have a single domain with a molecular weight of 11,000 and lack carbohydrate moieties. Family II has two disulfide bonds (e.g., chicken egg cystatin) (27). Members of family II have a single domain with a molecular weight of 13,000 and also lack carbohydrates. Family III includes blood plasma kininogens. They are large (60–120 kDa) secreted proteins with multiple disulfide bonds. Kininogens consist of three parts: a glycosylated amino-terminal heavy chain, the bradykinin moiety, and a carboxyl-terminal light chain (28). The heavy chain contains the inhibitory activity, which structure is a repeated cystatin-like domain. Two of them are capable of inhibiting cysteine proteinases.

Cystatin inhibitors are also present in plants. Oryzacystatin is a cysteine proteinase inhibitor from rice seeds that has no disulfide bonds like family I, but has identical amino acid residues to those in family II (29). The structures of the intron boundaries of oryzacystin are different from those found in humans. Abe et al. (30) cloned a cysteine inhibitor from corn. The deduced amino acid sequence showed homologies with the oryzacystin, no disulfide bonds, and sequences conserved with family II. A new family called phytocystatin has been proposed. However, ozyzacystatins lack signal peptides and the nucleotide sequence of corn cysteine inhibitor suggests that the protein exists as a precursor with a signal peptidelike human cystatin C. Potato multicystatin is also a proteinaceous inhibitor of cysteine proteases found in potato tubers. This inhibitor belongs to the cystatin superfamily. The inhibitor has a large size (85 kDa) because of the multiple cystatin units. The gene of potato multicystatin codes for eight stefinlike domains (31) and has a role in the plant's defense system according to gene expression studies.

C. Aspartyl Proteinases

The catalytic activity of aspartic proteinases (EC number 3.4.23.X) such as pepsin, renin, and cathepsin D depend on a pair of aspartic acid residues in the active site (Asp-32 and Asp-215). Most aspartic proteinases bind tightly to pepstatin, a microbial hexapeptide inhibitor (32). However, there are other sources of aspartic proteinase inhibitors, such as potato, yeast, the nematode *Ascaris,* and squash.

The potato cathepsin D inhibitor is a bifunctional inhibitor that also acts on serine proteinases. These inhibitors are similar to the soybean trypsin inhibitor family, which also possesses trypsin inhibitory activity (33). The aspartic proteinase inhibitors from potato form a multigene family of at least 10 members (34). This family is characterized by six cysteine residues and the active site responsible for the trypsin inhibitory activity is conserved at Arg-67.

An inhibitor from squash phloem exudate (35) has no similarity with any other known protein, which suggests that it belongs to a new inhibitor family. Squash, as yeast inhibitor, is an aspartic proteinase inhibitor that does not contain any disulfide bonds and there is no N-glycosylation site. The mass of this plant inhibitor is 10.5 kDa determined by deduced amino acid sequence, but the molecular mass obtained by gel filtration was almost twice this amount. This suggests that this proteinase inhibitor exists as dimer in its native form.

D. Metalloproteinases

Most metalloproteinases (EC 3.4.24.X) have zinc as the active metal in the catalytic site. Thermolysin is a bacterial metalloproteinase whose structure and sequence has been known since 1972 (36). In mammals, metalloproteinases are synthesized and secreted by connective tissues. They degrade different kinds of collagen and other connective tissue matrix macromolecules. These proteinases are important in the turnover of the connective tissue matrix and in the abnormal destruction of tissue proteins in some diseases (37).

Metalloproteinase are inhibited by metal chelators that remove the metal ion from the active site. However, some metalloproteinases inhibitors are involved in tissue remodelling. Tissue inhibitors of metalloproteinases (TIMP) are a family of secreted proteins that inhibit matrix metalloproteinases (38). TIMPs have several cysteine residues that form disulfide bonds and glycosylation is common in this group (39).

V. WHEN INHIBITION IS A PROBLEM (ANTINUTRITIONAL FACTOR)

A goal of the aquaculture industry is the replacement of expensive protein with less expensive protein sources. An alternative to using fish meal is the use of plant products. Legumes have been used for animal feeds because they have a better amino acid profile than other plant sources and the cost is less than fish or squid meal. However, seeds contain several antinutritive factors. For example, soybeans contain Kuniz and Bowman-Birk trypsin inhibitors. The Kuniz trypsin inhibitor has a molecular mass of 21 kDa and only inhibits trypsin. The Bowman-Birk trypsin inhibitor has a molecular mass of 8 KDa and inhibits both trypsin and chymotrypsin. Trypsin and chymotrypsin are the main serine proteases in the crustacean and fish digestive system (40) and therefore the nutritional impact of the protein source should be considered in diet formulation.

Feeding raw soybean meal inhibits growth, reduces protein digestibility, causes pancreatic enlargement, stimulates hypersecretion of digestive enzymes, and causes adverse physiological effects (41). The digestibility of protein and fat, weight gain, and trypsin activity in the intestinal contents of Atlantic salmon were reduced by increasing trypsin inhibitor (42).

Heat treatment of legume protein flour is one of the strategies to reduce these antinutritive factors. Heating inactivates proteinase inhibitors and improves the nutritive quality of soybeans. Shrimp fed with soy flour in which trypsin inhibitor content had been reduced had a higher weight gain than animals fed with no pretreated soy flour. Shrimp fed with meals having the lowest trypsin inhibitor content also had the highest body percentages of crude protein. These results correlated with those found by García-Carreño et al. (43,44), in which the proteolytic activity and degree of hydrolysis by shrimp extract of hepatopancreas, the organ that produces the digestive enzymes, increase if seed meal proteins are pretreated at 85°C.

VI. CONCLUSION

The identity and characterization of legume seed inhibitors are well known. However there are many other sources of proteinase inhibitors. Plants are the common source because of their defense system against predators. Other unexploited sources, including marine organisms, require further study. The purification and characterization of proteinase inhibitors and the cloning of genes will increase our knowledge of the use of these substances in the seafood industry.

REFERENCES

1. JR Whitaker. Principles of Enzymology for the Food Sciences, 2nd ed. New York: Marcel Dekker, 1994, pp 1–26.
2. G Salvesen, H Nagase. Inhibition of proteolytic enzymes. In: R Beynon, J Bond, eds. Proteolytic enzymes a practical approach. Oxford: IRL Press, 1989, pp 83–104.
3. FL García-Carreño. Proteinase inhibitors. TFST 7:197–204, 1996.

4. D Boulter, M Gatehouse, V Hilder. Use of cowpea trypsin inhibitor (Cpti) to protect plants against insect predation. Biotechnol Adv 7:489–497, 1989.

5. C Ryan. Defense response of plants. In: D Verma, T Hihn ed. Plant Gene Research: Genes Involved in Microbe Plant Interaction. New York: Springer-Verlag, 1984, pp 375–386.

6. FL García-Carreño. Protease inhibition in theory and practice. Biotechnol Ed 3:145–150, 1992.

7. FL García-Carreño, N Dimes, N Haard. Substrate–gel electrophoresis for composition and molecular weight of proteinases or proteinaceous proteinase inhibitors. Anal Biochem 214:65–69, 1993.

8. D. Panyam, A Kilara. Enhancing the functionality of food proteins by enzymatic modification. TFST 7:120–125, 1996.

9. S Arai, M Yamashita, M Fujimaki. Nutritional improvement of food proteins by means of the plastein reaction and its novel modification. Adv Exp Med Biol 105:663–680, 1978.

10. J Adler-Nissen. Limited enzymatic degradation of protein: a new approach in the industrial application of hydrolases. J Chem Technol Biotechnol 32:138, 1982.

11. A Gilberg. Enzymatic processing of marine raw materials. Proc Biochem 28:1–15, 1993.

12. R Lanier. Functional food protein ingredients from fish. In: Z Sikorski, B Sun Pan, F Shahidi, eds. Seafood Proteins. London: Chapman & Hall, 1994, pp 127–159.

13. Z Sikorski. The myofibrillar proteins in seafoods In: Z Sikorski, B Sun Pan, F Shahidi, ed. Seafood Proteins. London: Chapman & Hall, 1994, 57–40.

14. C Lee. Surimi process technology. Food Technol November: 69–80, 1994.

15. V Weerasinghe, M Morrissey, Y Chung, H An. Whey protein concentrate as a proteinase inhibitor in pacific whiting surimi. J Food Sci 61:367–371, 1996.

16. H An, V Weerasinghe, T Seymour, M Morrissey. Cathepsin degradation of Pacific whiting surimi proteins. J Food Sci 59:1013–1018, 1994.

17. ML Izquierdo, TA Haard, J Hung NF Haard. Oryzacystatin and other proteinase inhibitors in rice grain: potential use as a fish processing aid. J Agric Food Chem 42:616–622,1994.

18. B Lenarcic, M Krasovec, A Ritonja, I Olafsson, V Turk. Inactivation of human cystatin C and kininogen by human cathepsin D. FEBS Lett 280:211–215, 1991.

19. S Nakamura, M Ogawa, M Saito, S Nakai. Application of polymannosylated cystatin to surimi from roe-herring to prevent gel weakening. FEBS Lett 427:252–254, 1998.

20. H Toyohara, Y Makinodan. Comparison of calpain I and calpain II from carp muscle. Comp Biochem Physiol [B] 92:577–581, 1989.

21. H Toyohara, Y Makinodan, S Ikeda. Mutual inibitory effect of calpastains on calpains from carp muscle, carp erythrocytes and rat liver. Comp Biochem Physiol [B] 81:579–581, 1985.

22. M Laskowski, I Kato. Protein inhibitors of proteinases. Annu Rev Biochem 49:593–626, 1980.

23. R Read, M James. Introduction to the protein inhibitors: x-raycrystallography. In: AJ Barrett, G Salvesen, eds. Proteinase Inhibitors, vol 12. Cambridge: Elsevier, 1986, pp 515–569.

24. L Potempa, E Korzus, J Travis. The serpin superfamily of proteinase inhibitors: structure, function and regulation. J Biol Chem 269:15957–15960, 1994.

25. AJ Barret, ND Rawlings, ME Davies, W Machekeidt, G Salvesen, V Turk. Cysteine proteinase inhibitors of the cystatin superfamily. In: AJ Barrett, G Salvesen, eds. Proteinase Inhibitors, vol 12. Cambridge: Elsevier, 1986, pp 515–569.

26. A Ritonja, W Machleidt, A Barret. Amino acid sequence of the intracellular cysteine proteinase inhibitor cystatin B from human liver. Biochem Biophysic Res Commun 131:1187–1192, 1985.

27. R Colella, Y Sakaguchi, H Nagase, JWC Bird. Chicken egg white cystatin. J Biol Chem 264:17164–17169, 1989.

28. G Salvesen, C Parkes, M Abrahamson, A Grubb, A Barret. Human low Mr kininogen contains three copies of a cystatin sequence that are in structure and inhibitory activity for cysteine proteinases. Biochem J 234:429–434, 1986.

29. K Abe, Y Emori, H Kondo, K Susuki, S Arai. Molecular cloning of a cysteine proteinase inhibitor of rice (oryzacystatin). J Biol Chem 262:16793–16797, 1987.

30. M Abe, K Abe, M Kuroda, S Arai. Corn kernel cysteine proteinase inhibitor as a novel cystatin superfamily member of plant origin. Molecular cloning and expression studies. Eur J Biochem 209:933–937, 1992.

31. C Waldron, LM Wegrich, PA Merlo, TA Walsh. Characterization of a genomic sequence coding for potato multicystatin, an eight-domain cysteine proteinase inhibitor. Plant Mol Biol 23:801–812, 1993.

32. DH Rich. Inhibitors of aspatic proteinases. In: AJ Barrett, G Salvesen eds. Proteinase Inhibitors, vol 12. Cambridge: Elsevier, 1986, pp 179–217.

33. A Ritonja, I Krizaj, P Mesko, M Kopitar, P Lucovnik, B Strukelj, J Pungercar, DJ Buttle, AJ Barrett, V Turk. The amino acid sequence of a novel inhibitor of cathepsin D from potato. FEBS Lett 267:13–15, 1990.

34. B Strukelj, J Pungercar, P Mesko, D Barlic-Maganja, F Gubensek, I Kregar, V Turk. Characterization of aspartic proteinase inhibitors from potato at the gene, cDNA and protein levels. Biol Chem Hoppe-Seyler 373:477–482, 1992.

35. JT Christeller, PC Farley, RJ Ramsay, PA Sullivan WA Laing. Purification, characterization and cloning of an aspartic proteinase inhibitor from squash phloem exudate. Eur J Biochem 254:160–167, 1998.

36. PM Colman, JN Jansonius, BW Matthews. The structure of thermolysin: an electron density map at 2.3 Å resolution. J Mol Biol. 70:701–724, 1972.

37. TE Cawston. Protein inhibitors of metallo-proteinases. In: AJ Barrett, G Salvesen eds. Proteinase Inhibitors, vol 12. Cambridge: Elsevier, 1986.

38. KJ Leco, SS Apte, GT Taniguchi, SP Hawkes, R Khokha, GA Schultz, DR Edwards. Murine tissue inhibitor of metalloproteinases-4 (Timp-4): cDNA isolation and expression in adult mouse tissues. FEBS Lett 401:213–217, 1997.

39. RA Williamson, FA Marston, S Angal, P Koklitis, M Panico, HR Morris, AF Carne, BJ Smith, TJ Harris, RB Freedman. Disulphide bond assignment in human tissue inhibitor of metalloproteinases (TIMP). Biochem J. 268:267–274, 1990

40. MP Hernández-Cortés, JR Whitaker, FL García-Carreño. Purification and characterization of chymotrypsin from *Penaeus vannamei* (Crustacea: Decapoda). J Food Biochem 21:497–514, 1997.

41. D Sessa, C Lim. Efect of feeding soy products with varying trypsin inhibitor activities of growth of shrimp. JAOCS 69:209–212, 1992.

42. J Olli, K Hjelmeland, A Krogdahl. Soybean trypsin inhibitors in diets for Atlantic salmon (*Salmo salar,* L.): effect on nutrient digestibilities and trypsin in pyloric caeca homogenate and intestinal content. Comp Biochem Physiol 109A:923–928, 1994.

43. F García-Carreño, A Navarrete del Toro, M Díaz-López, M Hernández-Cortés, M Ezquerra. Proteinase inhibition of fish muscle enzymes using legume seed extracts. J Food Prot 59:312–318, 1996.

44. F García-Carreño, A Navarrete del Toro, M. Ezquerra. Digestive shrimp proteases for evaluation of protein digestibility in vitro. I. Effect of protease inhibitors in protein ingredients. J Marine Biotechnol 5:36–40, 1997.

20
Influence of High-Pressure Processing on Enzymes in Fish

Isaac N. A. Ashie
Novo Nordisk BioChem, North America, Inc., Franklinton, North Carolina

Tyre C. Lanier
North Carolina State University, Raleigh, North Carolina

I. INTRODUCTION

The myosystems of fresh fish and other muscle foods are replete with enzymes, the postmortem activity of which may be desirable or undesirable. In either case, their control can lead to maintaining or improving the product quality and shelf-life. High pressure as a food processing tool was initiated by Hite and his co-workers (1,2) who explored its possible use in the pasteurization of milk. During the same period, Bridgman (3) also demonstrated irreversible pressure-induced coagulation of ovalbumin. The primary focus has since been toward the application of high-pressure processing as an alternative to thermal processing for the control of microbial activity (4). However, in recent years there has been increasing interest in the use of high pressure for modification of proteins, enzymes, and other food macromolecules (5–9). Fish and other species living at varying depths within the aquatic milieu are somewhat adapted to different habitat pressures, which is manifested by their relative sensitivities to pressure treatment. Conceivably then, high pressure might be used in the control or enhancement of enzyme activity for the improvement of food quality and the development of new foods and food processes.

II. GENERAL EFFECTS OF PRESSURE ON PROTEINS
AND ENZYMES

The influence of pressure on proteins is primarily related to the volume changes induced. Reactions or biochemical processes that result in decreased volume or closer packing of molecules are enhanced by increasing pressure while those leading to increased volume are retarded, in accordance with Le Chaterlier's principle, which states that: *a system in equilibrium reacts in a manner to reduce the effect of any applied external stress.* In other words, if the volume of the system in the ground state differs from that in the activated or transition state, pressure will affect the progress of the reaction.

It may be expressed quantitatively as:

$$\delta \ln K/\delta P = -\Delta V/RT \tag{1}$$

where P is the applied pressure on the equilibrium constant (K) at a constant temperature (T). R is the gas constant and ΔV is the resulting change in volume.

Protein volume may be categorized into three components: the constitutive atomic volume; the void or conformational volume due to imperfect packing; and the solvation volume due to the electrostrictive effect of charged groups on the solvent and the rearrangement of solvent molecules around nonpolar groups (10). In order of increasing strength, the main types of interactions that maintain the native conformation and volume of proteins are the hydrophobic interactions (5–10 kJmol-1), hydrogen bonds (10–40 kJmol-1), electrostatic interactions (25–80 kJmol-1), and covalent interactions (200–400 kJmol-1) (11). Pressure treatment does not seem to have any effect on primary protein conformation since it cannot cleave the peptide (covalent) bonds (12). Hydrogen bond formation is slightly enhanced by high pressure due to the negative volume changes (–1 to –3 ml/mol) resulting from decreased interatomic distances upon formation of the bond. Formation of electrostatic or coulombic interactions is associated with positive volume changes (10 ± 10 ml/mol) and is therefore weakened at high pressures, while dissociation to separate charges is enhanced because of the negative volume changes associated with electrostriction of water around charged molecules (10,13). Thus the rate of an enzyme-catalyzed reaction involving transition state neutralization of ionic species will be reduced at high pressures while the formation of separate charges in the transition state will be enhanced. The effects of pressure on hydrophobic interactions are more complicated. These may be enhanced or inhibited depending on the nature of the nonpolar molecules involved in the interaction (10,12). Interactions between aliphatic groups are associated with positive volume changes while volume reduction occurs during stacking of aromatic groups.

Pressure-induced inactivation of enzymes was first investigated by Basset and Macheboeuf (14) and later by Curl and Jansen (15,16). The latter authors specifically studied proteases. Pressure effects on enzyme-catalyzed biochemical processes appear to be due to conformational changes associated with such processes as enzyme–substrate/ligand binding, modulator interactions, the activation event, subunit association or dissociation reactions, and denaturation (13,17,18). Substrate binding is considered to be most sensitive to pressure. Numerous studies investigating the effects of pressure on the Michaelis constant have shown that the volume changes associated with ligand binding vary greatly and thus enzyme activities may be enhanced, unaffected, or inhibited by pressure (18). For example, the electrostatically driven binding of proflavin to trypsin is inhibited at elevated pressures while binding to chymotrypsin is unaffected (19). Similarly, ligand binding to thermolysin from *Bacillus thermoproteolyticus* is enhanced by pressure while the homologous enzyme from *Bacillus subtilis* is inhibited (20).

Volume changes may also result from the rearrangement of water and other molecules in the surrounding medium, in response to pressure-induced removal or addition of amino acid residues and ligands from and to the surrounding medium (21). Hence enzyme activity may be greatly influenced by high pressure at various stages of the catalytic process.

According to the classic two-step model for enzyme inactivation (Eq. 2), the initial effect of an applied external stress causes the enzyme to undergo cooperative unfolding to form a reversibly denatured state that may then proceed to the irreversibly inactivated state depending on the conditions (22).

$$\text{Native} \Leftrightarrow \text{Reversibly Denatured} \Rightarrow \text{Inactive} \tag{2}$$

Pressure effects on proteins may either be reversible or irreversible depending on the amount of pressure applied, duration, and conditions of pressurization such as pH, ionic strength and type, temperature, and solvents. The native state of proteins is stable within relatively narrow zones of pressures that are determined also by these factors (23). In general however, reversible effects are often observed when pressures below 200 MPa are applied, while pressures in excess of 300 MPa tend to induce irreversible effects in proteins (12). Multimeric proteins are more sensitive to high pressure than monomeric proteins and the pressures required to alter protein structure increase from tertiary to secondary structure (12). Typical volume changes for dissociation of multimeric proteins are often large (-30 to -300 ml/mol) while the corresponding volume changes for denaturation of the most stable monomeric proteins range between -30 and -100 ml/mol (18,24,25). Penniston (26) showed that the activities of several multimeric enzymes were decreased primarily due to subunit dissociation induced by increasing pressure, while the monomeric enzymes showed varying levels of increased activity over the same pressure range. It is worth noting, however, that the activities of some multimeric enzymes can be enhanced by pressure (13).

III. EFFECTS OF HIGH PRESSURE ON SPECIFIC ENZYMES

A. ATPases

The negative charges on phosphates are complemented by positive charges at the enzyme-binding site, making electrostatic interactions the primary driving force in substrate binding during ATPase activity. Conversely, the transition state of phosphate hydrolysis involves a penta-coordinated configuration in which charge separation occurs. Thus while substrate binding may be reduced at high pressures, formation of the transition state is enhanced (27). However, it has been shown that substrate binding is generally the rate-limiting step in ATPase-catalyzed reactions such that enzyme activity is reduced under pressure (27). The effects of high pressure on the activities of Ca^{2+}- and Mg^{2+} -ATPase in fish myosystems have been widely investigated. Denaturation of actomyosin is often manifested as loss of Ca^{2+}-ATPase activity and this occurs very readily in fish muscle tissues subjected to high pressure. Milkfish (*Chanos chanos*) actomyosin pressurized to 200 MPa at 4 C lost about 80% of its original Ca^{2+}-ATPase activity and was almost completely inactivated upon pressurization at 300 MPa for 5 min (28). Flying fish (*Gypselurus opisthopus*) and sardine (*Sardinops melanostictus*) actomyosin ATPases were also inactivated, the activity in sardine being more sensitive than that of flying fish (29). Using fluorescence, solubility tests, and circular dichroism, Ishizaki et al. (30) demonstrated that the pressure-sensitive component responsible for the loss of actomyosin ATPase activity is the myosin subfragment-1 (S-1). This subfragment most readily unfolds, exposing hydrophobic groups with a concomitant loss of solubility and decreased helical content. Effects of pressure on ATPases are also influenced by such factors as temperature, amount and duration of pressurization, and myosin–actin interactions (28–30). Pressure sensitivity of myofibrillar ATPase activity is also related to the habitat temperature of fishes, as discussed below.

B. Proteolytic Enzymes

Kunugi (27) noted that changes in the reaction volumes of trypsin and chymotrypsin due to pressure increase were positive and negative, respectively, and would therefore result in the decrease and increase of their respective rates of reaction. This difference in transition state volume was attributed to the electrostatic nature of ligand binding to trypsin versus the hydrophobic nature of the substrate interaction with chymotrypsin. Yet at extremely high pressures (>300 MPa) chymotrypsin is inactivated (31). The stability of chymotrypsin was attributed to the compact globular structure of the active enzyme. Evidently the rupture at higher pressures of intramolecular hydrophobic interactions destabilizes the native state. Based on Raman spectroscopic analysis, however, Heremans and Heremans (32) have also attributed the volume and conformational changes under pressure to disruption of the salt bridge between Asp-194 and Ile-16, which is required for activity.

The activity of thermolysin, which displays a large negative apparent activation volume (-20 to -40 ml/mol), was increased 15-fold at 100 MPa while the corresponding activity for carboxypeptidase A, activation volume +20 to +35 ml/mol, was reduced to one-third of the control. This increased activity of thermolysin under pressure has been used to enhance the proteolysis of certain proteins. In this case, high pressure also acted as a so-called reagentless denaturant of the protein substrate, making it more accessible and thus susceptible to proteolysis (27).

Ashie and Simpson (33,34) investigated high-pressure effects on the activities of proteases endogenous to fish muscle that could cause its textural softening. The enzymes were inactivated to various extents depending on the amount of pressure applied and the duration of pressurization. Cathepsin C from bluefish (*Pomatomus saltatrix*) and sheepshead (*Calamus penna*) lost more than 80% of its original activity when pressurized at 300 MPa for 30 min. Under the same conditions, trypsin- and chymotrypsinlike enzymes retained 30–40% of their original activities. Activity of all enzymes continued to decline with storage at room temperature, but were partially recovered during storage at refrigerated temperatures. It is likely that autolysis prevailed over reactivation during storage at room temperature while reactivation predominated at low temperatures due to decreased proteolytic activity. Chung et al. (35) found that pressure treatment of Pacific whiting (*Merluccius productus*) surimi up to 240 MPa had seemingly no effect on the protease (cathepsin L) involved in gel softening during cooking.

In similar work with muscle of homeotherms, the activities of calpains I and II from rabbit skeletal muscle were unaffected when pressurized at 200 MPa for 10 min, but began to lose activity at higher pressures, becoming completely inactivated at 400 MPa (36). During investigation of the mechanism of meat tenderization by high pressure, Okamoto et al. (37) observed that pressurization of rabbit skeletal muscle to about 200 MPa caused significant disruption of the sarcoplasmic reticulum, resulting in the release of Ca^{2+} ions into the myofibrillar region. The increased levels of Ca^{2+} ions increased the activities of calpains, resulting in acceleration of meat conditioning and tenderization. Pressurization of beef muscle beyond 200 MPa also causes significant losses of aminopeptidase B and cathepsin B, D, H, and L activities (38,39). At lower pressures, however, Kurth (40) reported an increase in cathepsin B1 activity. Dufour et al. (41) reported that the rate of collagen hydrolysis by cathepsin B and collagenase was reduced by high-pressure treatment up to 300 MPa.

C. Transglutaminases

Transglutaminase (TGase) activity endogenous to muscle has been shown to be responsible for the low-temperature gelation (setting) of surimi pastes made

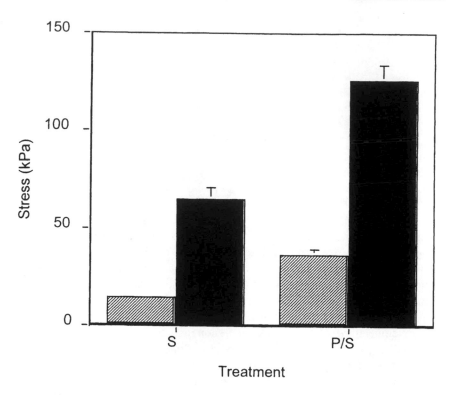

Figure 1 Effect of high-pressure treatment (200 MPa for 15 min) on TGase-induced gelation of Alaska pollock (*Theragra chalcogramma*) surimi. (From Ref. 47.)

from many fish species (42–44; see Chap. 6). Shoji et al. (45) found that a crude extract of the enzyme prepared from walleye pollock (*Theragra chalcogramma*) surimi retained about 75% of its original activity after being pressurized at 200 MPa for 10 min, but was completely inactivated by pressure in excess of 300 MPa. Contrary to this, the author and others found that TGase from both pollack and microbial sources survived these pressures, retaining their maximum activity (46–48). Pressurization of Alaska pollock surimi pastes at 300 MPa for 15 min prior to setting (2 h at 25°C) resulted in a threefold increase in gel strength compared to gels formed by setting alone (Fig. 1). This suggested not only that TGase activity survived the pressure treatment, which was confirmed by direct assay (Fig. 2), but also that the higher gel strength obtained by prior pressurization resulted from pressure-induced denaturation of the muscle proteins, rendering them more accessible to the enzyme for crosslinking. Nonaka et al. (48) similarly reported survival of microbial TGase (MTGase) activity after pressure treatments as high as 600 MPa. They also noted that bovine serum albumin and

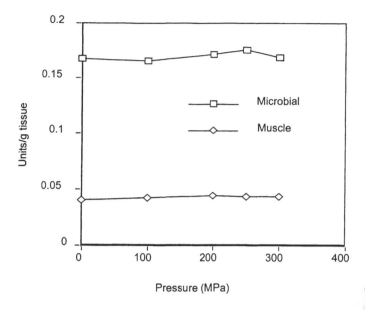

Figure 2 Effect of high-pressure treatment on microbial and Alaska pollock (*Theragra chalcogramma*) surimi TGases. (From Ref. 47.)

ovalbumin, which typically require denaturation by dithiothreitol to be suitable substrates for MTGase, showed good reactivity once they had been subjected to high-pressure treatment. Thus the reagentless denaturant aspect of high-pressure treatment (27) can apply equally well to improving the reactivity of both crosslinking and hydrolyzing enzymes, and could be applicable in other enzyme-catalyzed reactions to enhance the reactivity of substrates.

D. Other Enzymes

The key enzymes involved in regulation of glycogenolysis have shown differing sensitivities to high pressure (49). Phosphorylase kinase and phosphorylase phosphatase were readily inactivated when pressurized at 150 MPa for 5 min, while phosphorylase retained high levels of activity. Rapid inactivation of phosphatase by pressure resulted in the relatively high level of phosphorylase no longer being regulated by the phosphorylation–dephosphorylation cycle, with a resulting increase in the rate of glycogenolysis.

Enzymatic hydrolysis of fish muscle lipids during refrigerated storage often results in decreased phospholipid levels and an increase in free fatty acids. These are in turn very susceptible to autoxidation and result in oxidative rancidity, which may cause changes in flavor and texture of the fish (50,51). Following pressurization of cod (*Gadus macrocephalus*) and mackerel (*Scomber japonicus*) muscles in excess of 400 MPa, fatty acid accumulation was reduced while phospholipid levels were not significantly affected, indicating that the enzymatic degradation of phospholipids was inhibited (52).

IV. PRESSURE AND TEMPERATURE ADAPTATION

The thermal stability of myofibrillar Ca^{2+}-ATPase from whale, rabbit, and 40 fish species is strongly related to body temperature, which for poikilothermic fish species is essentially equivalent to the habitat temperature (53; see also Chap. 3). Thus myosin of cold-adapted species like krill and Alaska pollock is more readily denatured by heat than that of warm water fish (e.g., tuna) or warm-blooded animals. In a similar manner, we recently found that pressure-induced denaturation of myofibrillar Ca^{2+}-ATPase correlates with habitat/body temperatures of animal species (Fig. 3). Actomyosin Ca^{2+}-ATPase from cold-adapted Alaska pollock (*Theragra chalcogramma*) was almost completely inactivated on application of 200 MPa pressure, while that from tilapia (*Oreochromis aureus*), a tropical freshwater fish, was unaffected and more similar in stability to bovine myofibrillar Ca^{2+}-ATPase.

This relationship between stability and habitat temperature is not limited to myofibrillar ATPase. Simpson and Haard (55) reported a strong correlation between inactivation temperatures of pyruvate kinase, trypsin, and pepsin from various fish species and their habitat temperatures. Endogenous proteases from bluefish (*Pomatomus saltatrix*), sheepshead (*Calamus penna*), and bovine sources are more sensitive to pressure-induced inactivation than their bovine counterparts (34).

Thus it would seem that adaptations in the tertiary structure of myofibrillar proteins for high enthalpy environments can also stabilize the molecule at high pressure. Presumably enzymes or proteins from cold-adapted species are more flexible and highly compressible. These less ordered molecular structures would permit conformational changes that ensure reasonable levels of activity at the low habitat temperatures, while the more stable warm-adapted enzymes are more compact and of lower compressibility (56–58). For example, the CD spectra of Greenland cod (*Gadus ogac*) and bovine trypsins show significant differences in their internal structures. Under identical conditions, Greenland cod trypsin contained 7.3–7.8% α-helicity and 92.2–92.7% random coil, while bovine trypsin was comprised of 11.5–12% α-helix and 88–89.5% random coil (59). Other observations by these authors included the following. First, that Greenland cod trypsin with eight cysteine residues could form only four disulfide linkages compared to the six possible

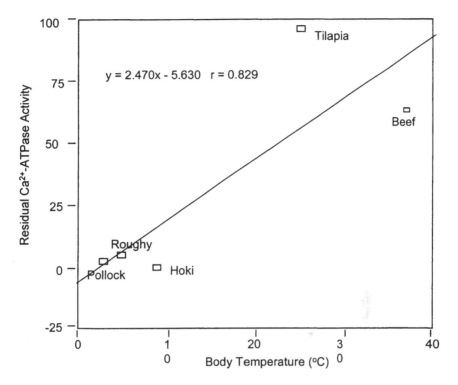

Figure 3 Relationship between body temperature and residual activity of Ca^{2+}-ATPase from myofibrillar proteins of bovine and different fish species after pressurization at 200 MPa for 15 min. (From Ref. 54.)

disulfide pairs in bovine trypsin. Second, the average hydrophobicity of Greenland cod trypsin (0.86 kcal/residue) is lower than that of bovine trypsin (1.04 kcal/residue). Comparative studies of enzymes from some thermophilic and mesophilic organisms have further shown that in many cases additional interactions may be found in thermophilic enzymes while amino acid residues sensitive to structure scrambling (Cys), deamidation (Asn, Gln), and oxidative damage (Met) have been replaced to stabilize the native state (60–63; see also Chap. 1).

The oceans cover about 70% of the earth's surface with an average depth of 3,800 m and a corresponding average pressure of 38 MPa. The Challenger Deep in the Marianas Trough with depth of 11,000 m is the deepest part of the oceans. Organisms that inhabit the deeper aquatic milieu may be subject to relatively high pressures and must have enzymes that are structurally stable to this stress. For instance, actin from fish species living at depths below 2000–3000 m were found to be more thermostable than those from shallow-living species adapted to low temperatures, and even showed greater stability than actin from rabbit and chicken (64,65). Analysis of the enthalpies, entropies, and volume changes of these pressure-adapted actins showed a reduced dependence on hydrophobic interactions that are unstable at the low temperatures and high pressures prevailing at such depths (31,32,66).

The membranes of various tissues of deep living fishes are also more stable to high pressure. 2-Deoxy-ATP binding by adenylyl cyclase of the shallow-living fish species *Sebastolobus alascanus* was more barosensitive than the deep-living *Sebastolobus altivelis,* with the corresponding apparent volume changes being 72.3 and 42.4 ml/mol, respectively (67). Elevated pressures also reduced the peak amplitude of the compound action potential of the vagus nerve of shallow-living fishes by 50%, while those of two deep sea (4000–4200 m) species were unaffected (68). Membranes in these species evidently have an inherently higher degree of fluidity than the shallow-living species, which enables them to function adequately under the low temperature–high pressure conditions of the abyssal habitat (69,70). This phenomenon, referred to as homeoviscous adaptation, has been demonstrated in the membranes of various marine species (69,71). One of the ways in which homeoviscous adaptation could regulate high-pressure effects on membrane-based function in the deep-sea is *via* the membrane-bound lipoprotein enzyme Na^+/K^+-ATPase, which has been shown to be more barotolerant than the homologous enzyme from surface-living species (71). The relative barostability of the deep sea enzyme has been attributed to its interaction with specific membrane lipids essential for modulating its activity and response to pressure (72). However, the reduced sensitivity to high pressure is limited to species occurring below 2000 m. For instance, the enzyme from the rattail fish, *Coryphaenoides acrolepis,* with maximal depth near 2000 m is as sensitive to pressure as that from cold-adapted surface fishes, but the deeper-living (5,000 m) *Coryphaenoides armatus* is much less sensitive (72). Similar increase in stability with increasing vertical distribution has been observed with M4-lactate dehydrogenase from different fish species (73). Hochachka (74) also showed that disruption of acetylcholine binding to actylcholinesterase by high pressure was more severe in shallow-living fish than in deep sea species.

Gross and Jaenicke (56) concluded in a recent review that adaptation to deep sea is dominated by accommodation to low temperature rather than to high pressure. For instance, lowering the reaction temperature from 20°C to 2°C, representative of temperatures at sea level and the ocean floor at 2000 m, respectively, leads to a 4–10-fold reduction in the reaction rate of lactate dehydrogenase, while only a 12% reduction in reaction rate is observed as a result of a 200 atm rise in pressure, corresponding to pressure at this depth (75,76).

V. OTHER FACTORS AFFECTING PRESSURE RESPONSE

As previously discussed, the effect of high pressure on enzyme activity and stability is greatly influenced by a number of environmental factors, including temperature, pH, modulator or substrate concentration, ionic composition and strength, and sugars and polyols. This section provides a closer look at the influence of these factors.

A. Temperature

Temperature variation affects a number of enzymic processes involving volume changes such as protein conformation, activation, and denaturation. The influence of temperature on pressure response of enzymes is largely dependent on the pressurization temperature in relation to the melting point (Tm) of the protein (77). Enzymes pressurized ($\geq$ 300 MPa) at low temperatures often show large negative changes in volume and are therefore inactivated or denatured, while those pressurized at higher temperatures close to Tm may be stabilized or destabilized depending

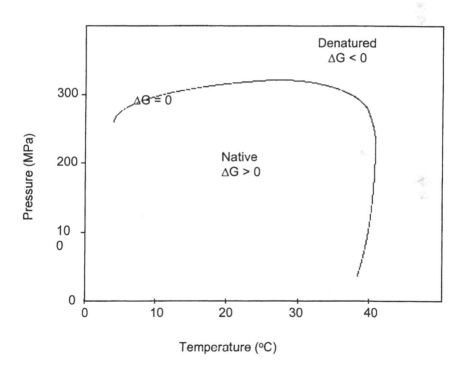

Figure 4 Typical pressure–temperature denaturation boundary for proteins. (From Ref. 77.)

on the enzyme (Fig. 4). This has been demonstrated for chymotrypsinogen, ribonuclease, aspartase, malate dehydrogenase, amyloglucosidase, and α-chymotrypsin (31, 77–80). Aspartase and α-chymotrypsin were inactivated at low pressurization temperatures (20°C and 45°C, respectively), but showed increased stability and activity when pressurized at higher temperatures. According to Nickerson (81), the stabilizing effect of pressure at high temperatures in such cases may be due to the influence of temperature on the highly ordered structure of electrostricted water, which often causes significant volume changes in enzymatic reactions. At high temperatures, the electrostriction effect on protein structure and stability is significantly reduced and may even be completely eliminated, thereby reducing the destabilizing effect of pressure on electrostatic interactions that are essential for protein activity and stability. The increase in enzyme stability at high temperatures further suggests that the volume change associated with denaturation becomes positive or less negative with increasing temperature. At extremely high pressures, however, any stabilizing effect of high temperature is minimal and enzymes are generally inactivated (Fig. 4). Thus far, the stabilizing effect of high pressure against thermal inactivation or denaturation seems to be a general characteristic of most enzymes stuied. It is worthy of note, however, that a study of pressurized adenylate kinase from some deep-sea thermophilic microorganisms did not show such stabilization (82).

Conversely, moderate or physiological pressures can enhance the activity of some enzymes at elevated temperature. For example, application of 0.7 MPa pressure increased pyrophosphatase activity 25-fold while pressures between 130 and 170 MPa shifted its upper temperature limit of activity from below 100°C at atmospheric pressure to above 105°C (83).

B. pH

Changes in the ionization constants of weak acids (and thus pH of buffering systems) have been shown to vary with pressure (84,85). For example, the pH of acetate, cacodylate, or phosphate buffers is changed by 1–2 pH units over a pressure range of about 600 MPa, while tris buffers are almost pressure insensitive with pH reduction of less than 0.2 units. These variations in pressure effects have been attributed to volume changes associated with the ionization equilibrium of the buffering species. Fructose diphosphatase shows no change in activation volume at pH 7.0, but exhibits a large negative volume change at pH 9.0 (86). Similar change in enzyme volume by pH alteration has been demonstrated in chymotrypsin, pepsin, ribonuclease, and other proteins (87). Pressurization of these enzymes would therefore elicit different responses at the different pH values. Zipp and Kauzmann (85) reported that changes in pH affect protein barostability in such a way that, at any pH from 4 to 13, a range of pressures may be found for which the protein undergoes changes from denatured to native and then back to a denatured state.

C. Modulator/Substrate Concentration

Molecules such as coenzymes, inhibitors, and activators, which bind specifically to native proteins, may stabilize them against external stress as was demonstrated for substrate-binding by lysozyme (88). Fructose diphosphate, a positive modulator of pyruvate kinase, is also capable of blocking the ATP inhibition of the enzyme under high-pressure conditions. Pressurization of actomyosin from yellowfin tuna (*Thunnus albacores*) showed that not only was the protein stabilized in the presence of lecithin, but ATPase activity was also even enhanced at certain pressures (89,90). Additionally, Schade et al. (91) found that coenzyme (NADH) binding of M4-lactate dehydrogenase shifted the dissociation transition to higher pressures. Pace and McGrath (88) attributed the stabilizing effects of these interactions to preferential substrate binding to the native conformation of the active site, resulting in a shift of the unfolding equilibrium (Eq. 2) in favor of the native state.

Concentration of the enzyme can also alter pressure effects. Mitochondrial ATPase and pepsin activity, at different concentrations, was more readily inactivated at lower concentrations (16, 92). On the other hand, trypsin and chymotrypsin activities were reduced with increasing enzyme concentration (15).

D. Ionic Composition and Strength

Ionic strength and composition have variable effects on volume changes in proteins (94). Addition of 0.1M sodium sulfate or ammonium sulfate stabilized hexokinase against pressure dissociation by increasing the pressure required for 50% dissociation by 0.4 MPa (94). The barosensitivity of pyruvate kinase from different fish species was inversely proportional to ionic strength (17). This has led to the suggestion that high ionic strength may facilitate stabilization to pressure in organisms living in the deep-sea environment. Indeed, higher solute concentrations have been demonstrated in the plasma of some deep-sea fishes (17,95).

Changes in activation volumes of pyruvate kinase from the fish species *Trematomus borchgrevinki* and *Scorpaena gutatta* were smaller in the presence of potassium and ammonium ions than for sodium ions (17). Replacement of Tris-HCl buffer with phosphate buffer of equal pH and ionic strength also increased the stability of M_4-lactate dehydrogenase (91). These effects have been found to depend on the type of salt and its position in the Hoffmeister series (18,96). For example, the effects of both anions and cations on M_4-lactate dehydrogenase from halibut (*Hippoglossus stenolepis*) followed the ranking of the Hoffmeister series (97). The salting-in ions have the tendency to inhibit the enzyme and increase the change in activation volume while salting-out ions reduce the reaction velocity to a lower extent and decrease the change in activation volume. Low and Somero (21) attributed the salt effects on activation volume to "hydration density" effects due to changes in exposure of water-density-modifying groups to solvent, and structural or conformational volume changes of the protein molecule itself, which is related to the habitat temperature of the species. Salting-in ions, which increase protein solubility, therefore enhance enzyme hydration such that peptide groups or polar side chains that are normally hydrated only in the transition state are largely hydrated. Thus, SCN⁻, Br⁻ and I⁻ will reduce or eliminate the negative contribution to the change in activation volume whereas salting-out ions increase the volume of the enzyme-substrate complex (96).

The salt effects on enzymes are in turn influenced by temperature. Greaney and Somero (96) showed that, at 10 C, 1 M KI inhibited alkaline phosphatase activity by 53%, whereas the same concentration caused only a 35% reduction in activity at 20°C. These effects are accompanied by changes in activation volume as a result of changes in "hydration density" by salting-in ions (96).

E. Sugars and Polyols

O'Sullivan and Tompson (cited in ref. 18) observed in the last century that the thermal stability of invertase was increased in the presence of cane sugar. The stabilization of seafood muscle proteins against freeze and heat denaturation by sugars and polyols has been well demonstrated in various studies (98,99). During surimi manufacturing, sucrose and/or sorbitol is therefore routinely added to stabilize the fish proteins in order to prevent loss of functionality. Recent reports indicate that these compounds could also serve as baroprotectants of proteins (54, 100,101). An evaluation of the baroprotective potential of some sugars and polyols on fish proteins showed that all the polyols studied imparted varying levels of baroprotection to fish (*Theragra chalcogramma*) muscle Ca^{2+}-ATPase activity. Sorbitol was the most effective and maltodextrins were least effective when subjected to 200 MPa pressure (Fig. 5). The order of effectiveness showed no correlation between number of equatorial hydroxyl groups in the sugars or polyols and baroprotection, as previously suggested by Fujii et al. (102). While the mechanism of baroprotection by these compounds is not yet clearly understood, we have proposed that barostabilization is likely to be attributable to solute exclusion or preferential hydration (103). This proposal is based on the greater baroprotection afforded by the lower-molecular-weight carbohydrates (Fig. 5), which also markedly affect the surface tension of water with increasing concentration (98).

These studies further showed that the barostabilizing capacity of the additives is greatly reduced at low temperatures, as may be encountered during pressure-shift freezing (54). For example, Atlantic flounder Ca^{2+}-ATPase, which

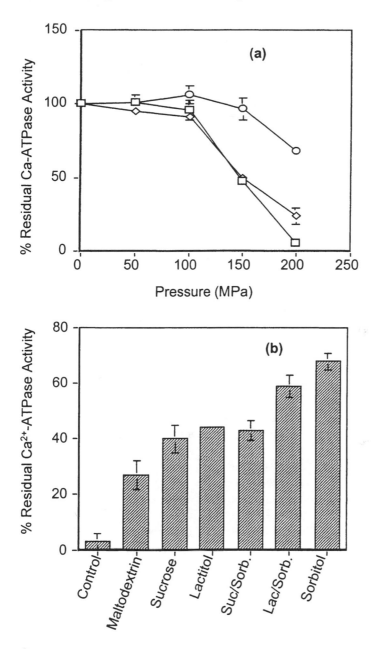

Figure 5 Baroprotective effect of sorbitol (a) on Alaska pollock Ca^{2+}-ATPase activity; and (b) relative baroprotective effects of various sugars and polyols on pollock muscle Ca^{2+}-ATPase activity. (From Ref. 54.)

retained full activity after pressurization at 5°C and 200 MPa in the presence of 8% sorbitol, lost about 50% activity during pressure-shift freezing. Considering the known tendency of proteins to undergo cold denaturation (104) and the observation that the pressure required for protein denaturation decreases when the temperature falls below freezing (77), the reduced protection under pressure may be due to a synergism of pressure and low temperature. Since hydrophobic bonds are weakened by both low temperature and high pressure, the combination would especially favor weakening of intramolecular hydrophobic bonds that stabilize enzyme conformation.

VI. FUTURE TRENDS AND PROSPECTS FOR HIGH-PRESSURE PROCESSING

High-pressure processing, compared to heating, better retains the natural taste and flavor of many foods. Thus it may prove useful as an alternative means of controlling the activity of seafood enzymes that affect seafood quality. Ashie and Simpson (34) demonstrated improved firmness and elasticity of fish muscle due to inactivation of endogenous protease. Pressure-induced seafood gels retain better color and flavor, and are glossier, smoother, and more elastic than heat-induced gels (105). Unlike heat treatment, which depends on a temperature gradient for transmission, high-pressure transmission is instantaneous and independent of product volume. This ensures a uniform effect on product quality throughout the product, and in a relatively short process time.

Substrate modification by pressure treatment (47) could also be exploited to enhance reaction rates and thus improve the efficiency of seafood enzyme-catalyzed processes (106), particularly for substrates that are relatively unreactive without prior pressure treatment.

Thus far the seafood industry has been rather apprehensive toward incorporating this technology into food processing operations. Pressure treatment remains largely a batch operation, and there are potential operational hazards and high equipment costs at present. Both academia and industry acknowledge the potential of the technology: a university–industry consortium has been established to address some of these issues. As a result of this collaboration, at least one company (Avomex Inc., High Pressure Research, Inc.) has taken the initiative to introduce several new food products processed by high pressure into the North American market. A number of high-pressure equipment manufacturers (ABB Autoclave Systems, Engineered Pressure Systems, Flow International Corp.) have also emerged recently; some have even developed semicontinuous high-pressure systems for fluid foods, and all are working to enhance the feasibility of large-scale high-pressure equipment for food processing.

REFERENCES

1. BH Hite. The effect of pressure in the preservation of milk. W Va Univ Agric Exp. Sta, Morgantown Bull 58: 15–35, 1899.
2. BH Hite, NJ Giddings, CE Weakly. The effects of pressure on certain microrganisms encountered in the preservation of fruits and vegetables. W Va Univ Agric Exp Sta, Morgantown Bull 146: 1–67, 1914.
3. PW Bridgman. The coagulation of albumen by pressure. J Biol Chem 19:511–512, 1914.
4. DG Hoover, C Metrick, AM Papineau, DF Farkas, D Knorr. Biological effects of high pressure on food microorganisms. Food Technol 43(3): 99–107, 1989.
5. K Ohmiya, K Fukami, S Shimizu, K Gekko. Milk curdling by rennet under high pressure. J Food Sci 52: 84–87, 1987.
6. K Ohmiya, T Kajino, S Shimizu, K Gekko. Effect of pressure on the association states of enzyme-treated caseins. Agric Biol Chem 53: 1–8, 1989.
7. EA Elgasim, WH Kennick. Effect of pressurization of pre-rigor beef muscles on protein quality. J Food Sci 45: 1122–1124, 1980.
8. R Hayashi. Application of high pressure to food processing and preservation: philosophy and development. In: WEL Spiess, H Schubert, eds. Engineering and Food. London: Elsevier Applied Science, 2: 815–826, 1989.
9. R Hayashi, A Hayashida. Increased amylase digestibility of pressure-treated starch. Agric Biol Chem 53: 2543–2544, 1989.
10. K Heremans. High pressure effects on proteins and other biomolecules. Annu Rev Biophys Bioeng 11:1–21, 1982.
11. E Dickinson. Enzymic crosslinking as a tool for food colloid rheology control and interfacial stabilization. Trends Food Sci Technol 8: 334–339, 1997.
12. C Balny, P Masson. Effects of high pressure on proteins. Food Rev Int 9: 611–628, 1993.
13. E Morild. The theory of pressure effects on enzymes. Adv Prot Chem 34: 93–166, 1981.
14. J Basset, MA Macheboeuf. Biological effects of ultra-high pressures. CR Acad Sci 196: 1431, 1932.
15. LA Curl, EF Jansen. Effect of high pressures on trypsin and chymotrypsin. J Biol Chem 184: 45–54, 1950.
16. LA Curl, EF Jansen. The effect of high pressures on pepsin and chymotrypsinogen. J Biol Chem 185: 713–723, 1950.
17. PS Low, GN Somero. Pressure effects on enzyme structure and function in vitro and under simulated in vivo conditions. Comp Biochem Physiol 52B: 67–74, 1975.
18. PC Michels, D Hei, DS Clark. Pressure effects on enzyme activity and stability at high temperatures. Adv Prot Chem 48: 341–376, 1996.
19. KA Heremans, J Snauwaert, HA Vandersypen, YV Nuland. In: J Osugi, ed. Proceedings of the Fourth International Conference on High Pressure. Kyoto, Japan: Physicochemical Society of Japan, pp 623–626, 1974.
20. M Fukuda, S Kunugi. Pressure dependence of thermolysin catalysis. Eur J Biochem 142: 565–570, 1984.

21. PS Low, GN Somero. Activation volumes in enzymic catalysis: their sources and modification by low molecular weight solutes. Proc Natl Acad Sci USA 72: 3014–3018, 1975b.

22. SE Zale, AM Klibanov. On the role of reversible denaturation (unfolding) in the irreversible thermal inactivation of enzymes. Biotech Bioeng 25: 2221–2230, 1983.

23. P Masson Pressure denaturation of proteins. In: C Balny, R Hayashi, K Heremans, P Masson, eds. High Pressure and Biotechnology. Colloque INSERM/John Libbey Eurotext Ltd 224: 89–99, 1992.

24. K Heremans. Biophysical chemistry at high temperatures. Rev Phys Chem Jpn 50: 256–273, 1980.

25. JC Cheftel. Effects of high hydrostatic pressure on food constituents: an overview. In: C Balny, R Hayashi, K Heremans, P Masson, eds. High Pressure and Biotechnology. Colloque INSERM/John Libbey Eurotext Ltd. 224:195–209, 1992.

26. JT Penniston. High hydrostatic pressure and enzymic activity: inhibition of multimeric enzymes by dissociation. Arch Biochem Biophys 142: 322–332, 1971.

27. S Kunugi. Effect of pressure on activity and specificity of some hydrolytic enzymes. In: C Balny, R Hayashi, K Heremans, P Masson, eds. High Pressure and Biotechnology. Colloque INSERM/John Libbey Eurotext Ltd. 224: 129–137, 1992.

28. WC Ko. Effect of high pressure on gelation of meat paste and inactivation of actomyosin Ca^{2+}-ATPase prepared from milkfish. Fisheries Sci 62:101–104, 1996.

29. WC Ko, M Tanaka, Y Nagashima, T Taguchi, K Amano. Effect of pressure treatment on actomyosin ATPases from flying fish and Sardine muscles. J Food Sci 56(2): 338–340, 1991.

30. S Ishizaki, M Tanaka, R Takai, T Taguchi. Stability of fish myosins and their fragments to high hydrostatic pressure. Fisheries Sci 61: 989–992, 1995a.

31. Y Taniguchi, K Suzuki. Pressure inactivation of α-chymotrypsin. J Phys Chem 87: 5185–5193, 1983.

32. L Heremans, K Heremans. Raman spectroscopic study of the changes in secondary structure of chymotrypsin: effect of pH and pressure on the salt bridge. Biochimica et Biophysica Acta 999: 192–197, 1989.

33. INA Ashie, BK Simpson. Effects of hydrostatic pressure on α_2-macroglobulin and selected proteases. J Food Biochem 18: 377–391, 1995.

34. INA Ashie, BK Simpson. Application of high hydrostatic pressure to control enzyme related fresh seafood texture deterioration. Food Res Int 29: 569–575, 1996.

35. YC Chung, A Gebrehiwot, DF Farkas, MT Morrissey. Gelation of surimi by high hydrostatic pressure. J Food Sci 59: 523–524, 543, 1994.

36. O Deschamps, P Cottin, A Largeteau, G Demazeau, A Ducastaing. Incidence of high pressures on the kinetic parameters of the Ca^{2+}-dependent thiol proteases (calpains) from rabbit skeletal muscle. In: C Balny, R Hayashi, K Heremans, P Masson, eds. High Pressure and Biotechnology. Colloque INSERM/John Libbey Eurotext Ltd 224: 129–137, 1992.

37. A Okamoto, A Suzuki, Y Ikeuchi, MSaito. Effects of high pressure treatment on Ca^{2+} release and Ca^{2+} uptake of sarcoplasmic reticulum. Biosci Biotech Biochem 59: 266–270, 1995.

38. N Homma, Y Ikeuchi, A Suzuki. Effects of high pressure treatment on the proteolytic enzymes in meat. Meat Sci 38: 219–228, 1994.

39. T Ohmori, T Shigehisa, S Taji, R Hayashi. Effect of high pressure on protease activities in meat. Agric Biol Chem 55: 357–361, 1991.

40. LB Kurth. Effect of pressure-heat treatments on cathepsin B1 activity. J Food Sci 51: 663–664, 667, 1986.

41. E Dufour, M Dalgalarrondo, G Herve, R Goutefongea, T Haertle. Proteolysis of type III collagen by collagenase and cathepsin B under high hydrostatic pressure. Meat Sci 42: 261–269, 1996.

42. GG. Kamath, TC Lanier, EA Foegeding, DD Hamann. Nondisulfide covalent crosslinking of myosin heavy chain in setting of Alaska pollock and Atlantic croaker surimi. J Food Biochem 16: 151–172, 1992.

43. H Araki, N Seki. Comparison of reactivity of transglutaminase to various fish actomyosins. Nippon Suisan Gakk 59: 711–716, 1993.

44. Y Tsukamasa, Y Shimizu. Factors affecting the transglutaminase-associated setting phenomenon in fish meat sol. Nippon Suisan Gakk 57: 535–540, 1991.

45. T Shoji, H Saeki, A Wakameda, M Nonaka. Influence of ammonium salt on the formation of pressure-induced gel from walleye pollack surimi. Nippon Suisan Gakk 60: 101–109, 1994.

46. GM Gilleland, TC Lanier, DD Hamann. Covalent bonding in pressure-induced fish protein gels. J Food Sci 62: 713–716, 1997.

47. INA Ashie, TC Lanier. High pressure effects on gelation of surimi and turkey breast muscle enhanced by microbial transglutaminase. J Food Sci 64: 704–708.

48. M Nonaka, R Ito, A Sawa, M Motoki, N Nio. Modification of several proteins by using Ca^{2+}-independent microbial transglutaminase with high-pressure treatment. Food Hydrocolloids 11 (3): 351–353, 1997.

49. DJ Horgan, R Kuypers. Effect of high pressure on the regulation of phosphorylase activity in pre-rigor rabbit muscle. Meat Sci 8: 65–77, 1983.

50. AJ deKoning, TH Mol. Rates of free fatty acid formation from phospholipids and neutral lipids in frozen cape hake (*Merluccius spp.*) mince at various temperatures. J Sci Food Agric 50: 391–398, 1990.

51. AA Fazal, LN Srikar. Effect of accumulated free fatty acids on reduction of salt soluble proteins of pombret and seer fish during frozen storage. J Food Sci Technol 26: 269–271, 1989.

52. T Ohshima, H Ushio, C Koizumi. High pressure processing of fish and fish products. Trends Food Sci Technol 4(11): 370–375, 1993.

53. A Hashimoto, A Kobayashi, K Arai. Thermostability of fish myofibrillar Ca^{2+}-ATPase and adaptation to environmental temperature. Bull Jpn Soc Sci Fish 48: 671–684, 1982.

54. INA Ashie, TC Lanier. Cryo- and baro-protection of proteins during pressure-assisted freezing of Seafoods. Proceedings of the Joint Meeting of Seafood Science

and Technology and Atlantic Fisheries and Technological Conference. Clearwater, Florida, 1996.

55. BK Simpson, NF Haard. Cold-adapted enzymes from fish. In: D Knorr, ed Food Biotechnology. New York: Marcel Dekker, 1987, pp 495–527.

56. M Gross, R Jaenicke. Proteins under pressure: the influence of high hydrostatic pressure on structure, function, and assembly of proteins and protein complexes. Eur J Biochem 221: 617–630, 1994.

57. GN Somero. Temperature as a selective force in protein evolution: the adaptational strategy of compromise. J Exp Zool 194: 175–188, 1975.

58. I A Johnston, NJ Walesby Molecular mechanisms of temperature adaptation in fish myofibrillar adenosine triphosphatases. J Comp Physiol 119: 195–206, 1977.

59. BK Simpson, NF Haard. Trypsin from Greenland cod, *Gadus ogac.* Isolation and comparative properties. Comp Biochem Physiol 79B: 613–622, 1984.

60. HM Chen, C Ford, PJ Reilly. Substitution of asparagine residues in Aspergillus awamori glucoamylase by site-directed mutagenesis to eliminate N-glycosylation and inactivation by deamidation. Biochem J 301: 275–281, 1994.

61. Y Suzuki, N Ito, T Yuuki, H Yamagata, S Udaka. J Biol Chem 264: 18933–18936, 1989. Amino acid residues stabilizing a Bacillus alpha-amylase against irreversible thermoinactivation.

62. MN Gupta. Thermostabilization of proteins. Biotechnol Appl Biochem 14: 1–11, 1991.

63. TJ Ahern, JI Casal, GA Petsko, AM Klibanov. Control of oligomeric enzyme thermostability by protein engineering. Proc Natl Acad Sci USA 84: 675–679, 1987.

64. RR Swezey, GN Somero. Polymerization thermodynamics and structural stabilities of skeletal muscle actins from vertebrates adapted to different temperatures and pressures. Biochemistry 21: 4496–4503, 1982.

65. RR Swezey, GN Somero. Pressure effects on actin self-assembly: interspecific differences in the equilibrium and kinetics of the G to F transformation. Biochemistry 24: 852–860, 1985.

66. Y Shibauchi, H Yamamoto, Y Segara. Conformational change of casein micelles by high pressure treatment. In: C Balny, R Hayashi, K Heremans, P Masson, eds. High Pressure and Biotechnology. Colloque INSERM/John Libbey Eurotext Ltd. 224: 239–242, 1992.

67. J F Siebenaller, AF Hagar, TF Murray. The effects of hydrostatic pressure on A1 adenosine receptor signal transduction in brain membranes of two congeneric marine fishes. J Exp Biol 159: 23–43, 1991.

68. AA Harper, AG Macdonald, CS Wardle, JP Pennec. The pressure tolerance of deep-sea fish axons: results of Challenger cruise 6B/85. Comp Biochem Physiol 88A: 647–653, 1987.

69. GN Somero. Adaptations to high hydrostatic pressure. Annu Rev Physiol 54: 557–577, 1992.

70. AR Cossins, AG Macdonald. The adaptation of biological membranes to temperature and pressure: fish from the deep and cold. J Bioenerg Biomembr 21: 115–135, 1989.

71. A Gibbs, GN Somero. Pressure adaptation of teleost gill Na^+/K^+-adenosine triphos-

phatase: role of the lipid and protein moieties. J Comp Physiol 160B: 431–439, 1990.

72. A Gibbs, GN Somero. Pressure adaptation of Na+/K+-ATPase in gills of marine teleost fishes. J Exp Biol 143: 475–492, 1989.

73. JP Hennessey Jr, JF Siebenaller. Pressure inactivation of tetrameric lactate dehydrogenase of confamilial deep-living fishes. J Comp Physiol 155B: 647–652, 1985.

74. PW Hochachka. Temperature and pressure adaptation of the binding site of acetylcholinesterase. Biochem J 143: 535–539, 1974.

75. PS Low, JL Bada, GN Somero. Temperature adaptation of enzymes: roles of the energy, the enthalpy, and the entropy of activation. Proc Natl Acad Sci USA. 70: 430–432, 1973.

76. R Jaenicke. Enzymes under extremes of physical conditions. Annu Rev Biophys Bioeng 10: 1–67, 1981.

77. SA Hawley. Reversible pressure-temperature denaturation of chymotrypsinogen. Biochemistry 10: 2436–2442, 1971.

78. RD Haight, RY Morita. Interaction between the parameters of hydrostatic pressure and temperature on aspartase of *Escherichia coli*. J Bacteriol 83: 112–120, 1962.

79. RY Morita, RD Haight. Malic dehydrogenase activity at 101°C under hydrostatic pressure. J Bacteriol 83: 1341–1346, 1974.

80. RP Rohrbach, MJ Mallarik. Increasing the stability of amyloglucosidase. United States Patent #4,415,656, 1983.

81. KW Nickerson. A hypothesis on the role of pressure in the origin of life. J Theor Biol 110: 487–499, 1984.

82. J Konisky, PC Michels, DS Clark. Pressure stabilization is not a general property of thermophilic enzymes: the adenylate kinases of *M. voltae, M. maripaludis, M. thermolithotrophicus,* and *M. jannaschii.* Appl Environ Microbiol 61: 2762–2764, 1995.

83. RY Morita, PF Mathemeier. Temperature-hydrostatic pressure studies on partially purified inorganic pyrophosphatase activity. J Bacteriol 88: 1667–1671, 1964.

84. RC Neuman Jr, W Kauzmann, A Zipp. Pressure dependence of a weak acid ionization in aqueous buffers. J Phys Chem 77: 2687–2691, 1973.

85. A Zipp, W Kauzmann. Pressure denaturation of metmyoglobin. Biochemistry 12: 4217–4228, 1973.

86. PW Hochachka, HW Behrisch, F Marcus. Pressure effects on catalysis and control of catalysis by liver fructose diphosphatase from an off-shore benthic fish. Am Zool 11: 437–449, 1971.

87. J Rasper, W Kauzmann. Volume changes in protein reactions. I. Ionization reactions of proteins. J Am Chem Soc 84: 1771–1777, 1962.

88. GN Pace, T McGrath. Substrate stabilization of lysozyme to thermal and guanidine hydrochloride denaturation. J Biol Chem 255: 3862–3865, 1980.

89. T Mustafa, TW Moon, PW Hochachka. Effects of pressure and temperature on the catalytic and regulatory properties of muscle pyruvate kinase from an off-shore benthic fish. Am Zool 11: 451–466, 1971.

90. S Ishizaki, T Hayashi, M Tanaka, T Taguchi. High pressure effects on lecithin-added yellowfin tuna actomyosin ATPase activities. Fisheries Sci 61: 703–705, 1995.

91. BC Schade, R Rudolph, HD Ludemann, R Jaenicke. Reversible high pressure dissociation of lactic dehydrogenase from pig muscle. Biochemistry 19: 1121–1130, 1980.

92. G Dreyfus, H Guimaraes-Motta, JL Silva. Effect of hydrostatic pressure on the mitochondrial ATP synthase. Biochemistry 27: 6704–6710, 1988.

93. PC Michels, DS Clark. Pressure dependence of enzyme catalysis. In: M W W Adams, R M Kelly, eds. Biocatalysis at Extreme Temperatures: Enzyme Systems Near and Above 100°C. ACS Symposium Series. American Chemical Society, Washington DC 498: 108–121, 1992.

94. K Ruan, G Weber. Dissociation of yeast hexokinase by hydrostatic pressure. Biochemistry 27: 3295–3301, 1988.

95. JHS Blaxter CS Wardle, BL Roberts. Aspects of the circulatory physiology and muscle systems of deepsea fish. J Marine Biol Assoc UK 51: 991–1006, 1971.

96. GS Greaney, GN Somero. Effects of anions on the activation thermodynamics and fluorescence emission spectrum of alkaline phosphatase: evidence for enzyme hydration changes during catalysis. Biochemistry 18: 5322–5332, 1979.

97. GN Somero, M Neubauer, PS Low. Neutral salt effects on the velocity and activation volume of the lactate dehydrogenase reaction: evidence for enzyme hydration changes during catalysis. Arch Biochem Biophys 181: 438–446, 1977.

98. GA MacDonald, TC Lanier, HE Swaisgood, DD Hamann. Mechanism for stabilization of fish actomyosin by sodium lactate. J Agric Food Chem 44: 106–112, 1996.

99. JW Park, TC Lanier, DP Green. Cryoprotective effects of sugar, polyols, and/or phosphates on Alaska pollack surimi. J Food Sci 53: 1–3, 1988.

100. EM Dumay, MT Kalichevsky, JC Cheftel. High pressure unfolding and aggregation of β-lactoglobulin and the baroprotective effects of sucrose. J Agric Food Chem 42: 1861–1868, 1994.

101. V Athes, D Combes. Influence of additives on high pressure stability of β-galactosidase from *Kluyveromyces lactis* and invertase from *Saccharomyces cerevisiae*. Enzyme Microb Technol 22: 532–537, 1998.

102. S Fujii, K Obuchi, H Iwahashi, T Fujii, Y Komatsu. Saccharides that protect yeast against hydrostatic pressure stress correlated to the mean number of equatorial OH groups. Biosci Biotechnol Biochem 60: 476–478, 1996.

103. T Arakawa, SJ Prestrelski, WC Kenney, JF Carpenter. Factors affecting short-term and long-term stabilities of proteins. Adv Drug Deliv Rev 10: 1–28, 1993.

104. PL Privalov. Cold denaturation of proteins. Crit Rev Biochem Mol Biol 25: 281–305, 1990.

105. M Okamoto, Y Kawamura, R Hayashi. Application of high pressure to food processing: textural comparison of pressure- and heat-induced gels of food proteins. Agric Biol Chem 54: 183–189, 1990.

106. NF Haard. Specialty enzymes from marine organisms. Food Technol 52:64–67, 1998.

21

Applications of Fish and Shellfish Enzymes in Food and Feed Products

Manuel Díaz-López
University of Almería, Almería, Spain

Fernando Luis García-Carreño
CIBNOR, La Paz, BCS, Mexico

I. INTRODUCTION

Because enzymes are molecules with biotransformation capabilities, they are important tools of biotechnology. Either working alone or as part of a cell, "Use of enzymes to accomplish specific desirable changes in foods has been practiced for centuries. The techniques have been handed down from generation to generation without any knowledge, until recently" (1). The importance of enzymes as tools in biotechnology and food or feed processing is increasing (2, 3). The demand for enzymes with specific properties is high, and various enzyme sources are being investigated. Enzymes are important components of food and feed for many reasons: growth, maturation, production, processing, storage, and spoilage; consumer preferences and selection; safety and control of predators; food intake, digestion, and assimilation and disease; and as analytical tools (4). The marine environment contains the largest pool of diversified genetic material and, hence, represents an enormous potential source of enzymes. The application of seafood enzymes to food and feed products has been developed during the last 30 years. Early work revealed the presence in the marine environment of several enzymes with unique properties (5, 6). This chapter will review some of the uses of such enzymes.

II. CHOICE OF ENZYME

Food and feed processing involves many different operational variables, such as temperature, pressure, flow rate, density, pH, viscosity, chemical composition, enzyme inhibitors, activators, or inactivators. The choice of an enzyme that works efficiently in such conditions is of paramount importance. Moreover, most enzyme-driven processes have a limited endpoint. The ability to restrain the enzyme activity after the work has been done, without affecting the product quality, should also be considered.

Enzyme stability and reaction rate under diverse process variables are required to achieve catalyst compatibility with industrial processes. An enzyme should be thoroughly evaluated in the laboratory, under the intended operating conditions, prior to scaling up to an industrial process. Purity of the intended enzyme is important because of the price/activity ratio.

Of particular interest for the food and feed industry is the use of enzymes that are active at low temperature. Most aquatic organisms are poikilotherms: their body temperature is similar to that of their environment. Poikilotherms living in polar, deep-sea, or any constantly low temperature environments, produce enzymes adapted in a variety of ways. These include increased enzyme concentration, change in the type of enzyme present, and adaptability of a homologous enzyme. Cold-adapted enzymes may manifest alterations in isoenzyme distribution, substrate binding, substrate turnover rate, thermal stability, physiological efficiency, and thermodynamic properties. Digestive proteolytic enzymes from cold-adapted aquatic organisms possess unique properties compared to mammalian proteases. Two key properties of cold-adapted enzymes useful in food and feed processing are high molecular activity at low reaction temperature and thermal instability (fast denaturation at moderate temperature). As industrial enzymes, these properties are advantageous because the process can be run at low temperature, thus minimizing bacterial activity and other interfering reactions, and the enzyme may be inactivated by moderate heat treatment after reaction.

Table 1 Tips for Developing an In-House Assay for Comparing Enzymes for Strength and Price

---

Identify the conditions for the proposed application: pH, temperature, moisture level, etc.
Consult the methods cited by the supplier
Select a substrate that is the same or similar to the material to be modified in your
    application
Integrate the application conditions and the proposed substrate with the published
    procedure
A functional property assay may prove to be more useful
Express the observed activity as specific activity to compare enzymes

---

This is very important in certain food process operations in which residual protease activity is undesirable (7–10).

Comparing enzymes from different suppliers for specific activity is a commonsense approach. Developing an in-house assay to compare enzymes in the lab is an easy task yielding economic and processing benefits. Tips for developing an in-house assay were developed by Boyce in 1986 (11) and are summarized in Table 1.

III. CURRENT APPLICATIONS OF FISH AND SHELLFISH ENZYMES IN SEAFOOD PRODUCTION

The traditional applications of enzymes in the seafood industry have been limited to very few products (fish protein hydrolysate, fish sauce, or cured herring). These processes are based on endogenous proteases in the fish (5). Recently, additional applications of fish or shellfish enzymes in the seafood industry have emerged. These include the improvement of the traditional applications by using exogenous enzymes to accelerate the process, production of other products (i.e., polyunsaturated fatty acids [PUFA]-enriched fish oils) (12), and the use of fish and shellfish enzymes to improve the production processes or as alternative processes in seafood production (selective removal of skin, fish-scale removal, or riddling process in cured roe production)(13–17).

A. Specialty Products

A product whose manufacture is made possible, directly or indirectly, by the use of enzymes is called a "specialty product." These products are differentiated from "conventional products," such as fresh, frozen, or canned fish or shellfish flesh, in which use of enzymes is not necessary for production, even though it may be advantageous (18). The main specialty products and the role of both the endogenous and exogenous enzymes in seafood processing are described below.

1. Mince Products

The term surimi is derived from the Japanese verb *suru,* which means "to mince" and it refers to minced and water-washed fish muscle tissues. The term is commonly identified with crab leg analogues, a surimi-based product, in Europe and in North America. Surimi is primarily used in the manufacture of various types of Japanese heat-gelled products such as *itatsuki, kamaboko, chikuwa, hanpen,* and *satsuma-age.* Essentially, surimi is a bland, preferably white, myofibrillar protein concentrate used for manufacturing different gelled or emulsion-type fishery products or as an ingredient in processing other foods. Surimi is regarded as a valuable, functional proteinaceous ingredient, similar to the use of soybean proteins concentrates, in a variety of food products (19, 20).

Alaska pollock (*Theragra chalcogramma*) has been the most frequently used raw material for surimi processing (21). However, other fish species, such as Pacific whiting (*Merluccius productus*) (22), arrowtooth flounder (*Atheresthes stomias*)(23), Atlantic menhaden (*Brevoorti tyrannus*)(24), and sardine (*Sardina pilchardus*)(25) are also being used.

Surimi is minced fish flesh that has been refined by leaching the soluble fraction with repeated washing with water, straining to remove any remaining connective tissue elements and certain factors that accelerate protein denaturation, and the separating of lipids and other undesirable components. The product is then dewatered, stabilized, and mixed with cryoprotectants, such as sugar, sorbitol, or polyphosphates, and frozen. Surimi forms thermoirreversible gels upon heating, constituted by a three-dimensional protein network formed mainly of actomyosin (complex of actin, tropomyosin, troponin, and myosin filaments) (19, 21).

The characteristic rheological properties of surimi are largely based on the gel-forming ability of the fish myofibrillar proteins, myosin being the most important. The textural characteristics developed during gelation are expressed as gel strength, which is the primary determinant for surimi quality and price (26). The quality of surimi products is highly affected by the species-determined properties of proteins and by the activity of the endogenous heat-stable proteinases, which have a gel-softening effect. These proteases are very important as a manufacturing variable (24, 27–28; see also Chap. 19).

An important step in the formation of a high-quality gel from surimi, as occurs in the production of *kamaboko,* is the formation of ε-γ–glutamyl-lysine[GL] crosslinks in the fish proteins (Fig. 1) (29). GL crosslinks of proteins contribute to gel strength enhancement in surimi (30; see also Chap. 6). The endogenous enzyme responsible for the GL cross-link formation in fish flesh is transglutaminase. Transglutaminase (protein-glutamine-γ–glutamyltransferase, EC 2.3.2.13) is an enzyme capable of catalyzing acyl-transfer reactions in which the γ-carboxyamide group of peptide-bound glutamine serves as the acyl donor. This reaction introduce covalent crosslinks between proteins as well as peptides and various primary amines (31). When the ε-amino groups of lysine residues in proteins act as acyl acceptors, ε-(γ-Glu)-Lys bonds are formed between proteins, both intra- and intermolecularly. Without primary amines in the reaction system, water becomes the acyl acceptor and the γ-carboxyamide groups of glutamine residues are deaminated, becoming glutamic acid residues (Fig. 1)(32).

Transglutaminase-catalyzed reactions can be used to modify the functional properties of food proteins, including surimi production. Transglutaminase has been used to catalyze the crosslinking of a number of proteins, such as whey proteins, meat proteins (myosin and actomyosin), soybean proteins, and gluten. The

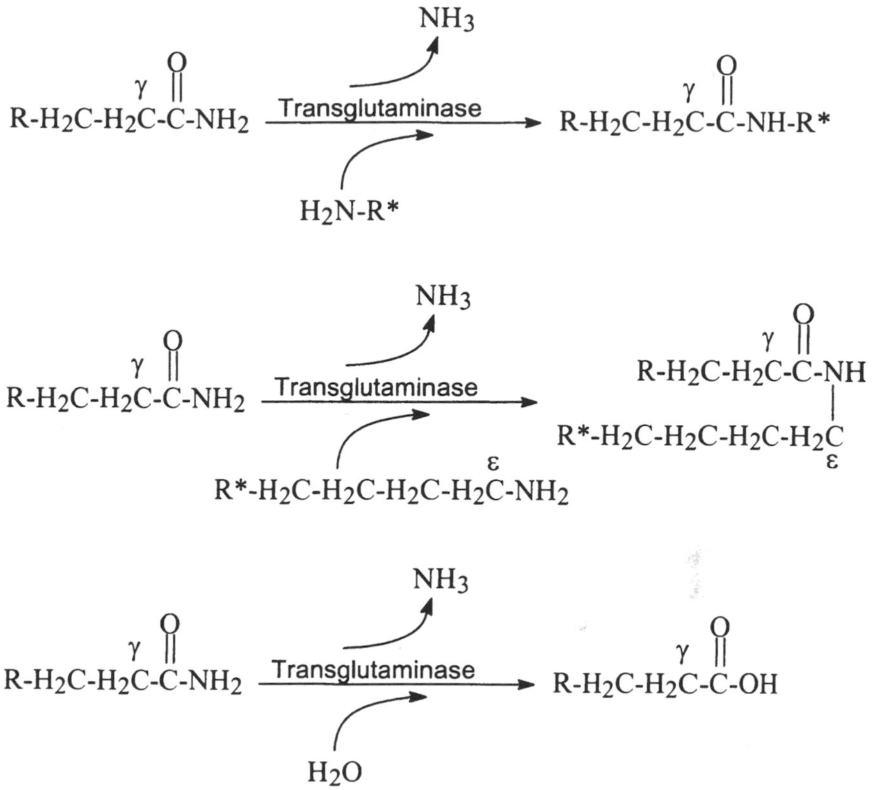

$$R\text{-}H_2C\text{-}H_2C\text{-}\overset{\gamma}{\overset{\displaystyle O}{\overset{\|}{C}}}\text{-}NH_2 \xrightarrow[\text{H}_2\text{N-R*}]{\text{Transglutaminase}} \overset{\text{NH}_3}{\nearrow} R\text{-}H_2C\text{-}H_2C\text{-}\overset{\gamma}{\overset{\displaystyle O}{\overset{\|}{C}}}\text{-}NH\text{-}R*$$

$$R\text{-}H_2C\text{-}H_2C\text{-}\overset{\gamma}{\overset{\displaystyle O}{\overset{\|}{C}}}\text{-}NH_2 \xrightarrow[R*\text{-}H_2C\text{-}H_2C\text{-}H_2C\text{-}H_2C\text{-}NH_2]{\text{Transglutaminase}} \overset{\text{NH}_3}{\nearrow}$$

$$R\text{-}H_2C\text{-}H_2C\text{-}\overset{\gamma}{\overset{\displaystyle O}{\overset{\|}{C}}}\text{-}NH$$
$$R*\text{-}H_2C\text{-}H_2C\text{-}H_2C\text{-}H_2C$$

$$R\text{-}H_2C\text{-}H_2C\text{-}\overset{\gamma}{\overset{\displaystyle O}{\overset{\|}{C}}}\text{-}NH_2 \xrightarrow[\text{H}_2\text{O}]{\text{Transglutaminase}} \overset{\text{NH}_3}{\nearrow} R\text{-}H_2C\text{-}H_2C\text{-}\overset{\gamma}{\overset{\displaystyle O}{\overset{\|}{C}}}\text{-}OH$$

Figure 1 Transglutaminase-catalyzed reactions. When the ε-amino group of peptide-bound lysine act as acyl acceptor, ε-(γ-glutamil)-lysine bonds are formed between the proteins, resulting in their crosslinking (middle reaction). $R\text{-}H_2C^{\gamma}\text{-}H_2C\ CONH_2$ = γ-carboxamide group of peptide-bound glutamine = acyl donor; $R*\text{-}NH_2$ = primary amine group = acyl acceptor; $R*\text{-}H_2C\text{-}H_2C\text{-}H_2C\text{-}H_2C^{\epsilon}NH_2$ = ε-amino group of peptide-bound lysine = acyl acceptor. (from Ref. 53.)

modification of food proteins by transglutaminase may lead to textured products (hamburger, meatballs, canned meat, frozen meat, molded meat, surimi-fish paste, krill paste, baked foods, protein powders from plants, etc.), help to protect lysine in food proteins from various chemical reactions, encapsulate lipids and lipid-soluble materials, avoid heat treatment for gelation, improve elasticity and water-holding capacity, modify solubility and functional properties, and produce food proteins of higher nutritive value by crosslinking of different proteins containing complementary limiting essential amino acids. Transglutaminase use in these products has usually been from exogenous origin (33–35).

For surimi production, the role of transglutaminase and some other endogenous enzymes has been reviewed (36). The activity of endogenous fish transglutaminase decreases rapidly after catch and is almost completely destroyed by freezing. Therefore, manufacturing at sea has hitherto been required for high-quality surimi, but the cost of producing surimi at sea is considerably higher than onshore. The solution to this problem could be the addition of exogenous transglutaminase to facilitate glutamyl-lysine crosslinking (36–37).

Transglutaminase has been found in animal and plant tissues or organs (31). In the 1960s, the purification, characterization, and application of transglutaminase started with material of animal origin. Guinea pig liver has been the sole source of commercial transglutaminase for decades (38, 39). Bovine transglutaminase-enriched plasma fraction has positively been used to enhance gel strength of Pacific whiting surimi (40). The scarcity of the source and the difficulties of isolation and purification procedures for obtaining tissue transglutaminase from animals have resulted in an extremely high price of the enzyme: about $80 for one unit. It is thus not possible to use such tissue transglutaminase in food processing on an industrial scale. Fish tissue has been the source of transglutaminase in the laboratory, but not on a commercial scale. Transglutaminase from Alaska pollock was isolated and found to induce gelation in minced fish, under the same conditions as mammalian transglutaminase (Ca^{2+} dependent) (41). Recently, red sea bream (*Pagrus major*) transglutaminase was purified but the process is longer than other purification protocols from different sources and is more expensive (42). Transglutaminase from plant tissue has also been separated and purified, but is not commercialized (43). Both sources of transglutaminase are still in their infancy. Currently, extracellular transglutaminase from microbial sources has been purified from the culture filtrate of *Streptomyces* sp. (44) and *Streptoverticillium* sp. (45, 46). Microbial fermentation makes it possible to achieve mass production of transglutaminase from cheap substrates and therefore its subsequent commercial use.

The production of transglutaminase from microorganisms makes it possible to use this enzyme in a variety of food processes. Two patents report methods for manufacturing a traditional Japanese fish paste, *kamaboko,* using transglutaminase. Fish paste products are manufactured from material containing fish meat as the main ingredient and 0.1–700 U transglutaminase/g fish meat protein. A mixture of 100 parts dehydrated walleye pollock, with 3 parts NaCl, 5 parts potato starch, 10 parts water, 0.5 part monosodium glutamate, and 0.01 part transglutaminase was packed in a film, heated at 60°C for 30 min and at 90°C for 20 min, and cooled to manufacture a surimi with acceptable texture and whiteness (47, 48). The onshore manufacture of surimi can be more technically challenging than processing at sea. It is difficult owing to the inactivation of transglutaminase after harvesting. In the laboratory experiments, an extracellular microbial transglutaminase, produced by *Streptoverticillium mobaraense*, has been added to a surimi preparation from Alaska pollock stored on ice for 2 days after catch, to obtain *kamaboko* gels. The results showed that gel strength was considerably improved through the formation of GL crosslinks by the addition of microbial transglutaminase in suwari gels (49).

In meat processing, methods for producing minced-meat products containing transglutaminase have been developed (50, 51). Minced meat and other food ingredients are mixed with transglutaminase, shaped, packed in pressure-resistant containers, and retorted to manufacture meat products such as hamburgers, meatballs, stuffed dumplings, and shao-mai (a typical Chinese food). These foods show improved elasticity, texture, taste, and flavor. Minced beef and pork, flour, onion, skim-milk powder, and condiments were mixed with water and microbial transglutaminase, packed with sauce in bags, and retorted to make raw hamburgers. Similar methods for meat and meat products treated with transglutaminase can be found in the literature (52, 53).

2. PUFA-Enriched Fish Oils

The health aspects of marine oils (lipids from fish or from marine mammals) were discovered by the Danish physicians Bang and Dyerberg in the 1970s by clinical examinations of Greenland seal hunters (54). Recently, medical and nutritional aspects of marine oils have received widespread attention, because they have been shown to have a preventive action against coronary heart diseases and to be necessary for normal brain and nervous tissue development in animal and humans (mainly young children) (55). Fish oils contain triacylglycerides and variable amounts of phospholipids, glycerol ethers, and wax esters. They contain a wide range of highly unsaturated long-chain fatty acids, with the number of carbon atoms ranging from 14 to 22 (56).

Polyunsaturated fatty acids (PUFAs) are essential components in human nutrition. For beneficial effect, the main PUFAs are the n-3 and n-6 families (also called ω3 and ω6). It is now recognized that the n-3 and n-6 fatty acids have distinct and sometimes opposing roles in human metabolism (57). Because n-3 and n-6 fatty acids are not interconverted in humans, they must be present in the diet. An appropriate ratio of n-3/n-6 PUFAs is 1:4. Nutritionists believe that the diet of many developed countries contains adequate quantities of the n-6 family, but a deficiency of the n-3 fatty acids. The dietary requirement for humans is about 1 g per day (58), but the Food and Agriculture Organization's (FAO's) Food Balance Sheet (1984–1986) indicates an average global consumption of only two-thirds of the minimum requirements. This had led to recommendations from expert committees that the intake of n-3 fatty acids be increased (59), particularly the long-chain fatty acids docosahexaenoic acid (C22:6, cis-4,7,10,13,16,19-docosahexaenoic acid [DHA]) and eicosapentaenoic acid (C20:5, cis-5,8,11,14,17-eicosapentaenoic acid, [EPA]). These desirable fatty acids are bound into triacylglycerides, which is the natural form of fatty acids in fish oil (12). Both DHA and EPA are immediate precursors of biologically active molecules such as prostaglandins and thromboxanes, which participate in controlling a wide range of biological functions. Although DHA and EPA can be synthesized in the body by elongation and desaturation of α-linolenic acid, ingestion of the preformed molecules is usually more effective, especially for the very young or the elderly (60).

Further medical research has led to the conclusion that the beneficial effects of increasing dietary EPA and DHA can be classified into two main areas. First, these agents sustain normal healthy life through the reduction of blood pressure and plasma triacylglycerides and cholesterol, and increased blood coagulation time, because platelet aggregation decreases (61). Second, they alleviate certain diseases: blood vessel disorders and inflammatory diseases, and control of an overactive immune function resulting in alleviation of autoimmune disease, such as arthritis and some types of dermatitis (62). DHA is an important structural component in the membranes of the brain, nervous tissue, and eye, and of particular importance in the development of the fetus and the young child (63).

Dietary supplementation of long-chain polyunsaturated n-3 fatty acids is hampered because these acids are present in low concentrations in the commonly available edible oils. Microbial oils that contain high levels of either EPA and DHA can be produced by fermentation, but at a relatively high cost. Most fish oils contain only moderate levels of DHA and EPA. Cod liver oil is a well-known source of n-3 PUFAs used in the pharmaceutical industry (64). Cod liver oil is a complex mixture of more than 50 different fatty acids forming triacylglycerols, of which there is usually 8–9% each of EPA and DHA, and 22–24% of all n-3 PUFAs (12). However, fish and fish dishes currently provide an average intake of 0.2 g n-3 PUFAs per day, mainly as EPA and DHA (according to the FAO's above-cited report). To reach the proposed health requirement of 1 g per day in human nutrition requires a much greater consumption of fish or the use of fish oil in foods (apart from the possibility of using pharmaceutical concentrations). It is possible to increase the PUFA content in fish oil, or marine oils in general and use the oils in foods. There is a great commercial interest in preparing PUFA-enriched fish oil (56).

Preparation of oils in which such polyunsaturated fatty acids are enriched to high concentrations is done by various chemical methods, including using enzymic ones. It is possible to prepare triacylglycerides containing up to 30% EPA and DHA directly from fish oils without splitting the fat. Several methods are available, such as winterization, molecular distillation, and solvent crystallization. Moreover, it is possible to increase the oil concentration of n-3 PUFAs by first generating free fatty acids or esters totaling above 30%, which can then be fractionated by a variety of chemical methods, including supercritical fluid extraction, complexing with urea, chromatography, or high-performance liquid chromatography (HPLC), obtaining 65–85% EPA + DHA levels according to the method (12). The PUFA concentrate can then be resynthesized into triacylglycerols using lipases from microorganisms such as yeast, fungi, and bacteria, under conditions that favor the reesterification process for triacylglycerol synthesis (12, 65–67).

The n-3 PUFAs are highly labile and may be destroyed by oxidation or *cis-trans* isomerization during processes that involve extreme pHs and high temperatures. Traditional chemical modification processes for fats and oils generally involve quite drastic conditions. Mild enzymatic modification processes are therefore necessary when heat-sensitive and labile fatty acids are involved. Lipase (triacylglycerol acylhydrolase, EC 3.1.1.3) is an enzyme that catalyzes the hydrolysis of a specific type of ester bond: triacylglycerides. They can be non-region-specific or n-1,3-specific towards triacylglycerides and they can possess specificity towards particular types of fatty acids (12). Moreover, lipases catalyze the reverse reaction: ester synthesis (from free fatty acids and alcohols) and interesterifications. There are several variations of the interesterification reaction: transesterification, in which ester–ester interchanges occur (Fig. 2), such as between triacylglycerides, triacylglycerides and monoester, monoesters, etc.; acidolysis, in which fatty acid exchange reactions occur, such as between triacylglycerides and free fatty acids; and alcoholysis, in which triacylglyceride is

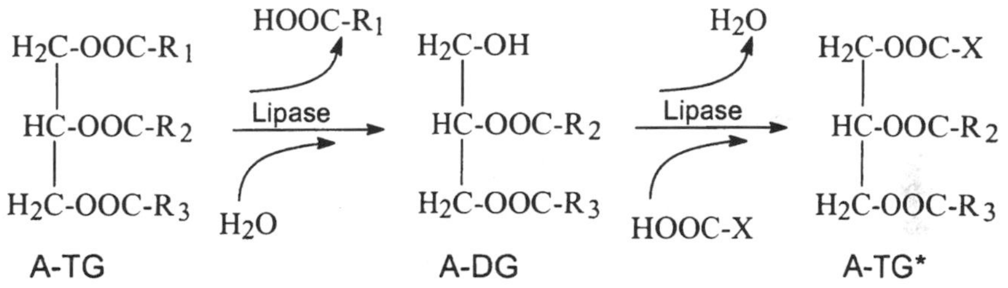

Figure 2 Schematic representation of a lipase-catalyzed transesterification to yield an acyltriglyceride enriched in EPA and/or DHA. This interesterification reaction involves sequential execution of the hydrolysis and reesterification reactions. The same enzyme catalyzes the deacylation/acylation step with different affinity by acyl positions and/or free fatty acids (PUFAs). A-TG = acyltriglyceride with native and arbitrary fatty acid residues (R_1, R_2, and R_3); A-DG = acyldiglyceride (also could be monoglyceride or glycerol, depending on the lipase specificity); HOOC-R_1 = product fatty acid; HOOC-X = new fatty acid to be incorporated. It is from a mixed of fatty acids in the reaction media, with a high proportion of EPA and/or DHA; A-TG* = acyltriglyceride with a incorporated fatty acid residue (OOC-X), which will be EPA or DHA depending on its concentration in the reaction media and the lipase specificity in the reesterification reaction. (from Ref. 224.)

allowed to react with a simple alcohol or glycerol (glycerolysis) leading to a mixture of mono- and diglycerides (12, 68).

Haraldsson surveyed the application of lipases to modify marine oils for preparing PUFA-enriched fish oil and predicted that this discovery might prove important for the marine oil industry (12), but there are only a few reports in the literature on such biotechnological applications (69). Apparently n-3 PUFAs, especially docosahexaenoic acid, are not good substrates for most commercially available lipases. This is simply a matter of fatty acid specificity and might change when lipases from fish or microorganisms, with a high affinity for long-chain polyunsaturated fatty acids, are isolated and made commercially available (68).

The discrimination of the *Candida rugosa* lipase to triacylglycerides containing different fatty acids has been reported. The C14–C18 saturated and monounsaturated fatty acids were preferentially hydrolyzed in triglycerides from capelin oil, whereas the long-chain monoenes, 20:1 and 22:1, and particularly, the PUFAs 18:4, 20:5, and 22:6, were resistant to the hydrolysis (70). Lipolytic enzymes, isolated from the Atlantic cod (*Gadus morhua*), have been shown to have preference for hydrolysis of the PUFAs over the shorter-chain acids. This specificity is completely opposite to that of the lipases commercially available at that time (mainly porcine pancreas lipase) (71–73). The specificity of purified cod lipase is in good agreement with the crude pancreas lipase of leopard shark (74) and digestive lipases of four marine fish species (75). Comparative studies of cod and human milk lipase showed that the cod enzyme has a higher reactivity for hydrolysis of very long chain polyunsaturated fatty acyl esters (76). A bile-salt-activated lipase (carboxyl ester lipase) has recently been purified and characterized from the extract of the delipidated powder of red sea bream (*Pagrus major*) hepatopancreas. The enzyme efficiently hydrolyzed ethyl esters of polyunsaturated fatty acids such as arachidonic acid and eicosapentaenoic acid, which were resistant to porcine pancreatic lipase (77). Thus, lipases from marine fish, such as red sea bream and cod, efficiently hydrolyze esters of highly unsaturated fatty acids, such as C20:4n-6 and C20:5n-3. This suggests that the fatty acid specificity of fish lipases might be a crucial factor when considering the use of lipolytic enzymes to obtain PUFA-enriched marine oils.

Different methods have been described for preparing concentrates of n-3 PUFAs, mainly EPA and DHA. However, all process described use lipases from microorganisms (*Pseudomonas* spp., *Chromobacterium viscosum, Candida rugosa* [*C. cylindracea*], *Candida antarctica, Aspergillus rhizopus, Rhizomucor miehei, Rhizopus niveus,* etc.), but not from fish, in spite of the more appropriate specificity of fish-derived enzymes. This could be because microbial lipases are less expensive and the n-3 PUFA specificity of commercially available lipases is adequate for production. These processes produce marine oils with high EPA and DHA concentration, reaching 85% of the two fatty acids (78–80). Porcine pancreatic phospholipase A_2 has also been used for extracting n-3 PUFA-enriched oil from cod roe and gave 24% EPA and 40% DHA in the free fatty acid fraction. In laboratory experiments, the recovery was 60%, corresponding to a yield of 6 g polyunsaturated fatty acids per kilogram cod roe (81). Although, with these results, one expects that there are industrial processes to obtain n-3 PUFAs via lipase, to the authors' knowledge there are no patents for these processes.

3. Caviar and Other Roe Production

One of the major delicatessen-type salted-fish products is caviar. The word "caviar" refers only to the riddled and cured roe of the sturgeon (*Acipenser* spp. and *Huso huso*), although in recent years it has been used to describe similarly treated roe of other, less expensive, fish species. In this way, black caviar is prepared from the eggs of sturgeon and red caviar from Salmonidae. Less expensive caviar is made *from the eggs of several fish: cod, catfish (Parasilurus astus), herring (Clupea harengus)*, capelin (*Mallotus villosus*), lumpfish (*Protopterus aethiopicus*), and some freshwater species (20). Caviar substitute is produced by coloring lumpsucker roe (*Cyclopterus lumpus*) to mimic sturgeon caviar. The preparation of caviar starts with the riddling process, which consists of separating the roe from the roe sack (ovaries). These should be taken from the fish at the moment of slaughter. The eggs are immediately washed with cold water; salted with dry, fine salt, 3–5% of the egg weight, or brined for 8–18 min in saturated brine at egg/brine ratio of 1:3; drained; and packed in cans or jars. Some assortments of caviar are pasteurized in sealed cans at 60–70°C for 2.5–3 h. The shelf-life of pasteurized caviar at 10–18°C is 3–4 months and at -2°C it is 6–8 months (82).

The riddling process is a somewhat laborious task, which is either carried out manually or mechanically. One problem in the conventional process is that it is difficult to release the roe particles from the supportive connective tissue of the roe sac without destroying a large amount of the roe. Sometimes the yield of intact roe is as low as 50% (83). This problem has been solved using enzymes. A US patent describes a method by which proteinases (acid, neutral, or alkaline conditions) are used to release salmon roe from the connective tissue (84).

To ease the riddling process in caviar production, fish enzymes are used to achieve a gentle separation of roe from the connective tissue. Enzymatic roe rinsing is a very delicate process demanding strict control of all the variables to avoid damage of the roe cell wall. Whole or split-roe sacs are immersed in a water bath containing the egg-releasing enzyme. Several enzymes have been used. Satisfactory results have been obtained with cold-active Atlantic cod pepsin (83, 85–88) and New Zealand orange roughy acid proteinase (*Hopolostethus atlanticus*) (89). That these enzymes have high activity at low temperatures gives them unique application possibilities (e.g., it is important to run the enzymatic process at a low temperature to keep bacterial contamination at a low level). Moreover, fish pepsins apparently split the linkages between the egg cells and the roe sack without damaging the eggs (15). After the enzyme treatment, the roe may be separated from the connective tissue by sedimentation in a flotation tank. The connective tissues are removed by flotation. By accurate adjustment of salt concentration and flotation conditions, it is also possible to remove damaged roe together with the connective tissue. The caviar yield from rainbow trout (*Oncorhynchus mykiss*) roe is about 90% compared to 70% with conventional mechanical methods (88, 90).

There is a small commercial production of cold-adapted pepsins from fish viscera (Atlantic cod) by a Norwegian company (Marine Biochemicals, Tromsø). This type of production is also beginning in Iceland at the Icelandic Fisheries Laboratories (IFL) (13, 87). Commercial production is also being started in New Zealand (89). Industrial production of caviar with these enzymes has been started in Canada, the United States, Australia, and several Scandinavian countries. Process equipment for enzymatic rinsing of roe in caviar production has been developed by Biotec-Mackzymal and by Trio Industries in Norway. So far, mainly salmon and trout caviar are produced by enzymatic methods, and the annual production is about 50 tons (14). The enzymatic production of shrimp (*Pandalus borealis*) roe caviar has also been investigated. Tests are being made at IFL (13).

Enzymatic methods, based on cold-adapted fish enzymes, can be used for other fish roe processing operations: to remove the mucoprotein layer on newly spawned eggs from walleye and catfish to improve hatching yield and survival of larvae, and the enzymatic removal of the chorion from fish eggs to facilitate observation of embryogenesis and microinjection of naked genes in research experiments (14).

4. Preparation of Cured Fish Products

Preparations of proteolytic enzymes may be used as processing aids in preparing ripened, salted fish. The annual world production of cured fish products amounts to several hundred thousand tons. Most of these are cured herring (*Clupea harengus*) (e.g., Dutch "matjes" made from immature, fatty, feeding herring) and fermented squid (*Illex illecebrosus*) (13, 91, 92). The cured products are made by a maturing or ripening process of raw materials. The sensory characteristics of cured fish result from the salting process and enzymatic changes in protein, lipids, and carbohydrates, from different interactions of the products of these enzymatic reactions, and from the added spices (20, 93).

In traditional cured fish products, a partial proteolysis occurs, catalyzed by endogenous fish enzymes during fish processing. This is caused by the activity of muscle proteases, which play an important role in muscle protein metabolism and in tissue degradation during postmortem tenderization (82, 94–96). The activity of these muscle proteases, however, is small compared to the proteolytic activity of the gastrointestinal tract (97), especially in herring and cod (98). These fish proteases play a decisive role in the maturing or ripening process of fish meat. The ripening of these products appears to depend on endogenous digestive enzymes producing the characteristic texture and flavor (15). The possibility that components from the feed also contribute to the characteristic flavor, either directly or by participating in some reactions with the muscle components, should not be ruled out (20).

The production of herring products probably originates from Scotland in the 8th Century, and is practiced in all countries that harvest fish in the North Atlantic and the North Sea. Traditionally, whole herring, with or without the head, are mixed with salt and stored cool (5–10°C) in barrels for a few months before they are washed, filleted and packed. In many countries, concentrated salt brine is added before the barrels are closed. During storage, enzymes from the digestive tract leak out and, together with other endogenous fish enzymes, cause partial digestion of the muscle proteins. The fish meat becomes smooth and pliable and attains a pleasant rich flavor. The final salt concentration can vary from approximately 4 to 18%, depending on the recipe used. The high salt concentration and low temperature restrict excessive proteolysis, mainly due to gastrointestinal proteases, and the ripening process usually takes from 2 to 7 months. This traditional maturing procedure is time-consuming and demands large storage capacity (13, 14).

Trypsin-type enzymes from the pyloric caecum are the major contributors to endopeptidase activity in the ripening process. The concentration of these digestive enzymes is greatly affected by seasonal changes in feeding intensity, the highest being before spawning (see Chaps. 1 and 9). The North Sea herring is the only fish caught in the period from May to August that contains sufficient digestive proteolytic enzyme activity to promote satisfactory ripening (13, 99).

Another problem related to storage capacity is the use of whole herring in the ripening process. When whole fish is salted, the filleting waste becomes too salty to be used as animal feed. By filleting before salting and ripening, two advantages are obtained: the storage capacity is reduced, and the filleting waste can be used as animal feed. However, endogenous fish enzymes are necessary to achieve maturation in herring fillets or gutted fish (14). To avoid the seasonal variations and induce or accelerate the ripening process, experiments involving the addition of exogenous enzymes to the brine have been done since the early 1970s (91, 99–102). Several methods for artificial ripening of herring involving enzyme addition have been developed and patented. The nature of enzyme addition has been different. Pancreatic enzymes from pig or cattle have been used to obtain good quality matjes herring from fish caught during nonfeeding period (99), or a proteolytic mixture containing mainly trypsin or chymotrypsin (from beef organs) has been used for the preparation of matjes herring from which head and viscera are removed (103). However, the main exogenous enzyme preparations used in the ripening process are of marine origin. By addition of exogenous fish enzymes the duration of the ripening process has been reduced and a quality similar to matjes herring obtained by traditional methods has been maintained. Nevertheless, if the enzyme preparations contain high lipase activity, the product attains an inferior flavor (99).

The first fish-enzyme addition used in artificial ripening was minced pyloric cecum from herring caught in midsummer. This preparation was added with salt to spent herring caught in October and a good ripening was obtained (100). Another artificial ripening method has been patented in which herring fillets are matured in brine containing digestive enzymes from herring. One part herring fillets is immersed in 1.3 parts brine containing 12% salt, 6% sugar, a spice mixture, and 5% minced herring pyloric cecum. After only 5 days at 3°C, the fillets attain the smooth consistency and flavor that characterize traditionally matured products. The final salt concentration is only 6%, and extensive washing before consumption is not necessary. The amount of pyloric cecum used in this method corresponds to the natural content in herring before spawning (summer-caught)(101–102). However, the industrial use of adding herring viscera enzymes to control the ripening does not appear to be widely practiced, perhaps because enzyme preparations from herring viscera are not available as commercial industrial products.

With the enzyme added as a crude extract to accelerate the ripening of herring fillets, an acceptable flavor has been obtained, but the stability of the final product is poor because of residual enzyme activity. This problem does not arise in the traditional long-term storage methods because the enzyme activity gradually declines during storage, giving a quite stable final product with little residual enzyme activity. Thus, to obtain stable products with rapid ripening methods, it is necessary to reduce the residual activity of the added enzyme as much as possible. In this sense, trypsin isolated from the pyloric cecum of Greenland cod (*Gadus ogac*) has been used in ripening experiments with herring and squid. The maturing was achieved mixing trypsin (25.5 mg) and eviscerated herring or squid mantle (1.5 kg), packed and pickled in brine, for 40 days at 10°C. The cod enzyme was more efficient than the bovine pancreas trypsin used as control (the protein solubilization rate was double). In addition, the residual trypsin activity in the brine decreased more rapidly for the cod enzyme than it did for the bovine enzyme, because of end product inhibition and enzyme instability (91). Cod trypsin has better catalytic properties than cattle trypsin at the temperature used in artificial ripening.

Squid liver has been investigated as a source of enzymes that may be used for supplementation of brined Atlantic short-finned squid (*Illex illecebrosus*) to accelerate the autolytic fermentation (92) and to tenderize squid

(*Illex argenticus*) (104). Squid liver extract has a comparatively high proteolytic activity against hemoglobin and casein at pH 2–8 and high proteolytic activity at low temperatures. It is rich in cathepsin C, an enzyme having dipeptidyl transferase and dipeptidyl hydrolase activity, optimal at pH 6–6.5, which improves the formation of tasty amino acids (glutamine, alanine, leucine, serine, lysine, arginine, and proline) in squid meat. The sweet delicious taste of brined squid may be enhanced by supplementing the brine with pure Greenland rock cod trypsin or with a neutral protease fraction from squid liver possessing cathepsin C activity during the fermentation process. Fresh squid mantles (less than 12 h postmortem) were skinned and sliced. The squid (500 g) was mixed with 350 ml brine solution including purified Greenland cod trypsin (26.5 mg, 7,000 BAPNA units) and a crude preparation of cathepsin C from squid liver (105 mg). The squid was incubated at 6°C and pH 6 for up to 120 days. The result was a brined squid with better sweetness and delicious taste than control. Cod trypsin acted to accelerate free amino acid accumulation during the first stage of fermentation, whereas cathepsin C activity resulted in a more sustained enhancement of free amino acid accumulation, mainly tasty amino acids (92).

Furthermore, most of the proteases of squid liver extract are also collagenolytic enzymes. At pH 7, the proteins of the squid sarcoplasmic and myofibrillar fractions, as well as squid collagen, are hydrolyzed to low-molecular-weight fractions after 24 h at 4°C. The high activity of squid liver enzymes, at neutral pH range and at refrigeration temperature, has been used to tenderize squid mantle meat before cooking. By soaking skinned squid mantle meat in liver extract (1:3) for 24 h at 20°C or 4°C prior to cooking 45 min in water, a significant hydrolysis of the proteins (65%) was obtained. A 40% decrease in toughness of the cooked product was obtained compared to the toughness of untreated cooked samples. The mantle cooked directly in enzyme preparations was not more tender than that of untreated samples (104).

Although artificial maturing methods have obvious advantages (reduction of storage capacity requirement and improved use of the raw material), the industrial applications of such methods still appear to be limited. One reason is that appropriate enzyme preparations are not commercially available. Their availability is complicated, especially for fish or shellfish enzymes, since both the source and purification process are expensive (15). Also, poor stability is obtained in the final products using the rapid methods as opposed to the traditional methods (14). Furthermore, the enzyme preparations should have low lipase activity, otherwise an undesirable fatty acid taste can occur in the cured product (99). The solution will be the use of purified fish or shellfish digestive enzymes, in particular cold-active proteases, because they are more suitable tools in accelerated maturing than mammalian enzymes (105).

5. Protein Hydrolysates from Fish and Shellfish

Fish protein hydrolysates (FPH) are a mixture of proteinaceous fragments and are prepared extensively by digestion of whole fish or other aquatic animals or parts thereof using proteolytic enzymes (endogenous and exogenous) at the optimal temperature and pH required by enzymes. The process of hydrolysis breaks down the protein into smaller peptides to obtain a water-soluble product (106, 107). FPH production is a means of transforming a large proportion of total landed fish, which remains underused (by-catch pelagic fish species and unconventional species such as krill), to a hydrosoluble protein concentrate with food applications (5, 107).

Liquid fish hydrolysates, prepared from small fish, were known in ancient Rome. Detailed description of how to prepare such a "liquemen" appears in *Roman Cookery Book*. Similar liquid products are popular in Southeast Asia (108). The real development of FPH started in Canada in the 1940s after World War II. Methods for the enzymatic preparation of protein hydrolysates from fish meat were developed (109). The first intended application of this amino nitrogen source was as peptone in growth media for microorganisms. Peptone was prepared by autolysis of fish viscera (110) and by a two-step enzymatic hydrolysis of whole fish, imitating in vivo digestion. The first digestion step was an autolysis of acidified minced fish with endogenous pepsins and peptidases. The second enzymatic hydrolysis step was done by proteinases added under alkaline conditions, prior to removal of the digested proteins (111). These peptones supported bacterial growth better than hydrolysates prepared by chemical hydrolysis and comparable to the best meat peptones on the market (112). At present, industrial fish peptones are an excellent substrate for biomass production in solid and submerged fermentation and are produced commercially in Norway and Japan (14).

However, the main use of fish and shellfish protein hydrolysates has been as food or feed ingredients (113). The production of FPH has been described in several reviews (113–120). The general procedure for fish and shellfish protein hydrolysate preparation could be detailed as follows (21).

1. Substrate. The raw materials are minced and mixed with an equal amount of water. The pH is adjusted to the desired level. Raw materials may include by products from filleting (waste, viscera, etc.), by-catch fish from trawlers, small underused lean fish species or krill (a collective name used for five or so species of shrimplike creatures, the most abundant of which is *Euphausia superba;* see Chap. 18).

2. Hydrolysis. Proteolysis is carried out by autolysis and/or by addition of a preparation of hydrolytic enzymes. Depending on the concentration of endogenous proteinase activity in the raw material, enzyme addition improves the hydrolysate yield. The enzymes are added at optimal hydrolytic temperature under

stirring. At the end of the incubation period, the enzymes are inactivated, and the material is pasteurized by heating to above 80–90°C.

Considering that the hydrolytic process is aimed essentially at increasing protein solubility, industrial operations use proteases with broad specificity that collectively achieve peptide cleavage at random. Commercial proteases are usually from animal, plant, or microbial source. Animal (mammalian) proteases include mostly pepsin (121), although chymotrypsin and trypsin have also been tested (122–124). Bromelain, papain, and ficin are the plant enzymes most widely used (114, 125) and patented (14). Microbial proteases of bacterial origin and low price have been evaluated (119, 123, 126). The use of proteolytic enzymes from fish itself has also been suggested (5).

3. Separation. Bones and scales are removed by sieving. Oil, sediment, and hydrolysate are subsequently separated by a three-phase separator.

4. Concentration. The hydrolysate is filtered, evaporated, and dried by a vacuum or spray-drier to prepare soluble FPH.

Type of raw material, hydrolytic power of enzyme, temperature, and incubation time are the main factors determining the protein hydrolysate yield. In enzymatic production of protein hydrolysates for food use, processes such as sterilizing the protein source, use of a bacteriostat during hydrolysis, and pasteurization of the hydrolyzed protein are necessary for good quality control (127).

Although most FPHs are produced by direct addition of an enzyme obtained from commercial sources, endogenous enzymes from the raw material are also used to manufacture different protein hydrolysate products. Thus, FPH from whole Caspian sprat (*Sprattus spracttus*), based on partial hydrolysis involving endogenous proteases, has been developed (108, 128). Krill is an underused marine shellfish with growing world production (107). It is very rich in proteases and undergoes rapid autolysis. During autolysis most of the krill protein is hydrolyzed to free amino acids after 48 h at 20°C. Because of these properties, an industrial process for production of free amino acids has been developed (129). Moreover, free amino acids are recovered as a byproduct in bulk production of krill proteolytic enzymes (a mixture of endopeptidases and exopeptidases) (130, 131), commercialized by a Swedish company (Pharmacia, Upsala) and efficiently used in debridement of necrotic animal wounds (132, 133).

The main use of FPH in food is in dietetic foods as a source of small peptides and amino acids, in products such as soufflés, meringues, macaroni, or bread, and in fish soup, fish paste, and shellfish analogues as flavoring compounds. High dispensability of such products render them suitable as a substitute for milk proteins (14, 107). The major feed applications are as milk replacers for calves (134, 135) and weaning pigs and as protein and attractants in fish feed (14). Hydrolyzed krill is now being investigated as an ingredient in starter diets for fish larvae and juveniles (136). Moreover, stimulation of the immune system has been demonstrated in animal feeding with FPH because of certain peptide compounds (137).

The choice of enzyme preparation and the process conditions determine yield (profitability), properties (solubility, emulsifying or gelling, flavor, nutritional value, etc.), and use (food or feed) of the product (21). Bitterness is a common problem in preparation of FPHs and for this reason they are unsuitable as food ingredients (138). This is caused by the formation of peptides containing bulky hydrophobic groups toward their C-terminal. The intensity of bitterness depends on the degree of hydrolysis and the specificity of protease (i.e., the size of peptides and the kind of peptide bonds cleaved) (106, 107).

Several methods have been suggested to mask the bitterness of FPHs. These include incorporation of glutamic acid or glutamyl-rich peptides, polyphosphates, gelatin, or glycine into the products (139). Bitterness can also be eliminated by selective separation (extraction or chromatography) and by enzymatic treatment, which consists of using exopeptidases to carry out the plastein reaction. The "plastein" is resynthesized from peptides and amino acids by transpeptidation (108, 140). Plastein is a high-molecular-weight proteinlike substance with different properties from those of the original protein and devoid of contaminants having flavor. Thus far, plastein formation has been facilitated with proteolytic enzymes (trypsin, chymotrypsin, pepsin, etc.) at 37–50°C, producing a hydrophobic product (141). Cold-adapted trypsins, chymotrypsins, and pepsins from fish have been proposed to facilitate the formation of hydrophilic plasteins at low temperature to maintain the water solubility of the protein hydrolysate (15). Another method of avoiding the bitterness in FPHs is to reduce the formation of bitter-tasting peptides and excessive protein hydrolysis to peptides and amino acids. Cold-adapted enzymes from fish have a relatively narrow specificity for peptide bonds and relatively high molecular activity (8). Addition of these enzymes during the hydrolysis process has been suggested, since this improves the amount of pleasant-tasting active amino acids and reduces the percentage of TCA-soluble proteins, thus preventing the bitterness problems in protein hydrolysates (5).

6.   Seafood Flavorings

Flavor enhancers or flavor potentiators are know to be chemical substances that have little flavor of their own but, when mixed with food products, have the ability to enhance the flavor of the food. Enhancers used extensively worldwide include monosodium glutamate (MSG), inosine-5′-monophosphate (5′-IMP), and guanosine-5′-monophosphate (5′-GMP). These compounds accentuate meaty flavor and have found applications as flavor potentiators in soups, sauces, gravies, and many other savory products (142).

Flavor potentiators occur naturally in sources such as sea tangle (*Lami-*

naria japonica), dried bonito tuna, and black mushroom (*Lentinus edodus*). However, flavoring agents have usually been produced by enzymatic transformations. Certain fresh seafood is rich in 5′-AMP (adenosine-5′-monophosphate), which may be converted to 5′-IMP by the action of endogenous enzymes released during processing. Several important enzymes that have applications in flavor-enhancing nucleotide and amino acid production have been identified in naturally occurring products of plant, animal, or microbial origin. These enzymes predominantly belong to the general group of hydrolases. Commercial production of extracts rich in flavoring agents is mainly from yeast extracts and hydrolyzed vegetables, both of which are rich in glutamic acid. Yeast extract is a concentrate of the soluble fraction of yeast generally made by autolysis. Hydrolyzed vegetable protein is a product composed of amino acids or peptide savory flavoring, obtained from soy bean, wheat, and other plant substances by a partial proteolysis followed by acid hydrolysis (142).

Seafood flavors are in high demand for use as additives in products such as kamaboko, artificial crab, and fish sausage, and cereal-based extrusion products such as shrimp chips (143). Seafood flavoring from various sources of raw material can be produced by enzymatic hydrolysis. Exogenous and endogenous proteolytic enzymes can aid the extraction of flavor compounds from fish and shellfish byproducts, although other types of enzymes may also be involved (5). A similar process to FPH manufacturing has been developed by a French company (Isnard-Lyraz) to recover seafood flavoring compounds from fish, shellfish, or mollusks, using whole animals or their byproducts (144). The process consists of liquefaction of the raw material by enzymatic hydrolysis, thermal inactivation of enzymes, separation of bones and shells, filtration or centrifugation, and concentration of the flavors and flavor enhancers naturally occurring in the raw materials. The purpose of the enzymatic hydrolysis is to liquefy and allow the separation of bones and shells, also facilitating concentration or drying. Moreover, enzymatic hydrolysis permits the nucleotide and protein transformation to flavor enhancers without producing additional flavor or taste. The final products consist of pastes or powders and are natural extracts with flavoring properties, but they are not flavors. Exogenous enzymes are used but endogenous enzymes are also of importance for the hydrolysis. No more details are available of this patented process (144).

Various kinds of byproducts have been used to produce marketable seafood flavoring. The production is mainly based on autohydrolysis by their endogenous enzymes. Thus the flavor from oysters (145), shrimp (146), and clam (147, 148) have been recovered.

7. Fish Sauces

Fish sauce has long been a traditional fermented fish product and is an important source of protein in Southeast Asia. It is also consumed in Europe and North

America. This seafood is a liquid product made by storing heavily salt-preserved fish material at tropical temperatures until it is solubilized by endogenous enzymes (21). Typical examples of such products are the *nuoc-mam* produced in Vietnam and Cambodia, *nam-pla* in Thailand, *patis* in the Philippines, *uwo-shoyu* in Japan, or *ngapi* in Burma (149).

The production of fish sauce has recently been reviewed (150). Traditional fish fermentation is actually a combination of salting, enzyme hydrolysis, and bacterial fermentation. The general procedure for fish sauce preparation is summarized as follows.

1. Salting fish. Small fish, both marine and fresh water, whole or minced, are mixed with sun-dried marine salt in a ratio of 3:1 for marine fish or 4:1 for fresh water fish (20–40% salt, minimum 15%). The mixture is then placed into fermentation tanks between two layers of salt.

2. Hydrolysis and fermentation. Closed tanks are stored at ambient tropical temperatures. Tissue solubilization occurs as a result of autolytic action by fish digestive enzymes. This hydrolysis provides the necessary nutrients for halophilic bacterial fermentation to begin, which plays an important role in flavor development. Protein hydrolysis is caused by trypsinlike enzymes (152) together with fish-gut peptidases. The pH of fish sauce ranges from neutral to slightly acid. Because of nonoptimal pH, partial inhibition of high salt concentration, and salt impurities, the activity of digestive proteases is at a minimum and the rate of autolysis is low (14). Under these conditions, the time needed for the full flavor of fish sauce to develop varies from 8 to 18 months. When fish sauce is made from cod viscera, mixed with salt (25% w/w) and stored at 22–27°C, only a 25 day fermentation period is needed. Moreover, this process permits the recovery of a tryptic enzyme concentrate by ultrafiltration during the fermentation process. Spray-dried enzyme concentrate has been used in the ripening of salting herring fillets (10 g kg^{-1}), obtaining a satisfactory ripening more rapidly than by the traditional ripening of ungutted herring (151).

3. First-quality fish sauce. After fermentation is completed, the liquid is drained off and saturated brine is added to the residue to extract the leftover soluble matter. The two liquids are then combined, filtered, and bottled to produce first-quality fish sauce. This amber protein hydrolysate has a high nutritional value, being rich in the essential amino acids lysine and methionine. However, the high salt concentration may be a health risk depending on the daily intake and the rate of perspiration.

4. Second-quality fish sauce. The remaining partly digested fraction is extracted several times with saturated brine until most of the proteinaceous material is recovered to obtain second-quality fish sauce. This product is cheaper and more accessible to poor people, but has an inferior nutritional value.

The long production time means that storage tanks of large capacity are required. From an economic point of view, to reduce this need it is desirable to speed up the solubilization (21). Several methods have been proposed to reduce the production time. The first is to raise the initial temperature. A couple of weeks at 45°C in the initial storage phase reduces the total production time from 1 year to 2 months (14). The second is addition of acid combined with reduced salt content. Pepsin is inhibited by salt. Thus, an initial phase of rapid autolysis is carried out at pH 4 (obtained by addition of hydrochloric acid) and low salt concentration, at 27±2°C. After this phase (5 days), samples are neutralized with sodium hydroxide solution, and salt is added to the normal level (250 g kg⁻¹). By this method, an acceptable flavor could be achieved after 2 months. However, some endogenous enzymes that are important for flavor development during storage at neutral conditions are denatured at low pH (153–154). The third is initial alkalization at low salt concentration. If the initial pH is raised to 11 by addition of sodium hydroxide at a moderate salt concentration, alkaline digestive proteases will be very active, whereas endogenous trypsin and chymotrypsin inhibitors will be denatured. The final product has a limited storage stability and needs to be neutralized with hydrochloric acid and salt addition (to 250 g kg⁻¹) to achieve the same level and flavor as traditional fish sauce. These conditions shorten the process to 2 months (155).

These accelerated methods have achieved an improved hydrolytic rate. However, inferior flavor has also been produced, the problem in most cases being that the taste is either too weak or too bitter. Bitterness appears for the same reasons commented on in the discussion of FPHs. However, in China a method based on initial acidification is apparently used by some fish sauce manufacturers (21).

Another method used to accelerate the protein hydrolysis in traditional fish sauce is the addition of enzyme-rich components. Plant enzymes were the first to be used (156). Papain from unripe papaya, bromelain from pineapple stems, and ficin from figs have all been tested. These enzymes are cysteine proteases most active under weak acid conditions. Fish sauce recovery was obtained after 2–3 weeks of the fermentation process. However, the characteristic flavor of the finished product was inferior to the traditional, although the best results were obtained by using bromelain preparations. This accelerated method is used today in commercial fish production in Thailand (14).

The use of squid hepatopancreas tissue to accelerate traditional methods in fish sauce production has been reported (5, 15). Capelin (*Mallotus villosus*) fish sauce was prepared by mixing minced fish with salt (25% w/w) and supplementing with squid hepatopancreas tissue (SHP) (2.5% w/w). After 11 months at 20–25°C, the fish sauce was recovered by filtration and stored at ambient temperature. Fish sauce was analyzed and evaluated by a sensory panel 2–3 months after filtration. The free amino acid content and hydrolysis rate during the first month were significantly higher than for the control fish sauce or for other proteolytic enzyme supplements (fungal protease, pronase, trypsin, chymotrypsin, or squid protease fraction). Preference analysis of capelin fish sauce showed that the product supplemented with squid hepatopancreas was highly accepted and preferred to a commercial product from the Philippines (157). The fermentation time of SHP-supplemented fish sauce can be shortened to 6 months, resulting in a free amino acid content and sensory evaluation score significantly higher than the control fish sauce. The better acceptability scores of SHP-supplemented fish sauce are caused by a higher content of free amino acids and peptides, ranging from 100 to more than 1,300 Da and rich in aspartic acid, serine, glutamic acid, and leucine. However, typical flavor has also been correlated with large peptides (the fraction greater than 10,000 Da) (158). Squid hepatopancreas proteases produce an acceptable proportion of taste-influencing free amino acids and peptides to yield fish sauce with more flavor and less fermentation time than in the traditional process.

B. Enzymes as Fish-Processing Aids

Over the last decades significant research and development have been done in seafood processing on biotechnological methods using enzymes as an alternative to conventional processing methods. This challenging task has been in progress in Iceland (13, 159) and Norway (88, 90) since the 1980s.

Mechanical processes have obvious limitations regarding selective and gentle treatment of the raw materials. In many situations, mechanical processing implies low yield and quality reduction. In some cases technical problems are not solved by mechanical means. Hydrolytic enzymes digest certain tissue structures, leaving others intact. Some normally resistant tissue structures are sensitive to enzyme digestion when correct conditions of pH, temperature, or the salt concentration are met. Because of this, enzymes can be used as specific tools in food processing, acting as gentle knives, where the aim is a selective removal or modification of certain tissue structures (14).

1. Deskinning

The common method of removing skin from fish fillets is purely mechanical: an automated machine in effect tears the skin off the flesh. The ease of deskinning varies greatly among fish species (18). The biochemical method of deskinning works because skin differs fundamentally from muscle in chemical composition and structure, and enzymes are generally very specific in their action (88). Enzymatic deskinning can be done by cold-adapted fish pepsin at low reaction temperatures (5, 15). In addition, fish pepsin at low pH breaks down fish skin rapidly but breaks down muscle protein relatively slowly (160). Proteases are usually mixed with carbohydrases to facilitate skin removal (16).

In Norway, an enzymatic skin removal method has been developed for herring. The whole herring is treated in 5% acetic acid at 10°C, to denature the skin collagen, and is immediately transferred to the enzyme bath containing cold-active fish pepsin and 0.5% acetic acid. After 1–2 h at 20°C, the skin is partially solubilized and can be washed off. Differences in skin structure and thickness on different parts of the body make it difficult to achieve uniform deskinning (14, 83, 161).

Proteolytic enzymes from squid intestines may also be used for removing the double-layered skin of squid. The approximate collagen content in the mantle muscle and skin of squid is 4.6 and 21.9% per dry tissue and 5.4 and 28.4% per crude protein. The collagen content per crude protein of the skin is about five times as much as that of the muscle (162). The conventional mechanical deskinning method removes only the pigmented outer skin, leaving the rubbery inner membrane intact. The enzymatic deskinning process involves soaking the squid tubes and tentacles in a weak salt solution containing a squid intestine extract (including liver) at 45°C for a short time, which selectively attacks the rubbery membrane without degrading the muscle tissue (13, 163). Most of the proteases of squid offal extract are also collagenolytic enzymes (164). A complete production line for a similar process of squid enzymatic deskinning has been developed by Biotec-Mackzymal in Norway and produced by Carnitech in Denmark (14). A company from Russia and Liechtenstein (Seatec) is producing a collagenase preparation from crab hepatopancreas that may be used in deskinning of squid (20).

Starry ray (*Raja radiata*) is the most common skate species in Iceland. It is caught in a by-catch/landed-catch proportion of 16:1 every year and it is underused. Some of the skate catch is used to manufacture and export skinned skate wing (fresh or frozen). The conventional deskinning method is both laborious and difficult, since the skate wing skin is partly covered with sharp spikes. Mechanical or manual skinning leaves residual skin fragments that must be removed when the process is repeated. An enzymatic method for deskinning skate wings has been developed to yield improved product. The skin collagen is denatured by a gentle and rapid treatment, followed by incubating the skate wings at low temperature (0–10°C) in a enzyme bath for a few h (or overnight at 0°C). The final step is to rinse the dissolved skin and remaining spikes from the skate wings (13). The enzyme preparation used contains cold-active proteinases with high collagenolytic activity, possibly from fish or shellfish digestive system, and carbohydrases, that may be acting by loosening the collagenous layer or by making the access to the denatured collagen easier for the proteinase enzymes.

2. Descaling

Descaling of fish species such as redfish (*Sebastes marinus*) or haddock (*Melanogrammus aeglefinus*) by mechanical means can be a problem because the scaling treatment is harsh and tends to damage the skin and lower fillet yield. Redfish is descaled before filleting so that the scales do not contaminate the fish flesh or the knives of the filleting machines during processing. The scaling is usually done in large rotating cylinder tunnels with a ribbed base: the scales are scraped from the skin and washed away by water. Often the skin is torn in the scaling treatment. The descaling of haddock can be difficult to accomplish without damaging the fish (especially during the summer months when the fish flesh is soft), and this process is essential to produce flesh fillets, with the skin on. Haddock is much more liable to damage than redfish and therefore it cannot be descaled in rotary cylinder tunnels. Usually haddock is descaled individually in special machines. If the fish is gutted before descaling stage, the fish flesh may be damaged, which can lead to a low yield (13, 88).

In Iceland, this mechanical process has been analyzed for replacement by a biotechnological application, using a cold-adapted fish enzyme solution for a more gentle descaling (13). The scales from redfish or haddock may be removed by the help of fish enzymes. The scales can be gently removed without affecting the skin or flesh after incubating the fish in an enzyme solution at 0°C followed by spraying with water. However, the research is still at a preliminary stage and the method is being tested as a laboratory method and has not been compared with the mechanical descaling. It is too early to predict whether enzymes can become an alternative to mechanical descaling during processing (15).

3. Membrane Removal

Salted-cod swim bladders are a product manufactured in small quantities (annual production of 30–50 tons) in Iceland for markets in Southern Europe; Italy being the major user. The fresh cod swim bladder is enclosed in a thin black membrane, but the market demands an almost white salted product. Thus, the black membrane has to be removed before the swim bladders are exported. The undesirable membrane is tightly bound to the fresh bladder and is impossible to remove efficiently by manual or mechanical means. This problem has been solved by subjecting the salted raw material to a simple enzymatic process in which the black membrane is hydrolyzed for about 20 min. This process is currently used by all salted-cod swim-bladder processors in Iceland. No details are available on this process or the origin of the hydrolytic enzyme (13). However, the source of enzyme may be from marine organisms (pepsin from cod or collagenase from crab hepatopancreas) because of the specific characteristics demanded by the process (3).

Membrane removal is a problem during the production of canned cod liver. This product is made mainly in Iceland with an annual production of about 200 tons and the principal markets are in Eastern Europe. The cod liver is sometimes infested with the seal-worm (*Phocanema decipiens*), which has to be removed before canning. It is necessary to remove a thin collagenous membrane that surrounds the liver, because the seal-worm is mainly contained in this membrane. Suitable machinery is not available for this removal. Thus, the membrane is removed by hand, but only with moderate success and considerable labor cost. This produces a bottleneck in the factory during the canning of cod liver. Recently, an enzymatic method has been developed to dissolve the cod liver collagenous membrane using fish proteases. The method has been tested in canning plants and compared to the manual methods. The throughput of liver during the membrane removal stage can be increased by 20–30% when using the enzymatic method. No additional details are available about the process (13).

IV. APPLICATIONS IN OTHER FOOD SECTORS

The main use of fish and shellfish enzymes in other sectors of the food industry is in dairy technology. However, other applications have been suggested such as meat tenderizing for specific fish collagenases, and the enzymatic clarification of fruit juice. Trypsin from stomachless fish is useful because of its special ability to digest native proteins substrates compared to its counterparts derived from species with a functional stomach (5, 15, 165, 166).

A. Enzymatic Milk Coagulation

The major use of enzymes in the dairy industry is in the coagulation of milk to make cheese. The Food and Nutrition Board of the United States National Research Council uses the term "rennet" to describe all milk-clotting enzyme preparations (except porcine pepsin) used for cheese-making. The same board defines rennet as aqucous extracts made from the fourth stomach of calves, kids, or lambs, and bovine rennet as aqueous extracts made from the fourth stomach of bovine animals, sheep, or goats. Microbial rennet followed by the name of the organism is the approved nomenclature for milk-clotting preparations derived from microorganisms. Milk-clotting enzymes generally recognized as safe (GRAS) are rennet and bovine rennet. However, the United States Standard of Identity for Cheddar cheese allows the use of rennet and other clotting enzymes of animal, plant, or microbial origin (167).

The milk coagulation process has two phases: enzymatic and nonenzymatic. Milk clotting begins with enzymatic cleavage at low pH of the chymosin-sensitive bond of k-caseins (Phe-105—Met-106) to form para-k-casein and a macropeptide. The hydrolysis of k-casein destroys its ability to stabilize casein micelles, rendering them susceptible to coagulation in the presence of calcium. The process is done at a temperature below 10°C. Enzymatic cleavage is followed by a nonenzymatic aggregation of the altered casein micelles into a firm gel structure (the curd) and the whey is drained off. This phase is done at 30–39°C. Both phases of milk clotting overlap (168–170).

There has been a continued interest in the search for rennet substitutes ever since a rennet shortage was anticipated in the 1960s because of a decline in the number of calves slaughtered and an increase in demand for cheese (171). Many proteolytic enzymes clot milk under appropriate conditions, and the importance of these proteases results not only from their ability to coagulate of milk but also from the relation between milk-clotting ability and the general proteolysis the enzyme may produce. Not all proteolytic enzymes are suitable for making acceptable cheese because their excessive general proteolysis leads to lower curd yield caused by excessive loss of fat and protein to the whey, and development of undesirable changes in texture (softening) and flavor (bitter off taste) during cheese aging (171–173). A few enzymes are found to be adequate when the above deficiencies are considered. Chymosin (EC 3.4.23.4) or rennin is an acid protease, from abomasa of suckling ruminants, with optimum pH, stability at pH near 7.0 (5.3–6.3), and narrow substrate specificity. This is the enzyme of choice for milk-clotting and the standard against which all others are evaluated. Chymosin converts the colloidal milk casein into a curd to give high yields and reduce proteolysis, which contributes to cheese aging (174–176).

Other proteases from microbial, plant, or animal origin have been evaluated as rennet substitutes. *Mucor miehei* rennet is the most common fungal protease preparation accepted by the cheese industry. However, fungal rennet has not been totally satisfactory since it may have a relatively broad specificity, be heat stable, cause bitterness in the cheese, and remain active in the whey (167, 177, 178). Plant proteases have been employed as rennet substitutes and appear to give rise to a softer curd than calf rennet. Milk-clotting factor from the flowers of Cardon (*Cyanara cardunculus*) is traditionally used in Portugal for making soft cheese (Serra cheese) from sheep milk, but its high proteolytic activity is not suitable in the manufacture of Edam and Roquefort cheese (179). Porcine pepsin is used as a rennet substitute but its ability to clot milk diminishes quickly above pH 6.5. This enzyme is mixed with bovine rennet and chymosin, obtaining a less expensive coagulant and avoiding low curd yield (175). Chicken pepsin has also been employed as a rennet substitute, but Cheddar cheese prepared with this enzyme can have intense off flavors (180).

Cold-adapted gastric proteases from fish and chymosinlike enzymes from marine mammals have several characteristics that make them suitable for both milk-clotting phases and that avoid the major problems with rennet substitutes pointed out above (8, 5, 15). Atlantic cod pepsin (181–182) and tuna (Atlantic tuna, *Thunnus obesus*) gastric enzyme (179) have been purified and proposed as milk-coagulating enzymes. A semipurified aminopeptidase preparation from squid hepatopancreas is currently under trial in Cheddar cheese ripening because it has potential to reduce bitterness and enhance the flavor of cheese effectively (183).

Cold-adapted fish pepsin (from Atlantic cod, Greenland cod, or Polar cod) has a lower temperature coefficient (i.e., lower Arrhenius activation energy) for milk clotting (1.4–1.7) compared to calf rennet and microbial rennet (2.0–2.8). Clotting can be done with a lower enzyme concentration, thereby conserving rennet and minimizing the presence of residual curd proteolysis (8). This pepsin also has a lower temperature optimum for hydrolysis. Moreover, gastric proteases from marine organisms are unstable at temperatures above 30°C, which make it possible to clot milk and subsequently heat-denature rennet during curd formation to inactivate proteases, and avoid softening problems and off-flavors (8, 181). This pepsin also has a high molecular activity at low reaction temperatures, which produces a cheese with much lower levels of free amino acids and bitter peptides during ripening (8).

These properties make cold-adapted pepsin an excellent rennet substitute. Atlantic cod pepsin has been used as a rennet substitute in the preparation of Cheddar cheese. The product, when aged less than 6 months, was judged acceptable by sensory panels. However, when the conventional cheddar cheese process was employed, this enzyme results in high loss of protein and fat to the whey during cheese making, and the product develops bitterness and a pasty consistency after prolonged aging (181). These problems are prevented by raising the temperature to 39°C in the initiation of the nonenzymatic phase during milk clotting (182).

Chymosinlike protease from the gastric mucosa of harp seal (*Pagophilus groenlandicus*) has catalytic properties similar to calf chymosin, and has been used successfully in Cheddar cheese preparation. Seal gastric protease is more stable than pepsin at neutral to alkaline pH, and needs lower pH (6.6) than calf rennet (pH 6.8–7) to clot milk. Cheddar cheese prepared with seal gastric proteases gave significantly higher sensory scores, and less free and peptide-bound amino acids than cheese made with calf rennet (178). Gastric mucosa from harp seals have four zymogens of acidic proteases (A, B, C, and D). Zymogen A, named pepsin A, has been isolated and is similar to calf chymosin in several physical and catalytic properties (184). Seal pepsin A appears to make a major contribution to the excellent cheese-making characteristic of the crude seal pepsins. However, the best results have been obtained with crude seal gastric proteases because the other components of the crude extract are probably responsible for the accelerated aging of Cheddar cheese made with this rennet substitute (185).

The extraction of milk-clotting enzymes from fish stomach mucosa or shellfish hepatopancreas for cheese manufacture would provide an inexpensive alternative to rennet substitutes, and could become a new food-related industry. Atlantic cod pepsin is now produced commercially in Norway (87, 88). In addition, the enzyme extraction would address a very important pollution and disposal problem, as a means of minimizing the waste associated with processed fish and shellfish, obtaining an additional value (179, 181, 183). The need for rennet substitutes decreased considerably with the commercial introduction of recombinant bovine chymosin.

B. Prevention of Oxidized Flavor in Milk

Spontaneous development of oxidized flavors is common in dairy products. Several native milk enzymes are implicated in the process. Peroxidase (EC 1.11.1.7), which is stable at pasteurization temperatures, catalyzes oxidative reactions in milk and dairy products. Xanthine oxidase (EC 1.2.3.2), which catalyzes oxidation of xanthine or hypoxanthine to uric acid and hydrogen peroxide, also causes oxidative flavors (167).

Inhibition of milk oxidation by enzymatic treatment has been studied. Bovine trypsin can prevent oxidized flavor in milk (186). However, milk treated with bovine trypsin before packing retained certain residual tryptic activity because the enzyme was not completely inactivated by the pasteurization process. The problem is the subsequent hydrolysis of milk protein, which decreases milk quality. Trypsin from cold-adapted fish (e.g., Greenland cod) has a lower free energy of activation at reaction temperatures below 30°C and more thermal instability than bovine trypsin (7). The prevention of oxidized flavors in milk by Greenland cod trypsin has been demonstrated. When milk is treated with pure enzyme (0.005% w/v) and kept at 4°C for 4 h, no retained trypsin was present after pasteurization (at 70°C for 45 min). Concentrations greater than 0.0013% of cold-adapted fish trypsin prevent the oxidation of lipids in raw milk (8, 91).

C. Preparation of Infant Milk

Cow's milk can be "humanized" by addition of lysozyme, making it suitable as an infant milk. Lysozyme acts as a preservative by reducing bacterial counts in the milk without affecting *Lactobacillus bifius* (187).

Lysozymes are widely distributed in nature and have been found in fish, shellfish, and other animals or plants, playing a role as a defense mechanism against infectious diseases (188).

Lysozyme from Arctic scallop (*Chlamys islandica*) has been recovered and purified from scallop waste generated in shellfish factories in Norway (88, 188). This lysozyme has a particularly low activation energy at low temperature, and is active at 0°C. The specific activity of the purified scallop viscera lysozyme is nearly 300% higher than hen egg white lysozyme (widely used as a antimicrobial enzyme in the food industry) (188). Both properties, catalytic and specificity, suggest that lysozyme could be applied as a bacteriostatic and prophylactic agent in food preservation and specialty products (e.g., infant milk) without the problems caused by use of antibiotics.

V. APPLICATIONS OF FISH AND SHELLFISH ENZYMES IN FEED PRODUCTS

A. Fish Silage

Fish material not for human consumption is used for the production of fish meal, which has a world market of considerable size. However, new ways of using small pelagic fish, fish waste, and fish viscera in industry or animal feed are constantly being sought (107). Distance from a fish meal plant is a factor because of transportation cost. Pelagic fish may be caught periodically in quantities exceeding the local fish-meal processing or freezing capacities. By-catch is often thrown overboard because of its low selling price. The production of fish silage offers a convenient way of using these resources.

The product of the process of preserving and storing wet fodder in a silo is called silage. The traditional use of the word has been in conjunction with green forage, preserved either by adding acid or by the anaerobic production of lactic acid by bacteria. Fish silage has been adopted for analogous products of whole fish or fish parts (189, 190). Fish silage may be described as a liquid product made from whole fish (by-catch fish) or fish waste plus acid or, less frequently, alkali. Liquefaction is caused by the action of enzymes naturally present in the fish and is accelerated by the acid, creating the right conditions for the enzymes that are active at low pH to hydrolyze quickly most of the protein. This process yields an aqueous solution and limits the growth of spoilage bacteria (191).

Fish silage was first produced in Scandinavia in the 1930s by Edin, who treated different types of fish and fish waste with a mixture of sulfuric and hydrochloric acids to preserve and liquefy them. Production of acid fish silage on an industrial scale started in Denmark in 1948, and in 1951 its annual production was about 15,000 tons (190).

There are two methods for the production of fish silage (190). The first is acid-preserved silage. This is produced by the addition of inorganic or organic acid, which lowers the pH below 3–4, which is enough to avoid microbial spoilage. Silages made with inorganic acids require a lower pH. This pH is optimum for enzymes naturally present in the raw material used. Endogenous enzymes, mainly pepsins and cathepsins, hydrolyze the tissue structures producing an amber liquid, which is an aqueous phase rich in small peptides and free amino acids. Since no salt is added during silage, the autolysis is much faster than in fish sauce, and a high recovery of aqueous phase and an oil-rich fraction are normally obtained after a few days of storage (depending on the kind or raw material). The aqueous phase has a bitter taste and is not suitable for human consumption. This method is the most widely used.

The second method of production is fermented silage. A bacterial fermentation is initiated by mixing minced or chopped raw material with a fermentable sugar, which favors growth of lactic acid bacteria. These bacteria are usually added as a starter culture. The lactic acid bacteria produce acids and antibiotics, which together destroy competing spoilage bacteria, and the low pH achieves the tissue hydrolysis.

The silage method is an alternative to fish meal production in animal feeding. The main advantages of fish silage production are the simple technology and low investment costs compared with fish meal production. Fish silage can be made in fishing vessels and in small isolated places where fish meal plants cannot be operated economically (191). Moreover, the energy requirements of silage production are very low compared with fish meal. Oil separation in the silage process is achieved by enzymatic hydrolysis instead of by heating and pressing as in fish meal production. The final product in fish silage is more stable and resistant to spoilage, putrefaction, development of pathogens, and fly infestation than fish meal in tropical areas, because of acid preservation. The main disadvantages are the high water content (70–80%), resulting in high transportation costs, variable storage stability, and variable chemical composition. Fish meal is less bulky and thus cheaper to transport and store. However, different considerations may determine whether fuel should be used to evaporate water or whether the product should be transported in the form of liquid silage (21, 190).

With these considerations in mind, fish meal factories are usually settled at fishing ports, where it is convenient to use the readily available waste material for fish meal. In small and isolated ports and fishing communities, supplies of fish waste may be small and irregular. Therefore it is generally not economical to produce fish meal in these places, whereas production of fish silage is a feasible option (191).

Acid-preserved silage is used commercially in Scandinavia, Denmark, and Poland. During the 1970s, the annual production in Denmark reached a level of about 60,000 tons (192). Commercial silage production has been limited because of the increasing demand for standardized feeds in modern husbandry. However, renewed interest in this technology is arising, because it is being recognized as the most useful method for solving problems with waste handling in the aquaculture industry. Moreover, new methods for silage fractionation and processing are being developed to achieve a standardized product. Recently, pilot-scale and small-industrial-scale production has been introduced in many countries without approaching the same size of the fish meal industry. The annual production of fish silage is about 120,000 tons, produced mainly by the use of formic acid, acetic acid, and mineral acids (190, 191). As yet there is no commercial production of fish silage by the fermentation method, but the method is in partial operation on a small scale to use local fish waste in developing countries (190, 193, 194). Its application in fish feeding has been demonstrated (195).

Industrial-scale processing equipment for acid-preserved silage was developed during the 1980s in Norway. The raw materials for silage production are pelagic fish, trash fish, fish wastes, viscera, and byproducts. The production process is simple and has been developed in depth (190). Recently, the process has been reviewed to respond to several questions raised by the fish silage industry (191). The raw materials are minced into small particles (3–4 mm in diameter), to distribute the enzymes throughout the mass of fish and also to ensure thorough blending of the acid to avoid pockets of untreated fish where bacterial growth may continue. The minced fish is then mixed with acid preservatives.

The choice of acids for preservation is between mineral acids (hydrochloric and sulfuric), organic acids (acetic, propionic, and formic), or a mixture of both. Mineral acid can lower the pH to 2, but this silage is an unfinished product and must be neutralized before feeding. However, the high salt level resulting from the neutralization is nutritionally undesirable and produces an unstable product. The organic acids are more expensive than the mineral acids but their use gives stabilization at higher pH (around 4.0). Silage produced with organic acids can be used in feed without neutralization (189). Formic acid is the most used acidulant for the production of fish silage.

The mixture is stirred constantly in the first steps of digestion and the temperature is chosen for the right hydrolysis rate. During the mixing stage, silage gradually liquefies because of the hydrolysis activity of fish endogenous proteases. These enzymes have an optimal pH range of 2–4 and their activity decreases sharply above pH 4. At pH 3–4 the process is catalyzed by the exo- and endopeptidases of the muscles and digestive organs and results in a large accumulation of amino acids. At pH 2, the activity is restricted to pepsin-type proteinases, and the liberation of short peptides and amino acids is reduced (196). Hydrolysis activity is dependent on temperature, with maximum activity at 45–50°C. Approximately 80% of the protein in acid-preserved silage becomes liquified after 1 week at temperatures between 23 and 30°C (190).

After the liquefaction step, the silage is stored in tanks to precipitate insoluble tissue fragments, bone, sand, and other heavy particles to the bottom. Silage is stored for long periods, because of the acid content, but antioxidants need to be added to prevent oxidation of the fat (197). During storage it is also necessary to consider the metal corrosion in the tanks where silage is placed and the other metallic parts: a normal commercial steel tank will corrode at about 0.7 mm/year if the silage is kept at 36°C using formic acid as preservative. All silage-processing equipment should be designed in stainless steel (corrosion rate of 0.004 mm/year) (191).

During the storage, silage can be separated into three phases: a lipid–protein emulsion on top, an aqueous phase containing soluble nitrogenous compounds in the middle, and a small insoluble fraction at the bottom. Hydrolysis is terminated by pasteurization at 85°C for 15 min to inactivate proteases and lipases. After heating, the silage is deoiled by decantation and centrifugation. The liquefied protein phase is acidified to pH 4 to prevent spoilage. To reduce the transport and storage costs of fish silage, it is possible to produce concentrated silage by vacuum evaporation to reduce the bulk by about 50%. Concentrated silage has a syrupy consistency and may be used as an additive in pellets for animal feeding (191).

There is extensive literature describing conditions in the mixing step to obtain good-quality silage in a short time. In Poland, minced fish offal is treated with 0.2% sulfuric acid, 0.2% hydrochloric acid, and 2% formic acid to decrease the pH of the pulp to 3.5–3.8. After 1 day of maturation at 40°C, the product takes on a liquid consistency and has a pleasant fishy smell. The product can be stored for months at room temperature and has a high nutritional value for poultry (149). In Norway, cod-viscera silage is prepared by adding 0.75% propionic acid and 0.75% formic acid to minced viscera. The product has a pH of 4.3 and after 17 days at 27°C, about 85% of the protein is solubilized and no further solubilization occurs. Silage retains its fresh acidic smell for at least 1 year at this temperature (198). In Iceland, viscera silage without liver was produced with 3% formic acid and pH 3.5–3.8, at 35±2°C for 6–7 days (191).

For raw materials having low acid–protease activity, it is necessary to add exogenous proteolytic enzymes or enzyme-rich raw material (14–15). This practice is common in France and the United States. The hydrolysis of minced fish offal in the presence of added enzymes take places in few hours at optimum process variables. The result is the solubilization of about 80% of the total nitrogen. Most of the soluble fraction is peptides and amino acids. A variety of proteinases of animal, plant, or microbial origin, with maximum activity at pH ranging from 2 to 8.5, and with broad range of temperatures, are commercially available. The accelerated method may be rather complicated and requires expensive equipment and accurate control (149, 191). Addition of commercial enzymes in fermented silage production to accelerate the liquefaction process has been used. Addition of bromelain (0.7–0.9% w/w) and *Lactobacillus plantarum* to minced whole fish with 15% molasses increased the proteolysis rate and decreased the liquefaction time from 15 days to 12 h (199).

Fish silage usually is better digested than fish meal and is used to feed immature (poorly developed digestive system) domestic animals, poultry (203), fish (204), and as milk replacers for young animals (weaning calves and piglets)(85, 205). When mature ruminants or fish are fed on highly concentrated feeds based on fish silage, the animal production and growth are reduced. This is probably caused by adverse effects of highly hydrolyzed protein in the digestive metabolism of these animals (206, 207). The substitution of 5–10% of the feed protein by silage protein is the highest recommended dose. This is advisable as there are indications that health, fertility, and general appearance are improved when some fish silage protein is included, for the same reasons as for fish protein hydrolysates (21, 190). Recently, fish silage has been investigated as a raw material for plastein synthesis. Fish silage with a 65–70% degree of hydrolysis is the optimum source for plastein reaction, with pepsin at pH 5.0 as the most productive enzyme (208).

B. Carotenoid Pigments

One of the outstanding features of the salmonid fishes (Atlantic salmon, *Salmo salar*) is the salmon-pink color of their flesh. This color is, in the consumer's mind, closely connected with the quality of the fish, and a correct coloring of the flesh is of great importance to the salmonid aquaculture industry. Farmed salmon with a different color that is natural for the species will have a low classification and a low price in the market. The color of the flesh of salmonids is caused by astaxanthin, which belongs to a large group of compounds named carotenoids. Salmonids are not able to synthesize astaxanthin and depend on an adequate supply through their feed to obtain the color. Wild salmonids obtain astaxanthin from small crustaceans such as krill and shrimps. In salmonid farmed production, carotenoid pigments are used in the feed for improving the attractive pinkish–red color of the fish meat. For rainbow trout (*Oncorhynchus mykiss*), this is done to produce a flesh color similar to that of salmon (209, 210).

An important source of astaxanthin in industrial production is the yeast *Phaffia rhodozyma*. A good review about the yeast culture, pigment extraction, and its use in salmonid farming has been published (210, 211). Astaxanthin is also the major carotenoid pigment in shrimp and lobster (14). Proteolytic enzyme treatment of shrimp (212, 213), snow-crab shell (214), and shellfish (215) wastes allows the recovery of the carotenoid pigment along with the protein, since about one-third of the dry matter in crustacean shell waste is protein. By this method, the carotenoid pigment is recovered in the form of a protein–carotenoid complex, which is more resistant to oxidation and gives better results than free astaxanthin in the coloring of farmed rainbow trout (5).

The proteolytic enzymes used to aid the extraction of carotenoprotein have been trypsin type proteases. Pure bovine trypsin has been used with shrimp waste. The waste was soaked in the extraction buffer 0.5 M trisodium ethylenediaminetetraacetate (Na_3 EDTA) in a proportion 1:3 w/v, mixed at 4°C, and the homogenate was added with 0.1% (w/w) bovine trypsin. The digestion was done at pH 7.7 for 24 h at 4°C. After filtration of the homogenate, the filtrate was precipitated with ammonium sulfate, recovering carotenoprotein after centrifugation. The extraction process recovered a carotenoprotein fraction containing about 80% of the protein and carotenoid pigments present in shrimp offal. The long-term stability of the astaxanthin associated with the carotenoprotein was improved by addition of protease inhibitor and antioxidant to the product. Composition and properties of this product allow its use as feed supplement for pen-reared salmonids, coloring the flesh. High temperatures speed up the yield of carotenoprotein and do not require EDTA, but the odor and taste of the product are negatively affected (212).

Because better results have been obtained at low temperature, experiments in carotenoprotein extraction from shrimp process waste have been made that replace bovine trypsin with Atlantic cod trypsin (213). Cod trypsin is a more efficient catalyst than bovine trypsin at low reaction temperatures (7). Under identical conditions and in the same extraction buffer as above, using the same enzyme concentration, 64 % of astaxanthin and 81% of shrimp waste protein was recovered as carotenoprotein with purified Atlantic cod trypsin. However, using pure bovine pancreatic trypsin, the carotenoprotein recovered was 49% of the astaxanthin and 65% of the waste protein. Pure cod trypsin is necessary for effective extraction, because when a semipurified cod trypsin fraction has been used, a poor yield of intact carotenoprotein is recovered. This is presumably caused by other proteolytic or lipolytic enzymes, which are present in semipurified extract and act to degrade the carotenoprotein. However, at the moment the use of pure cod trypsin as an extraction aid is not feasible for commercial preparations of carotenoprotein, because purified enzyme is very expensive (213). The solution will be to find other cheaper ways of obtaining cod trypsin without residual activity, or to use cold-adapted trypsin from other cheap psychrophilic organism sources. The autolysis of crustacean waste in acid ensilaging prior to pigment extraction may be used to increase the recovery yield of astaxanthin (216).

VI. FUTURE OF FISH AND SHELLFISH ENZYMES APPLICATIONS

Exploitation of fish and shellfish enzymes for biotechnological applications in the food and feed industry is complicated by variable availability of raw material and expensive production because of the comparatively low enzyme concentration. It seems that profitable processes can only be established when the fish or shellfish enzymes have unique properties (chemical conditions, specificity, etc.) that cannot be imitated by less costly enzymes of plant or microbial origin. In the future, some of these enzymes may be produced more profitably by recombinant DNA or gene technology. With this technology, it is possible to cut out a small segment or gene from any chromosome (DNA molecules) and recombine this gene with genes from another chromosome. Genes for enzymes produced in minute quantities or from obscure sources may be transferred to high-yielding microorganisms (2, 16, 20, 88).

Recombinant DNA technology has been employed to clone the calf gene responsible for rennet activity into *Escherichia coli* and *Saccharomyces cerevisiae* (217–218). Cheddar-cheese-making trials comparing recombinant chymosin with calf rennet have found no significant differences between the two. Recombinant enzymes are now approved for commercial use, and recombinant chymosin meets some religious dietary requirements not met by calf rennet (167, 187). The amino-acid sequence, cloning, and cDNA encoding of fish and shellfish enzymes with importance as food-processing aids have been described:

transglutaminase from salmon (*Onchorhynchus keta*) liver (219) and trypsin from crayfish (*Pacifastacus leniusculus*) hepatopancreas (220).

Another approach to enzyme engineering is site-directed mutagenesis. This methodology can be used to make small modifications in the nucleotide sequence of a gene. Recently, modified genes have been used to produce enzymes with slightly different amino acid composition/sequence in determined positions, but with markedly different catalytic and stability properties. High methionine content appears as a general property of cold-adapted serine proteinases, and one position (Met-134) is present in the cod chymotrypsins as well as in cod salmon, and dogfish trypsin. The inherent mobility in methionine side-chains may contribute to the maintenance of flexibility at low temperatures. This methodology could transforms proteases (obtained by cheaper biotechnological processes) into other proteases with kinetic and thermodynamic properties similar to cold-adapted proteases (221).

Extreme environmental conditions are found in the oceans. Pressures from 1 atm at sea level up to several hundred at the sea bottom (1 atm every 10 m depth); temperature below 0°C in the Arctic and Antarctic oceans and exceeding 100°C in the hydrothermal vents in the ocean bottom; salinities from fresh water in the salt marshes up to 6N NaCl in the precipitation pits (salt mines); from almost zero concentration of organic substances to eutrophic areas are frequent. Because of the diverse habitats in the seas, there is an immense biodiversity of marine organisms and an immense genetic diversity. Each organism has a metabolism adapted to such conditions. Because metabolism is driven by enzymes, a vast diversity of enzymes with particular kinetic capabilities is expected to be discovered.

Cold-adapted enzymes from fish living at temperatures about the freezing point of seawater and thermoresistant enzymes from organisms, including crustacea, living in the hydrothermal vents have been reported. Enzymes working better at 2–4 M NaCl from organisms living in salt mines are hot topics in marine biology. Each time an enzyme with special kinetic abilities is discovered, a new potential application for food technology arises.

Enzymes help in a variety of processes, and in some examples they have become an important and indispensable part of the processes used by the modern food industry to produce a large and diversified range of products for humans and animals. The foremost advantage of enzymatic processes are the enzymes' specificity of both substrate and reaction, transforming only one molecule in a complex mixture of analogous molecules, rendering almost 100% transformation without byproducts. The transformation is done under mild conditions of temperature and pH. However, although enzymes are usually considered catalysts working at mild conditions, new discoveries show that enzymes can work at extreme conditions. In some cases, enzyme technology competes with and even substitutes for traditional engineering technology (222).

The search for new and improved enzymes is an approach appearing somewhat empirical to the new technologies such as protein engineering, gene cloning, synzymes, and reaction in supercritical carbon dioxide. However, it is a proven approach that has withstood the test of time. The screening of microorganisms and plants has become standard procedure and screening of marine organisms is at its earliest stage. Food technology is an evolving activity; new processes and ingredients will appear, so there will be a rapid growth in industrial applications of enzymes. The challenge to the enzyme technologists in food processing is to find new enzymes and to understand the kinetic and molecular basis of enzyme transformation to improve the functional and nutritional properties of foodstuff (223).

REFERENCES

1. JR Whitaker. Principles of Enzymology for the Food Science, 2nd ed. New York: Marcel Dekker, 1994, pp 1–27.
2. NF Haard. Enzymes as food processing aids. Proceeding of Yenching International Symposium on Critical Issues in the Food Industry in Nineties, Beijing, China, 1994, pp 1–12.
3. NF Haard. Specialty enzymes from marine organisms. Biotecnologia 2:78–85, 1997.
4. G Reed. Introduction. In: T Nagodawithana, G Reed, eds. Enzymes in Food. Processing, 3rd ed. San Diego: Academic Press, 1993, pp 1–5.
5. NF Haard. A review of proteolytic enzymes from marine organisms and their application in the food industry. J Aquat Food Product Tech 1:17–35, 1992.
6. S de Vecchi, Z Coppes. Marine fish digestive proteases—relevance to food industry and the South-West Atlantic region—a review. J Food Biochem 20:193–214, 1996.
7. BK Simpson, NF Haard. Purification and characterization of trypsin from Greenland cod (*Gadus ogac*). 1. Kinetic and thermodynamic characteristics. Can J Biochem Cell Biol 62:894–900, 1984.
8. BK Simpson, NF Haard. Cold-adapted enzymes from fish. In: D Knorr, ed. Food Biotechnology. New York: Marcel Dekker, 1987, pp 495–527.
9. BU Dittrich. Life under extreme coditions: aspects of evolutionary adaptation to temperature in crustacean proteases. Polar Biol 12:269–274, 1992.
10. G Feller, E Narinx, JL Arpigny, M Aittaleb, E Baise, S Genicot, C Gerday. Enzymes from psychrophilic organisms. FEMS Microbiol Rev 18:189–202, 1996.
11. COL Boyce. Novo's Handbook of Practical Biotechnology. Bagsvaerd, Denmark: Novo Industri A/S, 1986, pp 19–27.
12. GG Haraldsson. The applications of lipases for modification of facts and oils, including marine oils. In: MN Voigt, JR Botta, eds. Advances in Fisheries Technology and Biotechnology for Increased Profitability. Lancaster, PA: Technomic Publishing, 1990, pp 337–357.

13. G Stefánsson, U Steingrímsdóttir. Application of enzymes for fish processing in Iceland—present and future aspects. In: MN Voigt, JR Botta, eds. Advances in Fisheries Technology and Biotechnology for Increased Profitability. Lancaster, PA: Technomic Publishing, 1990, pp 237–250.

14. A Gildberg. Enzymic processing of marine raw materials. Process Biochem 28:1–15, 1993.

15. NF Haard, BK Simpson. Proteases from aquatic organisms and their uses in the seafood industry. In: AM Martin, ed. Fisheries Processing: Biotechnological Applications. London: Chapman & Hall, 1994, pp 132–154.

16. NF Haard, BK Simpson, ZE Sikorski. Biotechnological applications and seafood proteins and other nitrogenous compounds. In: ZE Sikorski, BS Pan, F Shahidi, eds. Seafood Proteins. London: Chapman & Hall, 1994, pp 194–216.

17. I Kolodziejska, ZE Sikorski. The properties and utilization of proteases of marine fish and invertebrates. Pol J Food Nutr Sci 4/45:5–12, 1995.

18. O Vilhelmsson. The state of enzyme biotechnology in the fish processing industry. Trends Food Sci Technol 8:266–270, 1997.

19. T Ohshima, T Suzuki, C Koizumi. New developments in surimi technology. Trends Food Sci Technol 4:157–163, 1993.

20. TC Lanier. Functional food protein ingredients from fish. In: ZE Sikorski, BS Pan, F Shahidi, eds. Seafood Proteins. London: Chapman & Hall, 1994, pp 127–159.

21. ZE Sikorski, A Gildberg, A Ruiter. Fish products. In: A Ruiter ed. Fish and Fishery Products—Composition, Nutritive Properties and Stability. Wallingford: CAB International, 1995, pp 315–346.

22. MT Morrissey, PS Hartley, H An. Proteolytic activity in Pacific whiting and effects of surimi processing. J Aquat Food Product Tech 4:17–35, 1995.

23. RW Porter, B Koury, G Kudo. Inhibition of protease activity in muscle extracts and surimi from Pacific whiting, *Merluccius productus,* and arrowtooth flounder, *Atheresthes stomias* Mar Fish Rev 55:10–15, 1993.

24. SW Boye, TC Lanier. Effects of heat-stable alkaline protease activity of Atlantic menhaden (*Brevoorti tyrannus*) on surimi gels. J Food Sci 53:1340, 1988.

25. M Gómez-Guillén, MA Martí de Castro, P Montero. Rheological and microstructural changes in gels made from high and low quality sardine mince with added egg white during frozen storage. Z Lebensm Unters Forsch A 205:419–428, 1997.

26. TC Lanier. Measurement of surimi composition and functional properties. In: TC Lanier, CM Lee, eds. Surimi Technology. New York: Marcel Dekker, 1992, pp 123–163.

27. TA Seymour, MT Morrissey, MY Peters, H An. Purification and characterization of Pacific whiting proteases. J Agric Food Chem 42:2421–2427, 1994.

28. H An, V Weerasinghe, TA Seymour, MT Morrissey. Cathepsin degradation of Pacific whiting surimi proteins. J Food Sci 59:1013, 1994.

29. Y Kumazawa, T Numazawa, K Seguro, M Motoki. Suppression of surimi gel setting by transglutaminase inhibitor. J Food Sci 60:715–717, 726, 1995.

30. I Kimura, M Sugimoto, K Toyoda, N Seki, K Arai, T Fujita. A study on the crosslinking reaction of myosin in kamaboko "suwari" gels. Nippon Suisan Gakk 57:1389–1396, 1991.

31. JE Folk. Transglutaminases. Annu Rev Biochem 49:517–531, 1980.

32. H Ando, M Adachi, K Umeda, A Matsura, M Nonaka, R Uchio, H Tanaka, M Motoki. Purification and characteristic of a novel transglutaminase derived from microorganisms. Agric Biol Chem 53:2613–2617, 1989.

33. G Matheis, JR Whitaker. A review: enzymatic cross-linking of proteins applicable to foods. J Food Biochem 11:309–327, 1987.

34. N Kitabatake, E Doi. Improvement of protein gel by physical and enzymatic treatment. Food Rev Int 9:445–71, 1993.

35. M Motoki, K Seguro. Trends in Japanese soy protein research. Inform 5:308–313, 1994.

36. HJ An, MY Peters, TA Seymour. Roles of endogenous enzymes in surimi gelation. Trends Food Sci Technol 7:321–327, 1996.

37. K Seguro, Y Kumazawa, T Ohtsuka, S Toiguchi, M Motoki. Microbial transglutaminase and epsilon-(gamma-glutamyl)lysine crosslink effects on elastic properties of kamaboko gels. J Food Sci 60:305–311, 1995.

38. JM Connellan, SI Chung, NK Whetzel, LM Bradley, JE Folk. Structural properties of guinea pig liver transglutaminase. J Biol Chem 246:1093–1098, 1971.

39. PP Brookhart, PL MaMahon, M Takahashi. Purification of guinea pig liver transglutaminase using a phenylalanine–Sepharose 4B affinity column. Anal Biochem 128:202–205, 1983.

40. TA Seymour, MY Peters, MT Morrissey, H An. Surimi gel enhancement by bovine plasma proteins. J Agric Food Chem 45:2919–2923, 1997.

41. N Seki, H Uno, NH Lee, I Kimura, K Toyoda, T Fujita, K Arai. Transglutaminase activity in Alaska pollack muscle and surimi, and its reaction with myosin B. Nippon Suisan Gakk 56:125–132, 1990.

42. H Yasueda, Y Kumazawa, M Motoki. Purification and characterization of a tissue-type transglutaminase from Red Sea bream (*Pagrus major*). Biosci Biotechnol Biochem 58:2041–2045, 1994.

43. P Falcone, D Serafini-Fracassini, S Del Duca. Comparative studies of transglutaminase activity and substrates in different organs of *Helianthus tuberosus*. J Plant Physiol 142:263–273, 1993.

44. H Ando, A Marsura, H Susumu. Manufacture of transglutaminase with *Streptomyces*. Jpn Kokai Tokkyo Koho JP 04108381, 1990.

45. U Gerber, U Jucknischke, S Putzien, HL Fuchsbauer. A rapid and simple method for the purification of transglutaminase from *Streptoverticillium mobaraense*. Biochem J 299:825–829, 1994.

46. GJ Tsai, SM Lin, ST Jiang. Transglutaminase from *Streptoverticillium ladakanum* and application to minced fish product. J Food Sci 61:1234–1238, 1996.

47. Y Ichihara, A Wakameda, M Motoki. Fish meat paste products containing transglutaminase and their manufacture. Jpn Kokai Tokkyo Koho JP 02186961, 1990.

48. A Wakameda, Y Ichihara, S Toiguchi, M Motoki. Manufacture of fish meat paste with transglutaminase as phosphate substitute. Jpn Kokai Tokkyo Koho JP 02100653, 1990.

49. H Sakamoto, Y Kumazawa, S Toiguchi, K Seguro, T Soeda, M Motoki. Gel

strength enhancement by addition of microbial transglutaminase during onshore surimi manufacture. J Food Sci 60:300–304, 1995.

50. Y Takagaki, K Narukawa. Manufacture of frozen meat paste containing transglutaminase. Jpn Kokai Tokkyo Koho JP 02100651, 1990.

51. H Sakamoto, T Soeda. Minced meat products containing transglutaminase. Jpn Kokai Tokkyo Koho JP 03175929, 1991.

52. T Soeda. Production of coagulated foods using transglutaminase. Gekkan Fudo Kemikaru 8:108–113, 1992.

53. Y Zhu, J Bol, A Rinzema, J Tramper. Microbial transglutaminase—a review of its production and application in food processing. Appl Microbiol Biotechnol 44:277–282, 1995.

54. HO Bang, J Dyerberg. Lipid metabolism and ischemic heart disease in Greenland eskimos. Adv Nutr Res 3:1–21, 1986.

55. SM Barlow, FVK Young, IF Duthie. Nutritional recommendations of n-3 polyunsaturated fatty acids and the challenge to the food industry. Proc Nutr. Soc 49:13–21, 1990.

56. W Schmidtsdorff. Fish meal and fish oil - not only by-poducts. In: A Ruiter, ed. Fish and Fishery Products—Composition, Nutritive Properties and Stability. Wallingford: CAB International, 1995, pp 347–376.

57. TAB Sanders. Marine oils: metabolic effects and role in human nutrition. Proc Nutr Soc 52:47–472, 1993.

58. KS Bjerve, L Thoresen, K Bønaa, T Vik, H Johnsen, AM Brubakk. Clinical studies with alpha-linolenic and long chain n-3 fatty acids. Nutrition 8:130–135, 1992.

59. N Ashwell ed. Task force on unsaturated fatty acids: nutritional and physiological significance. British Nutrition Foundation, London: Chapman & Hall, 1992.

60. SE Carlson. The role of PUFA in infant nutrition. Inform 6:940–946, 1995.

61. PRC Howe. Can we recommend fish oil for hypertension? Clin Exp Pharmacol Physiol 22:199–203, 1995.

62. DE Hughes. Fish oil and the immune system. Nutr Food Sci 2:12–16, 1995.

63. YY Linko, K Hayakawa. Docosahexaenoic acid: a valuable neutraceutical? Trends Food Sci Technol 7:59–63, 1996.

64. H Breivik, KH Dahl. Production and quality control of n-3 fatty acids. In: JC Frölich, C von Schacky, eds. Clinical Pharmacology, Vol 5. Fish, Fish Oil and Human Health. New York: W. Zuckschwerdt Verlag, 1992, pp 25–39.

65. A Valenzuela, S Nieto. Technological innovation applicable to marine oils rich in n-3 fatty acids to allow its nutritional and pharmaceutical use: a challenge for the current decade [Spanish]. Arch Latinoam Nutr 44:223–231, 1994.

66. Y Tanaka, J Hirano, T Funada. Synthesis of docosahexaenoic acid-rich triglyceride with immobilized *Chromobacterium viscosum* lipase. J Am Oil Chem Soc 71:331–334, 1994.

67. SR Moore, GP McNeill. Production of triglycerides enriched in long-chain n-3 polyunsaturated fatty acids from fish oil. J Am Oil Chem Soc 73:1409–1414, 1996.

68. GG Haraldsson, PA Höskuldsson, ST Sigurdsson, F Thorsteinsson, S Gudbjarnason. The preparation of triglycerides highly enriched with Ω-3 polyunsaturated

fatty acids via lipase catalyzed interesterification. Tetrahed Lett 30:1671–1674, 1989.

69. GG Haraldsson, B Hjaltason. Using biotechnology to modify marine lipids. Inform 3:626–629, 1992.

70. Ø Lie, G Lambertsen. Fatty acid specificity of *Candida cylindracea* lipase. Fette Seifen Anstrich 88:365–369, 1986.

71. Ø Lie, G Lambertsen. Digestive lipolytic enzymes in cod (*Gadus morrhua*): fatty acid specificity. Comp Biochem Physiol 80B:447–450, 1985.

72. DR Gjellesvik. Fatty acid specificities of bile salt-dependent lipase: enzyme recognition and super-substrate effects. Biochim Biophys Acta 1086:167–172, 1991.

73. DR Gjellesvik, D Lombardo, BT Wather. Pancreatic bile salt dependent lipase from cod (*Gadus morhua*): purification and properties. Biochim Biophys Acta 1124:123–134, 1992.

74. JS Patton, TG Warner, AA Benson. Partial characterization of the bile salt dependent triacylglycerol lipase from the leopard shark pancreas. Biochim Biophys Acta 486:322–330, 1977.

75. JS Patton, JC Nevenzel, AA Benson. Specificity of digestive lipases in hydrolysis of wax esters and triglycerides studied in anchovy and others selected fish. Lipids 10:575–583, 1975.

76. DR Gjellesvik, AJ Raae, BT Wather. Partial purification and characterization of a triacylglycerol lipase from cod (*Gadus morhua*). Aquaculture 79:177–184. 1989.

77. N Iijima, S Tanaka, Y Ota. Purification and characterization of bile salt-activated lipase from the hepatopancreas of red sea bream, *Pagrus major.* Fish Physiol Biochem 18:59–69, 1998.

78. GP McNeill, RG Ackman, SR Moore. Lipase-catalyzed enrichment of long-chain polyunsaturated fatty acids. J Am Oil Chem Soc 73:1403–1407, 1996.

79. GG Haraldsson, B Kristinsson, R Sigurdardottir, GG Gudmundsson, H Breivik. The preparation of concentrates of EPA and DHA by lipase-catalyzed transesterification of fish oil with ethanol. J Am Oil Chem Soc 74:1419–1424, 1997.

80. H Breivik, GG Haraldsson, B Kristinsson. Preparation of highly purified concentrates of eicosapentaenoic acid and docosahexaenoic acid. J Am Oil Chem Soc 74:1425–1429, 1997.

81. DR Tocher, A Webster, JR Sargent. Utilization of porcine pancreatic phospholipase A_2 for the preparation of a marine fish oil enriched in (n-3) polyunsaturated fatty acids. Biotechnol Appl Biochem 8:83–95, 1986.

82. VI Shenderyuk, PJ Bykowski. Salting and marinating of fish. In: ZE Sikorski, ed. Seafood: Resources, Nutritional Composition, and Preservation. Boca Raton, FL: CRC Press, 1990, pp 147–162.

83. J Raa. Modern biotechnology: impact on aquaculture and the fish processing industry. Paper presented at the 5th World Productivity Congress, Jakarta, Indonesia, 13–16 April 1986.

84. T Sugihara, C Yashima, H Tamura, M Kawasaki, S Shimizu. Process for preparation of ikura (salmon egg). US Patent No 3759718, 1973.

85. A Gildberg, KA Almås. Utilization of fish viscera. In: M Le Maguer, P Jelen, eds.

Food Engineering and Process Applications—Vol. 2, Unit Operations. London: Elsevier Science, 1986, pp 388–393.

86. A Gildberg. Aspartic proteinases in fishes and aquatic invertebrates. Comp Biochem Physiol 91B:425–435, 1988.

87. KA Almås. Utilization of marine biomass for production of microbial growth media and biochemicals. In: MN Voigt, JR Botta, eds. Advances in Fisheries Technology and Biotechnology for Increased Profitability. Lancaster, PA: Technomic Publishing, 1990, pp 361–372.

88. J Raa. Biotechnology in aquaculture and the fish processing industry: a success story in Norway. In: MN Voigt, JR Botta, eds. Advances in Fisheries Technology and Biotechnology for Increased Profitability. Lancaster, PA: Technomic Publishing, 1990, pp 509–524.

89. RA Xu, RJ Wong, ML Rogers, GC Fletcher. Purification and characterization of acidic proteases from the stomach of the deepwater finfish orange roughy (*Hoplostethus atlanticus*). J Food Biochem 20:31–48, 1996.

90. T Wray. Fish processing: new uses for enzymes. Food Manuf 63:48–49, 1988.

91. BK Simpson, NF Haard. Trypsin from Greenland cod as a food-processing aid. J Appl Biochem 6:135–143, 1984.

92. YZ Lee, BK Simpson, NF Haard. Supplementation of squid fermentation with proteolytic enzymes. J. Food Biochem 6:127–134, 1982.

93. V Venugopal, F Shahidi. Traditional methods to process underutilized fish species for human consumption. Food Rev Int 14:35–97, 1998.

94. NF Haard. Enzymes in food myosystems. J Muscle Foods 1:293–338, 1990.

95. INA Ashie, BK Simpson. Proteolysis in food myosystems—a review. J Food Biochem 21:91–123, 1997.

96. I Kolodziejska, ZE Sikorski. Neutral and alkaline muscle proteases of marine fish and invertebrates—a review. J Food Biochem 20:349–363, 1996.

97. I Kolodziejska, ZE Sikorski. The digestive proteases of marine fish and invertebrates. Bull Sea Fish Inst 1(137): 51–56, 1996.

98. I Stoknes, T Rustad, V Mohr. Comparative studies of the proteolytic activity of tissue extracts from cod (*Gadus morhua*) and herring (*Clupea harengus*). Comp Biochem Physiol 106B:613–619, 1993.

99. TM Ritskes. Artificial ripening of maatjes-cured herring with the aid of proteolytic enzyme preparations. Fish Bull 69:647–654, 1971.

100. A Ruiter. Substitution of proteases in the enzymatic ripening of herring. Ann Technol Agric 21:597–605, 1972.

101. C Eriksson. Method of controlling the ripening process of herring. Canadian Patent No 969419, 1975.

102. K Opshaug. Procedure for accelerated enzymatic ripening of herring [Norwegian]. Norwegian Patent No 148207, 1983.

103. Anonymous. Production process for salted herrings (Matjes). British Patent No 1403221, 1975.

104. I Kolodziejska, J Pacana, ZE Sikorski. Effect of squid liver extract on proteins and on the texture of cooked squid mantle. J Food Biochem 16:141–150, 1992.

105. B Asgeirsson, JW Fox, JB Bjarnason. Purification and characterization of trypsin from the poikilotherm *Gadus morhua*. Eur J Bichem 180: 85–94, 1989.

106. E Bárzana, M García-Garibay. Production of fish protein concentrates. In: AM Martin, ed. Fisheries Processing: Biotechnological Applications. London: Chapman & Hall, 1994, pp 206–222.

107. V Venugopal, F Shahidi. Value-added products from underutilized fish species. Crit Rev Food Sci Nutr 35:431–453, 1995.

108. ZE Sikorski, M Naczk. Modification of technological properties of fish protein concentrates. CRC Critical Rev Food Sci Nutr 14:201–230, 1981.

109. HLA Tarr. Possibilities in developing fisheries by-products. Food Technol 2:268–277, 1948.

110. J Raa, A Gildberg. Autolysis and proteolytic activity of cod viscera. J Food Technol 11:619–628, 1976.

111. A Gildberg, I Batista, E Strøm. Preparation and characterization of peptones obtained by a two-step enzymatic hydrolysis of whole fish. Biotechnol Appl Biochem 11:413–423, 1989.

112. SE Vecht-Lifshitz, KA Almås, E Zomer. Microbial growth on peptones from fish industrial wastes. Lett Appl Microbiol 10:183–186, 1990.

113. IM Mackie. Fish protein hydrolysates. Process Biochem 17:26–31, 1982.

114. P Hevia, JR Whitaker, HS Olcott. Solubilization of fish protein concentrate with proteolytic enzymes. J Agric Food Chem 24:383–385, 1976.

115. IM Mackie. General review of fish protein hydrolysates. Anim Feed Sci Technol 7:113–124, 1982.

116. JD Owens, LS Mendoza. Enzymatically hydrolised and bacterially fermented fishery products. J Food Technol 20:73–293, 1985.

117. A Kilara. Enzyme-modified protein food ingredients. Process Biochem 20:149–158, 1985.

118. J Adler-Nissen. Enzymic hydrolysis of food proteins. New York: Elsevier, 1986, pp 263–313.

119. BD Rebeca, MT Peña-Vera, M Díaz-Castañeda. Production of fish protein hydrolysates with bacterial proteases; yield and nutritional value. J Food Sci 56:309–314, 1991.

120. V Venugopal. Production of fish protein hydrolyzates by microorganisms. In: AM Martin, ed. Fisheries Processing: Biotechnological Applications. London: Chapman & Hall, 1994, pp 223–243.

121. LL Lin, GM Pigott. Preparation and use of inexpensive crude pepsin for enzyme hydrolysis of fish J Food Sci 46:1569–1572, 1981.

122. J Montecalvo, SM Constantinides, CST Yang. Enzymatic modification of fish frame protein isolate. J Food Sci 49:1305–1309, 1984.

123. HH Baek, KR Cadwallader. Enzymatic hydrolysis of crayfish processing byproducts. J Food Sci 60:929–935, 1995.

124. BK Simpson, G Nayeri, V Yaylayan, INA Ashie. Enzymatic hydrolysis of shrimp meat. Food Chem 61:131–138, 1998.

125. GB Quaglia, E Orban. Enzymatic solubilisation of proteins of sardine (*Sardina pilchardus*) by commercial proteases. J Sci Food Agric 38:263–269, 1987.

126. GB Quaglia, E Orban. Influence of the enzymatic hydrolysis on structure and emulsifying properties of sardine (*Sardina pilchardus*) protein hydrolysates. J Food Sci 55:1571, 1990.

127. WJ Lahl, SD Braun. Enzymatic production of protein hydrolysates for food use. Food Technol 48:68–71, 1994.

128. ZE Sikorski, M Naczk. Changes in functional properties in fish protein preparations induced by hydrolysis. Acta Alim Pol 8:35–42, 1982.

129. T Ellingsen, V Mohr. A new process for utilization of Antarctic krill. Process Biochem 14:14–19, 1979.

130. KK Osnes, T Ellingsen, V Mohr. Hydrolysis of proteins by peptide hydrolases of Atlantic krill, *Euphausia superba*. Comp Biochem Physiol 83B:801–805, 1986.

131. B Karlstam, J Vincent, B Johansson, C Brynö. A simple purification method of squeezed krill for obtaining high levels of hydrolytic enzymes. Pre. Biochem 21:237–256, 1991.

132. L Hellgren, B Karlstam, V Mohr, J Vincent. Krill enzymes. A new concept for efficient debridement of necrotic ulcers. Int J Dermatol 30:102–103, 1991.

133. JR Mekkes, IC Le Poole, PK Das, JD Bos, W Westerhof. Efficient debridement of necrotic wounds using proteolytic enzymes derived from Atlantic krill: a double-blind, placebo-controlled study in a standardized animal wound model. Wound Rep Reg 6:50–58, 1998.

134. M Díaz-Castañeda, GJ Brisson. Replacement of skimmed milk with hydrolyzed fish protein and nixtamal in milk substitutes for dairy calves. J Dairy Sci 70:130–140, 1987.

135. M Díaz-Castañeda, GJ Brisson. Blood responses of calves fed milk substitutes containing hydrolyzed fish protein and lime-treated corn flour. J Dairy Sci 72:2095–2106, 1989.

136. F Kubitza, LL Lovshin. The use of freeze-dried krill to feed train largemouth bass (*Micropterus salmonides*): feeds and training strategies. Aquaculture 148:299–312, 1997.

137. C Vinot, P Bouchez, P Durand. Extraction and purification of peptides from fish protein hydrolysates. In: S Miyachi, I Karube, Y Ishida eds. Current Topics in Marine Biotechnology. Tokyo: Fuji Technology Press, 1989, pp 361–364.

138. V Mohr. Enzymes technology in the meat and fish industries. Process Biochem 15:18, 1980.

139. G Roy. Bitterness: reduction and inhibition. Trends Food Sci Technol 3:85–91, 1992.

140. B Pedersen. Removing bitterness from protein hydrolysates. Food Technol 48:96–98, 1994.

141. S Arai, M Fujimaki. The plastein reaction. Theoretical basis. Ann Nutr Aliment 32:701–707, 1978.

142. T Nagodawithana. Enzymes associated with savory flavor enhancement. In: T Nagodawithana, G Reed, eds. Enzymes in Food Processing, 3rd ed. San Diego: Academic Press, 1993, pp 401–421.

143. T Kawai. Fish flavor. Crit Rev Food Sci Nutr 36:257–298, 1996.

144. T In. Seafood flavourants produced by enzymatic hydrolysis. In: MN Voigt, JR

Botta, eds. Advances in Fisheries Technology and Biotechnology for Increased Profitability. Lancaster, PA: Technomic Publishing, 1990, pp 425–436.

145. CY Shiau, T Chai. Characterization of oyster shucking liquid wastes and their utilization as oyster soup. J Food Sci 55:374–378, 1990.

146. HY Chen, CF Li. Isolation, partial purification, and application of proteases from grass shrimp heads. Food Sci (China) 15:230–243, 1988.

147. Y Joh, LF Hood. Preparation and properties of dehydrated clam flavor from clam processing wash water. J Food Sci 44:1612, 1979.

148. HR Kim, HH Baek, SP Meyers, KR Cadwallader, JS Godber. Crayfish hepatopancreatic extract improves flavor extractability from crab processing by-product. J Food Sci 59:91–96, 1994.

149. ZE Sikorski, A Ruiter. Changes in proteins and nonprotein nitrogen compounds in cured, fermented, and dried seafoods. In: ZE Sikorski, BS Pan, F Shahidi, eds. Seafood Proteins. London: Chapman & Hall, 1994, pp 113–126.

150. P Saisithi. Traditional fermented fish: fish sauce production. In: AM Martin, ed. Fisheries Processing: Biotechnological Applications. London: Chapman & Hall, 1994, pp 111–131.

151. A Gildberg, S Xian-Quan. Recovery of tryptic enzymes from fish sauce. Process Biochem 29:151–155, 1994.

152. F Magno-Orejana, J Liston. Agents of proteolysis and its inhibition in patis (fish sauce) fermentation. J Food Sci 47:198–203, 1982.

153. CG Beddows, AG Ardeshir. The production of soluble fish protein solution for use in fish sauce manufacture. II. The use of acids at ambient temperature. J Food Technol 14:613–623, 1979.

154. A Gilgberg, J Espejo-Hermes, F Magno-Orejana. Acceleration of autolysis during fish sauce fermentation by adding acid and reducing the salt content. J Sci Food Agric 35:1363–1369, 1984.

155. A Gilgberg. Accelerated fish sauce fermentation by initial alkalification at low salt concentration. In: S Miyachi, I Karube, Y Ishida, eds. Current Topics In Marine Biotechnology. Tokyo: Fuji Technology Press, 1989, pp 101–104.

156. CG Beddows, AG Ardeshir. The production of soluble fish protein solution for use in fish sauce manufacture. I. The use of added enzymes. J Food Technol 14:603–612, 1979.

157. N Raksakulthai, YZ Lee, NF Haard. Effect of enzyme supplements on the production of fish sauce from male capelin (*Mallotus villosus*). Can Inst Food Sci Techenol J 19:28–33, 1986.

158. N Raksakulthai, NF Haard. Correlation between the concentration of peptides and amino acids and the flavour of fish sauce. ASEAN Food J 7:86–90, 1992.

159. G Stefánsson. Enzymes in the fishing industry. Food Technol 42:64–65, 1988.

160. A Gildberg, J Raa. Solubility and enzymatic solubilization of muscle and skin of capelin (*Mallotus villosus*) at different pH and temperature. Comp Bichem Physiol 63B:309–314, 1979.

161. KG Joakimsson. Enzymatic deskinning of herring (*Clupea harengus*). PhD thesis, Institute of Fisheries, University of Tromsø, Norway, 1984.

162. S Mizuta, R Yoshinaka, M Sato, M Sakaguchi. Isolation and partial characteriza-

tion of two distinct types of collagen in the squid *Todarodes pacificus.* Fisheries Sci 60:467–471, 1994.

163. DH Buisson, DK O'Donnel, DN Scott, SC Ting. Squid processing options for New Zealand. Fish Proc Bull 6:18–44, 1985.

164. JL Leuba, I Meyer. Skinning of squid. Fifth International Congress on Engineering and Food, Cologne, 1989, p. 240.

165. BK Simpson, NF Haard. Trypsin and a trypsin-like enzyme from the stomachless cunner. Catalytic and other physical characteristics. J Agric Food Chem 35:652–656, 1987.

166. BK Simpson, MV Simpson, NF Haard. On the mechanism of enzyme action: digestive proteases from selected marine organisms. Biotechnol Appl Biochem 11:226–234, 1989.

167. RJ Brown. Dairy products. In: T Nagodawithana, G Reed, eds. Enzymes in Food Processing, 3rd ed. San Diego: Academic Press, 1993, pp 347–361.

168. ML Green, SV Morant. Mechanism of aggregation of casein micelles in rennet-treated milk. J Dairy Res 48:57–63, 1981.

169. DJ McMahon, RJ Brown. Composition, structure, and integrity of casein micelles: a review. J Dairy Sci 67:499–512, 1984.

170. DJ McMahon, RJ Brown. Enzymatic coagulation of casein micelles: a review. J Dairy Sci 67:919–929, 1984.

171. PJ de Koning. Coagulating enzymes in cheese making. Dairy Ind Int 43:7–12, 1978.

172. ML Green. Review of the progress of dairy science: milk coagulants. J Dairy Res 44:159–188, 1977.

173. PF Fox. Proteolysis during cheese manufacture and ripening. J Dairy Sci 72:1379–1400, 1989.

174. S Visser. Proteolytic enzymes, and their action on milk proteins. A review. Neth Milk Dairy J 35:65–88, 1981.

175. DJ McMahon, RJ Brown. Effects of enzyme type on milk coagulation. J Dairy Sci 68:628–632, 1985.

176. B Manji, Y Kakuda. The role of protein denaturation, extent of proteolysis, and storage temperature on the mechanism of age gelation in a model system. J Dairy Sci 71:1455–1463, 1988.

177. JL Sardinas. Microbial rennets. Adv Appl Microbiol 15:39–73, 1972.

178. K Shamsuzzaman, NF Haard. Evaluation of harp seal gastric protease as a rennet substitute for Cheddar cheese. J Food Sci 48:179–182, 1983.

179. JFP Tavares, JAB Baptista, MF Marcone. Milk-coagulating enzymes of tuna fish waste as a rennet substitute. Int J Food Sci Nutr 48:169–176, 1997.

180. S Gordin, I Rosenthal. Efficacy of chicken pepsin as a milk clotting enzyme. J Food Protection 41:684–688, 1978.

181. P Brewer, N Helbig, NF Haard. Atlantic cod pepsin—characterization and use as a rennet substitute. Can Inst Food Sci Technol J 17:38–43, 1984.

182. NF Haard. Atlantic cod gastric protease. Characterization with casein and milk substrate and influence of sepharose immobilization on salt activation, temperature characteristics ad milk clotting reaction. J Food Sci 51:313, 1986.

183. FL García-Carreño, R Raksakulthai, NF Haard. Processing wastes. Exopeptidases from shellfish. In: A Bremmer, C Davis, B Austin, eds. Making the Most of the Catch. Hamilton, Queensland: AUSEAS, 1997, pp 37–43.

184. K Shamsuzzaman, NF Haard. Purification and characterization of a chymosinlike protease from the gastric mucosa of harp seal (*Pagophilus groenlandicus*). Can J Biochem Cell Biol 62:699–708, 1984.

185. K Shamsuzzaman, NF Haard. Milk clotting and cheese making properties of a chymosin-like enzyme from harp seal mucosa. J Food Biochem 9:173–192, 1985.

186. D Lim, WF Shipe. Proposed mechanism for the anti-oxygenic action of trypsin in milk. J Dairy Sci 55:753–758, 1972.

187. RFH Dekker. Enzymes in food and beverage processing. 1. Food Austral 46:36–139, 1994.

188. B Myrnes, A Johansen. Recovery of lysozyme from scallop waste. Prep Biochem 24:69–80, 1994.

189. IN Tatterson, ML Windsor. Fish silage. J Sci Food Agric 25:369–379, 1974.

190. J Raa, A Gildberg. Fish silage: a review. CRC Crit. Rev Food Sci Nutr 16:383–419, 1982.

191. S Arason. Production of fish silage. In: AM Martin, ed. Fisheries Processing: Biotechnological Applications. London: Chapman & Hall, 1994, pp 244–272.

192. J Wignall, I Tatterson. Fish silage. Process Biochem 11:17–19, 1976.

193. O Fagbenro, K Jauncey. Chemical and nutritional quality of raw, cooked and salted fish silages. Food Chem 48:331–335, 1993.

194. O Fagbenro. Preparation, properties and preservation of lactic acid fermented shrimp heads. Food Res Int 29:595–599, 1996.

195. O Fagbenro, K Jauncey. Chemical and nutritional quality of dried fermented fish silages and their nutritive value for tilapia (*Oreochromis niloticus*). Animal Feed Sci Technol 45:167–176, 1994.

196. MR Raghunath, AR McCurdy. Influence of pH on the proteinase complement and proteolytic products in rainbow trout viscera silage. J Agric Food Chem 38:45–50, 1990.

197. NF Haard, N Kariel, G Herzberg, LAW Feltham. Stabilization of protein and oil in fish silage for use as ruminant feed supplement. J Sci Food Agric 36:229–241, 1985.

198. A Gildberg, J Raa. Properties of a propionic acid/formic acid preserved silage of cod viscera. J Sci Food Agric 27:647–653, 1977.

199. E Tomé, BA Levy, RA Bello. Proteolytic activity control in fish silage [Spanish]. Arch Latinoam Nutr 45:317–321, 1995.

200. Å Krogdahl. Fish viscera silage as a protein source for poultry. I. Experiments with layer-type chicks and hens. Acta Agric Scand 35:3–23, 1985.

201. Å Krogdahl. Fish viscera silage as a protein source for poultry. II. Experiments with meat-type chickens and ducks. Acta Agric Scand 35:24–32, 1985.

202. AJ Jackson, AK Kerr, CB Cowey. Fish silage as a dietary ingredient for salmon. I. Nutritional and storage characteristics. Aquaculture 38:211–220, 1984.

203. FE Stone, RW Hardy. Nutritional value of acid stabilised silage and liquefied fish protein. J Sci Food Agric 37:797–803, 1986.

204. SP Lall. Nutritional value of fish silage in salmonid diets. Fish Silage Workshop, J Delabbie ed., Nova Scotia, Canada, 1991, pp 63–74.

205. IN Tatterson. Fish silage—preparation, properties and uses. Animal Feed Sci Technol 7:153–159, 1982.

206. F Johnsen, A Skrede. Evaluation of fish viscera silage as a feed resource. Acta Agric Scand 31:21–27, 1981.

207. RW Hardy, KD Shearer, J Spinelli. The nutritional properties of co-dried fish silage in rainbow trout (*Salmo gairdneri*) diets. Aquaculture 38:25–44, 1984.

208. MR Raghunath, AR McCurdy. Synthesis of plasteins from fish silage. J Sci Food Agric 54:655–658, 1991.

209. SJ de Groot. Edible species. In: A Ruiter ed. Fish and Fishery Products—Composition, Nutritive Properties and Stability. Wallingford: CAB International, 1995, pp 31–76.

210. A Tangerås, E Slinde. Coloring of salmonids in aquaculture: the yeast *Phaffia rhodozyma* as a source of astaxanthin. In: AM Martin, ed. Fisheries Processing: Biotechnological Applications. London: Chapman & Hall, 1994, pp 391–431.

211. JD Fontana, MB Chocial, M Baron, MF Guimaraes, M Maraschin, C Ulhoa, JA Florencio, TMB Bonfim. Astaxanthinogenesis in the yeast *Phaffia rhodozyma*—optimization of low-cost culture media and yeast cell-wall lysis. Appl Biochem Biotechnol 63:305–314, 1997.

212. BK Simpson, NF Haard. The use of proteolytic enzymes to extract carotenoproteins from shrimp waste. J Appl Biochem 7:212–222, 1985.

213. A Cano-López, BK Simpson, NF Haard. Extraction of carotenoprotein from shrimp process wastes with aid of trypsin from Atlantic cod. J Food Sci 52:503–504, 506, 1987.

214. W Manu-Tawiah, NF Haard. Recovery of carotenoprotein from the exoskeleton of snow crab, *Chionectes opilio*. Can Inst Food Sci Technol J 20:31–33, 1987.

215. T Ya, BK Simpson, H Ramaswamy, V Yaylayan, JP Smith, C Hudon. Carotenoproteins from lobster waste as a potential feed supplement for cultured salmonids. Food Biotechnol 5:87–93, 1991.

216. HM Chen, SP Meyers. Ensilage treatment of crawfish waste for improvement of astaxanthin pigment extraction. J Food Sci 48:1516, 1983.

217. M Teuber. Production of chymosin (EC 3.4.23.4) by microorganisms, and its use for cheesemaking. Bull Int Dairy Fed 251:3–15, 1990.

218. DT Moir, JI Mao, MJ Duncan, RA Smith, T Kohno. Production of calf chymosin by the yeast *S. cerevisiae*. Dev Ind Microbiol 26:75–85, 1985.

219. K Sano, K Nakanishi, N Nakamura, M Motoki, H Yasueda H. Cloning and sequence analysis of a cDNA encoding salmon (*Onchorhynchus keta*) liver transglutaminase. Biosci Biotechnol Biochem 60:1790–1794, 1996.

220. M Hernández-Cortes, L Cerenius, F García-Carreño, K Söderhäll. Purification and cDNA cloning of trypsin from *Pacifastacus leniusculus* hepatopancreas. Biol Chem, 380: 499–501, 1999.

221. R Leth-Larsen, B Ásgeirsson, M Thórólfsson, M Nørregaard-Madsen, P Højrup. Structure of chymotrypsin variant B from Atlantic cod, *Gadus marhua*. Biochim Biophys Acta 1297:49–56, 1996.

222. FM Christensen. Enzyme technology versus engineering technology in the food industry. Biotechnol Appl Biochem 11:249–265, 1989.

223. BP Wasserman. Evolution of enzyme technology: progress and prospects. Food Technol 44:118–122, 1990.

224. HR Reyes, CG Hill. Kinetic modeling of interesterification reactions catalyzed by immobilized lipase. Biotechnol Bioeng 43:171–182, 1994.

22
Uses of Enzymes from Marine Organisms

Asbjørn Gildberg
Norwegian Institute of Fisheries and Aquaculture, Ltd., Tromsø, Norway

Benjamin K. Simpson
McGill University, Ste. Anne de Bellevue, Quebec, Canada

Norman F. Haard
University of California, Davis, California

I. INTRODUCTION

In fish the digestive tract is the richest source of hydrolytic enzymes. Crude or partly purified preparations of such enzymes are utilized as biotechnological tools in the production of fishery products such as caviar, descaled skin-on fillets, and in the maturation of various fish delicacy products. These and many other applications will be discussed in this chapter, but initially two important examples of the utilization of autolytic digestion in product manufacturing will be described. These are fish sauce and fish silage production, in which hydrolytic enzymes from the fish itself play a key role in the solubilization and degradation of the tissue proteins.

II. FISH SAUCE

Fish sauce is a salt-preserved solution of enzymatically solubilized and digested fish protein, which is used as a food ingredient and condiment on vegetable dishes. Fish sauce is mainly used in Southeast Asia where the annual production is about 300,000 tons (1, 2). In Europe fish sauce was used in the ancient Greece and Roman cultures, but here the tradition almost vanished (3). Only special fish sauce products made from liver and fish fry are still produced in Greece and southern France, respectively (4).

Fish sauce is usually made from small pelagic species such as anchovies and sardines. In the traditional production the fish is mixed with solar-dried sea salt in a weight ratio about 1:3 and stored at tropical temperatures in earthern jars or concrete tanks. After 6–12 months of storage the fish sauce can be drained off as an amber aqueous solution in which salt and digested fish protein are the major ingredients. It has been a matter of discussion whether endogenous fish enzymes or microbial enzymes are the most important factors during fish sauce fermentation (5–8). Due to the very high salt content in fish sauce (20–25% w/w), the microbial activity is generally low. Hamm and Clague (9) showed that the total bacterial count was reduced from an initial level of 6.5×10^6/g to only 300/g during the first 3 weeks of anchovy sauce fermentation. Raksakulthai and Haard (10) concluded that the microbial activity did not contribute to the protein hydrolysis during fermentation of Arctic capelin (*Mallotus villosus*). There is no doubt, however, that bacterial methabolites such as free fatty acids, amines, and various other nitrogen-containing compounds are major contributors to fish sauce flavor and aroma (11–13). Also the endogenous hydrolytic fish enzymes are very much inhibited by the high salt concentration. The activity of crude pepsin from Atlantic salmon (*Salmo salar*) was very low at 10% and completely inhibited at 15% salt concentration (14). Also the tryptic enzymes are significantly inhibited by salt, but this inhibition is less pronounced and varies considerably with different species. Whereas the activity of salmon trypsins is reduced to about 65% at 15% salt, bovine trypsin still expresses about 70% of the total activity at 25% salt (6, 14). Orejana and Liston (6) showed that protein digestion during fish sauce fermentation was drastically reduced when tryptic enzymes were inhibited by addition of soybean trypsin inhibitor. This indicates that trypsins and chymotrypsins are of vital importance for tissue solubilization and protein digestion, even if their activity is partly inhibited by the salt. Comparison of tryptic enzymes from anchovy indicates that chymotrypsin most probably is more important than trypsin during fish sauce fermentation since chymotrypsin is more active at neutral and weak acid conditions (15).

The normal pH of fish sauce is about 6 (5, 10, 16, 17). This is below the optimal range for activity of trypsins and chymotrypsins, but they still express about 20 and 40%, respectively, of maximal activity at this pH (15, 18, 19). The enzymatic hydrolysis is relatively fast during the first few days of fermentation, but levels off rapidly after a few weeks (20, 21). This is partly due to a reduced proteolytic activity, which may be caused both by enzyme inactivation and product inhibition. During fish sauce fermentation of tropical species, the protease activity was reduced to only 1% after 5 weeks of fermentation (8), whereas no significant activities of either trypsin, chymotrypsin, or elastase were detected after 4 weeks of fermentation of Atlantic cod viscera (22). During fermentation of tropical anchovies a less pronounced activity reduction was recorded. Here the tryptic activity was reduced to about 20% after 4–5 months of storage at 37°C (6). This may possibly explain why anchovies are regarded as the most suitable raw material for fish sauce production in Southeast Asia.

Although the tryptic enzymes obviously are of vital importance during fish sauce fermentation, a number of other proteolytic enzymes, such as exopeptidases, cathepsins, and other tissue enzymes, may contribute significantly to the protein digestion. The aspartic proteinase, cathepsin D, is probably the most abundant muscle proteinase in fish (23). This enzyme, however, is inactivated by high salt concentrations and cannot contribute to tissue solubilization during fish sauce fermentation (24, 25). Raksakulthai and Haard (10) found that the protein hydrolysis during sauce fermentation of eviscerated capelin was almost equally good as when whole fish was used as a raw material. They concluded that thiol proteases including cathepsin C may play an important role since such enzymes were only moderately inhibited by 20–25% salt.

Although present in low concentrations, elastases and collagenases may play a role in the initial digestion of connective tissues (26, 27). These are serine proteinases active from weak acid to alkaline pH, and at least elastases may be very active even at high salt concentrations (26).

During the last decades two groups of proteases associated with fish muscle proteins have been given considerable attention: the calpains and the multicatalytic proteinases. Calpains are calcium-activated cysteine proteinases with optimal activity at neutral pH (28), but which are apparently strongly inhibited by high salt concentrations (29). Hence, the effect of calpains during fish sauce fermentation is probably insignificant. The multicatalytic proteinases are huge enzyme complexes with various tryptic activities. In in vitro systems these enzymes normally require quite high temperature for activation, but they may also be activated by free fatty acids (30). Although these enzymes requires at least 1% salt to be active, it is not known whether they are active at the very high salt concentration occuring during fish sauce fermentation (31).

Anchovies are the species most frequently used for fish sauce production. Two kinds of neutral serine proteinases have recently been purified from the muscle of salted anchovy (32). These enzymes still express some activity at above 25% salt and are probably active during sauce fermentation. The cysteine proteinase cathepsin L has also been detected in anchovy muscle. This enzyme is active at the weak acid conditions normally occuring in fish sauce (33).

Fish sauce fermentation is a time-consuming process, and several methods for speeding up the process have been tested out. Generally it is important to choose raw materials with a high content of proteolytic enzymes. Such activities vary among various species and within one species according to the seasons and feeding situation. Apparently fresh raw materials liquefy faster than freeze stored raw material even if the protease activity is the same (21). Although the initial proteolytic activity may be high, the tissue solubilization and protein digestion are significantly inhibited by the high salt content. To avoid this problem, initial fermentation at lower salt concentration has been tried. In fish sauce fermentation experiments with anchovies, it has been shown that the liquefaction process can be considerably speeded up if the initial fermentation is run at acid conditions and at a reduced salt content (34, 35). In one experiment pepsins and other acid proteinases were allowed to act at almost optimal conditions (pH 4 and 5% NaCl) for a few days before the fermentation mixture was neutralized and salt was added to the normal level. The problem with this method is that the peptic digestion is different from the tryptic digestion normally occuring during fish sauce fermentation, and the tryptic enzymes initially present in the raw material are susceptible to irreversible inactivation during the initial acidification (18). Both these factors contribute to inferior flavor development in the final product. Hence, initial alkalification at low salt concentration has been considered as a more convenient method for speeding up the liquefaction (14). So far this alternative has not been thoroughly tested out. One objection to alkali treatment is the possible formation of harmful compounds such as lysinoalanine (36).

Fish sauce fermentation normally occurs at about 30°C. Mabesa et al., (37) reported that the production time for fish sauce was reduced from 12 to 2 months if the fermentation mixture was heated to 45–50°C during the last 10 days of the 2 month fermentation period.

The fermentation process can also be accelerated by adding exogenous enzymes. Several crude preparations of plant enzymes have been tried with variable results. Most of them improve the liquefaction speed, but the aroma and flavor generally become too weak (34, 38). The best results were achieved with bromelain, which is a cysteine proteinase obtained from pinapple waste. The most logical step would be to add more fish enzyme, but so far little has been done to investigate this possibility. Raksakulthai et al. (20) compared the effect of adding squid hepatopancreas enzymes and various fungal and mammalian enzymes including pronase, trypsin, and chymotrypsin during sauce fermentation of capelin. They found that adding 2.5% squid hepatopancreas to minced male capelin gave a significantly higher protein digestion than in the control sample and in samples to which 0.003% of the various pure commercial enzymes were added. The sensory evaluation showed that fish sauce from capelin supplemented with squid hepatopancreas enzymes scored higher than all the other samples, including a commercial fish sauce sample from Southeast Asia. Based on these results a commercial production by which good-quality fish sauce can be produced within only 4 months has been established (39).

III. FISH SILAGE

Fish silage is a product related to fish sauce. It is made from chemically preserved minced fish, or fish waste, which is digested by endogenous hydrolytic enzymes. Unlike fish sauce, however, it is not preserved by salt but by acid, and it is not used for human cosumption but for fish or animal feed production.

Fish silage is a Scandinavian invention first used in Sweden during World War II (40). Since then industrial production has taken place in Poland and Denmark, but today Norway is the major producer of fish silage, with an annual production of about 200,000 tons. The main sources of raw material are fish waste, by-catch fish from the marine fisheries, and filleting waste from the salmon aquaculture. Whereas the silage made from marine fish basically serves as a raw material for fish feed production, the salmon silage is used in the production of feed for domestic animals. Various acids, both organic and inorganic, have been used for preservation. Generally organic acids are better preservatives than inorganic, and although the former are more expensive, formic acid has been found to be the most suitable (41). Today most of the silage production is based on preservation with formic acid (2–3%) combined with a commercial antioxidant such as ethoxyquin (200 ppm).

Originally fish silage was used directly as a wet suspension of digested protein and oil, which was mixed with other feed components into a moist feed product. Such feeds were well accepted and gave good results, particularly with pigs and poultry, as long as the fish oil content was kept below a level of about 1% (42). The problem with variable oil content was eliminated when the silage processing method was developed (43). This method is based on utilization of the endogenous enzymes of the raw material to digest and liquefy the fish tissue to make mechanical oil separation possible.

In principle, fish silage can be made from any fish raw material that is not deteriorated, by mincing and acid preservation. To achieve an efficient solubilization, however, it is of vital importance that the fish stomach is a part of the raw material. To obtain full preservation the silage must be acidified to pH about 4 or below, and at this pH the stomach pepsins are the most important protein digestive agents. Fish pepsins have optimal activity in the pH-range 2.5–4, which coincides exactly with the most suitable silage pH (42, 43). Lysozomal muscle enzymes such as cathepsins are active at acid conditions and may also contribute to the protein digestion. Among these the aspartic proteinase cathepsin D is probably most important since it is very abundant in fish muscle and has a pH optimum at about pH 4 (23, 44). The cysteine protease cathepsin B has been suggested to play a role in the digestion of collagenous connective tissues during the autolysis of fish silage (45).

At room temperature and up to about 45°C there is a rapid initial protein digestion. The speed of autolysis declines fast after a few days, even though a significant proteolytic activity can still be measured after long time storage (46, 47). An insolubile fraction always remains even after extensive autolysis (48). This fraction is rich in aromatic amino acids and cystine, but contains no detectable hydroxyproline (43). Collagen always contains hydroxyproline, and the absence of this amino acid in the sediment shows that collagenous connective tissues are efficiently digested in fish silage. This is as expected, since it is well known that collagen is efficiently solubilized and partly digested by fish pepsins at low pH (49).

Normally fish silage is made and stored at outdoor ambient temperatures which is often too low to achieve an efficient autolysis (41, 43) Experiments with cod viscera silage showed that optimal autolysis was obtained at pH slightly below 4 and about 37 C, and that the yield of soluble matter did not increase significantly after a few days of incubation at this temperature (43). Based on these results heating to 30–35°C for 3 days with stirring was chosen as the conditions for the industrial scale processing. Then the temperature is raised to 95°C to facilitate oil separation from sludge and the aqueous solution of digested protein. This final heat treatment also serves as an inactivation of all enzymes present in the raw material and as an efficient pasteurization of the products. After decanter centrifugation to remove sludge, the oil and aqueous solution is separated by a commercial oil separator, leaving only 0.1–0.3% residual oil in the aqueous solution of digested protein. Finally the protein solution is vacuum evaporated from an initial dry matter content of about 12% to about 40% in the final concentrate. Oil and protein concentrate are stable standardized products, which is highly demanded by the modern feed industry because of their excellent properties for both growth and performance for the animals. Supplementing 5–10% fish silage protein seems to be optimal in most cases. This is probably due to the fact that a combination of intact and digested protein in the feed provides optimal protein utilization in both fishes and many domestic animals (50).

Although feed production is by far the most important utilization of fish silage, more sophisticated biotechnological applications should also be mentioned. After an extensive autolysis most proteins are degraded to free amino acids and small peptides. However, if the autolysis is performed at a moderate temperature (about 25°C), a major part of the pepsins can be recovered by ultrafiltration on a 10 kDa membrane (21, 51). The enzyme concentrate may easily be freeze dried with only a moderate activity loss and applied in various biotechnological proccesses and as a digestive aid in fish feed (52). The permeate containing peptides and free amino acids can be vacuum evaporated, neutralized, and spray dried to yield a peptone with excellent properties as a nitrogen source in microbial growth media (51, 53). It should also be mentioned that certain peptides present in such a peptone made from fish stomach may improve the disease resistance in fishes by stimulating the nonspecific immune system (54, 55).

IV. APPLICATION OF FISH ENZYMES IN FISH PROCESSING

Fish and warm-blooded animals generally have a fairly similar set of enzymes, but there are of course major differences in temperature characteristics. Whereas enzymes in Arctic fish often have very flexible molecules with low activation energy, some tropical fish species have enzymes with temperature characteristics comparable to enzymes from mammals.

There is a considerable demand for enzymes with low temperature characteristics, both because they work more efficiently at low temperatures and because they can be easily inactivated by moderate heating. Such properties are very useful in the processing of fresh foods and particularly many fishery products that are easily damaged by heat. Several enzymatic fish processing methods have been developed as alternatives to mechanical methods, and the common purpose has been to improve product quality and yield. The biochemical principle in this kind of enzymatic processing is that the enzymes should act as biotechnological knifes with the capacity for selective degradation of specific tissue fractions.

A. Skin Removal

Mechanical removal of fish skin is often technically difficult and always involves some reduction in quality and yield due to damage of the fillets.. By a combination of chemical and enzymatic treatment it is possible to solubilize the collagenous skin tissue without degrading the muscle tissues.

Deskinning of Atlantic herring can be achieved by first exposing the fish to 5% acetic acid (5 min) to loosen the scales and denature the skin collagen. Then the skin is incubated with fish pepsin in an acidic solution (0.2–0.5% acetic acid) at 15–20°C. The skin is degraded and can be washed off in cold water after an incubation period from 30 to 120 min, depending on the size of the fish and the season of capture (56). Collagenous skin tissues in herring are tougher and thicker during the prespawning winter season than in the summer (57, 58). This method has proved to work well in pilot scale, although some problems with uneven deskinning may occur due to different skin thickness and toughness at different parts of the body.

In a method developed for enzymatic removal of the skin from skate wings, a short initial warm water treatment is used to make the skin degradable by a mixture of proteases and carbohydrases. The enzyme incubation is performed either for 4 h at 25°C or overnight at 5°C (59). It is not known whether the enzyme preparation is of marine origin, but according to Stefansson (60), enzymes from fish viscera can be used for deskinning skate wings.

An industrial method for deskinning squid has been in operation since the mid-1980s (61). In this method an enzyme of plant origin is used (52), but deskinning can also be obtained by activating endogenous enzymes during incubation of the squid in a salt solution at 45°C (62). It is not known which enzymes are taking part in the autolytic skin degradation, but both cathepsin D- and cathepsin B-like proteases have been detected in squid mantle muscle (63). More recently, it has been shown that proteolytic enzymes from squid liver can degrade squid mantle collagen even at low temperature and neutral pH (64). It is also reported that collagenases obtained from crab hepatopancreas can be used for removal of squid skin (65).

B. Scale Removal

There is an international market for fresh or frozen skin-on fillets from species such as haddock, hake, ocean perch, redfish, and salmon, but such products are only acceptable if the scales have been thoroughly removed. A mechanical device for scale removal has been available for a long time, but the mechanical descaling gives the fish a rough treatment that often results in great losses and quality reduction. Hence, enzymatic descaling methods have been developed (66, 67). The three basic elements in enzymatic descaling are mild acid treatment to denature and loosen the mucus layer, enzymatic degradation of the outer skin structures attaching the scales to the skin, and washing off scales and enzymes by water jets. Gentle and efficient descaling can be obtained even at low temperatures if a special mixture of fish digestive enzymes is used (65). To achieve optimal results it is important that the incubation conditions be carefully tested out for each fish species. An automatic descaling machine has been developed, and the method is promoted worldwide by the Norwegian company Biotec ASA (67).

C. Production of Caviar and Spermary Extracts

Proteases, including fish pepsins and collagenolytic enzymes from crab hepatopancreas, can be used to release fish roe from the connective tissues of the roe sac (51, 52, 65, 68,). A method for enzymatic production of caviar from salmon and trout roe has been used industrially for several years, and about 50 tons of caviar have been produced annually in Canada and Scandinavia (69). This method involves enzyme treatment of the roe sacs in a stirred tank, where degraded connective tissues and empty eggs are removed by floatation on top, whereas the sedimenting eggs are collected from the bottom, stabilized and drained to make a high-quality caviar. A 90% yield of caviar has been obtained from roe of rainbow trout compared to about 70% with the conventional mechanical method (52). Enzymatic caviar production is a very delicate process demanding strict control with all the condition parameters to avoid damage of the roe cell wall, which may cause leakage during storage.

In Japan spermary extracts from fish milt are used for human consumption. Spermary extracts can be made from skipjack tuna by autolysis, but a more efficient hydrolysis, and a better recovery, is obtained when enzymes from the skipjack pyloric caecum are added. A sensory evalutation revealed that incubation for 4 h at pH 6.4 and 37°C was most convenient for extract manufacturing (70).

D. Preparation of Carotenoprotein Complexes from Crustacean Waste

Carotenoprotein complexes were prepared from lobster shells using a semipurified extract from the pyloric caecum and intestines of Atlantic cod (71). The cod enzyme extract recovered a relatively lower yield of the products (about 14–46% less) than was recovered with bovine trypsin. Also, the carotenoid pigment content of the product recovered with the cod enzyme extract was 17–46% lower than the product prepared with bovine trypsin. However, the cod enzyme achieved a similar level of demineralization as the bovine enzyme. It was suggested that the lower product yield as well as lower pigment content of the product obtained with the cod enzyme was possibly due to the presence of other hydrolytic enzymes (e.g., proteases and lipases) that caused excessive degradation of proteins and lipids in the final product.

E. Supplementation of Squid and Herring Fermentation with Greenland Cod Trypsin

Greenland cod trypsin was used to facilitate the low-temperature (10°C) fermentation of herring and squid, and the performance of the fish enzyme was compared with that of bovine trypsin. The fermentation (as measured by the release of free amino acids and soluble protein in the fermentation brines) was more rapid with fish enzyme at the low temperature (10°C) than with the bovine enzymes (72). However, the reaction by the cod enzyme slowed down much more quickly than that by the bovine enzyme, such that the degree of hydrolysis by the two enzymes was identical.

F. Prevention of Copper-Induced Off Flavors in Raw Milk

Raw milk samples were dosed with Cu^{++} and then treated with either Grenland cod or bovine trypsin to evaluate the capacity of the trypsins to prevent off flavor development in the raw milk. While both trypsins were able to curtail oxidized flavors in raw milk samples, it was only the cod enzyme that was inactivated after heating at 70°C for 45 min (72). This illustrates the relative advantage of the lower thermal stability of the fish enzyme: after the desired transformation is achieved in the product, the enzyme can be inactivated by mild treatments to prevent continued proteolysis in the product to cause undesirable texture and flavor changes.

G. Preparation of Shrimp Flavor Extract with Turbot Protease

An ammonium sulfate protease extract was prepared from turbot pancreas and intestines. The semipurified fish enzyme extract exhibited trypsin activity mostly, and chymotrypsin activity to a smaller extent. The turbot extract was used to supplement the fermentation of shrimp heads. The product was analyzed for protein, fat, ash, pigments, and free amino acids. The flavor extract had higher levels of alanine, arginine, glycine, and proline (73). For the product prepared by cold water extraction the levels of the amino acids were 31.4% (arg), 16.7% (pro), 14.2% (gly), 9.5% (ala). For the product prepared by hot water extraction, the levels of the main amino acids were 33.9% (gly), 19.3% (pro), 14.1% (arg), and 8.3% (ala). The major free sugar in the extracts was ribose, and it made up 76% and 72% of the free sugars in the cold water and hot water extracted products, respectively.

V. UTILIZATION OF ENZYMES FROM MARINE INVERTEBRATES

Various enzymes from marine invertebrates have already been mentioned previously in connection with various applications in the processing of fishery products. In this section the possibilities of utilizing such enzymes as tools within other biotechnological fields will be discussed. The marine invertebrates represent an immensely diversified pool of biological material. Although only a very limited number of enzymes from such animals have been characterized, many of them have proved to have special properties that can be utilized biotechnologically.

A. Enzymes from Shrimp and Scallop Processing Waste

The international shrimp industry produces large amounts of solid and soluble waste. The solid waste has always been regarded as commercially interesting due to its content of chitin and carotenoproteins. Until recently, however, the solubles represented only a waste water pollution problem. Normally the shrimp processing factories receive the raw material in frozen blocks that must be thawed before processing. Olsen et al. (74) showed that the thawing water from Northern shrimp (*Pandalus borealis*) contained considerable amounts of a number of biotechnologically interesting enzymes including hyaluronidase, β-N-acetyl-glucosaminidase, chitinases, alkaline phosphatase, and several nucleases.

A simple method for pilot-scale recovery of these enzymes was developed. After clarification by ferric chloride precipitation, the enzymes in the thawing water were concentrated 50 times by ultrafiltration on a 10 kDa cut-off ultrafiltration membrane. After freeze drying the concentrate, a yield higher than 60% was obtained with all the enzyme activities measured.

The alkaline phosphatase from Northern shrimp has been purified to homogenity by chromatographic methods, and it proved to have a psychrophilic character making it very interesting as a tool in genetic engineering (75). The enzyme expresses about 40% of maximal activity at 10°C and can be more easily temperature inactivated than the corresponding calf enzyme normally used for the same purposes. Today the enzyme is produced commercially and highly prized on the international biotechnology market. Similar forms of temperature-labile alkaline phosphatases have also been purified from hepatopancreas of the shrimp *Penaeus monodon* (76).

At present the possibilities of utilizing more of the enzymes present in shrimp thawing water are being evaluated. Some of the nucleases seem to be the most interesting candidates for commercial production.

Extracts of scallop visceral waste have an extremely high storage stability even at room temperature. The reason for this is probably the combined effect of low-molecular-weight antibacterial compounds and enzymes disrupting the bacterial cell wall. Lysozyme activity corresponding to about 0.1 mg hen egg white lysozyme per ml was detected in the press liquid from the waste of Arctic scallop *(Chlamys islandica),* and three different lysozymes were purified from this liquid (77). The lysozymes were shown to be very psychrophilic, with optimal activity at about 37°C and 55% of maximal activity at 4°C. They had a specific activity three times as high as hen egg white lysozyme and a pH optimum about 6.

Both temperature and pH characteristics reveal that lysozymes from scallop can have interesting properties as preservatives during cold storage of fresh foods. It is not realistic, however, that such enzymes will be used for food preservation and can be produced commercially from the scallop waste itself. Recently the gene coding for one of the enzymes has been determined, and recombinant production is now being considered (78).

B. Enzymes from Squid-Processing Waste

The midgut gland or hepatopancreas of squid is a rich source of proteolytic enzymes. Extracts of the hepatopancreas from short-finned squid, *Illex illecebrosus,* have been experimentally used in a number of food processing applications including preparation of capelin fish sauce; salted, fermented squid mantle; matjes herring; squid tenderization; and Cheddar cheese ripening (79–88). Short-finned squid hepatopancreas represents almost 20% of round squid weight and is readily available as a byproduct from frozen tubes and dried squid. Fish sauce, with squid hepatopancreas as a fermentation aid, has been commercially produced in Canada for more than 10 years. The proteinases in squid hepatopancreas are primarily cysteine proteases, while the peptidases include cysteine, metallo-, and serine proteases (81). Early work showed that dipeptidyl amino peptidase is a major active component of squid hepatopancreas (82) and its products contribute to the delicious taste of fish sauce made with squid hepatopancreas (84,85).

More than 10 amino peptidases were characterized (81) and several carboxypeptidases have also been identified in squid hepatopancreas (89). The aminopeptidases appear to contribute to the delicious taste of aged Cheddar cheese, while Phe-X dipeptidases contribute to removal of bitter peptides. The utilization of squid peptidases to enhance cheese ripening requires removal of the proteinases from the extracts to obtain a peptidase enriched fraction (86,87,89) since the endoproteases cause excessive texture deterioration and formation of bitter peptides during ripening. Treatment of hepatopancreas extracts with Zn salts is effective in increasing the ratio of aminopeptidase to proteinase activity (81). Zn serves to inhibit cysteine proteinases while activating metalloproteases such as the aminopeptidases.

VI. UTILIZATION OF MARINE ENZYMES IN AQUACULTURE

A. Use of Enzymes to Evaluate Feed Quality

High-quality fish meal is in limited supply and is the most costly part of formulated diets for farmed fish. Accordingly, there is demand for alternative sources of protein in fish feed. The currently used in vitro method for evaluating the digestibility of feed protein is the solubility of N after digestion of the feed with porcine or bovine pepsin. Recent studies have shown that an in vitro digestibility method using digestive enzymes from fish gives a better correlation with in vivo protein digestibility than when enzymes from mammals are employed in the assay (90–95). The method has been successfuly applied to evaluating proteins for salmonids (rainbow trout, coho salmon, chinook salmon) (90–93) and white shrimp (94,95).

Although the spectrum of digestive enzymes in fish is not particularly distinctive from those of land animals (see Chap. 8), the homologues exhibit several different properties. This may explain why their use better predicts protein digestibility than commercially available digestive proteases from mammals. For example, cod pepsins are activated by NaCl, whereas porcine pepsins are inhibited (96,97); salmonid trypsins are several orders of magnitude more sensitive to soy bean trypsin inhibitor than the enzyme from mammals (93).

B. Use of Enzymes to Improve Feed Quality

Enzymes have been directly added to feed for farmed fish to act as "digestive aids." The results of such trials have been mixed. More recent approaches have been to use enzymes to upgrade poor-quality raw materials and/or to pretreat the feed material with enzymes rather than including the enzyme in the diet formulation. The feed value of canola meal for juvenile prawns (*Penaeus monodon*) was improved by adding a mixture of enzymes ("Porzyme," Finnfeeds International) to the diet (98). Enzyme addition increased the liveweight gain with a diet containing 64% canola meal by 28%. The addition of enzyme to diets using canola as a protein source also gave significant improvement in feed conversion ratios. Papaya leaf meal was also tested as an alternative protein and as a source of enzymes in juvenile tiger shrimp diets (99). Papaya leaf meal could replace 10% of fish meal protein without affecting survival, weight gain, and feed conversion ratio. Whether or not papaya enzymes (papain) contributed as digestive aids requires further study. However, pretreatment of soybean residue with papain yielded significantly better carp growth and reduced water turbidity of the fish tanks (100).

Pretreatment of soybean meal with phytase increased dietary phosphorous availability for rainbow trout (101). Convened phytin phosphorous replaced supplemental phosphorous in the diet and is effectively utilized by trout. The use of phytase would reduce phosphorous pollution from fish farm effluents.

VII. USE OF ENZYMES IN BIOSENSORS

In recent years there has been increased interest in the use of biosensors to evaluate fish quality. While this work has not advanced to the use of marine enzymes in biosensors, this is a likely direction for future developments in this field given their variety of distinctive properties (39).

A. Time–Temperature Integration

Time–temperature integrators are devices to monitor temperature exposure history and relate it to food shelf life. Enzymes exhibit a characteristic relationship between reaction rate and temperature as shown by the Arrhenius energy of activation (Ea). Recent studies have shown that enzyme–substrate systems can accurately reflect the dynamic nature of environmental temperature and predict the shelf life of chilled fish after storage temperature abuse (102).

B. Biologically Active Amines

Histamine and other biologically active amines are hazardous microbial products that sometimes accumulate in temperature-abused fish, notably in the scombroid family. The enzyme diamine oxidase, from different sources, has been immobilized on a surface electrode and used to monitor total amine content in fish during storage (103–106). An electrode porcine enzyme was stable at 5°C for 2 months and usable for at least 60 assays. The diamine oxidase used by Lopez-Sabater et al. (106) appears to be very selective for histamine and result of high-performance liquid chromatographic (HPLC) and biosensor analysis of 18 tuna samples gave a regression coefficient of 0.99 (p<0.001).

C. Hypoxanthine and K Value

The use of ATP degradation products as indices of fish freshness is reviewed in Chapters 2 and 12. A xanthine oxidase biosensor could detect less than 1 μM hypoxanthine and was stable for 6 weeks or 400 assays of fish tissue (107,108). A multienzyme electrode containing immobilized 5'-nucleosidase, nucleoside phosphorylase and xanthine oxidase was developed to assess the K-1 value (see Chaps. 2 and 12) of fish samples (109,110). The quantities of hypoxanthine, inosine, and inosine 5'-monophosphate can be determined with this electrode.

D. ATPase Activity

ATPase activity has also been used as an index of fish freshness and surimi quality (see Chap. 12). An enzyme sensor for ATPase activity consisted of an immobilized membrane of two enzymes—purine nucleoside phosphorylase and xanthine oxidase—with an oxygen electrode (111). Assay time with the electrode is only 3 mins. The correlation coefficient for the biosensor and a conventional colorimetric assay of ATPase in fish was 0.99.

Future studies might be directed at using homologous cold-adapted enzymes from fish that might better allow direct measurement of freshness indices at the chill storage temperature of the seafood product.

VIII. CONCLUSIONS

Enzyme technology in the seafood and aquaculture industry has undergone considerable growth in recent years. Early processes utilizing endogenous enzymes, such as fish sauce and fish silage, continue to be used in various parts of the world. Moreover, isolation, characterization, and application of marine enzymes has led to development of improved value of seafood products (e.g., skin and membrane removal), improved value of other food products (e.g., accelerated cheese ripening), utilization of processing waste materials for valuable byproducts (e.g., recovery of carotenoprotein from shellfish waste), basic biochemical applications (e.g., use of Arctic shrimp phosphatase in rDNA kits), diagnosis of the nutritive value of fish feed (e.g., assays for in vitro protein digestibility of fish feed), and improving the nutritional value fish feed ingredients. Future developments involving the use of rDNA technology will include expansion of the above uses and the application of marine enzymes in biosensors.

REFERENCES

1. K Amano. The influence of fermentation on the nutritive value of fish with special reference to fermented fish products in South-East Asia. In: E Heen, R Kreuzer, eds. Fish in Nutrition. London: Fishing News Books, 1962, pp 180–197.
2. P Saisithi Traditional fermented fish: fish sauce production In: AM Martin, ed Fisheries Processing: Biotechnological Applications. London: Chapman & Hall, 1994, pp 111–131.
3. TH Corcoran. Roman fish sauces. Classic J 58: 204–210, 1963.
4. AM Martin, TR Patel. Bioconversion of waste from marine organisms. In: AM Martin, ed. Bioconversion of Waste Materials to Industrial Products. London: Elsevier Applied Science, 1991, pp 417–420.
5. C Thongthai, M Siriwongpairat. Changes in the viable bacterial population, pH, and chloride concentration during the first month of nam pla (fish sauce) fermentation. J Sci Soc Thailand 4: 73–78, 1978.
6. FM Orejana, J Liston. Agents of proteolysis and its inhibition in patis (fish sauce) fermentation. J Food Sci 47: 198–203, 1981.
7. CH Lee. Fish fermentation technology—a review. In: PJA Reilly, RWH Parry, LE Barile, eds. Post-Harvest Technology Preservation and Quality of Fish in Southeast Asia. Manila: Echanis Press, 1990, pp 1–13.
8. C Thongthai, W Panbangred, C Khoprasert, S Dhaveetiyanond. Protease activities in the traditional process of fish sauce fermentation. In: PJA Reilly, RWH Parry, LE Barile, eds. Post-Harvest Technology Preservation and Quality of Fish in Southeast Asia. Manila: Echanis Press, 1990, pp 61–65.
9. WS Hamm, JA Clague. Temperature and salt purity effects on the manufacture of fish paste and sauce. Fish and Wildlife Service, Research rep. 24, US Government Printing Office, Washington DC, 1950, 11 p.
10. N Raksakulthai, NF Haard. Fish sauce from capelin (*Mallotus villosus*): contribution of cathepsin C to the fermentation. ASEAN Food J 7: 147–151. 1992.
11. J Dougan, GE Howard. Some flavouring constituents of fermented fish sauces. J Sci Food Agric 26: 887–894, 1975.
12. K Yatsunami, T Takenaka. Changes in nitrogenous components and protease activity of fermented sardine with rice-bran. Fisheries Sci 62: 790–795, 1996.
13. RR Peralta, M Shimoda, Y Osajima. Further identification of volatile compounds in fish sauce. J Agric Food Chem 44: 3606–3610, 1996.
14. A Gildberg. Accelerated fish sauce fermentation by initial alkalification at low salt concentration. In: S Miyachi, I Karube, Y Ishida, eds. Tokyo: Fuji Technol Press Ltd, 1989, pp 101–104.

15. MS Heu, HR Kim, JH Pyeun. Comparison of trypsin and chymotrypsin from the viscera of anchovy, Engraulis japonica. Comp Biochem Physiol 112B: 557–567, 1995.
16. J Nonaka, LM Dieu, C Koizumi. Studies of volatile constituents of fish sauces, nuoc-nam and shottsuru. J Tokyo Univ Fisheries 62: 1–10, 1975.
17. A Gildberg, C Thongthai. The effect of reduced salt content and addition of halophilic lactic acid bacteria on quality and composition of fish sauce made from sprat. J Sci Food Agric, submitted, 1998.
18. H Outzen, GJ Berglund, AO Smalås, NP Willassen. Temperature and pH sensitivity of trypsins from Atlantic salmon (*Salmo salar*) in comparison with bovine and porcine trypsin. Comp Biochem Physiol 115B: 33–45, 1996.
19. MM Kristiasson, HH Nielsen. Purification and characterization of two chymotrypsin-like proteases from the pyloric caeca of rainbow trout (*Oncorhychus mykiss*). Comp Biochem Physiol 101B: 247–253, 1992.
20. N Raksakulthai, YZ Lee, NF Haard. Effect of enzyme supplements on the production of fish sauce from male capelin (*Mallotus villosus*). Can Inst Food Sci Technol J 19: 28, 1986.
21. A Gildberg. Recovery of proteinases and protein hydrolysates from fish viscera. Bioresource Technol 39: 271–276, 1992.
22. A Gildberg, XQ Shi. Recovery of tryptic enzymes from fish sauce. Process Biochem 29: 151–155, 1994.
23. MJ Bonete, A Manjon, F Llorca, JL Iborra. Acid proteinase activity in fish I. Comparative study of extraction of cathepsins B and D from *Mujil auratus*. Comp Biochem Physiol 78B: 203–206, 1984.
24. PK Reddi, SM Constantinides, HA Dymsza. Catheptic activity in fish muscle. J Food Sci 37: 643–648, 1972.
25. SN Doke, V Ninjoor, GB Nadkarni. Characterization of cathepsin D from the skeletal muscle of fresh water fish, Tilapia mossambica. Agric Biol Chem 44: 1521–1528, 1980.
26. A Gildberg, K Øverbø. Purification and characterization of pancreatic elastase from Atlantic cod (*Gadus morhua*). Comp Biochem Physiol 97B: 775–782, 1990.
27. MM Kristjansson, S Gudmundsdottir, JW Fox, JB Bjarnason. Characterization of collagenolytic serine proteinase from Atlantic cod (*Gadus morhua*). Comp Biochem Physiol 110B: 707–717, 1995.
28. H Toyohara, Y Makinodan. Comparison of calpain I and calpain II from carp muscle. Comp Biochem Physiol 92B: 577–581, 1989.
29. JH Wang, ST Jiang properties of calpain II from tilapia muscle (*Tilapia nilotica* × *Tilapia aurea*). Agric Biol Chem 55: 339–345, 1991.
30. JJ Sanchez, EJ Folco, L Busconi, CB Martone, C Studdert, CA Casalongue. Multicatalytic proteinase in fish muscle. Mol Biol Rep 21: 63–69, 1995.
31. M Kinoshita, H Toyohara, Y Shimizu. Characterization of two distinct latent proteinases associated with myofibrils of crucian carp (*Carassius auratus* cuvieri). Comp Biochem Physiol 97B: 315–319, 1990.
32. M Ishida, N Sugiyama, M Sato, F Nagayama. Two kinds of neutral serine proteinases in salted muscle of anchovy, *Engraulis japonica*. Biosci Biotech Biochem 59: 1107–1112, 1995.

33. MS Heu, HR Kim, DM Cho, JS Godber, JH Pyeun. Purification and characterization of cathepsin L-like enzyme from the muscle of anchovy, *Engraulis Japonica*. Comp Biochem Physiol 118B: 523–529, 1997.

34. CG Beddows, AG Ardeshir. The production of soluble fish protein solution for use in fish sauce manufacture II. The use of scids at ambient temperature. J Food Technol 14: 613–623.

35. A Gildberg, JE Hermes, FM Orejana. Acceleration of autolysis during fish sauce fermentation by adding acid and reducing the salt content. J Sci Food Agric 35: 1363–1369, 1984.

36. AS Nashef, DT Osuga, HS Lee, AI Ahmed, JR Whitaker, RE Feeney. Effect of alkali on proteins. Disulfides and their products. Agric Food Chem 25: 245–251, 1977.

37. RC Mabesa, EV Carpio, LB Mabesa. An accelerated process for fish sauce (patis) production. In: PJA Reilly, RWH Parry, LE Barile, eds. Post-Harvest Technology, Preservation and Quality of Fish in Southeast Asia. Manila: Echanis Press, 1990, pp 45–49.

38. CG Beddows, M Ismail, KH Steinkraus. The use of bromelain in the hydrolysis of mackerel and the investigation of fermented fish aroma. J Food Technol 11: 379–388, 1976.

39. NF Haard. Speciality enzymes from marine organisms. Food Technol 52: 64–67, 1998.

40. H Edin. Investigations concerning the lack of protein caused by the import closure. Nord Jordbr Forsk 22: 142–158, 1940 (in Swedish).

41. IN Tatterson, ML Windsor. Fish silage. J Sci Food Agric 25: 369–379, 1974.

42. J Raa, A Gildberg. Fish silage—a review. CRC Crit Rev Food Sci Nutr 16: 383–419, 1982.

43. J Raa, A Gildberg. Autolysis and proteolytic activity of cod viscera. J Food Technol 11: 619–628, 1976.

44. A Gildberg. Aspartic proteinases in fishes and aquatic invertebrates. Comp Biochem Physiol 91B: 425–435, 1988.

45. GM Hall, D Keeble DA Ledward, RA Lawrie. Silage from tropical fish 1. Proteolysis. J Food Technol 20, 561–572, 1985.

46. A Gildberg, J Raa. Properties of a propionic acid/formic acid preserved silage of cod viscera. J Sci Food Agric 28: 647–653, 1977.

47. NF Haard, N Kariel, G Herzberg, LAW Feltham, K Winter. Stabilisation of protein and oil in fish silage for use as a ruminant feed supplement. 36: 229–241, 1985.

48. MR Raghunath, AR McCurdy. Autolysis-resistant sediment in fish silage. Biol Wastes 20: 227–239, 1987.

49. MLR Harkness, ED Harkness, MF Venn. Digestion of native collagen in the gut. Gut 19: 240–243, 1978.

50. FE Stone, RW Hardy. Nutritional value of acid stabilised silage and liquefied fish protein. J Sci Food Agric 37: 797–803, 1986

51. KA Almås. Utilization of marine biomass for production of microbial growth media and biochemicals. In: KN Voigt, JR Botta eds. Advances in Fisheries Technology and Biotechnology for Increased Profitability. Lancaster: Technomic Publishing, 1990, pp 361–372.

52. J Raa. Biotechnology in aquaculture and the fish processing industry: A success story in Norway. In: KN Voigt, JR Botta, eds. Advances in Fisheries Technology and Biotechnology for Increased Profitability. Lancaster: Technomic Publishing, 1990, pp 509–524.

53. SE Vecht-Lifshitz, KA Almås, E Zomer. Microbial growth on peptones from fish industrial wastes. Lett Appl Microbiol 10: 183–186, 1990.

54. A Gildberg, J Bøgwald, A Johansen, E Stenberg. Isolation of peptide fractions from a fish protein hydrolysate with strong stimulatory effect on Atlantic salmon (*Salmo salar*) head kidney leucocytes. Comp Biochem Physiol 114B: 97–101, 1996.

55. A Gildberg, H Mikkelsen, Effects of supplementing the feed to Atlantic cod (*Gadus morhua*) fry with lactic acid bacteria and immuno-stimulating peptides during a challenge trial with Vibrio anguillarum. Aquaculture 167: 103–113, 1998.

56. K Joakimsson. Enzymatic deskinning of herring. Thesis (in Norwegian), University of Tromsø, Tromsø, Norway, 1984.

57. JR McBride, RA MacLeod, DR Idler. Seasonal variations in the collagen content of Pacific herring tissues. J Fish Res Bd Can 17: 913–918, 1960.

58. RB Hughes. Chemical studies on the herring (*Clupea harengus*). VII.—Collagen and cohesiveness in heat-processed herring, and observations on a seasonal variation in collagen content. J Sci Food Agric 14: 432–441, 1963.

59. G Stefansson. Enzymes in the fishing industry. Food Technol 42(3): 64–65, 1988.

60. Anon. Enzymes skin the spiky fish. New Scand Technol 1: 20–21, 1991.

61. K Nilsen, MT Viana, J Raa. Biotechnology in squid processing: removing skins enzymatically. Infofish Int 2/89: 27–28, 1989.

62. J Raa, K Nilsen. A method for removal of connective skins from squid. Norwegian patent 150304, 1984 (in Norwegian).

63. J Sakai, Y Sakaguchi, JJ Matsumoto. Acid proteinase activity of squid mantle muscle: Some properties and subcellular distribution. Comp Biochem Physiol 70B: 791–794, 1981.

64. I Kolodziejska, J Pacana, ZE Sikorski. Effect of squid liver extract on proteins and on the texture of cooked squid mantle. J Food Biochem 16: 141–150, 1992.

65. ZE Sikorski, A Gildberg, A Ruiter. Fish products. In: A Ruiter, ed. Fish and Fishery Products. Wallingford: CAB International, 1995, pp 315–346.

66. G Stefansson, U Steingrisdottir. Application of enzymes for fish processing in Iceland—present and future aspects. In: MN Voigt, JR Botta, eds. Advances in Fisheries Technology and Biotechnology for Increased Profitability. Lancaster: Technomic Publising, 1990, pp 237–250.

67. R Svenning, E Stenberg, A Gildberg, K Nilsen. Biotechnological descaling of fish. Infofish Int 6/93: 30–31, 1993.

68. A Gildberg. Enzymic processing of marine raw materials. Process Biochem 28: 1–15, 1993.

69. NF Haard, BK Simpson. Proteases from aquatic organisms and their uses in the seafood industry. In: AM Martin, ed. Fisheries Processing: Biotechnological Applications. London: Chapman & Hall, 1994, pp 132–153.

70. Y Murata, T Hayashi, E Watanabe, K Toyama. Preparation of spermary extract by enzymolysis. Nippon Suisan Gakk 57: 1127–1132, 1991.

71. BK Simpson, L Dauphin, JP Smith. Recovery and characterization of carotenoprotein from lobster (*Homarus americanus*) waste. J. Aquat Food Prod Dev 1: 129–146, 1992.

72. BK Simpson, NF Haard. Trypsin from Greenland cod as a food processing aid. J Appl Biochem 6: 135–143, 1984.

73. BK Simpson, V Awafo, H Ramaswamy. Biotechnological approaches for the production of value-added ingredients from by-products of seafood harvesting. 4th International Food Conference, Mysore (India), Nov. 1998.

74. RL Olsen, A Johansen, B Myrnes. Recovery of enzymes from shrimp waste. Process Biochem 25: 67–68, 1990.

75. RL Olsen, K Øverbø, B Myrnes. Alkaline phosphatase from the hepatopancreas of shrimp (*Pandalus borealis*): a dimeric enzyme with catalytically active subunits. Comp Biochem Physiol 99B: 755–761, 1991.

76. AC Lee, NN Chuang. Characterization of different molecular forms of alkaline phosphatase in the hepatopanreas from the shrimp *Penaeus monodon* (Crustacea: decapoda). Comp Biochem Physiol 99B: 845–850, 1991.

77. B Myrnes, A Johansen. Recovery of lysozyme from scallop waste. Prep Biochem 24: 69–80, 1994.

78. B Myrnes. New componds revealed in scallop. Annual report: The Norwegian Institute of Fisheries and Aquaculture, Tromsø, Norway, p 27, 1997 (in Norwegian).

79. YZ Lee, BK Simpson, NF Haard. Supplementation of squid fermentation with proteolytic enzymes. J. Food Biochem 6:127–134, 1982.

80. I Kolodziejska, J Pacana, ZE Sikorski. Effect of squid liver extract on proteins and on the texture of cooked squid mantle. J Food Biochem 16:141–150, 1992.

81. R Raksakulthai and NF Haard Purification and characterization of aminopeptidase fractions from squid (*Illex illecebrosus*) hepatopancreas. J. Food Biochem. 23(2):, 1999.

82. KS Hameed, NF Haard. Isolation and characterization of cathepsin C from Atlantic short finned squid, *Illex illecebrosus*. Comp. Biochem. Physiol. 82B: 241–246, 1985.

83. N Raksakulthai, YZ Lee, NF Haard. Influence of mincing and fermentation aids on fish sauce prepared from male, inshore capelin, *Mallotus villosus*. Can. Inst. Food Sci. Technol. J. 19: 28–33, 1986.

84. N Raksakulthai, NF Haard. Contribution of cathepsin C to the fermentation of fish sauce (*Mallotus villosus*). In: S Otwell, ed. Proceedings, 12th Tropical and Subtropical Fisheries Technology Conference and 32nd Annual Atlantic Fisheries Technology Conference, 1988, pp. 658–670.

85. N Raksakulthai, NF Haard. Fish sauce from capelin (*Mallotus villosus*): contribution of cathepsin C to the fermentation. ASEAN Food J 7(3) 147–151, 1992.

86. FL Garcia-Carreno, NF Haard. Preparation of an exopeptidase enriched fraction from the hepatopancreas of decapods. Process Biochem 29:663–670, 1994.

87. NF Haard, FL Garcia-Carreno, LE Dimes. Exopeptidases from shellfish viscera. In: S Otwell, ed. Third Joint Meeting of the Atlantic Fisheries technological Society and the Tropical and Subtropical Fisheries Technology Society, Gainesville, FL: Florida Seagrant, 1994, pp. 351–367.

88. FL Garcia-Carreno, R Raksakulthai, NF Haard. Processing Wastes: Exopeptidases from shellfish, In: A Bremner, C Davis, B Austin, eds. Making the Most of the Catch. Brisbane Australia: AUSEAS, 1997, pp. 37–43.

89. N Raksakulthai. Accleration of Cheddar ripening with squid peptidases. Ph.D thesis, University of California, Davis (in preparation), 1999.

90. LE Dimes, NF Haard. Estimation of protein digestibility: I. Development of an in vitro method for estimating protein digestibility in salmonids. Comp Biochem Physiol 108A(2/3): 349–362, 1994.

91. LE Dimes, NF Haard, FM Dong, BA Rasco, IP Forster, WT Fairgraive, R Arndt, RW Hardy, FT Barrows, DA Higgs. Estimation of protein digestibility: II. In vitro assay of protein in salmonid feeds. Comp Biochem Physiol 108A(2/3): 363–370, 1994.

92. NF Haard. Digestibility and in vitro evaluation of vegetable proteins for salmonid feed. In: (C. Lim and D.J. Sessa, eds.) Nutrition and Utilization Technology in Aquaculture. Champaign, IL;, AOCS Press, 1995, pp. 199–219.

93. NF Haard, R Arndt, FM Dong. Estmation of protein digestibility IV. Properties of pyloric caeca enzymes from Coho salmon fed soybean meal. Comp Biochem Physiol 115B(4):533–540, 1996.

94. JM Ezquerra, FL Garcia-Carreno, NF Haard. Digestive proteinases from the hepatopancreas of white shrimp (*Penaeus vannamei*) fed with different diets. J Food Biochem 21(5): 401–419, 1997.

95. JM Ezquerra, FL Garcia-Carreno, R Civera, NF Haard. pH-stat method to predict protein digestibility in white shrimp (*Penaeus vannamei*). Aquaculture 157 (3/4):251–262, 1997.

96. J Squires, NF Haard, LAW Feltham, L. Pepsin isozymes from Greenland cod, *Gadus ogac*. 1. Purification and physical properties. Can. J. Biochem. Cell Biol 65: 205–209, 1986.

97. J Squires, NF Haard, LAW Feltham. Pepsin isozymes from Greenland cod, *Gadus ogac*. 2. Substrate specificity and kinetic properties. Can. J. Biochem. Cell Biol 65: 210–214, 1986.

98. J Buchanan, HZ Sarach, D Poppi, RT Cowan. Effects of enzyme addition to canola meal in prawn diets. Aquaculture 151:29–35, 1997.

99. VD Penaflorida, Growth and survival of juvenile tiger shrimp fed diets where fish meal is partially replaced with papaya *Carica papaya* L. or camote Ipomea batatas Lam. leaf meal. Israel J Aquacul Bamidgeh 47(1): 25–33, 1995.

100. MH Wong, LY Tang, FSL Kwok. The use of enzyme digested soybean residue for feeding common carp. Biomed Environ Sci 9(4): 418–423, 1996.

101. KD Cain, DL Garling. Pretreatment of soybean meal with phytase for salmonid diets to reduce phosphorous concentrations in hatchery effluents. Prog Fish Culturist 57(2): 114–119, 1995.

102. S Tsoka, PS Taoukis, P Christakopoulos, D Kekos, BJ Macris. Time temperature integration for chilled food shelf life monitoring using enzyme-substrate systems. Food Biotechnol 12(1–2): 139–155, 1998.

103. R Draischi, G Volpe, L Lucentina, A Cecilia, R Federico, G Palleschi. Determination of biogenic amines with an electrochemical biosensor and its application to salted anchovies. Food Chem 62(2): 225–232, 1998.

104. KB Male, P Bouvrette, JHT Luong, BF Gibbs. Amperometric biosensor for total histamine, putrescine and cadaverine using diamine oxidase. J. Food Sci 61 (5): 1012–1016, 1996.

105. M Ohashi, F Nomura, M Suzuki, O Adachi, N Arakawa. Oxygen-sensor-based simple assay of histamine in fish using purified amine oxidase. J Food Sci 59(3): 519–522, 1994.

106. EI Lopez-Sabater, JJ Rodriguez-Jerez, AX Roig-Sagues, MT Mora-Ventura. Determination of histamine in fish using an enzymic method. Food Addit Contam 10(5): 593–602, 1993.

107. C Qiong, P Tuzhi, Y Liju. Silk fibroin-cellulose acetate membrane electrodes incorporating xanthine oxidase for the determination of fish freshness. Anal Chim Acta 369(3): 245–251, 1998.

108. M-A Carsol, G Volpe, M Mascini. Amperometric detection of uric acid and hypoxanthine with xanthine oxidase immobilized and carbon based screen-printed electrode. Applications for fish freshness determination. Talanta 44(11): 2151–2159, 1997.

109. S Hu, C-C Liu. Amperometric sensor for fish freshness based on immobilized multi-enzyme modified electrode. Electroanalysis 9(16): 1229–1233, 1997.

110. Y-C Su, S-F Chou, C-Y Chen. Determination of fish-freshness with a multi-enzyme system. J Chin Agric Chem Soc 31(6): 717–731, 1993.

111. B Cheun, H Endo, T Hayashi, E Watanabe. Development of a sensor for ATPase activity. Fisheries Sci Tokyo 62(6): 950–954, 1996.

23

Recovery of Enzymes from Seafood-Processing Wastes

Haejung An
Oregon State University, Astoria, Oregon

Wonnop Visessanguan
National Center for Genetic Engineering and Biotechnology, Bangkok, Thailand

I. INTRODUCTION

Enzymes are biological catalysts capable of speeding up chemical reactions and are biological tools for improving food quality or food processing operations. Use of an enzyme as a processing aid has a number of advantages over use of chemicals, including high specificity, efficiency of catalysis at moderate temperatures, and being environmentally friendly. Among the enzymes used in food processing, hydrolases consist of the largest proportion (1). Overall, hydrolases dominate the global industrial enzyme sales, accounting for 97% of the total. Within the hydrolase class, proteases make up 24.0% of the market as the second largest group following carbohydrases. Most proteases are used in the food industry for a variety of products including baked foods, beer, wine, cereal, milk, and dairy products including cheese, chocolate, eggs and egg products, meat and fish products, legumes, and for production of protein hydrolysates and flavor extract (2). Among the potential applications, however, the enzymes recovered from fish have been most successfully used as seafood processing aids. Examples of such applications include accelerating fermentation of fish sauce or fermented matjes herring. For details on such applications the reader is referred to Chapters 21 and 22. Although most industrial enzymes for the food industry are derived from animal by-products, plants, and microbial sources, enzymes from aquatic microorganisms are likely to establish a significant part of the industrial enzyme market, in the future, due to their unique characteristics (3).

II. MARINE ENZYMES

Marine species are highly diversified and adapted to a variety of habitat conditions in the marine environment. Genetic variations within species as well as adaptation to different environmental conditions have resulted in a wide array of enzymes with unique properties that open new opportunities for a variety of applications. Marine enzymes are unique compared to those of land animals in that marine animals live in habitat where temperatures are often below 4°C and in some cases even below 0°C, as in the Arctic. Marine enzymes, therefore, are often highly cold adapted to compensate for such harsh living environment. Cold-adapted enzymes from marine organisms have relatively high molecular activity, physiological efficiency, and relatively low temperature optimum and stability as well as high salt tolerance if the enzymes are from stomachless fish (3). Low-temperature processing could provide various benefits including lower thermal energy requirements, protection of substrates or products from degradation, minimization of side reactions, and prevention of the destruction of other substances associated with raw material (4, 5).

III. SEAFOOD-PROCESSING WASTES

Wastes generated from seafood processing are one of the largest untapped protein resources. Recovery of edible portions in seafood processing has traditionally been low, ranging only from 20 to 30% (6). Until recently, such wastes have been processed into low-value products, such as fish and animal feeds and natural fertilizers. Most of the waste, not utilized as an edible product, consist of viscera, head, skin, shell, and bone, and the solids make up the largest proportion of wastes generated from seafood processing.

A. Disposal of Wastes

There are two ways to handle the large volumes of waste being produced: disposal or recovery as processing byproducts. The first choice is highly discouraged with the increased awareness of environmental impact. The Environmental Protection Agency's regulations for the disposal of waste at landfills or by dumping into oceans or rivers have become more restrictive. Thus, there has been a growing effort to convert seafood waste materials to valuable products. Handling process wastes through utilization and development of marketable byproducts has become of increasing interest.

B. Liquid and Solid Wastes

Seafood wastes generated from seafood processing differ in amount and composition depending on the specie and the process method employed. With more

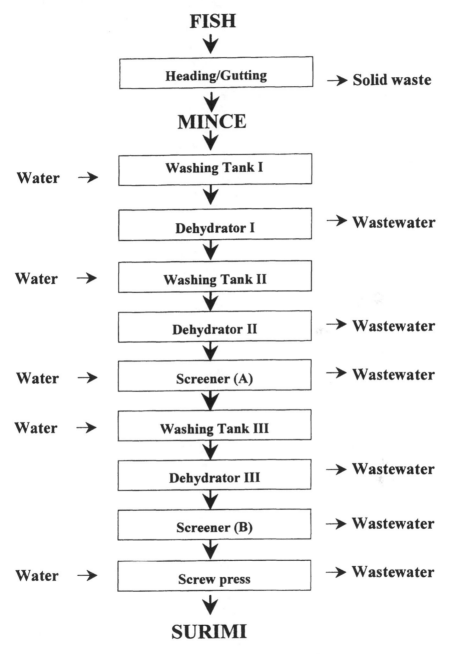

Figure 1 Schematic diagram of surimi processing and sources of liquid and solid processing waste. (From Ref. 8.)

processing steps there is normally a larger volume of waste generated. In most seafood processing operations, a large volume of liquid waste is generated by using clean water to wash, thaw, transport, cook, formulate, and/or package seafood products. The major part of solid waste is from inedible parts of animals. An average recovery of 25% is obtained when fish are processed into fillets, while the remaining 75% forms waste material (7). The wastes are primarily composed of heads, fins, tails, and viscera. The head, accounting for 18–24% of the dressed fish weight, is the primary waste. In a typical salmon cannery, production wastes can account for 31–38% of landed weight. In freezer operations, processing waste may range from 10 to 25% of landed weight depending on product form. Surimi processing generates a large volume of both liquid and solid waste (Fig. 1) (8–10). Recoveries of surimi from fish mince range only 15–22%. Conversely, 78–85% of the whole fish is disposed of as waste. The bulk of the waste, approximately 60%, is generated by the filleting operation and consists of heads, backbones, and viscera. Other significant wastes include skin and bones removed by the mincing/deboning process (10–12%), soluble proteins leached during the washing process (17–20%), and connective tissue and impurities removed by the refining operation (5–10%).

Shellfish processing also generates a considerable volume of waste. When cooked crab meat is produced, the amount of waste can be as high as 75–83% of the animal weight (11). The bulk of the waste material is carapace with some viscera and blood. Shrimp processing produces a large volume of wastes both as solids and liquids. The solid wastes are generated from heads and shells, while a large volume of processing water is discarded after use as a thawing medium for frozen shrimp or as a lubricant in automatic shrimp peelers.

IV. BIOCHEMICAL PROPERTIES OF MARINE ENZYMES

A. Digestive Enzymes

Marine animals contain high levels of proteases both in viscera and muscle. Fish viscera is a relatively large portion of the animal round weight: approximately 5% (12). In addition to its adequate nutritional value, it contains high levels of digestive enzymes, making it a suitable source for recovering proteases for food application. Digestive proteinases have been studied in several species of fish (13) and decapods (14–16). Proteinases found in the intestine of fish include trypsin, chymotrypsin, collagenase, elastase, carboxypeptidase and carboxyl esterase, and they are normally secreted from the pyloric caeca and pancreas (17). Some properties of the digestive proteases from marine organisms are shown in Table 1. Pepsin and trypsin are two main groups of proteinases found in fish viscera. Pepsin is found in fish stomach and is active at acid conditions (18), while trypsin is concentrated in pyloric cecum and active at neutral and alkaline conditions (19).

Table 1 Properties of Proteases Recovered from Fish

Enzyme	MW (kDa)	Optimum pH	Optimum T(°C)	Species	Reference
Pepsin					
Protease I	37.0	4.0	55	*Sardionos melanostica* (sardine)	33
Protease II	33.0	2.0	40		
Protease I	23.0	3.7	38	*Mallotus villosus* (Arctic fish capelin)	30
Protease II	27.0	2.5	43		
Pepsin A	42.0	2.0	32	*Boreogadus saida* (Polar cod)	81
Pepsin B	40.0	2.0	32		
Protease I	36.4	3.5		*Gadus ogac* (Greenland cod)	31
Protease II	35.9	2.5–3.0	30		
Protease III	37.9	2.5–3.0			
Pepsin I	35.5	3.5	40	*Gadus morhua* (Atlantic cod)	82–84
Pepsin IIa	34.0	3.0	40		
Pepsin IIb	34.0	3.0			
Trypsin					
Proteinase I	22.9	10.0	45	*Sardionos melanostica*	85
Proteinase II	28.9	10.0	45		
Proteinase III	27.0	10.0	45		
Trypsin I	28.0	8.0	45	*Mallotus villosus*	22
Trypsin II	28.0	7.5	45		
Trypsin	23.5	7.5	45	*Gadus ogac*	25, 86
Trypsin	24.0		35	*Tautogolabrus adspersus* (Cunner)	87–88
Trypsin isozymes	24.2–24.8	7.5–8.0	40	*Gadus morhua*	18, 86
Chymotrypsin					
Chy a	26.0	7.8	52	*Gadus morhua*	89
Chy b	26.0	7.8	52		
Elastase					
Pancreatic elastase	28.0	8.5	43	*Gadus morhua*	82
Intestinal elastase	24.8	8–8.5	40	*Gadus morhua*	90

Serine proteinase, mainly trypsin and chymotrypsin, play a major role in protein digestion (20). Fish trypsins are generally stable at alkaline pH (21). Purified trypsin from hybrid tilapia (*Tilapia nilotica/aurea*) intestines showed an optimum at pH 9.0 and 40°C against Nα-benzoyl-DL-arginine p-nitroanilide (BAPNA) (22). Two trypsinlike enzymes, enzyme I and II, isolated from the gut of capelin showed optimum pH at 8–9 against BAPNA with the optimum temperature at 42°C (23). Simpson et al. (24) reported that Atlantic cod trypsin was most active at pH 7.5 and 40°C against BAPNA. In Dover sole, the activity at pH 7.0–8.0 was due to trypsin and chymotrypsinlike enzymes, while the optimum activity at pH 9.5–10.5 was due to elastase (25). The optimum pH for hydrolysis of casein by Greenland cod trypsin was 9–9.5 in contrast to a pH optimum of 8.0 by bovine pancreas trypsin (26). Hjelmeland and Raa (23) found two trypsins from Arctic fish capelin with M_r of about 28,000. For the Greenland cod trypsin molecular weight, 23,500 Da was reported by Simpson and Haard (26). Trypsin A and B from anchovy had molecular weights of 27,000 and 28,000 Da (27). Cohen et al. (28) reported that carp trypsin molecular weight was 25,000 Da. A trypsinlike enzyme was reported to be the major form of protease in the digestive organs of Pacific whiting based on the molecular weight, the inhibition by N-tosyl-L-phenylalanine chloromethyl ketone (TLCK), and the activity towards specific substrates (29).

An aspartic proteinase, pepsin, has an extracellular function as the major gastric proteinase. Pepsin, secreted as a zymogen (pepsinogen), is activated by the acid in stomach to an active form (25). Pepsinlike protease with an optimum pH value of 1.7 was reported to be predominant in the stomach of Dover sole (25). Haard (30) reported that the initial rate of hemoglobin digestion was maximum at 35 C, pH 1.9 by Atlantic cod pepsin. Fish pepsins were shown to hydrolyze hemoglobin much faster than casein, myofibrillar proteins or sarcoplasmic proteins (31, 32). Guerard and Le Gal (33) reported that a hexapeptide is the smallest substrate to be hydrolyzed by fish pepsins. Most fish species contain two or three major pepsins with an optimum for hemoglobin digestion at pH between 2 and 4 (31, 34, 35). Gildberg et al. (18) reported that the affinity of the cod pepsin, especially pepsin I, was lower at pH 2 than at pH 3.5 towards hemoglobin. Furthermore, pH optimum was highly dependent on substrate concentration. Pepsins I and II showed similar pH optima at pH 3.0 at high concentrations of hemoglobin; whereas, pepsin I had a maximum activity at pH 3.5–4 at low substrate concentration. The reader is referred to Chapter 8 for more information on digestive proteinases.

B. Muscle Proteinases

Marine animals contain various proteases in muscle tissue that engage in protein turnover in vivo. Muscle proteases are located mainly in lysosomes, in the sarcoplasm, and in the extracellular matrix of the connective tissue surrounding each cell (36). Fish muscle of some species, which are prone to heat-induced tissue softening and textural degradation, contain relatively high levels of proteolytic activity. Among the muscle proteinases, cathepsins and alkaline proteinases have been identified as the most active proteinases (37). Surimi processing generates a large volume of both solid and liquid wastes. Typically the product yield is less than 20% of whole fish. During surimi processing, at least three volumes of water are used per weight of fish mince to remove water-soluble matter, such as sarcoplasmic proteins, blood, and pigments, thus generating a large volume of wastewater (10). Since only ground mince meat is used as a starting material to manufacture surimi, most enzymes originating from the muscle are washed out in the waste stream. The major enzymes identified in surimi wash water were cysteine lysozomal cathepsins, such as cathepsins B and L. Cathepsin B was shown to be the most active enzyme in fish mince, showing at least twice as much activity as cathepsin L (38). However, cathepsin L was identified as the most predominant proteinase in surimi wash water (39). No evidence of calpain or cathepsin H activity was found in surimi wash water presumably due to the instability of the enzymes and low temperature optimum.

Enzymes recovered from muscle were shown to be more susceptible to denaturation than enzymes recovered from viscera. Crude proteases recovered from seabass and gizzard-shad muscles almost lost their activities during storage while those from the viscera of fish were relatively stable (40). Cathepsin L was recovered from Pacific whiting surimi wash water in a freeze-dried powder form but was unstable, showing only 18.5% of the original activity after 7 week storage at 4°C (41), necessitating use of enzyme stabilizers to retain the activity of the recovered enzyme during storage.

C. Other Enzymes

Alkaline phosphatase is often used as a chemical reagent for detecting antigen–antibody reactions. It is widely distributed in nature and found to be rich in a number of crustaceans: crayfish (42), spiny lobster (43), and hermit crab (44). Three alkaline phosphatase isozymes, APase-I, II, and III, have been isolated from the cephalothoraxes of big-head shrimp (*Solenocera melantho*) (45). The enzyme is present in the digestive system, especially in the hepatopancreas. Olsen et al. (46) recovered alkaline phosphatase from the wastewater of the commercial processing of Northern shrimp (*Pandalus borealis*). This enzyme is a unique dimeric protein with a molecular weight of 155 kDa that may be dissociated into subunits that retain enzyme activity (47). The lower thermal stability of this enzyme has found application in gene cloning, since it can be more readily inactivated without disrupting the DNA structure (48). In addition to alkaline phosphatase, other carbohydrate-degrading enzymes, such as, hyaluronidase, β-N-acetyl glucosaminidase, and chitinase, have also been reported as valuable enzymes that can be recovered from shrimp wastewater (46).

V. METHODS USED FOR ENZYME RECOVERY

A. Ensilage

Silage generally refers to the process of preserving wet fodder (i.e., whole fish or parts of fish etc.) by adding acid or by the anaerobic production of lactic acid by bacteria (49). Silage methods have been used as an effective way to prepare raw materials for enzyme recovery. Reece (50) pointed out that the major drawback of the large-scale recovery of enzymes from fish waste is the geographical diversity of offal sources (i.e., at sea, on shore, or on aquaculture farms). The investigator first produced silage from thawed viscera by mixing with formic/sulfuric acid or formic/propionic acid to lower the pH to 3.0–4.2. The silage product showed relatively constant total activity of acid proteases for salmon and cod up to 20 days of storage, although a significant drop in the activity was observed with mackerel in 13 days of storage. Although the process was most effective for recovering acid protease activity from the fish offal of salmon, cod, and mackerel, the investigator was also able to coextract an alkaline protease from salmon due to the stability of the enzyme at acidic pH condition. A large-scale process for production of cod pepsin concentrate has been developed by Gildberg and Almås (51) and Almås (52). Cod stomach manually separated from intestines was ground and mixed with formic acid at 1.7% (v/w) and kept at 25°C to prepare an acid hydrolysate (12). Autolysis occurred rapidly in the first 3 days. The amount of pepsin released was approximately 2 g active enzyme/L of aqueous phase. The aqueous phase was recovered at 70% of the total volume and passed through ultrafiltration with the average permeate flux rate 7 L/m/h. The recovered enzyme was concentrated six times. By this procedure, about 75% of the pepsin activity was recovered in the concentrate and the activity corresponded to about 9 g pepsin/L. The enzyme retained 98% of its initial activity after 10 days, 74% in 2 months, and 30% in 2 years of storage with the pH still below 4.

B. Membrane Technology

Membrane technology involves separating components from fluid or gaseous streams by forcing the stream to flow under pressure over the surface of membrane. The technology is actually a family of processes that include reverse osmosis, nanofiltration, ultrafiltration, and microfiltration, which can be differentiated by the separation range (Fig. 2) (53). Among them, ultrafiltration is the most widely used process for protein and enzyme recovery. Ultrafiltration is

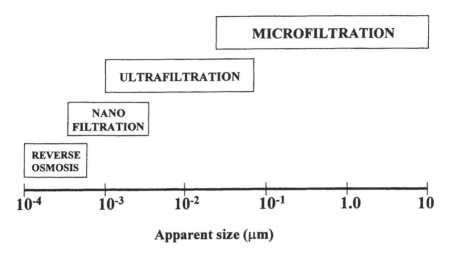

Figure 2 Comparison of apparent size of separation of various cross-flow filter technologies. (From Ref. 53.)

a pressure-driven membrane process that generally uses membranes with pore sizes in the range of 0.05 μm to 1 nm at pressures from 10 to 100 psi to remove both ionic species and low-molecular-weight solutes from the stream being processed (53). Ultrafiltration can offer a great advantage to recover and concentrate a small amount of substances from a large volume of liquid phase. Since positive pressure is employed as a driving force to push the liquid phase through the membrane pores without involving the phase change, ultrafiltration is especially suitable for the separation of sensitive biological substances with activity, such as enzymes. While water and particles smaller than membrane pores pass through, larger molecules (i.e., colloids, emulsion droplets, and particulates) are retained and concentrated.

Ultrafiltration processes have three distinctive characteristics from the conventional filtration process: they are crossflow systems in which the solution flows parallel to the membrane surface, clearing away any particles accumulated on the membrane surface; they are critically dependent on membrane materials and their nominal molecular weight cutoff (MWCO); and they depend on membrane geometry in the actual equipment, which highly affects the efficiency of the entire process (54). The mechanism of separation is complex and is influenced by numerous factors, such as method of membrane manufacture, composition of the membrane, chemical interactions between the feed stream and the membrane, fluid dynamics of the membrane, pressure, temperature, and velocity of the feed stream (55). The main problem in many ultrafiltration processes is fouling or the buildup of the layer on the membrane surface until the retained mass offers hydrodynamic resistance and interferes with flux (56). Fouling dramatically reduces the efficiency of the membrane filtration process and accelerates membrane deterioration. Membranes available today have characteristics that make them capable of performing most of the separations required economically (57).

Ultrafiltration has been widely used in the food and beverage industry (55). In dairy processing, ultrafiltration is used to concentrate skim or whole milk and to fractionate whey, a byproduct of cheese processing. Ultrafiltration is used to remove undesirable components including microorganisms in juice, wine, and beer, to increase clarity and stability of the products. Ultrafiltration has been successfully applied to recover amylolytic enzymes from spent culture medium of *Bacillus subtilis* (58) and pectinolytic enzymes from citric acid fermentation broth of *Aspergillus niger* (59). More than 80% of enzymic activity was recovered in the retentate without any changes in enzymic properties and all enzymes recovered exhibited better quality than the commercial enzymes. However, in seafood processing, ultrafiltration has been applied primarily to reduce the organic load in the wastewater stream from the processing plants (60–62) and to recover enzymes from processing solid-waste silage (50,63–65) and processing wastewater (46).

Fish digestive enzymes were recovered as byproducts from fish-processing wastes of cod and mackerel (50). By using ultrafiltration and ion exchange chromatography, mackerel and cod proteases were obtained at approximately 1g from 1 kg of offal with about 11 and 13-fold purification factor, respectively. A concentrate of pepsin can be recovered from an acid aqueous autolysate of cod stomach silage preserved with formic acid by ultrafiltration and the ultrafiltration concentrate contained about four times as much activity when whole cod viscera was used as raw material (64). In a very high salt concentration (20–25%), a concentrate of tryptic enzymes can be recovered by ultrafiltration from fish sauce, the liquid product developed during fermentation of heavily salted fish materials (65). A large-scale process for the recovery of alkaline phosphatase in wastewater from the shrimp processing was described (46). By using a ultrafiltration unit fitted with two hollow-fiber cartridges with a nominal MWCO 1×10^4 dalton, clarified wastewater (435 L) obtained after removing the insoluble matter and shrimp debris from shrimp processing was concentrated 50 times to 8.5 L retentate that contained more than 99% of soluble proteins with high activity of alkaline phosphatase. The retentate was freeze-dried to 580 g powder and could be stored for at least 1 month at 4°C without any losses of activity.

C. Ohmic Heating

Ohmic heating is a novel method to deliver thermal energy to the heating medium. When an alternating current passes through a material to be heated, its temperature rises due to the electrical resistance, thus converting electrical energy to the thermal energy (66). Thus, the heating rate is extremely high and it can be applied to a large volume of materials. This technique has been used to process food particulates aseptically (67), maximize gel formation of Pacific whiting surimi (68), and remove high-molecular-weight proteins in surimi wash water (SWW) (69). The ohmic heating process was successfully applied to recover heat-stable cathepsins from SWW. The process served as a good preliminary method to remove contaminating proteins, mainly of myofibrillar proteins. The ohmic heating process alone recovered higher total activity, (193%, of the cathepsin) from the starting material by removing the naturally occurring proteinase inhibitor, presumed to be a cystatin compound (70). However, the subsequent step (ultrafiltration) used to recover the cathepsin contributed to a 40% loss of the enzyme activity due to fouling of membrane and protein denaturation.

VI. ALTERNATIVE METHODS FOR ENZYME RECOVERY

A. Precipitation

The recovery of enzymes can be facilitated by selective denaturation of contaminating proteins. The simplest way to achieve selective denaturation is by careful heating. Shifting the pH of the homogenate can also be regarded as a very simple process that requires only minor amounts of chemicals. To achieve optimum purification effects, both methods can be combined. Heat treatment and pH-shift can also result in agglomeration of particles, thus improving the sedimentation of cell debris. Scaling up of this process in large batches can be difficult because heating and cooling rates decrease as the process volume increases. Büntemeyer et al. (71) reported that carefully designed continuous pilot plant with sufficient residence time flexibility and heat exchangers suitable for treatment of media with varying viscosities eliminated the disadvantage of a large-scale batch process. When the investigators conducted testing using commercial baker's yeast, they were able to achieve 100% recovery of invertase with the 73 purification factor by combining the two steps:

1. Acidification with 85% phosphoric acid or 100% acetic acid (pH 4) showed the highest purification fold with heat-treated samples, while pH 3 was more desirable with unheat-treated materials).
2. Heat treatment at 50°C for 10 min.

Schnell and Kula (72) reported the purification of intracellular enzymes by continuous heat treatment after cell disruption. For yeast alcohol dehydrogenase, the authors achieved a twofold higher purification compared to the untreated crude extract. The enzyme recovery was about 80%.

B. Aqueous Two-Phase Systems

Aqueous two-phase systems (ATPS) have been widely used for their potential to improve both the yield and purification potential of downstream processing (73). ATPS form readily upon mixing aqueous solutions of two hydrophilic polymers, or of a polymer and a salt, above a certain threshold concentration. ATP systems are differentiated from traditional liquid–liquid extraction systems because they are prepared using two water-soluble but mutually incompatible materials, not preexisting immiscible organic and aqueous phases (74).

ATPS are ideally suited to macromolecular polyelectrolytes such as proteins and enzymes, because they can be optimized for partition on the basis of charge, hydrophobicity, and molecular weight. The ability to combine these three independent separation bases into a single step gives ATPS a powerful advantage over classic chromatography. The separation potential of ATPS increases with the size or molecular weight of the materials, based on interfacial tension effects. Thus, enzymes with molecular weights generally in the range of 20–100 kDa partition more effectively than peptides, amino acids, or salts (74).

An ATPS forms a specific environment suitable for maintaining enzymes in their native structure and for concentration/purification by means of selective partitioning of the enzyme to one of the phases (73). Extraction using ATPS in many cases offers a better alternative to existing methods in the early processing stages, especially with regards to scale of operations, space-time yield, enrichment of production, and continuous operation for the separation and purification of desired enzyme/proteins from a complex mixture. Furthermore, ATPS can remove byproducts such as other undesirable enzymes and proteins, unidentified polysaccharides, and pigments that are present in the system. Technical advances on the polymer/salt and the polymer/polymer phase systems have enhanced the chances that this technique can be used even on a large scale. In most applications, the phase system is made up by polyethylene glycol (PEG) and a salt. The product is usually collected in the PEG-rich top phase, while cells disintegrate and nucleic acids are displaced to the salt-rich bottom phase (75).

An ATPS was applied to remove yeast cell debris and concentrate β-galactosidase simultaneously (76). Figure 3 shows that cell debris was partitioned to the bottom phase, as PEG concentration was increased and phosphate concentration decreased. Most adequate enzyme recovery and cell debris elimination was achieved with low salt concentration (6–9%) and PEG concentration below the binodal curve (25–29%). With these conditions, 60–87% of the enzyme was recovered in the top phase, while cell debris remained in the bottom (76). Using the large-scale fermentation and purification process, Stranberg et al. (75) achieved purification of the fusion protein (AGβgal) composed of the five IgG-binding regions of staphylococcal protein A, two IgG-binding

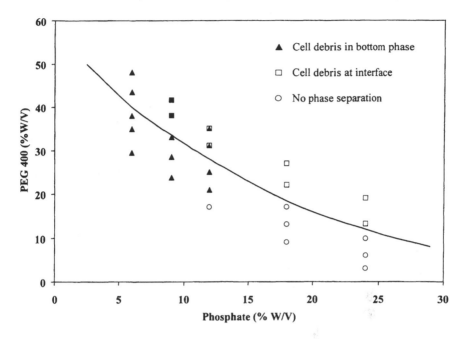

Figure 3 Optimization of aqueous two-phase system using PEG400 and phosphate salt for elimination of cell debris in downstream process. (From Ref. 76.)

regions from streptococcal protein G, and β-galactosidase from *E. coli* using the aqueous two-phase extraction. When the recovery was estimated by the β-galactosidase activity fused with the protein, it was shown that top phase containing PEG retained 66% of the fusion protein, providing a 2.4-fold purification. PEG present in the extract was removed by diafiltration in the subsequent step. The PEG concentration decreased effectively from 43 g/L, initially found in the retentate, to 0.7 g/L with the complete diafiltration.

A process was developed for recovering microbially produced recombinant chymosin from the fermentation broth to be used for coagulation of milk proteins (74, 77). After the fermentation, the broth was acidified to inactivate the culture, and 4% PEG 8000 and 10.5% anhydrous sodium sulfate was added. The chymosin-rich PEG phase was collected for further processing, while the salt-rich raffinate containing fungal cells and contaminant proteins was discarded. The final volume of the extract was approximately 28% that of the original fermentation broth. In addition, the ATP step resulted in an extremely selective purification for chymosin, effectively removing α-amylase, glycoamylase, acid phosphatase, and leucine amino peptidase, which could not have been easily achieved by using conventional chromatographic methods. To render the product food-grade, the contaminating PEG was removed by ion exchange using IBF Spherodex SP resin.

C. Chromatography

Chromatography is a powerful technique to achieve high degrees of purity. It has been well established in biotechnological industries as a production-scale unit operation and as an analytical tool to monitor the quality of raw materials and end products and the purification efficiency of sequential downstream operations. Chromatography offers a variety of methods to separate proteins based on charge, hydrophobicity, size, and molecular recognition. Purification of proteins or enzymes from tissues or cell culture may be performed via different separation techniques used separately or in combination.

1. Size Exclusion Chromatography

Size exclusion chromatography, often referred to as gel filtration, offers the great advantage of being one of the least denaturing techniques, an advantage that sometime is decisive when enzymes are to be purified. Size exclusion chromatography is based on the principle of separating proteins according to their hydrodynamic volume. Their molecular weight can, therefore, be used if the size of the molecule to be separated is presumably different from that of the contaminant proteins (78). The limitation of this method is that the largest sample volume suitable for an optimal separation is relatively small with regards to the column size; as a consequence, the protein solution usually has to be concentrated prior to application onto column or after protein elution. Gel filtration is irreplaceable when the specific ligand for the protein to be separated is not known or not easily available, and affinity chromatography cannot be used (79). An example was made with interleukin 2 (IL2) purification. It is a protein secreted by T lymphocytes. Its relative molecular weight is about 30,000 and its isoelectric point is pH 4–5. No ligand has been identified for this protein. Therefore, its purification was achieved by using gel fitration on ULTROGEL AcA 54 (80) to separate it from its closely related contaminating colony-stimulating factors. Although other steps were used (i.e., precipitation and ion exchange), gel filtration was the most important step in the purification scheme of IL-2. This method has been successfully used for separation of other biomolecules, such as growth factors, tissue plasminogen activator (TPA), and monoclonal antibodies.

2. Ion Exchange Chromatography

Ion-exchange chromatography has been often applied to fractionate proteins. It is dependent on the electrostatic interactions between the protein and charged groups on the exchangers (78). When adsorbate is introduced into an ion exchange column, it will attach to the oppositely charged adsorbent because of the electrostatic attraction between the charged surface groups of the protein and the oppositely charged groups of the ion exchange adsorbent. The nature of the ion exchange groups can be classified as weak when they are derived from a weak acid, such as carboxymethyl (CM), or weak base diethylaminoethyl (DEAE) or strong when derived from strong acid, such as sulfonate (S) and sulfopropyl (SP), or strong base, such as quaternary amine (Q) and quaternary aminoethyl (QAE). The strength of adsorption between a protein and an ion exchanger is also dependent on the pH of the solution. It is desirable to carry out adsorption at a pH close to neutrality to prevent denaturation of proteins by extremes of pH. The bound proteins are usually eluted with an ionic strength gradient or a pH gradient for strongly bound proteins. These procedures reduce the electrical attraction between the protein and support, thereby promoting elution (81).

Ion exchange chromatography provides a unique separation method for simplified purification of recombinant proteins produced by genetic engineering of plants or "biofarming." In order to evaluate suitability of canola as a recombinant protein production host and the simplicity of purification scheme, mutants of T4 lysozyme of varying charges were tested (82). Among the mutants, single point mutation was most effective in shifting elution pattern of the lysozyme. The point mutation introduced the increase of +1 charge on the lysozyme, requiring the increase in conductivity of the eluent by 0.068 mS/cm (27.8 mM NaCl). The change in charge shifted the point of elution into a valley between two major native canola protein peaks, thus making it possible to enrich the protein on a single step.

Hirudin is a potent thrombin-specific inhibitor isolated from the salivary gland of the leech that shows the potential to prevent a variety of types of thrombotic diseases. When recombinant hirudin expressed in *Saccaromyces cerevisiae* was purified, several types of column were compared for their efficiency to concentrate hirudin. Q-Sepharose column showed the highest yield of the protein at 97%, although the purification factor was only 5.3 (83). In comparison, the metal affinity chromatography step using Cu(II)-chelate IDA-Sepharose 6B showed the purification factor of 58 with the yield of 85%. On the ion exchange step, the protein was loaded at pH 7.0 and was eluted with a linear salt gradient from 0 to 0.5 M NaCl in the wash buffer.

3. Affinity Chromatography

Affinity chromatography is based on specific recognition between two relevant biomolecules. It is a most attractive method to purify specifically a dilute protein secreted in the culture medium. Affinity chromatography offers two great advantages: its conceptual simplicity and its high specificity. The ligands used can be small or macromolecules, such as proteins or dye. Their mechanism of action may be either specific or general. Generally, it is used as the last step of the purification scheme, since overloading or a great diversity of proteins can reduce the ligand specificity and the support adsorption capacity and leads to destruction of the ligand structure. The associated cost with affinity chromatography may be high if separation of a specific support calls for long and complex chemical reactions or ligands are rare and expensive. However, affinity chromatography can produce products with high purity in a single step, thus keeping the overall cost reasonable. A high correlation has been shown between the product purity and the value of the products (Fig. 4), making it possible to get a higher return in the long run.

Natural or Synthethic Inhibitor Ligands. Highly effective purification procedures for serine proteases have been reported using immobilized protein inhibitors (85–87). The most popular inhibitors are aprotinin and soybean

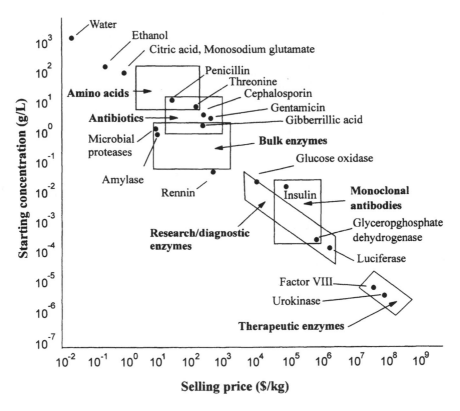

Figure 4   Inverse relation of feed concentration and product price. (From Ref. 84.)

trypsin inhibitor (STI). These are small proteins available at low cost and are very stable. They inhibit proteases by tight interaction at their active site through a peptidic loop that fits into the enzyme active site. The inhibitor peptidic loop may be cleaved by the protease but also the broken bond may be resynthesized. Proteases purified by chromatography on such immobilized inhibitors are usually applied at neutral pH and eluted through pH decrease. A trypsinlike enzyme from a pathogenic nematode *Anisakis simplex* was purified from larval extract by adsorption on STI agarose following an anion-exchange chromatography step (88). Active peak from the ion-exchange step was adsorbed onto the STI agarose gel in a batch mode, and the serine protease was eluted using eletroeluter. The purification factor was 65 with a yield of 12%. A recombinant catalytic domain of bovine enterokinase (RtEK) expressed in *Escherichia coli* as a fusion protein to the C-terminus of DsbA (an analogue of thioredoxin) was purified using an affinity chromatography on immobilized STI. RtEK liberated from the fusion protein by autocatalytic cleavage was purified by chromatography on STI agarose equilibrated in a pH 8 buffer and elution by lowing the pH to 3. Only correctly folded active proteins were bound to the immobilized inhibitor.

A strain of *Flavobacterium odoratum* producing elastolytic enzyme was isolated from soil and cultured with a medium containing poultry feather meal as a source of nitrogen (89). In order to purify the enzyme, the elastin–cellulose column was prepared for affinity chromatography. Using the extract precipitated by ammonium sulfate, the enzyme was concentrated 48-fold with 51% yield including the ammonium sulfate precipitation step. The purification factor was comparable to and the yield was superior to the three-step purification scheme composed of ammonium sulfate, ion change chromatography on DEAE column, and size exclusion chromatography on Sephadex G-75, which showed the purification factor of 49 with 12% yield.

Benzamidine derivatives have been widely used as immobilized ligands for purification of trypsin and trypsinlike enzymes (90, 91). The interaction sites are shown to be the guanidium moiety of benzamidine and an aspartyl side chain situated at the bottom of the specificity pocket (92, 93). However, the interaction between the benzamidine-based supports and proteases are not exclusively affinity for active trypsinlike enzymes at their active site, since the ionic and hydrophobic character of benzamidine derivatives makes them prone to nonspecific interactions. Thrombins were bound to immobilized amidines that were N,N'-substituted with alkyl substituents, although these amidine derivatives cannot accommodate protease specificity pockets (94).

Dye Binding. The use of dye molecules as affinity ligands introduced a pseudoaffinity chromatography in which the structure of the dye molecule resembles a biospecific ligand (95). It was first developed by the serendipitous observation that pyruvate kinase bound to the chromophore (Cinacron Blu F3GA) of soluble Blue Dextran used as a void volume marker in size-exclusion chromatography (96). Dyes used for purification procedures are aromatic polycyclic molecules on which are grated polar groups such as sulfonates, hydroxyls, amines, or amides.

The dyes are able to bind to the active site of a protein or enzyme by virtue of its similarity to a naturally occurring substrate. Several synthetic textile dyes are successfully used as immobilized ligands in affinity chromatography. Two major classes of dye molecules useful for affinity-based separations belong to the cibacron and procion families. The most commonly used dyes are Cibacron Blue F3GA and Procion Blue MX-3G, and MX-R. These dyes are characterized by reactive triazone rings with one or two replaceable chlorine atoms. The active triazine ring provides convenient chemistry for immobilization to hydroxyl-containing supports, forming an ether bond between the dye and matrix. The dye portion of the molecules usually consists of anthraquinone or naphthalene derivatives, often containing one or more sites of substitution with amines and/or sulfate groups. Many of the dyes also contain azo linkages between aromatic components of their structures.

Chromatography on immobilized Cibacron Blue was used to purify IL-1B-converting enzyme, a thiol protease. The yield was 89%, and the purification factor 5 (97). A matrix metalloprotease was also purified by using both Cibacron Blue and Green A. The yield was 100% with the purification factor of 6.5. These proteases were retained on the immobilized dye at low ionic strength and eluted by increasing an ionic strength. Immobilized Reactive Red 120 was used for purification of calpain from chicken gizzard smooth muscle (98). Partially purified protcin was loaded on a column equilibrated in 20 mM MOPS buffer pH 7.2 containing 0.5 M NaCl. The bound protein was eluted using the same buffer but without NaCl. This procedure resulted in a electrophoretically homogenous protein. The purification factor was 7.3 with the yield of 80%. A calcium-activated protease, the hatching enzyme from sea urchin embryo, was also purified using immobilized Reactive Red 120. The enzyme preparation was concentrated first by diafiltration. After application of the extract onto the Reactive Red 120 column, the bound protein was eluted by the gradient using 0–60% ethylene glycol. This procedure allowed recovery of electrophoretically pure enzyme by the single chromatography step with the yield of 67% and the purification factor of 29.4 (99).

VII. CONCLUSIONS

There is increasing interest in recovering valuable byproducts, such as enzymes and pharmaceuticals, from fish-processing wastes. Filleting wastes and processing wash water are a source of various enzyme byproducts. Process operations that have been used for the isolation of enzymes from fish processing waste include acidification, membrane filtration, and rapid heating. Additional research is needed to develop other scale-up methods, such as chromatography, for the industrial production of enzyme byproducts from fish-processing offal.

ACKNOWLEDGMENT

We thank Dr. Ajoy Velayudhan, Department of Bioresource Engineering, Oregon State University, for his helpful discussion and references.

REFERENCES

1. Anonymous. Industrial enzymes to top $1.8 billion. Appl Genet News 19: 3, 1998.
2. BK Simpson, NF Haard. Trypsin and a trypsin-like enzymes from the stomachless cunner. J Agric Food Chem 35: 652–654, 1987.
3. NF Haard. Specialty enzymes from marine organisms. Food Tech 52 (7): 64–67, 1998.
4. HO Hultin. Enzymes from organisms acclimated to low temperatures. In: JP Danehy, B Wolnak eds. Enzymes, The Interface Between Technology and Economics. New York: Marcel Dekker, 1978, pp 161–178.
5. BK Simpson, NF Haard. Cold-adapted enzymes from fish. In D Knorr, ed. Food Biotechnology. New York: Marcel Dekker, 1987, pp 495–572.
6. C Crapo, B Paust, J Babbitt. Recoveries and yields from Pacific fish and shellfish. Marine Advisory Bulletin No. 37. Fairbanks: Alaska Sea Grant College Program, University of Alaska, 1988.
7. JA Dassow. Product yields from various Alaska fish species. Seattle, WA: National Marine Fisheries Service, 1979.
8. J Lin, JW Park, MT Morrissey. Recovered protein and reconditioned water from surimi processing waste. J Food Sci 60: 4–9. 1995
9. Anonymous. Characterization of Alaska Seafood Wastes. A Report to Alaska Fisheries Development Foundation. Kodiak: University of Alaska. Fishery Industrial Technology Center, 1988.
10. TM Lin, JW Park, MT Morrissey. Recovered protein and reconditioned water from surimi processing waste. J Food Sci 60: 4–9, 1995.
11. DT Hoopes. Alaska's fisheries resources—the Dungeness crab. Fisheries Facts No. 6. Seattle, WA: National Marine Fisheries Service, 1973.
12. A Gildberg. Recovery of proteinase and protein hydrolysate from fish viscera. Bioresource Tech 39: 271–276, 1992.
13. SD Vecchi, Z Coppes. Marine fish digestive proteases—relevance to food industry and the south-west Atlantic region—a review. J Food Biochem 20: 193–214, 1998.

14. FL García-Carreño, NF Haard. Characterization of proteinase classes in langostilla and crayfish extracts. J Food Biochem 17:97–113, 1993.

15. FL García-Carreño, NF Haard. Characterization of proteinase classes in Langostilla (*Pleuroncodes planipes*) and crayfish (*Pacififastacus astacus*) extracts. J Food Biochem 17:97–113, 1993.

16. FL García-Carreño, MP Hernández-Cortés, NF Haard. Enzymes with peptidase and proteinase activity from the digestive systems of a freshwater and a marine decapod. J Agric Food Chem 42: 1456–1461, 1994.

17. NF Haard. Protein hydrolysis in seafood. In: F Shahidi, JR Botta eds. Seafood: Chemistry, Processing Technology and Quality. Glasgow: Blackie Academic and Professional/Chapman and Hall Publishing Co., 1994, pp 10–33.

18. A Gildberg, RL Olsen, JB Bjannasson. Catalytic properties and chemical composition of pepsin from Atlantic cod (*Gadus morhua*). Comp Biochem Physiol 69B: 323–330, 1990.

19. B Asgeirsson, JW Fox, J Bjarnason. Purification and characterization of trypsin from the poikilotherm *Gadus morhua*. Eur J Biochem 180: 85–94, 1989.

20. A Martinez, JL Serra. Proteolytic activities in the digestive tract of anchovy (*Engraulis encrasicholus*). Comp Biochem Physiol 93B: 61–66, 1989.

21. AJ Vithayathill, F Buck, M Bier, FF Nord. On the mechanism of enzyme action. LXXII. Comparative studies of trypsins of various origins. Arch Biochem Biophys 92: 532–540, 1961.

22. MG El-Shemy, RE Levin. Characterization of affinity-purified trypsin from hybrid tilapia (*Tilapia milotic/aurea*). J Food Biochem 21: 163–175, 1997.

23. K Hjelmeland, J Raa. Characteristics of two trypsin type isozymes isolated from the Arctic fish capelin (*Mallotus villosus*). Comp Biochem Physiol 71B(4): 557–562, 1982.

24. BK Simpson, MV Simpson, NF Haard. Properties of trypsin from the pyloric ceaca of Atlantic cod (*Godus morhua*). J Food Sci 55: 959–961, 1990.

25. J Clarks, NL Macdonald, JR Stark. Metabolism in marine flatfish. III Measurement of elastase activity in the digestive tract of Dover sole (*Solea solea* L). Comp Biochem Physiol 81B: 695–700, 1985.

26. BK Simpson, NF Haard. Purification and characterization of trypsin from the Greenland cod (*Gadus ogac*). I. Kinetic and thermodynamic characteristics. Can J Biochem Cell Biol 62(9): 894–900, 1984.

27. A Martinez, RL Olsen, J Serra. Purification and characterization of trypsin-like enzymes from the digestive tract of anchovy, *Engraulis encrasicholus*. Comp Biochem Physiol 91B:677–684, 1988.

28. T Cohen, A Gertler, A, Y Birk. Pancreatic proteolytic enzymes from carp (*Cyprinus carpio*)-I. Purification and physical properties of trypsin, chymotrypsin, elastase and carboxypeptidase B. Comp Biochem Physiol 69B:639–646, 1981.

29. S Benjakul, MT Morrissey, TA Seymour, H An. Proteolytic activities in solid wastes and digestive organs of Pacific whiting. OSU-Seafood Laboratory, Astoria, OR, 1999.

30. NF Haard. Atlantic cod protease. 1. Characterization with casein and milk substrate and influence of Sepharose immobilization on salt activation, temperature characteristics and milk clotting reaction. J Food Sci 51 (2): 313–316, 326, 1986.

31. A Gildberg, J Raa. Purification and characterization of pepsins from the Arctic fish capelin (*Mallotus villosus*). Comp Biochem Physiol 75A: 337–342, 1983.

32. EJ Squires, NF Haard, LAW Feltham. Gastric proteases of the Greenland cod (*Gadus ogac*). Isolation and kinetic properties. Can J Biochem Cell Biol 64:205–214, 1984.

33. F Guerard, Y LeGal. Characterization of a chymosin-like pepsin from the dogfish (*Scyliorhinus canicula*). Comp Biochem Physiol 88B: 823–827, 1987.

34. M Noda, K Murakami. Studies on proteinases from the digestive organs of sardine. II. Purification and characterization of two acid proteinases from the stomach. Biochem Biophys Acta 65B: 27–34, 1981.

35. SS Twinning, PA Alexander, DM Glick. A pepsinogen from rainbow trout. Comp Biochem Physiol 75B: 109–112, 1983.

36. H Kirshke, AJ Barrett. Chemistry of lysosomal proteases. In: H Gaumann, FJ Ballard, eds. Lysosomes: Their Role in Protein Breakdown. New York: Academic Press, 1987, pp 193–238.

37. H An, MY Peters, TA Seymour. Roles of endogenous enzymes on surimi gelation. Trends Food Sci Technol 7: 321–327, 1996.

38. H An, V Weerasinghe, TA Seymour, MT Morrissey. Degradation of Pacific whiting surimi proteins by cathepsins. J Food Sci 59: 1013–1017, 1033, 1994.

39. S Benjakul, TA Seymour, MT Morrissey, H An. Proteinases from Pacific whiting surimi wash water: Identification and characterization. J Food Sci 61: 1165–1170, 1996.

40. DS Lee, MS Heu, DS Kim, JH Pyeun. Some properties of the crude proteases from fish for application in seafood fermentation industry. J Korean Fish Soc 29(3): 309–319, 1996.

41. S Benjakul, TA Seymour, TMT Morrissey, H An. Characterization of proteinase recovered from Pacific whiting surimi wash water. J Food Biochem 22: 1–16, 1998.

42. JM Denucé. Phosphatase and esterases in the digestive gland of the crayfish *Orconectes virilis*. Arch Int Physiol Biochem 75: 159–160, 1967.

43. DF Travis. The molting cycle of the spiny lobster, *Panulirus argus* latreille. II Preecdysial histological and histochemical changes in the hepatopancreas and integumental tissues. Biol Bull 108: 88–112, 1955.

44. S Chockalingam. Studies on enzymes associated with calcification of the cuticle of the hermit crab, *Clibanarius olivaceous*. Mar Biol 10: 169–182, 1971.

45. JF Shaw, WC Chen. Isozymes of bighead shrimp alkaline phosphatase. Biosci Biotech Biochem 58: 28–31, 1994.

46. RL Olsen, A Johansen, B Myrnes. Recovery of enzyme from shrimp waste. Process Biochem April: 67–68, 1990.

47. RL Olsen, KO Kersti, B Myrnes. Alkaline phosphatase from the hepatopancreas of shrimp: a dimeric enzyme with catalytically active subunits. Comp Biochem Physiol 99B: 755–761, 1991.

48. J Raa. New commercial products based on waste from the fish processing industry. In: A Bremner, C Davis, B Austin eds, Making the Most of the Catch, Brisbane, Australia: AUSEAS, 1997, pp 1–4.

49. J Raa, A Gildberg. Fish silage: a review. CRC Crit Rev Food Sci Nutr 16: 383–419, 1982.

50. P Reece. Recovery of proteinase from fish waste. Process Biochem June: 62–66, 1988.
51. A Gildberg, KA Almås. Utilization of fish viscera. In: ML Maguer, P Jelen, eds. Food Engineering and Process Applications, vol. 2. New York: Elsevier Applied Science 1986, p 383.
52. KA Almås. Utilization of marine biomass for production of microbial growth media and biochemicals. In: MNVoigt, JR Botta, eds. Advances in Fisheries Technology and Biotechnology for Increased Profitability. Lancaster, PA: Technomic Pub. Co., 1990, pp 361–372.
53. J Short. Membrane separation in food processing. In: RK Singh, SSH Rizvi, eds. Bioseparation Processes in Foods. New York: Marcel Dekker, 1995, pp 333–350.
54. PA Belter, EL Cussler, WS Hu. Ultrafiltration and electrophoresis. In: PA Belter, EL Cussler, WS Hu, eds. Bioseparation, A New York: Wiley-Interscience, 1988, pp 237–270.
55. JD Dziezak. Membrane separation technology offers processors ultimate potential. Food Tech 9: 108–113, 1990.
56. DJ Paulson, RL Wilson. Crossflow membrane technology: its use in the food industry, recent innovations. In: M Kroger, R Shapiro, eds. Changing Food Technology. Lancaster, PA: Technomic Pub. Co., 1987, p 85.
57. SN Gaeta. The industrial development of polymeric membranes and membrane modules for reverse osmosis and ultrafiltration. In: A Caetano, MN De Pinho, E Drioli, H Muntau, eds. Membrane Technology: Applications to Industrial Wastewater Treatment. The Netherlands: Kluwer Academic Publishers, 1995, pp 47–62.
58. J Bohdziewic. Ultrafiltration of technical amylolytic enzymes. Process Biochem 31: 185–191, 1996.
59. J Bohdziewic, M Bodzek. Ultrafiltration preparation of pectinolytic enzymes from citric acid fermentation broth. Process Biochem 29: 99–107, 1994.
60. Y Miyata. Concentration of protein from the wash water of red meat fish by ultrafiltration membrane. Bull Jpn Soc Sci Fish 50: 659–663, 1984.
61. D Green, L Tzoo, AC Chao, TC Lanier. Strategies of handling soluble wastes generated during minced fish (surimi) production. Proceedings, 39th Industry Conference, Purdue University, 1984, pp 565–571.
62. R Ninomiya, T Okawa, T Tsuchiya, JJ Matsunoto. Recovery of water soluble protein in waste wash water of fish processing plants of ultrafiltration. Bull Jpn Soc Sci Fish 51: 1133–1158, 1985.
63. NF Haard. Enzymes from food myosystems. J Muscle Food 1 (14): 293–338, 1990.
64. A Gildberg. Recovery of proteinase and protein hydrolysates from fish viscera. Bioresource Tech 39: 271–276, 1992.
65. A Guilberg, XQ Shi. Recovery of tryptic enzymes from fish sauce. Process Biochem 29: 151–155, 1994.
66. CH Biss, SA Coombes, PJ Skudder. The development and application of ohmic heating for the continuous heating particulate foodstuffs. In: RW Field, JA Howell, eds. Processing Engineering in the Food Industry: Developments and Opportunities. Inc., New York: Elsevier Science Publishing Co., 1989, pp 259–272.
67. DL Parrott. Use of ohmic heating for aseptic processing of food particulates. Food Tech 46 (12): 68–72, 1992.

68. J Yongsawasdigul, JW Park, E Kolbe, YA Dagga, MT Morrissey. Ohmic heating maximizes gel functionality of Pacific whiting surimi. J Food Sci 60: 10–14, 1995.
69. L Huang, Y Chen, MT Morrissey. Coagulation of fish proteins from frozen fish mince wash water by ohmic heating. Astoria, OR: OSU-Seafood Laboratory, 1996.
70. S Benjakul, MT Morrissey, TA Seymour, H An. Recovery of proteinase from Pacific whiting surimi wash water. J Food Biochem 21: 431–444, 1997.
71. K Büntemeyer, KH Kroner, H Hustedt, WD Deckwer. Process for large-scale recovery of intracellular yeast invertase based on heat and pH-shift treatment. Process Biochem December: 212–216, 1989.
72. J Schnell, MR Kula, Investigations of heat treatment to improve the isolation of intracellular enzymes. Bioproc Eng 4: 129, 1989.
73. S Tanuja, ND Srinivas, KSMS Raghava-Rao, MK Gowthaman, Aqueous two phase extraction for downstream processing of amyloglucosidase. Process Biochem 32: 635–641, 1997.
74. T Becker. 1995. Separation of purification processed for recovery of industrial enzymes. In: RK Singh, SSH Rizvi, eds. Bioseparation Processes in Foods. New York: Marcel Dekker, Inc, 1995.
75. L Strandberg, K Kohler, SO Enfors. 1991. Large scale fermentation and purification of a recombinant protein from *Escherichia coli*. Process Biochem 26: 225–234, 1991.
76. M Gonzalez, C Peña, LT Casas. Partial purification of β-galactosidase from yeast by an aqueous two-phase system method. Process Biochem Intl. 157–161, 1990.
77. HG Heinsohn, JD Lorch, KJ Hayenga, RE Arnold. Process for the recovery of microbially produced chymosin. U.S. Patent 5,1139,943, Aug. 18, 1992.
78. RK Scope. Protein Purification: Principles and Practice, 3rd ed. New York: Springer-Verlag, 1984.
79. C Sene, E Boschetti. Place of chromatography in the separation and purification of proteins produced from cultured cells. In: A Mizrahi, ed. Downstream Process: Equipment and Techniques. Advances in Biotechnological Processes, vol 8. New York: Alan R. Liss, 1988, p 241–314.
80. MB Frank, J Watson, D Mochizuki, S Gillis. Biochemical and biological characterization of lymphocyte regulatory molecules. VIII. Purification of interleukin 2 from a human T cell leukemia. J Immunol 127: 2361–2365, 1981.
81. HA Chase. Adsorption separation processes for protein purification. In: A Mizrahi, ed. Downstream Process: Equipment and Techniques. Advances in Biotechnological Processes, vol 8. New York: Alan R. Liss, 1988, pp 163–204.
82. C Zhang, CE Glatz. Process engineering strategy for recombinant protein recovery from canola by cation exchange chromatography. Biotechnol Prog 15: 12–18, 1999.
83. JH Sohn, ES Choi, BH Chung, DJ Youn, JH Seo, SK Rhee. Process development for the production of recombinant hirudin in *Saccharomyces cerevisiae:* from upstream to downstream. Process Biochem 30: 653–660, 1995.
84. EL Cussler, H Ding. Bioseparation, especially using hollow fibers. 1 In: RK Singh, SSH Rizvi, eds. Bioseparation Processes in Foods. New York: Marcel Dekker, 1995, pp 1–15.
85. O Ibrahim-Granet, O Bertrand. Separation of proteases: old and new approaches. J Chromatography 684: 239–263, 1996.

86. NC Robinson, RW Tye, H Neurath, KA Walsh. Isolation of trypsins by affinity chromatography. Biochemistry 10: 2743–2747, 1971.
87. J Turkova, O Hubalkova, M Krivakova, J Coupek. Affinity chromatography on hydroxyalkyl methacrylate gels. I. Preparation of immobilized chymotrypsin and its use in the isolation of proteolytic inhibitors. Biochim Biophys Acta 322: 1–9, 1973.
88. SR Morris, JA Sakanari. Characterization of the serine protease and serine protease inhibitor from the tissue-penetrating nematode *Anisakis simplex*. J Biol Chem 269: 27650–27656, 1994.
89. Zhi-ying, Y., Guo-xiong, G, Wei-qin Z., and Zhe-fu, L. Elastolytic activity from *Flavobacterium odoratum:* microbial screening and cultuvation, enzyme production and purification. Process Biochem 29: 427–436, 1994.
90. G Schmer. The purification of bovine thrombin by affinity chromatography on benzamidine-agarose. Hoppe Seylers Z Physiol Chem 353 (5): 810–4, 1972.
91. HF Hixson, AH Nishikawa. Affinity chromatography: purification of bovine trypsin and thrombin. Arch Biochem Biophys 154: 501–509, 1973.
92. W Bode, P Schwager. The refined crystal structure of bovine beta-trypsin at 1.8 Å resolution. II. Crystallographic refinement, calcium binding site, benzamidine binding site and active site at pH 7.0. J Mol Biol 98: 693–717, 1975.
93. W Bode, D Turk, J Sturzebecher. Geometry of binding of the benzamidine- and arginine-based inhibitors N alpha-(2-naphthyl-sulphonyl-glycyl)-DL-p-amidinophenylalanyl-pipe ridine (NAPAP) and (2R,4R)-4-methyl-1-[N alpha-(3-methyl-1,2,3,4-tetrahydro-8-quinolinesulphonyl)-L-arginyl]-2-piperidine carboxylic acid (MQPA) to human alpha-thrombin. X-ray crystallographic determination of the NAPAP-trypsin complex and modeling of NAPAP-thrombin and MQPA-thrombin. Eur J Biochem 193: 175–182, 1990.
94. S Khamlichi, D Muller, R Fuks, J Jozefonvicz. Specific adsorption of serine proteases on coated silica beads substituted with amidine derivatives. J Chromatogr 510: 123–132, 1990.
95. SS Deshpande. Affinity chromatography. In: RK Singh, SSH Rizvi, eds. Bioseparation Processes in Foods. New York: Marcel Dekker, 1995.
96. Haeckel, B Hess, W Lauterborn, KH Wüster. Purification and allosteric properties of yeast pyruvate kinase. Hoppe Seylers Z Physiol Chem 349: 699–714, 1968.
97. SR Kronheim, A Mumma, T Greenstreet, PJ Glackin, K van Ness, CJ March, RA Black. Purification of interleukin-1 beta converting enzyme, the protease that cleaves the interleukin-1 beta precursor. Arch Biochem Biophys 296: 698–703, 1992.
98. DR Hathaway, DK Werth, JR Haeberle. Limited autolysis reduces the Ca^{2+} requirement of a smooth muscle Ca^{2+}-activated protease. J Biol Chem 257: 9072–9077, 1982.
99. T Lepage, C Gache. Purification and characterization of the sea urchin embryo hatching enzyme J Biol Chem 264: 4787–4793, 1989.

Index